Inorganic Chemistry

Inorganic Chemistry

Second Edition

James E. House

Illinois Wesleyan University
and Illinois State University

AMSTERDAM • WALTHAM • HEIDELBERG • LONDON • NEW YORK • OXFORD
PARIS • SAN DIEGO • SAN FRANCISCO • SINGAPORE • SYDNEY • TOKYO
Academic Press is an Imprint of Elsevier

Academic Press is an Imprint of Elsevier
225 Wyman Street, Waltham, MA 02451, USA
The Boulevard, Langford Lane, Kidlington, Oxford OX5 1GB, UK

First edition 2008
Second edition 2013

Library of Congress Cataloging-in-Publication Data
House, J. E.
  Inorganic chemistry / James House. − 2nd ed.
    p. cm.
  Includes bibliographical references and index.
1. Chemistry, Inorganic − Textbooks. I. Title.
  QD151.5.H68 2013
  546−dc23

                              2012017867

British Library Cataloging-in-Publication Data
A catalogue record for this book is available from the British Library

ISBN: 978-0-12-385110-9

For information on all Elsevier publications
visit our website at www.store.elsevier.com

Contents

Preface to the Second Edition

The development of inorganic continues at an astonishing pace, and the number of journal pages devoted annually to the field continues to increase. Since the publication of the first edition of this book, exciting results have been obtained in many areas, especially in the application of inorganic compounds in medicine.

The present edition continues the same format as the first with five major areas that constitute atomic and molecular structure; condensed phases; acids, bases, and solvents; chemistry of the elements; and chemistry of coordination compounds. Although the structure of the book has not been changed, there are some significant modifications. First, the coverage of superacids in Chapter 10 has been expanded in order to reflect the increasing utilization of these materials in inorganic chemistry. Second, a new chapter on bioinorganic chemistry has been added, with numerous illustrations from the recent literature that describe the use of metal complexes in medicine. Third, the coverage of catalysis by complexes has been expanded with new material on mechanisms. Fourth, the entire manuscript has been edited to enhance clarity of presentation. The judicious use of color illustrations is also new to this edition. Finally, a significant number of additional end of chapter questions and problems have been added.

Although there have been substantial changes in the coverage from the first edition, flexibility in the order of presentation remains. After the introductory chapters on atomic and molecular structure, the nature of inorganic condensed phases is described with a theme of understanding properties of such materials. Chapter 8 deals with dynamic processes in solids because of the importance of the topic in many industrial processes.

Explanations do not exist for all observations in the behavior of inorganic compounds. Consequently, throughout the text there are instances mentioned which illustrate unanswered questions in science. It is hoped that some of these might generate an interest that would cause the reader to pursue further study in inorganic chemistry.

The author wishes to thank all of those students who have used the first edition and pointed out errors. Their contribution to the preparation of the second edition is gratefully acknowledged. The author also wishes to thank his wife, Kathleen, for enduring yet another long project.

Preface to the First Edition

No single volume, certainly not a textbook, can come close to including all of the important topics in inorganic chemistry. The field is simply too broad in scope and it is growing at a rapid pace. Inorganic chemistry textbooks reflect a great deal of work and the results of the many choices that authors must make as to what to include and what to leave out. Writers of textbooks in chemistry bring to the task backgrounds that reflect their research interests, the schools they attended, and their personalities. In their writing, authors are really saying "this is the field as I see it." In these regards, this book is similar to others.

When teaching a course in inorganic chemistry, certain core topics are almost universally included. In addition, there are numerous peripheral areas that may be included at certain schools but not at others depending on the interests and specialization of the person teaching the course. The course content may even change from one semester to the next. The effort to produce a textbook that presents coverage of a wide range of optional material in addition to the essential topics can result in a textbook for a one semester course that contains a thousand pages. Even a "concise" inorganic chemistry book can be nearly this long. This book is not a survey of the literature or a research monograph. It is a textbook that is intended to provide the background necessary for the reader to move on to those more advanced resources.

In writing this book, I have attempted to produce a concise *textbook* that meets several objectives. First, the topics included were selected in order to provide essential information in the major areas of inorganic chemistry (molecular structure, acid–base chemistry, coordination chemistry, ligand field theory, solid-state chemistry, etc.). These topics form the basis for competency in inorganic chemistry at a level commensurate with the one semester course taught at most colleges and universities.

When painting a wall, better coverage is assured when the roller passes over the same area several times from different directions. It is the opinion of the author that this technique works well in teaching chemistry. Therefore, a second objective has been to stress fundamental principles in the discussion of several topics. For example, the hard–soft interaction principle is employed in discussion of acid–base chemistry, stability of complexes, solubility, and predicting reaction products. Third, the presentation of topics is made with an effort to be clear and concise so that the book is portable and user friendly. This book is meant to present in convenient form a readable account of the essentials of inorganic chemistry that can serve as both as a textbook for a one semester course, upper-level course and as a guide for self study. It is a *textbook* not a review of the literature or a research monograph. There are few references to the original literature, but many of the advanced books and monographs are cited.

Although the material contained in this book is arranged in a progressive way, there is flexibility in the order of presentation. For students, who have a good grasp of the basic principles of quantum mechanics and atomic structure, Chapters 1 and 2 can be given a cursory reading but not included in the required course material. The chapters are included to provide a resource for review and self study. Chapter 4 presents an overview of structural chemistry early so the reader can become familiar with many types of inorganic structures before taking up the study of symmetry or chemistry of specific elements. Structures of inorganic solids are discussed in Chapter 7, but that material could easily be studied before Chapters 5 or 6. Chapter 6 contains material dealing with intermolecular forces and polarity of molecules because of the importance of these topics when interpreting properties of substances and their chemical behavior. In view of the importance of the topic, especially in industrial chemistry, this book includes material on rate processes involving inorganic compounds in the solid state (Chapter 8). The chapter begins with an overview of some of the important aspects of reactions in solids before considering phase transitions and reactions of solid coordination compounds.

It should be an acknowledged fact that no single volume can present the descriptive chemistry of all the elements. Some of the volumes that attempt to do so are enormous. In this book, the presentation of descriptive chemistry of the elements is kept brief with the emphasis placed on types of reactions and structures that summarize the behavior of many compounds. The attempt is to present an overview of descriptive chemistry that will show the important classes of compounds and their reactions without becoming laborious in its detail. Many schools offer a descriptive inorganic chemistry course at an intermediate level that covers a great deal of the chemistry of the elements. Part of the rationale for offering such a course is that the upper-level course typically concentrates more heavily on principles of inorganic chemistry. Recognizing that an increasing fraction of the students in the upper-level inorganic chemistry course will have already had a course that deals primarily with descriptive chemistry, this book is devoted to a presentation of the principles of inorganic chemistry while giving a brief overview of descriptive chemistry in Chapters 12−15, although many topics that are primarily descriptive in nature are included in other sections. Chapter 16 provides a survey of the chemistry of coordination compounds and that is followed by Chapters 17−23 that deal with structures, bonding, spectra, and reactions of coordination compounds. The material included in this text should provide the basis for the successful study of a variety of special topics.

Doubtless, the teacher of inorganic chemistry will include some topics and examples of current or personal interest that are not included in any textbook. That has always been my practice, and it provides an opportunity to show how the field is developing and new relationships.

Most textbooks are an outgrowth of the author's teaching. In the preface, the author should convey to the reader some of the underlying pedagogical philosophy which resulted in the design of his or her book. It is unavoidable that a different teacher will have somewhat different philosophy and methodology. As a result, no single book will be completely congruent with the practices and motivations of all teachers. A teacher who writes the textbook for his or her course should find all of the needed topics in the book. However, it is unlikely that a book written by someone else will ever contain exactly the right topics presented in exactly the right way.

The author has taught several hundred students in inorganic chemistry courses at Illinois State University, Illinois Wesleyan University, University of Illinois, and Western Kentucky University using the materials and approaches set forth in this book. Among that number are many who have gone onto graduate school, and virtually all of that group have performed well (in many cases very well!) on registration and entrance examinations in inorganic chemistry at some of the most prestigious institutions. Although it is not possible to name all of those students, they have provided the inspiration to see this project to completion with the hope that students at other universities may find this book useful in their study of inorganic chemistry. It is a pleasure to acknowledge and give thanks to Derek Coleman for his encouragement and consideration as this project progressed. Finally, I would like to thank my wife, Kathleen, for reading most of the manuscript and making many helpful suggestions. Her constant encouragement and support have been needed at many times as this project was underway.

Structure of Atoms and Molecules

CHAPTER 1

Light, Electrons, and Nuclei

The study of inorganic chemistry involves interpreting, correlating, and predicting the properties and structures of an enormous range of materials. Sulfuric acid is the chemical produced in the largest tonnage of any compound. A greater number of tons of concrete is produced, but it is a mixture rather than a single compound. Accordingly, sulfuric acid is an inorganic compound of enormous importance. On the other hand, inorganic chemists study compounds such as hexaaminecobalt(III) chloride, $[Co(NH_3)_6]Cl_3$, and Zeise's salt, $K[Pt(C_2H_4)Cl_3]$. Such compounds are known as coordination compounds or coordination complexes. Inorganic chemistry also includes areas of study such as nonaqueous solvents and acid–base chemistry. Organometallic compounds, structures and properties of solids, and the chemistry of elements other than carbon comprise areas of inorganic chemistry. However, many compounds of carbon (e.g. CO_2 and Na_2CO_3) are also inorganic compounds. The range of materials studied in inorganic chemistry is enormous, and a great many of the compounds and processes are of industrial importance. Moreover, inorganic chemistry is a body of knowledge that is expanding at a very rapid rate, and knowledge of the behavior of inorganic materials is fundamental to the study of the other areas of chemistry.

Because inorganic chemistry is concerned with structures and properties as well as the synthesis of materials, the study of inorganic chemistry requires familiarity with a certain amount of information that is normally considered to be physical chemistry. As a result, physical chemistry is normally a prerequisite for taking a comprehensive course in inorganic chemistry. There is, of course, a great deal of overlap of some areas of inorganic chemistry with the related areas in other branches of chemistry. Knowledge of atomic structure and properties of atoms is essential for describing both ionic and covalent bonding. Because of the importance of atomic structure to several areas of inorganic chemistry, it is appropriate to begin our study of inorganic chemistry with a brief review of atomic structure and how our ideas about atoms were developed.

1.1 SOME EARLY EXPERIMENTS IN ATOMIC PHYSICS

It is appropriate at the beginning of a review of atomic structure to ask the question, "How do we know what we know?" In other words, "What crucial experiments have been performed and what do the results tell us about the structure of atoms?" Although it is not necessary to consider all of the early experiments in atomic physics, we should describe some of them and explain the results. The first of these experiments was that of J.J. Thompson from 1898 to 1903, which dealt with cathode rays. In the experiment, an evacuated tube that contains two electrodes has a large potential difference generated between the electrodes as shown in Figure 1.1.

3

Inorganic Chemistry. DOI: http://dx.doi.org/10.1016/B978-0-12-385110-9.00001-7

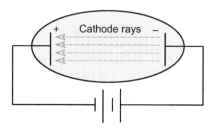

FIGURE 1.1
Design of a cathode ray tube.

Under the influence of a high electric field, the gas in the tube emits light. The glow is the result of electrons colliding with the molecules of gas that are still present in the tube even though the pressure has been reduced to a few torr. The light that is emitted is found to consist of the spectral lines characteristic of the gas inside the tube. Neutral molecules of the gas are ionized by the electrons streaming from the cathode, which is followed by recombination of electrons with charged species. Energy (in the form of light) is emitted as this process occurs. As a result of the high electric field, negative ions are accelerated toward the anode and positive ions are accelerated toward the cathode. When the pressure inside the tube is very low (perhaps 0.001 torr), the mean free path is long enough that some of the positive ions strike the cathode, which emits rays. Rays emanating from the cathode stream toward the anode. Because they are emitted from the cathode, they are known as *cathode rays*.

Cathode rays have some very interesting properties. First, their path can be bent by placing a magnet near the cathode ray tube. Second, placing an electric charge near the stream of rays also causes the path they follow to exhibit curvature. From these observations, we conclude that the rays are electrically charged. The cathode rays were shown to carry a negative charge because they were attracted to a positively charged plate and repelled by the one that carried a negative charge.

The behavior of cathode rays in a magnetic field is explained by recalling that a moving beam of charged particles (they were not known to be electrons at the time) generates a magnetic field. The same principle is illustrated by passing an electric current through a wire that is wound around a compass. In this case, the magnetic field generated by the flowing current interacts with the magnetized needle of the compass causing it to point in a different direction. Because the cathode rays are negatively charged particles, their motion generates a magnetic field that interacts with the external magnetic field. In fact, some important information about the nature of the charged particles in cathode rays can be obtained by studying the curvature of their path in a magnetic field of known strength.

Consider the following situation.Suppose a cross wind of 10 mph is blowing across a tennis court. If a tennis ball is moving perpendicular to the direction the wind is blowing, the ball will follow a curved path. It is easy to rationalize that if a second ball had a cross sectional area that was twice that of the tennis ball but the same mass, it would follow a more curved path because the wind pressure on it would be greater. On the other hand, if a third ball having twice the cross sectional area and twice the mass of the tennis ball were moving perpendicular to the wind direction, it would follow a path with the same curvature as the tennis ball. The third ball would experience twice as much wind pressure as the tennis ball, but it would have twice the mass, which tends to cause the ball to move in a straight line (inertia). Therefore, if the path of a ball is being studied when it is subjected to wind pressure applied perpendicular to its motion, an analysis of the curvature of the path could be used to determine the ratio of the cross sectional area to the mass of the ball, but neither property alone.

A similar situation exists for a charged particle moving under the influence of a magnetic field. The greater the mass is, the greater will be the tendency of the particle to travel in a straight line. On the other hand, the higher its charge is, the greater will be its tendency to travel in a curved path in the magnetic field. If a particle has two units of charge and two units of mass, it will follow the same path as one that has one unit of charge and one unit of mass. From the study of the behavior of cathode rays in a magnetic field, Thompson was able to determine the charge to mass ratio for cathode rays, but not the charge or the mass alone. The negative particles in cathode rays are electrons, and Thompson is credited with the discovery of the electron. From his experiments with cathode rays, Thompson determined the charge to mass ratio of the electron to be -1.76×10^8 coulomb/g.

It was apparent to Thompson that if atoms in the metal electrode contained negative particles (electrons), then they must also contain positive charges because atoms are electrically neutral. Thompson proposed a model for the atom in which positive and negative particles were embedded in some sort of matrix. The model became known as the plum pudding model because it resembled plums embedded in a pudding. Somehow, an equal number of positive and negative particles were held in this material. Of course we now know that this is an incorrect view of the atom, but the model did account for several features of atomic structure.

The second experiment in atomic physics that increased our understanding of atomic structure was conducted by Robert A. Millikan in 1908. This experiment has become known as the Millikan Oil Drop experiment because of the way in which oil droplets were used. In the experiment, oil droplets (made up of organic molecules) were sprayed into a chamber where a beam of X-rays was directed on them. The X-rays ionized molecules by removing one or more electrons producing cations. As a result, some of the oil droplets carried an overall positive charge. The entire apparatus was arranged in such a way that a negative metal plate, the charge of which could be varied, was at the top of the chamber. By varying the (known) charge on the plate, the attraction between the plate and a specific droplet could be varied until it exactly equaled the gravitational force on the droplet. Under this condition, the droplet could be suspended with an electrostatic force pulling the drop upward that equaled the gravitational force pulling downward on the droplet. Knowing the density of the oil and having measured the diameter of the droplet, the mass of the droplet was calculated. It was a simple matter to calculate the charge on the droplet because the charge on the negative plate with which the droplet interacted was known. Although some droplets may have had two or three electrons removed, the calculated charges on the oil droplets were always a multiple of the smallest charge measured. Assuming that the smallest measured charge corresponded to that of a single electron, the charge on the electron was determined. That charge is -1.602×10^{-19} coulomb or -4.80×10^{-10} esu (electrostatic units: 1 esu $= 1$ g$^{1/2}$ cm$^{3/2}$ s^{-1}). Because the charge to mass ratio was already known, it was now possible to calculate the mass of the electron, which is 9.11×10^{-31} kg or 9.11×10^{-28} g.

The third experiment that is crucial to understanding atomic structure was carried out by Ernest Rutherford in 1911 and is known as Rutherford's experiment. It consists of bombarding a thin metal foil with alpha (α) particles. Thin foils of metals, especially gold, can be made so thin that the thickness of the foil represents only a few atomic diameters. The experiment is shown diagrammatically in Figure 1.2.

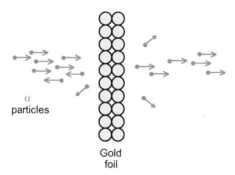

FIGURE 1.2
A representation of Rutherford's experiment.

It is reasonable to ask why such an experiment would be informative in this case. The answer lies in understanding what the Thompson plum pudding model implies. If atoms consist of equal numbers of positive and negative particles embedded in a neutral material, a charged particle such as an α particle (which is a helium nucleus) would be expected to travel near an equal number of positive and negative charges when it passes through an atom. As a result, there should be no *net* effect on the α particle, and it should pass directly through the atom or a foil that is only a few atoms in thickness.

A narrow beam of α particles impinging on a gold foil should pass directly through the foil because the particles have relatively high energies. What happened was that most of the α particles did just that, but some were deflected at large angles and some came essentially back toward the source! Rutherford described this result in terms of firing a 16-inch shell at a piece of tissue paper and having it bounce back at you. How could an α particle experience a force of repulsion great enough to cause it to change directions? The answer is that such a repulsion could result only when all of the positive charge in a gold atom is concentrated in a very small region of space. Without going into the details, calculations showed that the small positive region was approximately 10^{-13} cm in size. This could be calculated because it is rather easy on the basis of electrostatics to determine what force would be required to change the direction of an α particle with a +2 charge traveling with a known energy. Because the overall positive charge on an atom of gold was known (the atomic number), it was possible to determine the approximate size of the positive region.

Rutherford's experiment demonstrated that the total positive charge in an atom is localized in a very small region of space (the nucleus). Because the majority of α particles passed through the gold foil, it was indicated that they did not come near a nucleus. In other words, most of the atom is empty space. The diffuse cloud of electrons (which has a size on the order of 10^{-8} cm) did not exert enough force on the α particles to deflect them. The plum pudding model simply did not explain the observations from the experiment with α particles.

Although the work of Thompson and Rutherford had provided a view of atoms that was essentially correct, there was still the problem of what made up the remainder of the mass of atoms. It had been postulated that there must be an additional ingredient in the atomic nucleus, and it was demonstrated in 1932 by James Chadwick. In his experiments, a thin beryllium target was bombarded with α particles. Radiation having high penetrating power was emitted, and it was initially assumed that they were high-energy γ rays. From studies of the penetration of these rays in lead, it was concluded that the particles had an energy of approximately 7 MeV. Also, these rays were shown to eject protons having energies of approximately 5 MeV from paraffin. However, in order to explain some of the observations, it was shown that if the radiation were γ rays, they must have an energy that is approximately 55 MeV. If an α particle interacts with a beryllium nucleus so that it becomes captured, it is possible to show that the energy (based on mass difference between the products and reactants) is only about 15 MeV. Chadwick studied the recoil of nuclei that were bombarded by the radiation emitted from beryllium when it was a target for α particles and showed that if the radiation consists of γ rays, the energy must be a function of the mass of the recoiling nucleus, which leads to a violation of the conservation of momentum and energy. However, if the radiation emitted from the beryllium target is presumed to carry no charge and consist of particles having a mass approximately that of a proton, the observations could be explained satisfactorily. Such particles were called neutrons, and they result from the reaction

$$\,^{9}_{4}\mathrm{Be} + \,^{4}_{2}\mathrm{He} \rightarrow \left[\,^{13}_{6}\mathrm{C}\right]^{*} \rightarrow \,^{12}_{6}\mathrm{C} + \,^{1}_{0}n \tag{1.1}$$

Atoms consist of electrons and protons in equal numbers and in all cases except the hydrogen atom, some number of neutrons. Electrons and protons have equal but opposite charges, but greatly different masses. The mass of a proton is 1.67×10^{-24} g. In atoms that have many electrons, the electrons are not all held with the same energy so we will discuss later the shell structure of electrons in atoms. At this point, we see that the early experiments in atomic physics have provided a general view of the structures of atoms.

1.2 THE NATURE OF LIGHT

From the early days of physics, a controversy had existed regarding the nature of light. Some prominent physicists, such as Isaac Newton, had believed that light consisted of particles or "corpuscles." Other scientists of that time believed that light was wave-like in its character. In 1807, a crucial experiment was conducted by T. Young in which light showed a diffraction pattern when a beam of light was passed through two slits. Such behavior showed the wave character of light. Other work by A. Fresnel and F. Arago had dealt with interference, which also depended on light having a wave character.

The nature of light and the nature of matter are intimately related. It was from the study of light emitted when matter (atoms and molecules) was excited by some energy source or the absorption of light by matter that much information was obtained. In fact, most of what we know about the structure of atoms and molecules has been obtained by studying the interaction of electromagnetic radiation with matter or electromagnetic radiation emitted from matter. These types of interactions form the basis of several types of spectroscopy, techniques that are very important in studying atoms and molecules.

In 1864, Maxwell showed that electromagnetic radiation consists of transverse electric and magnetic fields that travel through space at the speed of light (3.00×10^{8} m sec^{-1}). The electromagnetic spectrum consists of the several types of waves (visible light, radio waves, infrared radiation, etc.) that form a continuum as shown in Figure 1.3. In 1887, Hertz produced electromagnetic waves by means of an apparatus that generated an oscillating electric charge (an antenna). This discovery led to the development of radio.

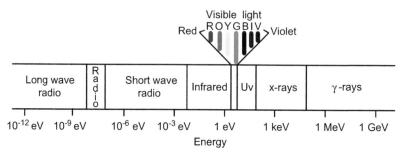

FIGURE 1.3
The electromagnetic spectrum.

Although all of the developments that have been discussed are important to our understanding of the nature of matter, there are other phenomena that provide additional insight. One of them concerns the emission of light from a sample of hydrogen gas through which a high voltage is placed. The basic experiment is shown in Figure 1.4. In 1885, Balmer studied the visible light emitted from the gas by passing it through a prism that separates the light into its components.

The four lines observed are as follows.

$$H_\alpha = 656.28 \text{ nm} = 6562.8 \text{ Å}$$

$$H_\beta = 486.13 \text{ nm} = 4861.3 \text{ Å}$$

$$H_\gamma = 434.05 \text{ nm} = 4340.5 \text{ Å}$$

$$H_\delta = 410.17 \text{ nm} = 4101.7 \text{ Å}$$

This series of spectral lines for hydrogen became known as the Balmer Series, and the wavelengths of these four spectral lines were found to obey the relationship

$$\frac{1}{\lambda} = R_H \left(\frac{1}{2^2} - \frac{1}{n^2} \right) \tag{1.2}$$

where λ is the wavelength of the line, n is an integer larger than 2, and R_H is a constant known as Rydberg's constant that has the value 109,677.76 cm^{-1}. The quantity $1/\lambda$ is known as the *wave number* (the number of complete waves per centimeter), which is written as $\bar{\nu}$ ("nu bar"). From Eqn (1.2) it can be seen that as n assumes larger values, the lines become more closely spaced, but when n equals infinity, there is a limit reached. That limit is known as the *series limit* for the Balmer Series. Keep in mind that these spectral lines, the first to be discovered for hydrogen, were in the visible region of the electromagnetic spectrum. Detectors for visible light (human eyes and photographic plates) were available at an earlier time than were detectors for other types of electromagnetic radiation.

Eventually, other series of lines were found in other regions of the electromagnetic spectrum. The Lyman Series was observed in the ultraviolet region, whereas the Paschen, Brackett, and Pfund Series

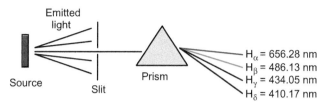

FIGURE 1.4
Separation of spectral lines due to refraction in a prism spectroscope.

were observed in the infrared region of the spectrum. All of these lines were observed as they were *emitted* from excited atoms, so together they constitute the *emission spectrum* or *line spectrum* of hydrogen atoms.

Another of the great developments in atomic physics involved the light emitted from a device known as a black body. Because black is the best absorber of all wavelengths of visible light, it should also be the best emitter. Consequently, a metal sphere, the interior of which is coated with lampblack, emits radiation (black body radiation) having a range of wavelengths from an opening in the sphere when it is heated to incandescence. One of the thorny problems in atomic physics was trying to predict the intensity of the radiation as a function of wavelength. In 1900, Max Planck arrived at a satisfactory relationship by making an assumption that was radical at that time. Planck assumed that absorption and emission of radiation arises from oscillators that change frequency. However, Planck assumed that the frequencies were not continuous but rather that only certain frequencies were allowed. In other words, the frequency is *quantized*. The permissible frequencies were multiples of some fundamental frequency, v_0. A change in an oscillator from a lower frequency to a higher one involves the absorption of energy whereas energy is emitted as the frequency of an oscillator decreases. Planck expressed the energy in terms of the frequency by means of the relationship

$$E = hv \tag{1.3}$$

where E is the energy, v is the frequency, and h is a constant (known as Planck's constant, 6.63×10^{-27} erg s $= 6.63 \times 10^{-34}$ J s). Because light is a transverse wave (the direction the wave is moving is perpendicular to the displacement), it obeys the relationship

$$\lambda v = c \tag{1.4}$$

where λ is the wavelength, v is the frequency, and c is the velocity of light (3.00×10^{10} cm sec^{-1}). By making these assumptions, Plank arrived at an equation that satisfactorily related the intensity and frequency of the emitted black body radiation.

The importance of the idea that energy is quantized is impossible to overstate. It applies to all types of energies related to atoms and molecules. It forms the basis of the various experimental techniques for studying the structure of atoms and molecules. The energy levels may be electronic, vibrational, or rotational depending on the type of experiment conducted.

In the 1800s, it was observed that when light was shone on a metal plate contained in an evacuated tube an interesting phenomenon occurs. The arrangement of the apparatus is shown in Figure 1.5.

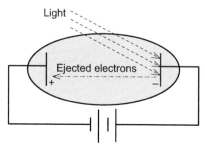

FIGURE 1.5
Apparatus for demonstrating the photoelectric effect.

When light was shone on the metal plate, an electric current flows. Because light and electricity are involved, the phenomenon became known as the *photoelectric effect*. Somehow, light is responsible for the generation of the electric current. Around 1900, there was ample evidence that light behaved as a wave, but it was impossible to account for some of the observations on the photoelectric effect by considering light in that way. Observations on the photoelectric effect include the following:

1. The incident light must have some minimum frequency (the *threshold frequency*) in order for electrons to be ejected.
2. The current flow is instantaneous when the light strikes the metal plate.
3. The current is proportional to the intensity of the incident light.

In 1905, Albert Einstein provided an explanation of the photoelectric effect by assuming that the incident light acts as particles. This allowed for instantaneous collisions of light particles (*photons*) with electrons (called photoelectrons), which resulted in the electrons being ejected from the surface of the metal. Some minimum energy of the photons was required because the electrons are bound to the metal surface with some specific binding energy that depends on the type of metal. The energy required to remove an electron from the surface of a metal is known as the *work function* (w_0) of the metal. Ionization potential (which corresponds to removal of an electron from a gaseous atom) is not the same as the work function. If an incident photon has an energy that is greater than the work function of the metal, the ejected electron will carry away part of the energy as kinetic energy. In other words, the kinetic energy of the ejected electron will be the difference between the energy of the incident photon and the energy required to remove the electron from the metal. This can be expressed by the equation

$$1/2\, mv^2 = h\nu - w_0 \tag{1.5}$$

By increasing the negative charge on the plate to which the ejected electrons move, it is possible to stop the electrons and thereby stop the current flow. The voltage necessary to stop the electrons is known as the *stopping potential*. Under these conditions, what is actually being determined is the kinetic energy of the ejected electrons. If the experiment is repeated using incident radiation with a different frequency, the kinetic energy of the ejected electrons can again be determined. By using light having several known incident frequencies, it is possible to determine the kinetic energy of the electrons corresponding to each frequency and make a graph of the kinetic energy of the electrons vs ν. As can be seen from Eqn (1.5), the relationship should be linear with the slope of the line being h, Planck's constant, and the intercept is $-w_0$. There are some similarities between the photoelectric effect described here and photoelectron spectroscopy of molecules that is described in Section 3.4.

Although Einstein made use of the assumption that light behaves as a particle, there is no denying the validity of the experiments that show that light behaves as a wave. Actually, light has characteristics of both waves and particles, the so-called *particle–wave duality*. Whether it behaves as a wave or a particle depends on the type of experiment to which it is being subjected. In the study of atomic and molecular structure, it is necessary to use both concepts to explain the results of experiments.

1.3 THE BOHR MODEL

Although the experiments dealing with light and atomic spectroscopy had revealed a great deal about the structure of atoms, even the line spectrum of hydrogen presented a formidable problem to the physics of that time. One of the major obstacles was that energy was not emitted continuously as the electron moves about the nucleus. After all, velocity is a vector quantity that has both a magnitude and a direction. A change in direction constitutes a change in velocity (acceleration) and an accelerated electric charge should emit electromagnetic radiation according to Maxwell's theory. If the moving electron lost energy continuously, it would slowly spiral in toward the nucleus and the atom would "run down." Somehow, the laws of classical physics were not capable of dealing with this situation, which is illustrated in Figure 1.6.

Following Rutherford's experiments in 1911, Neils Bohr proposed in 1913 a dynamic model of the hydrogen atom that was based on certain assumptions. The first of these assumptions was that there were certain "allowed" orbits in which the electron could move without radiating electromagnetic energy. Further, these were orbits in which the angular momentum of the electron (which for a rotating object is expressed as mvr) is a multiple of $h/2\pi$ (which is also written as $\hbar$),

$$mvr = \frac{nh}{2\pi} = n\hbar \tag{1.6}$$

where m is the mass of the electron, v is its velocity, r is the radius of the orbit, and n is an integer that can take on the values 1, 2, 3, …, and $\hbar$ is $h/2\pi$. The integer n is known as a *quantum number*, or more specifically, the *principal* quantum number.

Bohr also assumed that electromagnetic energy was emitted as the electron moved from a higher orbital (larger n value) to a lower one and absorbed in the reverse process. This accounts for the fact that

FIGURE 1.6
As the electron moves around the nucleus, it is constantly changing direction.

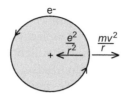

FIGURE 1.7
Forces acting on an electron moving in a hydrogen atom.

the line spectrum of hydrogen shows only lines having certain wavelengths. In order for the electron to move in a stable orbit, the electrostatic attraction between the electron and the proton must be balanced by the centrifugal force that results from its circular motion. As shown in Figure 1.7, the forces are actually in opposite directions, so we equate only the *magnitudes* of the forces.

The electrostatic force is given by the coulombic force as e^2/r^2, and the centrifugal force on the electron is mv^2/r. Therefore, we can write

$$\frac{mv^2}{r} = \frac{e^2}{r^2} \tag{1.7}$$

From Eqn (1.7) we can calculate the velocity of the electron as

$$v = \sqrt{\frac{e^2}{mr}} \tag{1.8}$$

We can also solve Eqn (1.6) for v to obtain

$$v = \frac{nh}{2\pi mr} \tag{1.9}$$

Because the moving electron has only one velocity, the values for v given in Eqns (1.8) and (1.9) must be equal.

$$\sqrt{\frac{e^2}{mr}} = \frac{nh}{2\pi mr} \tag{1.10}$$

We can now solve for r to obtain

$$r = \frac{n^2h^2}{4\pi^2me^2} \tag{1.11}$$

In Eqn (1.11), only r and n are variables. From the nature of this equation, we see that the value of r, the radius of the orbit, increases as the square of n. For the orbit with $n = 2$, the radius is four times that when $n = 1$, etc. Dimensionally, Eqn (1.11) leads to a value of r that is given in cm if the constants are assigned their values in the cm-g-s system of units (only h, m, and e have units).

$$[(\text{g cm}^2/\text{sec}^2)\text{sec}]^2/[\text{g}(\text{g}^{1/2} \text{ cm}^{3/2}/\text{sec})^2] = \text{cm}. \tag{1.12}$$

From Eqn (1.7), we see that

$$mv^2 = \frac{e^2}{r}$$

(1.13)

Multiplying both sides of the equation by ½ we obtain

$$\frac{1}{2}mv^2 = \frac{e^2}{2r}$$

(1.14)

where the left hand side is simply the kinetic energy of the electron. The total energy of the electron is the sum of the kinetic energy and the electrostatic potential energy, $-e^2/r$.

$$E = \frac{1}{2}mv^2 - \frac{e^2}{r} = \frac{e^2}{2r} - \frac{e^2}{r} = -\frac{e^2}{2r}$$

(1.15)

Substituting the value for r from Eqn (1.11) into Eqn (1.15) we obtain

$$E = -\frac{e^2}{2r} = -\frac{2\pi^2 me^4}{n^2 h^2}$$

(1.16)

from which we see that there is an inverse relationship between the energy and the square of the value n. The lowest value of E (and it is negative!) is for $n = 1$, and $E = 0$ when n has an infinitely large value that corresponds to complete removal of the electron. If the constants are assigned values in the cm-g-s system of units, the energy calculated will be in ergs. Of course $1 J = 10^7$ erg and 1 cal $= 4.184$ J.

By assigning various values to n, we can evaluate the corresponding energy of the electron in the orbits of the hydrogen atom. When this is done, we find the energies of several orbits as follows.

$n = 1$,	$E = -21.7 \times 10^{-12}$ erg
$n = 2$,	$E = -5.43 \times 10^{-12}$ erg
$n = 3$,	$E = -2.41 \times 10^{-12}$ erg
$n = 4$,	$E = -1.36 \times 10^{-24}$ erg
$n = 5$,	$E = -0.87 \times 10^{-12}$ erg
$n = 6$,	$E = -0.63 \times 10^{-12}$ erg
$n = \infty$,	$E = 0$

These energies can be used to prepare an energy level diagram such as that shown in Figure 1.8. Note that the binding energy of the electron is lowest when $n = 1$, and the binding energy is 0 when $n = \infty$.

Although the Bohr model successfully accounted for the line spectrum of the hydrogen atom, it could not explain the line spectrum of any other atom. It could be used to predict the wavelengths of spectral lines of other species that had only one electron such as He^+, Li^{2+}, Be^{3+}, etc. Also, the model was based on assumptions regarding the nature of the allowed orbits that had no basis in classical physics. An additional problem is also encountered when the Heisenberg Uncertainty Principle is

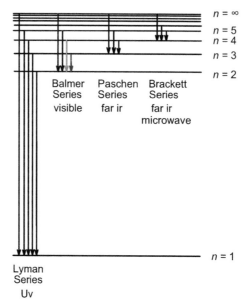

FIGURE 1.8
An energy level diagram for the hydrogen atom.

considered. According to this principle, it is impossible to know exactly the position and momentum of a particle simultaneously. Being able to describe an orbit of an electron in a hydrogen atom is equivalent to knowing its momentum and position. The Heisenberg Uncertainty Principle places a limit on the accuracy to which these variables can be known simultaneously. That relationship is

$$\Delta x \times \Delta(mv) \geq \hbar \tag{1.17}$$

where Δ is read as the uncertainty in the variable that follows. Planck's constant is known as the fundamental unit of action (it has units of energy multiplied by time), but the product of momentum multiplied by distance has the same dimensions. The essentially classical Bohr model explained the line spectrum of hydrogen, but it did not provide a theoretical framework for understanding atomic structure.

1.4 PARTICLE–WAVE DUALITY

The debate concerning the particle and wave nature of light had been lively for many years when in 1924 a young French doctoral student, Louis V. de Broglie, developed a hypothesis regarding the nature of particles. In this case, the particles were "real" particles such as electrons. De Broglie realized that for electromagnetic radiation, the energy could be described by the Planck equation

$$E = h\nu = \frac{hc}{\lambda} \tag{1.18}$$

However, one of the consequences of Einstein's special theory of relativity (in 1905) is that a photon has an energy that can be expressed as

$$E = mc^2 \tag{1.19}$$

This famous equation expresses the relationship between mass and energy, and its validity has been amply demonstrated. This equation does not indicate that a photon has a mass. It does signify that because a photon has energy, its *energy* is *equivalent* to some mass. However, for a given photon, there is only one energy. So

$$mc^2 = \frac{hc}{\lambda} \tag{1.20}$$

Rearranging this equation leads to

$$\lambda = \frac{h}{mc} \tag{1.21}$$

Having developed the relationship shown in Eqn (1.21) for photons, de Broglie considered the fact that photons have characteristics of both particles and waves, as we have discovered earlier in this chapter. He reasoned that if a "real" particle such as an electron could exhibit properties of both particles and waves, the wavelength for the particle would be given by an equation that is equivalent to Eqn (1.21) except for the velocity of light being replaced by the velocity of the particle.

$$\lambda = \frac{h}{mv} \tag{1.22}$$

In 1924, this was a result that had not been experimentally verified, but the verification was not long in coming. In 1927, C. J. Davisson and L. H. Germer conducted the experiments at Bell Laboratories in Murray Hill, New Jersey. A beam of electrons accelerated by a known voltage has a known velocity. When such a beam impinges on a crystal of nickel metal, a diffraction pattern is observed! Moreover, because the spacing between atoms in a nickel crystal is known, it is possible to calculate the wavelength of the moving electrons, and the value corresponds exactly to the wavelength predicted by the de Broglie equation. Since this pioneering work, electron diffraction has become one of the standard experimental techniques for studying molecular structure.

De Broglie's work clearly shows that a moving electron can be considered as a wave. If it behaves in that way, a stable orbit in a hydrogen atom must contain a whole number of wavelengths or otherwise there would be interference that would lead to cancellation (destructive interference). This condition can be expressed as

$$mvr = \frac{nh}{2\pi} \tag{1.23}$$

This is precisely the relationship that was required when Bohr assumed that the angular momentum of the electron is quantized for the allowed orbits.

Having now demonstrated that a moving electron can be considered as a wave, it remained for an equation to be developed to incorporate this revolutionary idea. Such an equation was obtained and solved by Erwin Schrödinger in 1926 when he made use of the particle–wave duality ideas of de Broglie even before experimental verification had been made. We will describe this new branch of science, wave mechanics, in Chapter 2.

1.5 ELECTRONIC PROPERTIES OF ATOMS

Although we have not yet described the modern methods of dealing with theoretical chemistry (quantum mechanics), it is possible to describe many of the properties of atoms. For example, the energy necessary to remove an electron (the *ionization energy* or *ionization potential*) from a hydrogen atom is the energy that is equivalent to the series limit of the Lyman Series. Therefore, atomic spectroscopy is one way to determine ionization potentials for atoms.

If we examine the relationship between the first ionization potentials for atoms and their atomic numbers, the result can be shown graphically as in Figure 1.9. Numerical values for ionization potentials are shown in Appendix A.

Several facts are apparent from this graph. Although we have not yet dealt with the topic of electron configuration of atoms, you should be somewhat familiar with this topic from earlier chemistry courses. We will make use of some of the ideas that deal with electron shells here but delay presenting the details until later.

1. The helium atom has the highest ionization potential of any atom. It has a nuclear charge of $+2$, and the electrons reside in the lowest energy level close to the nucleus.
2. The noble gases have the highest ionization potentials of any atoms in their respective periods. Electrons in these atoms are held in shells that are completely filled.
3. The Group IA elements have the lowest ionization potentials of any atoms in their respective periods. As you probably already know, these atoms have a single electron that resides in a shell outside of other shells that are filled.
4. The ionization potentials within a period generally increase as you go to the right in that period. For example, $B < C < O < F$, etc. However, in the case of nitrogen and oxygen, the situation is reversed. Nitrogen, which has a half-filled shell, has a higher ionization potential than oxygen, which has one electron more than a half-filled shell. There is some repulsion between the two electrons that reside in the same orbital in an oxygen atom, which makes it easier to remove one of them.

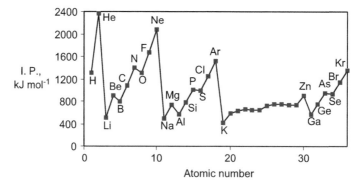

FIGURE 1.9
The relationship between first ionization potential and atomic number.

5. In general, the ionization potential decreases for the atoms in a given group going down in the group. For example, Li > Na > K > Rb > Cs and F > Cl > Br > I. The outer electrons are farther from the nucleus for the larger atoms, and there are more filled shells of electrons between the nucleus and the outermost electron.

6. Even for the atom having the *lowest* ionization potential, Cs, the ionization potential is approximately 374 kJ mol^{-1}.

These are some of the general trends that relate the ionization potentials of atoms with regard to their positions in the periodic table. We will have opportunities to discuss additional properties of atoms later.

A second property of atoms that is vital to understanding their chemistry is the energy released when an electron is added to a gaseous atom,

$$X(g) + e^-(g) \rightarrow X^-(g) \quad \Delta E = \text{electron addition energy} \tag{1.24}$$

For most atoms, the addition of an electron occurs with the *release* of energy so the value of ΔE is *negative*. There are some exceptions, most notably the noble gases and Group IIA metals. These atoms have completely filled shells so any additional electrons would have to be added in a new empty shell. Nitrogen also has virtually no tendency to accept an additional electron because of the stability of the half-filled outer shell.

After an electron is added to an atom, the "affinity" that it has for the electron is known as the *electron affinity*. Because energy is released when an electron is added to most atoms, it follows that energy would be required to remove the electron so the electron affinity is positive for most atoms. The electron affinities for most of the main group elements are shown in Table 1.1. It is useful to remember that 1 eV per atom is equal to 96.48 kJ mol^{-1}.

Several facts are apparent when the data shown in Table 1.1 are considered. In order to see some of the specific results more clearly, Figure 1.10 has been prepared to show how the electron affinity varies

Table 1.1 Electron Affinities of Atoms in kJ mol^{-1}

H 72.8							
Li 59.6	Be −18		B 26.7	C 121.9	N −7	O[a] 141	F 328
Na 52.9	Mg −21		Al 44	Si 134	P 72	S[b] 200	Cl 349
K 48.4	Ca −186	Sc ... Zn 18 ... 9	Ga 30	Ge 116	As 78	Se 195	Br 325
Rb 47	Sr −146	Y ... Cd 30 ... −26	In 30	Sn 116	Sb 101	Te 190	I 295
Cs 46	Ba −46	La ... Hg 50 ... −18	Tl 20	Pb 35	Bi 91	Po 183	At 270

[a] −845 kJ mol^{-1} for addition of two electrons.
[b] −531 kJ mol^{-1} for addition of two electrons.

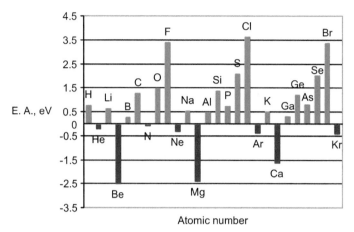

FIGURE 1.10
Electron affinity as a function of atomic number.

with position in the periodic table (and therefore orbital population). From studying Figure 1.10 and the data shown in Table 1.1, the following relationships emerge.

1. The electron affinities for the halogens are the highest of any group of elements.
2. The electron affinity generally increases in going from left to right in a given period. In general, the electrons have been added to the atoms in the same outer shell. Because the nuclear charge increases in going to the right in a period, the attraction for the outer electron increases accordingly.
3. In general, the electron affinity decreases going downward for atoms in a given group.
4. The electron affinity of nitrogen is out of line with those of other atoms in the same period because it has a stable half-filled shell.
5. Whereas nitrogen has an electron affinity that is approximately zero, phosphorus has a value greater than zero even though it also has a half-filled outer shell. The effect of a half-filled shell decreases for larger atoms because that shell has more filled shells separating it from the nucleus.
6. In the case of halogens (Group VIIA), the electron affinity of fluorine is *lower* than that of chlorine. This is because the fluorine atom is small and the outer electrons are close together and repelling each other. Adding another electron to an F atom, although very favorable energetically, is not as favorable as it is for chlorine, which has the highest electron affinity of any atom. For Cl, Br, and I, the trend is in accord with the general relationship.
7. Hydrogen has a substantial electron affinity, which shows that we might expect compounds containing H^- to be formed.
8. The elements in Group IIA have negative electron affinities showing that the addition of an electron to those atoms is not energetically favorable. These atoms have two electrons in the outer shell, which can hold only two electrons.
9. The elements in Group IA can add an electron with the release of energy (a small amount) because their singly occupied outer shells can hold two electrons.

Table 1.2 Atomic Radii in pm

H 37							
Li 134	Be 113		B 83	C 77	N 71	O 72	F 71
Na 154	Mg 138		Al 126	Si 117	P 110	S 104	Cl 99
K 227	Ca 197	Sc ... Zn 161 ... 133	Ga 122	Ge 123	As 125	Se 117	Br 114
Rb 248	Sr 215	Y ... Cd 181 ... 149	In 163	Sn 140	Sb 141	Te 143	I 133
Cs 265	Ba 217	La ... Hg 188 ... 160	Tl 170	Pb 175	Bi 15	Po 167	At —

As is the case with ionization potential, electron affinity is a useful property when considering the chemical behavior of atoms, especially when describing ionic bonding, which involves electron transfer.

In the study of inorganic chemistry, it is important to understand how atoms vary in size. The relative sizes of atoms determine to some extent the molecular structures that are possible. Table 1.2 shows the sizes of atoms in relationship to the periodic table.

Some of the important trends in the sizes of atoms can be summarized as follows.

1. The sizes of atoms in a given group increase as one progresses down the group. For example, the covalent radii for Li, Na, K, Rb, and Cs are 134, 154, 227, 248, and 265 pm, respectively. For F, Cl, Br, and I the covalent radii are 71, 99, 114, and 133 pm, respectively.
2. The sizes of atoms decrease in progressing across a given period. Nuclear charge increases in such a progression as long as electrons in the outer shell are contained in the same type of shell. Therefore, the higher the nuclear charge is (farther to the right in the period), the greater will be the attraction for the electrons and the closer to the nucleus they will reside. For example, the radii for the first long row of atoms are as follows.

Atom	Li	Be	B	C	N	O	F
Radius, pm	134	113	83	77	71	72	71

Other rows in the periodic table follow a similar trend. However, for the third long row, there is a general decrease in radius except for the last two or three elements in the transition series. The covalent radii of Fe, Co, Ni, Cu, and Zn are 126, 125, 124, 128, and 133 pm, respectively. This effect is a manifestation of the fact that the $3d$ orbitals shrink in size as the nuclear charge increases (going to the right), and the additional electrons populating these orbitals experience greater repulsion. As a result, the size decreases to a point (at Co and Ni), but after that the increase in repulsion produces an increase in size (Cu and Zn are larger than Co and Ni).

3. The largest atoms in the various periods are the Group IA metals. The outermost electron resides in a shell that is outside other completed shells (the noble gas configurations), so it is loosely held (low ionization potential) and relatively far from the nucleus.

An interesting effect of nuclear charge can be seen by examining the radius of a series of species that have the same nuclear charge but different numbers of electrons. One such series are the ions that have 10 electrons (the neon configuration). The ions include Al^{3+}, Mg^{2+}, Na^+, F^-, O^{2-}, and N^{3-}, for which the nuclear charge varies from 13 to 7. Figure 1.11 shows the variation in size of these species with nuclear charge.

Note that the N^{3-} ion (radius 171 pm) is much larger than the nitrogen atom, for which the covalent radius is only 71 pm. The oxygen atom (radius 72 pm) is approximately half the size of the oxide ion (radius 140 pm). Anions are always larger than the atoms from which they are formed. On the other hand, the radius of Na^+ (95 pm) is much smaller than the covalent radius of the Na atom (radius 154 pm). Cations are always smaller than the atoms from which they are formed.

Of particular interest in the series of ions is the Al^{3+} ion, which has a radius of only 50 pm, whereas the atom has a radius of 126 pm. As will be described in more detail later (see Chapter 7), the small size and high charge of the Al^{3+} ion causes it (and similar ions with high charge to size ratio or *charge density*) to have some very interesting properties. It has a great affinity for the negative ends of polar water molecules so that when an aluminum compound is dissolved in water, evaporating the water does not remove the water molecules that are bonded directly to the cation. The original aluminum compound is not recovered.

Because inorganic chemistry is concerned with the properties and reactions of compounds that may contain any element, understanding the relationships between properties of atoms is important. This topic will be revisited numerous times in later chapters, but the remainder of this chapter will be devoted to a brief discussion of the nuclear portion of the atom and nuclear transformations. We now know that it is not possible to express the weights of atoms as whole numbers that represent multiples of the mass of a hydrogen atom as had been surmised about two centuries ago. Although Dalton's atomic theory was based on the notion that all atoms of a given element were identical, we now know that this is not correct. As students in even elementary courses now know, the atomic masses represent

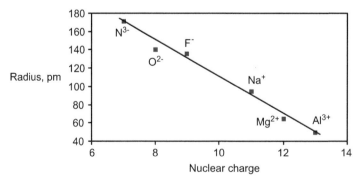

FIGURE 1.11
Radii of ions having the neon configuration.

averages resulting from most elements existing in several isotopes. The application of mass spectroscopy techniques has been of considerable importance in this type of study.

1.6 NUCLEAR BINDING ENERGY

There are at present 116 known chemical elements. However, there are well over 2000 known nuclear species as a result of several isotopes being known for each element. About three-fourths of the nuclear species are unstable and undergo radioactive decay. Protons and neutrons are the particles that are found in the nucleus. For many purposes, it is desirable to describe the total number of nuclear particles without regard to whether they are protons or neutrons. The term *nucleon* is used to denote both of these types of nuclear particles. In general, the radii of nuclides increase as the mass number increases, with the usual relationship being expressed as

$$R = r_0 A^{1/3} \tag{1.25}$$

where A is the mass number and r_0 is a constant that is approximately 1.2×10^{-13}.

Any nuclear species is referred to as a *nuclide*. Thus, $^{1}_{1}H$, $^{23}_{11}Na$, $^{12}_{6}C$, and $^{238}_{92}U$ are different recognizable species or nuclides. A nuclide is denoted by the symbol for the atom with the mass number written to the upper left, the atomic number written to the lower left, and any charge on the species, $q\pm$ to the upper right. For example,

$$^{A}_{Z}X^{q\pm}$$

As was described earlier in this chapter, the model of the atom consists shells of electrons surrounding the nucleus, which contains protons and, except for the isotope ^{1}H, a certain number of neutrons. Each type of atom is designated by the atomic number, Z, and a symbol derived from the name of the element. The mass number, A, is the whole number nearest to the mass of that species. For example, the mass number of $^{1}_{1}H$ is 1 although the actual mass of this isotope is 1.00794 atomic mass units (amu). Because protons and neutrons have masses that are essentially the same (both are approximately 1 atomic mass unit, amu), the mass number of the species minus the atomic number gives the number of neutrons, which is denoted as N. Thus, for $^{15}_{7}N$, the nucleus contains seven protons and eight neutrons.

When atoms are considered to be composed of their constituent particles, it is found that the atoms have lower masses than the sum of the masses of the particles. For example, $^{4}_{2}He$ contains two electrons, two protons, and two neutrons. These particles have masses of 0.0005486, 1.00728, and 1.00866 amu, respectively, which gives a total mass of 4.032977 amu for the particles. However, the actual mass of $^{4}_{2}He$ is 4.00260 amu so there is a *mass defect* of 0.030377 amu. That "disappearance" of mass occurs because the particles are held together with an energy that can be expressed in terms of the Einstein equation,

$$E = mc^2 \tag{1.26}$$

If 1 g of mass is converted to energy, the energy released is

$$E = mc^2 = 1\,g \times \left(3.00 \times 10^{10}\ cm/s\right)^2 = 9.00 \times 10^{20}\ erg$$

When the mass being converted to energy is 1 amu (1.66054×10^{-24} g), the amount of energy released is 1.49×10^{-3} erg. This energy can be converted to electron volts by making use of the conversion that $1\,\text{eV} = 1.60 \times 10^{-12}$ erg. Therefore, 1.49×10^{-3} erg/1.60×10^{-12} erg/eV is 9.31×10^{8} eV. When dealing with energies associated with nuclear transformations, energies are ordinarily expressed in MeV with 1 MeV being 10^{6} eV. Consequently, the energy equivalent to 1 amu is 931 MeV. When the mass defect of 0.030377 amu found for $^{4}_{2}\text{He}$ is converted to energy, the result is 28.3 MeV. In order to make a comparison between the stability of various nuclides, the total binding energy is usually divided by the number of nucleons, which in this case is 4. Therefore, the *binding energy per nucleon* is 7.07 MeV.

As an aside issue, it may have been noted that we neglected the attraction energy between the electrons and the nucleus. The first ionization energy for He is 24.6 eV and the second is 54.4 eV. Thus, the total binding energy of the electrons to the nucleus in He is only 79.9 eV, which is only 0.000079 MeV and is totally insignificant compared to the 28.3 MeV represented by the total binding energy. Attractions between nucleons are enormous compared to binding energies of electrons in atoms. Neutral atoms have the same number of electrons and protons, the combined mass of which is almost exactly the same as that of a hydrogen atom. Therefore, no great error is introduced when calculating mass defects by adding the mass of an appropriate number of hydrogen atoms to that of the number of neutrons. For example, the mass of $^{16}_{8}\text{O}$ can be approximated as the mass of eight hydrogen atoms and eight neutrons. The binding energy of the electrons in the eight hydrogen atoms is ignored.

When similar calculations are performed for many other nuclides, it is found that the binding energy per nucleon differs considerably. The value for $^{16}_{8}\text{O}$ is 7.98 MeV, and the highest value is approximately 8.79 MeV for $^{56}_{26}\text{Fe}$. This suggests that for a very large number of nucleons, the most stable arrangement is for them to make $^{56}_{26}\text{Fe}$, which is actually abundant in nature. Figure 1.12 shows the binding energy per nucleon as a function of mass number of the nuclides.

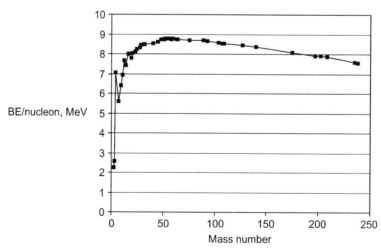

FIGURE 1.12
The average binding energy per nucleon as a function of mass number.

With the highest binding energy per nucleon being for species such as $^{56}_{26}$Fe, we can see that the *fusion* of lighter species to produce nuclides that are more stable should release energy. Because very heavy elements have lower binding energy per nucleon than do nuclides having mass number from about 50 to 80, fission of heavy nuclides is energetically favorable. One such nuclide is $^{235}_{92}$U, which undergoes fission when bombarded with low energy neutrons.

$$^{235}_{92}U + ^{1}_{0}n \rightarrow ^{92}_{36}Kr + ^{141}_{56}Ba + 3^{1}_{0}n \qquad (1.27)$$

When $^{235}_{92}$U undergoes fission, many different products are obtained because there is not a great deal of difference in the binding energy per nucleon for nuclides having a rather wide range of mass numbers. If the abundances of the products are plotted against the mass numbers, a double humped curve is obtained, and the so-called symmetric split of the $^{235}_{92}$U is not the most probable event. Fission products having atomic numbers in the ranges of 30−40 and 50−60 are much more common than two $_{46}$Pd isotopes.

1.7 NUCLEAR STABILITY

The atomic number, Z, is the number of protons in the nucleus. Both the proton and neutron have masses that are approximately 1 atomic mass unit, amu. The electron has a mass of only about 1/1837 of the proton or neutron so almost all of the mass of the atoms is made up by the protons and neutrons. Therefore, adding the number of protons to the number of neutrons gives the approximate mass of the nuclide in amu. That number is called the mass number and is given the symbol A. The number of neutrons is found by subtracting the atomic number, Z, from the mass number, A. Frequently, the number of neutrons is designated as N and $(A - Z) = N$. In describing a nuclide, the atomic number and mass number are included with the symbol for the atom. This is shown for an isotope of X as $^{A}_{Z}$X.

Although the details will not be presented here, there is a series of energy levels or shells where the nuclear particles reside. There are separate levels for the protons and neutrons. For electrons, the numbers 2, 10, 18, 36, 54, and 86 represent the closed shell arrangements (the noble gas arrangements). For nucleons, the closed shell arrangements correspond the numbers of 2, 8, 20, 28, 50, and 82 with a separate series for protons and neutrons. It was known early in the development of nuclear science that these numbers of nucleons represented stable arrangements although it was not known why these numbers of nucleons were stable. Consequently, they were referred to as *magic numbers*.

Another difference between nucleons and electrons is that nucleons pair whenever possible. Thus, even if a particular energy level can hold more than two particles, two particles will pair when they are present. Thus, for two particles in degenerate levels, we show two particles as ↑↓ rather than ↑ ↑. As a result of this preference for pairing, nuclei with even numbers of protons and neutrons have all paired particles. This results in nuclei that are more stable than those which have unpaired particles. The least stable nuclei are those in which *both* the number of neutrons and the number of protons is odd. This difference in stability manifests itself in the number of stable nuclei of each type. Table 1.3 shows the numbers of stable nuclei that occur.

Figure 1.13 shows graphically the relationship between the number of neutrons and the number of protons for the stable nuclei. The data show that there does not seem to be any appreciable difference

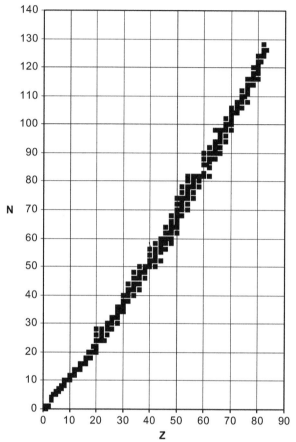

FIGURE 1.13
The relationship between the number of neutrons and protons for stable nuclei.

Table 1.3 Numbers of Stable Nuclides Having Different Arrangements of Nucleons

Z	N	Number of Stable Nuclides
Even	Even	162
Even	Odd	55
Odd	Even	49
Odd	Odd	4

in stability when the number of protons or neutrons is even whereas the other is odd (the even–odd and odd–even cases). The small number of nuclides that have odd Z and odd N (so-called odd–odd nuclides) is very small, which indicates that there is an inherent instability in such an arrangement. The most common stable nucleus which is of the odd–odd type is $^{14}_{7}N$.

1.8 TYPES OF NUCLEAR DECAY

We have already stated that the majority of known nuclides are unstable and undergo some type of decay to produce another nuclide. The starting nuclide is known as the parent and the nuclide produced is known as the daughter. The most common types of decay processes will now be described.

When the number of neutrons is compared to the number of protons that are present in all stable nuclei, it is found that they are approximately equal up to atomic number 20. For example, in $^{40}_{20}$Ca it is seen that $Z = N$. Above atomic number 20, the number of neutrons is generally greater than the number of protons. For $^{235}_{92}$U, $Z = 92$, but $N = 143$. In Figure 1.13, each small square represents a stable nuclide. It can be seen that there is a rather narrow band of stable nuclei with respect to Z and N, and that the band gets farther away from the line representing $Z = N$ as the atomic number increases. When a nuclide lies outside the band of stability, radioactive decay occurs in a manner that brings the daughter into or closer to the band of stability.

1. Beta ($-$) decay (β^-): When we consider $^{14}_{6}$C, we see that the nucleus contains six protons and eight neutrons. This is somewhat "rich" in neutrons so the nucleus is unstable. Decay takes place in a manner that decreases the number of neutrons and increases the number of protons. The type of decay that accomplishes this is the emission of a β^- particle as a neutron in the nucleus is converted into a proton. The β^- particle is simply an electron. The beta particle that is emitted is an electron that is produced as a result of a neutron in the nucleus being transformed into a proton, which remains in the nucleus.

$$n \rightarrow p^+ + e^- \tag{1.28}$$

The ejected electron did not exist before the decay, and it is not an electron from an orbital. One common species that undergoes β^- decay is $^{14}_{6}$C,

$$^{14}_{6}C \rightarrow ^{14}_{7}N + e^- \tag{1.29}$$

In this decay process, the mass number stays the same because the electron has a mass that is only 1/1837 of the mass of the proton or neutron. However, the nuclear charge increases by 1 unit as the number of protons is increased by one. As we shall see later, this type of decay process takes place when the number of neutrons is somewhat greater than the number of protons.

Nuclear decay processes are often shown by means of diagrams that resemble energy level diagrams with the levels displaced to show the change in atomic number. The parent nucleus is shown at a higher energy than the daughter. The x-axis is really the value of Z with no values indicated. The decay of $^{14}_{6}$C can be shown as follows.

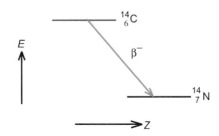

2. Beta (+) or positron emission (β^+): This type of decay is one that occurs when a nucleus has a greater number of protons than neutrons. In this process, a proton is converted into a neutron by emitting a positive particle known as a β^+ particle or *positron*. The positron is a particle having the mass of an electron but carrying a positive charge. It is sometimes called the antielectron and shown as e^+. The reaction can be shown as

$$p^+ \rightarrow n + e^+ \tag{1.30}$$

One nuclide that undergoes β^+ decay is $^{14}_8O$,

$$^{14}_8O \rightarrow {}^{14}_7N + e^+ \tag{1.31}$$

In β^+ decay, the mass number remains the same but the number of protons decreases by one although the number of neutrons increases by one. The decay scheme for this process is shown as follows.

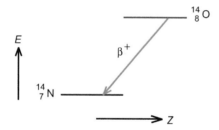

In this case, the daughter is written to the left of the parent because the nuclear charge is decreasing.

3. Electron capture (EC): In this type of decay, an electron from outside the nucleus is captured by the nucleus. Such a decay mode occurs when there is a greater number of protons than neutrons in the nucleus.

$$^{64}_{29}Cu \xrightarrow{E.C.} {}^{i64}_{28}Ni \tag{1.32}$$

In electron capture, the nuclear charge decreases by 1 because what happens is that a proton in the nucleus interacts with the electron to produce a neutron.

$$\underset{\substack{\text{inside} \\ \text{nucleus}}}{p^+} + \underset{\substack{\text{outside} \\ \text{nucleus}}}{e^-} \rightarrow \underset{\substack{\text{inside} \\ \text{nucleus}}}{n} \tag{1.33}$$

In order for this to occur, the orbital electron must be very close to the nucleus. Therefore, electron capture is generally observed when the nucleus has a charge of $Z \approx 30$. However, a few cases are known in which the nucleus has a considerably smaller charge than this. Because the electron that is captured is one in the shell closest to the nucleus, the process is sometimes called K-capture. Note that electron capture and β^+ decay accomplish the same changes in the nucleus. Therefore, they are sometimes competing processes and the same nuclide may decay simultaneously by both processes.

4. Alpha (α) decay: As we shall see later, the alpha particle, which is a helium nucleus, is a stable particle. For some unstable heavy nuclei, the emission of this particle occurs. Because the α particle contains a magic number of both protons and neutrons (2), there is a tendency for this particular combination of particles to be the one emitted rather than some other combination, such as ^6_3Li, etc. In alpha decay, the mass number decreases by 4 units, the number of protons decreases by two and the number of neutrons decreases by two. An example of alpha decay is the following.

$$^{235}_{92}\text{U} \rightarrow {}^{231}_{90}\text{Th} + \alpha \qquad (1.34)$$

5. Gamma emission (γ): Gamma rays are high-energy photons that are emitted as an atomic nucleus undergoes de-excitation. This situation is entirely equivalent to the spectral lines emitted by atoms as electrons fall from higher energy states to lower ones. In the case of gamma emission, the de-excitation occurs as a proton or neutron that is in an excited state falls to a lower nuclear energy state. However, the question naturally arises as to how the nuclide attains the higher energy state. The usual process is that the excited nuclear state results from some other event. For example, $^{38}_{17}\text{Cl}$ decays by β^- emission to $^{38}_{18}\text{Ar}$, but this nuclide exists in an excited state. Therefore, it is designated as $^{38}_{18}\text{Ar}^*$ and it relaxes by the emission of gamma rays. A simplified decay scheme can be shown as follows.

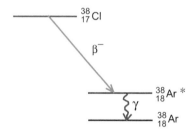

Gamma emission almost always follows some other decay process that results in an excited state in the daughter nucleus due to a nucleon being in a state above the ground state.

The decay of $^{226}_{88}\text{Ra}$ to $^{222}_{86}\text{Rn}$ can occur to either the ground state of the daughter or to an excited state that is followed by emission of a γ ray. The equation and energy diagram for this process are shown as follows.

$$^{226}_{88}\text{Ra} \rightarrow {}^{222}_{86}\text{Rn} + \alpha \qquad (1.35)$$

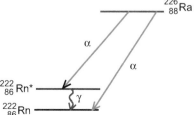

1.9 PREDICTING DECAY MODES

For light nuclei, there is a strong tendency for the number of protons to be approximately equal to the number of neutrons. In many stable nuclides, the numbers are exactly equal. For example, ^4_2He, $^{12}_6C$, $^{16}_8O$, $^{20}_{10}Ne$, and $^{40}_{20}Ca$ are all stable nuclides. In the case of the heavier stable nuclides, the number of neutrons is greater than the number of protons. Nuclides such as $^{64}_{30}Zn$, $^{208}_{82}Pb$, and $^{235}_{92}U$ all have a larger number of neutrons than protons with the difference increasing as the number of protons increases. If a graph is made of the number of protons vs the number of neutrons and all the stable nuclides are located on the graph, it is seen that the stable nuclides fall in a rather narrow band. This band is sometimes referred to as the band of stability. Figure 1.13 shows this relationship. If a nucleus has a number of either type of nuclide that places it outside this band, the nuclide will undergo a type of decay that will bring it into the band. For example, $^{14}_6C$ has six protons and eight neutrons. This excess of neutrons over protons can be corrected by a decay process that transforms a neutron into a proton. Such a decay scheme can be summarized as

$$n \rightarrow p^+ + e^- \tag{1.36}$$

Therefore, $^{14}_6C$ undergoes radioactive decay by β^- emission,

$$^{14}_6C \rightarrow ^{14}_7N + e^- \tag{1.37}$$

On the other hand, $^{14}_8O$ has eight protons but only six neutrons. This imbalance of protons and neutrons can be corrected if a proton is transformed into a neutron as is summarized by the equation

$$p^+ \rightarrow n + e^+ \tag{1.38}$$

Thus, $^{14}_8O$ undergoes decay by positron emission,

$$^{14}_8O \rightarrow ^{14}_7N + e^+ \tag{1.39}$$

Electron capture accomplishes the same end result as positron emission, but because the nuclear charge is low, positron emission is the expected decay mode. Generally, electron capture is not a competing process unless $Z \approx 30$ or so.

Figure 1.14 shows how the transformations shown in Eqns (1.37) and (1.39) are related to the relationship between numbers of protons and neutrons for the two decay processes. The point labeled a on the graph and the arrow starting from that point show the decay of $^{14}_6C$. Point b on the graph represents $^{14}_8O$, and the decay is indicated by the arrow.

Although the use of the band of stability to predict stability is straightforward, there are further applications of the principles discussed that are useful also. For example, consider the following cases.

$^{34}_{14}Si$	$t_{1/2} = 2.8$ s
$^{33}_{14}Si$	$t_{1/2} = 6.2$ s
$^{32}_{14}Si$	$t_{1/2} = 100$ yr

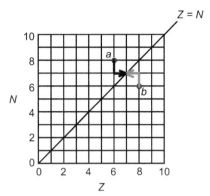

FIGURE 1.14

Predicting decay mode from the relative number of protons and neutrons.

Although all three of these isotopes of silicon are radioactive, the heaviest of them, ^{34}Si, lies farthest from the band of stability and it has the shortest half-life. Generally, the farther a nuclide lies from the band of stability, the shorter its half-life is. There are numerous exceptions to this general rule, and we will discuss some of them here. First, consider these cases:

(even–odd)	$^{27}_{12}\text{Mg}$	$t_{1/2} = 9.45$ min
(even–even)	$^{28}_{12}\text{Mg}$	$t_{1/2} = 21.0$ h

Although ^{28}Mg is farther from the band of stability than is ^{27}Mg, the former is an even–even nuclide whereas the latter is an even–odd nuclide. As we have seen earlier, even–even nuclides tend to be more stable. Consequently, the even–even effects here outweigh the fact that ^{28}Mg is farther from the band of stability. Another interesting case is shown by considering these isotopes of chlorine.

(odd–odd)	$^{38}_{17}\text{Cl}$	$t_{1/2} = 37.2$ min
(odd–even)	$^{39}_{17}\text{Cl}$	$t_{1/2} = 55.7$ min

In this case, the $^{38}_{17}\text{Cl}$ is an odd–odd nucleus whereas $^{39}_{17}\text{Cl}$ is an odd–even nucleus. Thus, even though $^{39}_{17}\text{Cl}$ is farther away from the band of stability, it has a slightly longer half-life. Finally, let us consider two cases where both of the nuclei are similar in terms of numbers of nucleons. Such cases are the following:

(odd–even)	$^{39}_{17}\text{Cl}$	$t_{1/2} = 37.2$ min
(even–odd)	$^{39}_{18}\text{Ar}$	$t_{1/2} = 259$ yrs

In this case, there is no real difference with respect to the even/odd character. The large difference in half-life is related to the fact that $^{39}_{17}\text{Cl}$ is farther from the band of stability than is $^{39}_{18}\text{Ar}$. This is in accord with the general principle stated earlier. Although specific cases might not follow the general trend, it is generally true that the farther a nuclide is from the band of stability, the shorter its half-life will be.   In some cases, a nuclide may undergo decay by more than one process at the same time. For example, ^{64}Cu undergoes decay by three processes simultaneously.

$$^{64}_{28}\text{Ni} \text{ by electron capture, 19\%}$$
$$\nearrow$$
$$^{64}_{29}\text{Cu} \longrightarrow \ ^{64}_{28}\text{Ni by } \beta^+ \text{ emission, 42\%}$$
$$\searrow$$
$$^{64}_{30}\text{Zn} \text{ by } \beta^- \text{ emission, 39\%}$$

The rate of disappearance of ^{64}Cu is by all three processes, but by making use of different types of counting methods, it is possible to separate the rates of the processes.

There are three naturally occurring radioactivity series that consist of a sequence of steps that involve α and β decay until a stable nuclide results. The uranium series involves the decay of $^{238}_{92}$U in a series of steps that eventually produces $^{206}_{82}$Pb. Another series involves $^{235}_{92}$U that decays in a series of steps that ends in $^{207}_{82}$Pb, which is stable. In the thorium series, $^{232}_{90}$Th is converted into $^{208}_{82}$Pb. Although there are other individual nuclides that are radioactive, these are the three prominent decay series.

The introduction to characteristics of nuclei presented here is intimately related to how certain species are important to chemistry (such as dating materials by determining their carbon-14 content). Also, the application of isotopic tracers is a useful technique that will be illustrated in later chapters.

References for Further Study

Blinder, S.M., 2004. *Introduction to Quantum Mechanics in Chemistry, Materials Science, and Biology*, Academic Press, San Diego. A good survey book that shows the applications of quantum mechanics to many areas of study.

Emsley, J., 1998. *The Elements*, 3rd ed., Oxford University Press, New York. This book presents a wealth of data on properties of atoms.

House, J.E., 2004. *Fundamentals of Quantum Chemistry*. Elsevier, New York. An introduction to quantum mechanical methods at an elementary level that includes mathematical details.

Krane, K., 1995. *Modern Physics*, 2nd ed., Wiley, New York. A good introductory book that described developments in atomic physics.

Loveland, W.D., Morrissey, D., Seaborg, G.T., 2006. *Modern Nuclear Chemistry*, Wiley, New York.

Serway, R.E., 2000. *Physics for Scientists and Engineers*, 5th ed., Saunders (Thompson Learning), Philadelphia. An outstanding physics text that presents an excellent treatment of atomic physics.

Sharpe, A.G., 1992. *Inorganic Chemistry*, Longman, New York. Chapter 2 presents a good account of the development of the quantum mechanical way of doing things in chemistry.

Warren, W.S., 2000. *The Physical Basis of Chemistry*, 2nd ed., Academic Press, San Diego, CA. Chapter 5 presents the results of some early experiments in atomic physics.

QUESTIONS AND PROBLEMS

1. If a short-wave radio station broadcasts on a frequency of 9.065 megahertz (MHz), what is the wavelength of the radio waves?

2. Calculate the wavelength and frequency of the first three lines in the Balmer Series. What would be the values for the series limit?

3. One of the lines in the line spectrum of mercury has a wavelength of 435.8 nm. (a) What is the frequency of this line? (b) What is the wave number for the radiation? (c) What energy (in kJ mol^{-1}) is associated with this radiation?

4. The ionization potential for the NO molecule is 9.25 eV. What is the wavelength of a photon that would just ionize NO with the ejected electron having no kinetic energy?

5. What is the de Broglie wavelength of an electron (mass 9.1×10^{-28} g) moving at 1.5% of the velocity of light?

6. What energy is associated with an energy change in a molecule that results in an absorption at 2100 cm^{-1}? (a) in ergs; (b) in joules; (c) in kJ mol^{-1}.

7. What wavelength of light will just eject an electron from the surface of a metal that has a work function of 2.75 eV?

8. If light having a wavelength of 2537 Å falls on the surface of a copper plate, the ejected electrons have an energy of 0.20 eV. What is the longest wavelength of light that could eject electrons from copper?

9. If an electron in a hydrogen atom falls from the state with $n = 5$ to that where $n = 3$, what is the wavelength of the photon emitted?

10. If a moving electron has a kinetic energy of 2.35×10^{-12} erg, what would be its de Broglie wavelength?

11. The work function for barium is 2.48 eV. If light having a wavelength of 400 nm shines on a barium cathode, what is the maximum velocity of the ejected electrons?

12. If a moving electron has a velocity of 3.55×10^5 m sec^{-1}, what is its de Broglie wavelength?

13. What is the velocity of the electron in the first Bohr orbit?

14. In each of the following pairs, select the one that has the highest first ionization potential. (a) Li or Be; (b) Al or F; (c) Ca or P; (d) Zn or Ga

15. In each of the following pairs, which species is larger? (a) Li$^+$ or Be^{2+}; (b) Al^{3+} or F$^-$; (c) Na$^+$ or Mg^{2+}; (d) S^{2-} or F$^-$.

16. In each of the following pairs, which atom releases a greater amount of energy when an electron is added? (a) P or C; (b) N or Na; (c) H or I; (d) S or Si.

17. The bond energy in H$_2^+$ is 256 kJ mol^{-1}. What wavelength of electromagnetic radiation would have enough energy to dissociate H$_2^+$?

18. For the HCl molecule, the first excited vibrational state is 2886 cm^{-1} above the ground state. How much energy is this in erg/molecule? in kJ mol^{-1}?

19. The ionization potential for the PCl$_3$ molecule is 9.91 eV. What is the frequency of a photon that will just remove an electron from a PCl$_3$ molecule? In what spectral region would such a photon be found? From which atom in the molecule is the electron removed?

20. Arrange the following in the order of increasing first ionization potential: B, Ne, N, O, P.

21. Explain why the first ionization potentials for P and S differ by only 12 kJ mol^{-1} (1012 and 1000 kJ mol^{-1}, respectively) whereas those for N and O differ by 88 kJ mol^{-1} (1402 and 1314 kJ mol^{-1}, respectively).

22. Arrange the following in the order of increasing first ionization potential: H, Li, C, F, O, N.

23. Arrange the following in the order of increasing first ionization potential: Ne, Li, Na, F, S, Mg.

24. Arrange the following in the order of decreasing amount of energy released when an electron is added to the atom: O, F, N, Cl, S, Br

25. Arrange the following in the order of decreasing size: Cl, O, I$^-$, O^{2-}, Mg^{2+}, F$^-$.

26. Calculate the binding energy per nucleon for the following: $^{18}_{8}$O; $^{23}_{11}$Na; $^{40}_{20}$Ca.

27. Predict the decay mode for the following and write the reaction for the predicted decay mode. (a) $^{35}_{16}$S; (b) $^{17}_{9}$F; (c) $^{43}_{20}$Ca.

28. How much energy (in MeV) would be released by the fusion of three $^{4}_{2}$He nuclei to produce $^{12}_{6}$C?

CHAPTER 2

Basic Quantum Mechanics and Atomic Structure

In the previous chapter, we saw that the energies of electron orbits are quantized. It was also mentioned that dealing with an electron in an atom would require considering the wave character of the moving particle. The question remains as to how we proceed when formulating and solving such a problem. The procedures and methods employed constitute the branch of science known as quantum mechanics or wave mechanics. In this chapter, we will present only a very brief sketch of this important topic because it is assumed that readers of this book will have studied quantum mechanics in a physical chemistry course. The coverage here is meant to provide an introduction to the terminology and basic ideas of quantum mechanics or, preferably, to provide a review.

2.1 THE POSTULATES

To systematize the procedures and basic premises of quantum mechanics, a set of postulates have been developed that provides the usual starting point for studying the topic. Most books on quantum mechanics give a precise set of rules and interpretations, some of which are not necessary for the study of inorganic chemistry at this level. In this section, we will present the postulates of quantum mechanics and provide some interpretation of them, but for complete coverage of this topic, the reader should consult a quantum mechanics text, such as those listed in the references at the end of this chapter.

Postulate I: For any possible state of a system, there exists a wave function, Ψ, that is a function of the coordinates of the parts of the system and time that completely describes the system.

This postulate establishes that the description of the system will be in the form of a mathematical function. If the coordinates used to describe the system are Cartesian coordinates, the function Ψ will contain these coordinates and the time as variables. For a very simple system that consists of only a single particle, the function Ψ, known as the wave function, can be written as

$$\Psi = \Psi(x, y, z, t). \tag{2.1}$$

If the system consists of two particles, the coordinates must be specified for each of the particles, which results in a wave function that is written as

$$\Psi = \Psi(x_1, y_1, z_1, x_2, y_2, z_2, t). \tag{2.2}$$

As a general form of the wave function, we write

$$\Psi = \Psi(q_i, t), \tag{2.3}$$

33

Inorganic Chemistry. DOI: http://dx.doi.org/10.1016/B978-0-12-385110-9.00002-9

where the q_i are appropriate coordinates for the particular system. With the specific form of the coordinates not being specified, the q_i are referred to as *generalized* coordinates. Because Ψ describes the system in some particular state, the state is known as the *quantum state* and Ψ is called the *state function* or *complete wave function*.

There needs to be some physical interpretation of the wave function and its relationship to the state of the system. One interpretation is that the square of the wave function, Ψ^2, is proportional to the probability of finding the parts of the system in a specified region of space. For some problems in quantum mechanics, differential equations arise that can have solutions that are complex (contain $(-1)^{1/2} = i$). In such a case, we use $\Psi^*\Psi$, where Ψ^* is the complex conjugate of Ψ. The complex conjugate of a function is the function that results when i is replaced by $-i$. Suppose we square the function $(a + ib)$:

$$(a + ib)^2 = a^2 + 2aib + i^2b^2 = a^2 + 2aib - b^2. \tag{2.4}$$

Because the expression obtained contains i, it is still a complex function. Suppose, however, that instead of squaring $(a + ib)$ we multiply by its complex conjugate, $(a - ib)$:

$$(a + ib)(a - ib) = a^2 - i^2b^2 = a^2 + b^2. \tag{2.5}$$

The expression obtained by this procedure is a real function. Thus, in many instances, we will use the product $\Psi^*\Psi$ instead of Ψ^2; although if Ψ is real, the two are equivalent.

For a system of particles, there is complete certainty that the particles are *somewhere* in the system. The probability of finding a particle in a volume element, $d\tau$, is given by $\Psi^*\Psi\, d\tau$ so that the total probability is obtained from the integration

$$\int \Psi^* \Psi\, d\tau. \tag{2.6}$$

An event that is impossible has a probability of zero, and a "sure thing" has a probability of 1. For a given particle in the system, the probability of finding the particle in all the volume elements that make up all space must add up to 1. Of course, the way of summing the volume elements is by performing an integration. Therefore, we know that

$$\int_{\text{All space}} \Psi^* \Psi\, d\tau = 1. \tag{2.7}$$

When this condition is met, we say that the wave function, Ψ, is *normalized*. In fact, this is the definition of a normalized wave function.

However, there are other requirements on Ψ that lead to it being a "well-behaved" wave function. For example, the above-mentioned integral must equal 1.00 ... so the wave function cannot be infinite. As a result, we say that Ψ must be *finite*. Another restriction on Ψ relates to the fact that there can be only one probability of finding a particle in a particular place. As an example, there is only one probability of finding an electron at a particular distance from the nucleus in a hydrogen atom. Therefore, we say that the wave function must be *single valued* so that there results only one value for the probability. Finally, we must take into account the fact that the probability does not vary in an abrupt way. Increasing the distance by 1% should not cause a 50% change in probability. The requirement is expressed by saying that Ψ must be *continuous*. Probability varies in some continuous manner, not abruptly. A wave function is said to be well behaved if it has the characteristics of being finite, single valued, and continuous.

Another concept that is important when considering wave functions is that of *orthogonality*. If the functions ϕ_1 and ϕ_2 are related so that

$$\int \phi_1^* \phi_2 d\tau = 0 \quad \text{or} \quad \int \phi_1 \phi_2^* d\tau = 0 \tag{2.8}$$

the functions are said to be *orthogonal*. In this case, changing the limits of integration may determine whether such a relationship exists. For Cartesian coordinates, the limits of integration are $-\infty$ to $+\infty$ for the coordinates x, y, and z. For a system described in terms of polar coordinates (r, θ, and ϕ), the integration limits are the ranges of those variables, $0 \to \infty$, $0 \to \pi$, and $0 \to 2\pi$, respectively.

Postulate II: For every dynamical variable (also known as a classical observable), there exists a corresponding operator.

Quantum mechanics is concerned with operators. An *operator* is a symbol that indicates that some mathematical action must be performed. For example, some operators that are familiar include $(x)^{1/2}$ (taking the square root of x), $(x)^2$ (squaring x), dy/dx (taking the derivative of y with respect to x), and so on. Physical quantities such as momentum, angular momentum, position coordinates, and energy are known as dynamical variables (classical observables for a system), each of which has a corresponding operator in quantum mechanics. Coordinates are identical in both operator and classical forms. For example, the coordinate r is simply r in either case. On the other hand, the operator for momentum in the x direction (p_x) is $(\hbar/i)/(d/dx)$. The operator for the z component of angular momentum (in polar coordinates) is $(\hbar/i)/(d/d\phi)$. Kinetic energy, $\frac{1}{2}mv^2$, can be written in terms of momentum, p, as $p^2/2m$ so it is possible to arrive at the operator for kinetic energy in terms of the operator for momentum. Table 2.1 shows some of the operators that are necessary for an introductory study of quantum mechanics. Note that the operator for kinetic energy is obtained from the momentum operator because the kinetic energy, T, can be expressed as $p^2/2m$. Note also that the operator for potential energy is expressed in terms of the generalized coordinates, q_i, because the form of the potential energy depends on the system. For example, the electron in a hydrogen atom has a potential energy, $-e^2/r$, where e is the charge on the electron. Therefore, the operator for the potential energy is simply $-e^2/r$ that is unchanged from the classical form.

Operators have properties that can be expressed in mathematical terms. If an operator is *linear*, it means that

$$\alpha(\phi_1 + \phi_2) = \alpha\phi_1 + \alpha\phi_2, \tag{2.9}$$

where ϕ_1 and ϕ_2 are functions that are being operated on by the operator α. Another property of the operators that is often useful in quantum mechanics is that when C is a constant,

$$\alpha(C\phi) = C(\alpha\phi). \tag{2.10}$$

An operator is Hermitian if

$$\int \phi_1^* \alpha \phi_2 \, d\tau = \int \phi_2 \alpha^* \phi_1 \, d\tau. \tag{2.11}$$

It can be shown that if an operator, α, meets this condition, the quantities calculated will be real rather than complex or imaginary. Although it is stated without proof, all of the operators to be discussed subsequently meet these conditions.

Table 2.1 Some Common Operators in Quantum Mechanics

Quantity	Symbol Used	Operator Form
Coordinates	x, y, z, r	x, y, z, r
Momentum		
x	p_x	$\dfrac{\hbar}{i}\dfrac{\partial}{\partial x}$
y	p_y	$\dfrac{\hbar}{i}\dfrac{\partial}{\partial y}$
z	p_z	$\dfrac{\hbar}{i}\dfrac{\partial}{\partial z}$
Kinetic energy	$\dfrac{p^2}{2m}$	$-\dfrac{\hbar^2}{2m}\left(\dfrac{\partial^2}{\partial^2 x}+\dfrac{\partial^2}{\partial^2 y}+\dfrac{\partial^2}{\partial^2 z}\right)$
Kinetic energy	T	$-\dfrac{\hbar}{i}\dfrac{\partial}{\partial t}$
Potential energy	V	$V(q_i)$
Angular momentum		
L_z (Cartesian)		$\dfrac{\hbar}{i}\left(x\dfrac{\partial}{\partial y}-y\dfrac{\partial}{\partial x}\right)$
L_z (polar)		$\dfrac{\hbar}{i}\dfrac{\partial}{\partial \phi}$

Postulate III: The permissible values that a dynamical variable may have are those given by $\alpha\phi = a\phi$, where α is an operator corresponding to the dynamical variable whose permissible values are a and ϕ is an eigenfunction of the operator α.

When reduced to an equation, Postulate III can be written as:

$$\underset{\substack{\text{operator}}}{\alpha}\ \underset{\substack{\text{wave}\\\text{function}}}{\phi}\ =\ \underset{\substack{\text{constant}\\\text{(eigenvalue)}}}{a}\ \underset{\substack{\text{wave}\\\text{function}}}{\phi}\ . \qquad (2.12)$$

When an operator operates on a wave function to produce a constant times the original wave function, the function is said to be an *eigenfunction* of that operator. In terms of the equation above, the operator α operating on ϕ yields a constant a times the original wave function. Therefore, ϕ is an eigenfunction of the operator α with an *eigenvalue* of a. We can use several examples to illustrate these ideas.

Suppose $\phi = e^{ax}$, where a is a constant and the operator is $\alpha = d/dx$. Then

$$d\phi/dx = a\,e^{ax} = (\text{constant})\,e^{ax}. \qquad (2.13)$$

Thus, we see that the function e^{ax} is an eigenfunction of the operator d/dx with an eigenvalue of a. If we consider the operator $(\ )^2$ operating on the same function, we find

$$(e^{ax})^2 = e^{2ax}, \qquad (2.14)$$

which does not represent the original function multiplied by a constant. Therefore, e^{ax} is not an eigenfunction of the operator $(\ )^2$. When we consider the function $e^{in\phi}$ (where n is a constant) being operated on by the operator for the z component of angular momentum, $(\hbar/i)(d/d\phi)$, where $h/2\pi$ is represented as $\hbar$ (read as "h bar"), we find

$$\frac{\hbar}{i}\left[\frac{d(e^{in\phi})}{d\phi}\right] = n\frac{\hbar}{i}e^{in\phi} \tag{2.15}$$

which shows that the function $e^{in\phi}$ is an eigenfunction of the operator for the z component of angular momentum.

One of the most important techniques in quantum mechanics is known as the variation method. That method provides a way of starting with a wave function and calculating a value for a property (dynamical variable or classical observable) by making use of the operator for that variable. It begins with the equation $\alpha\phi = a\phi$. Multiplying both sides of the equation by ϕ^*, we obtain

$$\phi^*\alpha\phi = \phi^*a\phi. \tag{2.16}$$

Because α is an *operator*, it is not necessarily true that $\phi^*\alpha\phi$ is equal to $\phi a\phi^*$ so the order of writing the symbols is preserved. We now perform integration using the equation in the form

$$\int_{\text{All space}} \phi^*\alpha\phi\, d\tau = \int_{\text{All space}} \phi^*a\phi\, d\tau. \tag{2.17}$$

However, the eigenvalue a is a constant so it can be removed from the integral on the right-hand side of the equation. Then, solving for a gives

$$\langle a \rangle = \frac{\int \phi^*\alpha\phi\, d\tau}{\int \phi^*\phi\, d\tau}. \tag{2.18}$$

The order of the quantities in the numerator must be preserved because α is an operator. For example, if an operator is d/dx, it is easy to see that $(2x)(d/dx)(2x) = (2x)(2) = 4x$ is not the same as $(d/dx)(2x)(2x) = (d/dx)(4x^2) = 8x$.

If the wave function ϕ is normalized, the denominator in Eqn (2.18) is equal to 1. Therefore, the value of a is given by the relationship

$$\langle a \rangle = \int \phi^*\alpha\phi\, d\tau \tag{2.19}$$

The value of a calculated in this way is known as the *average* or *expectation* value, and it is indicated by $\bar{a}$ or $<a>$. The operator to be used is the one that corresponds to the variable being calculated.

The utilization of the procedure outlined above will be illustrated by considering an example that is important as we study the atomic structure. In polar coordinates, the normalized wave function for the electron in the $1s$ state of a hydrogen atom is

$$\psi_{1s} = \frac{1}{\sqrt{\pi}}\frac{1}{a_0^{3/2}}e^{-r/a_0} = \psi_{1s}^* \tag{2.20}$$

where a_0 is a constant known as the first Bohr radius. Note that in this case the wave function is real (meaning that it does not contain i) so ψ and ψ^* are identical. We can now calculate the average value for the radius of the $1s$ orbit by making use of

$$\langle r \rangle = \int \psi^*(\text{operator})\psi \, d\tau, \tag{2.21}$$

where the operator r is the same in operator or classical form. The volume element, $d\tau$, in polar coordinates is $d\tau = r^2 \sin\theta \, dr \, d\theta \, d\phi$. Making the substitutions for $d\tau$ and the operator, we obtain

$$\langle r \rangle = \int_0^\infty \int_0^\pi \int_0^{2\pi} \frac{1}{\sqrt{\pi}} \left(\frac{1}{a_0}\right)^{3/2} e^{-r/a_0} (r) \frac{1}{\sqrt{\pi}} \left(\frac{1}{a_0}\right)^{3/2} e^{-r/a_0} r^2 \sin\theta \, dr \, d\theta \, d\phi. \tag{2.22}$$

Although this integral looks as if it might be very difficult to evaluate, it becomes much simpler when some factors are combined. For example, the r that is the operator is the same as the r in the volume element. Combining these gives a factor of r^3. Next, the factors involving π and a_0 can be combined. When these simplifications are made, the integral can be written as

$$\langle r \rangle = \int_0^\infty \int_0^\pi \int_0^{2\pi} \frac{1}{\pi a_0^3} e^{-2r/a_0} r^3 \sin\theta \, dr \, d\theta \, d\phi \tag{2.23}$$

We can further simplify the problem by recalling the relationship from calculus that

$$\int \int f(x)g(y)dx \, dy = \int f(x)dx \int g(y)dy \tag{2.24}$$

allows us to write the integral above in the form

$$\langle r \rangle = \int_0^\infty \frac{1}{\pi a_0^3} e^{-2r/a_0} r^3 dr \int_0^\pi \int_0^{2\pi} \sin\theta \, d\theta d\phi \tag{2.25}$$

From a table of integrals, it is easy to verify that the integral involving the angular coordinates can be evaluated to give 4π. Also from a table of integrals, it can be found that the exponential integral can be evaluated by making use of a standard form,

$$\int_0^\infty x^n e^{-bx} \, dx = \frac{n!}{b^{n+1}} \tag{2.26}$$

For the integral being evaluated here, $b = 2/a_0$ and $n = 3$ so that the exponential integral can be written as

$$\int_0^\infty r^3 e^{-2r/a_0} dr = \frac{3!}{\left(\dfrac{2}{a_0}\right)^4} \tag{2.27}$$

Making the substitutions for the integrals and simplifying, we find that

$$\langle r \rangle = (3/2)a_0, \tag{2.28}$$

where a_0, the first Bohr radius, is 0.529 Å. The expectation value for the radius of the $1s$ orbit is 1.5 times that distance. Note that a complicated looking problem actually turns out to be much simpler than might have been supposed. In this case, looking up two integrals in a table removes the necessity for integrating the functions by brute force. A great deal of elementary quantum mechanics can be handled in this way.

It may have been supposed that the average value for the radius of the first orbit in a hydrogen atom should evaluate to 0.529 Å. The answer to why it does not lies in the fact that the probability of finding the electron as a function of distance from the nucleus in a hydrogen atom can be represented as shown in Figure 2.1.

The *average* distance is that point where there is an equal probability of finding the electron on either side of that distance. It is that distance that we have just calculated by the above-mentioned procedure. On the other hand, the probability as a function of distance is represented by a function that goes through a maximum. Where the probability function has its maximum value is the *most probable* distance and that distance for an electron in the $1s$ state of a hydrogen atom is a_0.

The applications of the principles just discussed are many and varied. It is possible to calculate the expectation values for a variety of properties by precisely the procedures illustrated. Do not be intimidated by such calculations. Proceed in a stepwise fashion, look up any needed integrals in a table, and handle the algebra in a clear, orderly manner.

In addition to the postulates stated earlier, one additional postulate is normally included in the required list.

Postulate IV: The state function, Ψ, results from the solution of the equation

$$\hat{H}\psi = E\psi,$$

where $\hat{H}$ is the operator for total energy, the Hamiltonian operator.

For a variety of problems in quantum mechanics, the first step in the formulation is to write the equation

$$\hat{H}\psi = E\psi$$

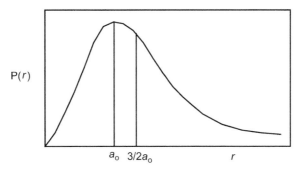

FIGURE 2.1
The probability of finding the electron as a function of distance from the nucleus.

and then substitute the appropriate function for the Hamiltonian operator. In classical mechanics, Hamilton's function is the sum of the kinetic (translational) and potential energies. This can be written as

$$H = T + V, \tag{2.29}$$

where T is the kinetic energy, V is the potential energy, and H is Hamilton's function. When written in *operator* form, the equation becomes

$$\hat{H}\psi = E\psi \tag{2.30}$$

For some systems, the potential energy is some function of a coordinate. For example, the potential energy of an electron bound to the nucleus of a hydrogen atom is given by $-e^2/r$, where e is the charge on the electron and r is a coordinate. Therefore, when this potential function is placed in operator form, it is the same as in classical form, $-e^2/r$.

To put the kinetic energy in operator form, we make use of the fact that the kinetic energy can be written in terms of the momentum as

$$T = \frac{1}{2}mv^2 = \frac{(mv)^2}{2m} = \frac{p^2}{2m}. \tag{2.31}$$

Because momentum has x, y, and z components, the total momentum can be represented as

$$T = \frac{p_x^2}{2m} + \frac{p_y^2}{2m} + \frac{p_z^2}{2m}. \tag{2.32}$$

Earlier it was shown that the operator for the x component of momentum can be written as $(\hbar/i)(d/dx)$. The operators for the y and z components have the same form except for the derivatives being with respect to those variables. Because the momentum in each direction is squared, the operator must be correspondingly used twice:

$$\left(\frac{\hbar}{i}\frac{\partial}{\partial x}\right)^2 = \frac{\hbar^2}{i^2}\frac{\partial^2}{\partial x^2} = -\hbar^2\frac{\partial^2}{\partial x^2}. \tag{2.33}$$

As a result, the operator for the total kinetic energy is

$$T = -\frac{\hbar^2}{2m}\left(\frac{\partial^2}{\partial x^2} + \frac{\partial^2}{\partial y^2} + \frac{\partial^2}{\partial z^2}\right) = -\frac{\hbar^2}{2m}\nabla^2, \tag{2.34}$$

where ∇^2 is the *Laplacian operator* that is often referred to as simply the *Laplacian*.

2.2 THE HYDROGEN ATOM

There are several introductory problems that can be solved exactly by quantum mechanical methods. These include the particle in a one-dimensional box, the particle in a three-dimensional box, the rigid rotor, the harmonic oscillator, and barrier penetration. All these models provide additional insight to the methods of quantum mechanics, and the interested reader should consult a quantum mechanics text, such as those listed in the references at the end of this chapter. Because of the nature of this book, we will progress directly to the problem of the hydrogen atom, which was solved in 1926 by Erwin Schrödinger. His starting point was a three-dimensional wave equation that had been developed earlier by physicists who were dealing with the so-called flooded planet problem. In this model, a sphere was assumed to be covered with water, and the problem was to deal with the wave motion that would result if the surface were disturbed. Schrödinger did not derive a wave equation. He adapted one that already existed. His adaptation consisted of representing the wave motion of an electron by means of the de Broglie relationship that had been established only 2 years earlier. Physics was progressing at a rapid pace in that time period.

We can begin directly by writing the equation

$$\hat{H}\psi = E\psi \tag{2.35}$$

and then determine the correct form for the Hamiltonian operator. We will assume that the nucleus remains stationary with the electron revolving around it (known as the Born–Oppenheimer approximation) and will deal only with the motion of the electron. The electron has a kinetic energy of $\frac{1}{2}mv^2$, which can be written as $p^2/2m$. Eqn (2.34) shows the operator for kinetic energy.

The interaction between an electron and a nucleus in a hydrogen atom gives rise to a potential energy that can be described by the relationship $-e^2/r$. Therefore, using the Hamiltonian operator and Postulate IV, the wave equation can be written as

$$-\frac{\hbar^2}{2m}\nabla^2\psi - \frac{e^2}{r}\psi = E\psi. \tag{2.36}$$

Rearranging the equation and representing the potential energy as V gives

$$\nabla^2\psi + \frac{2m}{\hbar^2}(E - V) = 0. \tag{2.37}$$

The difficulty in solving this equation is that when the Laplacian is written in terms of Cartesian coordinates, we find that

$$r = \sqrt{x^2 + y^2 + z^2}. \tag{2.38}$$

The wave equation is a second-order partial differential equation in three variables. The usual technique for solving such an equation is to use a procedure known as the separation of variables. However, with r expressed as the square root of the sum of the squares of the three variables, it is impossible to separate the variables. To circumvent this problem, a change of coordinates to polar coordinates is made. After that is done, the Laplacian must be transformed into polar coordinates, which is a tedious task. When the transformation is made, the variables can be separated so that three second-order differential equations each containing one coordinate as the variable are obtained. Even after this is done, the resulting equations are quite complex and the solution of two of the three equations requires the use of series techniques to solve them. The solutions are described in detail in most quantum mechanics books so it is not necessary to solve the equations here (see suggested readings at the end of this chapter), but Table 2.2 shows the wave functions. These wave functions are referred to as hydrogen-like wave functions because they apply to any one-electron system (e.g. He^+, Li^{2+}).

From the mathematical restrictions on the solution of the equations comes a set of constraints known as *quantum numbers*. The first of these n, the principal quantum number, is restricted to integer values (1, 2, 3, …). The second quantum number is l, the orbital angular momentum quantum number, and it

Table 2.2 Complete Normalized Hydrogen-Like Wave Functions

$$\psi_{1s} = \frac{1}{\pi^{1/2}}\left(\frac{Z}{a}\right)^{3/2} e^{-Zr/a}$$

$$\psi_{2s} = \frac{1}{4(2\pi)^{1/2}}\left(\frac{Z}{a}\right)^{3/2}\left(2 - \frac{Zr}{a}\right)e^{-Zr/2a}$$

$$\psi_{2p_z} = \frac{1}{4(2\pi)^{1/2}}\left(\frac{Z}{a}\right)^{5/2} r e^{-Zr/2a}\cos\theta$$

$$\psi_{2p_x} = \frac{1}{4(2\pi)^{1/2}}\left(\frac{Z}{a}\right)^{5/2} r e^{-Zr/2a}\sin\theta\cos\phi$$

$$\psi_{2p_y} = \frac{1}{4(2\pi)^{1/2}}\left(\frac{Z}{a}\right)^{5/2} r e^{-Zr/2a}\sin\theta\cos\phi$$

$$\psi_{3s} = \frac{1}{81(3\pi)^{1/2}}\left(\frac{Z}{a}\right)^{3/2}\left(27 - 18\frac{Zr}{a} + 2\frac{Z^2 r^2}{a^2}\right)e^{-Zr/3a}$$

$$\psi_{3p_z} = \frac{2^{1/2}}{81\pi^{1/2}}\left(\frac{Z}{a}\right)^{5/2}\left(6 - \frac{Zr}{a}\right)r e^{-Zr/3a}\cos\theta$$

$$\psi_{3p_z} = \frac{2^{1/2}}{81\pi^{1/2}}\left(\frac{Z}{a}\right)^{5/2}\left(6 - \frac{Zr}{a}\right)r e^{-Zr/3a}\sin\theta\cos\phi$$

$$\psi_{3p_Y} = \frac{2^{1/2}}{81\pi^{1/2}}\left(\frac{Z}{a}\right)^{5/2}\left(6 - \frac{Zr}{a}\right)r e^{-Zr/3a}\sin\theta\sin\phi$$

$$\psi_{3d_{xy}} = \frac{1}{81(2\pi)^{1/2}}\left(\frac{Z}{a}\right)^{7/2} r^2 e^{-Zr/3a}\sin^2\theta\sin 2\phi$$

$$\psi_{3d_{xz}} = \frac{2^{1/2}}{81\pi^{1/2}}\left(\frac{Z}{a}\right)^{7/2} r^2 e^{-Zr/3a}\sin\theta\cos\theta\cos\phi$$

$$\psi_{3d_{yz}} = \frac{2^{1/2}}{81\pi^{1/2}}\left(\frac{Z}{a}\right)^{7/2} r^2 e^{-Zr/3a}\sin\theta\cos\theta\sin\phi$$

$$\psi_{3d_{x^2-y^2}} = \frac{1}{81(2\pi)^{1/2}}\left(\frac{Z}{a}\right)^{7/2} r^2 e^{-Zr/3a}\sin\theta\cos 2\theta$$

$$\psi_{3d_{z^2}} = \frac{1}{81(6\pi)^{1/2}}\left(\frac{Z}{a}\right)^{7/2} r^2 e^{-Zr/3a}(3\cos^2\theta - 1)$$

must also be an integer such that it can be at most $(n - 1)$. The third quantum number is m, the magnetic quantum number which gives the projection of the l vector on the z-axis as shown in Figure 2.2.

The three quantum numbers that arise as mathematical restraints on the differential equations (boundary conditions) can be summarized as follows.

$n =$ principal quantum number $= 1, 2, 3, \ldots$
$l =$ orbital angular quantum number $= 0, 1, 2, \ldots, (n-1)$
$m =$ magnetic quantum number $= 0, +1, +2, \ldots, + l$.

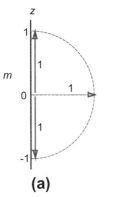
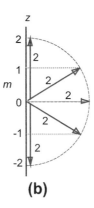

FIGURE 2.2
Projections of vectors $l = 1$ and $l = 2$ on the z-axis.

Note that from the solution of a problem involving three dimensions, three quantum numbers result, unlike the Bohr approach that specified only one. The quantum number n is essentially equivalent to the n that was assumed in the Bohr model of hydrogen.

A spinning electron also has a spin quantum number, which is expressed as $\pm\frac{1}{2}$ in units of $\hbar$. However, that quantum number does not arise from the solution of a differential equation in Schrödinger's solution of the hydrogen atom problem. It arises because as in the case of other fundamental particles, the electron has an intrinsic spin that is half integer in units of $\hbar$, the quantum of angular momentum. As a result, four quantum numbers are required to completely specify the state of the electron in an atom. The Pauli Exclusion Principle states that *"no two electrons in the same atom can have identical sets of four quantum numbers."* We will illustrate this principle later.

The lowest energy state is the one characterized by $n = 1$ requires that $l = 0$ and $m = 0$. A state for $l = 0$ is designated as an s state so the lowest energy state is known as the $1s$ state because states are designated by the value of n followed by a lowercase letter to represent the l value. The values of l are denoted by letters as follows:

Value of l:	0	1	2	3
State designation:	**s**harp	**p**rincipal	**d**iffuse	**f**undamental

The words sharp, principal, diffuse, and fundamental were used to describe certain lines in atomic spectra, and it is the first letter of each word that is used to name s, p, d, and f states.

For hydrogen, the notation $1s^1$ is used where the superscript denotes a single electron in the $1s$ state. Because an electron can have a spin quantum number of $+\frac{1}{2}$ or $-\frac{1}{2}$, two electrons having opposite spins can occupy the $1s$ state. The helium atom, having two electrons, has the configuration $1s^2$, with the electrons having spins of $+\frac{1}{2}$ and $-\frac{1}{2}$.

As we have seen in earlier sections, wave functions can be used to perform useful calculations to determine values for dynamical variables. Table 2.2 shows the normalized wave functions in which the nuclear charge is given as Z ($Z = 1$ for hydrogen) for one-electron species (H, He$^+$, etc.). One of the

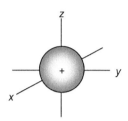

FIGURE 2.3
The three-dimensional surface representing the probability region of an *s* orbital.

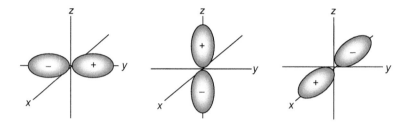

FIGURE 2.4
The three-dimensional surfaces of *p* orbitals.

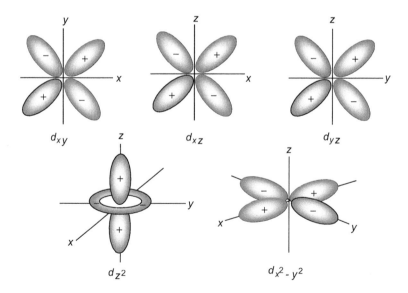

FIGURE 2.5
Three-dimensional representations of the five *d* orbitals.

results that can be obtained by making use of wave functions is that it is possible to determine the shapes of the surfaces that encompass the region where the electron can be found some fraction (perhaps 95%) of the time. Such drawings result in the orbital contours that are shown in Figures 2.3–2.5.

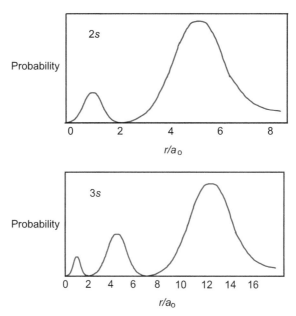

FIGURE 2.6
Radial distribution plots for 2s and 3s wave functions.

It is interesting to examine the probability of finding an electron as a function of distance when s orbitals having different n values are considered. Figure 2.6 shows the radial probability plots for the 2s and 3s orbitals. Note that the plot for the 2s orbital has one node (where the probability goes to 0), whereas that for the 3s orbital has two nodes. It can be shown that the nodes occur at $r = 2a_0$ for the 2s orbital and at $1.90a_0$ and $7.10a_0$ for the 3s orbital. It is a general characteristic that ns orbitals have $(n - 1)$ nodes. It is also noted that the distance at which the maximum probability occurs increases as the value of n increases. In other words, a 3s orbital is "larger" than a 2s orbital. Although not exactly identical, this is in accord with the idea from the Bohr model that the sizes of allowed orbits increase with increasing n value.

2.3 THE HELIUM ATOM

In previous sections, the basis for applying quantum mechanical principles has been illustrated. Although it is possible to solve exactly several types of problems, it should not be inferred that this is always the case. For example, it is easy to formulate wave equations for numerous systems, but generally they cannot be solved exactly. Consider the case of the helium atom illustrated in Figure 2.7 to show the coordinates of the parts of the system.

We know that the general form of the wave equation is

$$\hat{H}\psi = E\psi \tag{2.39}$$

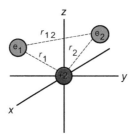

FIGURE 2.7
Coordinate system for the helium atom.

in which the Hamiltonian operator takes on the form appropriate to the particular system. In the case of the helium atom, there are two electrons, each of which has kinetic energy and is attracted to the +2 nucleus. However, there is also the repulsion between the two electrons. With reference to Figure 2.7, the attraction terms can be written as $-2e^2/r_1$ and $-2e^2/r_2$. The kinetic energies represented as $\frac{1}{2}mv_1^2$ and $\frac{1}{2}mv_2^2$ in operator form can be written as $(-\hbar^2/2m)\nabla_1^2$ and $(-\hbar^2/2m)\nabla_2^2$. However, we now must include the repulsion between the two electrons that gives rise to the term $+e^2/r_{12}$ in the Hamiltonian. When the complete Hamiltonian is written out, the result is

$$\hat{H} = -\frac{\hbar^2}{2m}\nabla_1^2 - \frac{\hbar^2}{2m}\nabla_2^2 - \frac{2e^2}{r_1} - \frac{2e^2}{r_2} + \frac{e^2}{r_{12}} \tag{2.40}$$

from which we obtain the wave equation for the helium atom,

$$\hat{H}\psi = E\psi \tag{2.41}$$

or

$$\left(-\frac{\hbar^2}{2m}\nabla_1^2 - \frac{\hbar^2}{2m}\nabla_2^2 - \frac{2e^2}{r_1} - \frac{2e^2}{r_2} + \frac{e^2}{r_{12}}\right)\psi = E\psi. \tag{2.42}$$

To solve the wave equation for the hydrogen atom, it is necessary to transform the Laplacian into polar coordinates. That transformation allows the distance of the electron from the nucleus to be expressed in terms of r, θ, and ϕ, which in turn allows the separation of variables technique to be used. Examination of Eqn (2.40) shows that the first and third terms in the Hamiltonian are exactly like the two terms in the operator for the hydrogen atom. Likewise, the second and fourth terms are also equivalent to those for a hydrogen atom. However, the last term, e^2/r_{12}, is the troublesome part of the Hamiltonian. In fact, even after polar coordinates are employed, that term prevents the separation of variables from being accomplished. Not being able to separate the variables to obtain three simpler equations prevents an exact solution of Eqn (2.40) from being carried out.

When an equation describes a system exactly but the equation cannot be solved, there are two general approaches that are followed. First, if the exact equation cannot be solved exactly, it may be possible to obtain approximate solutions. Second, the equation that describes the system exactly may be modified to produce a different equation that now describes the system only approximately but can be solved exactly. These are the approaches to solve the wave equation for the helium atom.

Because the other terms in the Hamiltonian essentially describe two hydrogen atoms except for the nuclear charge of +2, we could simply neglect the repulsion between the two electrons to obtain an equation that can be solved exactly. In other words, we have "approximated" the system as two hydrogen atoms meaning that the binding energy of an electron to the helium nucleus should be 27.2 eV, twice the value of 13.6 eV for the hydrogen atom. However, the actual value for the first ionization potential of helium is 24.6 eV because of the repulsion between the two electrons. Clearly, the approximate wave equation does not lead to a correct value for the binding energy of the electrons in a helium atom. This is equivalent to saying that an electron in a helium atom does not experience the effect of being attracted by a nucleus having a +2 charge, but an attraction is less than that value because of the repulsion between electrons. If this approach is taken, it turns out that the effective nuclear charge is $27/16 = 1.688$ instead of exactly 2.

If the problem is approached from the standpoint of considering the repulsion between the electron as being a minor irregularity or perturbation in an otherwise solvable problem, the Hamiltonian can be modified to take into account this perturbation in a form that allows the problem to be solved. When this is done, the calculated value for the first ionization potential is 24.58 eV.

Although we cannot solve the wave equation for the helium atom exactly, the approaches described provide some insight in regard to how we might proceed in cases where approximations must be made. The two major approximation methods are known as the variation and perturbation methods. For details of these methods as applied to the wave equation for the helium atom, see the quantum mechanics books listed in the suggested readings at the end of this chapter. A detailed treatment of these methods is beyond the scope of this book.

2.4 SLATER WAVE FUNCTIONS

We have just explained that the wave equation for the helium atom cannot be solved exactly because of the term involving $1/r_{12}$. If the repulsion between two electrons prevents a wave equation from being solved, it should be clear that when there are more than two electrons, the situation is worse. If there are three electrons present (as in the lithium atom), there will be repulsion terms involving $1/r_{12}$, $1/r_{13}$, and $1/r_{23}$. Although there are a number of types of calculations that can be performed (particularly the self-consistent field calculations), they will not be described here. Fortunately, for some situations, it is not necessary to have an exact wave function that is obtained from the exact solution of a wave equation. In many cases, an approximate wave function is sufficient. The most commonly used approximate wave functions for one electron are those given by J. C. Slater, and they are known as Slater wave functions or *Slater-type orbitals* (usually referred to as STO orbitals).

Slater wave functions have the mathematical form

$$\psi_{n,l,m} = R_{n,l}(r)e^{-Zr/a_0 n}Y_{l,m}(\theta, \phi).$$

(2.43)

When the radial function $R_{n,l}(r)$ is approximated, the wave functions can be written as

$$\psi_{n,l,m} = r^{n^*-1}e^{-(Z-s)r/a_0 n^*}Y_{l,m}(\theta, \phi),$$

(2.44)

where s is a constant known as a screening constant, n^* is an effective quantum number that is related to n, and $Y_{l,m}(\theta,\phi)$ is a spherical harmonic that gives the angular dependence of the wave function.

The spherical harmonics are functions that depend on the values of l and m as indicated by the subscripts. The quantity $(Z-s)$ is sometimes referred to as the effective nuclear charge, Z^*. The screening constant is calculated according to a set of rules that are based on the effectiveness of electrons in shells to screen the electron being considered from the effect of the nucleus.

The calculation of the screening constant for a specific electron is as follows.

1. The electrons are written in groups as follows:

$$1s \mid 2s\ 2p \mid 3s\ 3p \mid 3d \mid 4s\ 4p \mid 4d \mid 4f \mid 5s\ 5p \mid 5d \mid \ldots$$

2. Electrons residing outside the shell in which the electron being described resides do not contribute to the screening constant.
3. A contribution of 0.30 is assigned for an electron in the $1s$ level, but for other groups, 0.35 is added for each electron in that group.
4. A contribution of 0.85 is added for each electron in an s or p orbital for which the principle quantum number is one less than that for the electron being described. For electrons in s or p orbitals that have an n value of 2 or more lower than that of the orbital for the electron being considered, a contribution of 1.00 is added for each electron.
5. For electrons in d and f orbitals, a contribution of 1.00 is added for each electron in orbitals having lower n than the one where the electron being considered resides.
6. The value of n^* is determined from n as based on the following table.

$n = 1$	2	3	4	5	6
$n^* = 1$	2	3	3.7	4.0	4.2

To illustrate the use of the rules above, we will write the Slater wave function for an electron in an oxygen atom. The electron resides in a $2p$ orbital so $n = 2$ means that $n^* = 2$. There are four electrons in the $2p$ orbitals in an oxygen atom so the fourth electron is screened by three others. However, an electron in a $2p$ orbital is also screened by the two electrons in the $1s$ orbital and two in the $2s$ orbital. Two electrons in the $1s$ level give screening that can be written as $2(0.85) = 1.70$. The screening constant is the same for electrons in the $2s$ and $2p$ states, and the electron in the $2p$ state that is being considered has five other electrons involved in screening. Therefore, those five electrons give a contribution to the screening constant of $5(0.35) = 1.75$. Adding the contributions gives a total screening constant of $1.70 + 1.75 = 3.45$ meaning that the effective nuclear charge is $8 - 3.45 = 4.55$. Using this value gives $(Z - s)/n^* = 2.28$ and a Slater wave function that can be written as

$$\psi = re^{-2.28r/a_0 n^*} Y_{2,m}(\theta, \phi). \tag{2.45}$$

The important aspect of this approach is that it is now possible to arrive at an approximate one-electron wave function that can be used in other calculations. For example, Slater-type orbitals form the basis of many of the high-level molecular orbital calculations using self-consistent field theory and other approaches. However, in most cases, the Slater-type orbitals are not used directly. Quantum mechanical calculations on molecules involve the evaluation of a large number of integrals, and exponential integrals of the STO are much less efficient in calculations. In practice, the STO functions are represented as a series of functions known as Gaussian functions, which are of the form $a \exp(-br^2)$. The set of functions to be used, known as the *basis set*, is then constructed as a series of Gaussian functions representing each STO. When a three-term Gaussian is used, the orbitals are known as a STO-3G basis set. The result of this transformation is that the computations are completed much more quickly because Gaussian integrals are much easier to compute. For a more complete discussion of this advanced topic, see the book, *Quantum Chemistry* by J. P. Lowe listed in the references at the end of this chapter.

2.5 ELECTRON CONFIGURATIONS

For $n = 1$, the only value possible for both l and m is 0. Therefore, there is only a single state possible, the one for $n = 1$, $l = 0$, and $m = 0$. This state is denoted as the 1s state. If $n = 2$, it is possible for l to be 0 or 1. For the $l = 0$ state, the $n = 2$ and $l = 0$ combination gives rise to the 2s state. Again, $m = 0$ is the only possibility because of the restriction on the values of m being 0 up to $\pm l$. Two electrons can be accommodated in the 2s state that is filled in the beryllium atom so the electron configuration is $1s^2 2s^2$.

For the quantum state of $n = 2$ and $l = 1$, we find that there are three values of m possible, $+1$, 0, and -1. Therefore, each value of m can be used with spin quantum numbers of $+\frac{1}{2}$ and $-\frac{1}{2}$. This results in six combinations of quantum numbers that obey the restrictions on their values given above. The six possible sets of quantum numbers can be shown as follows:

Electron 1	Electron 2	Electron 3	Electron 4	Electron 5	Electron 6
$n = 2$	$n = 2$	$n = 2$	$n = 2$	$n = 2$	$n = 2$
$l = 1$	$l = 1$	$l = 1$	$l = 1$	$l = 1$	$l = 1$
$m = +1$	$m = 0$	$m = -1$	$m = +1$	$m = 0$	$m = -1$
$s = +\frac{1}{2}$	$s = +\frac{1}{2}$	$s = +\frac{1}{2}$	$s = -\frac{1}{2}$	$s = -\frac{1}{2}$	$s = -\frac{1}{2}$

A state for $l = 1$ is known as a p state so the six sets of quantum numbers above belong to the 2p state. Population of the 2p state is started with boron and completed with neon in the first long period of the periodic table. However, each m value denotes an orbital so that there are three orbitals where electrons can reside. *Electrons remain unpaired as long as possible* when populating a set of orbitals. For convenience, we will assume that **the orbitals fill by starting with the highest positive value of m first and then going to successive lower values.** Table 2.3 shows the maximum population of states based on the value for l.

Table 2.3 Maximum Orbital Populations			
l Value	State	*m* Values Possible	Maximum Population
0	s	0	2
1	p	$0, \pm 1$	6
2	d	$0, \pm 1, \pm 2$	10
3	f	$0, \pm 1, \pm 2, \pm 3$	14
4	g	$0, \pm 1, \pm 2, \pm 3, \pm 4$	18

According to the Bohr model, the quantum number n determines the energies of the allowed states in a hydrogen atom. We now know that both n and l are factors that determine the energy of a set of orbitals. In a general way, it is the sum $(n + l)$ that determines the energy, and the energy increases as the sum of n and l. However, when there are two or more ways to get a particular sum of $(n + l)$, the combination with lower n is normally used first. For example, the $2p$ state has $(n + l) = (2 + 1) = 3$ as does the $3s$ state where $(n + l) = (3 + 0) = 3$. In this case, the $2p$ level fills first because of the greater importance of the lower n value. Table 2.4 shows the scheme used to arrange the orbitals in the order in which they are normally filled. As a result, the general order of filling shells is as follows with the sum $(n + l)$ being given directly below for each type of shell.

State	$1s$	$2s$	$2p$	$3s$	$3p$	$4s$	$3d$	$4p$	$5s$	$4d$	$5p$	$6s$
$(n + l)$	1	2	3	3	4	4	5	5	5	6	6	6

By following the procedures described, we find that for the first 10 elements, the results are as follows.

H	$1s^1$
He	$1s^2$
Li	$1s^2 2s^1$
Be	$1s^2 2s^2$
B	$1s^2 2s^2 2p^1$
C	$1s^2 2s^2 2p^2$
N	$1s^2 2s^2 2p^3$
O	$1s^2 2s^2 2p^4$
F	$1s^2 2s^2 2p^5$
Ne	$1s^2 2s^2 2p^6$

For some of the atoms, this is an oversimplification because an electron configuration such as $1s^2 2s^2 2p^2$ does give the complete picture. Each m value denotes an orbital in which two electrons can reside with opposite spins. When $l = 1$, m can have the values of $+1$, 0, and -1, which denote three orbitals. Therefore, the two electrons could be paired in one of the three orbitals or they could be unpaired and reside in different orbitals. We have already stated that **we will start with the highest positive value of m first and with the positive value of s.** For a $2p^2$ configuration, two possible arrangements are as follows:

$$m = \quad \frac{\uparrow\downarrow}{+1} \ \frac{}{0} \ \frac{}{-1} \quad \text{or} \quad \frac{\uparrow}{+1} \ \frac{\uparrow}{0} \ \frac{}{-1}$$

Although the explanation will be provided later, the configuration on the right is lower in energy than that on the left. As we have already stated, electrons remain unpaired as long as possible. Therefore, the carbon atom that has the electron configuration $1s^2 2s^2 2p^2$ has two unpaired electrons. In a similar way, the nitrogen atom has three electrons in the $2p$ orbitals,

$$m = \quad \frac{\uparrow}{+1} \ \frac{\uparrow}{0} \ \frac{\uparrow}{-1}$$

Table 2.4 Filling of Orbitals According to Increasing $(n + l)$ Sum[a]

n	l	$(n + l)$	State
1	0	1	$1s$
2	0	2	$2s$
2	1	3	$2p$
3	0	3	$3s$
3	1	4	$3p$
4	0	4	$4s$
3	2	5	$3d$
4	1	5	$4p$
5	0	5	$5s$
4	2	6	$4d$
5	1	6	$5p$
6	0	6	$6s$
4	3	7	$4f$
5	2	7	$5d$
6	1	7	$6p$
7	0	7	$7s$

[a]*In general, energy increases in going down in the table.*

so it has three unpaired electrons. On the other hand, the oxygen atom has the configuration $1s^2\ 2s^2\ 2p^4$, which has only two unpaired electrons.

$$\begin{array}{ccc} \uparrow\downarrow & \uparrow & \uparrow \\ m = +1 & 0 & -1 \end{array}$$

It is important when discussing the chemistry of elements to keep in mind the actual arrangement of electrons, not just the overall configuration. For example, with the nitrogen atom having one electron in each of the three $2p$ orbitals, the addition of another electron would require it to be paired in one of the orbitals. Because of repulsion between electrons in the same orbital, there is little tendency of the nitrogen atom to add another electron. Therefore, the electron affinity of nitrogen is very close to 0. We have already mentioned in Chapter 1 that the ionization potential for the oxygen atom is lower than that for the nitrogen atom, even though the oxygen atom has a higher nuclear charge. We can now see why this is so. Because oxygen has the configuration $1s^2\ 2s^2\ 2p^4$, one of the $2p$ orbitals has a pair of electrons in it. Repulsion between the electrons in this pair reduces the energy with which they are bound to the nucleus so it is easier to remove one of them than it is to remove an electron from a nitrogen atom. There are numerous other instances of electron configurations that give a basis for interpreting properties of atoms.

We could follow the procedures illustrated above to write the electron configurations of elements 11 through 18 in which the $3s$ and $3p$ orbitals are being filled. However, we will not write all these out, but rather we will summarize the electron configuration of argon $1s^2\ 2s^2\ 2p^2\ 3s^2\ 3p^6$ as (Ar). When this is done, the next element, potassium, has the configuration (Ar) $4s^1$ and that of calcium is (Ar) $4s^2$. The sum $(n + l)$ is 4 for both the $3p$ and $4s$ levels, and the lower value of n is used first ($3p$). The next levels to be filled are those for $(n + l) = 5$, and these are the $3d$, $4p$, and $5s$. In this case, the $3d$ orbitals

IA 1	IIA 2	IIIB 3	IVB 4	VB 5	VIB 6	VIIB 7	VIIIB 8	VIIIB 9	VIIIB 10	IB 11	IIB 12	IIIA 13	IVA 14	VA 15	VIA 16	VIIA 17	VIIIA 18
1 H 1.0079																	2 He 4.0026
3 Li 6.941	4 Be 9.0122											5 B 10.81	6 C 12.011	7 N 14.0067	8 O 15.9994	9 F 18.9984	10 Ne 20.179
11 Na 22.9898	12 Mg 24.305											13 Al 26.9815	14 Si 28.0855	15 P 30.9738	16 S 32.06	17 Cl 35.453	18 Ar 39.948
19 K 39.0983	20 Ca 40.08	21 Sc 44.9559	22 Ti 47.88	23 V 50.9415	24 Cr 51.996	25 Mn 54.9380	26 Fe 55.847	27 Co 58.9332	28 Ni 58.69	29 Cu 63.546	30 Zn 65.38	31 Ga 69.72	32 Ge 72.59	33 As 74.9216	34 Se 78.96	35 Br 79.904	36 Kr 83.80
37 Rb 85.4678	38 Sr 87.62	39 Y 88.9059	40 Zr 91.22	41 Nb 92.9064	42 Mo 95.94	43 Tc (98)	44 Ru 101.07	45 Rh 102.906	46 Pd 106.42	47 Ag 107.868	48 Cd 112.41	49 In 114.82	50 Sn 118.69	51 Sb 121.75	52 Te 127.60	53 I 126.905	54 Xe 131.29
55 Cs 132.905	56 Ba 137.33	57 La* 138.906	72 Hf 178.48	73 Ta 180.948	74 W 183.85	75 Re 186.207	76 Os 190.2	77 Ir 192.22	78 Pt 195.09	79 Au 196.967	80 Hg 200.59	81 Tl 204.383	82 Pb 207.2	83 Bi 208.980	84 Po (209)	85 At (210)	86 Rn (222)
87 Fr (223)	88 Ra 226.025	89 Ac* 227.028	104 Rf (257)	105 Db (260)	106 Sg (263)	107 Bh (262)	108 Hs (265)	109 Mt (266)	110 Ds (271)	111 Rg (272)							

*Lanthanide Series	58 Ce 140.12	59 Pr 140.908	60 Nd 144.24	61 Pm (145)	62 Sm 150.36	63 Eu 151.96	64 Gd 157.25	65 Tb 158.925	66 Dy 162.50	67 Ho 164.930	68 Er 167.26	69 Tm 168.934	70 Yb 173.04	71 Lu 174.967
*Actinide Series	90 Th 232.038	91 Pa 231.036	92 U 238.029	93 Np 237.048	94 Pu (244)	95 Am (243)	96 Cm (247)	97 Bk (247)	98 Cf (251)	99 Es (252)	100 Fm (257)	101 Md (258)	102 No (259)	103 Lr (260)

FIGURE 2.8
The periodic table of the elements.

have the lower n so that set of orbitals is filled next. Therefore, the configurations for the next few elements are as follows:

$$\text{Sc(Ar)}4s^2\ 3d^1 \qquad \text{Ti(Ar)}4s^2\ 3d^2 \qquad \text{V(Ar)}4s^2\ 3d^3.$$

Following this pattern, we expect chromium to have the configuration (Ar) $4s^2\ 3d^4$, but it is actually (Ar) $4s^1\ 3d^5$. The reason for this will be explained later, but we will mention here that it has to do with the interaction of electrons having the same spin by means of coupling of their angular momenta. The next element, Mn, has the configuration (Ar) $4s^2\ 3d^5$, and filling of the $3d$ shell is regular until copper is reached. There, instead of the configuration being (Ar) $4s^2\ 3d^9$, it is (Ar) $4s^1\ 3d^{10}$, but the configuration of zinc is the expected $4s^2\ 3d^{10}$.

It is not necessary here to list all the irregularities that occur in electron configurations. We should point out that irregularities of the type just discussed occur only when the orbitals involved have energies that differ only slightly. There is no case where an atom such as carbon is found with the ground state configuration of $1s^2\ 2s^1\ 2p^3$ or $1s^2\ 2s^2\ 2p^1\ 3s^1$ instead of $1s^2\ 2s^2\ 2p^2$. The difference in energy between the $2s$ and $2p$ states and that between the $2p$ and $3s$ states is simply too great for an electron to reside in the higher state unless there has been electron excitation by some means. Electron configurations for atoms can now be written by making use of the procedures described above and referring to the periodic table shown in Figure 2.8. Table 2.5 gives the ground state electron configurations for all the atoms.

Table 2.5 Electronic Configurations of Atoms

Z	Symbol	Configuration	Z	Symbol	Configuration
1	H	$1s^1$	36	Kr	(Ar) $3d^{10}\,4s^2\,4p^6$
2	He	$1s^2$	37	Rb	(Kr) $5s^1$
3	Li	$1s^2\,2s^1$	38	Sr	(Kr) $5s^2$
4	Be	$1s^2\,2s^2$	39	Y	(Kr) $4d^1\,5s^2$
5	B	$1s^2\,2s^2\,2p^1$	40	Zr	(Kr) $4d^2\,5s^2$
6	C	$1s^2\,2s^2\,2p^2$	41	Nb	(Kr) $4d^4\,5s^1$
7	N	$1s^2\,2s^2\,2p^3$	42	Mo	(Kr) $4d^5\,5s^1$
8	O	$1s^2\,2s^2\,2p^4$	43	Tc	(Kr) $4d^5\,5s^2$
9	F	$1s^2\,2s^2\,2p^5$	44	Ru	(Kr) $4d^7\,5s^1$
10	Ne	$1s^2\,2s^2\,2p^6$	45	Rh	(Kr) $4d^8\,5s^1$
11	Na	(Ne) $3s^1$	46	Pd	(Kr) $4d^{10}$
12	Mg	(Ne) $3s^2$	47	Ag	(Kr) $4d^{10}\,5s^1$
13	Al	(Ne) $3s^2\,3p^1$	48	Cd	(Kr) $4d^{10}\,5s^2$
14	Si	(Ne) $3s^2\,3p^2$	49	In	(Kr) $4d^{10}\,5s^2\,5p^1$
15	P	(Ne) $3s^2\,3p^3$	50	Sn	(Kr) $4d^{10}\,5s^2\,5p^2$
16	S	(Ne) $3s^2\,3p^4$	51	Sb	(Kr) $4d^{10}\,5s^2\,5p^3$
17	Cl	(Ne) $3s^2\,3p^5$	52	Te	(Kr) $4d^{10}\,5s^2\,5p^4$
18	Ar	(Ne) $3s^2\,3p^6$	53	I	(Kr) $4d^{10}\,5s^2\,5p^5$
19	K	(Ar) $4s^1$	54	Xe	(Kr) $4d^{10}\,5s^2\,5p^6$
20	Ca	(Ar) $4s^2$	55	Cs	(Xe) $6s^1$
21	Sc	(Ar) $3d^1\,4s^2$	56	Ba	(Xe) $6s^2$
22	Ti	(Ar) $3d^2\,4s^2$	57	La	(Xe) $5d^1\,6s^2$
23	V	(Ar) $3d^3\,4s^2$	58	Ce	(Xe) $4f^2\,6s^2$
24	Cr	(Ar) $3d^5\,4s^1$	59	Pr	(Xe) $4f^3\,6s^2$
25	Mn	(Ar) $3d^5\,4s^2$	60	Nd	(Xe) $4f^4\,6s^2$
26	Fe	(Ar) $3d^6\,4s^2$	61	Pm	(Xe) $4f^5\,6s^2$
27	Co	(Ar) $3d^7\,4s^2$	62	Sm	(Xe) $4f^6\,6s^2$
28	Ni	(Ar) $3d^8\,4s^2$	63	Eu	(Xe) $4f^7\,6s^2$
29	Cu	(Ar) $3d^{10}\,4s^1$	64	Gd	(Xe) $4f^7\,5d^1\,6s^2$
30	Zn	(Ar) $3d^{10}\,4s^2$	65	Tb	(Xe) $4f^9\,6s^2$
31	Ga	(Ar) $3d^{10}\,4s^2\,4p^1$	66	Dy	(Xe) $4f^{11}\,6s^2$
32	Ge	(Ar) $3d^{10}\,4s^2\,4p^2$	67	Ho	(Xe) $4f^{11}\,6s^2$
33	As	(Ar) $3d^{10}\,4s^2\,4p^3$	68	Er	(Xe) $4f^{12}\,6s^2$
34	Se	(Ar) $3d^{10}\,4s^2\,4p^4$	69	Tm	(Xe) $4f^{13}\,6s^2$
35	Br	(Ar) $3d^{10}\,4s^2\,4p^5$	70	Yb	(Xe) $4f^{14}\,6s^2$

(Continued)

Table 2.5 Electronic Configurations of Atoms *(continued)*

Z	Symbol	Configuration	Z	Symbol	Configuration
71	Lu	(Xe) $4f^{14}\,5d^1\,6s^2$	91	Pa	(Rn) $5f^2\,6d^1\,7s^2$
72	Hf	(Xe) $4f^{14}\,5d^2\,6s^2$	92	U	(Rn) $5f^3\,6d^1\,7s^2$
73	Ta	(Xe) $4f^{14}\,5d^3\,6s^2$	93	Np	(Rn) $5f^5\,7s^2$
74	W	(Xe) $4f^{14}\,5d^4\,6s^2$	94	Pu	(Rn) $5f^6\,7s^2$
75	Re	(Xe) $4f^{14}\,5d^5\,6s^2$	95	Am	(Rn) $5f^7\,7s^2$
76	Os	(Xe) $4f^{14}\,5d^6\,6s^2$	96	Cm	(Rn) $5f^7\,6d^1\,7s^2$
77	Ir	(Xe) $4f^{14}\,5d^7\,6s^2$	97	Bk	(Rn) $5f^8\,6d^1\,7s^2$
78	Pt	(Xe) $4f^{14}\,5d^9\,6s^1$	98	Cf	(Rn) $5f^{10}\,7s^2$
79	Au	(Xe) $4f^{14}\,5d^{10}\,6s^1$	99	Es	(Rn) $5f^{11}\,7s^2$
80	Hg	(Xe) $4f^{14}\,5d^{10}\,6s^2$	100	Fm	(Rn) $5f^{12}\,7s^2$
81	Tl	(Xe) $4f^{14}\,5d^{10}\,6s^2\,6p^1$	101	Md	(Rn) $5f^{13}\,7s^2$
82	Pb	(Xe) $4f^{14}\,5d^{10}\,6s^2\,6p^2$	102	No	(Rn) $5f^{14}\,7s^2$
83	Bi	(Xe) $4f^{14}\,5d^{10}\,6s^2\,6p^3$	103	Lr	(Rn) $5f^{14}\,6d^1\,7s^2$
84	Po	(Xe) $4f^{14}\,5d^{10}\,6s^2\,6p^4$	104	Rf	(Rn) $5f^{14}\,6d^2\,7s^2$
85	At	(Xe) $4f^{14}\,5d^{10}\,6s^2\,6p^5$	105	Db	(Rn) $5f^{14}\,6d^3\,7s^2$
86	Rn	(Xe) $4f^{14}\,5d^{10}\,6s^2\,6p^6$	106	Sg	(Rn) $5f^{14}\,6d^4\,7s^2$
87	Fr	(Rn) $7s^1$	107	Bh	(Rn) $5f^{14}\,6d^5\,7s^2$
88	Ra	(Rn) $7s^2$	108	Hs	(Rn) $5f^{14}\,6d^6\,7s^2$
89	Ac	(Rn) $6d^1\,7s^2$	109	Mt	(Rn) $5f^{14}\,6d^7\,7s^2$
90	Th	(Rn) $6d^2\,7s^2$	110	Ds	(Rn) $5f^{14}6d^8\,7s^2$
			111	Rg	(Rn) $5f^{14}6d^9\,7s^2$

2.6 SPECTROSCOPIC STATES

We have already alluded to the fact that within an overall electron configuration of an atom, there is interaction of the electrons by means of coupling of angular momenta. This results from the fact that a spinning electron that is moving in an orbit has angular momentum resulting from its spin as well as from its orbital movement. These are vector quantities that couple in accord with quantum mechanical rules. In one of the coupling schemes, the individual spin angular momenta of the electrons couple to give an overall spin of S. In addition, the orbital angular momenta of the electrons couple to give an overall orbital angular momentum, L. Then, these resulting vector quantities couple to give the total angular momentum vector, J, for the atom. Coupling occurs by this scheme, which is known as $L-S$ or Russell–Saunders coupling, for atoms in approximately the top half of the periodic table.

In a manner similar to that by which the atomic states were designated as s, p, d, or f, the letters S, P, D, and F correspond to the values of 0, 1, 2, and 3, respectively, for the angular momentum vector, L. After the values of the vectors L, S, and J have been determined, the overall angular momentum is described by a symbol known as a *term symbol* or *spectroscopic state*. This symbol is constructed as $^{(2S+1)}L_J$ where the appropriate letter is used for the L value as listed above, and the quantity $(2S+1)$ is known as the *multiplicity*. For one unpaired electron, $(2S+1)=2$, and a multiplicity of 2 gives rise to a doublet. For two unpaired electrons, the multiplicity is 3, and the state is called a triplet state.

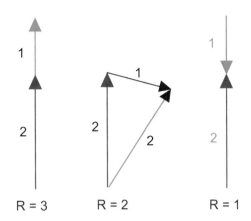

FIGURE 2.9
Vector sums for two vectors having lengths of 2 and 1 units. The resultant can have values of 3, 2, or 1.

Figure 2.9 shows how two vectors can couple according to the quantum mechanical restrictions. Note that vectors l_1 and l_2 having lengths of 1 and 2 units can couple to give resultants that are 3, 2, or 1 units in length. Therefore, the resultant, R, for these combinations can be written as $|l_1 + l_2|$, $|l_1 + l_2 - 1|$ or $|l_1 - l_2|$.

In heavier atoms, a different type of coupling scheme is sometimes followed. In that scheme, the orbital angular momentum, l, couples with the spin angular momentum, s, to give a resultant, j for a single electron. Then, these j values are coupled to give the overall angular momentum for the atom, J. Coupling of angular momenta by this means, known as $j-j$ coupling, occurs for heavy atoms, but we will not consider this type of coupling further.

In $L-S$ coupling, we need to determine the following sums to deduce the spectroscopic state of an atom:

$$L = \sum l_i$$
$$S = \sum s_i$$
$$M = \sum m_i = L,\ L-1,\ L-2,\ \cdots,\ 0,\ \cdots,\ -L$$
$$J = |L+S|,\ \cdots,\ |L-S|.$$

Note that if all the electrons are paired, the sum of spins is 0 so a singlet state results. Also, if all the orbitals in a set are filled, for each electron with a positive value of m, there is also one having a negative m value. Therefore, the sum of the m values is 0 (means that $L = 0$) leads to the conclusion that for an electron configuration that represents all filled shells the spectroscopic state is $^1S_0$. The subscript 0 arises because the combination $J = L + S = 0 + 0 = 0$. This is the case for the noble gas atoms that have all filled shells. We will now illustrate the application of these principles by showing how to determine the spectroscopic states of several atoms.

For the hydrogen atom, the single electron resides in the $1s$ state so the l and m values for that orbital are 0 (this means that L is also 0), and the single electron has a spin of $\frac{1}{2}$. Because $L = 0$, the spectroscopic state is an S state, and the multiplicity is 2 because the sum of spins is $\frac{1}{2}$. Therefore, the spectroscopic state of the hydrogen atom is ^2S. Note that the state being denoted as an S state is not related to the sum of spins vector, S. The value of J in this case is $|0 + \frac{1}{2}| = \frac{1}{2}$, which means that the spectroscopic state for the hydrogen atom is written as $^2S_{1/2}$. Any electron in an s orbital has l and m values of 0 regardless of the n value. As a result, any atom, such as Li, Na, K, and so on, having all closed shells and an ns^1 electron outside the closed shells have a spectroscopic state of $^2S_{1/2}$.

In the simple example above, only one spectroscopic state is possible. In many cases, more than one spectroscopic state can result from a given electron configuration because the electrons can be arranged in different ways. For example, the electron configuration np^2 could be arranged as

$$m = \quad \frac{\uparrow\downarrow}{+1} \ \frac{}{0} \ \frac{}{-1} \quad \text{or} \quad \frac{\uparrow}{+1} \ \frac{\uparrow}{0} \ \frac{}{-1}$$

that results in different spectroscopic states. With the electrons paired as shown in the diagram on the left, the state will be a singlet, whereas that on the right will give rise to a triplet state. Although we will not show the details here, the np^2 configuration gives rise to several spectroscopic states, but only one of them is the state of lowest energy, the spectroscopic *ground state*. Fortunately, there is a set of rules, known as Hund's rules, that permits us to determine the ground state (which usually concerns us most) very easily. Hund's rules can be stated as follows:

1. The state with the highest multiplicity gives the lowest energy for equivalent electrons.
2. For states having the highest multiplicity, the state with highest L is lowest in energy.
3. For shells that are less than half filled, the state with the lowest J lies lowest in energy, but for shells that are more than half filled, the lowest energy state has the highest J value.

In accord with the first rule, the electrons remain unpaired as long as possible when filling a set of orbitals because that is how the maximum multiplicity is achieved. With regard to the third rule, if a state is exactly half filled, the sum of the m values that gives the L vector is 0 and $|L + S|$ and $|L - S|$ are identical so only one J value is possible.

Suppose we wish to find the spectroscopic ground state for the carbon atom. We need not consider the filled $1s$ and $2s$ shells because both give $S = 0$ and $L = 0$. If we place the two electrons in the $2p$ level in the two orbitals that have $m = +1$ and $m = 0$, we will find that the sum of spins is 1, the maximum value possible. Moreover, the sum of the m values will give $L = 1$, which leads to a P state. With $S = 1$ and $L = 1$, the J values possible are 2, 1, and 0. Therefore, by application of Rule 3, the spectroscopic ground state for the carbon atom is $^3P_0$, but there are two other states, $^3P_1$ and $^3P_2$ that have energies that are only slightly higher (16.5 and 43.5 cm^{-1}, respectively). It can be shown that there are also states designated as $^1D_2$ and $^1S_0$ possible for the np^2 configuration. These have energies that are 10,193.7 and 21,648.4 cm^{-1} (122 and 259 kJ mol^{-1}, respectively) above the ground state. Note that these singlet states, which correspond to arrangements in the two electrons have been forced to pair, are significantly higher in energy than the ground state, and it is correct to say that the electrons remain unpaired as long as possible.

As has already been mentioned, the ground state is generally the only one that is of concern. If we need to determine the ground state for Cr^{3+}, which has the outer electron arrangement of $3d^3$ (for transition metals, the electrons are lost from the $4s$ level first), we proceed as before by placing the

Table 2.6 Spectroscopic States Arising for Equivalent Electrons

Electron Configuration	Spectroscopic States	Electron Configuration	Spectroscopic States
s^1	^2S	d^2	$^3F, \, ^3P, \, ^1G, \, ^1D, \, ^1S$
s^2	^1S	d^3	^4F (ground state)
p^1	^2P	d^4	^5D (ground state)
p^2	$^3P, \, ^1D, \, ^1S$	d^5	^6S (ground state)
p^3	$^4S, \, ^2D, \, ^2P$	d^6	^5D (ground state)
p^4	$^3P, \, ^1D, \, ^1S$	d^7	^4F (ground state)
p^5	^2P	d^8	^3F (ground state)
p^6	^1S	d^9	^2D
d^1	^2D	d^{10}	^1S

electrons in a set d orbitals beginning with the highest m value and working downward while keeping the electrons unpaired to give the highest multiplicity:

$$m = \frac{\uparrow}{+2} \; \frac{\uparrow}{+1} \; \frac{\uparrow}{0} \; \frac{}{-1} \; \frac{}{-2}.$$

For this arrangement, the sum of spins is $3/2$ and the L value is 3. These values give rise to the J values of $|3 + 3/2|$, $|3 + 3/2 - 1|$,..., $|3 - 3/2|$, which are $9/2$, $7/2$, $5/2$, and $3/2$. Because the set of orbitals is less than half filled, the lowest J corresponds to the lowest energy, and the spectroscopic ground state for Cr^{3+} is $^4F_{3/2}$. The spectroscopic states can be worked out for various electron configurations using the procedures described above. Table 2.6 shows a summary of the spectroscopic states that arise from various electron configurations.

Although we have not given a complete coverage to the topic of spectroscopic states, the discussion here is adequate for the purposes described in this book. In Chapter 18, it will be necessary to describe what happens to the spectroscopic states of transition metal ions when these ions are surrounded by other groups when coordination compounds form.

In this chapter, a brief review of quantum mechanical methods and the arrangement of electrons in atoms has been presented. These topics form the basis for understanding how quantum mechanics is applied to problems in molecular structure and the chemical behavior of the elements. The properties of atoms discussed in Chapter 1 are directly related to how the electrons are arranged in atoms. Although the presentation in this chapter is not exhaustive, it provides an adequate basis for the study of topics in inorganic chemistry. Further details can be found in the references.

References for Further Study

Silbey, R.J., Alberty, R.A., Bawendi, M.G., 2005. *Physical Chemistry*, 4th ed. Wiley, New York. Excellent coverage of early experiments that led to development of quantum mechanics.

DeKock, R.L., Gray, H.B., 1980. *Chemical Bonding and Structure*, Benjamin-Cummings, Menlo Park, CA. One of the best introductions to bonding concepts available.

Emsley, J., 1998. *The Elements*, 3rd ed. Oxford University Press, New York. A good source for an enormous amount of data on atomic properties.

Gray, H.B., 1965. *Electrons and Chemical Bonding*. Benjamin, New York. An elementary presentation of bonding theory that is both readable and well illustrated.

Harris, D.C., Bertolucci, M.D., 1989. *Symmetry and Spectroscopy*, Dover Publications, New York. Chapter 2 presents a very good introduction to quantum mechanics.

House, J.E., 2003. *Fundamentals of Quantum Chemistry*. Elsevier, New York. An introduction to quantum mechanical methods at an elementary level that includes mathematical details.

Lowe, J.P., 1993. *Quantum Chemistry*, 2nd ed. Academic Press, New York. An excellent treatment of advanced applications of quantum mechanics to chemistry.

Mortimer, R.G., 2008. *Physical Chemistry*, 3rd ed. Academic Press, San Diego, CA. A physical chemistry text that described the important experiments in atomic physics.

Sharpe, A.G., 1992. *Inorganic Chemistry*, 3rd ed. Longman, New York. An excellent book that gives a good summary of quantum mechanics at an elementary level.

Warren, W.S., 2000. *The Physical Basis of Chemistry*, 2nd ed. Academic Press, San Diego, CA. Chapter 6 presents a good tutorial on elementary quantum mechanics.

QUESTIONS AND PROBLEMS

1. Determine which of the following are eigenfunctions of the operator d/dx (a and b are constants). (a) e^{-ax}; (b) $x\,e^{-bx}$; (c) $(1 + e^{bx})$.

2. Is the function $\sin e^{ax}$ an eigenfunction of the operator d^2/dx^2?

3. Normalize the function e^{-2x} in the interval zero to infinity.

4. Normalize the function e^{-ax} in the interval zero to infinity.

5. Write the complete Hamiltonian operator for the lithium atom. Explain why the wave equation for Li cannot be solved exactly.

6. Write a set of four quantum numbers for the "last" electron in each of the following: (a) Ti; (b) S; (c) Sr; (d) Co; (e) Al.

7. Write a set of four quantum numbers for the "last" electron in each of the following: (a) Sc; (b) Ne; (c) Se; (d) Ga; (e) Si.

8. Draw vector coupling diagrams to show all possible ways in which vectors $L = 3$ and $S = 5/2$ can couple.

9. Write a set of four quantum numbers for the "last" electron in each of the following: (a) Si; (b) Se; (c) As; (d) Ga; (e) Ar.

10. Write complete electron configurations for the following: (a) Si; (b) S^{2-}; (c) K^+; (d) Cr^{2+}; (e) Fe^{2+}; (f) Zn.

11. Determine the spectroscopic ground states for the following: (a) P; (b) Sc; (c) Si; (d) Ni^{2+}.

12. The difference in energy between the $^3P_0$ and $^3P_1$ states of the carbon atom is 16.4 cm^{-1}. How much energy is this in $kJ \text{ mol}^{-1}$?

13. Determine the spectroscopic ground state for each of the following: (a) Be; (b) P; (c) F^-; (d) Al; (e) Sc.

14. Determine the spectroscopic ground state for each of the following: (a) Ti^{3+}; (b) Fe; (c) Co^{2+}; (d) Cl; (e) Cr^{2+}.

15. The spectroscopic ground state for a certain first row transition metal is $^6S_{5/2}$. (a) Which metal is it? (b) What would be the ground state spectroscopic state of the +2 ion of the metal described in (a)? (c) What would be the spectroscopic ground state for the +3 ion of the metal described in (a)?

16. What are all the types of atomic orbitals possible for $n = 5$? How many electrons can be held in orbitals that have $n = 5$? What would be the atomic number of the atom that would have all the shells with $n = 5$ filled?

17. Calculate the expectation value for $1/r$ for the electron in the $1s$ state of a hydrogen atom.

18. Calculate the expectation value for r^2 for the electron in the $1s$ state of a hydrogen atom.

19. The ionization potential for the hydrogen atom is 13.6 eV. Even though the nuclear charge in He is twice that of the hydrogen atom, the ionization potential is not twice the value of 13.6 eV. Explain why this is so. Look up the actual ionization potential of He. Using that value and the actual ionization potential, calculate the repulsion energy between the two electrons in a helium atom. What distance of separation of the electrons does this correspond to?

20. If a particle moves in a one-dimensional box of length a in which the potential energy is 0, show that the average position of the particle is $a/2$. The wave function for the lowest energy level is

$$\psi = \sqrt{\frac{a}{2}} \sin \sqrt{a}x.$$

CHAPTER 3

Covalent Bonding in Diatomic Molecules

Although the first two chapters were devoted to presenting the basic principles of quantum mechanics and their application to atomic structure, we must also be concerned with providing information about the structure of molecules. In fact, the structures of molecules constitute the basis for their chemical behavior. The observation that SF_4 reacts very rapidly and vigorously with water whereas SF_6 does not exhibit this behavior is related to the difference in the structures of the molecules. A good understanding of molecular structure is necessary to interpret differences in chemical behavior of inorganic species. Although the chemical formulas for CO_2 and NO_2 may not look much different, the chemistry of these compounds is greatly different. In this chapter, a description of the covalent bond will be presented as it relates to diatomic molecules and their properties. In the next two chapters, bonding in more complex molecules will be described, and the topic of molecular symmetry will be addressed.

3.1 THE BASIC IDEAS OF MOLECULAR ORBITAL METHODS

The formalism that applies to the molecular orbital method will be illustrated by considering a hydrogen molecule, but a more detailed description of both H_2^+ and H_2 will be given in the next section. To begin our description of diatomic molecules, let us imagine that two hydrogen atoms that are separated by a relatively large distance are being brought closer together. As the atoms approach each other, there is an attraction between them that gets greater the shorter the distance between them becomes. Eventually, the atoms reach a distance of separation that represents the most favorable (minimum energy) distance, the bond length in the H_2 molecule (74 pm).

As the distance between the atoms decreases, the nuclei begin to repel each other as do the two electrons. However, there are forces of attraction between the nucleus in atom 1 and the electron in atom 2 and between the nucleus in atom 2 and the electron in atom 1. We can illustrate the interactions involved as shown in Figure 3.1.

We know that for each atom the ionization potential is 13.6 eV, the bond energy for the H_2 molecule is 4.51 eV (432 kJ mol^{-1}), and the bond length is 76 pm. Keep in mind that bond energies are expressed as the energy necessary to *break* the bond and are therefore *positive* quantities. If the bond *forms*, energy equivalent to the bond energy is released so it is a *negative* quantity.

The fact that the nuclei do not get closer together does not mean that the forces of attraction and repulsion are equal. The minimum distance is that distance where the total energy (attraction and repulsion) is most favorable. Because the molecule has some vibrational energy, the internuclear distance is not constant, but the equilibrium distance is R_o. Figure 3.2 shows how the energy of interaction between two hydrogen atoms varies with internuclear distance.

61

Inorganic Chemistry. DOI: http://dx.doi.org/10.1016/B978-0-12-385110-9.00003-0

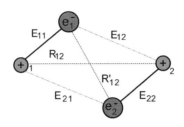

FIGURE 3.1

Interactions within a hydrogen molecule. The quantities represented as R are the repulsions and those as E are the attraction energies. The subscripts indicate the nuclei and electrons involved.

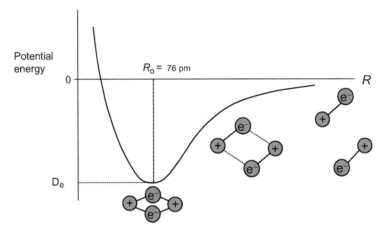

FIGURE 3.2

The interaction of two hydrogen atoms that form a molecule as they come together.

In order to describe the hydrogen molecule by quantum mechanical methods, it is necessary to make use of the principles given in Chapter 2. It was shown that a wave function provided the starting point for application of the methods that permitted the calculation of values for the dynamical variables. It is with a *wave function* that we must again begin our treatment of the H_2 molecule by the molecular orbital method. But what wave function do we need? The answer is that we need a wave function for the H_2 *molecule*, and that wave function is constructed from the *atomic* wave functions. The technique used to construct molecular wave functions is known as the linear combination of atomic orbitals (LCAO-MO). The linear combination of atomic orbitals can be written mathematically as:

$$\psi = \sum a_i \phi_i \tag{3.1}$$

In this equation, ψ is the molecular wave function, ϕ is an atomic wave function, and a is a weighting coefficient, which gives the relative weight in the "mix" of the atomic wave functions. The summation is over i, the number of atomic wave functions (number of atoms) being combined. If a diatomic molecule is being described, there are only two atoms involved so the equation becomes

$$\psi = a_1 \phi_1 + a_2 \phi_2 \tag{3.2}$$

Although the combination has been written as a sum, the difference is also an acceptable linear combination. The weighting coefficients are variables that must be determined.

In Chapter 2, it was shown that in order to calculate the average value for a dynamical variable a whose operator is α, it is necessary to make use of the relationship

$$\langle a \rangle = \frac{\int \psi^* \alpha \psi \, d\tau}{\int \psi^* \psi \, d\tau} \tag{3.3}$$

If the property we wish to determine is the energy, this equation becomes

$$E = \frac{\int \psi^* \hat{H} \psi \, d\tau}{\int \psi^* \psi \, d\tau} \tag{3.4}$$

where $\hat{H}$ is the Hamiltonian operator, the operator for total energy. The expression shown in Eqn (3.2) is substituted for ψ in the equation above, which gives

$$E = \frac{\int (a_1 \phi_1{}^* + a_2 \phi_2{}^*) \hat{H} (a_1 \phi_1 + a_2 \phi_2) d\tau}{\int (a_1 \phi_1{}^* + a_2 \phi_2{}^*)(a_1 \phi_1 + a_2 \phi_2) d\tau} \tag{3.5}$$

When the multiplications are carried out and the constants are removed from the integrals, we obtain

$$E = \frac{a_1^2 \int \phi_1{}^* \hat{H} \phi_1 d\tau + 2 a_1 a_2 \int \phi_1{}^* \hat{H} \phi_2 d\tau + a_2^2 \int \phi_2{}^* \hat{H} \phi_2 d\tau}{a_1^2 \int \phi_1{}^* \phi_1 d\tau + 2 a_1 a_2 \int \phi_1{}^* \phi_2 d\tau + a_2^2 \int \phi_2{}^* \phi_2 d\tau} \tag{3.6}$$

In writing this equation, it was assumed that

$$\int \phi_1{}^* \hat{H} \phi_2 d\tau = \int \phi_2{}^* \hat{H} \phi_1 d\tau \tag{3.7}$$

and that

$$\int \phi_1{}^* \phi_2 d\tau = \int \phi_2{}^* \phi_1 d\tau \tag{3.8}$$

These assumptions are valid for a diatomic molecule composed of identical atoms (*homonuclear diatomic*) because ϕ_1 and ϕ_2 are identical and real in this case. In working with the quantities in equations such as Eqn (3.6), certain elements are frequently encountered. For simplicity, the definitions that will be adopted are as follows.

$$H_{11} = \int \phi_1{}^* \hat{H} \phi_1 d\tau \tag{3.9}$$

$$H_{12} = \int \phi_1{}^* \hat{H} \phi_2 d\tau \tag{3.10}$$

Because $\hat{H}$ is the operator for total energy, H_{11} represents the binding energy of an electron in atom 1 to its nucleus. If the subscripts on the wave functions are both 2, the binding energy of electron 2 to its nucleus is indicated. Such integrals represent the energy as an electrostatic interaction so they are known as *Coulomb integrals*. Integrals of the type shown in Eqn (3.10) indicate the energy of the interaction of the electron in atom 1 with the nucleus in atom 2. Therefore, they are known as *exchange integrals*. As a result of the Hamiltonian being an operator for energy, both types of integrals represent energies. Furthermore, because these integrals represent favorable interactions, they are both negative (representing attractions) in sign.

Because the integral shown in Eqn (3.9) represents the energy with which electron 1 is bound to nucleus 1, it is simply the binding energy of the electron in atom 1. The binding energy of an electron is the reverse (with respect to sign) of the ionization potential. Therefore, it is customary to represent these Coulomb integrals in terms of the ionization potentials by reversing the sign. Although it will not be shown here, the validity of this approximation lies in a principle known as Koopmans' theorem. The valence state ionization potential (VSIP) is generally used to give the value of the Coulomb integral. This assumes that the orbitals are identical in the ion and the neutral atom. However, this relationship is not strictly correct. Suppose an electron is being removed from a carbon atom that has a configuration of $2p^2$. With there being two electrons in a set of three orbitals, there are 15 microstates that represent the possible permutations in placing the electrons in the orbitals. There is an exchange energy that is associated with this configuration because of the interchangeability of the electrons in the orbitals, and we say that the electrons are *correlated*. When one electron is removed, the single electron remaining in the $2p$ orbitals has a different exchange energy, so the measured ionization potential also has associated with it other energy terms related to the difference in exchange energy. Such energies are small compared to the ionization potential so the VSIP energies are normally used to represent the Coulomb integrals.

The exchange integrals (also knows as resonance integrals) represent the interaction of nucleus 1 with electron 2 and the interaction of nucleus 2 with electron 1. Interactions of this type must be related to the distance separating the nuclei so the value of an exchange integral can be expressed in terms of the distance separating the atomic nuclei.

In addition to the integrals that represent energies, there are integrals of a type in which no operator occurs. These are represented as

$$S_{11} = \int \phi_1^* \phi_1 d\tau \tag{3.11}$$

$$S_{12} = \int \phi_1^* \phi_2 d\tau \tag{3.12}$$

Integrals of this type are known as *overlap integrals* and in a general way, they represent effectiveness with which the orbitals overlap in a region of space. If the subscripts are identical, orbitals on the same atom are indicated, and if the atomic wave functions are normalized, the value of such an integral is 1. As a result, we can write

$$S_{11} = \int \phi_1^* \phi_1 d\tau = S_{22} = \int \phi_2^* \phi_2 d\tau = 1 \tag{3.13}$$

On the other hand, the integrals of the type

$$S_{12} = \int \phi_1^* \phi_2 d\tau = S_{21} = \int \phi_2^* \phi_1 d\tau = 1 \tag{3.14}$$

are related to the degree of overlap of an orbital on atom 1 with an orbital on atom 2. If the two atoms are separated by a large distance, the overlap integral approaches 0. However, if the atoms are closer together, there is some overlap of the orbitals and $S > 0$. If the atoms were forced together in such a way that the two nuclei coincided (internuclear distance is 0), we would expect $S = 1$ because the orbitals would be congruent. Clearly, the value of an overlap integral such as those shown in Eqn (3.14) must be somewhere between 0 and 1, and it must be a function of the internuclear distance. With the exchange integrals and overlap integrals both being functions of the internuclear distance, it should be possible to express one in terms of the other. We will return to this point later.

The appearance of Eqn (3.6) can be simplified greatly by using the notation described above. When the substitutions are made, the result is

$$E = \frac{a_1^2 H_{11} + 2a_1 a_2 H_{12} + a_2^2 H_{22}}{a_1^2 + 2a_1 a_2 S_{12} + a_2^2} \tag{3.15}$$

in which we have assumed that S_{11} and S_{22} are both equal to 1 because of the atomic wave functions being normalized. We now seek to find values of the weighting coefficients that make the energy a minimum. To find a minimum in the energy expression, we take the partial derivatives with respect to a_1 and a_2 and set them equal to 0.

$$\left(\frac{\partial E}{\partial a_1}\right)_{a_2} = 0 \quad \left(\frac{\partial E}{\partial a_2}\right)_{a_1} = 0 \tag{3.16}$$

When the differentiations are carried out with respect to a_1 and a_2 in turn while keeping the other constant, we obtain two equations that after simplification can be written as

$$a_1(H_{11} - E) + a_2(H_{12} - S_{12}E) = 0 \tag{3.17}$$

$$a_1(H_{21} - S_{21}E) + a_2(H_{22} - E) = 0 \tag{3.18}$$

These equations are known as the *secular equations* and in them the weighting coefficients a_1 and a_2 are the unknowns. These equations constitute a pair of linear equations that can be written in the form

$$ax + by = 0 \quad \text{and} \quad cx + dy = 0 \tag{3.19}$$

It can be shown that a nontrivial solution for a pair of linear equations requires that the determinant of the coefficients must be equal to 0. This means that

$$\begin{vmatrix} H_{11} - E & H_{12} - S_{12}E \\ H_{21} - S_{21}E & H_{22} - E \end{vmatrix} = 0 \tag{3.20}$$

The molecule being described is a homonuclear diatomic, so $H_{12} = H_{21}$ and $S_{12} = S_{21}$. If we represent S_{12} and S_{21} by S and let $H_{11} = H_{22}$, the expansion of the determinant yields

$$(H_{11} - E)^2 - (H_{12} - SE)^2 = 0 \tag{3.21}$$

By equating the two terms on the left hand side of Eqn (3.21), taking the square root gives

$$H_{11} - E = \pm(H_{12} - SE) \tag{3.22}$$

from which we find two values for E (denoted as E_b and E_a)

$$E_b = \frac{H_{11} + H_{12}}{1 + S} \quad \text{and} \quad E_a = \frac{H_{11} - H_{12}}{1 - S} \tag{3.23}$$

The energy state labeled E_b is known as the *bonding* or *symmetric* state, whereas that designated as E_a is called the *antibonding* or *asymmetric* state. Because both H_{11} and H_{12} are *negative* (binding) energies, E_b represents the state of lower energy. Figure 3.3 shows the qualitative energy diagram for the bonding and antibonding molecular orbitals relative to the 1s atomic orbital.

Figure 3.4 shows a more correctly scaled energy level diagram that results for the hydrogen molecule. Note that the energy for the 1s atomic orbital of a hydrogen atom is at -1312 kJ mol^{-1} because the ionization potential is 1312 kJ mol^{-1} (13.6 eV). Note also that the bonding molecular orbital has an energy of -1528 kJ mol^{-1}, which is lower than that of the 1s state.

If a hydrogen molecule is separated into the two constituent atoms, the result is equivalent to taking the two electrons in the bonding molecular orbital and placing them back in the atomic orbitals.

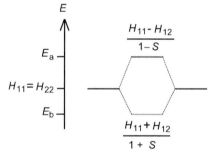

FIGURE 3.3
Combination of two *s* orbitals to produce bonding and antibonding orbitals.

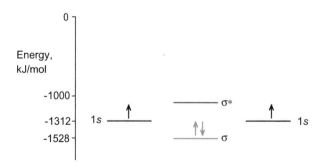

FIGURE 3.4
The energy level diagram for the H_2 molecule.

Because there are two electrons, that energy would be $2(1528 - 1312) = 432$ kJ mol^{-1}, the bond energy of the H_2 molecule. From the molecular orbital diagram, it can be seen that although the energy of the antibonding state is higher than that of the hydrogen atom, it is still very *negative*. An energy of 0 does not result because even when the atoms are completely separated the energy of the system is the sum of the binding energies in the atoms, which is 2 (-1312) kJ mol^{-1}. The bonding and antibonding states are "split" below and above the energy state of the electron in an atom, not below and above an energy of 0. However, the antibonding state is raised a greater amount than the bonding state is lowered relative to the atomic orbital energy. This can be seen from the relationships shown in Eqns (3.23) because in the first case the denominator is $(1 + S)$ but in the other it is $(1 - S)$.

If we substitute the values for the energy as shown in Eqn (3.23) into the secular equations, we find that

$$a_1 = a_2 (\text{for the } symmetric \text{ state}) \quad \text{and} \quad a_1 = -a_2 (\text{for the } antisymmetric \text{ state})$$

When these relationships between the weighting coefficients are used, it is found that

$$\psi_b = a_1\phi_1 + a_2\phi_2 = \frac{1}{\sqrt{2 + 2S}}(\phi_1 + \phi_2) \tag{3.24}$$

$$\psi_a = a_1\phi_1 - a_2\phi_2 = \frac{1}{\sqrt{2 - 2S}}(\phi_1 - \phi_2) \tag{3.25}$$

If we let A represent the normalization constant, the condition for normalization is that

$$1 = \int A^2(\phi_1 + \phi_2)d\tau = A^2\left[\int \phi_1^2 d\tau + \int \phi_2^2 d\tau + 2\int \phi_1\phi_2 d\tau\right] \tag{3.26}$$

The first and second integrals on the right hand side of this equation evaluate to 1 because the atomic wave functions are assumed to be normalized. Therefore, the right hand side of the equation reduces to

$$1 = A^2[1 + 1 + 2S] \tag{3.27}$$

and we find that the normalization constant is given by

$$A = \frac{1}{\sqrt{2 + 2S}} \tag{3.28}$$

and the wave functions can be represented as shown in Eqns (3.24) and (3.25)

Although we have dealt with a diatomic molecule consisting of two hydrogen atoms, the procedure is exactly the same if the molecule is Li_2 except that the atomic wave functions are 2s wave functions and the energies involved are those appropriate to lithium atoms. The VSIP for lithium is only 513 kJ mol^{-1} rather than 1312 kJ mol^{-1} as it is for hydrogen.

When performing molecular orbital calculations, overlap and exchange integrals must be evaluated. With modern computing techniques, overlap integrals are most often evaluated as part of the calculation. The wave functions are of the Slater type (see Section 2.4), and the overlap integrals can be evaluated for varying bond lengths and angles. Many years ago, it was common for the values of overlap integrals to be looked up in a massive set of tables, which presented the values of the overlap integrals for the various combinations of atomic orbitals and internuclear distances. These tables, known as the Mulliken tables, were prepared by R. A. Mulliken and coworkers, and they were essential for giving part of the data needed to perform molecular orbital calculations.

The exchange integrals, H_{ij}, are evaluated by representing them as functions of the Coulomb integrals, H_{ii}, and the overlap integrals. One such approximation is known as the Wolfsberg–Helmholtz approximation, which is written as

$$H_{12} = -KS\left(\frac{H_{11} + H_{22}}{1 + S}\right) \tag{3.29}$$

where H_{11} and H_{22} are the Coulomb integrals for the two atoms, S is the overlap integral, and K is a constant having a numerical value of approximately 1.75. Because the overlap integral is a function of the bond length, so is the exchange integral.

The quantity $(H_{11} + H_{22})$ may not be the best way to combine Coulomb integrals for cases where the atoms have greatly different ionization potentials. In such cases, it is preferable to use the Ballhausen–Gray approximation,

$$H_{12} = -KS(H_{11}H_{22})^{1/2} \tag{3.30}$$

Another useful approximation for the H_{12} integral is that known as the Cusachs approximation, and it is written as

$$H_{12} = \frac{1}{2}S(K - |S|)(H_{11} + H_{22}) \tag{3.31}$$

Although the energy of a chemical bond as a function of internuclear distance can be represented by the potential energy curve shown in Figure 3.2, neither the Wolfsberg–Helmholtz nor the Ballhausen–Gray approximation is a function that possesses a minimum. However, the approximation of Cusachs is a mathematical expression that does pass through a minimum.

It has already been shown that the energy of the bonding molecular orbital can be written as $E_b = (H_{11} + H_{12})/(1 + S)$. Suppose there are two electrons in a bonding orbital having this energy and the bond is being broken. If the bond breaking is homolytic (one electron ends up on each atom), the electrons will reside in atomic orbitals having an energy of H_{11} (which is the same as H_{22} if a homonuclear molecule is being disrupted). Before the bond is broken, the two electrons have a total energy that is given by the expression $2[(H_{11} + H_{12})/(1 + S)]$, and after the bond is broken the binding energy for the two electrons is $2H_{11}$. Therefore, the bond energy (BE) can be expressed as

$$BE = 2H_{11} - 2\left(\frac{H_{11} + H_{12}}{1 + S}\right) \tag{3.32}$$

In order to use this equation to calculate a bond energy, it is necessary to have the values for H_{12} (the values for the H_{ii} integrals are usually available by approximating from the ionization potentials) and S. To a rough approximation (known as neglecting the overlap), the value of S can be assumed to be 0 because the value is small (in the range 0.1–0.4) in many cases.

3.2 THE H$_2^+$ AND H$_2$ MOLECULES

The simplest diatomic molecule consists of two nuclei and a single electron. That species, H$_2^+$, has properties, some of which are well known. For example, in H$_2^+$, the internuclear distance is 104 pm and the bond energy is 268 kJ mol^{-1}. Proceeding as illustrated in the previous section, the wave function for the bonding molecular orbital can be written as

$$\psi_b = a_1\phi_1 + a_2\phi_2 = \frac{1}{\sqrt{2+2S}}(\phi_1 + \phi_2) \tag{3.33}$$

This wave function describes a bonding orbital of the σ type that arises from the combination of two 1s wave functions for atoms 1 and 2. To make that point clear, the wave function could be written as

$$\psi_b(\sigma) = a_1\phi_{1(1s)} + a_2\phi_{2(1s)} = \frac{1}{\sqrt{2+2S}}\left(\phi_{1(1s)} + \phi_{2(1s)}\right) \tag{3.34}$$

The expression above is actually a one-electron wave function, which is adequate in this case but not for the H$_2$ molecule. The energy associated with this molecular orbital can be calculated as shown in Eqn 3.4 from which we obtain

$$E[\psi_b(\sigma)] = \int [\psi_b(\sigma)] \; \hat{H} \; [\psi_b(\sigma)] \; d\tau \tag{3.35}$$

When the approximation is made that the overlap can be neglected, $S = 0$, and the normalization constant is $1/2^{1/2}$. Therefore, after substituting the results shown in Eqn (3.34) for $\psi_b(\sigma)$, the expression for the energy of the molecular orbital can be written as

$$E[\psi_b(\sigma)] = \frac{1}{2}\int \left(\phi_{1(1s)} + \phi_{2(1s)}\right) \hat{H}\left(\phi_{1(1s)} + \phi_{2(1s)}\right) d\tau \tag{3.36}$$

Separating the integral gives

$$E[\psi_b(\sigma)] = \frac{1}{2}\int \phi_{1(1s)} \hat{H} \phi_{1(1s)}d\tau + \frac{1}{2}\int \phi_{2(1s)}\hat{H} \phi_{2(1s)}d\tau + \frac{1}{2}\int \phi_{1(1s)}\hat{H}\phi_{2(1s)}d\tau$$
$$+ \frac{1}{2}\int \phi_{2(1s)}\hat{H}\phi_{1(1s)}d\tau \tag{3.37}$$

As we saw earlier, the first two terms on the right hand side of this equation represent the electron binding energies in atoms 1 and 2, respectively, which are H_{11} and H_{22}, the Coulomb integrals. The last two terms represent the exchange integrals, H_{12} and H_{21}. In this case, $H_{11} = H_{22}$ and $H_{12} = H_{21}$ because the nuclei are identical. Therefore, the energy of the orbital is

$$E[\psi_b(\sigma)] = \frac{1}{2}H_{11} + \frac{1}{2}H_{11} + \frac{1}{2}H_{12} + \frac{1}{2}H_{12} = H_{11} + H_{12} \tag{3.38}$$

Although the development will not be shown here, the energy of the antibonding orbital can be written as

$$E[\psi_a(\sigma)] = \frac{1}{2}H_{11} + \frac{1}{2}H_{11} - \frac{1}{2}H_{12} - \frac{1}{2}H_{12} = H_{11} - H_{12} \tag{3.39}$$

We have already noted that in the case where $S = 0$ the molecular orbitals reside above and below the atomic states by an amount H_{12}. Using this approach, the calculated values for the bond energy and internuclear distance in H_2^+ do not agree well with the experimental values. Improved values are obtained when an adjustment in the total positive charge of the molecule is made as was done in the case of the helium atom. For H_2^+, it turns out that the electron is acted on as if the total nuclear charge was about 1.24 rather than 2. This is essentially the same approach as that taken in the case of the helium atom (see Section 2.3). Also, the molecular orbital wave function was constructed by taking a linear combination of $1s$ atomic wave functions. A better approach is to take an atomic wave function that contains not only s character but also a contribution from the $2p_x$ orbital that lies along the internuclear axis. When these changes are made, the agreement between the calculated and experimental properties of H_2^+ is much better.

The wave functions described above are one-electron wave functions, but the H_2 molecule has two electrons to be dealt with. In the methods of molecular orbital theory, a wave function for the two electrons in the hydrogen molecule is formed by taking the product of the two one-electron wave functions. Therefore, the wave function for the bonding molecular orbital in H_2 is written in terms of the atomic wave functions, ϕ, as

$$\psi_{b,1} \psi_{b,2} = \left[\phi_{A,1} + \phi_{B,1}\right]\left[\phi_{A,2} + \phi_{B,2}\right] \tag{3.40}$$

In this case, the subscript b indicates the bonding (σ) orbital, A and B subscripts denote the two nuclei, and subscripts 1 and 2 denote electrons 1 and 2, respectively. Expanding the expression on the right hand side of Eqn (3.40) gives

$$\psi_{b,1} \psi_{b,2} = \phi_{A,1}\phi_{B,2} + \phi_{A,2}\phi_{B,1} + \phi_{A,1}\phi_{A,2} + \phi_{B,1}\phi_{B,2} \tag{3.41}$$

In this expression, the term $\phi_{A,1} \phi_{B,2}$ essentially represents the interaction of two $1s$ orbitals for hydrogen atoms A and B. The term $\phi_{A,2} \phi_{B,1}$ represents the same type of interaction with the electrons interchanged. However, the term $\phi_{A,1} \phi_{A,2}$ represents *both* electrons 1 and 2 interacting with nucleus A. That means the structure described by the wave function is ionic, $H_A^- H_B^+$. In an analogous way, the term $\phi_{B,1} \phi_{B,2}$ represents both electrons interacting with nucleus B, which corresponds to the structure $H_A^+ H_B^-$. Therefore, what we have devised for a molecular wave function actually describes the hydrogen molecule as a "hybrid" (a valence bond term that is applied incorrectly) of

$$H_A : H_B \leftrightarrow H_A^- H_B^+ \leftrightarrow H_A^+ H_B^-$$

As in the case of the H_2^+ molecule, the calculated properties (bond energy and bond length) of the H_2 molecule do not agree well with the experimental values when this wave function is utilized. An improvement is made by allowing the nuclear charge to be a variable with the optimum value being about 1.20. Additional improvement in calculated and experimental values is achieved when the atomic wave functions are not considered to be pure $1s$ orbitals but rather allow some mixing of the $2p$ orbitals. Also, the wave function shown in Eqn (3.41) makes no distinction in the weighting given to the covalent and ionic structures. Our experience tells us that for identical atoms that have the same electronegativity, an ionic structure would not be nearly as significant as a covalent one. Therefore, weighting parameters should be introduced that adjust the contributions of the two types of structures to reflect the chemical nature of the molecule.

The discussion above is presented in order to show how the basic ideas of the molecular orbital approach are employed. It is also intended to show how to approach getting improved results after the basic ideas are used to generate molecular wave functions. For the purposes here, it is sufficient to indicate the nature of the changes rather than presenting quantitative results of the calculations.

A simple interpretation of the nature of a covalent bond can be seen by considering some simple adaptations of the wave function. For example, it is ψ^2 that is related to probability of finding the electrons. When we write the wave function for a bonding molecular orbital as ψ_b, that means that because $\psi_b = \phi_A + \phi_B$

$$\psi_b^2 = (\phi_A + \phi_B)^2 = \phi_A^2 + \phi_B^2 + 2\phi_A\phi_B \tag{3.42}$$

Because the last term (when integration is performed over all space) becomes

$$\int \phi_A \phi_B \, d\tau$$

this is actually an overlap integral. The expression in Eqn (3.42) indicates that there is an *increased* probability of finding the electrons between the two nuclei as a result of orbital overlap. This is, of course, for the *bonding* molecular orbital. For the antibonding orbital, the combination of atomic wave functions is written as

$$\psi_a^2 = (\phi_A - \phi_B)^2 = \phi_A^2 + \phi_B^2 - 2\phi_A\phi_B \tag{3.43}$$

This expression indicates that there is a decreased probability (indicated by the term $-2\phi_A\phi_B$) of finding the electrons in the region between the two nuclei. In fact, there is a nodal plane between the positive and negative (with respect to algebraic sign) of the two regions of the molecular orbital. As a simple definition, we can describe a covalent bond as the increased probability of finding electrons between two nuclei or an increase in electron density between the two nuclei when compared to the probability or density that would exist simply because of the presence of two atoms.

3.3 DIATOMIC MOLECULES OF SECOND-ROW ELEMENTS

The basic principles dealing with the molecular orbital description of the bonding in diatomic molecules have been presented in the previous section. However, somewhat different considerations are involved when second-row elements are involved in the bonding because of the differences between s and p orbitals. When the orbitals being combined are p orbitals, the lobes can combine in such a way that the overlap is symmetric around the internuclear axis. Overlap in this way gives rise to a σ bond. This type of overlap involves p orbitals for which the overlap is essentially "end on" as shown in Figure 3.5. For reasons that will become clear later, it will be assumed that the p_z orbital is the one used in this type of combination.

The essential idea is that orbital lobes of the same mathematical sign can lead to favorable overlap (the overlap integral has a value >0). This can occur between orbitals of different types in several ways. Figure 3.6 shows a few of the types of orbital overlap that lead to bonding. As we shall see in later chapters, some of these types are quite important.

Representing the p_z orbitals on atoms 1 and 2 by z_1 and z_2, the combinations of atomic wave functions can be shown as

$$\psi(\sigma_z) = \frac{1}{\sqrt{2 + 2S}} [\phi(z_1) + \phi(z_2)] \tag{3.44}$$

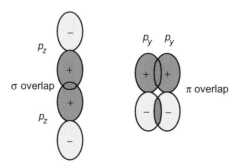

FIGURE 3.5
Possible ways for overlap of p orbitals to occur.

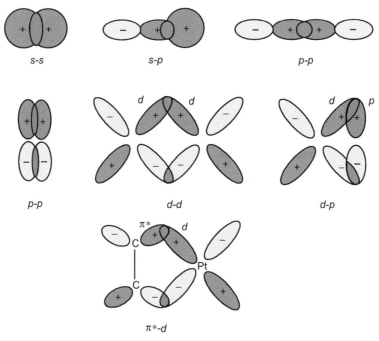

FIGURE 3.6
Some types of orbital overlap that lead to energetically favorable interactions (overlap integral >0).

$$\psi(\sigma_z)^* = \frac{1}{\sqrt{2-2S}}[\phi(z_1) - \phi(z_2)] \tag{3.45}$$

After the σ bond has formed, further interaction of the p orbitals on the two atoms is restricted to the p_x and p_y orbitals, which are perpendicular to the p_z orbital. When these orbitals interact, the region of orbital overlap is not symmetrical around the internuclear axis but rather on either side of the internuclear axis, and a π bond results. Orbital overlap of this type is also shown in Figures 3.5 and 3.6. The combinations of wave functions for the bonding π orbitals can be written as

$$\psi(\pi_x) = \frac{1}{\sqrt{2 + 2S}} [\phi(x_1) + \phi(x_2)] \tag{3.46}$$

$$\psi(\pi_y) = \frac{1}{\sqrt{2 + 2S}} [\phi(x_1) + \phi(x_2)] \tag{3.47}$$

The two bonding π orbitals represented by these wave functions are degenerate. The wave functions for the antibonding states are identical in form except that negative signs are used in the combination of atomic wave functions and in the normalization constants.

The combination of three p orbitals on one atom with three p orbitals on another leads to the formation of one σ and two π bonding molecular orbitals. The order in which the orbitals are populated is

$$\sigma(2s)\ \sigma^*(2s)\ \sigma(2p_z)\ \pi(2p_x)\ \pi(2p_y)\ldots$$

It might be assumed that the $\sigma(2p_z)$ orbital would always have a lower energy than the two π orbitals, but that is not necessarily the case. Orbitals of similar energy interact best when combining as a result of hybridization. Mixing of the 2s and $2p_z$ orbitals is allowed in terms of symmetry (see Chapter 5), but the combination of 2s and $2p_x$ or $2p_y$ results in zero overlap because they are orthogonal. For the elements early in the second period where the nuclear charge is low, the 2s and 2p orbitals are similar in energy so it is possible for them to hybridize extensively. For the later members of the group (N, O, and F), the higher nuclear charge causes the difference in energy between the 2s and 2p orbitals to be great enough that they cannot hybridize effectively. One result of hybridizing orbitals is that their energies are changed, and in this case that results in the order of filling the σ and π orbitals to be reversed so that for B_2 and C_2, there is experimental evidence to show that the π orbitals lie lower in energy than the σ orbital. For the atoms later in the second period, the extent of hybridization of the 2s and 2p orbitals is slight, which results in the σ orbital lying lower in energy than the two π orbitals. The order of filling the molecular orbitals for the early elements is

$$\sigma(2s)\ \sigma^*(2s)\ \pi(2p_x)\ \pi(2p_y)\ \sigma(2p_z)\ldots$$

Figure 3.7 shows both of the molecular orbital energy diagrams that result for diatomic molecules of second-row elements.

The fact that the B_2 molecule is paramagnetic shows that the highest occupied molecular orbitals (HOMO) are the degenerate π orbitals each of which is occupied by one electron. Further evidence that this is the correct energy level scheme to be used for C_2 comes from the fact that the molecule is diamagnetic. The molecular orbital configurations for these molecules can be written as

$$B_2 \quad (\sigma)^2 (\sigma *)^2 (\pi)^1 (\pi)^1$$

$$C_2 \quad (\sigma)^2 (\sigma *)^2 (\pi)^2 (\pi)^2$$

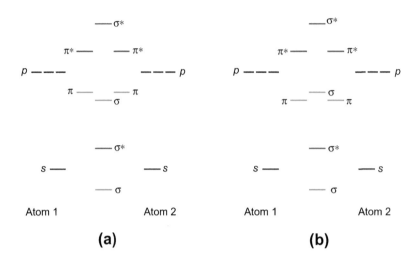

FIGURE 3.7
Energy level diagrams for diatomic molecules of second-row elements. Early members of the series follow the diagram shown in (b), whereas later members follow (a).

Sometimes, these configurations are shown with the atomic orbitals indicated from which the molecular orbitals arise. For example, B_2 could be described as

$$B_2 \quad [\sigma(2s)]^2 [\sigma^*(2s)]^2 [\pi(2p_x)]^1 [\pi(2p_y)]^1$$

Although the subject of symmetry has not yet been discussed in this book, it is true that σ orbitals that are bonding in character have "g" symmetry because the wave functions are symmetric with respect to the center of the bond. Essentially, this means that if $\psi(x,y,z)$ is equal to $\psi(-x,-y,-z)$, the function is said to be an even function or to have *even parity*. This is signified by "g", which comes from the German word *gerade* meaning even. If $\psi(x,y,z)$ is equal to $-\psi(-x,-y,-z)$, the function has *odd parity* and it is indicated by "u", which comes from the word *ungerade* meaning uneven. An atomic s orbital is g, whereas a p orbital is u in symmetry. Although a bonding s orbital is g, π bonding orbitals have "u" symmetry because they are antisymmetric with respect to the internuclear axis. Antibonding molecular orbitals of each type have the symmetry labels reversed. Sometimes the symmetry character of the molecular orbital is indicated by means of a subscript. When this is done, the representation for B_2 is

$$B_2 \quad (\sigma_g)^2 (\sigma_u)^2 (\pi_u)^1 (\pi_u)^1$$

Molecular orbitals are sometimes given numerical prefixes to show the order in which the orbitals having those type and symmetry designations are encountered. When this is done, the order of filling the molecular orbitals for the second-row elements is shown as

$$1\sigma_g \ 1\sigma_u \ 2\sigma_g \ 1\pi_u \ 1\pi_u \ 1\pi_g \ 1\pi_g \ 2\sigma_u$$

In this case, "1" indicates the first instance where an orbital of that type is encountered. A "2" indicates the second time an orbital having the designation following the number is encountered. The various ways to identify molecular orbitals are shown here because different schemes are sometimes followed by different authors.

In writing these configurations for diatomic molecules of second-row elements, we have omitted the electrons from the $1s$ orbitals because they are not part of the valence shells of the atoms. When considering the oxygen molecule, for which the σ orbital arising from the combinations of the $2p_z$ orbitals lies lower in energy than the π orbitals, we find that the electron configuration is

$$O_2 \quad (\sigma)^2 \, (\sigma^*)^2 \, (\sigma)^2 \, (\pi)^2 \, (\pi^*)^1 \, (\pi^*)^1$$

Note that there are two unpaired electrons in the degenerate π^* orbitals, and as a result, the oxygen molecule is paramagnetic. Figure 3.8 shows the orbital energy diagrams for the diatomic molecules of the second-row elements.

A concept that is important when considering bonds between atoms is the *bond order*, B. The bond order is a measure of the *net* number of electron pairs used in bonding. It is related to the number of electrons in bonding orbitals (N_b) and the number in antibonding orbitals (N_a) by the equation

$$B = \frac{1}{2}(N_b - N_a) \tag{3.48}$$

The bond order for each diatomic molecule is given in Figure 3.8 as is the bond energy. Note that there is a general increase in bond energy as the bond order increases. This fact makes it possible to see why certain species behave as they do. For example, the bond order for the O_2 molecule is $(8 - 4)/2 = 2$, and

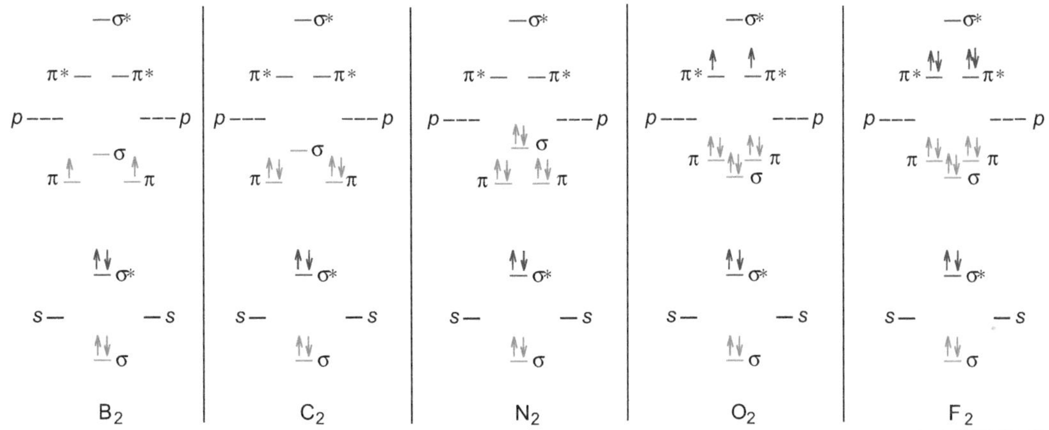

	B$_2$	C$_2$	N$_2$	O$_2$	F$_2$
B.O.	1	2	3	2	1
R, pm	159	131	109	121	142
B.E., eV	3.0	5.9	9.8	5.1	1.6

FIGURE 3.8
Molecular orbital diagrams for second-row homonuclear diatomic molecules. Keep in mind that in these molecular orbital diagrams, the atomic orbitals are not all at the same energy so neither do the molecular orbitals of the same type have the same energy for different molecules.

we say that a bond order of 2 is equivalent to a double bond. If an electron is removed from an oxygen molecule, the species O_2^+ results, and the electron removed comes from the highest occupied orbital, which is a π^* (antibonding) orbital. The bond order for O_2^+ is $(8 - 3)/2 = 2.5$, which is higher than that of the O_2 molecule. Because of this, it is not at all unreasonable to expect there to be some reactions in which an oxygen molecule reacts by losing an electron to form O_2^+, the dioxygenyl cation. Of course, such a reaction would require the reaction of oxygen with a very strong oxidizing agent. One such oxidizing agent is PtF_6 that contains platinum in the +6 oxidation state. The reaction with oxygen can be written as

$$PtF_6 + O_2 \rightarrow O_2^+ + PtF_6^-$$ (3.49)

Although this reaction shows the formation of O_2^+, it is also possible to add one electron to the O_2 molecule to produce O_2^-, the superoxide ion, or two electrons to form O_2^{2-}, the peroxide ion. In each case, the electrons are added to the antibonding π^* orbitals, which reduces the bond order from the value of 2 in the O_2 molecule. For O_2^-, the bond order is 1.5, but it is only 1 for O_2^{2-}, the peroxide ion. The $O-O$ bond energy in the peroxide ion has a strength of only 142 kJ mol^{-1} and, as expected, most peroxides are very reactive compounds. The superoxide ion is produced by the reaction

$$K + O_2 \rightarrow KO_2$$ (3.50)

In addition to the homonuclear molecules, the elements of the second period form numerous important and interesting heteronuclear species, both neutral molecules and diatomic ions. The molecular orbital diagrams for several of these species are shown in Figure 3.9. Keep in mind that the energies of the molecular orbitals having the same designations are not equal for these species. The diagrams are only qualitatively correct.

It is interesting to note that both CO and CN$^-$ are isoelectronic with the N_2 molecule. That is, they have the same number and arrangement of electrons as the N_2 molecule. However, as we will see later,

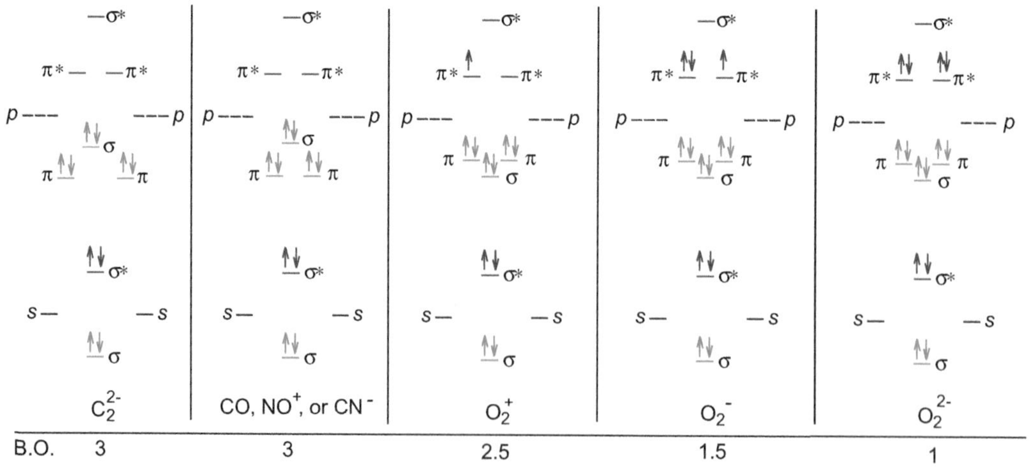

FIGURE 3.9
Molecular orbital diagrams for some heteronuclear molecules and ions of second-row elements.

Table 3.1 Characteristics of Some Diatomic Species

Species	N_b	N_a	B^a	R, pm	DE^b, eV
H_2^+	1	0	0.5	106	2.65
H_2	2	0	1	74	4.75
He_2^+	2	1	0.5	108	3.1
Li_2	2	0	1	262	1.03
B_2	4	2	1	159	3.0
C_2	6	2	2	131	5.9
N_2	8	2	3	109	9.76
O_2	8	4	2	121	5.08
F_2	8	6	1	142	1.6
Na_2	2	0	1	308	0.75
Rb_2	2	0	1	–	0.49
S_2	8	4	2	189	4.37
Se_2	8	4	2	217	3.37
Te_2	8	4	2	256	2.70
N_2^+	7	2	2.5	112	8.67
O_2^+	8	3	2.5	112	6.46
BN	6	2	2	128	4.0
BO	7	2	2.5	120	8.0
CN	7	2	2.5	118	8.15
CO	8	2	3	113	11.1
NO	8	3	2.5	115	7.02
NO^+	8	2	3	106	–
SO	8	4	2	149	5.16
PN	8	2	3	149	5.98
SiO	8	2	3	151	8.02
LiH	2	0	1	160	2.5
NaH	2	0	1	189	2.0
PO	8	3	2.5	145	5.42

^{a} B is the bond order, $(N_b - N_a)/2$.
^{b} DE is the dissociation energy (1 eV = 96.48 kJ mol^{-1}).

these species are quite different from N_2 in their chemical behavior. The properties of many homonuclear and heteronuclear molecules and ions are presented in Table 3.1.

3.4 PHOTOELECTRON SPECTROSCOPY

Most of what we know about the structure of atoms and molecules has been obtained by studying the interaction of electromagnetic radiation with matter. Line spectra reveal the existence of shells of different energy where electrons are held in atoms. From the study of molecules by means of infrared spectroscopy, we obtain information about vibrational and rotational states of molecules. The types of bonds present, the geometry of the molecule, and even bond lengths may be determined in specific cases. The spectroscopic technique known as photoelectron spectroscopy (PES) has been of enormous importance in determining how electrons are bound in molecules. This technique provides direct information on the energies of molecular orbitals in molecules.

In PES, high-energy photons are directed to the target from which electrons are ejected. The photon source that is frequently employed is the He(I) source that emits photons having an energy of 21.22 eV as the excited state $2s^1 2p^1$ relaxes to the $1s^2$ ground state. The ionization potential for the hydrogen atom is 13.6 eV, and the first ionization potential for many molecules is of comparable magnitude. The principle on which PES works is that a photon striking an electron causes the electron to be ejected. The kinetic energy of the ejected electron will by

$$\frac{1}{2}mv^2 = hv - I \tag{3.51}$$

where hv is the energy of the incident photon and I is the ionization potential for the electron. This situation is somewhat analogous to the photoelectric effect (see Section 1.2). A molecule, M, is ionized by a photon,

$$hv + M \rightarrow M^+ + e^- \tag{3.52}$$

Electrons that are ejected are passed through an analyzer and by means of a variable voltage, electrons having different energies can be detected. The number of electrons having specific energies is counted, and a spectrum showing the number of electrons emitted (intensity) vs. energy is produced. In most cases, when an electron is removed during ionization, most molecules are in their lowest vibrational state. Spectra for diatomic molecules show a series of closely spaced peaks that correspond to ionization that leads to ions that are in excited vibrational states. If ionization takes place with the molecule in its lowest vibrational state to produce the ion in its lowest vibrational state, the transition is known as an *adiabatic ionization*. When a diatomic molecule is ionized, the most intense absorption corresponds to ionization with the molecule and the resulting ion having the same bond length (see Section 13.6). This is known as the *vertical ionization*, and it leads to the ion being produced in excited vibrational states. In general, the molecule and the ion have nearly identical bond lengths when the electron is ejected from a nonbonding orbital.

Applications of the PES technique to molecules have yielded an enormous amount of information regarding molecular orbital energy levels. For example, PES has shown that the bonding π orbitals in oxygen are higher in energy than the σ orbital arising from the combination of the $2p$ wave functions. For nitrogen, the reverse order of orbitals is found. When electrons are ejected from the bonding σ_{2p} orbital of O_2, two absorption bands are observed. There are two electrons populating that orbital, one with a spin of $+\frac{1}{2}$ and the other with a spin of $-\frac{1}{2}$. If the electron removed has a spin of $-\frac{1}{2}$, the electron having a spin of $+\frac{1}{2}$ remains, and it can interact with the two electrons in the π^* orbitals that have spins of $+\frac{1}{2}$. This can be shown as follows where $(\sigma)^{1(+\frac{1}{2})}$ means that there is one electron having a spin of $+\frac{1}{2}$ in the σ orbital, etc.

$$O_2^+ \quad (\sigma)^2(\sigma^*)^2(\sigma)^{1\left(+\frac{1}{2}\right)}(\pi)^2(\pi)^2(\pi^*)^{1\left(+\frac{1}{2}\right)}(\pi^*)^{1\left(+\frac{1}{2}\right)}$$

If the electron removed from the s orbital has a spin of $+\frac{1}{2}$, the resulting O_2^+ ion is

$$O_2^+ \quad (\sigma)^2(\sigma^*)^2(\sigma)^{1\left(-\frac{1}{2}\right)}(\pi)^2(\pi)^2(\pi^*)^{1\left(+\frac{1}{2}\right)}(\pi^*)^{1\left(+\frac{1}{2}\right)}$$

These two O_2^+ ions have slightly different energies as is exhibited by their photoelectron spectra. Studies such as these have contributed greatly to our understanding of molecular orbital energy diagrams. We will not describe the technique further, but more complete details of the method and its use can be found in the references at the end of this chapter.

3.5 HETERONUCLEAR DIATOMIC MOLECULES

Atoms do not all have the same ability to attract electrons. When two different types of atoms form a covalent bond by sharing a pair of electrons, the shared pair of electrons will spend more time in the vicinity of the atom that has the greater ability to attract them. In other words, the electron pair is *shared*, but it is not shared *equally*. The ability of an atom in a molecule to attract electrons to it is expressed as the *electronegativity* of the atom. Earlier, for a homonuclear diatomic molecule, we wrote the combination of two atomic wave functions as

$$\psi = a_1\phi_1 + a_2\phi_2 \tag{3.53}$$

where we did not have to take into account the difference in the ability of two atoms to attract electrons. For two different types of atoms, we can write the wave function for the bonding molecular orbital as

$$\psi = \phi_1 + \lambda\phi_2 \tag{3.54}$$

where the parameter λ is a weighting coefficient. Actually, a weighting coefficient for the wave function of one atom is assumed to be 1, and a different weighting factor, λ, is assigned for the other atom depending on its electronegativity.

When two atoms share electrons unequally, it means that the bond between them is polar. Another way to describe this is to say that the bond has *partial ionic character*. For the molecule AB, this is equivalent to drawing two structures, one of which is covalent and the other ionic. However, there are actually three structures that can be drawn,

$$\begin{array}{ccc} A:B \leftrightarrow A^+B^- \leftrightarrow A^-B^+ \\ I \qquad II \qquad III \end{array} \tag{3.55}$$

If we write a wave function for the molecule to show a combination of these structures, it is written as

$$\psi_{molecule} = a\psi_I + b\psi_{II} + c\psi_{III} \tag{3.56}$$

where a, b, and c are constants and ψ_I, ψ_{II}, and ψ_{III} are wave functions that correspond to the structures I, II, and III, respectively. Generally, we have some information about the magnitudes of a, b, and c. For example, if the molecule being considered is HF, the resonance structure $H^- F^+$ will contribute very little to the actual structure of the molecule. It is contrary to the chemical nature of the H and F atoms to have a structure with a negative charge on H and a positive charge on F. Accordingly, the weighting coefficient for structure III must be approximately 0. For molecules that are predominantly covalent in nature, even structure II will make a smaller contribution than will structure I.

The dipole moment, μ, for a diatomic molecule (the situation for polyatomic molecules that have several bonds is more complex) can be expressed as

$$\mu = q \times r \tag{3.57}$$

where q is the quantity of charge separated and r is the distance of separation. If an electron were completely transferred from one atom to the other, the quantity of charge separated would be e, the charge on an electron. For bonds in which an electron pair is shared unequally, q is less than e, and if the sharing is equal, there is no charge separation, $q = 0$, and the molecule is nonpolar. For a polar molecule, there is only one bond length, r. Therefore, the ratio of the actual or observed dipole moment (μ_{obs}) to that assuming complete transfer of the electron (μ_{ionic}) will give the ratio of the amount of charge separated to the charge of an electron.

$$\frac{\mu_{obs}}{\mu_{ionic}} = \frac{q \cdot r}{e \cdot r} = \frac{q}{e} \tag{3.58}$$

The ratio q/e gives the fraction of an electron that appears to be transferred from one atom to another. This ratio can also be considered as the partial ionic character of the bond between the atoms. It follows that the percent of ionic character is 100 times the fraction of ionic character. Therefore,

$$\% \text{ Ionic character} = \frac{100\mu_{obs}}{\mu_{ionic}} \tag{3.59}$$

The actual structure of HF can be represented as a composite of the covalent structure H—F in which there is equal sharing of the bonding electron pair and the ionic structure $H^+ F^-$ where there is complete transfer of an electron from H to F. Therefore, the wave function for the HF molecule wave function can be written in terms of the wave functions for those structures as

$$\psi_{molecule} = \psi_{covalent} + \lambda\psi_{ionic} \tag{3.60}$$

The squares of the coefficients in a wave function are related to probability. Therefore, the total contribution from the two structures is $1^2 + \lambda^2$, whereas the contribution from the ionic structure is given by λ^2. As a result, $\lambda^2/(1^2 + \lambda^2)$ gives the fraction of ionic character to the bond and

$$\% \text{ Ionic character} = \frac{100\lambda^2}{(1 + \lambda^2)} \tag{3.61}$$

because $1^2 = 1$. Therefore,

$$\frac{\mu_{obs}}{\mu_{ionic}} = \frac{\lambda^2}{(1 + \lambda^2)} \tag{3.62}$$

For the HF molecule, the bond length is 0.92 Å (0.92×10^{-8} cm $= 0.92 \times 10^{-10}$ m), and the measured dipole moment is 1.91 Debye or 1.91×10^{-18} esu cm. If an electron were completely transferred from H to F, the dipole moment (μ_{ionic}) would be

$$\mu_{ionic} = 4.80 \times 10^{-10} \text{esu} \times 0.92 \times 10^{-8} \text{cm} = 4.41 \times 10^{-18} \text{esu cm} = 4.41 \text{ D}$$

Therefore, the ratio μ_{obs}/μ_{ionic} is 0.43, which means that

$$0.43 = \frac{\lambda^2}{(1 + \lambda^2)} \tag{3.63}$$

from which we find that $\lambda = 0.87$. Therefore, the wave function for the HF molecule can be written as

$$\psi_{molecule} = \psi_{covalent} + 0.87\psi_{ionic} \tag{3.64}$$

From the analysis above, it appears that we can consider the polar HF molecule as consisting of a hybrid made from a purely covalent structure contributing 57% and an ionic structure contributing 43% to the actual structure.

$$H : F \leftrightarrow H^+F^-$$
$$57\% \qquad 43\%$$

Of course, HF is actually a polar covalent molecule, but from the extent of the polarity, it behaves *as if it were* composed of the two structures shown above. A similar analysis can be carried out for all of the hydrogen halides, and the results are shown in Table 3.2.

A simple interpretation of the effect of two atoms in a diatomic molecule is seen from the molecular orbital description of the bonding. Different atoms have different ionization potentials, which results in the values for the Coulomb integrals used in a molecular orbital calculation being different. In fact, according to Koopmans' theorem, the ionization potential with the sign changed gives the value for the Coulomb integral. In terms of a molecular orbital energy level diagram, the atomic states of the two atoms are different and the bonding molecular orbital will be closer in energy to that of the atom having the higher ionization potential. For example, in the HF molecule, there is a single σ bond between the two atoms. The ionization potential for H is 1312 kJ mol^{-1} (13.6 eV), whereas that for F is 1680 kJ mol^{-1} (17.41 eV). When the wave functions for the hydrogen 1s and fluorine 2p orbital are combined, the resulting molecular orbital will have an energy that is closer to that of the fluorine orbital than to that of the hydrogen orbital. In simple terms, this means that the bonding molecular orbital is more like a fluorine orbital than a hydrogen orbital. This is loosely equivalent to saying that the electron spends more time around the fluorine atom as we did in describing bonding in HF in valence bond terms.

Table 3.2 Parameters for Hydrogen Halide Molecules, HX

Molecule	r, pm	μ_{obs}, D	μ_{ionic}, D	% Ionic character	
				$100\mu_{obs}/\mu_{ionic}$	$\chi_x - \chi_H$
HF	92	1.91	4.41	43	1.9
HCl	128	1.03	6.07	17	0.9
HBr	143	0.78	6.82	11	0.8
HI	162	0.38	7.74	5	0.4

1 Debye 10^{-18} esu cm. The electronegativities of atoms H and X are χ_A and χ_B.

The bonding in hetronuclear species can be considered as the mixing of atomic states to generate molecular orbitals with the resulting molecular orbitals having a larger contribution from the more electronegative atom. For example, the ionization potential for Li is 520 kJ mol^{-1} (5.39 eV), whereas that of hydrogen is 1312 kJ mol^{-1} (13.6 eV). Therefore, the bonding orbital in the LiH molecule will have a great deal more of the character of the hydrogen $1s$ orbital. In fact, the compound LiH is substantially ionic, and we normally consider the hydrides of the Group IA metals to be ionic. When we consider the compound LiF, the ionization potentials of the two atoms (energy of the atomic states for which the wave functions are being combined) are so different that the resulting "molecular orbital" is essentially the same as an atomic orbital on the fluorine atom. This means that in the compound, the electron is essentially transferred to the F atom when the bond forms. Accordingly, we consider LiF to be an ionic compound in which the species present are Li$^+$ and F$^-$.

3.6 ELECTRONEGATIVITY

As has just been described, when a covalent bond forms between two atoms, there is no reason to assume that the pair of electrons is shared *equally* between the atoms. What is needed is some sort of way to provide a relative index of the ability of an atom to attract electrons. Linus Pauling developed an approach to this problem by describing a property now known as the *electronegativity* of an atom. This property gives a measure of the tendency of an atom in a molecule to attract electrons. Pauling devised a way to give numerical values to describe this property that makes use of the fact that the covalent bonds between atoms of different electronegativity are more stable than if they were purely covalent (with equal sharing of the electron pair). For a diatomic molecule AB, the actual bond energy, D_{AB}, is written as

$$D_{AB} = \frac{1}{2}[D_{AA} + D_{BB}] + \Delta_{AB} \qquad (3.65)$$

where D_{AA} and D_{BB} are the bond energies in the purely covalent diatomic species A$_2$ and B$_2$, respectively. Because the actual bond between A and B is stronger than if the bond were purely covalent, the term Δ_{AB} corrects for the additional stability. The degree to which the sharing of the electron pair is unequal depends on the property known as electronegativity. Pauling related the additional stability of the bond to the tendency of the atoms to attract electrons by means of the equation

$$\Delta_{AB} = 96.48|\chi_A - \chi_B|^2 \qquad (3.66)$$

In this equation, χ_A and χ_B are values that describe the electron attracting ability (electronegativity) of atoms A and B, respectively. The constant 96.48 appears so that the value of Δ_{AB} will be given in kJ mol^{-1}. If the constant is 23.06, the value of Δ_{AB} will be in kcal mol^{-1}. Note that it is the *difference* between the values for the two atoms that is related to the additional stability of the bond. With the values of Δ_{AB} known for many types of bonds, it is possible to assign values for χ_A and χ_B, but only when there is a value known for at least one atom. Pauling solved this problem by *assigning* a value of 4.0 for the electronegativity of fluorine. In that way, the electronegativities of all other atoms are positive values between 0 and 4.0. Based on more recent bond energy values, the value of 3.98 is sometimes used. It would not have made any difference if the fluorine atom had been assigned a value of 100 because other atoms would then have electronegativities between 96 and 100.

With the electronegativity of the fluorine atom being assigned a value of 4.0, it was now possible to determine a value for hydrogen because the H−H and F−F bond energies were known as was the bond energy for the H−F molecule. Using those bond energies, the electronegativity of H is found to be about 2.2. Keep in mind that it is only the *difference* in electronegativity that is related to the additional stability of the bond, not the actual values. Pauling electronegativity values for many atoms are shown in Table 3.3.

Whereas the approach described above is based on the average bond energy of A_2 and B_2 as described by the arithmetic mean, $\frac{1}{2}(D_{AA} + D_{BB})$, a different approach is based on the average bond energy being given by $(D_{AA} \times D_{BB})^{\frac{1}{2}}$. This is a geometric mean, which gives a value for the additional stability of the molecule as

$$\Delta' = D_{AB} - (D_{AA} \times D_{BB})^{\frac{1}{2}} \tag{3.67}$$

For molecules that are highly polar, this equation gives better agreement with the electronegativity difference between the atoms and the additional stability of the bond than does Eqn (3.65).

Pauling based electronegativity values on bond energies between atoms, but that is not the only way to approach the problem of the ability of atoms in a molecule to attract electrons. For example, the ease of removing an electron from an atom, the ionization potential, is related to its ability to attract electrons to itself. The electron affinity also gives a measure of the ability of an atom to hold on to an electron that it has gained. These *atomic* properties should therefore be related to the ability of an atom in a molecule to attract electrons. It is natural to make use of these properties in an equation to express the electronegativity of an atom. Such an approach was taken by Mulliken who proposed that the electronegativity, χ, of an atom A could be expressed as

$$\chi_A = \frac{1}{2}[I_A + E_A] \tag{3.68}$$

In this equation, I_A is the ionization potential and E_A is the electron affinity for the atom and it is the average of these two properties that Mulliken proposes to use as the electronegativity of the atom. When

Table 3.3 Pauling Electronegativities of Atoms

H							
2.2							
Li	Be		B	C	N	O	F
1.0	1.6		2.0	2.6	3.0	3.4	4.0
Na	Mg		Al	Si	P	S	Cl
1.0	1.3		1.6	1.9	2.2	2.6	3.2
K	Ca	Sc … Zn	Ga	Ge	As	Se	Br
0.8	1.0	1.2 … 1.7	1.8	2.0	2.2	2.6	3.0
Rb	Sr	Y … Cd	In	Sn	Sb	Te	I
0.8	0.9	1.1 … 1.5	1.8	2.0	2.1	2.1	2.7
Cs	Ba	La … Hg	Tl	Pb	Bi	Po	At
0.8	0.9	1.1 … 1.5	1.4	1.6	1.7	1.8	2.0

the energies are expressed in electron volts, the Mulliken electronegativity for the fluorine atom is 3.91 rather than the value of 4.0 assigned by Pauling. In general, the electronegativity values on the two scales do not differ by very much.

If a property is as important as is electronegativity, it is not surprising that a large number of approaches have been taken to provide measures of the property. Although we have already described two approaches, we should also mention one additional method. Allred and Rochow made use of the equation

$$\chi_A = 0.359\left(\frac{Z^*}{r^2}\right) + 0.744 \tag{3.69}$$

In this equation, Z^* is the *effective* nuclear charge, which takes into account the fact that an outer electron is screened from experiencing the effect of the *actual* nuclear charge by the electrons that are closer to the nucleus (see Section 2.4). In principle, the Allred–Rochow electronegativity scale is based on the electrostatic interaction between valence shell electrons and the nucleus.

Probably the most important use of electronegativity values is in predicting bond polarities. For example, in the H–F bond, the shared electron pair will reside closer to the fluorine atom because it has an electronegativity of 4.0 and that of the hydrogen atom is 2.2. In other words, the electron pair is shared, but not equally. If we consider the HCl molecule, the shared electron pair will reside closer to the chlorine atom, which has an electronegativity of 3.2, but the electron pair will be shared more nearly equally than is the case for HF because the difference in electronegativity is smaller for HCl. We will have many opportunities to use this principle when describing structures of inorganic compounds.

Having shown that the weighting coefficient (λ) of the term giving the contribution of an ionic structure to the molecular wave function is related to the dipole moment of the molecule, it is logical to expect that equations could be developed that relate the ionic character of a bond to the electronegativities of the atoms. Two such equations that give the percent ionic character of the bond in terms of the electronegativities of the atoms are

$$\% \text{ Ionic character} = 16|\chi_A - \chi_B| + 3.5|\chi_A - \chi_B|^2 \tag{3.70}$$

$$\% \text{ Ionic character} = 18|\chi_A - \chi_B|^{1.4} \tag{3.71}$$

Although the equations look very different, the calculated values for the percent ionic character are approximately equal for many types of bonds. If the difference in electronegativity is 1.0, Eqn (3.70) predicts 19.5% ionic character, whereas Eqn (3.71) gives a value of 18%. This difference is insignificant for most purposes. After one of these equations is used to estimate the percent ionic character, Eqn (3.61) can be used to determine the coefficient λ in the molecular wave function. Figure 3.10 shows how percent ionic character varies with the difference in electronegativity.

When the electrons in a covalent bond are shared equally, the length of the bond between the atoms can be approximated as the sum of the covalent radii. However, when the bond is polar, the bond is not only stronger than if it were purely covalent but also shorter. As shown earlier, the amount by which a polar bond between two atoms is stronger than if it were purely covalent is related to the difference in electronegativity between the two atoms. It follows that the amount by which the bond is shorter than

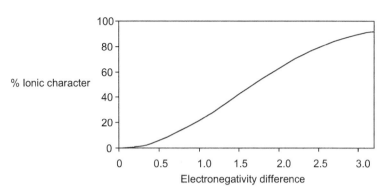

FIGURE 3.10
The variation in percent ionic character to a bond and the difference in the electronegativities of the atoms.

the sum of the covalent radii should also be related to the difference in electronegativity. An equation that expresses the bond length in terms of atomic radii and the difference in electronegativity is the Schomaker–Stevenson equation. That equation can be written as

$$r_{AB} = r_A + r_B - 9.0|\chi_A - \chi_B| \tag{3.72}$$

where χ_A and χ_B are the electronegativities of atoms A and B, respectively, and r_A and r_B are their covalent radii expressed in pm. This equation provides a good approximation to bond lengths. When the correction for the difference in electronegativity is applied to polar molecules, the calculated bond lengths agree considerably better with experimental values.

In this chapter, the basic ideas related to the molecular orbital approach to covalent bonds have been presented. Other applications of the molecular orbital method will be discussed in Chapters 5 and 15.

3.7 SPECTROSCOPIC STATES FOR MOLECULES

For diatomic molecules, there is coupling of spin and orbital angular momenta by a coupling scheme that is similar to the Russell–Saunders procedure described for atoms. When the electrons are in a specific molecular orbital, they have the same orbital angular momentum as designated by the m_l value. As in the case of atoms, the m_l value depends on the type of orbital. When the internuclear axis is the z-axis, the orbitals that form σ bonds (which are symmetric around the internuclear axis) are the s, p_z, and d_{z^2} orbitals. Those that form π bonds are the p_x, p_y, d_{xz}, and d_{yz} orbitals. The $d_{x^2-y^2}$ and d_{xy} can overlap in a "sideways" fashion with one stacked above the other, and the bond would be a δ bond. For these types of molecular orbitals, the corresponding m_l values are:

$$
\begin{aligned}
\sigma : & \quad m_l = 0 \\
\pi : & \quad m_l = \pm 1 \\
\delta : & \quad m_l = \pm 2
\end{aligned}
$$

As in the case of atoms, the molecular term symbol is written as ^{2S+1}L where L is the absolute value of M_L (the highest positive value). The molecular states are designated as for atoms except for the use of capital Greek letters.

$$M_L = 0 \text{ the spectroscopic state is } \Sigma$$

$$M_L = 1 \text{ the spectroscopic state is } \Pi$$

$$M_L = 2 \text{ the spectroscopic state is } \Delta$$

After writing the molecular orbital configuration, the vector sums are obtained. For example, in H_2, the two bonding electrons reside in a σ orbital, and they are paired so $S = +\frac{1}{2} + (-\frac{1}{2}) = 0$. As shown above, for a σ orbital, the m_l is 0 so the two electrons combined have $M_L = 0$. Therefore, the ground state for the H_2 molecule is $^1\Sigma$. As in the case of atoms, all filled shells have $\Sigma s_i = 0$, which results in a $^1\Sigma$ state.

The N_2 molecule has the configuration $(\sigma)^2 (\sigma^*)^2 (\sigma)^2 (\pi)^2 (\pi)^2$ so all of the populated orbitals are filled. Therefore, the spectroscopic state is $^1\Sigma$. For O_2, the unfilled orbitals are $(\pi_x^*)^1 (\pi_y^*)^1$, and the filled orbitals do not determine the spectroscopic state. For a π orbital, $m_l = \pm 1$. These vectors could be combined with spin vectors of $\pm \frac{1}{2}$. If both spins have the same sign, $|S| = 1$ and the state will be a triplet. If the spins are opposite, $|S| = 0$ and the state is a singlet. Because $M_l = \Sigma m_l$ and the m_l values for the π orbitals are ± 1, the possible values for M_L are 2, 0, and -2. Possible ways to combine M_S and M_L are shown below when the values are (m_l, s):

M_L		M_S	
	1	0	-1
2		$(1,\frac{1}{2}), (1,-\frac{1}{2})$	
0	$(1,\frac{1}{2}), (-1,\frac{1}{2})$	$(1,\frac{1}{2}), (-1,-\frac{1}{2})$	$(1,-\frac{1}{2}), (-1,-\frac{1}{2})$
		$(1,-\frac{1}{2}), (-1,\frac{1}{2})$	
-2		$(-1,\frac{1}{2}), (-1,-\frac{1}{2})$	

Those cases for $M_L = 2$ result when the spins are *opposed* and, therefore, represent a $^1\Delta$ state. There is one combination where $M_L = 0$ with the S vector having values of $+1$, 0, and -1, which corresponds to $^3\Sigma$. The remaining combinations correspond to the $^1\Sigma$ term. Of these states ($^1\Delta$, $^1\Sigma$, and $^3\Sigma$), the one having the highest multiplicity lies lowest in energy so the ground state of the O_2 molecule is $^3\Sigma$. The *ground state* could be identified quickly by simply placing the electrons in separate π orbitals with parallel spins and obtaining the M_L and M_S values.

For the CN molecule, the configuration is $(\sigma)^2 (\sigma_z)^2 (\pi_x)^2 (\pi_y)^2 (\sigma_z)^1$. The single electron in the σ_z orbital gives $M_L = 0$ and $S = \frac{1}{2}$ so the ground state is $^2\Sigma$. Several species such as N_2, CO, NO^+, and CN^- have the configuration $(\sigma)^2 (\sigma_z)^2 (\pi_x)^2 (\pi_y)^2 (\sigma_z)^2$, which is a closed shell arrangement. Therefore, the ground state for these species is $^1\Sigma$. The NO molecule has the configuration $(\sigma)^2 (\sigma_z)^2 (\pi_x)^2 (\pi_y)^2 (\sigma_z)^2 (\pi_x^*)^1$, which gives rise to $S = \frac{1}{2}$ and $M_L = 1$. These values give rise to a ground state that is $^2\Pi$.

References for Further Study

Cotton, F.A., Wilkinson, G., Murillo, C.A., Bochmann, M., 1999. *Advanced Inorganic Chemistry*, 6th ed. John Wiley, New York. Almost 1400 pages devoted to all phases of inorganic chemistry. An excellent reference text.

DeKock, R.L., Gray, H.B., 1980. *Chemical Bonding and Structure*. Benjamin Cummings, Menlo Park. CA. One of the best introductions to bonding available. Highly recommended.

Greenwood, N.N., Earnshaw, A., 1997. *Chemistry of the Elements*, 2nd ed. Butterworth Heinemann, New York. Although this is a standard reference text on descriptive chemistry, it contains an enormous body of information on bonding.

House, J.E., 2003. *Fundamentals of Quantum Chemistry*. Elsevier, New York. An introduction to quantum mechanical methods at an elementary level that includes mathematical details.

Lide, D.R. (Ed.), 2003. *CRC Handbook of Chemistry and Physics*, 84th ed. CRC Press, Boca Raton, FL Various sections in this massive handbook contain a large amount of data on molecular parameters.

Lowe, J.P., Peterson, K., 2005. *Quantum Chemistry*, 3rd ed. Academic Press, New York. This is an excellent book for studying molecular orbital methods at a higher level.

Mackay, K., Mackay, R.A., Henderson, W., 2002. *Introduction to Modern Inorganic Chemistry*, 6th ed. Nelson Thornes, Cheltenham, UK. One of the very successful standard texts in inorganic chemistry.

Mulliken, R.S., Rieke, A., Orloff, D., Orloff, H., 1949. Overlap integrals and chemical binding. *J. Chem. Phys.* 17, 510. and "Formulas and Numerical Tables for Overlap Integrals", *J. Chem. Phys.* 17, 1248−1267. These two papers present the basis for calculating overlap integrals and show the extensive tables of calculated values.

Pauling, L., 1960. *The Nature of the Chemical Bond*, 3rd ed. Cornell University Press, Ithaca, New York. Although somewhat dated in some areas, this is a true classic in bonding theory.

Sharpe, A.G., 1992. *Inorganic Chemistry*, 3rd ed. Longman, New York. Excellent coverage of bonding concepts in inorganic molecules.

QUESTIONS AND PROBLEMS

1. For each of the following, draw a molecular orbital energy level diagram and give the bond order. Tell whether the species would be more or less stable after gaining an electron. (a) O_2^+; (b) CN; (c) S_2; (d) NO; (e) Be_2^+.

2. Explain in terms of molecular orbitals why Li_2 is stable but Be_2 is not.

3. Which has the greater bond energy, NO or C_2? Explain by making appropriate drawings.

4. Numerical data are given below for the BN and BO molecules. Match the properties to these molecules and explain your answers.
   Data: 120 pm, 128 pm, 8.0 eV, 4.0 eV.

5. If the H−H and S−S bond energies are 266 and 432 kJ mol^{-1}, respectively, what would be the H−S bond energy?

6. The stretching vibration for NO is found at 1876 cm^{-1}, whereas that for NO^+ is at 2300 cm^{-1}. Explain this difference.

7. What is the ClF bond length if the covalent radii of Cl and F are 99 and 71 pm, respectively? Explain your answer in terms of resonance.

8. Consider a diatomic molecule A_2 in which there is a single σ bond. Excitation of an electron to the σ^* state gives rise to an absorption at 15,000 cm^{-1}. The binding energy of an electron in the valence shell of atom A is −9.5 eV.
   (a) If the overlap integral has a value of 0.12, determine the value of the exchange integral, H_{12}.
   (b) Calculate the actual values of the bonding and antibonding molecular orbitals for the A_2 molecule.
   (c) What is the single bond energy in the A_2 molecule?

9. Arrange the species O_2^{2-}, O_2^+, O_2, and O_2^- in the order of decreasing bond length. Explain this order in terms of molecular orbital populations.

10. Explain why the electron affinity of the NO molecule is 88 kJ mol^{-1} but that of the CN molecule is 368 kJ mol^{-1}.

11. The electron affinity of the NO molecule is about 88 kJ mol^{-1}, whereas that for the C_2 molecule is about 341 kJ/mol. Explain this difference in terms of the molecular orbital diagrams for the molecules.

12. In the spectrum of the CN molecule, an absorption band centered around 9000 cm^{-1} appears. Explain the possible origin of this band in terms of the molecular orbitals in this molecule. What type of transition is involved?

13. Consider the Li_2 molecule that has a dissociation energy of 1.03 eV. The first ionization potential for the Li atom is 5.30 eV. Describe the bonding in Li_2 in terms of a molecular orbital energy diagram. If a value of 0.12 is appropriate for the overlap integral, what is the value of the exchange integral?

14. Sketch a molecular orbital energy level diagram for HF. Using that diagram as a basis, describe the polar nature of the HF molecule.

15. For a molecule XY, the molecular wave function can be written as

$$\psi_{molecule} = \psi_{covalent} + 0.70\psi_{ionic}$$

Calculate the percent of ionic character in the X−Y bond. If the X−Y bond length is 142 pm, what is the dipole moment of XY?

16. What Pauling electronegativity is predicted for an element X if the H−X bond energy is 402 kJ mol^{-1}? The H−H bond energy is 432 kJ mol^{-1} and the X−X bond energy is 335 kJ mol^{-1}. What would be the percent ionic character of the H−X bond? If the molecular wave function is written as

$$\psi_{molecule} = \psi_{covalent} + \lambda\psi_{ionic}$$

what is the value of λ?

17. Suppose the bond energies of A_2 and X_2 are 210 and 345 kJ mol^{-1}, respectively. If the electronegativities of A and X are 2.0 and 3.1, respectively, what will be the strength of the A−X bond? What will be the dipole moment if the internuclear distance is 125 pm?

18. For a molecule XY, the molecular wave function can be written as

$$\psi_{molecule} = \psi_{covalent} + 0.50\psi_{ionic}$$

Calculate the percent ionic character of the X−Y bond. If the bond length is 148 pm, what is the dipole moment of XY?

19. Determine the spectroscopic ground states for the following diatomic molecules:
    (a) BN; (b) C_2^+; (c) LiH; (d) CN$^-$; (e) C_2^-

CHAPTER 4

A Survey of Inorganic Structures and Bonding

Molecular structure is the foundation on which chemistry, the study of matter and the changes it undergoes, rests. Much of chemistry is concerned with changes at the molecular level as structures are elucidated and chemical reactions occur. This is true not only in inorganic chemistry but also in all areas of chemical science. Consequently, this chapter is devoted to an overview of some of the basic ideas concerning bonding and the structure of molecules. Although other aspects of bonding will be discussed in later chapters, this chapter is intended to provide an introduction to structural inorganic chemistry early in the study of the subject. More details concerning the structure of specific inorganic materials will be presented in later chapters as most of the structures discussed here will be revisited in the context of the chemistry of the compounds. It should be kept in mind that for many purposes, a theoretical approach to bonding is not necessary. Accordingly, this chapter provides a nonmathematical view of molecular structure that is useful and adequate for many uses in inorganic chemistry. Because some of the principles are different for molecules that contain only single bonds, this topic will be introduced first.

In this chapter, the descriptions of molecular structure will be primarily in terms of a valence bond approach, but the molecular orbital method will be discussed in Chapter 5. As we shall see, construction of molecular orbital diagrams for polyatomic species is simplified by making use of symmetry, which will also be discussed in Chapter 5.

4.1 STRUCTURES OF MOLECULES HAVING SINGLE BONDS

One of the most important factors when describing molecules that have only single bonds is the repulsion that exists between electrons. Repulsion is related to the number of electron pairs both shared and unshared around the central atom. When only two pairs of electrons surround the central atom (as in BeH_2), the structure is almost always linear because that gives the configuration of lowest energy. When there are four pairs of electrons around the central atom (as in CH_4), the structure is tetrahedral. From your prior study of chemistry, the hybrid orbital types sp and sp^3 used to describe these cases are probably familiar. It is not unusual to hear someone say that CH_4 is tetrahedral because the carbon atom is sp^3 hybridized. However, CH_4 is tetrahedral because that structure represents the *configuration of lowest energy*, and our way of *describing* a set of orbitals that *matches* that geometry is by combining the wave functions for the $2s$ and three $2p$ orbitals. It can be shown that the four resulting orbitals point toward the corners of a tetrahedron.

Based on the requirement that repulsion should be minimized, idealized structures can be obtained based on the number of electrons surrounding the central atom. However, unshared pairs (sometimes called lone pairs) of electrons behave somewhat differently than shared pairs of electrons. A shared pair

89

Inorganic Chemistry. DOI: http://dx.doi.org/10.1016/B978-0-12-385110-9.00004-2

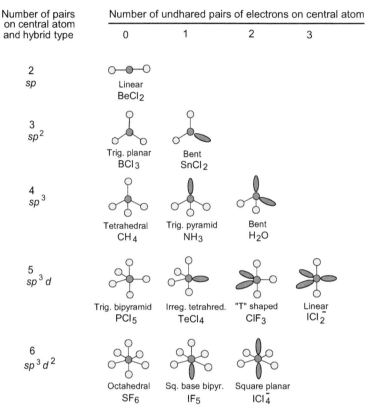

FIGURE 4.1
Molecular structure based on hybrid orbital type.

of electrons is essentially localized in the region of space between the two atoms sharing the pair. An unshared pair of electrons is bound only to the atom on which they reside, and as a result, they are able to move more freely than a shared pair of electrons so more space is required for an unshared pair. This has an effect on the structure of the molecule.

Figure 4.1 shows the common structural types that describe a very large number of inorganic molecules. Linear, trigonal planar, tetrahedral, trigonal bipyramid, and octahedral structures result when there are 2, 3, 4, 5, and 6 bonding pairs, respectively, but no unshared pairs on the central atom. The hybrid orbital types for these structures are sp, sp^2, sp^3, sp^3d, and sp^3d^2, respectively.

Rationalizing the probable structure for a molecule involves finding the number of electrons around the central atom and placing them in orbitals pointing in directions that minimize repulsion. However, there are complications when the details of the structure are considered. For example, a molecule such as BF_3 has only three pairs of electrons around the central atom (three valence shell electrons from B and one from each F atom). Therefore, the structure that gives the lowest energy is a trigonal plane with bond angles of 120°,

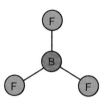

and the hybrid orbital type is sp^2. On the other hand, there are also six electrons around Sn in the gaseous $SnCl_2$ molecule (four valence electrons from Sn and one from each Cl atom). Molecules that have the same number of electrons are said to be *isoelectronic*. However, the hybrid orbital type may not be sp^2 because the bond angle certainly is not 120° in this case, and the structure of $SnCl_2$ is

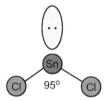

The unshared pair of electrons resides in an orbital that might be considered as sp^2, but as a result of the electrons being held to only one atom, there is more space required than there is for a shared pair. A shared pair of electrons is more restricted in its motion because of being attracted to two atoms simultaneously. As a result, the repulsion between the unshared pair and the shared pair is sufficient to force the bonding pairs closer together, which causes the bond angle to be much smaller than the expected 120°. In fact, the bond angle is much closer to that expected if p orbitals are used by Sn. On the other hand, the Sn—Cl bonds are quite polar so the bonding electron pairs reside much closer to the Cl atoms, which makes it easier for the bond angle to be small than if they were residing close to the Sn atom.

In some molecules that involve sp^2 hybrid orbitals on the central atom, the bond angles deviate considerably from 120°. For example in F_2CO, the bond angle is 108°.

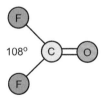

Various explanations have been proposed for the large deviation of the bonding from that expected on the basis of sp^2 hybrids. One simple approach to this problem is to consider that C—F bonds are quite polar with the shared electron pairs drawn closer to the fluorine atoms. Therefore, those bonding pairs are farther apart than they would be if they were being shared equally by C and F. There is less repulsion between the bonding pairs of electrons, and the effect of the C=O double bond is that the π orbitals give rise to some repulsion with the C—F bonding pairs forcing them closer together. Using this approach, we would expect the bond angle in phosgene, Cl_2CO, to be *larger* than it is in F_2CO because the electron pairs in the C—Cl bonds will reside closer to the carbon atom than if the bonds were C—F. Therefore, the bonding electron pairs would be closer to each other in Cl_2CO than in F_2CO. In either case, the carbon atom is at the positive end of the bond dipole because both F and Cl atoms have higher electronegativities than carbon. The structure of Cl_2CO,

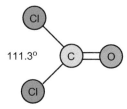

indicates that this interpretation is correct. Of course, the Cl atoms are larger than the F atoms so it is tempting to attribute the larger bond angle in Cl_2CO to that cause. The structure of formaldehyde, H_2CO, is useful in this connection because the H atom is *smaller* than either F or Cl. However, the structure,

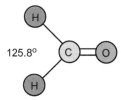

indicates that repulsion between the terminal atoms may not be significant. In this case, the H—C—H bond angle is *larger* than that expected for sp^2 hybrids on the central atom. When the polarity of the C—H bonds is considered, it is found that the carbon atom is at the *negative* end of the bond dipoles (see Chapter 6). Therefore, the C—H bonding pairs of electrons reside closer to the carbon atom (and hence closer to each other), and we should expect repulsion between them to cause the bond angle to be larger than 120°. The measured bond angle is in agreement with this rationale.

It is interesting to also see the difference in the bond angles in OF_2 and OCl_2, which have the structures

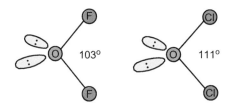

Although it might be tempting to ascribe this difference in bond angles to the difference in size between F and Cl, the location of *bonding* pairs of electrons is also important. The O—F bonds are polar with the bonding pairs closer to the fluorine atoms (and thus farther away from O and each other) allowing the bond angle to be smaller. The O—Cl bonds are polar, but with oxygen having the higher electronegativity, the shared pairs of electrons are closer to the oxygen atom (and to each other). As a result, there should be greater repulsion between the *bonding* pairs in OCl_2 than those in OF_2, and the OCl_2 bond angle is the larger of the two in agreement with this rationale. In both cases, there are two unshared pairs of electrons on the oxygen atom, and the bond angle deviates from the tetrahedral angle. If the situation were as simple as the effect produced by the unshared pairs, we would expect the bond angle to be

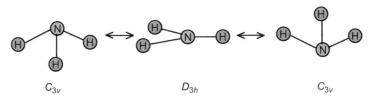

C_{3v} D_{3h} C_{3v}

FIGURE 4.2
The inversion of the ammonia molecule.

slightly larger in OCl_2 because the Cl atoms are larger than F. A difference of 8° probably indicates more repulsion between *bonding* pairs also.

In the CH_4 molecule, the bond angle is the expected value, 109°28′. There are eight electrons around the carbon atom (four valence shell electrons from C and one from each H atom), which results in a regular tetrahedral structure. In the ammonia molecule, the nitrogen atom has eight electrons around it (five from the N atom and one from each H atom), but one pair of electrons is an unshared pair.

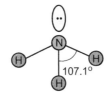

Although the hybrid orbital type used by N is sp^3, the bond angles in the NH_3 molecule are 107.1° rather than 109°28′ found in a regular tetrahedron. The reason for this difference is that the unshared pair requires more space and forces the bonding pairs slightly closer together. Although a structure such as that shown above displays a *static* model, the ammonia actually undergoes a vibrational motion known as *inversion*. In this vibration, the molecule passes from the structure shown above to that which is inverted by passing through a trigonal planar transition state (see Figure 4.2).

This vibration has a frequency that is approximately 1010 s^{-1}. The barrier height for inversion is 2076 cm^{-1}, but the difference between the first and second vibrational states is only 950 cm^{-1}, which is equivalent to 1.14 kJ mol^{-1}. Using the Boltzmann distribution law, we can calculate that the second vibrational state is populated only to the extent of 0.0105 so clearly there is not a sufficient amount of thermal energy available to cause the *rapid* inversion if the molecule must pass *over* a barrier that is 2076 cm^{-1} in height. In this case, the inversion involves *quantum mechanical tunneling*, which means that the molecule passes from one structure to the other without having to pass *over* the barrier.

The structure of the H_2O molecule shows the effect of two unshared pairs of electrons.

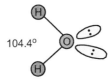

In this case, the two unshared pairs of electrons force the bonding pairs closer together so the observed bond angle is only 104.4°. Two unshared pairs in the H_2O molecule cause a greater effect than does one unshared pair in the NH_3 molecule. The effects that produce slight deviations from the bond angles expected for regular geometric structures are the result of a principle known as *valence shell electron pair repulsion* (VSEPR). The basis for this principle is that in terms of repulsion,

unshared pair – unshared pair > shared pair – unshared pair > shared pair – shared pair

When the effects of unshared pairs are considered according to this scheme, not only is the correct overall structure often deduced but also the slight deviations from regular bond angles are often predicted.

An interesting application of VSEPR is illustrated by the structure of SF_4. The sulfur atom has 10 electrons around it (six valence shell electrons from S and one from each of the four F atoms). We predict that the structure will be based on a trigonal bipyramid, but there are two possible structures.

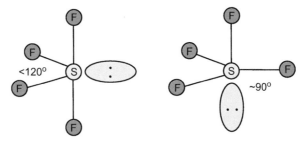

Only one of these structures is observed for the SF_4 molecule. In the structure on the left, the unshared pair of electrons is located at approximately 90° from two bonding pairs and approximately 120° from other two bonding pairs. In the structure on the right, the unshared pair is located at approximately 90° from three bonding pairs and 180° from the other bonding pair. These two possibilities may not look very different, but the repulsion between electron pairs is inversely related to the distance of separation raised to an exponent that may be as large as 6. A small difference in distance leads to a substantial difference in repulsion. As a result, the structure that has only two bonding electron pairs at 90° from the unshared pair is lower in energy, and the structure on the left is the correct one for SF_4. *In structures based on a trigonal bipyramid, unshared pairs of electrons are found in equatorial positions.*

Note that violations of the octet rule by the central atom occur with atoms such as sulfur and phosphorus. These are atoms that have *d* orbitals as part of their valence shells so they are not limited to a maximum of eight electrons. There is another interesting feature for molecules that are based on the trigonal bipyramid model for five bonds. If we consider the molecules PF_5 and PCl_5, there are 10 electrons around the phosphorus atom (five bonds), which point toward the corners of a trigonal bipyramid.

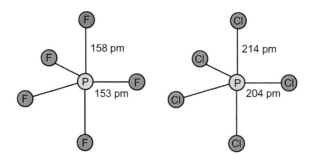

However, the bonds in the axial positions are slightly longer than those in the equatorial positions. For this type of structure, the hybrid bond type on the phosphorus atom is considered as sp^3d. However, the hybrid orbital type that gives three bonds in a trigonal plane is sp^2, and it can be shown that the dp combination is one that gives two orbitals directed at $180°$ from each other. Therefore, sp^3d hybrids can be considered as $sp^2 + dp$, and the bond lengths reflect the fact that the orbitals used in bonding three of the chlorine atoms are different from the other two. This is a general characteristic of molecules having five pairs of electrons around the central atom in a structure based on a trigonal bipyramid, and the axial bonds are usually longer than those in the equatorial plane.

One of the interesting aspects of the structure in which five pairs of electrons surround the central atom is that the equatorial positions make use of sp^2 hybrids, whereas the axial bonds are dp. As we have seen, any *unshared* pairs of electrons are found in *equatorial* positions. A further consequence of this is that peripheral atoms of high electronegativity bond best to orbitals of low s character and peripheral atoms of low electronegativity bond better to orbitals of high s character. The result of this preference is that if the mixed halide PCl_3F_2 is prepared, the fluorine atoms are found in axial positions. Also, atoms that can form multiple bonds (which usually have lower electronegativity) bond better to orbitals of higher s character (the sp^2 equatorial positions).

In view of the principles described above, the molecule PCl_2F_3 would be expected to have two fluorine atoms in the axial positions and two chlorine atoms and one fluorine atom in equatorial positions. However, at temperatures above $-22\,°C$, the NMR spectrum of PCl_2F_3 shows only one doublet, which results from the splitting of the fluorine resonance by the ^{31}P. When the NMR is taken at $-143\,°C$, the NMR spectrum is quite different and shows the presence of fluorine atoms bonded in more than one way. It is evident that at temperatures above $-22\,°C$, either all the fluorine atoms are equivalent or somehow they exchange rapidly so that only fluorine atoms in one environment are present. Earlier, the inversion vibration of the NH_3 molecule, which has a frequency that is on the order of $1010\,s^{-1}$, was described. The question arises as to what type of structural change could occur in PCl_2F_3 or PF_5 that would make the fluorine atoms in equatorial and axial positions appear equivalent on the time scale of NMR experiments. A mechanism that is believed to correctly describe this situation is known as the *Berry pseudorotation*, and it is illustrated in Figure 4.3.

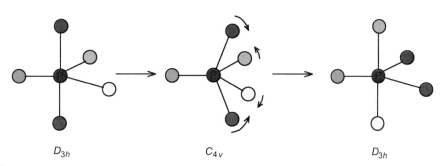

D_{3h} C_{4v} D_{3h}

FIGURE 4.3
The change in configuration in a trigonal bipyramid by a Berry pseudorotation. Although shown in different colors, all of the peripheral atoms may be identical.

In this process, the molecule passes through a square base pyramid configuration as the rotation of four of the groups occurs. This mechanism is similar to the inversion of the ammonia molecule except for the fact that the movement of atoms is by *rotational* motion. At very low temperature, thermal energy is low and the vibration occurs slowly enough that the fluorine resonance indicates fluorine atoms in two different environments (axial and equatorial). At higher temperatures, the structural change is rapid, and only one fluorine environment is indicated. As we saw for NH_3, not all molecules have static structures.

Perhaps no other pair of molecules exhibits the effect of molecular structure on reactivity more clearly than SF_4 (sp^3d orbitals) and SF_6 (sp^3d^2 orbitals) whose structures are

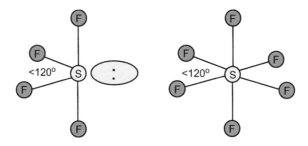

SF_6 is a remarkably inert compound. In fact, it is so unreactive that it is used as a gaseous dielectric material. Also, the gas can be mixed with oxygen to create a kind of synthetic atmosphere, and rats can breath the mixture for several hours with no ill effects. On the other hand, SF_4 is a very reactive molecule that reacts with H_2O rapidly and vigorously.

$$SF_4 + 3\,H_2O \rightarrow 4\,HF + H_2SO_3 \tag{4.1}$$

Because of the instability of H_2SO_3, this reaction can also be written as

$$SF_4 + 2\,H_2O \rightarrow 4\,HF + SO_2 \tag{4.2}$$

The fact that SF_6 does not react with water is not due to thermodynamic stability. Rather, it is because there is no low-energy pathway for the reaction to take place (kinetic stability). Six fluorine atoms surrounding the sulfur atom effectively prevent attack, and the sulfur atom has no unshared pairs of electrons where other molecules might attack. In SF_4, not only is there sufficient space for an attacking species to gain access to the sulfur atom but also the unshared pair is a reactive site. As a result of these structural differences, SF_6 is relatively inert, but SF_4 is very reactive.

There are several compounds that consist of two different halogens. These so-called *interhalogen* compounds have structures that contain only single bonds and unshared pairs of electrons. For example, in BrF_3, the bromine atom has 10 electrons surrounding it (seven valence shell electrons and one from each fluorine atom). The structure is drawn to place the unshared pairs of electrons in equatorial positions based on a trigonal bipyramid. Because of the effects of the unshared pairs of electrons, the axial bonds are forced closer together to give bond angles of 86°.

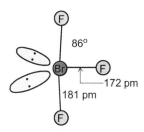

Except for slight differences in bond angles, this is also the structure for ClF_3 and IF_3. When IF_3 reacts with SbF_5, the reaction is

$$IF_3 + SbF_5 \rightarrow IF_2^+ + SbF_6^- \tag{4.3}$$

The structure of IF_2^+ can be deduced by recognizing that there are eight electrons around the iodine atom, seven valence electrons from I and one from each of the two fluorine atoms, but one electron has been removed giving the positive charge. The electrons will reside in four orbitals pointing toward the corners of a tetrahedron, but two pairs are unshared.

Note that the structures of many of these species are similar to the model structures shown in Figure 4.1. Although the hybrid orbital type is sp^3, the structure is characterized as bent or angular, not tetrahe-dral. On the other hand, the IF_2^- ion, which has 10 electrons around the iodine atom, has a linear structure.

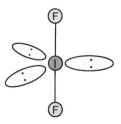

Note that in this case the unshared pairs of electrons are in equatorial positions, which results in a linear structure for IF_2^-, even though the hybrid orbital type is sp^3d. It is the arrangement of *atoms* not electrons that determines the structure for a molecule or ion. It is apparent that the simple procedures described in this section are adequate for determining the structures of many molecules and ions in which there are only single bonds and unshared pairs of electrons.

Xenon forms several compounds with fluorine, among which are XeF_2 and XeF_4. With filled s and p valence shell orbitals, the xenon atom provides eight electrons and each fluorine atom contributes one electron. Therefore, in XeF_2, there are 10 electrons around the Xe atom, which makes the XeF_2 molecule

Table 4.1 Average Bond Energies

Bond	Energy, kJ mol^{-1}	Bond	Energy, kJ mol^{-1}	Bond	Energy, kJ mol^{-1}
H–H	435	O=S	523	Ge–H	285
H–O	464	O–N	163	Ge–F	473
H–F	569	O–P	388	P–P	209
H–Cl	431	O–As	331	P–F	498
H–Br	368	O–C	360	P–Cl	331
H–I	297	O=C	745	P–Br	268
H–N	389	O≡C	1075	P–I	215
H–P	326	O–Si	464	P–O	368
H–As	297	O=Si	640	Si–F	598
H–Sb	255	O≡Si	803	Si–Cl	402
H–S	368	O–Ge	360	As–As	180
H–Se	305	S–S	264	As–O	331
H–Te	241	S=S	431	As–F	485
H–C	414	S–Cl	272	As–Cl	310
H–Si	319	S–Br	212	As–Br	255
H–Ge	285	S–C	259	As–H	297
H–Sn	251	N–F	280	C–C	347
H–Pb	180	N–Cl	188	C=C	611
H–B	331	N–N	159	C≡C	837
H–Mg	197	N=N	711	C–N	305
H–Li	238	N≡N	946	C–F	490
H–Na	201	N=O	594	C–Cl	326
H–K	184	N–Ge	255	C–Br	272
H–Rb	167	N–Si	335	O–F	213
O=O	498	O–Cl	205		
O–O	142	Ge–Ge	188		

isoelectronic with the IF$_2^-$ ion. Both XeF$_2$ and IF$_2^-$ are linear. In XeF$_4$, there are 12 electrons around the Xe atom, which results in the structure being

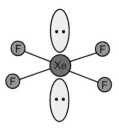

The molecule is square planar as is IF$_4^-$, which is isoelectronic with XeF$_4$.

Although we have described the structures of several molecules in terms of hybrid orbitals and VSEPR, not all structures are this simple. The structures of H$_2$O (bond angle 104.4°) and NH$_3$ (bond angles

$107.1°$) were described in terms of sp^3 hybridization of orbitals on the central atom and *comparatively small* deviations from the ideal bond angle of $109°28'$ caused by the effects of unshared pairs of electrons. If we consider the structures of H_2S and PH_3 in those terms, we have a problem. The reason is that the bond angle for H_2S is $92.3°$, and the bond angles in PH_3 are $93.7°$. Clearly, there is more than a minor deviation from the expected tetrahedral bond angle of $109°28'$ caused by the effect of unshared pairs of electrons!

If the bond angles in H_2S and PH_3 were $90°$, we would suspect that the orbitals used in bonding were the $3p$ valence shell orbitals. The sulfur atom has two of the p orbitals singly occupied, and overlap of hydrogen $1s$ orbital could produce two bonds at $90°$. Similarly, the phosphorus atom has the three $3p$ orbitals singly occupied, and overlap of three hydrogen $1s$ orbitals could lead to three bonds at $90°$. Although we were correct to assume that sp^3 orbital were used by the central atoms in H_2O and NH_3, it appears that we are not justified in doing so for H_2S and PH_3.

Why does hybridization occur when the central atom is oxygen or nitrogen but not when it is S or P? The answer lies in the fact that there are two major results of hybridization of orbitals. The first is that the orbitals are directed in space at different angles than are the unhybridized atomic orbitals. We have already seen the types of structures that result and how less repulsion results. However, the other result of hybridization is that the orbitals are *changed in size*. The hybridization of $3s$ and $3p$ orbitals on sulfur or phosphorus would produce more favorable bond angles with regard to repulsion, but the overlap between those orbitals and the hydrogen $1s$ orbitals is less effective. The hydrogen orbital can overlap better with a smaller unhybridized p orbital on sulfur or phosphorus. The result is that the orbitals used by the central atom have a very slight degree of hybridization but closely resemble pure p orbitals. Based on this analysis, we would expect that H_2Se and AsH_3 would have bond angles that deviate even more from the tetrahedral bond angle. In accord with this, the bond angles for these molecules are $91.0°$ and $91.8°$, respectively, indicating that the bonding orbitals on the central atoms are nearly pure p orbitals. The hydrogen compounds of the heavier members of Groups V and VI have bond angles that are even closer to right angles (H_2Te, $90°$; SbH_3, $91.3°$).

4.2 RESONANCE AND FORMAL CHARGE

For many species, the approach taken earlier with molecules that have only single bonds and unshared pairs of electrons is inadequate. For example, the molecule CO has only 10 valence shell electrons with which to achieve an octet around each atom. The structure $|C\equiv O|$ makes use of exactly 10 electrons and that makes it possible to place an octet (three shared pairs and one unshared pair) around each atom. A simple procedure for deciding how to place the electrons is as follows:

1. Determine the total number of valence shell electrons from all the atoms (N) that are available to be distributed in the structure.
2. Multiply the number of atoms present by eight to determine how many electrons would be required to give an octet around each atom (S).
3. The difference ($S - N$) gives the number of electrons that must be shared in the structure.
4. If possible, change the distribution of electrons to give favorable formal charges (discussed later in this chapter) on the atoms.

For CO, the total number of valence shell electrons is 10, and to give octets around two atoms, it would require 16 electrons. Therefore, $16 - 10 = 6$ electrons must be shared by the two atoms. Six electrons are equivalent to three pairs or three covalent bonds. Thus, we are led to the structure for CO that was shown earlier.

For a molecule such as SO_2, we find that the number of valence shell electrons is 18 and three atoms would require 24 electrons to make three octets. Therefore, $24 - 18 = 6$, the number of electrons that must be shared, which gives a total of three bonds between the sulfur atom and the two oxygen atoms. However, because we have already concluded that each atom in the molecule must have an octet of electrons around it, the sulfur atom must also have an unshared pair of electrons in addition to the three pairs that it is sharing. This can be shown as in the structure

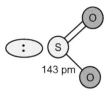

However, the same features are shown in the structure

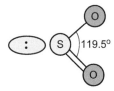

A situation in which the electrons can be arranged in more than one way constitutes *resonance*. In Chapter 3, the resonance structures H–F and $H^+ F^-$ were used to describe HF, but in the case of SO_2, neither of the resonance structures contains ions. The structures shown above involve different ways of arranging the electrons, which still conform to the octet rule. The true structure of SO_2 is one that lies half way between the two structures shown (is a hybrid made up of equal contributions from these two structures). It is *not* part of the time one structure and part of the time the other. The molecule is *all* of the time a *resonance hybrid* of the structures shown. In this case, the two structures contribute equally to the true structure, but this is not always the case. As a result of the unshared pair of electrons on the sulfur atom, the bond angle in SO_2 is 119.5°.

The double bond that is shown in each of the two structures above is not localized as is reflected by the two resonance structures. However, the two single bonds and the unshared pair are localized as a result of the hybrid orbitals in which they reside. The hybrid orbital type is sp^2, which accounts for the bond angle being 119.5°. There is one p orbital not used in the hybridization that is perpendicular to the plane of the molecule, which allows for the π bonding to the two oxygen atoms simultaneously. The π bond is described as being *delocalized*, and this can be shown as follows:

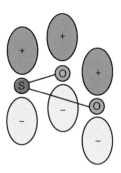

A single S—O bond has a length of approximately 150 pm, but as a result of the multiple bonding between sulfur and oxygen, the observed bond length in SO_2 where the bond order is 1.5 is 143 pm.

The following are rules that apply to drawing resonance structures. Remember that resonance relates to different ways of placing electrons in the structures, not ways of arranging the atoms themselves.

1. The atoms must be in the same relative positions in all structures drawn. For example, it can be demonstrated experimentally that the SO_2 molecule has a bent or angular structure. Structures showing the molecule with some other geometry (e.g., linear) are not permitted.
2. Structures that maximize the number of electrons used in bonding (consistent with the octet rule) contribute more to the true structure.
3. All resonance structures drawn must show the same number of unpaired electrons if there are any. A molecule or ion has a fixed number of unpaired electrons, and all resonance structures drawn for that species must show that number of unpaired electrons.
4. Negative formal charges normally reside on the atoms having higher electronegativity.

The NO_2 molecule illustrates the application of Rule 3. Because the NO_2 molecule has a total of 17 valence shell electrons, there are 8 pairs of electrons and 1 unpaired electron. Structures drawn for NO_2 must reflect this. Therefore, we draw the structure for NO_2 as shown below:

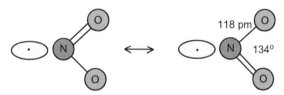

Note that the unpaired electron resides on the nitrogen atom giving it a total of seven electrons. Because the oxygen atom has higher electronegativity, the oxygen atoms are given complete octets. This is also consistent with the observation that NO_2 dimerizes by pairing of electrons on two molecules as shown by the equation,

$$2 \cdot NO_2 \rightleftarrows O_2N:NO_2 \tag{4.4}$$

Note that in the NO_2 molecule the bond angle is much larger than the 120° expected when the central atom is using sp^2 hybrid orbitals. In this case, the nonbonding orbital located on the nitrogen atom contains only one electron so repulsion between that orbital and the shared electron pairs is small. Therefore, the bond angle is larger because the repulsion between the bonding pairs is not balanced by the repulsion of the single electron in the nonbonding orbital. However, when the structure of NO_2^- is considered, there is an unshared *pair* of electrons on the nitrogen atom.

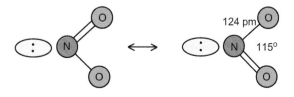

The repulsion between the unshared pair and the bonding electrons is much greater than that in NO_2 as is reflected by the bond angle being only 115°. The N–O bond length is 124 pm in NO_2^- because the nitrogen atom has an octet of electrons. There is less residual attraction for the bonding pairs of electrons, so the N–O bonds are longer than those in NO_2.

The concept of *formal* charges is a very useful one that is essentially a way of keeping track of electrons. To determine the formal charge on each atom in a structure, we must first apportion the electrons among the atoms. This is done according to the following procedure:

1. Any unshared pairs of electrons belong to the atom on which the electrons are located.
2. Shared electron pairs are divided equally between the atoms sharing them.
3. The total number of electrons on an atom in a structure is the sum from steps 1 and 2.
4. Compare the total number of electrons that appear to be on each atom to the number of valence shell electrons that it normally has. If the number of electrons in the valence shell is greater than that indicated in Step 3, the atom *appears* to have lost one or more electrons and has a *positive* formal charge. If the number indicated in Step 3 is larger than the number in the valence shell, the atom *appears* to have gained one or more electrons and has a *negative* formal charge.
5. Structures that have formal charges with the same sign on adjacent atoms will contribute little to the true structure.
6. The sum of formal charges on the atoms must total the overall charge on the species.

Earlier we showed the structure $|C\equiv O|$ for the carbon monoxide molecule. Each atom has one unshared pair of electrons, and there are three shared pairs. If the shared pairs are divided equally, each atom appears to have three electrons from among those shared. Therefore, each atom in the structure *appears* to have a total of five electrons. Carbon normally has four electrons in its valence shell so in the structure shown it has a formal charge of −1. An oxygen atom normally has six valence shell electrons so it *appears* that the oxygen atom has lost one electron giving it a formal charge of +1. The triple bond is only about 112.8 pm in length. Of course, the procedure is simply a bookkeeping procedure because no electrons have been lost or gained.

Formal charges can be used to predict the stable arrangement of atoms in many molecules. For example, nitrous oxide, N_2O, might have the structures shown as follows:

$$\bar{N} = O = \bar{N} \quad \text{and} \quad \bar{N} = N = \bar{O}$$

It might be tempting to assume that the structure on the left is correct, but not when the formal charges are considered. Note that the formal charges are circled to distinguish them from ionic charges. Following the procedure outlined above, the formal charges are as shown below:

$$\overset{\text{−1}}{\underset{}{\bar{N}}} = \overset{\text{+2}}{\underset{}{O}} = \overset{\text{−1}}{\underset{}{\bar{N}}}$$

A +2 formal charge on oxygen, the atom with the second highest electronegativity, is not in agreement with the rules for distributing formal charges. Therefore, the correct structure is the one shown on the right, which also accounts for the fact that N_2O can react as an oxidizing agent because of the oxygen atom being in a terminal position. *In general, the atom of lowest electronegativity will be found as the central atom.* Although we know that the arrangement of atoms is NNO, there is still the problem of resonance structures. For the N_2O molecule, three resonance structures can be shown as follows:

$$\underset{I}{\overset{(-1)\ (+1)\ (0)}{\underline{N}=N=\underline{\overset{..}{O}}}} \quad \longleftrightarrow \quad \underset{II}{\overset{(0)\quad(+1)\ (-1)}{|N\equiv N-\underline{\overset{..}{O}}|}} \quad \longleftrightarrow \quad \underset{III}{\overset{(-2)\quad(+1)\quad(+1)}{|\underline{N}-N\equiv O|}}$$

Structure III will contribute approximately 0% to the true structure because it has "like" formal charges on adjacent atoms, has a positive formal charge on the oxygen atom whereas a nitrogen atom has a -2 formal charge, and has higher formal charges overall. Deciding the relative contributions from Structures I and II is somewhat more difficult. Although Structure II has the negative formal charge on the oxygen atom, it also has a triple bond between the two nitrogen atoms, which results in six shared electrons in a small region of space. Two double bonds are generally preferred over a triple bond and a single bond. Structure I has two double bonds, even though it places a negative formal charge on a nitrogen atom. As a result of these factors, we suspect that Structures I and II would contribute about equally to the actual structure.

In this case, there is a simple experiment that will determine if this is correct. Structure I places a negative formal charge on the terminal nitrogen atom, but Structure II places a negative formal charge on the oxygen atom on the opposite end of the molecule. If the two structures contribute equally, these effects should cancel, which would result in a molecule that is not polar. In fact, the dipole moment of N_2O is only 0.17 D so Structures I and II must make approximately equal contributions.

Bond lengths are also useful when deciding contributions from resonance structures. Structure I shows a double bond between N and O, whereas Structure II shows the N−O bond as a single bond. If the structures contribute equally, the experimental N−O bond length should be approximately half way between the values for N−O and N=O, which is the case. Thus, we have an additional piece of evidence that indicates that Structures I and II contribute about equally to the actual structure. The observed bond lengths in the N_2O molecule are shown below (in picometers).

$$\text{N}\overset{112.6}{-\!\!-\!\!-\!\!-}\text{N}\overset{118.6}{-\!\!-\!\!-\!\!-}\text{O}$$

Known bond lengths for other molecules that contain bonds between N and O atoms are useful in assessing the contributions of resonance structures in this case. The N−N bond length is 110 pm, whereas that for the N=N bond is usually approximately 120−125 pm depending on the type of molecule. Likewise, the nitrogen to oxygen bond length in the NO molecule in which the bond order is 2.5 is 115 pm. On the other hand, in NO^+ (which has a bond order of 3), the bond length is 106 pm. From these values, it can be seen that the observed bond lengths in N_2O are consistent with the fact that the true structure is a hybrid of Structures I and II above.

We can also illustrate the application of these principles by means of other examples. Consider the cyanate ion, NCO^-. In this case, there are 16 valence shell electrons that need to be distributed. To provide a total of 3 octets, 24 electrons would be needed. Therefore, eight electrons must be shared, which means that there will be a total of four bonds, two from the central atom to each of the terminal atoms. Four bonds in the form of two double bonds in each direction can be expected to give a linear structure. Our first problem is how to arrange the atoms. Showing the total of 16 electrons, the arrangements can be illustrated as follows (the formal charges also indicated):

$$\overset{\overset{\text{-1}}{}}{\underline{\text{N}}} = \text{C} = \overset{\overset{\text{0}}{}}{\underline{\text{O}}} \qquad \overset{\overset{\text{-1}}{}}{\underline{\text{N}}} = \overset{\overset{\text{+2}}{}}{\text{O}} = \overset{\overset{\text{-2}}{}}{\underline{\text{C}}} \qquad \overset{\overset{\text{-2}}{}}{\underline{\text{C}}} = \overset{\overset{\text{+1}}{}}{\text{N}} = \overset{\overset{\text{0}}{}}{\underline{\text{O}}}$$

$$\text{I} \qquad\qquad\qquad\qquad \text{II} \qquad\qquad\qquad\qquad \text{III}$$

Deciding which arrangement of atoms is correct is based on the formal charge on the central atom. In the first structure, there are four shared pairs of electrons on the carbon atom, and dividing each pair equally leads to a total of four electrons around the carbon atom. Because a carbon atom normally has four valence shell electrons, the carbon atom in Structure I has a formal charge of 0. In Structure II, dividing each of the bonding pairs equally leads to accounting for four electrons on the oxygen atom that normally has six valence shell electrons. Therefore, in Structure II, the oxygen atom has a formal charge of +2. Of the three atoms in the structure, oxygen has the highest electronegativity so this structure is very unfavorable. In Structure III, dividing each bonding pair equally makes it appear that the nitrogen atom has four electrons on it, but there are five electrons in the valence shell of nitrogen. This results in a formal charge of +1 on the nitrogen atom and a −2 formal charge on the carbon atom.

Both Structures II and III have an arrangement of atoms that places a positive formal charge on atoms that are higher in electronegativity than carbon. Consequently, the most stable arrangement of atoms is as shown in Structure I. Some compounds containing the ion having Structure III (the fulminate ion) are known, but they are much less stable than the cyanates (Structure I). In fact, mercury fulminate has been used as a detonator.

As a general rule, we can see that for triatomic species containing 16 electrons, the fact that there must be four bonds to the central atom will result in a positive formal charge on that atom unless it is an atom that has only four valence shell electrons. Therefore, in cases where one of the three atoms is carbon, that atom is likely to be the central atom. A nitrogen atom in the central position would be forced to have a +1 formal charge, and an oxygen atom would have a +2 formal charge. By considering the structures of numerous 16-electron triatomic species, it will be generally found that the central atom is the one having the lowest electronegativity.

Now that we have determined that Structure I is correct for the cyanate ion, we still need to consider resonance structures. In keeping with the rules given earlier, the acceptable resonance structures that can be devised are

$$\overset{\overset{\text{-1}}{}}{\underline{\text{N}}} = \text{C} = \overset{\overset{\text{0}}{}}{\underline{\text{O}}} \longleftrightarrow |\text{N} \equiv \text{C} - \overset{\overset{\text{-1}}{}}{\underline{\text{O}}}| \longleftrightarrow |\overset{\overset{\text{-2}}{}}{\underline{\text{N}}} - \text{C} \equiv \overset{\overset{\text{+1}}{}}{\text{O}}|$$

$$\text{I} \qquad\qquad\qquad\qquad \text{II} \qquad\qquad\qquad\qquad \text{III}$$

In Structure I, the formal charges are −1, 0, and 0 on the nitrogen, carbon, and oxygen atoms, respectively. In Structure II, the corresponding formal charges are 0, 0, and −1. However, in Structure III, the formal charges on the atoms are nitrogen −2, carbon 0, and oxygen +1. Immediately, we see that the most electronegative atom, oxygen, has a positive formal charge, and we realize that if the actual structure is a resonance hybrid of these three structures, Structure III will contribute approximately 0%. This structure essentially represents removing electron density from an oxygen atom and placing it on a nitrogen atom. We now must estimate the contributions of the other two structures.

Although Structure II has a negative formal charge on the atom having the highest electronegativity, it also has a triple bond, which places a great deal of electron density in a small region of space. The repulsion that arises causes the bond to be less favorable than the -1 formal charge on the oxygen atom would suggest. On the other hand, the two double bonds in Structure I still provide a total of four bonds without as much repulsion as that resulting from a triple bond. Structure I also has a -1 formal charge on nitrogen, the second most electronegative of the three atoms. When all these factors are considered, we are led to the conclusion that Structures I and II probably contribute about equally to the true structure.

For CO_2, the structure contains two σ bonds and two π bonds, and it can be shown as

$$\bar{\underline{O}} = C = \bar{\underline{O}}$$

Two σ bonds are an indication of sp hybridization on the central atom, which leaves two p orbitals unhybridized. These orbitals are perpendicular to the molecular axis and can form π bonds with p orbitals on the oxygen atoms.

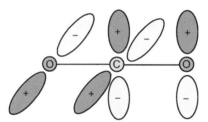

In CO_2, the C=O bond length is 116 pm, which is slightly shorter than the usual length of C=O bonds that are approximately 120 pm in length. Note that CO_2, NO_2^+, SCN^-, OCN^-, and N_2O are all triatomic molecules having 16 electrons and all are linear.

There are several important chemical species that consist of four atoms and have a total of 24 valence shell electrons. Some of the most common isoelectronic species of this type are CO_3^{2-}, NO_3^-, SO_3, and PO_3^- (known as the *meta*phosphate ion). Because four atoms would require a total of 32 electrons in order for each to have an octet, we conclude that 8 electrons must be shared in 4 bonds. With four bonds to the central atom, there can be no unshared pairs on that atom if the octet rule is to be obeyed. Therefore, we can draw the structure for CO_3^{2-} showing one double C=O bond and two single C—O bonds as

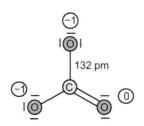

The double bond could also be shown in the other two positions so the true structure is a resonance hybrid of the three structures. The structure is a trigonal plane that has three identical bonds with each being an average of one double bond and two single bonds. The result is that there is a bond order of 1.33. Because of the structure being a trigonal plane, we understand that the hybrid orbital type used by carbon is sp^2. As a result, there is one additional p orbital on the central atom that is perpendicular to the plane and that orbital is empty. Therefore, a filled p orbital on an oxygen atom can overlap with the empty p orbital on the carbon atom to yield a π bond. This π bond is not restricted to one oxygen atom because the other two also have filled p orbitals that can be used in π bonding. The result is that the π bond is delocalized over the entire structure as is shown below.

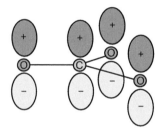

Note that because the carbon atom has a formal charge of 0, there is no necessity to draw structures having more than one double bond. The carbon atom has no valence shell orbitals other than $2s$ and $2p$ orbitals so only four pairs of electrons can be held in four valence shell orbitals. The structures of CO, CO_2, and CO_3^{2-} have been described in which the bond orders are 3, 2, and 4/3 and the bond lengths are 112.8, 116 (in CO_2, but approximately 120 in C=O bonds), and 132 pm, respectively. As expected, the bond length decreases as the bond order increases. A typical bond length for a C—O single bond is 143 pm so we have four C—O bond lengths that can be correlated with bond order. Figure 4.4 shows the relationship between the bond order and bond length for these types of C—O bonds.

A relationship such as this is useful in cases where a C—O bond length is known because from it the bond order can be inferred. On this basis, it is also sometimes possible to estimate the contributions of

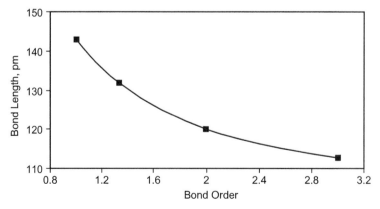

FIGURE 4.4
The relationship between bond order and bond length for bonds between carbon and oxygen.

various resonance structures. Pauling proposed an equation relating bond length relative to that of a single bond between the atoms that is written as

$$D_n = D_1 - 71 \log n \tag{4.5}$$

where D_n is the bond length for a bond of order n, D_1 is the length of a single bond, and n is the bond order. Using this equation, calculated bond lengths for C–O bonds having bond orders of 4/3, 2, and 3 are 134, 122, and 106 pm, respectively. The calculated length of a bond of 4/3 order is quite close to that in CO_3^{2-}. In many molecules, the length of a C=O bond is approximately 120 pm so the agreement is again satisfactory. In the case of the C≡O bond, the molecule that possesses this bond is carbon monoxide, which has some unusual characteristics because of ionic character (see Chapter 3) so the agreement is not as good between the experimental and calculated bond length. However, Eqn (4.5) is useful in many situations to provide approximate bond lengths.

For the SO_3 molecule, the structure we draw first by considering the number of valence shell electrons and the octet rule is

In this case, even with there being one double bond, the formal charge on the sulfur atom is +2 so structures that show *two* double bonds are possible and they can reduce further the positive formal charge. In drawing structures such as

for SO_3, the sulfur atom has 10 electrons around it, which means that the octet rule is not obeyed. However, in addition to the 3s and 3p valence orbitals, the sulfur atom also has empty 3d orbitals that can overlap with filled p orbitals on oxygen atoms. Therefore, unlike the carbon atom in CO_3^{2-}, the sulfur atom in SO_3 can accept additional electron density, and structures showing two double bonds are permissible. The S–O bond length in SO_2 where the bond order is 1.5 is 143 pm. That is almost exactly the S–O bond length in SO_3, which indicates that that a bond order of about 1.5 is correct for this molecule. Therefore, there must be some contributions from structures that show two double bonds because the bond order would be 4/3 if only one double bond is present.

The sulfate ion, SO_4^{2-}, exhibits some bonding aspects that deserve special consideration. First, there are five atoms so 40 electrons would be required to provide an octet of electrons around each atom. However, with only 32 valence shell electrons (including the two that give the -2 charge), there must be eight electrons shared. The four bonds will be directed toward the corners of a tetrahedron, which gives the structure

Although this structure agrees with several of our ideas about structure and bonding, there is at least one problem. If we determine the formal charges on the atoms, we find that there is a −1 formal charge on each oxygen atom but a +2 formal charge on the sulfur atom. Although sulfur has a lower electronegativity than oxygen, there is disparity in the electron densities on the atoms. This situation can be improved by taking an unshared pair on one of the oxygen atoms and making it a shared pair.

There is no reason that any particular oxygen atom should be chosen so there are four equivalent structures in which the double bond is shown to a different oxygen atom in each one. The question arises as to how this type of bonding occurs. When an oxygen atom is bonded to the sulfur atom by a single bond, there are three unshared pairs of electrons that reside in p orbital on the oxygen atom. Although the sulfur atom makes use of sp^3 hybrid orbitals (from $3s$ and $3p$ valence shell orbitals) in forming the four single bonds, the $3d$ orbitals are not greatly higher in energy, and they are empty. The symmetry (mathematical signs) of the filled p orbitals on an oxygen atom matches that of a d orbital. Therefore, electron density is shared between the oxygen and sulfur atoms, but the electrons come from filled orbitals on the oxygen atom. The result is that there is some double bond character to each S—O bond as a result of π bond formation, and the S—O bond lengths are shorter than expected for a single bond between the atoms.

In the H_2SO_4 molecule, there are two oxygen atoms that are bonded to hydrogen atoms and the sulfur atom. These oxygen atoms are unable to participate effectively in π bonding so the structure of the molecule is

This structure reflects that there is significant multiple bonding to two oxygen atoms but not to the two others that have hydrogen atoms attached. This behavior is also seen clearly in the structure of HSO_4^-,

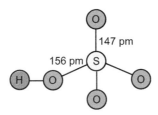

Note that the distance between the sulfur atom and the three oxygen atoms that do not have hydrogen attached is slightly greater than the corresponding distance in H_2SO_4. The reason is that the back donation is now spread over three terminal oxygen atoms rather than only two. The bond order of S—O bonds is slightly greater in H_2SO_4 than that in HSO_4^- when only terminal oxygen atoms are considered.

The PO_4^{3-} (known as the *ortho*phosphate ion) and ClO_4^- ions are isoelectronic with SO_4^{2-} ion, and their structures are

Although the phosphorus atom in PO_4^{3-} has a positive formal charge, it has a significantly lower electronegativity than oxygen. Therefore, there is not as much contribution from structures having double bonds. On the other hand, a +3 formal charge on the chlorine atom in ClO_4^- can be partially relieved by shifting some electron density from nonbonding orbitals on oxygen atoms to the empty d orbitals on chlorine atoms.

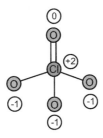

As in the case of the SO_4^{2-} ion, this is accomplished by overlap of filled p orbitals on oxygen atoms with empty d orbitals on the chlorine to give π bonds. The result of the contributions from structures in which there is some double bond character is that the bonds between chlorine and oxygen atoms are shorter than expected if the bonds were only single bonds. There is no reason why there cannot be some double bond character to more than one oxygen atom, which results in structures such as

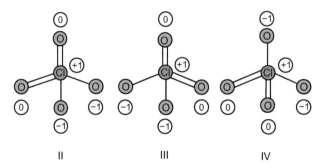

|| ||| IV

As expected, the extent to which the bonds in ClO_4^- are shortened from the single bond distance is rather large.

The structure of the H_3PO_4 molecule shown below is similar in many regards to that of H_2SO_4, but there are some significant differences.

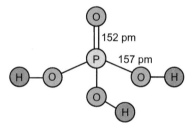

In sulfuric acid, the distance between the sulfur atom and the oxygen atoms without hydrogen atoms attached is 143 pm, the corresponding P—O distance in H_3PO_4 is 152 pm. This indicates that there is much less double bond character to the P—O bonds than there is to the S—O bonds. This is to be expected because of the +2 formal charge on sulfur when the structure is drawn showing only single bonds, whereas the phosphorus atom has only a +1 formal charge in a structure for H_3PO_4 showing only single bonds. Moreover, the phosphorus atom has a lower electronegativity than sulfur (2.2 compared with 2.6) so less double bonding to relieve a negative formal charge would be expected. The bond length of the HO—P single bond is 157 pm.

A structure for phosphorous acid, $(HO)_2HPO$, can be drawn as follows:

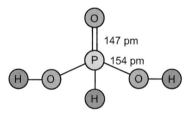

One of the hydrogen atoms is bound directly to the phosphorus atom and is not normally acidic. Note that in this case, the distance between the phosphorus atom and the oxygen atom without a hydrogen

atom attached is only 147 pm indicating more double bond character than there is in the corresponding bond in the H_3PO_4 molecule.

Another interesting structure is that of the dithionite ion, $S_2O_4^{2-}$, which is shown as

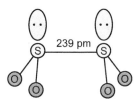

In several compounds that contain S—S single bonds, the bond length is approximately 205 pm. The very long S—S bond in $S_2O_4^{2-}$ is indicative of "loose" bonding, which is illustrated by the fact that when $^{35}SO_2$ is added to a solution containing $S_2O_4^{2-}$, some of the $^{35}SO_2$ is incorporated in $S_2O_4^{2-}$ ions. In contrast, the structure of dithionate, $S_2O_6^{2-}$,

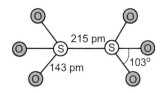

contains an S—S bond of more normal length, and the ion is more stable than is $S_2O_4^{2-}$. In $S_2O_4^{2-}$, the SO bonds are typical of SO bonds that have very little double bond character (151 pm). The fact that the S—O bonds in dithionate have substantial double bond character is indicated by the 143 pm bond length, which is equal to that in SO_2.

4.3 COMPLEX STRUCTURES: A PREVIEW OF COMING ATTRACTIONS

In addition to the structures discussed so far in this chapter, inorganic chemistry involves many others that can be considered to be chains, rings, or cages. In this section, several of the important structures will be described without resorting to theoretical interpretation. Some of the structures shown occur for several isoelectronic species so they represent structural *types*. In some cases, reactions are shown to illustrate processes that lead to products having such structures. These structures are often the result of an atom bonding to others of its own kind (which is known as *catenation*) or the formation of structures in which there are bridging atoms (especially oxygen because oxygen normally forms two bonds). An example of the latter type of structure is the pyrosulfate ion, $S_2O_7^{2-}$. This ion can be formed by adding SO_3 to sulfuric acid or by removing water (as in heating and thus the name *pyro*sulfate) from H_2SO_4 or a bisulfate.

$$H_2SO_4 + SO_3 \rightarrow H_2S_2O_7 \tag{4.6}$$

$$2\,NaHSO_4 \xrightarrow{\Delta} Na_2S_2O_7 + H_2O \tag{4.7}$$

The structure of the $S_2O_7^{2-}$ ion contains an oxygen bridge,

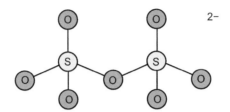

In this structure, as in the case of SO_4^{2-}, there is some double bond character to the bonds between the sulfur atoms and the terminal oxygen atoms. In addition to being the structure of $S_2O_7^{2-}$, this is the structure of the isoelectronic species $P_2O_7^{4-}$ (the *pyrophosphate* ion), $Si_2O_7^{6-}$ (pyrosilicate), and Cl_2O_7 (dichloride heptoxide). The peroxydisulfate ion, $S_2O_8^{2-}$, has a *peroxide* linkage between the two sulfur atoms.

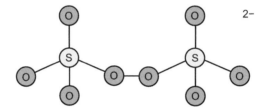

Dichloride hepatoxide, Cl_2O_7, results from the dehydration of $HClO_4$ with a strong dehydrating agent such as P_4O_{10}.

$$12\ HClO_4 + P_4O_{10} \rightarrow 6\ Cl_2O_7 + 4H_3PO_4 \tag{4.8}$$

The *di-* or *pyro*phosphate ion results from the partial dehydration of phosphoric acid,

$$2\ H_3PO_4 \rightarrow H_4P_2O_7 + H_2O \tag{4.9}$$

The reaction can be shown as a molecule of water being lost from two H_3PO_4 molecules as shown in Figure 4.5.

The polyphosphoric acids can be considered as arising from the reaction of $H_4P_2O_7$ with another molecule of H_3PO_4 by the loss of water. In the process shown below, the product is $H_5P_3O_{10}$, which is known as tripolyphosphoric acid.

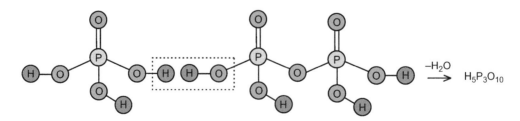

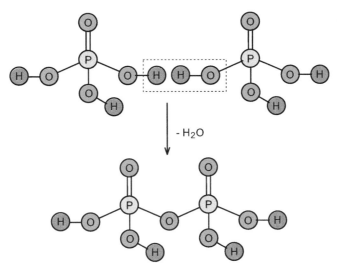

FIGURE 4.5
The partial dehydration of H_3PO_4.

The pyrophosphate ion also results from the dehydration of a salt such as Na_2HPO_4,

$$2\,Na_2HPO_4 \overset{\Delta}{\rightarrow} Na_4P_2O_7 + H_2O \qquad (4.11)$$

Complete hydration of P_4O_{10} in excess water produces orthophosphoric acid (H_3PO_4).

$$P_4O_{10} + 6\,H_2O \rightarrow 4\,H_3PO_4 \qquad (4.12)$$

but partial hydration of the oxide P_4O_{10} produces $H_4P_2O_7$.

$$P_4O_{10} + 4\,H_2O \rightarrow 2\,H_4P_2O_7 \qquad (4.13)$$

Elemental phosphorus is obtained on a large scale by the reduction of calcium phosphate with carbon in an electric furnace at $1200-1400\,°C$.

$$2\,Ca_3(PO_4)_2 + 6\,SiO_2 + 10\,C \rightarrow 6\,CaSiO_3 + 10\,CO + P_4 \qquad (4.14)$$

The element has several allotropic forms that are made up of tetrahedral P_4 molecules.

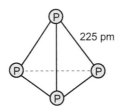

Combustion of phosphorus produces two oxides, P_4O_6 and P_4O_{10}, depending on the relative concentrations of the reactants.

$$P_4 + 3\,O_2 \;\rightarrow\; P_4O_6 \tag{4.15}$$

$$P_4 + 5\,O_2 \;\rightarrow\; P_4O_{10} \tag{4.16}$$

The structures of both P_4O_6 and P_4O_{10} are based on the P_4 tetrahedron. In the case of P_4O_6, there is an oxygen atom forming a bridge between each pair of phosphorus atoms along the edges of the tetrahedron resulting in a structure that can be shown as

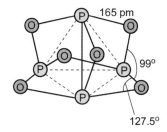

In this structure, the tetrahedron of phosphorus atoms is preserved. In keeping with the fact that when compared with P_4O_6, P_4O_{10} has four additional oxygen atoms, these oxygen atoms are bonded to the phosphorus atoms to give a structure that has not only the six bridging oxygen atoms but also one terminal oxygen atom bonded to each of the phosphorus atoms.

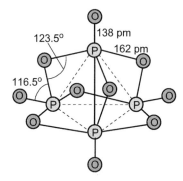

   Elemental phosphorus is only one of several elements whose structures are polyatomic species. Another is the structure of elemental sulfur that consists of puckered S_8 rings.

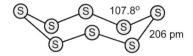

Although this eight-membered ring represents the structure of the molecule in the rhombic crystal phase, it is by no means the only sulfur molecule. Other ring structures have the formulas S_6, S_7, S_9, S_{10}, S_{12}, and S_{20}. Just as gaseous oxygen contains O_2 molecules, sulfur vapor contains S_2 molecules that are paramagnetic. Selenium also exists as Se_8 molecules, but catenation is less pronounced than in the case of sulfur, and tellurium shows even less tendency in this regard. Tellurium resembles metals in its chemistry more than does either sulfur or selenium.

Tetrasulfur tetranitride, S_4N_4, has a structure that can be considered as a hybrid of the two resonance structures shown below.

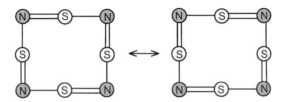

Although the structures above show positions of the bonds in the resonance structures, the geometric structure of the molecule is

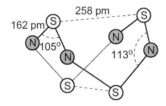

The distance between the sulfur atoms is considerably shorter than expected on the basis of radii of the isolated atoms. Therefore, there is believed to be a long weak bond between the sulfur atoms. A very large number of derivatives of S_4N_4 are known, and some of them will be described in Chapter 14.

Elemental boron exists as an icosahedral B_{12} molecule, the structure of which is shown in Figure 4.6. This structure has two staggered planes that contain five boron atoms in each as well as two boron atoms in apical positions.

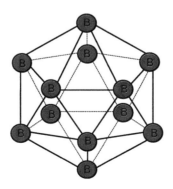

FIGURE 4.6
The structure of the B_{12} molecule.

No survey of polyatomic elements would be complete without showing the form of carbon that exists as C_{60}. Known as *buckminsterfullerene* (named after R. Buckminster Fuller, the designer of the geodesic dome), C_{60} has a cage structure that has 12 pentagons and 20 hexagons on the surface as shown in Figure 4.7(a). Each carbon atom makes use of sp^2 hybrid orbitals and is bonded by three σ bonds and one π bond with the π bonds being delocalized. A very large number of derivatives of C_{60} are known, and other forms of carbon have general formula C_x ($x \neq 60$). Carbon also exists as diamond and graphite that have the structures as shown in Figure 4.7(b) and (c).

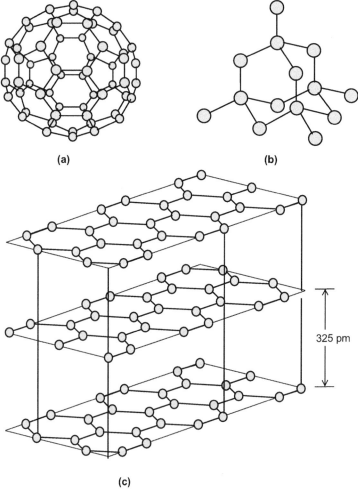

(a)

(b)

325 pm

(c)

FIGURE 4.7
Some of the structures of elemental carbon (see Chapter 13).

In addition to the structures of these elements, a great deal of structural inorganic chemistry is concerned with the silicates. These materials form a vast array of naturally occurring and synthetic solids whose structures are based on the tetrahedral unit SiO_4. There are structures that contain discrete SiO_4^{4-} ions as well as bridged structures such as $Si_2O_7^{6-}$. Because the Si atom has two fewer valence shell electrons than the S atom, the SiO_4^{4-} and SO_4^{2-} ions are isoelectronic as are the $Si_2O_7^{6-}$ and $S_2O_7^{2-}$ ions.

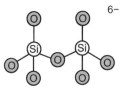

The SiO_4^{4-} ion, known as *orthosilicate*, occurs in minerals such as *zircon* ($ZrSiO_4$), *phenacite* (Be_2SiO_4), and *willemite* (Zn_2SiO_4). The SiO_3^{2-} ion is known as the *meta*silicate ion. Some of the minerals that contain the $Si_2O_7^{6-}$ ion are *thortveitite* ($Sc_2Si_2O_7$) and *hemimorphite* ($Zn_4(OH)_2Si_2O_7$).

Another important silicate structural type is based on a six-membered ring that contains alternating Si and O atoms and has the formula $Si_3O_9^{6-}$.

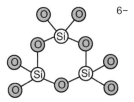

The $P_3O_9^{3-}$ ion (known as the trimetaphosphate ion) and $(SO_3)_3$, a trimer of SO_3, also have this structure. The trimetaphosphate ion can be considered as the anion of the acid $H_3P_3O_9$ (trimetaphosphoric acid), a trimeric form of HPO_3 (metaphosphoric acid). Note that this acid is formally related to $H_5P_3O_{10}$ (tripolyphosphoric acid) by the reaction

$$H_3P_3O_9 + H_2O \rightleftarrows H_5P_3O_{10} \tag{4.17}$$

A six-membered ring also is present in the anion of $Na_3B_3O_6$. It contains bridging oxygen atoms but has only one terminal oxygen atom on each boron atom.

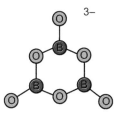

Although the complete description of its structure will not be shown here, boric acid, $B(OH)_3$, has a sheet structure in which each boron atom resides in a trigonal planar environment of oxygen atoms. There is hydrogen bonding between the OH groups in neighboring molecules.

Because there are numerous silicates whose structures are made up of repeating patterns based on the SiO_4 tetrahedron, a type of shorthand notation has been developed for drawing the structures. For example, the SiO_4 unit can be shown as follows:

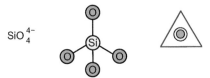

Then, the complex structures shown in Figure 4.8 are built up by combining these units. These structures are based on SiO_4 tetrahedra that share a corner or an edge. In the drawings, the solid circle

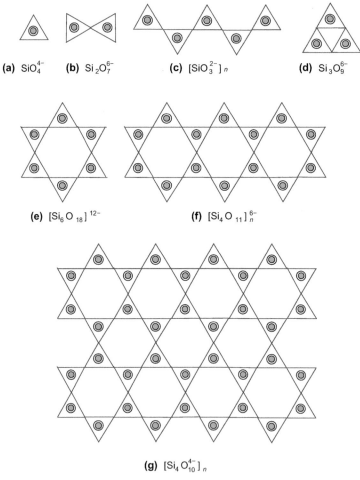

FIGURE 4.8
Structures of some common silicates.

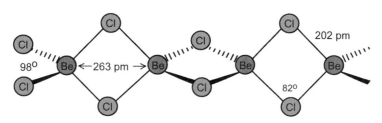

FIGURE 4.9
The structure of $BeCl_2$.

represents the silicon atom, and the open circle around it represents an oxygen atom that is directed upward out of the page. From combinations of the basic SiO_4 units, a wide range of complex structures result as shown in Figure 4.8.

Although the formula for beryllium chloride is $BeCl_2$, the compound exists in chains in the solid state. The bonding is covalent, and the environment around each Be is essentially tetrahedral with each Cl bridging between two Be atoms separated by 263 pm. The structure is shown in Figure 4.9. In the $BeCl_2$ monomer, there are only two bonds to Be and they contribute only four electrons around the atom. In the bridged structure, unshared pairs of electrons on Cl are donated to complete octets around Be atoms.

In this chapter, procedures for drawing molecular structures have been illustrated, and a brief overview of structural inorganic chemistry has been presented. The structures shown include a variety of types, but many others could have been included. The objective was to provide an introduction and review to the topics of VSEPR, hybrid orbitals, formal charge, and resonance. The principles discussed and types of structures shown will be seen later to apply to the structures of many other species.

4.4 ELECTRON-DEFICIENT MOLECULES

In this chapter, many of the basic principles related to structure and bonding in molecules have already been illustrated. However, there is another type of compound that is not satisfactorily described by the principles illustrated so far. The simplest molecule of this type is diborane, B_2H_6. The problem is that there are only 10 valence shell electrons available for use in describing the bonding in this molecule.

The BH_3 molecule is not stable as a separate entity. This molecule can be stabilized by combining it with another molecule that can donate a pair of electrons (indicates as :) to the boron atom to complete the octet (see Chapter 9). For example, the reaction between pyridine and B_2H_6 produces $C_5H_5N:BH_3$. Another stable adduct is carbonyl borane, $OC:BH_3$, in which a pair of electrons is donated from carbon monoxide, which stabilizes borane. In CO, the carbon atom has a negative formal charge so it is the "electron-rich" end of the molecule. Because the stable compound is B_2H_6, the bonding in that molecule should be explained.

The *framework* of B_2H_6 can be visualized by considering $B_2H_4^{2-}$, which is isoelectronic with C_2H_4. Starting with $B_2H_4^{2-}$ in which there would be a π bond like that in C_2H_4, the bonding can be represented as shown Figure 4.10.

The planar framework has σ bonds as shown in Figure 4.10, which involve sp^2 hybrid orbitals on the boron atoms. This leaves one unhybridized p orbital that is perpendicular to the plane. The B_2H_6 *molecule* can be considered being made by adding two H^+ ions to a hypothetical $B_2H_4^{2-}$ ion that is

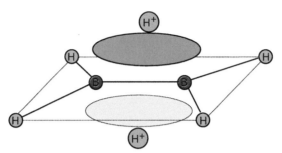

FIGURE 4.10
The structure resulting from the addition of two H^+ ions to $B_2H_4^{2-}$.

isoelectronic with C_2H_4 because each carbon atom has one more electron than does a boron atom. In the $B_2H_4^{2-}$ *ion*, the two additional electrons reside in a π bond that lies above and below the plane of the structure shown above. When two H^+ ions are added, they become attached to the lobes of the π bond to produce a structure the details of which are illustrated in Figure 4.11.

In each of the B—H—B bridges, only two electrons bond the three atoms together by having the orbitals on the boron atoms simultaneously overlap the hydrogen $1s$ orbital. A bond of this type is known as a *two-electron three-center bond*. In terms of molecular orbitals, the bonding can be described as the combination of two boron orbitals and one hydrogen orbital to produce three molecular orbitals, of which only the one of lowest energy is shown in Figure 4.12. Bonding of this type and other boron hydrides, which have three-center two-electron bonds with hydrogen bridges, will be discussed in Chapter 13.

Aluminum alkyls having the empirical formula AlR_3 are dimerized to give Al_2R_6 structures in which the two-electron three-center bonds involve alkyl groups as the bridging units. For example, $Al_2(CH_3)_6$ has the structure and dimensions as shown in Figure 4.13.

Other aluminum alkyls also exist as dimers as a result of bridging by the alkyl groups. In the structure shown in Figure 4.13, there are four CH_3 groups that are nonbridging and only two that are bridging. If an aluminum alkyl is prepared that contains more than one type of alkyl group, it is possible to determine which type of group forms the bridges. One such case is that of $Al_2(CH_3)_2(t\text{-}C_4H_9)_4$. In this case, it is found that the methyl groups are contained in the bridges, but none are found in the more numerous terminal positions. Therefore, we conclude that the methyl group forms stronger bridges between two aluminum atoms than does the t-butyl group. When other compounds of this type are prepared so that competition is set up between potential bridging groups, it is found that the strengths of bridges between aluminum atoms varies as $CH_3 > C_2H_5 > t\text{-}C_4H_9$. The associated nature of the

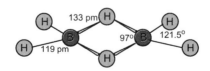

FIGURE 4.11
The structure of the B_2H_6 molecule.

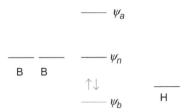

FIGURE 4.12
A simplified molecular orbital diagram for a three-center B—H—B bond.

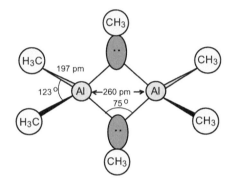

FIGURE 4.13
The structure of the $Al_2(CH_3)_6$ dimer.

aluminum alkyls will be discussed in greater detail in Chapter 6. Chloride ions form the bridges in the Al_2Cl_6 dimer. In fact, dimers exist when aluminum is bonded to numerous atoms and groups. The stability of the bridges formed varies in the order $H > Cl > Br > I > CH_3$.

A polymeric structure is exhibited by "beryllium dimethyl," which is actually $[Be(CH_3)_2]_n$ (see the structure of $(BeCl_2)_n$ shown in Figure 4.9), and $LiCH_3$ exists as a tetramer, $(LiCH_3)_4$. The structure of the tetramer involves a tetrahedron of Li atoms with a methyl group residing above each face of the tetrahedron. An orbital on the CH_3 group forms multicentered bonds to four Li atoms. There are numerous compounds for which the electron-deficient nature of the molecules leads to aggregation.

4.5 STRUCTURES HAVING UNSATURATED RINGS

In addition to the types of structures shown thus far, there are several others that are both interesting and important. One such type of structure contains unsaturated rings. Because $R-C\equiv N$ is called a nitrile, compounds containing the $-P\equiv N$ group were originally called phosphonitriles. An unstable molecule having the formula $:N-PH_2$ is known as phosphazine. Although this molecule is unstable, polymers containing this monomer unit with chlorine replacing the hydrogen atoms are well known. Heating a solution of PCl_5 and NH_4Cl in C_6H_5Cl or $HCl_2C-CHCl_2$ leads to the reaction

$$n\,NH_4Cl + n\,PCl_5 \rightarrow (NPCl_2)_n + 4n\,HCl \qquad (4.18)$$

Several materials having the formula $(NPCl_2)_n$ are known, but the most extensively studied of them is the cyclic trimer, $(NPCl_2)_3$, which has the structure

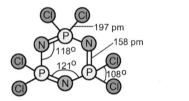

Compounds of this type are known as *phosphazines*, and they have structures that have a planar ring because of π bonding. In $(NPCl_2)_3$, the P—N distance is 158 pm, which is much shorter than the approximately 175 pm that is characteristic of a P—N single bond. Whether these compounds are strictly aromatic in character is not certain. The bonding in $(NPCl_2)_3$ is much more complex than that in benzene because unlike benzene in which the delocalized π orbital arises from overlap of unhybridized p orbitals on the carbon atoms, the P—N bonding involves p_N–d_P (the subscripts show the atoms involved) overlap to give p_π–d_π bonding. Delocalization does not appear to be as complete in the phosphazines as it is in benzene. Substitution of two groups on the phosphorus atoms in the ring can lead to three types of products. If the substituents are on the same phosphorus atom, the product is known as *geminal*, but if substitution is on different phosphorus atoms, the product may have *cis* or *trans* configuration depending on whether the two groups are on the same or opposite sides of the ring. These structures are shown in Figure 4.14. The chemistry of these compounds will be discussed in more detail in Chapter 13.

Carbon atoms have four valence shell electrons. Because boron and nitrogen atoms have three and five valence shell electrons, respectively, one boron and one nitrogen atom are formally equivalent to two carbon atoms. Therefore, a compound that contains an even number (n) of carbon atoms could have an analogous structure that contains $n/2$ atoms of boron and $n/2$ atoms of nitrogen. The most common case of this type is a compound that is analogous to benzene, C_6H_6, which has the formula $B_3N_3H_6$. This compound, sometimes referred to as "inorganic benzene", is actually *borazine* and has a structure that can be shown using resonance structures analogous to those used to describe benzene (Figure 4.15).

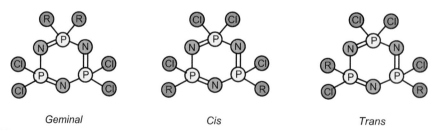

Geminal   Cis   Trans

FIGURE 4.14
The three isomers of $N_3P_3Cl_4R_2$.

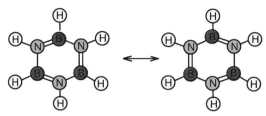

FIGURE 4.15
Resonance structures for borazine, $B_3N_3H_6$.

Therefore, the borazine molecule is considered to be aromatic. Borazine has many properties that are similar to those of benzene, although it is more reactive as a result of the B–N bonds being somewhat polar rather than purely covalent as in benzene.

The trichloro derivative having a chlorine atom bound to each of the boron atoms is known as *B*-trichloroborazine. Borazine can be prepared by several different methods, one of which is the reduction of the trichloro compound.

$$6\,NaBH_4 + 2\,Cl_3B_3N_3H_3 \rightarrow 2\,B_3N_3H_6 + 6\,NaCl + 3\,B_2H_6 \qquad (4.19)$$

Borazine can also be prepared directly by the reaction of diborane with ammonia.

$$3\,B_2H_6 + 6\,NH_3 \rightarrow 2\,B_3N_3H_6 + 12\,H_2 \qquad (4.20)$$

More of the chemistry of this interesting compound will be presented in Chapter 14.

4.6 BOND ENERGIES

Closely related to molecular structure are the energies associated with chemical bonds. It is frequently possible to make decisions regarding the stability of alternative structures based on the types of bonds present. However, because SF_4 has four S–F single bonds, it is not possible to determine whether the trigonal pyramidal or irregular tetrahedron is the stable structure because each of them has four bonds. However, for many situations, bond energies provide a useful tool.

Consider a compound such as $N(OH)_3$, which we would suspect is not the most stable arrangement for one nitrogen atom, three oxygen atoms, and three hydrogen atoms. Suppose one reaction of such a molecule can be written as

We can consider this reaction as taking place by breaking all the bonds in $N(OH)_3$ and then making all the bonds in HNO_2 (which is actually HONO) and H_2O. Breaking the bonds in $N(OH)_3$ means breaking 3 N—O bonds (163 kJ each) and 3 H—O bonds (464 kJ each). The energy required would be $3(163 \text{ kJ}) + 3(464 \text{ kJ}) = 1881 \text{ kJ}$. When the products form, the bonds formed are two H—O bonds in water and one H—O bond, one N—O bond (163 kJ), and one N=O bond (594 kJ) in HNO_2. These bonds give a total of 2149 kJ for the bonds *formed*, which means that this the amount of energy *released*. Therefore, for the entire process, the energy change is $(1881 \text{ kJ} - 2149 \text{ kJ}) = -268 \text{ kJ}$ so $N(OH)_3$ should be unstable with respect to the HNO_2 and H_2O products. Such calculations tell us nothing about the *rate* of a process because the rate depends on the pathway, whereas the thermodynamic stability depends only on the initial and final states. Even reactions that are energetically favorable may take place slowly (or not at all) because there may be no low-energy pathway.

An additional example of the use of bond energies will be considered. Earlier in this chapter, we considered the structure of the cyanate ion, OCN^-. Suppose the structures being considered are

$$\bar{\underline{O}} = C = \bar{\underline{N}} \qquad \bar{\underline{O}} = N = \bar{\underline{C}}$$
$$\text{I} \qquad\qquad\qquad \text{II}$$

and we wish to make a prediction regarding the relative stability of the structures. From the discussion presented earlier, we know that Structure I is more likely because in Structure II there is a positive formal charge on the nitrogen atom. The bond energies that are needed are as follows:

C = O 745 kJ	N = O 594 kJ
C = N 615 kJ	C = N 615 kJ

If we start with the atoms (and one additional electron) and form the bonds, Structure I would release a total energy of -1360 kJ, whereas Structure II would release -1209 kJ. Therefore, we predict (correctly) that the structure for the ion should be OCN^- rather than ONC^-. However, the structure

$$\bar{\underline{N}} = O = \bar{\underline{C}}$$

involves one N=O bond (594 kJ) and one C=O bond (745 kJ) so it appears from bond energies to be almost as stable as Structure I above (1360 vs. 1339 kJ). However, this structure places a +2 formal charge on the oxygen atom, which is contrary to the principles of bonding. It is best not to stretch the bond energy approach too far and to use it in conjunction with other information, such as formal charges and electronegativities when trying to decide issues related to stability.

The bond energy approach may not always be adequate because it is the free energy change, ΔG, that is related to equilibrium constants, and the free energy is given by

$$\Delta G = \Delta H - T\Delta S \qquad\qquad (4.22)$$

Accordingly, the difference in entropy for the structures in question may also be a factor in some cases. In spite of this, bond energies provide a basis for comparing structures. To use this approach, energies for many types of bonds are needed, and they are given in Table 4.1.

It should always be kept in mind that bond energies are usually *average* values based on the energies of bonds in several types of molecules. In a given molecule, a certain bond may have an energy that is somewhat different from the value given in the table. Therefore, in cases where a decision on the

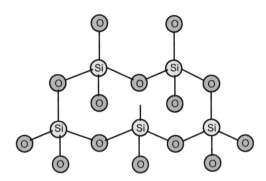

FIGURE 4.16
The network structure of SiO$_2$.

difference in stability between two structures is based on bond energies, regard small differences as being inconclusive.

One interesting and highly significant application of bond energies involves the enormous difference in the character of CO_2 and SiO_2. In the case of CO_2, the structure is a monomeric molecule that has two double bonds.

$$\bar{\underline{O}}=C=\bar{\underline{O}}$$

In the case of SiO_2, the structure is a *network* in which oxygen atoms form bridges between silicon atoms with each Si being surrounded by four oxygen atoms as shown in Figure 4.16.

The π bonding in CO_2 results in strong double bonds that actually are stronger than two C—O single bonds (C=O is 745 kJ mol^{-1}, whereas each C—O is 360 kJ mol^{-1}). Between Si and O, π bonding is not as strong because of the difference in size of the orbitals on the two atoms. As a result, a Si—O single bond is 464 kJ mol^{-1} (which is stronger than a C—O bond because of the greater polarity), whereas a Si=O bond is only 640 kJ mol^{-1}. Therefore, it is energetically more favorable for carbon to form two C=O double bonds, but four Si—O single bonds are more favorable energetically than two Si=O bonds. Carbon dioxide is a monomeric gas, whereas SiO$_2$ (which exists in several forms) is an extended solid that melts at over 1600 °C.

A compound that is less stable than some other compound into which it should transform with a release of energy may be locked in the less stable form because of the unfavorable kinetics for its transformation. Suppose the reaction

$$A \rightarrow B \qquad (4.23)$$

is thermodynamically favorable. If the pathway for the reaction is such that the rate is very slow, A may not react because of kinetic inertness rather than its thermodynamic stability. Such a situation is known

B (formed rapidly but less stable)

A

$\qquad (4.24)$

C (formed slowly but stable)

as *kinetic* stability rather than *thermodynamic* stability. A similar situation exists when the following system is considered.

In this case, the predominant product of the reaction will be B, even though it is less stable than C because of the difference in reaction rates. Sometimes, B is referred to as the *kinetic* product, whereas C is called the *thermodynamic* product.

In this chapter has been presented an elementary discussion of bonding principles that have broad applicability in the study of inorganic chemistry. These topics have been introduced by showing many of the important types of structures that are encountered in the study of inorganic materials. These structures will be revisited in later chapters, but the intent is to show many different types of structures and the relationships between many of them. Resonance, repulsion, electronegativity, and formal charge have been used to explain many of the aspects of bonding that are useful for understanding structures.

References for Further Study

Cotton, F.A., Wilkinson, G., Murillo, C.A., Bochmann, M., 1999. *Advanced Inorganic Chemistry*, 6th ed. John Wiley, New York. Almost 1400 pages devoted to all phases of inorganic chemistry. An excellent reference text.

DeKock, R.L., Gray, H.B., 1980. *Chemical Bonding and Structure*. Benjamin Cummings, Menlo Park, CA. An excellent introduction to many facets of bonding and structure of a wide range of molecules.

Douglas, B.E., McDaniel, D., Alexander, J., 1994. *Concepts and Models of Inorganic Chemistry*, 3rd ed. John Wiley, New York. A well-known text that provides a wealth of information on structures of inorganic materials.

Greenwood, N.N., Earnshaw, A., 1997. *Chemistry of the Elements*, 2nd ed. Butterworth Heinemann, New York. Although this is a standard reference text on descriptive chemistry, it contains an enormous body of information on structures of inorganic compounds.

Huheey, J.E., Keiter, E.A., Keiter, R.L., 1993. *Inorganic Chemistry: Principles of Structure and Reactivity*, 4th ed. Benjamin Cummings, New York. A popular text that has stood the test of time.

Lide, D.R. (Ed.), 2003. *CRC Handbook of Chemistry and Physics*, 84th ed. CRC Press, Boca Raton, FL Structural and thermodynamic data are included for an enormous number of inorganic molecules in this massive data source.

Mackay, K., Mackay, R.A., Henderson, W., 2002. *Introduction to Modern Inorganic Chemistry*, 6th ed. Nelson Thornes, Cheltenham, UK. One of the standard texts in inorganic chemistry.

Pauling, L., 1960. *The Nature of the Chemical Bond*, 3rd ed. Cornell University Press, Ithaca, NY. A true classic in bonding theory. Although somewhat dated, this book still contains a wealth of information on chemical bonding.

Sharpe, A.G., 1992. *Inorganic Chemistry*, 3rd ed. Longman, New York. Excellent coverage of bonding concepts in inorganic molecules and many other topics.

Atkins, P., Overton, T., Rourke, J., Weller, M., Armstrong, F., 2010. Shriver and Atkins Inorganic Chemistry, 5th ed. Oxford, New York. A highly regarded textbook in inorganic chemistry.

QUESTIONS AND PROBLEMS

1. Draw structures for the following showing correct geometry and all valence shell electrons: (a) OCS; (b) XeF_2; (c) H_2Te; (d) ICl_4^+; (e) $BrCl_2^+$; (f) PH_3.

2. Draw structures for the following showing correct geometry and all valence shell electrons: (a) SbF_4^+; (b) ClO_2^-; (c) CN_2^{2-}; (d) ClF_3; (e) $OPCl_3$; (f) SO_3^{2-}.

3. Suppose that a molecule is composed of two atoms of phosphorus and one of oxygen. Draw structures for two possible isomers of the molecule. For the more stable structure, draw the resonance structures. Which structure is least important?

4. Draw structures for the following showing correct geometry and all valence shell electrons: (a) Cl_2O; (b) ONF; (c) $S_2O_3^{2-}$; (d) PO_3^-; (e) ClO_3^-; (f) ONC^-.

5. The bond angle in ONCl is 116°. Explain what this means in terms of hybridization and how hybridization of orbitals on N allows a π bond to be present.

6. Explain why the bond length for two of the N to O bonds in HNO_3 are shorter than they are in NO_3^-.

7. The N—O bond length in NO_2^+ is 115 pm, but it is 120 pm in the NO_2 molecule. Explain this difference.

8. In H_3PO_4, one P—O bond has a different length than the other three. Would it be shorter or longer than the others? Why?

9. A P—N single bond normally has a length of approximately 176 pm. In $(PNF_2)_3$ (which contains a six-membered ring of alternating P and N atoms), a P—N bond length of 156 pm is found. Explain the difference in bond lengths.

10. The C≡O bond has a length of 113 pm, and it is the strongest bond found for a diatomic molecule. Why is it stronger than the bond in N_2 with which it is isoelectronic?

11. The reaction $CaC_2 + N_2 \rightarrow CaCN_2 + C$ produces calcium cyanamide, which has been widely used as a fertilizer. Draw the structure of the cyanamide ion and describe the bonding.

12. The O=O and S=S bond energies are 498 and 431 kJ mol^{-1}, respectively, and the O—O and S—S bond energies are 142 and 264 kJ mol^{-1}, respectively. Explain why it is reasonable to expect structures containing sulfur that have extensive catenation, but such structures are not expected for oxygen.

13. Why are there two different P—O bond lengths in P_4O_{10} and why do they differ by such a large extent?

14. Explain the slight difference in N—O bond lengths in NO_2^- (124 pm) and NO_3^- (122 pm).

15. In the compound ONF_3, the O—N bond length is 116 pm. The N—O single bond length is 121 pm. Draw resonance structures for ONF_3 and explain the short observed bond length.

16. Given the bond energies (in kJ mol^{-1}) C—O (360), C=O (745), Si—O (464), and Si=O (640), explain why extensive structures containing Si—O—Si linkages are stable, whereas similar structures containing C—O—C bonds are not.

17. The dehydration of malonic acid, $HO_2C—CH_2—CO_2H$, produces C_3O_2 (known as tricarbon dioxide or carbon suboxide). Draw the structure for C_3O_2 and describe the bonding in terms of resonance structures.

18. Explain why the N—O bond lengths decrease for the following species in the order $NO_2^- > NO_2 > NO_2^+$.

19. In the solid state, PBr_5 exists as $PBr_4^+Br^-$, but PCl_5 exists as $PCl_4^+PCl_6^-$. Explain this difference in behavior.

20. The stability of the oxy anions of the Group V elements decreases in the order $PO_4^{3-} > AsO_4^{3-} > SbO_4^{3-}$. Explain this trend in stability.

21. Although the electron affinity of fluorine is less than that of chlorine, F_2 is more reactive than Cl_2. Explain some of the reasons for this difference in reactivity.

22. Draw the structure for H_5IO_6. Explain why iodine can form H_5IO_6, but chlorine does not form H_5ClO_6.

23. Draw structures for $N(OH)_3$ and ONOH and show using bond energies why $N(OH)_3$ would not be expected to be stable.

24. By making use of bond energies, show that H_2CO_3 would be expected to decompose into CO_2 and H_2O.

25. Most *gem* diol compounds of carbon (containing two OH groups on the same carbon atom) are unstable. Using bond energies, show that this is expected.

26. The following species containing antimony are known: $SbCl_3$, $SbCl_4^-$, $SbCl_5$, $SbCl_5^{2-}$, and $SbCl_6^-$. Draw structures for each and predict bond angles using VSEPR. What type of hybrid orbitals is used by Sb in each case?

27. A reaction that produces NCN_3, known as cyanogen azide, is

$$BrCN + NaN_3 \rightarrow NaBr + NCN_3 (\text{cyanogenazide})$$

Draw the structure of cyanogen azide. Speculate on the stability of this molecule.

28. Draw two possible structures for thiocyanogen, $(SCN)_2$, and comment on the relative stability of the structures you draw.

29. Explain why the F—O—F bond angle in OF_2 is 102°, whereas the Cl—O—Cl angle in OCl_2 is 115°.

30. Although PF_5 exists as a molecule, "NF_5" exists as $NF_4^+F^-$. Explain this difference.

31. The P—O and Si—O single bond lengths are 175 and 177 pm, respectively. In PO_4^{3-} and SiO_4^{4-} ions, the bond lengths are 154 and 161 pm, respectively. Why are the bonds shorter than single bonds? Why is the P—O bond shortened more than is the Si—O bond?

CHAPTER 5

Symmetry and Molecular Orbitals

In the previous chapter, the structures of many molecules and ions were described by drawing structures showing how the electrons are distributed. However, there is another way in which the structures of molecules are described. That way uses different language and symbols to convey information about the structures in an efficient, unambiguous way. In this way, the structures of molecules and ions are described in terms of their *symmetry*. Symmetry has to do with the spatial arrangement of objects and the ways in which they are interrelated. For example, the letter "T" has a plane that bisects it along the "post," giving two halves that are identical in relationship to that plane. However, the letter "R" does not have such a plane that divides it into two identical parts. This simple example illustrates a symmetry characteristic that is known as a plane of symmetry. There is much more that can be done with symmetry in terms of molecular structure, so this chapter is devoted to this important topic.

5.1 SYMMETRY ELEMENTS

Understanding symmetry as related to molecular structure is learning to look at molecules in ways to see the spatial relationships of atoms to each other. Visualization of a molecule as a three-dimensional assembly is accomplished in terms of symmetry elements. Symmetry elements are lines, planes, and points that have a special relationship to a structure. In the discussion above, it was pointed out that the letter "T" has a plane that divides it into two identical fragments. That plane is known as a *plane of symmetry* or a *mirror plane* (which is designated as σ). The letter "H" also has such a plane that divides it into two identical parts, the plane perpendicular to the page that cuts the crossbar in half. There are also lines about which the letter H can be rotated to achieve an orientation that is identical to that shown. For example, a line in the plane of the page that passes through the midpoint of the cross bar is such a line. There is another that is perpendicular to the plane of the page that passes through the center of the crossbar and a third line about which H can be rotated to give the identical structure lies in the plane of the page and runs along the cross bar. Rotation by 180° around any of these three lines gives a structure that is identical to H.

Rotation of the H around either of the lines shown gives the letter in the orientation shown. The third line is perpendicular to the page and passes through the center of the cross bar.

Inorganic Chemistry. DOI: http://dx.doi.org/10.1016/B978-0-12-385110-9.00005-4

The lines that we have just described are known as lines of symmetry or *rotation axes* that are designated by the letter C. In this case, the angle of rotation required to give the same orientation of H is 180°, so the axes are C_2 axes. The subscript is an index that is obtained by dividing 360° by the angle through which the structure must be rotated to give an orientation that is indistinguishable from the original. Therefore, $360°/180° = 2$, so each of the axes is known as a C_2 axis. To be precise, the axes that we have described are known more correctly as *proper rotation axes*. It is important to distinguish between the proper rotation axis itself and the *operation* of actually rotating the molecule. Of course, any object can be rotated by 360° to give an orientation that is unchanged, so all objects have a C_1 axis. Rotations may be carried out sequentially, and the rotation of a molecule *m* times around an axis of degree of symmetry *n* is indicated as C_n^m.

If we consider the H_2O molecule, which has the arrangement of atoms (electrons are not localized and therefore are not considered when determining symmetry) shown as

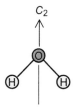

we see that there is a line through the oxygen atom that bisects the H—O—H bond angle about which rotation by 180° would leave the molecule unchanged. That line is a C_2 axis. Although any line through the structure is a C_1 axis, the C_2 axis is the axis of highest symmetry because a smaller rotation around it gives back the original structure. *The axis of highest symmetry in a structure is defined as the z-axis.*

From the structure drawn above for the water molecule (and that shown later in Figure 5.5), we can see that there are also two planes that divide the H_2O molecule into identical fragments. One is the plane of the page; the other is perpendicular to the page and bisects the oxygen atom, leaving a hydrogen atom on either side. A plane of symmetry is denoted as σ. Because we ordinarily take the z-axis to be the vertical direction, both of the planes of symmetry are vertical planes that contain the z-axis. They are designated as σ_v planes. The H_2O molecule thus has one C_2 axis and two vertical planes (σ_v), so its symmetry designation (also known as the *point group*) is C_{2v}. We will explain more about such designations later.

The structure of the ClF_3 molecule is based on there being 10 electrons around the central atom (seven valence electrons from Cl and one from each F atom). As we have seen earlier, unshared pairs of electrons are found in equatorial positions, so the structure can be shown as

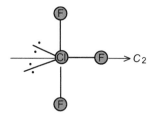

It is easy to see that the line passing through the chlorine atom and the fluorine atom in the equatorial position is a C_2 axis. Rotation about this axis by $180°$ leaves these two atoms unchanged, but the fluorine atoms in axial positions are interchanged. There are two planes that bisect the molecule. One of these is the plane of the page, which cuts all the atoms in half, and the other is the plane perpendicular to that plane that bisects the Cl atom and the fluorine atom in the equatorial position. This collection of symmetry elements (one C_2 axis and two vertical planes) means that the ClF_3 molecule can also be designated as a molecule having C_{2v} symmetry.

The formaldehyde molecule, H_2CO, has a structure that is shown as

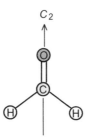

The line passing through the carbon and oxygen atoms is a C_2 axis and rotation of the molecule around that axis by $180°$ leaves those two atoms in their positions but interchanges the positions of the two hydrogen atoms. Moreover, there are two planes that bisect the molecule. One is perpendicular to the plane of the page that bisects the carbon and oxygen atoms, leaving one hydrogen on either side. The other plane is the plane of the page that bisects all four atoms. Therefore, the formaldehyde molecule also has C_{2v} symmetry. In each of the cases described above, we find that the molecule has one C_2 axis and two σ_v planes that intersect along the C_2 axis. These characteristics define the symmetry type known as C_{2v}, which is also the point group to which the molecule belongs.

The ammonia molecule has a structure that can be shown as

This pyramidal molecule has a C_3 axis that runs through the nitrogen atom and through the center of the triangular base formed by the three hydrogen atoms. Rotation around this axis by $120°$ leaves the position of the nitrogen atom unchanged but interchanges the hydrogen atoms. Viewed from the top looking down the C_3 axis, the ammonia molecule is seen as

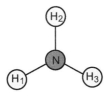

where the view is directly down the C_3 axis and the subscripts on the hydrogen atoms are to identify their positions. Clockwise rotation by 120° around the C_3 axis results in the molecule having the orientation

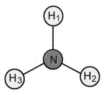

There are also three mirror planes that bisect the molecule along each N–H bond. Therefore, the NH_3 molecule has one C_3 axis and three σ_v planes, so the point group for the molecule is C_{3v}.

The BF_3 molecule has a planar structure, and it can be represented as (the subscripts are to identify the fluorine atoms)

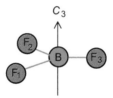

with a C_3 axis perpendicular to the plane of the molecule. Rotation of the molecule by 120° around that axis causes the fluorine atoms to interchange positions but gives an orientation that is identical.

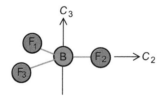

In addition to the C_3 axis (the z-axis), there are also three vertical planes of symmetry, which can be seen clearly in Figure 5.1. They are perpendicular to the plane of the molecule and bisect the molecule along each B–F bond. Because the molecule is planar, there is also a horizontal plane of symmetry (σ_h) that

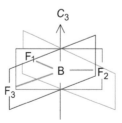

FIGURE 5.1
The BF_3 molecule showing planes of symmetry.

bisects all four atoms. Each B—F bond is also a C_2 axis because rotation around the bond would give the same orientation of the molecule except for interchanging the positions of the fluorine atoms. One of the C_2 axes is shown above, and rotation by $180°$ around that axis would produce the orientation

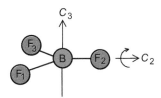

Note that each of the C_2 axes not only is coincident with a B—F bond but also is the line of intersection of the horizontal plane with one of the vertical planes. It is generally true that the intersection of a vertical plane of symmetry with a horizontal plane generates a C_2 axis. The list of symmetry elements that we have found for the BF_3 molecule includes one C_3 axis, three vertical planes (σ_v), three C_2 axes, and one horizontal plane (σ_h). A molecule possessing these symmetry elements, such as BF_3, SO_3, CO_3^{2-}, and NO_3^-, is said to have D_{3h} symmetry. In the cases of H_2O, ClF_3, H_2CO, and NH_3, the symmetry elements included only a C_n axis and n vertical planes. These molecules belong to the general symmetry type known as C_{nv}. Molecules that have a C_n axis and also n C_2 axes perpendicular to the C_n axis are known as D_n molecules.

The XeF_4 molecule has a planar structure with unshared pairs of electrons above and below the plane.

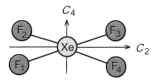

A line perpendicular to the plane of the molecule that passes through the Xe atom is a C_4 axis. There are four vertical mirror planes that intersect along the C_4 axis. Two of them cut the molecule along opposite Xe—F bonds; the other two bisect the molecule by bisecting opposite F—Xe—F angles. One of these planes cuts the horizontal plane of the molecule along the C_2 axis shown in the drawing above. There is, of course, a horizontal plane of symmetry, σ_h. The intersections of the four vertical planes with the horizontal plane generate four C_2 axes that are perpendicular to the C_4 axis.

However, the XeF_4 molecule has one additional symmetry element. The center of the Xe atom is a point through which each fluorine atom can be moved the same distance that it was originally from that point to achieve an orientation that is identical with the original. If this operation is carried out with the XeF_4 molecule oriented as above, the resulting orientation can be shown as

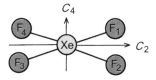

A center of symmetry (designated as i) is a point through which each atom can be moved a like distance to achieve an orientation that is identical to the original. The center of the Xe atom in XeF_4 is such

a point. Having four C_2 axes perpendicular to the C_4 (as a result of having a horizontal plane of symmetry), the XeF_4 belongs to the D_{4h} symmetry type.

Linear molecules belong to one of two symmetry types. The first is typified by HCN, which has the structure

$$H - C \equiv N \rightarrow C_\infty$$

Rotation of the molecule through any angle around the axis that lies along the bonds gives the same orientation. The rotation may even be by an infinitesimally small angle. Division of 360° by such a small angle gives a value approaching infinity, so the axis is known as a C_∞ axis. An infinite number of planes that intersect along the C_∞ axis bisect the molecule. Having a C_∞ axis and an infinite number of σ_v planes, the symmetry type is $C_{\infty v}$. Some other molecules and ions that have this symmetry are N_2O, OCS, CNO^-, SCN^-, and HCCH.

Linear molecules having a different symmetry type are typified by CO_2, which has the structure

$$\begin{array}{c} C_2 \\ \uparrow \\ O = C = O \longrightarrow C_\infty \end{array}$$

In this case, the C_∞ axis (the z-axis or vertical axis) functions as it did in the case of HCN, and there are an infinite number of σ_v planes that intersect along the C_∞ axis. However, the molecule also has a plane of symmetry that bisects the carbon atom leaving one oxygen atom on either side. Because that plane is perpendicular to the C_∞ axis, it is a horizontal plane. The intersection of an infinite number of vertical planes with a horizontal plane generates an infinite number of C_2 axes one of which is shown above. Moreover, the molecule has a center of symmetry, the center of the carbon atom. A center of symmetry is a point through which each atom can be moved the same distance it lies from the point initially to achieve an identical orientation of the molecule. The symmetry type (point group) that corresponds to one C_∞ axis, an infinite number of σ_v planes, one horizontal σ_h plane, an infinite number of C_2 axes perpendicular to the C_∞, and a center of symmetry is known as $D_{\infty h}$. Linear molecules that have a center of symmetry belong to this point group. In addition to CO_2, other examples of molecules of this type include XeF_2, ICl_2^-, CS_2, and BeF_2.

In addition to molecules having the symmetry types discussed above, there are a few special types. One of them is illustrated by the molecule ONCl. This molecule has the structure

This molecule has no rotation axis of higher symmetry than C_1. However, it does have one plane of symmetry, the one that bisects all three of the atoms. A molecule that has only a plane of symmetry is designated as C_s.

A tetrahedral molecule such as CH_4 or SiF_4 illustrates another of the special symmetry types. The structure of CH_4 is shown in Figure 5.2, which shows the four bonds as being directed toward opposite corners of a cube.

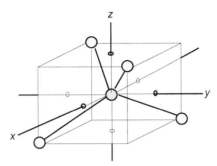

FIGURE 5.2
The tetrahedral CH₄ molecule shown in relationship to a cube.

It should be readily apparent that each of the C−H bonds constitutes a C_3 axis and that there are four such axes. Three mirror planes intersect along a C_3 axis, which suggests that there should be 12 such planes. However, each plane of this type also bisects the molecule along another C−H bond, so there are actually only six planes of symmetry. Although a tetrahedron has a geometric center, there is no center of symmetry. However, it is also apparent that each of the coordinate axes is a C_2 axis, and there are three such axes. These C_2 axes bisect pairs of H−C−H bond angles. When considering a tetrahedral structure, we encounter a type of symmetry element different from those we have so far.

Consider the z-axis in the tetrahedral structure shown in Figure 5.2. If the molecule is rotated by 90° clockwise around that axis and each atom is reflected through the x−y plane, the structure obtained is identical to that of the original molecule. The operation of rotating a molecule around an axis and then reflecting each atom through a plane perpendicular to the axis of rotation defines an *improper rotation axis*, which is designated as an *S* axis. Each of the three axes in the coordinate system for a tetrahedral molecule is an improper rotation axis. Because the amount of rotation required to produce an orientation exactly like the original is 90°, the axes are S_4 axes. Collectively, the symmetry elements present in a regular tetrahedral molecule consist of three S_4 axes, four C_3 axes, three C_2 axes (coincident with the S_4 axes), and six mirror planes. These symmetry elements define a point group known by the special symbol T_d.

It should be noted that if one hydrogen atom (on the z-axis) in the CH₄ structure is replaced by F, the resulting molecule, CH₃F, no longer has T_d symmetry. In fact, there is a C_3 axis that runs along the C−F bond and three vertical planes that intersect along that axis, so the symmetry is reduced to C_{3v}. We say that the symmetry is reduced because there are not as many symmetry elements present as there were in the original molecule.

The chair conformation of the cyclohexane molecule shown as

also illustrates the nature of an improper rotation axis. That structure can also be visualized as shown in Figure 5.3, with the z-axis pointing out of the page. Atoms represented as green circles lie above the plane of the page, whereas those represented by red circles lie below the page. The z-axis is a C_3 axis, but

FIGURE 5.3

A representation of the cyclohexane molecule. Green circles represent atoms above the plane of the page and red circles represent atoms below the page.

it is also an S_6 axis. Rotation around the z-axis by 360°/6 followed by reflection of each atom through the plane of the page (which is the x–y plane that is perpendicular to the axis about which the molecule is rotated) gives the identical orientation of the molecule. It should be apparent that in this case rotation around the z-axis by 120° accomplishes the same result as the S_6 operation.

From the foregoing discussion, it should be apparent that the S_6 axis and the operations performed can be described as

$$C_6 \cdot \sigma_{xy} = S_6$$

in which there is rotation around the z-axis by 60° followed by reflection of each atom through the x–y plane. Performing the operations twice would give

$$S_6^2 = C_6 \cdot \sigma_{xy} \cdot C_6 \cdot \sigma_{xy} = C_6^2 \cdot \sigma_{xy}^2 = C_3 = C_3 \cdot E$$

Another of the special symmetry types is the regular octahedron, which is illustrated by the structure of the SF_6 molecule. The lines that run along pairs of bonds at 180° to each other are C_4 axes, and there are three of them. A regular octahedron has four triangular faces on both the top and bottom halves of the structure. A line drawn through the center of a triangular face on the top half of the structure and out the center of the triangular face opposite it on the bottom half of the structure is an S_3 axis. There are four axes of this type. Lines drawn to bisect opposite pairs of bond angles are C_2 axes, and there are six of them. There are a total of nine mirror planes, and a molecule that has a regular octahedral structure also has a center of symmetry, i. All of these symmetry elements are typical of a symmetry type known as O_h.

In Chapter 4, the icosahedral structure of the B_{12} molecule was shown. Although all of the symmetry elements of a molecule having this structure will not be enumerated, the symmetry type is known as I_h.

Although not listed among the symmetry elements for a structure, there is also the identity operation, E. This operation leaves the orientation of the molecule unchanged from the original. This operation is essential when considering the properties that are associated with group theory. When a C_n operation is carried out n times, it returns the structure to its original orientation. Therefore, we can write

$$C_n^n = E$$

During the study of inorganic chemistry, the structures for a large number of molecules and ions will be encountered. Try to visualize the structures and think of them in terms of their symmetry. In that way, when you see that Pt^{2+} is found in the complex $PtCl_4^{2-}$ in an environment described as D_{4h}, you will know immediately what the structure of the complex is. This "shorthand" nomenclature is used to convey precise structural information in an efficient manner. Table 5.1 shows many common structural types for molecules along with the symmetry elements and point groups of those structures.

Table 5.1 Common Point Groups and Their Symmetry Elements

Point Group	Structure	Symmetry Elements	Examples
C_1	—	None	CHFClBr
C_s	—	One plane	ONCl, OSCl$_2$
C_2	—	One C_2 axis	H$_2$O$_2$
C_{2v}	Bent AB$_2$ or planar XAB$_2$	One C_2 axis and two σ_v planes at 90°	NO$_2$, H$_2$CO
C_{3v}	Pyramidal AB$_3$	One C_3 axis and three σ_v planes	NH$_3$, SO$_3^{2-}$, PH$_3$
C_{nv}	—	One C_n axis and n σ_v planes	BrF$_5$ (C_{4v})
$C_{\infty v}$	Linear ABC	One C_∞ axis and ∞ σ_v planes	OCS, HCN, HCCH
D_{2h}	Planar	Three C_2 axes, one σ_h, two σ_v planes, and i	C$_2$H$_4$, N$_2$O$_4$
D_{3h}	Planar AB$_3$ or AB$_5$ trig. bipy.	One C_3 axis, three C_2 axes, three σ_v, and one σ_h	BF$_3$, NO$_3^-$, CO$_3^{2-}$, PCl$_5$
D_{4h}	Planar AB$_4$	One C_4 axis, four C_2 axes, four σ_v, one σ_h, and i	XeF$_4$, IF$_4^-$, PtCl$_4^{2-}$
$D_{\infty h}$	Linear AB$_2$	One C_∞ axis, one σ_h, ∞ σ_v planes, and i	CO$_2$, XeF$_2$, NO$_2^+$
T_d	Tetrahedral AB$_4$	Four C_3, three C_2, and three S_4 and six σ_v planes	CH$_4$, BF$_4^-$, NH$_4^-$
O_h	Octahedral AB$_6$	Three C_4, four C_3, six C_2, and four S_6 axes, nine σ_v, and i	SF$_6$, PF$_6^-$, Cr(CO)$_6$
I_h	Icosahedral	Six C_5, 10 C_3, 15 C_2, and 20 S_6 axes and 15 planes	B$_{12}$, B$_{12}$H$_{12}^{2-}$

5.2 ORBITAL SYMMETRY

The branch of mathematics that gives the rules for manipulating groups is known as group theory. Before we can describe how to make use of symmetry to describe molecular orbitals and molecular structure, we will present a very brief introduction to the basic ideas related to group theory. A group consists of a set of symmetry elements and the operations that can be performed on the set. The group must obey a set of rules that will be presented later. At this point, we need only to use the designations as follows, which for the present can be regarded as definitions that will be amplified later.

A denotes a nondegenerate orbital or state that is symmetric around the principal axis.

B denotes a nondegenerate orbital or state that is antisymmetric around that axis.

E and *T* denote doubly and triply degenerate states, respectively. Subscripts 1 and 2 indicate symmetry or antisymmetry, respectively, with respect to a rotation axis other than the principal axis of symmetry.

In Chapter 3, the molecular orbital approach was used to describe the bonding in diatomic molecules. When considering more complicated molecules, the molecular orbital approach is more complicated, but the use of symmetry greatly simplifies the process of constructing the energy level diagram. One important aspect of the use of symmetry is that the symmetry character of the orbitals

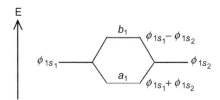

FIGURE 5.4
Two combinations of $1s$ wave functions that give different symmetry.

used in bonding by the central atom must match the symmetry of the orbitals on the peripheral atoms. For example, the combination of two hydrogen atom $1s$ wave functions, $\phi_{1s}(1) + \phi_{1s}(2)$, transforms as A_1 (or a_1 if molecular orbitals are described), but the combination $\phi_{1s}(1) - \phi_{1s}(2)$ transforms as B_1 (or b_1 for the molecular orbitals). As stated above, it can be seen that a singly degenerate state that is symmetric about the internuclear axis is designated as A_1. A singly degenerate state that is antisymmetric about the internuclear axis is designated as B_1. As shown in Chapter 3, the orbital combinations $\phi_{1s}(1) + \phi_{1s}(2)$ and $\phi_{1s}(1) - \phi_{1s}(2)$ represent the bonding and antibonding molecular orbitals for the H_2 molecule. Therefore, the qualitative molecular orbital diagram for the H_2 molecule can be constructed as shown in Figure 5.4.

Although it is stated here without proof, it can be shown that the irreducible representations of any group are *orthogonal*. *Only combinations of orbitals that have the same irreducible representation give nonzero elements in the secular determinant.* For the $2s$ orbital of the oxygen, any of the four operations on the C_2 group leaves the $2s$ orbital unchanged. Therefore, the $2s$ orbital transforms as A_1. It can also be seen that the signs of the p_x orbital are unchanged under the E and σ_{xy} operations, but the C_2 and σ_{yz} operations do change the signs of the orbital, which means that it transforms as B_1. The p_z orbital does not change signs during the C_2, E, σ_{xz}, or σ_{yz} operations, so it transforms as A_1. Following this procedure, it is found that the p_y orbital transforms as B_2. We can summarize the symmetry character of the valence shell orbitals on the oxygen atom as follows:

Orbital	Symmetry
$2s$	A_1
$2p_z$	A_1
$2p_x$	B_1
$2p_y$	B_2

Molecular orbitals are constructed in such a way that the combinations of atomic orbitals are the same as the irreducible representations of the groups that conform to the symmetry of the molecule. The character table for the point group to which the molecule belongs lists these combinations. Because H_2O is a C_{2v} molecule, the character table shown later in this chapter lists only A_1, A_2, B_1, and B_2 in accord with the symmetry of the oxygen orbitals listed above. The orbitals from both hydrogen atoms must be combined in such a way that the combination matches the symmetry of the oxygen orbitals. The combinations of hydrogen orbitals are known as *group orbitals*. Because they are combined in such

Table 5.2 Transformations for s and p Orbitals on a Central Atom in Different Symmetry

Point Group	Structure	s	p_x	p_y	p_z
C_{2v}	Bent triatomic	A_1	B_1	B_2	A_1
C_{3v}	Pyramidal	A_1	E	E	A_1
D_{3h}	Trigonal planar	A_1'	E'	E'	A_2''
C_{4v}	Pyramidal	A_1	E	E	A_1
D_{4h}	Square plane	A_1'	E_u	E_u	A_{2u}
T_d	Tetrahedral	A_1	T_2	T_2	T_2
O_h	Octahedral	A_1	T_{1u}	T_{1u}	T_{1u}
$D_{\infty h}$	Linear	Σ_g	Σ_u	Σ_u	Σ_g^+

a way that they match the symmetry of the orbitals on the central atom, they are sometimes referred to as *symmetry adapted linear combinations* (SALC).

The two combinations of hydrogen orbitals are $\phi_{1s}(1) + \phi_{1s}(2)$ and $\phi_{1s}(1) - \phi_{1s}(2)$, which have A_1 and B_1 symmetry (or a_1 and b_2 symmetry for orbitals). From the table above, it can be seen that the $2s$ and $2p_z$ oxygen orbitals have an A_1 symmetry designation, so their combinations with the hydrogen group orbitals produce a_1 and b_1 molecular orbitals. To deal with more complicated molecules, we need to know how s and p orbitals transform in environments of different symmetry. Table 5.2 shows the designations for the s and p orbitals on the central atom in several structural types.

5.3 A BRIEF LOOK AT GROUP THEORY

The mathematical apparatus for treating combinations of symmetry operations lies in the branch of mathematics known as group theory. A mathematical group behaves according to the following set of rules. A group is a set of elements and the operations that obey these rules.

1. The combination of any two members of a group must yield another member of the group (closure).
2. The group contains the identity, E, multiplication by which commutes with all other members of the group ($EA = AE$) (identity).
3. The associative law of multiplication must hold so that $(AB)C = A(BC) = (AC)B$ (associative).
4. Every member of the group has a reciprocal such that $B \cdot B^{-1} = B^{-1} \cdot B = E$ where the reciprocal is also a member of the group (inverse).

Let us illustrate the use of these rules by considering the structure of the water molecule shown in Figure 5.5. First, it is apparent that reflection through the $x–z$ plane, indicated by σ_{xz}, transforms H′ into H″. More precisely, we could say that H′ and H″ are interchanged by reflection. Because the z-axis contains a C_2 rotation axis, rotation about the z-axis of the molecule by 180° will take H′ into H″ and H″ into H′ but with the "halves" of each interchanged with respect to the $y–z$ plane. The same result would follow from reflection through the $x–z$ plane followed by reflection

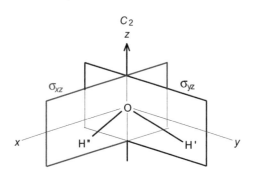

FIGURE 5.5
Symmetry elements of the water molecule. Both atoms are in the $y-z$ plane.

through the $y-z$ plane. Therefore, we can represent this series of symmetry operations in the following way:

$$\sigma_{xz} \cdot \sigma_{yz} = C_2 = \sigma_{yz} \cdot \sigma_{xz}$$

where C_2 is rotation around the z-axis by $360°/2$. This establishes that C_2 and σ_{yz} are both members of the group for this molecule. We see that in accord with rule 1, the combination of two members of the group has produced another member of the group, C_2. If reflection through the $x-z$ plane is followed by that operation, the molecule goes back to the arrangement shown in Figure 5.5. Symbolically, this combination of operations can be described as

$$\sigma_{xz} \cdot \sigma_{xz} = E$$

Also, from the figure it easy to see that

$$\sigma_{yz} \cdot \sigma_{yz} = E$$

and

$$C_2 \cdot C_2 = E$$

Further examination of Figure 5.5 shows that reflection through the $y-z$ plane, σ_{yz}, will cause the "halves" of the H′ and H″ atoms lying on either side of the $y-z$ plane to be interchanged. If we perform that operation and then rotate the molecule by $360°/2$ around the C_2 axis, we achieve exactly the same result as reflection through the $x-z$ plane produces. Thus,

$$\sigma_{yz} \cdot C_2 = \sigma_{xz} = C_2 \cdot \sigma_{yz}$$

In a similar way, it is easy to see that reflection through the $x-z$ plane followed by a C_2 operation gives the same result as σ_{yz}. Finally, it can be seen that the reflections σ_{xz} and σ_{yz} in either order leads to the same orientation that results from the C_2 operation.

$$\sigma_{xz} \cdot \sigma_{yz} = C_2 = \sigma_{yz} \cdot \sigma_{xz}$$

Table 5.3 Multiplication of Symmetry Operations for the H_2O (C_{2v}) Molecule

	E	C_2	σ_{xz}	σ_{yz}
E	E	C_2	σ_{xz}	σ_{yz}
C_2	C_2	E	σ_{yz}	σ_{xz}
σ_{xz}	σ_{xz}	σ_{yz}	E	C_2
σ_{yz}	σ_{yz}	σ_{xz}	C_2	E

Start with the operation in the left hand column and proceed to the desired operation at the top of a column. Then, read down that column to obtain the desired product.

The associative law, rule 3, has also been demonstrated here. Additional relationships are provided by the following:

$$E \cdot E = E$$

$$C_2 \cdot E = C_2 = E \cdot C_2$$

$$\sigma_{yz} \cdot E = \sigma_{yz} = E \cdot \sigma_{yz}, \text{etc.}$$

All these combinations of operations can be summarized in a *group multiplication table* arranged as shown in Table 5.3.

The multiplication table (Table 5.3) for the C_{2v} group is thus constructed so that the combination of operations follows the four rules presented at the beginning of this section. Obviously, a molecule having a different structure (symmetry elements and operations) would require a different table. To provide further illustrations of the use of symmetry elements and operations, the ammonia molecule, NH_3, will be considered (see Figure 5.6). Figure 5.6 shows that the NH_3 molecule has a C_3 axis through the nitrogen atom and three reflection planes containing that C_3 axis. The identity operation, E, and the C_3^2 operation complete the list of symmetry operations for the NH_3 molecule. It should be apparent that

$$C_3 \cdot C_3 = C_3^2$$

$$C_3^2 \cdot C_3 = C_3 \cdot C_3^2 = E$$

$$\sigma_1 \cdot \sigma_1 = E = \sigma_2 \cdot \sigma_2 = \sigma_3 \cdot \sigma_3$$

Reflection through σ_2 does not change H″, but it does interchange H′ and H‴. Reflection through σ_1 leaves H′ in the same position but interchanges H″ and H‴. We can summarize these operations as

$$H' \xleftrightarrow{\sigma_2} H'''$$

$$H''' \xleftrightarrow{\sigma_1} H''$$

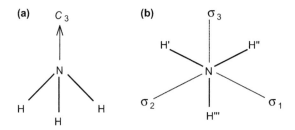

FIGURE 5.6
The pyramidal NH_3 molecule that has C_{3v} symmetry. In (b), the C_3 axis is directed upward perpendicular to the plane of the paper at the nitrogen atom.

However, C_3^2 would move H′ to H‴, H″ to H′, and H‴ to H″, which is exactly the same orientation as that σ_2 followed by σ_1 produced. It follows, therefore, that

$$\sigma_2 \cdot \sigma_1 = C_3^2$$

This process could be continued so that all the combinations of symmetry operations would be worked out. Table 5.4 shows the multiplication table for the C_{3v} point group to which a pyramidal molecule such as NH_3 belongs.

Multiplication tables can be constructed for the combination of symmetry operations for other point groups. However, it is usually not the multiplication table itself that is of interest. The multiplication table for the C_{2v} point group is shown in Table 5.3. If we replace E, C_2, σ_{xz}, and σ_{yz} by $+1$, we find that the numbers still obey the multiplication table. For example,

$$C_2 \cdot \sigma_{xz} = \sigma_{yz} = 1 \cdot 1 = 1$$

Thus, the values of the operations all being $+1$ satisfy the laws of the C_{2v} group. This set of four numbers (all $+1$) provides one representation of the group. Another is given by the relationships

$$E = 1, \quad C_2 = 1, \quad \sigma_{xz} = -1, \quad \sigma_{yz} = -1$$

Table 5.4 The Multiplication Table for the C_{3v} Point Group

	E	C_3	C_3^2	σ_1	σ_2	σ_3
E	E	C_3	C_3^2	σ_1	σ_2	σ_3
C_3	C_3	C_3	E	σ_3	σ_1	σ_2
C_3^2	C_3^2	E	C_3	σ_2	σ_3	σ_1
σ_1	σ_1	σ_2	σ_3	E	C_3	C_3^2
σ_2	σ_2	σ_3	σ_1	C_3^2	E	C_3
σ_3	σ_3	σ_1	σ_2	C_3	C_3^2	E

Table 5.5 Character Table for the C_{2v} Point Group

	E	C_2	σ_{xz}	σ_{yz}
A_1	1	1	1	1
A_2	1	1	−1	−1
B_1	1	−1	1	−1
B_2	1	−1	−1	1

which also obey the rules shown in the table. From other relationships, we know that the character table (see Table 5.5) summarizes the four *irreducible* representations for the C_{2v} point group.

The symbols at the left in Table 5.5 give the symmetry properties of the irreducible representation of the group. We will now briefly discuss what the symbols mean. Suppose we have a vector of unit length lying coincident with the x-axis as shown in Figure 5.7.

The identity operation does not change the orientation of the vector. Reflection in the x–z plane leaves the vector unchanged, but reflection through the y–z plane changes it to a vector of unit length in the $-x$ direction. Likewise, the C_2 operation around the z-axis changes the vector in the same way. Therefore, the vector is said to transform as $+1$ for the operations E and σ_{xz}, but it transforms as -1 for the operations C_2 and σ_{yz}. Table 5.5 shows the row containing these numbers ($+1, -1, +1,$ and -1 under the operations $E, C_2, \sigma_{xz},$ and σ_{yz}, respectively) labeled as B_1. It is easy to show how the other rows can be obtained in a similar manner. The four representations, $A_1, A_2, B_1,$ and B_2, are the irreducible representations of the C_{2v} group. It can be shown that these four irreducible representations cannot be separated or decomposed into other representations.

For a given molecule belonging to a particular point group, it is possible to consider the various symmetry species as indicating the behavior of the molecule under symmetry operations. As will be shown later, these species also determine the ways in which the atomic orbitals can combine to produce molecular orbitals because the combinations of atomic orbitals must satisfy the character table of the group. We need to give some meaning that is related to molecular structure for the species $A_1, B_2,$ etc.

The following conventions are used to label species in the character tables of the various point groups:

1. The symbol A is used to designate a nondegenerate species that is symmetric about the principal axis.

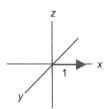

FIGURE 5.7
A unit vector lying along the x-axis.

2. The symbol B is used to designate a nondegenerate species that is antisymmetric about the principal axis.
3. The symbols E and T represent double and triply degenerate species, respectively.
4. If a molecule possesses a center of symmetry, the subscript "g" indicates symmetry with respect to that center and the subscript "u" indicates antisymmetry with respect to that center of symmetry.
5. For a molecule that has a rotation axis other than the principal one, symmetry or antisymmetry with respect to that axis is indicated by subscripts 1 or 2, respectively. When no rotation axis other than the principal one is present, these subscripts are sometimes used to indicate symmetry or antisymmetry with respect to a vertical plane, σ_v.
6. The marks $'$ and $''$ are sometimes used to indicate symmetry or antisymmetry with respect to a horizontal plane, σ_h. It should now be apparent how the species A_1, A_2, B_1, and B_2 arise.

Character tables have been worked out and are tabulated for all the common point groups. Presenting all the tables here would go beyond the scope of the discussion of symmetry and group theory as used in this book. Tables for some common point groups are shown in Appendix B.

We have barely scratched the surface of the important topic of symmetry. An introduction such as that presented here serves to introduce the concepts and the nomenclature as well as making one able to recognize the more important point groups. Thus, the symbol T_d or D_{4h} takes on precise meaning in the language of group theory. The applications of group theory include, among others, coordinate transformations, analysis of molecular vibrations, and the construction of molecular orbitals. Only the last of these uses will be illustrated here. For further details on the applications of group theory, see the books by Cotton and by Harris and Bertolucci listed in the references.

5.4 CONSTRUCTION OF MOLECULAR ORBITALS

The application of symmetry concepts and group theory greatly simplifies the construction of molecular orbitals. For example, it can be shown that the combination of two hydrogen $1s$ wave functions $\phi_{1s}(1) + \phi_{1s}(2)$ transforms as A_1 (usually written as a_1 when orbitals are considered) and the combination $\phi_{1s}(1) - \phi_{1s}(2)$ transforms as B_1 (sometimes written as b_1). According to the description of species in the character tables, we see that the A_1 combination is a singly degenerate state that is symmetric about the internuclear axis. Also, the B_1 combination represents a singly degenerate state that is antisymmetric about the internuclear axis. Therefore, the states described by the combinations $(\phi_{1s}(1) + \phi_{1s}(2))$ and $(\phi_{1s}(1) - \phi_{1s}(2))$ describe the bonding (a_1) and antibonding (b_1) molecular orbitals, respectively, in the H_2 molecule as shown in Figure 5.4.

For any group, the irreducible representations must be orthogonal. Therefore, only interactions of orbitals having the same irreducible representations lead to nonzero elements in the secular determinant. It remains, then, to determine how the various orbitals transform under different symmetry groups. For H_2O, the coordinate system is shown in Figure 5.5. Performing any of the four operations possible for the C_{2v} group leaves the $2s$ orbital unchanged. Therefore, that orbital transforms as A_1. Likewise, the p_x orbital does not change sign under E or σ_{xz} operations, but it does change signs under C_2 and σ_{yz} operations. This orbital thus transforms as B_1. In a like manner, we find that p_z transforms as A_1

(does not change signs under C_2, E, σ_{xz}, or σ_{yz} operations). Although it may not be readily apparent, the p_y orbital transforms as B_2.

The possible wave functions for the molecular orbitals for molecules are those constructed from the irreducible representations of the groups giving the symmetry of the molecule. These are readily found in the character table for the appropriate point group. For water, which has the point group C_{2v}, the character table (see Table 5.5) shows that only A_1, A_2, B_1, and B_2 representations occur for a *molecule* having C_{2v} symmetry. We can use this information to construct a qualitative molecular orbital scheme for the H_2O molecule as shown in Figure 5.8.

In doing this, we must recognize that there are two hydrogen $1s$ orbitals and the orbitals from the oxygen atom must interact with both of them. Therefore, it is not each hydrogen $1s$ orbital individually that is used but rather a combination of the two. These combinations are called *group orbitals*, and in this case the combinations can be written as $(\phi_{1s}(1) + \phi_{1s}(2))$ and $(\phi_{1s}(1) - \phi_{1s}(2))$. In this case, the $2s$ and $2p_z$ orbitals having A_1 symmetry mix with the combination of hydrogen $1s$ orbitals having A_1 symmetry to produce three molecular orbitals having A_1 symmetry (one bonding, one nonbonding, and one antibonding). The $2p_x$ orbital having B_1 symmetry combines with the combination of hydrogen orbitals having the same symmetry, $(\phi_{1s}(1) - \phi_{1s}(2))$. The $2p_y$ orbital remains uncombined as a $2p_y$ orbital that does not have the correct symmetry to interact with either of the combinations of hydrogen orbital. Therefore, it remains as a nonbonding π orbital of b designation. In the case of H_2O, the four orbitals of lowest energy will be populated because the atoms have a total of eight valence shell electrons. Therefore, the bonding can be represented as

$$(a_1)^2(b_1)^2(a_1^n)^2(b_2)^2$$

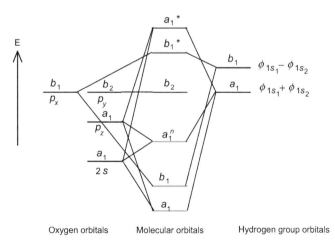

FIGURE 5.8
A molecular orbital diagram for the H_2O molecule.

As in the case of atomic orbitals and spectroscopic states (see Chapter 2), we use *lower case letters to denote orbitals or configurations* and *upper case letters to indicate states*. It should also be pointed out that the a_1 and b_1 orbitals are σ bonding orbitals, but the b_2 molecular orbital is a nonbonding π orbital.

Having considered the case of the H_2O molecule, we would like to be able to use the same procedures to construct the qualitative molecular orbital diagrams for molecules having other structures. To do this, it is required that we know how the orbitals of the central atom transform when the symmetry is different. Table 5.2 shows how the s and p orbitals are transformed, and more extensive tables can be found in the comprehensive books listed at the end of this chapter.

If we now consider a planar molecule such as BF_3 (D_{3h} symmetry), the z-axis is defined as the C_3 axis. One of the B—F bonds lies along the x-axis as shown in Figure 5.9. The symmetry elements present for this molecule include the C_3 axis, three C_2 axes (coincident with the B—F bonds and perpendicular to the C_3 axis), three mirror planes each containing a C_2 axis and the C_3 axis, and the identity. Thus, there are 12 symmetry operations that can be performed with this molecule. It can be shown that the p_x and p_y orbitals both transform as E' and the p_z orbital transforms as A_2''. The s orbital is A_1' (the prime indicating symmetry with respect to σ_h). Similarly, we could find that the fluorine p_z orbitals are A_1, E_1, and E_1. The qualitative molecular orbital diagram can then be constructed as shown in Figure 5.10.

It is readily apparent that the three σ bonds are capable of holding the six bonding electrons in the a_1' and e' molecular orbitals. The possibility of some π bonding is seen in the molecular orbital diagram as a result of the availability of the a_2'' orbital, and in fact there is some experimental evidence for this type of interaction. The sum of the covalent radii of boron and fluorine atoms is about 152 pm (1.52 Å), but the experimental B—F bond distance in BF_3 is about 129.5 pm (1.295 Å). Part of this "bond shortening" may be due to partial double bonds resulting from the π bonding. A way to show this is by means of the three resonance structures of the valence bond type that are represented in Figure 5.11.

From these resonance structures, we determine a bond order of 1.33 for the B—F bonds, which would predict the observed bond shortening. However, another explanation of the "short" B—F bonds is based on the fact that the difference in electronegativity between B and F is about 2.0 units, which causes the bonds to have substantial ionic character. In fact, calculations show that the positive charge on boron is probably as high as 2.5–2.6, indicating that the bonding is predominantly ionic. The B^{3+} ion is very small, and as will be shown in Chapter 7, the relative size of the cation and anion determines the number of anions that can be placed around a cation in a crystal. In the case of BF_3, only three F^- ions can surround B^{3+}, so there is no possibility of forming an extended network as is required in a crystal lattice. Therefore, it is likely that BF_3 is best considered as a monomer that is in reality an "ionic molecule".

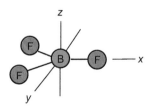

FIGURE 5.9

The coordinate system for the BF_3 molecule.

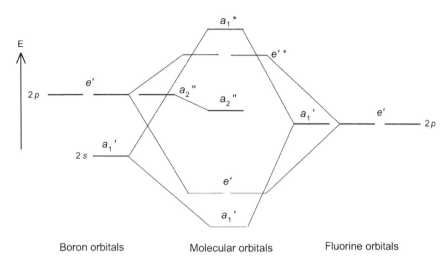

FIGURE 5.10
A molecular orbital diagram for the BF_3 molecule.

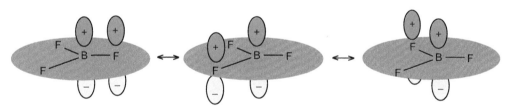

FIGURE 5.11
Three structures that taken together represent the BF_3 molecule.

Having seen the development of the molecular orbital diagram for AB_2 and AB_3 molecules, we will now consider tetrahedral molecules such as CH_4, SiH_4, or SiF_4. In this symmetry, the valence shell s orbital on the central atom transforms as A_1, whereas the p_x, p_y, and p_z orbitals transform as T_2 (see Table 5.2). For methane, the combination of hydrogen orbitals that transforms as A_1 is

$$\phi_{1s}(1) + \phi_{1s}(2) + \phi_{1s}(3) + \phi_{1s}(4)$$

and a combination that transforms as T_2 is

$$\phi_{1s}(1) - \phi_{1s}(2) + \phi_{1s}(3) - \phi_{1s}(4)$$

where the coordinate system is as shown in Figure 5.12.

Using the orbitals on the carbon atom and combining them with the group orbitals from the four hydrogen atoms (linear combination of orbitals having symmetry matching the carbon atom orbitals), we obtain the molecular orbital diagram shown in Figure 5.13.

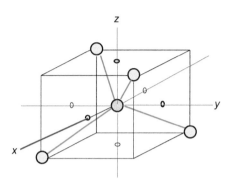

FIGURE 5.12
The regular tetrahedral structure.

The hydrogen group orbitals are referred to as *symmetry adapted linear combinations* (SALC). Although their development will not be shown here, the molecular orbital diagrams for other tetrahedral molecules are similar.

For an octahedral AB_6 molecule such as SF_6, the valence shell orbitals are considered to be the *s*, *p*, and *d* orbital of the central atom. It is easy to see that a regular octahedron has a center of symmetry so that "*g*" and "*u*" designations must be used on the symmetry species to designate symmetry or

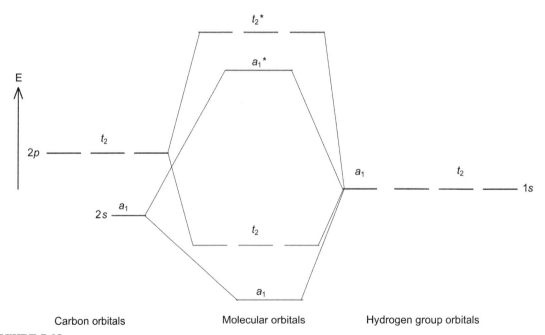

FIGURE 5.13
The molecular orbital diagram for a tetrahedral molecule such as CH_4.

asymmetry with respect to that center. Clearly the s orbital transforms as A_{1g}. The p orbitals being directed toward the corners of the octahedron are degenerate and change sign upon reflection through the center of symmetry. They thus constitute a T_{1u} set. Of the set of d orbitals, the d_{z^2} and $d_{x^2-y^2}$ orbitals are directed toward the corners of the octahedron, and they do not change sign upon inversion through the center of symmetry. Thus, these orbitals are designated as E_g. The remaining d_{xy}, d_{yz}, and d_{xz} orbitals form a triply degenerate set designated as T_{2g}.

If we consider only σ bonding, we find that T_{1u}, E_g, and A_{1g} orbitals are used by the six groups attached. The resulting energy level diagram is shown in Figure 5.14. In this section, we have seen how symmetry considerations are used to arrive at qualitative molecular orbital diagrams for molecules having several common structural types. The number of molecules and ions that have C_{2v}, C_{3v}, $C_{\infty v}$, $D_{\infty h}$, T_d, and O_h symmetry is indeed large. Energy level diagrams such as those shown in this section are widely used to describe structural, spectroscopic, and other properties of the molecules. We have not, however, set about to actually calculate anything. In Chapter 17, we will present an overview of the molecular orbital approach to the bonding in coordination compounds. The more sophisticated mathematical treatments of molecular orbital calculations are beyond the intended scope of this inorganic text, and they are not necessary for understanding the basic applications of symmetry to molecular orbital diagrams.

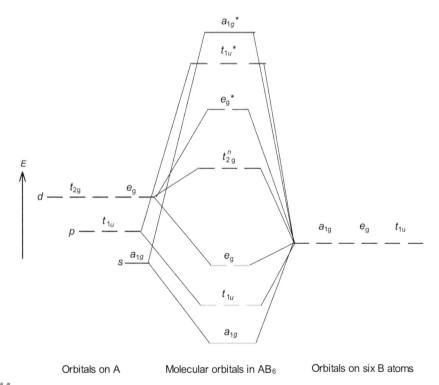

Orbitals on A   Molecular orbitals in AB_6 Orbitals on six B atoms

FIGURE 5.14
The molecular orbital diagram for an octahedral molecule.

5.5 ORBITALS AND ANGLES

Up to this point in this chapter, we have approached the description of the molecular orbitals for molecules with the structure already being known. Intuitively, we know that the H_2O molecule has an angular structure, whereas the BeH_2 molecule is linear. From prior experience, we know that there are eight electrons around the central atom in H_2O but only four around the Be atom in BeH_2. We now want to address the difference in structure using the molecular orbital approach.

One of the simplest approaches to comprehensive molecular orbital calculations is the Extended Hückel Method. This method was developed by Roald Hoffman in the 1960s, and it was applied to hydrocarbon molecules. From the discussion presented in Chapters 2 and 3, we know that one of the first things that has to be done is to choose the *atomic* wave functions that will be used in the calculations. One of the most widely used types of wave functions is that known as the Slater wave functions (see Section 2.4). In the Extended Hückel Method, the molecular wave functions are approximated as

$$\psi_i = \sum_j c_{ij}\phi_j \tag{5.1}$$

where $j = 1, 2, \ldots, n$. For a hydrocarbon molecule having the formula C_nH_m, there will be m hydrogen $1s$ orbitals, n carbon $2s$ orbitals, and $3n$ carbon $2p$ orbitals. Although the details will not be shown, this combination of orbitals leads to a secular determinant of dimension $4n + m$. Unlike the original method developed by Erich Hückel, which neglected all interactions except that between adjacent atoms, the Extended Hückel Method retains the off-diagonal elements, so it takes into account additional interactions between atoms in the molecule. As was described in Chapter 3, there are both coulomb and exchange integrals to evaluate, and the overlap integrals must be approximated.

Coulomb integrals, written as H_{ii}, represent the binding energy of an electron in atom i. Therefore, by Koopmans' theorem, these energies are equivalent in magnitude to the ionization of an electron from those orbitals. Accordingly, the values used (in eV) are as follows: H($1s$), -13.6; C($2s$), -21.4; and C($2p$), -11.4. Next, it is necessary to represent the exchange integrals, written as H_{ij}, and one of the most common ways is to use the Wolfsberg−Helmholtz approximation,

$$H_{ij} = 0.5\,K(H_{ii} + H_{jj})S_{ij} \tag{5.2}$$

where K is a constant having a value of approximately 1.75 and S_{ij} is the overlap integral for the wave functions for the orbitals on atoms i and j.

Although the overlap integral for two $1s$ wave functions is a function of internuclear distance, the situation is different when p orbitals are involved. Because of their angular character, the overlap of a p orbital with two hydrogen $1s$ orbitals will have a value that depends on the angle formed by the H−X−H bonds. Therefore, an adjustment that depends on the bond angle must be made to the value of the overlap integral. When the molecular orbital energies are calculated, it is found that they vary (as is expected) depending on the bond angle. In fact, the bond angle can be treated as an adjustable parameter, and the energies of the molecular orbitals can be plotted as a function of the bond angle as it is varied from 90° to 180°. It must be remembered that the molecular orbitals will have different designations that depend on symmetry. A linear H−X−H molecule will give rise to σ_g, σ_u, and two degenerate π_u molecular orbitals (that are perpendicular to the axis of the molecule). As was shown earlier in this chapter, if the bond angle is 90°, the molecular orbitals will be a_1, b_2, a_1, and b_1 (in increasing energy).

Knowing how the orbitals are arranged in terms of energy, we can make a graph (that is only qualitative) to show the energy of the orbitals as the bond angle is varied from 90° to 180°. A diagram of this type was prepared by Arthur D. Walsh over half a century ago, and it is appropriately known now as a *Walsh diagram*. Figure 5.15 shows the diagram of a triatomic molecule. When interpreting this diagram, it is essential to have a mental picture of how the two hydrogen $1s$ orbitals interact with the s and p orbitals on the central atom as the bond angle varies from 90° to 180°.

For two pairs of electrons around the central atom (as in BeH_2), the lowest energy is achieved when the two pairs of electrons occupy the a_1 and b_2 orbitals when the bond angle is 180°. For three pairs of electrons (as in BH_2^+ or CH_2^{2+}) occupying the three orbitals of lowest energy, a lower energy is obtained when the structure is bent. For four pairs of electrons (as in the H_2O molecule), the lowest energy is not 90°, but it is closer to that value than to 180°. Keep in mind that these are *qualitative* applications of

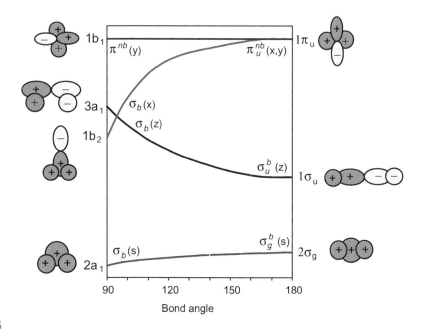

FIGURE 5.15

A Walsh diagram for a molecule having the formula XH_2. The energy levels are labeled with both of the commonly used types of symbols. In the drawing at the upper left, the bending causes the hydrogen orbitals to be nonbonding with regard to overlap with the p_y orbital of the central atom. In the diagram at the upper right, the combination of hydrogen orbitals is nonbonding with respect to the p_x and p_y orbital.

a graph. The results are, of course, in accord with what we know from simple valence bond (hybrid orbital) approaches described in Chapter 4. However, it is important to know that such a molecular orbital approach as that used by Walsh exists.

Although the complexity increases rapidly, there is no reason that Walsh diagrams cannot be constructed for XY_3 pyramidal, XY_4 tetrahedral, XY_6 octahedral, and other molecules. In fact, they have been prepared, but their applications will not be described here. Insofar as these diagrams are amenable to quantitative interpretation, the predictions are in accord with what we know from experimental evidence and valence bond methods.

5.6 SIMPLE CALCULATIONS USING THE HÜCKEL METHOD

Although we have presented some of the principles related to symmetry and its use in describing molecular orbital methods, it is still reassuring to be able to calculate something even at an elementary level. Such a simple approach was developed in the 1930s by Erich Hückel. It was developed to make molecular orbital calculations for organic molecules, and the method is now called the HMO method. In this connection, the interested reader should consult the classic book on the subject by John D. Roberts, *Notes on Molecular Orbital Calculations*. However, it is possible to extend the Hückel approach to include atoms other than carbon, so a brief description of the method will be presented and its use for some "inorganic" molecules will be illustrated. An essential idea is that the σ and π bonding can be separated, and that the energy is given by

$$E_{total} = E_\sigma + E_\pi \tag{5.3}$$

When dealing with carbon atoms, the coulomb integral (H_{ii}) is represented as α and the exchange integral (H_{ij}) is represented as β. It is also assumed that interaction between nonadjacent atoms can be ignored so if $|i - j| \geq 2$ the $H_{ij} = 0$. Finally, in simple HMO there is complete neglect of overlap so $S_{ij} = 0$.

If we begin with a simple molecule such as ethylene, it is apparent that the σ-bonded structure is represented as

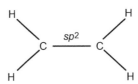

with the hybridization scheme being sp^2. This leaves a p orbital perpendicular to the plane of the molecule that can form a π bond between the carbon atoms. Although they are not included explicitly in the calculations, the wave functions for the σ bonds can be written as

$$\psi_{CH} = a_1\psi_1 + a_2\psi_2 \tag{5.4}$$

$$\psi_{CC(\sigma)} = a_3\psi_{sp^2(1)} + a_4\psi_{sp^2(2)} \tag{5.5}$$

The useful part of the calculation in the Hückel method is for the π bond,

$$\psi_{CC(\pi)} = a_5\psi_{p(1)} + a_6\psi_{p(2)} \tag{5.6}$$

As was shown in Chapter 3, the secular determinant can be written as

$$\begin{vmatrix} H_{11} - E & H_{12} - S_{12}E \\ H_{12} - S_{12}E & H_{22} - E \end{vmatrix} = 0 \tag{5.7}$$

Note that it has been assumed that $H_{12} = H_{21}$ and $S_{12} = S_{21}$, which means that the two bonded atoms are identical. When $S_{12} = S_{21} = 0$ as has been described above, after letting $H_{11} = H_{22} = \alpha$ and $H_{12} = H_{21} = \beta$, the secular determinant becomes

$$\begin{vmatrix} \alpha - E & \beta \\ \beta & \alpha - E \end{vmatrix} = 0 \tag{5.8}$$

Dividing each element in the determinant by β gives

$$\begin{vmatrix} \dfrac{\alpha - E}{\beta} & 1 \\ 1 & \dfrac{\alpha - E}{\beta} \end{vmatrix} = 0 \tag{5.9}$$

In order to simplify handling this expression, we let $x = (\alpha - E)/\beta$. Therefore, the determinant can be written as

$$\begin{vmatrix} x & 1 \\ 1 & x \end{vmatrix} = 0 \tag{5.10}$$

so that $x^2 - 1 = 0$ and $x^2 = -1$. This equation has roots of $x = 1$ and $x = -1$, which lead to

$$\frac{\alpha - E}{\beta} = 1 \text{ and } \frac{\alpha - E}{\beta} = -1 \tag{5.11}$$

These equations lead to the energy values $E = \alpha + \beta$ and $E = \alpha - \beta$. Both α and β represent negative quantities, and each carbon atom contributes one electron to the π bond, so the energy level diagram can be shown as illustrated in Figure 5.16. From the molecular orbital diagram, we predict that electronic transitions of the $\pi \rightarrow \pi^*$ should be possible. In fact, most hydrocarbon molecules that have an empty π^* orbital absorb in the ultraviolet region around 200–250 nm.

If the two electrons were residing in p orbitals on separate carbon atoms, their total energy would be 2α. However, when they are in a π molecular orbital, their energy is $2(\alpha + \beta)$. The difference

$$2(\alpha + \beta) - 2\alpha = 2\beta \tag{5.12}$$

represents the delocalization energy. Because a C—C bond has an energy of about 347 kJ mol^{-1} and C=C is about 619 kJ mol^{-1}, the additional stability (the energy of the π bond) must be approximately 272 kJ mol^{-1} so $\beta \approx 136$ kJ mol^{-1}. The ionization potential for a carbon atom is 1086 kJ mol^{-1}, so this is the value for α. It is often found that H_{12} is approximately 15% of H_{11}, so the values shown are in reasonable agreement with this estimate.

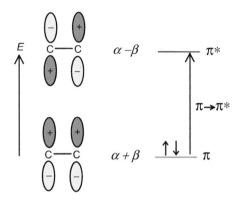

FIGURE 5.16

Molecular orbitals for ethylene. Promotion of an electron from the ground state to the excited state is known as a $\pi-\pi^*$ transition and is usually accompanied by an absorption of radiation in the ultraviolet region of the spectrum.

Despite its simplicity, the Hückel method enables other useful properties to be deduced. For example, the wave function for the bonding molecular orbital is

$$\psi_b = a_1\phi_1 + a_2\phi_2 \tag{5.13}$$

so we can evaluate the constants a_1 and a_2. We know that

$$\int \psi_b{}^2 \, d\tau = \int (a_1\phi_1 + a_2\phi_2) d\tau \tag{5.14}$$

and by using the abbreviations as shown in Section 3.1, we find that

$$a_1^2 S_{11} + a_2^2 S_{22} + 2a_1 a_2 S_{12} = 1 \tag{5.15}$$

Because we assume that $S_{11} = S_{22} = 1$ and $S_{12} = S_{21} = 0$, this equation reduces to

$$a_1^2 + a_2^2 = 1 \tag{5.16}$$

From the secular equations and the minimization of energy (see Eqns (3.17) and (3.18)),

$$a_1(\alpha - E) + a_2\beta = 0 \tag{5.17}$$

$$a_1\beta + a_2(\alpha - E) = 0 \tag{5.18}$$

Dividing by β and letting $x = (\alpha - E)/\beta$, we find that

$$a_1 x + a_2 = 0 \tag{5.19}$$

$$a_1 + a_2 x = 0 \tag{5.20}$$

For the bonding state, $x = -1$, so $a_1^2 = a_2^2 = 1 = 2a_1^2$. Therefore,

$$a_1 = \frac{1}{\sqrt{2}} = 0.707 = a_2 \tag{5.21}$$

and

$$\psi_b = 0.707\,\phi_1 + 0.707\,\phi_2 \tag{5.22}$$

Because $a_1^2 = a_2^2 = 1/2$, half of the bonding pair of electrons (one electron) resides on each atom. Therefore, the *electron density* (ED) is $2(\frac{1}{2}) = 1$. The *bond order* between two atoms is given by

$$B_{XY} = \sum_{i=1}^{n} a_x a_y p_i \tag{5.23}$$

where a is a weighting factor (as calculated above) and p_i is the population of orbital i. The summation is made over all n populated orbitals. For the ethylene molecule, the bond order between the carbon atoms is $B_{CC} = 2(0.707)(0.707) = 1$, so the order of the π bond is 1. When the σ bond is included, the total bond order between the carbon atoms is 2.

For a linear system consisting of three carbon atoms (which includes the allyl radical and the cation and anion derived from it), the coulomb integrals will be identical

$$H_{11} = H_{22} = H_{33} = \alpha \tag{5.24}$$

Because only interactions between *adjacent* atoms are considered,

$$H_{13} = H_{31} = 0 \tag{5.25}$$

The exchange integrals for interaction of adjacent atoms will be

$$H_{12} = H_{21} = H_{23} = H_{32} = \beta \tag{5.26}$$

As before, the overlap integrals are neglected and we can proceed directly to the secular determinant. After the substitutions are made and each element is divided by β, the result can be shown as follows (where $x = (\alpha - E)/\beta$).

$$\begin{vmatrix} H_{11} - E & H_{12} & 0 \\ H_{21} & H_{22} - E & H_{23} \\ 0 & H_{32} & H_{33} - E \end{vmatrix} = \begin{vmatrix} \alpha - E & \beta & 0 \\ \beta & \alpha - E & \beta \\ 0 & \beta & \alpha - E \end{vmatrix} = \begin{vmatrix} x & 1 & 0 \\ 1 & x & 1 \\ 0 & 1 & x \end{vmatrix} = 0 \tag{5.27}$$

By expanding the determinant, we obtain the characteristic equation

$$x^3 - 2x = 0 \tag{5.28}$$

for which the roots are $x = 0$, $x = -(2)^{1/2}$, and $x = 2^{1/2}$. Setting each of these values equal to $(\alpha - E)/\beta$, we obtain

$$\frac{\alpha - E}{\beta} = -\sqrt{2} \qquad \frac{\alpha - E}{\beta} = 0 \qquad \frac{\alpha - E}{\beta} = \sqrt{2}$$

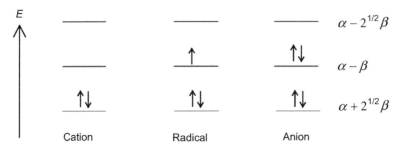

FIGURE 5.17
The energy level diagram for the allyl radical, cation, and anion species.

$$E = \alpha + \sqrt{2}\beta \quad E = \alpha \quad E = \alpha - \sqrt{2}\beta$$

The energy level diagram including electron populations for the allyl radical, cation, and anion can be shown as illustrated in Figure 5.17.

The orbital diagram and energy levels for the allyl system are shown in Figure 5.18. The arrangement of the molecular orbitals of the allyl species will be useful when discussing the bonding of this ligand in metal complexes (Chapters 16 and 21).

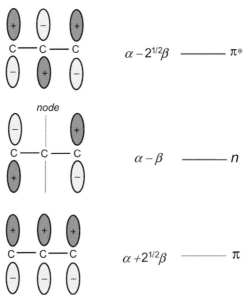

FIGURE 5.18
Molecular orbital diagram for the allyl species.

From the secular equations and determinant, we obtain

$$\begin{vmatrix} a_1 x & a_2 & 0 \\ a_1 & a_2 x & a_3 \\ 0 & a_2 & a_3 x \end{vmatrix} = 0 \tag{5.29}$$

$$a_1 x + a_2 = 0 \qquad a_1 + a_2 x + a_3 = 0 \qquad a_2 + a_3 x = 0$$

Starting with the $x = -2^{1/2}$ root, we find

$$-a_1\, 2^{1/2} + a_2 = 0 \qquad a_1 - a_2\, 2^{1/2} + a_3 = 0 \qquad a_2 - a_3\, 2^{1/2} = 0$$

From the first of these equations, we find that $a_2 = a_1\, 2^{1/2}$ and from the last $a_2 = a_3\, 2^{1/2}$. Therefore, $a_1 = a_3$ and substituting this into the second equation gives

$$a_1 - a_2\, 2^{1/2} + a_1 = 0 \tag{5.30}$$

However, $a_2 = a_1\, 2^{1/2}$ and we know that

$$a_1^2 + a_2^2 + a_3^2 = 1 \tag{5.31}$$

and when the values found above are substituted into this equation we find that

$$a_1^2 + 2a_1^2 + a_1^2 = 4a_1^2 = 1 \tag{5.32}$$

Therefore, $a_1^2 = 1/4$ and $a_1 = \tfrac{1}{2}$, which is also true for a_3, and $a_2 = 2^{1/2}/2 = 0.707$. The wave function for the bonding orbital is

$$\psi_b = 0.500\,\psi_1 + 0.707\,\psi_2 + 0.500\,\psi_3 \tag{5.33}$$

Using the root $x = 0$, a similar procedure allows us to evaluate the constants for the next molecular orbital (which is nonbonding) and gives the wave function

$$\psi_n = 0.707\,\psi_1 - 0.707\,\psi_3 \tag{5.34}$$

The root $x = 2^{1/2}$ leads to the wave function for the antibonding orbital,

$$\psi_b = 0.500\,\psi_1 - 0.707\,\psi_2 + 0.500\,\psi_3 \tag{5.35}$$

By making use of the weighting coefficients and the populations of the orbital, the electron density (ED) at each atom can be calculated as before. For the allyl radical,

$$ED_{C1} = 2(0.500)^2 + 1(0.707)^2 = 1.00 \tag{5.36}$$

$$ED_{C2} = 2(0.707)^2 + 1(0)^2 = 1.00 \tag{5.37}$$

$$ED_{C3} = 2(0.500)^2 + 1(-0.707)^2 = 1.00 \tag{5.38}$$

In a similar way, the electron densities can be found for the atoms in the cation and anion. The results can be summarized as follows:

Electron density at	C_1	C_2	C_3
Radical	1.00	1.00	1.00
Cation	0.500	1.00	0.500
Anion	1.50	1.00	1.50

Although the evaluations will not be shown, the bond orders between the atoms are identical because the difference in the orbital populations occurs in the *nonbonding* orbital. For the arrangement C=C–C, the π bond is localized between two atoms in the same way it is in ethylene. Therefore, the energy would be $2(\alpha + \beta)$. If the π bond is spread over the whole molecule, C⋯C⋯C, the energy would be $2(\alpha + 2^{1/2}\beta)$, which is lower than that of the previous structure by -0.828β. This energy represents the amount by which the structure with delocalized electron density is more stable than one in which the π bond is restricted to a location between two carbon atoms. This energy of stabilization is known as the *resonance energy*.

If we suppose that the carbon atoms form a ring structure, the problem is somewhat different because $H_{13} = H_{31} = \beta$. The secular determinant can be written as

$$\begin{vmatrix} x & 1 & 1 \\ 1 & x & 1 \\ 1 & 1 & x \end{vmatrix} = 0 \tag{5.39}$$

which leads to the equation

$$x^3 - 3x + 2 = 0 \tag{5.40}$$

and the roots are $x = -2$, $x = 1$, and $x = 1$. Therefore, $E = \alpha + 2\beta$ and $E = \alpha - \beta$. Because the latter energy occurs twice, *degenerate* orbitals are indicated. The energy level diagram for a three-membered ring system is shown in Figure 5.19.

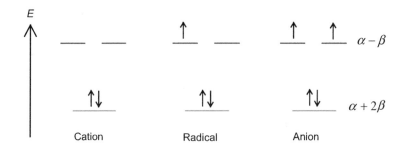

FIGURE 5.19
Energy level diagrams for species containing three carbon atoms in a ring.

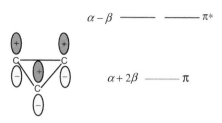

FIGURE 5.20
The molecular orbital diagram for cyclopropene.

For the cation, a localized π bond would lead to an energy of $2(\alpha + \beta)$, whereas a delocalized π bond would have an energy of $2(\alpha + 2\beta)$. The resonance energy is 2β. For the anion, the total energy is $2(\alpha + 2\beta) + 2(\alpha - \beta)$, which gives $4\alpha + 2\beta$. If there were a localized π bond and two electrons located on two carbon atoms, the energy would be $2(\alpha + \beta) + 2\alpha$, which is $4\alpha + 2\beta$. This is identical to the result for the case when the π bond is delocalized, so there is no resonance stabilization of the anion. Based on resonance stabilization, we predict (correctly) that the ring structure is more stable for the cation than the anion. The molecular orbital diagram for the cyclopropene ring is shown in Figure 5.20.

An interesting conjecture is in regard to whether H_3^+, which has been observed in gas discharge and mass spectrometry, would have a linear or ring structure. The energy level *diagrams* will be identical to those for the C_3 systems shown in Figures 5.17 and 5.19, although the *actual values* for α and β will be different. For the arrangement $[H-H-H]^+$, the energy levels are found to be $\alpha + 2^{1/2}\beta$, α, and $\alpha - 2^{1/2}\beta$. For the ring structure,

the energy levels are $\alpha + 2\beta$, $\alpha - \beta$, and $\alpha - \beta$. Therefore, for two electrons, the total energies are $E_L = 2(\alpha + 2^{1/2}\beta)$ for the linear structure and $E_R = 2(\alpha + 2\beta)$ for the ring structure. From this calculation, it is predicted that the ring structure is more stable by an amount -1.2β. This is expected because if H_3^+ is formed by the interaction of H_2 and H^+, the H^+ will attach at the region of highest electron density, which is the bond in H_2.

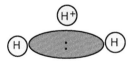

Thus, the cyclic structure for H_3^+ is indicated by both experimental evidence and calculations. Several species have been identified that can be represented as H_n^+ (with n equal to an odd number). They are derived from H_3^+ by adding H_2 molecules at the corners of the ring (presumably perpendicular to the ring). The species with n equal to an even number are less stable.

Over half a century ago, Arthur Frost and Boris Musulin (1953) published an interesting procedure for obtaining the energies of molecular orbitals of ring systems. The first step involves drawing a circle

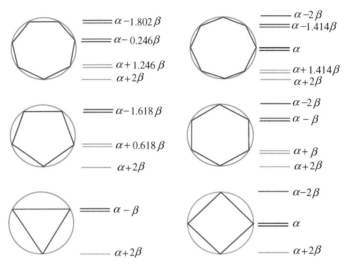

FIGURE 5.21
Frost—Musulin diagrams for cyclic systems having 3-, 4-, 5-, 6-, 7-, and 8-membered rings.

having a convenient radius that will be defined as 2β. Next, inscribe a regular polygon having a number of sides equal to the number of carbon atoms in the structure. Place one vertex of the polygon at the bottom of the circle. The height above the bottom where each vertex makes contact with the circle gives the energy of a molecular orbital. This is illustrated in Figure 5.21 for rings containing three to eight carbon atoms. Above the lowest level, the levels occur in degenerate pairs until the highest level is reached. From this simple procedure, the energy levels summarized in Table 5.6 are found for several cyclic systems.

Although the details will not be shown, it is easy to compute the resonance energies to determine the stabilities of rings with five carbon atoms (the cyclopentadiene, Cp, ring). When this is done, it is found

Table 5.6 Energy Levels For Cyclic Systems

	Orbital Energies					
No. Atoms =	3	4	5	6	7	8
E_8	–	–	–	–	–	$\alpha - 2\beta$
E_7	–	–	–	–	$\alpha - 1.802\beta$	$\alpha - 1.414\beta$
E_6	–	–	–	$\alpha - 2\beta$	$\alpha - 1.802\beta$	$\alpha - 1.414\beta$
E_5	–	–	$\alpha - 1.618\beta$	$\alpha - \beta$	$\alpha - 0.246\beta$	α
E_4	–	$\alpha - 2\beta$	$\alpha - 1.618\beta$	$\alpha - \beta$	$\alpha - 0.246\beta$	α
E_3	$\alpha - \beta$	α	$\alpha + 0.618\beta$	$\alpha + \beta$	$\alpha + 1.246\beta$	$\alpha + 1.414\beta$
E_2	$\alpha - \beta$	α	$\alpha + 0.618\beta$	$\alpha + \beta$	$\alpha + 1.246\beta$	$\alpha + 1.414\beta$
E_1	$\alpha + 2\beta$	$\alpha + 2\beta$	$\alpha + 2\beta$	$\alpha + 2\beta$	$\alpha + 2\beta$	$\alpha + 2\beta$

that $Cp^- > Cp > Cp^+$, which is in agreement with the fact that there is an extensive chemistry associated with the cyclopentadienyl anion.

The procedure of Frost and Musulin can be adapted to chain systems with π bonding in the following way. For a chain having m atoms, draw a polygon as before except that it must have $m + 2$ sides. Disregard the top and bottom vertices, and use only one side of the polygon where it makes contact with the circle to determine the energy levels.

Although the Hückel method is most often applied to organic molecules, the H_3^+ case discussed shows that it can also be applied to some inorganic species. Let us consider the pyrrole molecule

We can write the wave functions as linear combinations of atomic orbitals as before, but in this case, atom 1 is the nitrogen atom and it has a coulomb integral H_{11} that is *not* the same as that for a carbon atom. Therefore, H_{11} will be represented as α_N, which is often approximated in terms of the value for carbon. In this case, the nitrogen contributes two electrons to the π system, and the correction is made that results in the secular determinant having the form $\alpha_N = \alpha_C + (3/2)\beta$. The secular determinant becomes

$$\begin{vmatrix} \alpha + \dfrac{3}{2}\beta - E & \beta & 0 & 0 & \beta \\ \beta & \alpha - E & \beta & 0 & 0 \\ 0 & \beta & \alpha - E & \beta & 0 \\ 0 & 0 & \beta & \alpha - E & \beta \\ \beta & 0 & 0 & \beta & \alpha - E \end{vmatrix} = 0 \tag{5.41}$$

Letting $x = (\alpha - E)/\beta$ and simplifying gives the secular determinant

$$\begin{vmatrix} x + \dfrac{3}{2} & 1 & 0 & 0 & 1 \\ 1 & x & 1 & 0 & 0 \\ 0 & 1 & x & 1 & 0 \\ 0 & 0 & 1 & x & 1 \\ 1 & 0 & 0 & 1 & x \end{vmatrix} = 0 \tag{5.42}$$

which results in the polynomial equation

$$x^5 + \frac{3}{2}x^4 - 5x^3 - \frac{9}{2}x^2 + 5x + \frac{7}{2} = 0 \tag{5.43}$$

Equations such as this were normally solved by graphing before the days in which a calculator removed the need for such tedious techniques. Using numerical techniques, the roots can be found to be $x = -2.55$, -1.15, -0.618, 1.20, and 1.62. The three lowest energy states are populated with six electrons (nitrogen is presumed to contribute two electrons to the bonding). Therefore, the resonance energy is $6\alpha + 7.00\beta - (6\alpha + 8.64\beta) = -1.64\beta$. After the constants $a_1...a_5$ are evaluated, the wave functions can be shown to be

$$\psi_{MO(1)} = 0.749\,\psi_1 + 0.393\,\psi_2 + 0.254\,\psi_3 + 0.254\,\psi_4 + 0.393\,\psi_5 \tag{5.44}$$

$$\psi_{MO(2)} = 0.503\,\psi_1 - 0.089\,\psi_2 - 0.605\,\psi_3 - 0.605\,\psi_4 - 0.089\,\psi_5 \tag{5.45}$$

$$\psi_{MO(3)} = 0.602\,\psi_2 + 0.372\,\psi_3 - 0.372\,\psi_4 - 0.602\,\psi_5 \tag{5.46}$$

$$\psi_{MO(4)} = 0.430\,\psi_1 - 0.580\,\psi_2 + 0.267\,\psi_3 + 0.267\,\psi_4 + 0.580\,\psi_5 \tag{5.47}$$

$$\psi_{MO(5)} = 0.372\,\psi_2 - 0.602\,\psi_3 + 0.602\,\psi_4 - 0.372\,\psi_5 \tag{5.48}$$

Only the three lowest levels are populated. By following the procedure above, the electron densities at each position can be calculated (the nitrogen atom is in position 1).

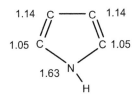

The electron densities total six, the number of electrons in the π system. Note that the highest electron density is at the nitrogen atom, which has the highest electronegativity.

As should be evident, part of the problem in dealing with structures that contain atoms other than carbon is what values to use for α and β. The values that have been suggested are based on correlating calculated properties with other known data. Because the Hückel method is not a quantitative scheme for calculating properties of molecules, we will not address the issue of correcting the values of α and β further.

If we perform an analysis of the N–C–N molecule where the nitrogen atoms are contributing one electron each, we let $\alpha_N = \alpha + \frac{1}{2}\beta$. Following the Hückel procedures, we arrive at the secular determinant

$$\begin{vmatrix} x + \dfrac{1}{2} & 1 & 0 \\ 1 & x & 1 \\ 0 & 1 & x + \dfrac{1}{2} \end{vmatrix} = 0 \tag{5.49}$$

from which it can be shown that the polynomial equation has roots of $x = -1.686$, -0.500, and 1.186. Therefore, the values calculated for the energies of the molecular orbitals are $\alpha + 1.686\beta$, $\alpha + 0.500\beta$, and $\alpha - 1.186\beta$. The first of these is doubly occupied, but the second is singly occupied. The wave functions for the first two levels are

$$\psi_{MO(1)} = 0.541\,\psi_1 + 0.643\,\psi_2 + 0.541\,\psi_3 \tag{5.50}$$

$$\psi_{MO(2)} = 0.707\,\psi_1 - 0.707\,\psi_3 \tag{5.51}$$

In earlier sections, we have seen that the energy of a molecular orbital can be expressed in terms of the coefficients a_i, and the coulomb and exchange integrals. For the second wave function, this can be expressed as

$$E_2 = a_1^2(\alpha + 1/_2\beta) + a_2^2(\alpha) + a_3^2(\alpha + 1/_2\beta) + 2a_1a_2\beta + 2a_2a_3\beta \tag{5.52}$$

The last two terms are zero, which after substituting for the coefficients gives

$$E_2 = 0.707^2(\alpha + 1/_2\beta) + 0(\alpha) + (-0.707)^2(\alpha + 1/_2\beta)$$

$$= 0.500\,\alpha + 0.250\,\beta + 0 + 0.500\,\alpha + 0.250\,\beta = \alpha + 0.500\,\beta \tag{5.53}$$

which is precisely the energy found for that orbital. The calculated electron densities are

$$\overset{\displaystyle 1.09 \qquad\quad 0.827 \qquad\quad 1.09}{N_1 \text{------} C \text{------} N_2}$$

Although no claim can be made that the calculation is quantitative, it is somewhat encouraging that the values obtained are in agreement with what we know about the nature and electronegativities of the atoms. Although it is a primitive method, the Hückel approach provides some interesting exercises for small inorganic species (such as the H_3^+ case discussed above). The insight that the method provides about orbitals in organic species (particularly alkenes) that function as ligands in coordination compounds will be useful in the discussion of these complexes in Chapters 16 and 21.

In this chapter, we have presented an overview of symmetry and its importance when applying molecular orbital methods to molecular structure. Although far from rigorous and complete, the principles described are sufficient for the study of inorganic chemistry at the undergraduate level. Additional details for readers seeking more advanced coverage are to be found in the references listed. It should also be mentioned that computer programs are available for carrying out molecular orbital calculations on several levels that include simple Hückel, extended Hückel, and more sophisticated types of calculations.

References for Further Study

Adamson, A.W., 1986. *A Textbook of Physical Chemistry*, 3rd ed. Academic Press College Division, Orlando. Chapter 17. One of the best treatments of symmetry available in a physical chemistry text.

Cotton, F.A., 1990. *Chemical Applications of Group Theory*, 3rd ed. Wiley, New York. The standard text on group theory for chemical applications.

DeKock, R.L., Gray, H.B., 1980. *Chemical Bonding and Structure*. Benjamin Cummings, Menlo Park, CA. An excellent introduction to bonding that makes use of group theory at an elementary level.

Drago, R.S., 1992. *Physical Methods for Chemists*. Saunders College Publishing, Philadelphia. Chapters 1 and 2 present a thorough foundation in group theory and its application to interpreting experimental techniques in chemistry. Highly recommended.

Fackler, J.P., 1971. *Symmetry in Coordination Chemistry*. Academic Press, New York. A clear introduction to symmetry.

Frost, A., Musulin, B., 1953. *J. Chem. Phys.* 21, 572. The original description showing energies of molecular orbitals by means of inscribed polygons.

Harris, D.C., Bertolucci, M.D., 1989. *Symmetry and Spectroscopy*. Dover, New York. Chapter 1 presents a good summary of symmetry and group theory.

QUESTIONS AND PROBLEMS

1. Make sketches of the following showing approximately correct geometry and all valence shell electrons. Identify all symmetry elements present and determine the point group for the species.
    (a) OCN^-; (b) IF_2^+; (c) ICl_4^-; (d) SO_3^{2-}; (e) SF_6; (f) IF_5; (g) ClF_3; (h) SO_3, (i) ClO_2^-; (j) NSF

2. Make sketches of the following showing approximately correct geometry and all valence shell electrons. Identify all symmetry elements present and determine the point group for the species.
    (a) CN_2^{2-}; (b) PH_3; (c) PO_3^-; (d) $B_3N_3H_6$; (e) SF_2; (f) ClO_3^-; (g) SF_4; (h) C_3O_2; (i) AlF_6^{3-}; (i) F_2O

3. Consider the molecule AX_3Y_2, which has no unshared electron pairs on the central atom. Sketch the structures for all possible isomers of this compound and determine the point group to which each belongs.

4. Match each characteristic listed on the left with the appropriate species from the list on the right that exhibits the characteristic.

Has three C_2 axes	OCN^-
Has $C_{\infty v}$ symmetry	BrO_3^-
Has one C_3 axis	SO_4^{2-}
Has only one mirror plane	XeF_4
Has a center of symmetry	$OSCl_2$

5. Tell which species fits the description or has the indicated symmetry element(s).

Has three S_4 axes	NF_3
Has a center of symmetry	OCS
Has C_{2v} symmetry	PO_4^{3-}
Has $C_{\infty v}$ symmetry	SCl_2
Has one C_3 axis	XeF_2

6. How many chlorine atoms in CCl_4 must be replaced by hydrogen atoms to give molecules having C_{3v} and C_{2v} symmetry, respectively?

7. Draw structures for the following showing correct geometry and identify all of the symmetry elements present in each.

(a) OCN^-; (b) SO_3^{2-}; (c) H_2S; (d) ICl_4^-; (e) ICl_3; (f) ClO_3^-; (g) NO_2^-; (h) IF_5

8. Draw structures for the following showing correct geometry and identify all of the symmetry elements present in each.

(a) C_6H_6; (b) SF_4; (c) ClO_2^-; (d) OF_2; (e) XeF_4; (f) SO_3; (g) Cl_2CO; (h) NF_3

9. Draw structures for the following showing correct geometry and identify all of the symmetry elements present in each.

(a) CH_3Cl; (b) SeF_4; (c) BrO_3^-; (d) SF_2; (e) SnF_2; (f) $S_2O_3^{2-}$; (g) H_2CO; (h) PCl_3

10. Use the symmetry of the atomic orbitals of the central atom to construct (using appropriate hydrogen group orbitals) the molecular orbital diagrams for the following.

(a) BeH_2; (b) HF_2^-; (c) CH_2; (d) H_2S

11. Use the symmetry of the atomic orbitals of the central atom to construct (using appropriate combinations of group orbitals on the peripheral atoms) the molecular orbital diagrams for the following.

(a) AlF_3; (b) BH_4^-; (c) SF_6; (d) NF_3

12. Consider the molecule Cl_2B-BCl_2.

(a) If the structure is planar, what is the point group of the molecule?

(b) Draw a structure for Cl_2B-BCl_2 that has an S_4 axis.

13. Use the procedure outlined in the text to obtain the multiplication table for the C_{4v} point group.

14. Follow the procedure used in the text in obtaining the character table for the C_{2v} point group and develop the character table for the C_{3v} point group.

15. Using the procedure shown in this chapter, calculate the electron density at each position in the pyrrole molecule.

16. By analogy to the carbon systems that contain three atoms, describe the structure of interhalogen species (see Chapter 15) such as I_3^-. Assuming that only p orbitals are used, describe the bonding in this species.

17. From the diagrams shown in Figure 5.21, would you expect the lowest energy spectral band to be at higher energy for the cyclopropene cation, cyclopentadiene anion, or benzene?

18. Use the Hückel method to determine whether H_3^- should have a linear or a ring structure. Calculate the electron density at each atom and the bond orders for the more stable structure.

19. Describe how you would carry out a Hückel calculation for the HFH^+ ion. What would you expect to find the most stable structure to be?

20. When the Hückel method is applied to cyclopentadiene, the secular determinant gives rise to the equation $x^5 - 5x^3 + 5x + 2 = 0$. Solve this equation and determine the energies of the five molecular orbitals. Use these energies to predict the relative stabilities of C_5H_5, $C_5H_5^+$, and $C_5H_5^-$. How do your predictions compare to the chemistry of these species (see Chapter 21).

2

Condensed Phases

CHAPTER 6

Dipole Moments and Intermolecular Interactions

Although the forces that hold molecules and solids together dominate the study of matter, there are other forces that affect chemical and physical properties. These are forces that arise as a result of the interactions between complete molecular units. Matter is composed of electrically charged particles so it is reasonable to expect that there exists *some* force between any two molecules in close proximity.

Forces between molecules are of several types. Some compounds consist of polar molecules that attract each other as a result of the electrical charges. Other compounds consist of nonpolar molecules, but the electrons in one molecule are weakly attracted to the nuclei in another as a result of instantaneous electron distributions that are not symmetrical. Still other molecules contain hydrogen atoms that are attached to other atoms having high electronegativity, which leaves the hydrogen with a residual positive charge. As a result, the hydrogen atom can become attracted to an unshared pair of electrons on an atom in the same or another molecule. This type of interaction is known as hydrogen bonding. Although the forces that exist between molecules may amount to only $10-20$ kJ mol^{-1}, they have a great influence on physical properties and in some cases chemical behavior. It is essential to have an understanding of these types of forces (sometimes called *nonchemical* or *nonvalence* forces) to predict and interpret the properties and behavior of inorganic compounds. This chapter is devoted to the subject of intermolecular interactions.

6.1 DIPOLE MOMENTS

Because atoms have different electronegativities, pairs of electrons that are shared in covalent bonds are not necessarily shared equally. The result is that the bond has a polarity with the center of negative charge generally residing on the atom having the higher electronegativity. For a covalent bond between two atoms, the dipole moment, μ, is expressed as

$$\mu = q \times r \tag{6.1}$$

where q is the quantity of charge separated and r is the distance of separation. In Chapter 3, the relationship between the dipole moment and the weighting coefficient of the ionic term in the molecular wave function for a diatomic molecule was determined. Several properties of molecules are related to their polarity, and it is a useful parameter for understanding molecular structure so it is appropriate to explore this topic in greater detail. Before doing so, a comment on units is appropriate. The charge on an electron is 1.6022×10^{-19} coulomb, and internuclear distances can be expressed in meters. As a result, **169**

Inorganic Chemistry. DOI: http://dx.doi.org/10.1016/B978-0-12-385110-9.00006-6

the units on dipole moments are coulomb meter (C m). A unit of polarity is defined as the *debye*, which is named after Peter Debye who did pioneering work on polar molecules. The relationship in SI units is

$$1 \text{ debye} = 1 \text{ D} = 3.33564 \times 10^{-30} \text{ C m}$$

Historically (as well as currently by many chemists), the quantity of charge separated is expressed in electrostatic units, esu, which is $g^{1/2} \text{ cm}^{3/2} \text{ s}^{-1}$. The charge on the electron is 4.80×10^{-10} esu, and when the internuclear distances are expressed in centimeters,

$$1 \text{ debye} = 1 \text{ D} = 10^{-18} \text{ esu cm}$$

Of course, the results are identical in either set of units, but the latter units are somewhat more convenient for some purposes and will be used in this discussion.

For molecules that have several polar bonds, a rough approximation of the overall dipole moment can be made by considering the *bond moments* as vectors and finding the vector sum. Consider the water molecule that has the structure

and for which the overall dipole moment is 1.85 D. If we consider that value to be the vector sum of the two O—H bond moments, we find that

$$1.85 \text{ D} = 2 \cos 52.25 \times \mu_{O-H} \tag{6.2}$$

Solving for μ_{OH}, we find a value of 1.51 D. We have another way to estimate the dipole moment of the O—H bond by making use of the equation

$$\% \text{ Ionic character} = 16|\chi_A - \chi_B| + 3.5|\chi_A - \chi_B|^2 \tag{6.3}$$

where χ_A and χ_B are the electronegativities of the atoms. By calculating the percent ionic character, we can determine the charge on the atoms. For an O—H bond,

$$\% \text{ Ionic character} = 16|3.5 - 2.1| + 3.5|3.5 - 2.1|^2 = 29.4\% \tag{6.4}$$

Therefore, because the length of the O—H bond is 1.10×10^{-8} cm (110 pm),

$$\mu_{O-H} = 0.294 \times 4.8 \times 10^{-10} \text{ esu} \times 1.10 \times 10^{-8} \text{ cm} = 1.58 \times 10^{-18} \text{ esu cm} = 1.58 \text{ D} \tag{6.5}$$

In this case, the agreement of the values calculated by the two methods is good, but it is not always so. One reason is that the simple vector approach ignores the effects of unshared pairs of electrons. Also, highly polar bonds can induce additional charge separation in bonds that might not otherwise be polar. In some cases, bonds may be essentially nonpolar as is the case for P—H and C—H bonds. Finally, many molecules are not adequately represented by a single structure due to resonance. As a result, the calculation of dipole moments for all but simple molecules is not a trivial problem.

The effect of molecular geometry can often be evaluated in a straightforward manner. Consider the tetrahedral CH_4 molecule, which will be shown as having one C—H bond pointing "up" and the other three forming a tripod-like base:

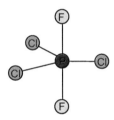

The bond pointing "up" constitutes one C—H bond in that direction, and the other three must exactly equal the effect of one C—H pointing "down." The "down" component of each of the three bonds can be obtained from $\cos(180 - 109° 28') = 1/3$. Therefore, the three bonds exactly equal the effect of the one bond pointing in the "up" direction. This would be true for any regular tetrahedral molecule so the dipole moment would be zero.

In Chapter 4, it was discussed that peripheral atoms of high electronegativity tend to bond to hybrid orbitals having a low degree of s character. In that connection, the molecule PCl_3F_2 is nonpolar indicating that the structure of the molecule is

The axial orbitals used by phosphorus in this molecule can be considered as dp in character (see Chapter 4), which have no s character, whereas the orbitals in equatorial positions are sp^2 hybrids. As expected, the fluorine atoms are found in axial positions and the molecule is nonpolar. This illustration shows the value of dipole moments in predicting the details of molecular structure. Table 6.1 shows dipole moments for a large number of inorganic molecules.

One of the interesting aspects of dipole moments for molecules is seen when the molecules NH_3 and NF_3 are considered:

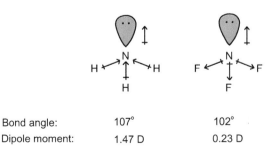

| Bond angle: | 107° | 102° |
| Dipole moment: | 1.47 D | 0.23 D |

For these two molecules, the structures are quite similar, and with the electronegativities of the atoms being $N = 3.0$, $H = 2.1$, and $F = 4.0$, even the polarities of the bonds are similar. The large difference

Table 6.1 Dipole Moments for Some Inorganic Molecules

Molecule	Dipole Moment, D	Molecule	Dipole Moment, D
H_2O	1.85	NH_3	1.47
PH_3	0.58	AsH_3	0.20
SbH_3	0.12	$AsCl_3$	1.59
AsF_3	2.59	HF	1.82
HCl	1.08	HBr	1.43
HI	0.44	$SOCl_2$	1.45
SO_2Cl_2	1.81	SO_2	1.63
PCl_3	0.78	F_2NH	1.92
OPF_3	1.76	SPF_3	0.64
SF_4	0.63	IF_5	2.18
HNO_3	2.17	H_2O_2	2.2
H_2S	0.97	N_2H_4	1.75
NO	0.15	NO_2	0.32
N_2O	0.16	$PFCl_4$	0.21
NF_3	0.23	ClF_3	0.60

in dipole moment is caused by the fact that in NH_3 there is a considerable effect produced by the unshared pair of electrons that causes the negative end of the dipole to lie in that direction. The N–H bonds are polar, with the positive ends lying in the direction of the hydrogen atoms. Therefore, the effect of the polar bonds adds to the effect of the unshared pair of electrons, giving rise to a large dipole moment. In NF_3, the unshared pair of electrons gives a negative charge to that region of the molecule, but the fluorine atoms having higher electronegativity than the nitrogen atom cause the negative ends of the polar N–F bonds to lie toward the fluorine atoms. Thus, the unshared pair of electrons and the polar N–F bonds act in opposition to each other, which results in a low dipole moment for the NF_3 molecule.

The dipole moments of ClF_3 and BrF_3 provide another interesting illustration of the effects of unshared pairs of electrons. The molecules can be shown as follows:

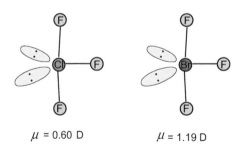

$\mu = 0.60$ D $\mu = 1.19$ D

Owing to the difference in electronegativities, the Br–F bonds are more polar than are the Cl–F bonds. However, the unshared pairs in ClF_3 are held closer to the Cl atom, and the polarity of the Cl–F bond is in the opposite direction as the resultant of the two unshared pairs of electrons. In BrF_3, the equatorial

Table 6.2 Bond Moments for Some Types of Polar Bonds

Bond	Moment, D	Bond	Moment, D
H–O	1.51	C–F	2.0
H–N	1.33	C–Cl	1.47
H–S	0.68	C–Br	1.4
H–P	0.36	C–O	0.74
H–C	0.40	C=O	2.3
P–Cl	0.81	C–N	0.22
P–Br	0.40	C=N	0.9
As–F	2.0	C≡N	3.5
As–Cl	1.6	As–Br	1.3

Br–F bond is slightly more polar than a Cl–F bond, but the unshared pairs reside farther away from the Br atom. Therefore, there is a greater effect produced by the two unshared pairs of electrons in BrF_3 that dominates the polarity of the equatorial Br–F bond. The result is that the dipole moment of BrF_3 is about twice as large as that of ClF_3.

In many cases, it is useful to have some way to approximate the polarity of a molecule or a portion of the molecule. For this purpose, knowing the polarity of specific bonds can provide an approach to the problem. Table 6.2 shows the bond moments for numerous types of bonds.

If the structure of a molecule and the bond moments are known, an approximate dipole moment can be obtained by treating the polar bonds as vectors. Figure 6.1 provides the vector diagram from which we obtain the resultant for two bonds as

$$\mu = \sqrt{\mu_1^2 + \mu_2^2 + 2\,\mu_1\,\mu_2\,\cos\,\theta} \tag{6.6}$$

For molecules where the angle is obtuse (indicated by the dotted lines in Figure 6.1), the relationship is written as

$$\mu = \sqrt{\mu_1^2 + \mu_2^2 + 2\,\mu_1\,\mu_2\,\cos\,(180 - \theta)} \tag{6.7}$$

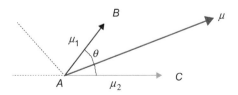

FIGURE 6.1
The vector model used to calculate dipole moments from bond moments.

If this approach is taken for the methyl chloride, CH_3Cl, molecule, which has the structure

the polarity of each C—H bond results in the negative end being toward the carbon atom. The resultant of the three C—H bonds added to the moment for the C—Cl bond gives the overall dipole moment for the molecule. It is found that each C—H bond is at $(180-109.5)° = 70.5°$ from a line directly opposite the C—Cl bond. Consequently, for three C—H bonds, the result is

$$3\,\mu_{CH} \times \cos 70.5 = 3\,\mu_{CH} \times 0.33 = 1\,\mu_{CH}$$

Therefore, the dipole moment for CH_3Cl is calculated to be

$$\mu_{molecule} = \mu_{CH} + \mu_{CCl} = 0.40 + 1.47 = 1.87\,D$$

For $CClF_3$, the polarity of the three C—F bonds is in opposition to the C—Cl bond so that

$$\mu_{molecule} = \mu_{CF} - \mu_{CCl} = 2.00 - 1.47 = 0.53\,D$$

which is very close to the measured dipole moment of 0.50 D.

6.2 DIPOLE—DIPOLE FORCES

When two atoms having different electronegativities share a pair of electrons, the electrons are not shared equally. As a result, the bond between the atoms is said to be polar because the electrons will reside closer to the atom of higher electronegativity giving it a negative charge. For a diatomic molecule, the dipole moment, μ, is given by

$$\mu = q \times r \tag{6.8}$$

and in Chapter 3, it was shown that HCl behaves as if 17% of the charge on an electron is transferred from H to Cl. For HF, the charge separation is 43% of the charge on the electron. When molecules having charge separations approach each other, there are electrostatic forces between them. Orientations such as those shown in Figure 6.2(a) and (b) that place opposite charges closer together represent lower energy (a negative value, E_A).

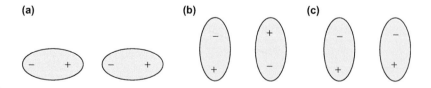

FIGURE 6.2
Arrangements of dipoles: (a) and (b) lead to attraction (interaction energy negative) and (c) leads to repulsion (interaction energy positive).

An arrangement such as that shown in Figure 6.2(c) leads to repulsion (the energy is positive, E_R). Although it might be assumed that such an arrangement could not occur, this is not exactly true. It represents a higher energy state than those in which the orientations place opposite charges in close proximity, but the population of a state of higher energy is governed by the Boltzmann Distribution Law. For two states having energies defined by E_A and E_R as described above, the populations of the states (n_A and n_R) can be related to the energy difference between them, ΔE, by

$$\frac{n_R}{n_A} = e^{-\Delta E/kT} \tag{6.9}$$

where k is Boltzmann's constant, T is the temperature (K), and the other quantities are defined above. Therefore, although the repulsive state is higher in energy, it can have a small population depending on the temperature and ΔE. Because the population of the attractive state is larger, there is a net attraction between two polar molecules. The net energy between two dipoles that are restricted in orientation, E_D, is

$$E_D = -\frac{\mu_1 \mu_2}{r^3} [2 \cos \theta_1 \cos \theta_2 - \sin \theta_1 \sin \theta_2 \cos(\phi_1 - \phi_2)] \tag{6.10}$$

In this equation, θ_1, θ_2, ϕ_1, and ϕ_2 are the angular coordinates that describe the orientations of the polar molecules 1 and 2, μ_1 and μ_2 are their dipole moments, and r is the average distance of separation of the molecules. It should be noted that if two dipoles are restricted to fixed orientations in a solid, the energy varies as $1/r^3$ as shown above and an energy expression containing that factor is frequently encountered. However, in a liquid, the orientations change, and all orientations from the antiparallel attractive arrangement to the parallel repulsive arrangement are possible. There is some average orientation that can be obtained by summing all the possible orientations. When this is taken into account and the average orientation is used, the energy varies as $1/r^6$ and is expressed as

$$E_D = -\frac{2\mu_1^2 \mu_2^2}{3r^6 \, kT} \tag{6.11}$$

If only one type of polar molecule is present, the interaction energy can be expressed as

$$E_D = -\frac{2\mu^4}{3r^6 kT} \tag{6.12}$$

On a molar basis, the energy of interaction is given by

$$E_D = -\frac{2\mu^4}{3r^6 RT} \tag{6.13}$$

Although the association of polar molecules is accompanied by an energy change of only approximately $2-5$ kJ mol^{-1}, the effect on physical properties is great.

It is important to keep in mind that the ability of dipoles to associate is influenced by their environment. Many studies on association of dipoles have been carried out in solutions containing polar molecules. If the molecules of the solvent are polar or can have polarity induced in them (see Section 6.3), the association of the solute molecules will be hindered. The solvent molecules will surround the polar solute molecules, which will inhibit their interaction with other solute molecules. The solute molecules must be at least partially "desolvated" before association can occur. If we represent a polar molecule as D, the association reaction to form dimers can be written as

$$2D \rightleftarrows D_2 \tag{6.14}$$

or in a more general form where aggregates may contain n molecules, the reaction is

$$nD \rightleftarrows D_n \tag{6.15}$$

The equilibrium constants for these reactions may differ by as much as a factor of $10-100$ depending on the nature of the solvent. If the solvent is a nonpolar one such as hexane, the interactions between the polar solute molecules are much stronger than are the interactions between solvent and solute. As a result, the equilibrium constant for dipole association will be large. On the other hand, if the solvent consists of polar molecules such as CH_3OH, the association of the solute molecules may be completely prevented owing to the interaction of polar solute molecules with the polar solvent molecules. The solute molecules must become partially "desolvated" to form dimers or larger aggregates. A solvent such as chlorobenzene or chloroform may not prevent association of polar molecules completely, but the equilibrium constant will almost always be smaller than it is when the solvent consists of nonpolar molecules, such as hexane or CCl_4.

Although it does not strictly involve dipole association, an interesting case that illustrates the principles described above involves the association of lithium alkyls. For $LiCH_3$, one of the stable aggregates is the hexamer, $(LiCH_3)_6$. In a solvent such as toluene, the hexamer units are maintained, but in a solvent like $(CH_3)_2NCH_2CH_2N(CH_3)_2$, which interacts strongly because of the unshared pairs of electrons on the nitrogen atoms, methyllithium exists as a solvated monomer. As will be discussed in Chapter 9, the association of electron donors and acceptors is also strongly affected by the ability of the solvent to interact with solute molecules.

6.3 DIPOLE-INDUCED DIPOLE FORCES

The electrons in molecules and atoms can be moved somewhat under the influence of a charge that generates an electrostatic force on the electrons. As a result, the electron cloud has some *polarizability*, which is represented as α. The total number of electrons may not be as important as the mobility of the electrons in determining the polarizability of a molecule. Consequently, molecules that have delocalized π electron systems generally have higher polarizabilities than do molecules having a similar number of electrons that are held in localized bonds. When a polarizable molecule having a spherical charge distribution approaches a polar molecule, a charge separation is induced in the molecule that was originally nonpolar. This interaction results in some force of attraction between the two species.

The polarizability of one molecule and the magnitude of the dipole moment of the other are the major factors that determine the strength of the interaction. The larger the dipole moment (μ) of the polar molecule and the higher the polarizability of the other molecule, the greater the strength of the interaction. Mathematically, the energy of the interaction of a dipole with a polarizable molecule can be expressed as

$$E_I = -\frac{2\alpha\mu^2}{r^6} \tag{6.16}$$

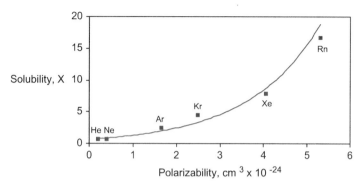

FIGURE 6.3
Solubility of noble gases in water (in mole fraction) as a function of polarizability.

The interaction by dipole-induced dipole forces exists in addition to the London forces that cause an attraction even between molecules that are nonpolar. One of the most striking manifestations of dipole-induced dipole forces is that shown by the solubility of the noble gases in water. The greater the energy of interaction between the solute and solvent, the greater the solubility of the gas. Polar water molecules induce a charge separation in the noble gas molecules that increases with the polarizability of the noble gas. Accordingly, it would be expected that helium would interact very weakly with water and radon should interact more strongly in accord with their polarizabilities. Consequently, the solubility of the noble gases in water decreases in the order $Rn > Xe > Kr > Ar > Ne > He$ as is shown in Figure 6.3.

Another important consequence of dipole-induced dipole interactions is the difference in the solubility of oxygen and nitrogen in water. Expressed as grams of gas dissolved per 100 g of water, the solubilities are 0.006945 and 0.002942, respectively, at 0 °C. Both are nonpolar molecules, but the O_2 molecule has a greater polarizability. As a result, polar H_2O molecules cause a greater charge to be induced in the oxygen molecules, which results in stronger interactions with the solvent leading to greater solubility.

An important extension of these ideas is to cases where an ion interacts with polar molecules (ion-dipole forces). In such cases, the polarity of the molecule is increased because of the inductive effect caused by the ion. Polar solvent molecules that surround an ion in the solvation sphere do not have the same polarity as do the molecules in the bulk solvent.

6.4 LONDON (DISPERSION) FORCES

In addition to the intermolecular forces that exist as a result of permanent charge separations in molecules, there must be some other type of force. Sometimes referred to as electronic van der Waals forces, they cause deviations from the ideal gas equation. These forces are not related to whether or not the molecules have a permanent dipole moment, but rather they exist between all molecules. Otherwise, it would be impossible to liquefy substances such as CH_4, O_2, N_2, and noble gases. The liquid and solid phases of many nonpolar compounds simply would not exist. We can see how such forces arise by considering two noble gas atoms that are in close proximity as shown in Figure 6.4. At some instant, the

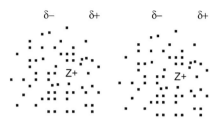

FIGURE 6.4

An instantaneous distribution of electrons that leads to polarity in two atoms. There will be a force of attraction between the atoms (or molecules), even though they do not have permanent polarity. The number of electrons and their ability to be moved will determine the magnitude of the attractive force.

majority of electrons in one atom might be located on one side of the atom leaving the other half with an instantaneous positive charge. That charge can attract the electrons in another atom so that there is a net force of attraction between them. The result is an electron arrangement that can be called an *instantaneous dipole*. In 1929, Fritz London studied the forces that arise from this type of interaction and they are called *London forces* or *dispersion forces*.

To arrive at a mathematical relationship to describe London forces, we will use an intuitive approach. First, the ability of the electrons to be moved within the molecule is involved. Atoms or molecules in which the electrons are highly localized cannot have instantaneous dipoles of any great magnitude induced in them. A measure of the ability of electrons in a molecule to be shifted is known as the electronic polarizability, α. In fact, each of the interacting molecules has a polarizability so the energy arising from London forces, E_L, is proportional to α^2. London forces are important only at short distances, which means that the distance of separation is in the denominator of the equation. In fact, unlike Coulomb's law, which has r^2 in the denominator, the expression for London forces involves r^6. Therefore, the energy of interaction as a result of London forces is expressed as

$$E_L = -\frac{3h\nu_0\alpha^2}{4r^6} \tag{6.17}$$

where α is the polarizability, ν_0 is the zero point vibration frequency, and r is the average distance of separation between the molecules. The quantity $h\nu_0$ is the ionization potential, I, for the molecule. Therefore, the London energy can be represented by

$$E_L = -\frac{3I\alpha^2}{4r^6} \tag{6.18}$$

It is interesting to note that many different types of molecules have ionization potentials that do not differ greatly. Table 6.3 shows typical values for molecular ionization potentials for a wide variety of substances.

Because the ionization potentials are similar in magnitude, I can be replaced by a constant with no great effect on the value of E_L. The values of α for helium and argon are 2.0×10^5 pm^3 and 1.6×10^6 pm^3, respectively. Calculations show that for helium atoms separated by a distance of

Table 6.3 Ionization Potentials (IP) for Selected Molecules

Molecule	IP, eV	Molecule	IP, eV
CH_3CN	12.2	$C_2H_5NH_2$	8.86
$(CH_3)_2NH$	8.24	$(CH_3)_3N$	7.82
NH_3	10.2	HCN	13.8
H_2O	12.6	H_2S	10.4
CH_4	12.6	CS_2	10.08
HF	15.77	SO_2	12.34
CH_3SH	9.44	C_6H_5SH	8.32
C_6H_5OH	8.51	CH_3OH	10.84
C_2H_5OH	10.49	BF_3	15.5
CCl_4	11.47	PCl_3	9.91
AsH_3	10.03	$AsCl_3$	11.7
$(CH_3)_2CO$	9.69	$Cr(CO)_6$	8.03
C_6H_6	9.24	1,4-Dioxane	9.13
$n\text{-}C_4H_{10}$	10.63	OF_2	13.6

1 eV is equivalent to 98.46 kJ mol^{-1}.

300 pm (3 Å), the energy of interaction is 76.2 J mol^{-1}, but the interaction of argon atoms at a distance of 400 pm (4 Å) is 1050 J mol^{-1}. In agreement with the difference in these energies, solid argon melts at 89 K, whereas solid helium is obtained at 1.76 K at a pressure of 29.4 atmospheres. If we consider two nonpolar molecules such as CCl_4 ($\alpha = 2.6 \times 10^7$ pm^3, b.p. 77 °C) and C_6H_6 ($\alpha = 2.5 \times 10^7$ pm^3, b.p. 80 °C), we find that the polarizabilities are very nearly equal as are the boiling points. For these molecules, the interaction is only by London forces so the comparison of boiling points and polarizabilities is valid.

In general, if the intermolecular forces are only of the London type, the boiling point (the temperature at which molecules of a liquid become separated from each other) will be higher the larger the molecule and the greater the number of electrons. For example, F_2 and Cl_2 are gases, Br_2 is a liquid, and I_2 is a solid at room temperature. The boiling points for $GeCl_4$ and $SnCl_4$ are 86.5 °C and 114.1 °C in accordance with the difference in numbers of electrons and hence the polarizabilities. The increase in boiling point of the hydrocarbon series, C_nH_{2n+2}, as n increases provides a familiar illustration of this principle.

If two different types of molecules having polarizabilities α_1 and α_2 are interacting, the London energy between them can be expressed as

$$E_L = -\frac{3h\alpha_1\alpha_2}{2r^6} \cdot \frac{\nu_1\nu_2}{\nu_1 + \nu_2} = -\frac{3\alpha_1\alpha_2 I_1 I_2}{2r^6(I_1 + I_2)} \tag{6.19}$$

In Chapter 7, it will be pointed out that the bonding in solid silver halides is somewhat covalent. This results from the ions being polarizable so that the cation and anion have induced charge separations. For AgI, the electrostatic attraction is 808 kJ mol^{-1}, but the London attraction is 130 kJ mol^{-1}. From the examples presented, it is clear that London forces are significant enough to greatly affect the physical properties of compounds.

One of the consequences of increasing the number of electrons that accompanies an increase in molecular weight is that the London forces between molecules increase. The boiling points of a series of compounds composed of nonpolar molecules should reflect the greater force of attraction between molecules. To illustrate this trend, the boiling points of a series of organic compounds such as the hydrocarbons are considered. However, the trend is also well illustrated by numerous series of inorganic compounds. Figure 6.5 shows the boiling points of some of the halogen compounds of Group IIIA and IV having the formulas EX_3 and EX_4.

For those compounds, most of which consist of covalent molecules, there is an expected increase in boiling point with increasing molecular weight. The trigonal planar BX_3 and tetrahedral SiX_4 and GeX_4 compounds follow the expected trend as X progresses from F to I. Except for AlF_3, the aluminum compounds are essentially covalent and exist as dimers as do the aluminum alkyls, $[AlR_3]_2$, the structures of which were described in Chapter 4 and will be discussed further later in this chapter. However, there is a marked difference when AlF_3 is considered. In this case, the compound is essentially ionic, which results in a boiling point of approximately 1300 °C. The total ionization potential to produce Al^{3+} is 5,139 kJ mol^{-1} so only in a case where there is a high lattice energy will the compound be ionic. Thus, the small size of the Al^{3+} and F^- allows them to form a lattice that is sufficiently stable to offset the high ionization energy required to produce Al^{3+}. The high boiling point for AlF_3 is a reflection of the different type of bonding present in that compound compared to the other aluminum halides. It is useful to remember that there is a continuum of bond character from covalent to ionic, and the bonding in AlF_3 is definitely toward the ionic end of the spectrum.

When a molecule dissolves in a liquid, some of the forces between the solvent molecules must be overcome. Unless an individual molecule of solute is considered, forces between the solute molecules must also be overcome. Effective solvation of solute molecules requires that the solute–solvent interactions lead to a favorable energy if the solute is to have significant solubility. Because nonpolar molecules interact due to the London forces between them, it should not be surprising to learn that compounds such as BI_3 that are nonpolar are soluble in nonpolar solvents, such as CCl_4 and CS_2. The

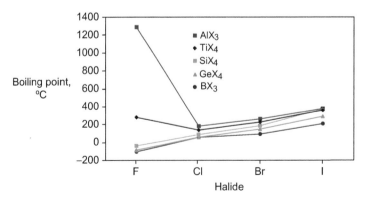

FIGURE 6.5
Boiling points of Group IIIA and Group IVA halides.

polarizability of BI_3 is substantial owing to the large number of electrons contained in three iodine atoms so the BI_3 molecules interact well with CCl_4 and CS_2, both of which have relatively high polarizability. Likewise, $AlBr_3$ and AlI_3 are soluble in alcohol, ether, and carbon disulfide, whereas AlF_3, which is composed of essentially nonpolarizable ions, is not.

6.5 THE VAN DER WAALS EQUATION

In 1873, J.D. van der Waals recognized deficiencies in the ideal gas equation and developed an equation to eliminate two problems. First, the volume of the container is not the *actual* volume available to the molecules of the gas because the molecules themselves occupy some volume. The first correction to the ideal gas equation was to subtract the volume of the molecules from V, the volume of the container, to give the net volume accessible to the molecules. When modified to include the number of moles, n, the corrected volume is $(V - nb)$, where b is a constant that depends on the type of molecule.

From the ideal gas equation, it is found that for one mole of gas, $PV/RT = 1$, which is known as the *compressibility factor*. For most real gases, there is a large deviation from the ideal value, especially at high pressure where the gas molecules are forced closer together. From the discussions in previous sections, it is apparent that the molecules of the gas do not exist independently from each other because of forces of attraction even between nonpolar molecules. Dipole–dipole, dipole-induced dipole, and London forces are sometimes *collectively* known as van der Waals forces because all these types of forces result in deviations from ideal gas behavior. Because forces of attraction between molecules reduce the pressure that the gas exerts on the walls of the container, van der Waals included a correction to the pressure to compensate for the "lost" pressure. That term is written as n^2a/V^2, where n is the number of moles, a is a constant that depends on the nature of the gas, and V is the volume of the container. The resulting equation of state for a real gas, known as van der Waals equation, is written as

$$\left(P + \frac{n^2a}{V^2}\right)(V - nb) = nRT \tag{6.20}$$

In van der Waals equation, it is the term n^2a/V^2 that is of interest in this discussion because that term gives information about intermolecular forces. Specifically, it is the parameter a that is related to intermolecular forces rather than the number of moles, n, or the volume, V. It should be expected that the parameter a would show a correlation with other properties that are related to the forces between both organic and inorganic molecules.

We will consider first a relatively simple case where the interactions between molecules are all of the same type (nonpolar molecules interacting as a result of London forces). For a liquid, the boiling point gives a measure of the strength of the forces between molecules in the liquid state because those forces must be overcome in order for the molecules to escape as a vapor. Figure 6.6 shows the boiling points of the noble gases and a few other substances as a function of the van der Waals a parameter.

It is apparent that for these nonpolar molecules the correlation is satisfactory. In this case, a characteristic of the liquid state (the boiling point) is correlated with a parameter from an equation that was developed to explain the behavior of gases. The liquid and gaseous states are referred to as *fluids*, and van der Waals equation can be considered as an equation that applies to fluids as well as to gases

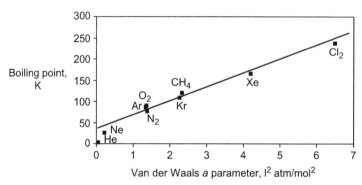

FIGURE 6.6
Variation in boiling points of nonpolar molecules with van der Waals a parameter.

through the use of the reduced variables (see references at the end of this chapter). Table 6.4 gives values for the van der Waals a parameter for molecules most of which are nonpolar.

Although a correlation between a property of a *liquid* and the a parameter in van der Waals equation might be expected, we should remember that for nonpolar molecules the solid phase is also held together by London forces. Of course, the energy holding the solid together is the lattice energy so we should attempt a correlation of the lattice energy of solids composed of nonpolar molecules with the van der Waals a parameter. Such a correlation is shown in Figure 6.7 where the lattice energies of the noble gases and a few other nonpolar substances are plotted against a. It is immediately apparent that a linear relationship results, even though a property of a *solid* is being considered as a function of the parameter that results from considering of the interaction of molecules of real gases.

Table 6.4 Values for the van der Waals a Parameter for Selected Molecules

Molecule	a, l^2 atm mol^{-2}	Molecule	a, l^2 atm mol^{-2}
He	0.03412	C_2H_6	5.489
H_2	0.2444	SO_2	6.714
Ne	0.2107	NH_3	4.170
Ar	1.345	PH_3	4.631
Kr	2.318	C_6H_6	18.00
Xe	2.318	CCl_4	20.39
N_2	1.390	SiH_4	4.320
O_2	1.360	SiF_4	4.195
CH_4	2.253	$SnCl_4$	26.91
Cl_2	6.493	C_2H_6	5.489
CO_2	3.592	N_2O	3.782
CS_2	11.62	$GeCl_4$	22.60

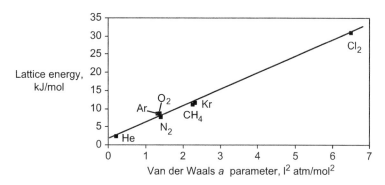

FIGURE 6.7
Variation in lattice energy of some nonpolar molecules with van der Waals a parameter.

The utility of the van der Waals a parameter should not be underestimated when physical properties of substances are being correlated and interpreted. Neither for that matter should the b parameter because it is related to effective molecular dimensions, but that is not our concern in this chapter.

6.6 HYDROGEN BONDING

Thousands of articles and several books have been published on the subject of hydrogen bonding. It is a phenomenon that pertains to many areas of the chemical sciences, and it is an important type of molecular interaction. The fact that it is referred to as *hydrogen* bonding suggests that hydrogen is unique in this ability, and so it is. Of all the atoms, only hydrogen leaves a completely bare nucleus exposed when it forms a single covalent bond to another atom. Even lithium has a filled $1s$ level around the nucleus after the single electron in the $2s$ level is used in covalent bonding. When a hydrogen atom is bonded to an element having an electronegativity of about 2.6 or greater (F, O, N, Cl, or S), the polarity of the bond is sufficient that the hydrogen carries a positive charge that enables it to be attracted to a pair of electrons on another atom. That attraction is known as a hydrogen bond (sometimes referred to as a hydrogen bridge). This type of interaction can be shown as

$$X-H\cdots\cdots:Y$$

Hydrogen bonding occurs in many situations in chemistry. Materials such as proteins, cellulose, starch, and leather have properties that are the result of hydrogen bonding. Even solid materials such as NH_4Cl, $NaHCO_3$, NH_4HF_2, and ice have strong hydrogen bonding between units. Water and other liquids that have OH groups on the molecules (e.g. alcohols) have extensive hydrogen bonding. There are two types of hydrogen bonds, which are illustrated by the following examples.

Intermolecular   Intramolecular

Both types of hydrogen bonds occur in pure liquids as well as in solutions. Many substances are associated at least partially in the vapor phase as a result of hydrogen bonding. For example, hydrogen cyanide is associated to give structures such as

$$\cdots HCN \cdots HCN \cdots HCN \cdots$$

The association of acetic acid in the vapor phase occurs so that the molecular weight of the gas indicates that it exists as dimers.

Studies have indicated that the association of HF in the gas phase leads predominantly to dimers or hexamers with small amounts of tetramers. Hydrogen bonding in liquids such as sulfuric and phosphoric acids is responsible for them being viscous liquids that have high boiling points.

Association of alcohols in the liquid state occurs with the formation of several types of species including chains,

in which the $O - H \cdots O$ bond distance is approximately 266 pm. The liquid also contains rings with the most predominant unit apparently being $(ROH)_6$, which can be shown as

The vapor of CH_3OH also contains some cyclic tetramer, $(CH_3OH)_4$, for which the heat of association has been found to be 94.4 kJ mol^{-1} leading to a value of 23.6 kJ mol^{-1} for each hydrogen bond. The equilibrium composition of alcohols in the vapor phase is temperature and pressure dependent. Boric acid, $B(OH)_3$, consists of sheet-like structures as a result of hydrogen bonding.

The body of information relating hydrogen bonding to physical properties is enormous. Only a brief summary will be presented, but more complete discussions can be found in the references cited at the end of this chapter. Perhaps the most familiar and elementary example of the effect of hydrogen bonding is reflected by the fact that the boiling point of water is 100 °C, whereas that of liquid H_2S is −61 °C. Figure 6.8 shows the boiling points of the hydrogen compounds of the elements in Groups IVA−VIIA.

There is no hydrogen bonding in the hydrogen compounds of the elements in Group IVA elements so CH_4, SiH_4, GeH_4, and SnH_4 show the expected increase in boiling point with increasing molecular weight. For the hydrogen compounds of the Group VA elements, only NH_3 exhibits significant hydrogen bonding so its boiling point (−33.4 °C) is clearly out of line with those of the other compounds (e.g. PH_3 has a boiling point of −85 °C). Water shows clearly the effect of strong hydrogen bonds (in fact, multiple hydrogen bonds), which results in a boiling point of 100 °C for a compound that has a molecular weight of only 18. With the high electronegativity of fluorine, the polar H−F bond is susceptible to forming strong hydrogen bonds as is clearly illustrated by the boiling point of HF being 19.4 °C, whereas HCl boils at −84.9 °C.

It is also interesting to note that although the compounds have almost identical formula weights, the boiling point of BF_3 is −101 °C, although boric acid, $B(OH)_3$, is a solid that decomposes at 185 °C. Dimethyl ether and ethanol have the formula $C_2H_6O_2$, but the boiling points are −25 and 78.5 °C, respectively. Hydrogen bonding between OH groups in alcohols leads to intermolecular forces that are not present in dimethyl ether.

When a liquid is changed into a vapor, the entropy of vaporization can defined as

$$\Delta S_{vap} = S_{vapor} - S_{liquid} \cong S_{vapor} \cong \Delta H_{vap}/T. \tag{6.21}$$

where T is the boiling point in K. If the liquid is one in which London forces give the only type of interaction between molecules and the vapor is completely random, the entropy of vaporization can be

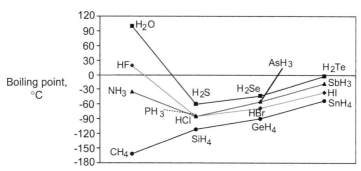

FIGURE 6.8
Boiling points of hydrides of Groups IVA, VA, VIA, and VIIA.

represented as $\Delta H_{vap}/T$. The entropy of a mole of a random gas is approximately 88 J mol^{-1} K^{-1}, and the constant value for the entropy of vaporization of a liquid is known as *Trouton's rule*. Table 6.5 shows the data for testing this rule with a wide range of liquids.

For CCl$_4$, the heat of vaporization is 30.0 kJ mol^{-1} and the boiling point is 76.1 °C, which gives a value of ΔS_{vap} of 86 J mol$^-$ K^{-1}, a value that is in good agreement with Trouton's rule. On the other hand, the heat of vaporization of CH$_3$OH is 35.3 kJ mol^{-1} and the boiling point is 64.7 °C. These values lead to a value for ΔS_{vap} of 104 J mol^{-1} K^{-1}. The deviation from Trouton's rule is caused by the fact that in the liquid state the molecules are strongly associated giving a structure to the liquid (lower entropy). Therefore, vaporization of CH$_3$OH leads to a larger entropy of vaporization than would be the case if the molecules of the liquid and vapor were arranged randomly.

Acetic acid provides a different situation. The boiling point of acetic acid is 118.2 °C and the heat of vaporization is 24.4 kJ mol^{-1}. These values yield an entropy of vaporization of only 62 J mol^{-1} K^{-1}. In this case, the liquid is associated to produce dimers as described above, but those dimers also exist in the vapor. Therefore, structure persists in the vapor so that the entropy of vaporization is much lower than would be the case if a vapor consisting of randomly arranged monomers were produced. It is interesting to note from the examples described above that a property such as the entropy of vaporization can provide insight as to the extent of molecular association.

Other properties are also affected by hydrogen bonding. For example, the solubility of *o-*, *m-*, and *p*-nitrophenols (NO$_2$C$_6$H$_4$OH) are greatly different as a result of hydrogen bonding. The solubility of *p*-nitrophenol (which can hydrogen bond to a solvent such as water) in water is greater than that of *o*-nitrophenol in which there is intramolecular hydrogen bonding. On the other hand, *o*-nitrophenol is much more soluble in benzene than is *p*-nitrophenol. The *ortho* isomer has intramolecular hydrogen bonds that allow the interaction of the solvent with the ring of the solute to be the dominant factor. As a result, *o*-nitrophenol is many times more soluble in benzene than is *p*-nitrophenol.

Table 6.5 Thermodynamic Data for the Vaporization of Several Liquids

Liquid	Boiling Point, °C	ΔH_{vap}, J mol^{-1}	ΔS_{vap}, J mol^{-1} K^{-1}
Butane	−1.5	22,260	83
Naphthalene	218	40,460	82
Methane	−164.4	9270	85
Cyclohexane	80.7	30,100	85
Carbon tetrachloride	76.7	30,000	86
Benzene	80.1	30,760	87
Chloroform	61.5	29,500	88
Ammonia	−33.4	23,260	97
Methanol	64.7	35,270	104
Water	100	40,650	109
Acetic acid	118.2	24,400	62

Hydrogen bond formation also leads to differences in chemical properties. For example, the enolization reaction of 2,4-pentadione (acetylacetone) is assisted by the formation of an intramolecular hydrogen bond.

$$
\begin{array}{cc}
:O: & :O: \\
\| & \| \\
CH_3-C-CH_2-C-CH_3
\end{array}
\rightleftharpoons
\begin{array}{cc}
:O:\cdots H-\overset{..}{O}: \\
\| & | \\
CH_3-C-CH=C-CH_3
\end{array}
\qquad (6.22)
$$

In the neat liquid, the enol form is dominant, and in solutions, the composition of the equilibrium mixture depends greatly on the solvent. For example, when the solvent is water, hydrogen bonding between the solvent and the two oxygen atoms can occur, which helps to stabilize the keto form that makes up 84% of the mixture. When the solvent is hexane, 92% of the acetylacetone exists in the enol form in which intramolecular hydrogen bonding stabilizes that structure. Acetone undergoes only an extremely small amount of enolization (estimated to be less than 10^{-5}%), which is at least partially the result of there being no possibility of hydrogen bonding in the enol form.

One of the most convenient ways to study hydrogen bonding experimentally is by means of infrared spectroscopy. When a hydrogen atom becomes attracted to an unshared pair of electrons on an atom in another molecule, the covalent bond holding the hydrogen atom becomes weakened slightly. As a result, the absorption band that corresponds to the stretching vibration of the bond is shifted to a position that may be *lower* by up to 400 cm^{-1}. Because of the attraction between the hydrogen atom and a pair of electrons in another molecule, the bending vibrations of the covalent bonds holding the hydrogen atom are hindered. Therefore, the bending vibrations are shifted to *higher* frequencies. Although the hydrogen bond itself is weak, there is a stretching vibration for that bond that does not exist prior to the formation of the hydrogen bond. Because hydrogen bonds are weak, the stretching vibration occurs at very low wave numbers (generally 100–200 cm^{-1}). All these vibrations and the regions where they are found in the infrared spectrum are summarized in Table 6.6.

Table 6.6 Infrared Spectral Features Associated With Hydrogen Bonding

Vibration	Assignment	Spectral Region, cm^{-1}
$\leftarrow \quad \rightarrow$ X—H $\cdots$ B /	ν_s, the X–H stretch	3500–2500
$\uparrow$ X—H $\cdots$ B / $\downarrow$	ν_b, the in plane bend[a]	1700–1000
$\oplus$ X—H $\cdots$ B / $\oplus$	ν_t, the out of plane bend[b]	400–300
$\leftarrow \quad \rightarrow$ X—H $\cdots$ B /	ν_σ, the H$\cdots$B stretch[c]	200–100

[a] Bending in the plane of the page. Hydrogen bonding causes higher ν_b.
[b] Bending perpendicular to the plane of the page. Hydrogen bonding causes higher ν_t.
[c] Stretching of the hydrogen bond to the donor atom. Increases with hydrogen bond strength.

In very dilute solutions of CH_3OH in CCl_4, the alcohol molecules are widely separated and the equilibrium

$$n\,CH_3OH \rightleftarrows (CH_3OH)_n \tag{6.23}$$

is shifted far to the left. The infrared spectrum of such a dilute solution shows a single band at 3642 cm^{-1} that corresponds to the "free" OH stretching vibration. As the concentration of alcohol is increased, other bands appear at 3504 and 3360 cm^{-1} that are due to higher aggregates that result from intermolecular hydrogen bonding between the OH groups as shown earlier. Figure 6.9 shows the spectra of 0.05, 0.15, and 0.25 M CH_3OH in CCl_4 solutions.

The spectra show that there are "free" OH groups at all concentrations, but the very broad peak at 3360 cm^{-1} shows a large fraction of the alcohol is bound in aggregates when the concentration is 0.25 M. In addition to the cyclic structures shown earlier, these aggregates are believed to have structures that can be represented as follows:

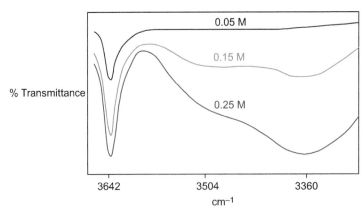

As described earlier, there is doubtless a complex equilibrium that involves several species of aggregates having both chain and ring structures.

The effect of the solvent on equilibria involving molecular aggregation has been discussed in connection with dipole association. However, the nature of the solvent also has an effect on the position where the OH stretching band is observed in the infrared spectrum, even though the OH group may not be involved in hydrogen bonding. The ability of the solvent to "solvate" the OH group affects vibrational energy levels, even though the interaction is not actually considered to be hydrogen bonding. The absorption band due to the stretching vibration of the O–H bond in CH_3OH in the vapor

FIGURE 6.9
Infrared spectra for solutions of CH_3OH in CCl_4.

phase is seen at 3687 cm^{-1}. In n-C$_7$H$_{16}$, CCl$_4$, and CS$_2$, the band is seen at 3649, 3642, and 3626 cm^{-1}, respectively. A hydrocarbon molecule has no unshared electron pairs to interact even weakly with the O—H bond so the stretching vibration is at the highest position found in any of these solvents when n-C$_7$H$_{16}$ is the solvent. The band appearing at 3626 cm^{-1} when the solvent is CS$_2$ is indicative of very weak hydrogen bonding to the solvent in this case. When the solvent is benzene, the position of the OH stretching vibration for CH$_3$OH in a dilute solution is at 3607 cm^{-1} indicating significant interaction of the OH group with the π electron system of the benzene ring. Benzene is known to form complexes with Lewis acids because of its ability to behave as a Lewis base (see Chapter 9). It is certainly not an "inert" solvent that should be chosen when hydrogen bonding studies are being carried out.

A relationship has been developed that relates the position of the O—H stretching band of an alcohol to the electronic character of the solvent. That equation is based on the assumption that an oscillating electric dipole is interacting with a solvent of dielectric constant ε. The equation can be written as

$$\frac{\nu_g - \nu_s}{\nu_g} = C\frac{\varepsilon - 1}{2\varepsilon + 1} \tag{6.24}$$

where ν_g and ν_s are the stretching frequencies in the gas phase and in solution and C is a constant. The dielectric constant at high frequency is usually approximated as the square of the index of refraction, n. When that is done, the shift in the position of the stretching band, $\Delta\nu = (\nu_g - \nu_s)$ is represented by the equation

$$\frac{\Delta\nu}{\nu_g} = C\frac{n^2 - 1}{2n^2 + 1} \tag{6.25}$$

which is known as the Kirkwood—Bauer equation. Figure 6.10 shows the correlation of the positions of the O—H stretching band of CH$_3$OH in heptane, CCl$_4$, CS$_2$, and benzene. The first three solvents appear

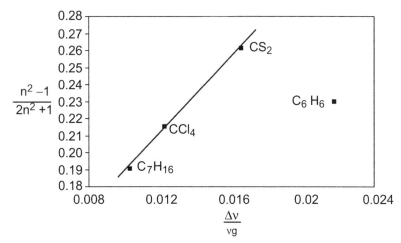

FIGURE 6.10
A Kirkwood—Bauer plot showing the effect of solvent on the O—H stretching band of methanol in different solvents.

to solvate CH_3OH in a "normal" manner and obey the Kirkwood–Bauer equation, but benzene is clearly interacting in a different way. As mentioned earlier, benzene is an electron donor that even forms complexes with metals. It is apparent from Figure 6.10 that benzene is by no means an "inert" solvent with regard to hydrogen bonding.

Hydrogen bond energies are sometimes described as weak, normal, or strong based on the strength of the bonds. Weak hydrogen bonds are those that are weaker than about 12 kJ mol^{-1} and are typical of the intramolecular hydrogen bond in 2-chlorophenol. Normal hydrogen bonds (which include the vast majority of cases) are those that have energies of perhaps 10–40 kJ mol^{-1}. Typical of this type of bond are those that occur between alcohols and amines. The strong hydrogen bonds found in the symmetric bifluoride ion, $[F\cdots H\cdots F]^-$, have a bond energy of approximately 142 kJ mol^{-1}. This ion has a distance between fluorine centers of 226 pm so each bond is 113 pm in accord with the bond order being ½. In this case, the strength of the hydrogen bonds is comparable to that of weak covalent bonds, such as F–F, I–I, and the O–O bond in O_2^{2-}.

Strengths of chemical bonds correspond the bond enthalpies for molecules in the gas phase, and it is desirable to measure enthalpies of hydrogen bonding for that type of interaction. However, most hydrogen-bonded systems are not stable enough to exist at the temperatures required to vaporize the donor and acceptor. Therefore, strengths of hydrogen bonds are usually determined calorimetrically by mixing solutions containing the donor and acceptor. When the influence of the solvent is considered, as illustrated by the relationship shown in Eqn (6.24), the question arises as to whether the measured enthalpy is actually that of the hydrogen bond. The situation can be illustrated by means of a thermochemical cycle in which B is an electron pair donor and $-X\cdots H$ is the species that forms the hydrogen bond.

$$
\begin{array}{ccccc}
\text{:B(gas)} & + & -X-H\text{(gas)} & \xrightarrow{\ \Delta H_{HB}\ } & [-X-H\cdots\text{:B] (gas)} \\[1mm]
\big\uparrow \Delta H_1 & & \big\uparrow \Delta H_2 & & \big\uparrow \Delta H_3 \\[1mm]
& & & \xrightarrow{\ \Delta H_{HB}'\ } & \\[1mm]
\text{:B(solv)} & + & -X-H\text{(solv)} & & [-X-H\cdots\text{:B] (solv)}
\end{array}
$$

The actual strength of the hydrogen bond, ΔH_{HB}, is not necessarily the same as that given by the enthalpy of the reaction measured in solution, $\Delta H_{HB}'$. An ideal solution is one in which the heat of mixing is 0 so the question arises as to whether $-X-H$ and B form ideal solutions with the solvent. If the solvent is one such as benzene, which forms weak hydrogen bonds to $-X-H$, those bonds must be broken before the hydrogen bond to B can form. Therefore, the enthalpy measured in solution will not be the same as that which corresponds to the gas phase reaction. The same situation exists if the solvent is one that interacts with B. Mathematically, the requirement for the gas and solution phase enthalpies to be equal is that $|\Delta H_1 + \Delta H_2| = |\Delta H_3|$. If the extent to which the solvent is a participant in the process is indicated by the position of the "free" O–H stretching band, it can be seen that heptane is the most nearly "inert" solvent of those discussed above. In fact, the "inertness" decreases in the order heptane > CCl_4 > CS_2 >> C_6H_6. This series is in accord with the trend illustrated in Figure 6.10. A good test for evaluating the role (if any) of the solvent in hydrogen bonding is to determine the enthalpy of the hydrogen bond formation in different solvents to see if the measured enthalpy is the same. Although the solvent most widely used in hydrogen bonding studies has probably been CCl_4, hexane or heptane is usually a better choice.

There have been many studies on the formation of hydrogen bonds between alcohols and a wide range of bases. If the bases are of similar type (e.g. all nitrogen donor atoms in amines), there is also frequently a rather good correlation between the shift of the O−H stretching band and other properties. For example, stretching frequency shifts of the OH bonds in alcohols have been correlated with base strength of the electron pair donor. As long as the bases have similar structure, the correlations are generally satisfactory. Figure 6.11 shows such a correlation for trimethylamine and triethylamine as well as a series of methyl-substituted pyridines. It is apparent that the correlation is quite good, and it can be expressed as

$$\Delta v_{OH} = a\, pK_b + b \tag{6.26}$$

where a and b are constants.

The availability of electrons on the donor atom in a base molecule determines not only its ability to bind H^+ but also its ability to attract a hydrogen atom in forming a hydrogen bond. Consequently, it is reasonable to expect some relationship to exist between base strength and hydrogen bonding ability.

Relationships between the stretching frequency shifts and hydrogen bond enthalpies have also been established. Such a correlation can be written in the form

$$-\Delta H = c\, \Delta v_{OH} + d \tag{6.27}$$

where c and d are constants. For bases of a different structural type, the constants may have different values. Many correlations of this type have been developed, and some of them are useful empirical relationships.

Hydrogen bonding is special type of acid−base interaction (see Chapter 9). Probably the most important equation relating hydrogen bond strengths is the equation known as the Drago four-parameter equation,

$$-\Delta H = C_A C_B + E_A E_B \tag{6.28}$$

which applies to many types of acid−base interactions. This equation is based on the assumption that a bond (including a hydrogen bond) is made up of a covalent part and an electrostatic part. The

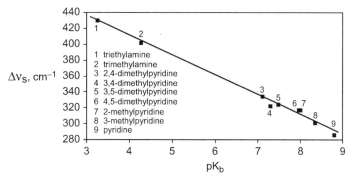

FIGURE 6.11
Shift of the OH stretching band for methanol hydrogen bonded to several bases as a function of base strength.

covalent contribution to the bond enthalpy is given by the product of parameters giving the covalent bonding ability of the acid and base (C_A and C_B), and the product of the electrostatic parameters (E_A and E_B) gives the ionic contribution to the bond. When tables are available that give the needed parameters, the calculated enthalpies of interaction agree remarkably well with the experimental values. Because of the wider use of the Drago equation in other types of acid–base interactions, it will be discussed in more detail in Chapter 9.

As the brief introduction to the subject presented here shows, hydrogen bonding is extremely important in all areas of chemistry. Additional topics including discussions of experimental methods for studying hydrogen bonding can be found in the references cited at the end of this chapter.

6.7 COHESION ENERGY AND SOLUBILITY PARAMETERS

Molecules have forces of attraction between them, and these intermolecular forces are responsible for many of the properties of liquids. There is a cohesion energy that holds the molecules together. The energy necessary to overcome these forces to vaporize a mole of liquid is known as the *cohesion energy* of the liquid or the *energy of vaporization*. It is related to the enthalpy of vaporization by the equation

$$\Delta H_{vap} = \Delta E_{vap} + \Delta(PV) \tag{6.29}$$

Therefore, because $\Delta(PV) = RT$ we can write

$$\Delta E_{vap} = E_C = \Delta H_{vap} - RT \tag{6.30}$$

where E_c is the cohesion energy of the liquid. The quantity E_c/V_m (where V_m is the molar volume of the liquid) is known as the cohesion energy density. There is a thermodynamic relationship

$$dE = T\, dS - P\, dV \tag{6.31}$$

This equation can be written as

$$\frac{\partial E}{\partial V} = T\left(\frac{\partial S}{\partial V}\right)_T - P = T\left(\frac{\partial P}{\partial T}\right)_V - P \tag{6.32}$$

where P is the *external* pressure. The internal pressure, P_i, is given by

$$P_i = T\left(\frac{\partial P}{\partial T}\right)_V \tag{6.33}$$

which can also be written as

$$P_i = \left(\frac{(\partial V/\partial T)_P}{(\partial V/\partial P)_T}\right) \tag{6.34}$$

The quantity $(\partial V/\partial T)_P$ is the coefficient of thermal expansion and $(\partial V/\partial P)_T$ is the coefficient of compressibility of the liquid. For many liquids, the internal pressure is in the range of 2000–8000 atmospheres. Because the internal pressure is so much greater than the external pressure,

$$E_c = P_i - P \approx P_i \tag{6.35}$$

The solubility parameter, δ, is expressed in terms of the cohesion energy per unit volume by the equation

$$\delta = \sqrt{\frac{E_c}{V_m}} \tag{6.36}$$

where V_m is the molar volume. The dimensions for δ are $(\text{energy/volume})^{1/2}$ with suitable units being $[\text{cal}/(\text{cm}^3/\text{mol})]^{1/2}$ or $\text{cal}^{1/2}\ \text{cm}^{-3/2}\ \text{mol}^{-1}$, a unit, h, knows as 1 Hildebrand in honor of Joel Hildebrand who did pioneering research on the liquid state for many years. Many of the standard tables that give solubility parameters in those units. Because 1 cal = 4.184 J, the conversion factor from $\text{cal}^{1/2}\ \text{cm}^{-3/2}\ \text{mol}^{-1}$ to $\text{J}^{1/2}\ \text{cm}^{-3/2}\ \text{mol}^{-1}$ is $4.184^{1/2} = 2.045$. Both sets of units are in common use (sometimes both are indiscriminately referred to as h), and most of the extensive tables found in the older literature give values in $\text{cal}^{1/2}\ \text{cm}^{-3/2}\ \text{mol}^{-1}$. The solubility parameters in $\text{J}^{1/2}\ \text{cm}^{-3/2}\ \text{mol}^{-1}$ for some common liquids are shown in Table 6.7.

The cohesion energies of liquids determine their mutual solubility. If two liquids have greatly differing cohesion energies, they will not mix because each liquid has a greater affinity for molecules of its own kind than for those of the other liquid. Water ($\delta = 53.2\ h$) and carbon tetrachloride ($\delta = 17.6\ h$)

Table 6.7 Solubility Parameters for Liquids (in $\text{J}^{1/2}\ \text{cm}^{-3/2}\ \text{mol}^{-1}$)

Liquid	Solubility Parameter, h	Liquid	Solubility Parameter, h
C_6H_{14}	14.9	CS_2	20.5
CCl_4	17.6	CH_3NO_2	25.8
C_6H_6	18.6	Br_2	23.5
$CHCl_3$	19.0	$HCON(CH_3)_2$[a]	24.7
$(CH_3)_2CO$	20.5	C_2H_5OH	26.0
$C_6H_5NO_2$	23.7	H_2O	53.2
$n\text{-}C_5H_{12}$	14.5	CH_3COOH	21.3
$C_6H_5CH_3$	18.2	CH_3OH	29.7
XeF_2	33.3	XeF_4	30.9
$(C_2H_5)_2O$	15.8	$n\text{-}C_8H_{18}$	15.3
$(C_2H_5)_3B$	15.4	$(C_2H_5)_2Zn$	18.2
$(CH_3)_3Al$[b]	20.8	$(C_2H_5)_3Al$[b]	23.7
$(n\text{-}C_3H_7)_3Al$[b]	17.0	$(i\text{-}C_4H_9)_3Al$[b]	15.7

[a] N,N-dimethylformamide.
[b] These compounds are extensively dimerized.

provide an example that illustrates this principle. In contrast, methanol ($\delta = 29.7\ h$) and ethanol ($\delta = 26.0\ h$) are completely miscible.

Solubility parameters for many liquids are available in extensive tables (see references at the end of this chapter). To determine solubility parameters for liquids, we need to know the heat of vaporization. Over a reasonable temperature range, the relationship between the vapor pressure of a liquid and temperature is given by

$$\ln p = -\frac{\Delta H_{vap}}{RT} + C \tag{6.37}$$

where p is the vapor pressure, ΔH_{vap} is the heat of vaporization, T is the temperature (K), and C is a constant. If the vapor pressure is available at several temperatures, one way in which the heat of vaporization is determined is from the slope of the line that results when natural logarithm of the vapor pressure is plotted against $1/T$. Therefore, it is possible to evaluate the solubility parameter for a liquid if vapor pressure data are available and if the density is known so that the molar volume can be calculated.

Although Eqn (6.37) is commonly used to represent the vapor pressure as a function of temperature, it is by no means the best equation for the purpose. For many compounds, a more accurate representation of the vapor pressure is given by the Antoine equation,

$$\log p = A - \left(\frac{B}{C + t}\right) \tag{6.38}$$

In this equation, A, B, and C are constants that are different for each liquid, and t is the temperature in °C. Numerical procedures exist for determining the values for A, B, and C for a liquid if the vapor pressure is known at several temperatures. Making use of the equation

$$E_C = \Delta H_{vap} - RT \tag{6.39}$$

after determining the heat of vaporization from a plot of $\ln p$ vs. $(C + t)$, the cohesion energy can be expressed as

$$E_c = RT\left(\frac{2.303\ BT}{(C + t)} - 1\right) \tag{6.40}$$

This equation is the one most often used to calculate the cohesion energy of a liquid. From the molar mass and density of the liquid, the molar volume can be determined, and by means of Eqn (6.36), the value of δ can be determined. The importance of the solubility parameter for interpreting several types of interactions will now be illustrated.

The solubility parameter provides a way to assess the degree of cohesion within a liquid. The values for nonpolar liquids in which there are only relatively weak intermolecular forces are typically in the range of 15–18 h ($J^{1/2}\ cm^{-3/2}\ mol^{-1}$). This includes compounds such as CCl_4, C_6H_6, and alkanes. Molecules of these liquids interact only by London forces so there is no strong association of molecules. For a given series of molecules (such as alkanes), it is to be expected that the value of δ should increase slightly as the molecular weight increases. This trend is observed with the solubility parameter for

n-pentane being 14.5 h, whereas that for n-octane is 15.3 h. On the other hand, molecules of CH_3OH and C_2H_5OH interact not only by London forces but also by dipole–dipole forces and hydrogen bonding. As a result, the solubility parameters for these compounds are in the range of 25–30 h. It is clear that the solubility parameter can provide useful insight to the nature of the intermolecular forces in liquids.

In addition to the obvious uses of solubility parameters in predicting physical properties, it is also possible in some cases to study other types of intermolecular interactions. For example, the solubility parameter for triethylboron, $(C_2H_5)_3B$, is 15.4 h, whereas that for triethylaluminum, $(C_2H_5)_3Al$, is 23.7 h. Triethylboron is known from other studies not to associate whereas triethylaluminum exists as dimers.

Another important use of solubility parameters is in interpreting the effects of different solvents on the rates of reactions. In a chemical reaction, it is the concentration of the transition state that determines the rate of the reaction. Depending on the characteristics of the transition state, the solvent used can either facilitate or hinder its formation. For example, a transition state that is large and has little charge separation is hindered in its formation by using a solvent that has a high value of δ. The volume of activation is usually positive for forming such a transition state, which requires expansion of the solvent. A reaction of this type is the esterification of acetic anhydride with ethyl alcohol:

$$(CH_3CO_2)O + C_2H_5OH \rightarrow CH_3C(O)OC_2H_5 + CH_3COOH \qquad (6.41)$$

Because the transition state is a large aggregate with low charge separation, the rate decreases with increasing δ of the solvent. The rate of the reaction is almost 100 times as great in hexane ($\delta = 14.9\ h$) as it is in nitrobenzene ($\delta = 23.7\ h$).

When two reacting species form a transition state in which ions are being formed, the volume of activation is often negative. The formation of such a transition state is assisted by a solvent that has a high solubility parameter. The reaction

$$(C_2H_5)_3N + C_2H_5I \rightarrow (C_2H_5)_4N^+I^- \qquad (6.42)$$

is of this type, and it passes through a transition state in which charge is being separated. The formation of such a transition state is assisted by a solvent having a large solubility parameter. For this reaction, the rate constant increases approximately as a linear function of δ for several solvents. In the reaction

$$CH_3I + Cl^- \rightarrow CH_3Cl + I^- \qquad (6.43)$$

the transition state can be shown as

in which the -1 charge is spread over a large structure. Consequently, a solvent of high δ inhibits the formation of the transition state, and it is found that the rate constant for the reaction when dimethylformamide, $HCON(CH_3)_2$ ($\delta = 24.7\ h$) is the solvent is over 10^6 times as great as when the solvent is CH_3OH ($\delta = 29.7\ h$).

The cases described above serve to illustrate two important principles that relate the rates of reactions to the solubility parameter of the solvent. First, solvents having large δ values assist in the formation of transition states having high polarity or charge separation. Second, solvents having large δ values hinder the formation of transition states that are large nonpolar structures. However, a large number of properties of solvents have been used to try to correlate and interpret how changing solvents alter the rates of reactions. It is clear that the solubility parameter is one important consideration when interpreting the role of the solvent in reaction kinetics or choosing a solvent as a reaction medium. It is not appropriate to discuss this topic further in a general book such as this, but the references listed below should be consulted for additional details.

6.8 SOLVATOCHROMISM

The characteristics of solutions of iodine in organic solvents illustrate an important aspect of solution chemistry. Iodine vapor has a deep purple color as a result of the absorption of light in the visible region of the spectrum. The maximum of the absorption band is observed at 538 nm. However, solutions that contain iodine have absorption maxima that vary in position with the nature of the solvent. When I_2 is dissolved in a solvent such as CCl_4 or heptane, the solution has a blue-purple color. If iodine is dissolved in benzene or an alcohol, the solution has a brown color. This difference in color is the result of the change in the relative energy of the π and π^* orbitals of the I_2 molecule as a result of their interaction with the solvent. The interaction is very weak for solvents such as CCl_4 and heptane, so in those solvents the absorption maximum is observed in about the same position as it is for gaseous I_2. However, I_2 is a Lewis acid that interacts with electron pair donors. As a result, molecules that have unshared pairs of electrons such as alcohols or accessible electrons as in the case of the π system in benzene interact with I_2 molecules to perturb the molecular orbitals. The peak maximum is shifted to lower energy the more strongly the solvent interacts with the I_2 molecules. A change in the absorption spectrum (and hence a change in color) produced by a solvent is referred to as *solvatochromism*. A similar change in color that results from a change in temperature is referred to as *thermochromism*.

The nature and magnitude of the solvent–solute interaction depend on the molecular structures of the species. However, it should be apparent that this type of interaction provides a way to assess the interaction between a solute and the solvent. This is an extremely important area of chemistry with regard to understanding the role of the solvent as it relates to effects on solubility, equilibria, spectra, and rates of reactions. As a result, several numerical scales have been devised to correlate the effects of solvent interactions, some of which are based on solvatochromic effects. However, in most cases, complex dyes have been utilized as the probe solutes, but it is interesting to note that iodine also exhibits solvatochromism. Some transition metal complexes also exhibit solvatochromism as a result of changes in structure that depend on the nature of the solvent (see Section 18.9). In some cases, this can involve a change from square planar to tetrahedral geometry.

References for Further Study

Atkins, P.W., de Paula, J., 2002. *Physical Chemistry*, 7th ed. Freeman, New York. Chapter 21 of this well-known physical chemistry text gives a good introduction to intermolecular forces.

Connors, K.A., 1990. *Chemical Kinetics: The Study of Reaction Rates in Solution*. Wiley, New York. A valuable resource for learning about solvent effects on reactions.

Dack, M.J.R., 1975. In: Weissberger, A. (Ed.), *Techniques of Chemistry. Solutions and Solubilities*, vol. VIII. Wiley, New York. Detailed discussions of solution theory and the effects of solvents on processes.

Hamilton, W.C., Ibers, J.A., 1968. *Hydrogen Bonding in Solids*. W.A. Benjamin, New York. This book shows how hydrogen bonding is an important factor in the structure of solids.

Hildebrand, J., Scott, R., 1962. *Regular Solutions*. Prentice Hall, Englewood Cliffs, NJ. One of the standard reference texts on theory of solutions.

Hildebrand, J., Scott, R., 1949. *Solubility of Non-Electrolytes*, 3rd ed. Reinhold, New York. The classic book on solution theory.

House, J.E., 2007. *Principles of Chemical Kinetics*, 2nd ed. Elsevier/Academic Press, San Diego. Chapters 5 and 9 contain discussions of factors affecting reactions in solution and the influence of solubility parameter of the solvent on reaction rates.

Israelachvili, J., 1991. *Intermolecular and Surface Forces*, 2nd ed. Academic Press, San Diego, CA. Good coverage of intermolecular forces.

Jeffrey, G.A., 1997. *An Introduction to Hydrogen Bonding*. Oxford University Press, New York. An excellent, modern treatment of hydrogen bonding and its effects. Highly recommended.

Joesten, M.D., Schaad, L.J., 1974. *Hydrogen Bonding*. Marcel Dekker, New York. A good survey of hydrogen bonding.

Parsegian, V.A., 2005. *Van der Waals Forces: A Handbook for Biologists, Chemists, Engineers, and Physicists*. Cambridge University Press, New York. A good reference on a topic that pervades all areas of chemistry.

Pauling, L., 1960. *The Nature of the Chemical Bond*, 3rd ed. Cornell University Press, Ithaca, NY. This classic monograph in bonding theory also presents a great deal of information on hydrogen bonding.

Pimentel, G.C., McClellan, A.L., 1960. *The Hydrogen Bond*. Freeman, New York. The classic book on all aspects of hydrogen bonding. It contains an exhaustive survey of the older literature.

Reid, R.C., Prausnitz, J.M., Sherwood, T.K., 1977. *The Properties of Gases and Liquids*. McGraw-Hill, New York. This book contains an incredible amount of information on the properties of gases and liquids. Highly recommended.

QUESTIONS AND PROBLEMS

1. For OF_2 and H_2O, the bond angles are $103°$ and $104.5°$, respectively. However, the dipole moments of these molecules are 0.30 and 1.85 D, respectively. Explain the factors that cause such a large difference in the dipole moments of the molecules.

2. Explain why the heat of interaction between C_6H_5OH and R_2O is considerably larger than that for C_6H_5OH interacting with R_2S.

3. Explain why m-$NO_2C_6H_4OH$ and p-$NO_2C_6H_4OH$ have different acid strengths. Which is stronger? Why?

4. In a certain solvent (A), the O–H stretching band of methanol is observed at 3642 cm^{-1}. In that solvent, the heat of reaction of methanol with pyridine is $-36.4 \text{ kJ mol}^{-1}$. In another solvent (B), the O–H stretching band is observed at 3620 cm^{-1} and the heat of reaction with pyridine is $-31.8 \text{ kJ mol}^{-1}$.

   (a) Write the equation for the interaction between methanol and pyridine.

   (b) Explain the thermodynamic data using a completely labeled thermochemical cycle as part of your discussion.

5. The boiling points of methanol and cyclohexane are 64.7 and $80.7 \,°C$, respectively, and their heats of vaporization are 34.9 and 30.1 kJ mol^{-1}. Determine the entropy of vaporization for these liquids and explain the difference between them.

6. The boiling point of CH_3OH is 64.7 °C and that of CH_3SH is 6 °C. Explain this difference.

7. The viscosities for three liquids at various temperatures are given below in centipoise (cp):

Temperature, °C	10	20	30	40	60
C_8H_{18}: η, cp	6.26	5.42	4.83	4.33	2.97
CH_3OH: η, cp	6.90	5.93	–	4.49	3.40
C_6H_6: η, cp	7.57	6.47	5.61	4.36 (50°)	–

(a) Explain why the viscosity of CH_3OH (formula weight 32) is approximately the same as that of octane (formula weight 114).

(b) Explain why the viscosity of C_6H_6 is higher than that of C_8H_{18} despite the difference in formula weights.

(c) Make an appropriate graph for each liquid to determine the activation energy of viscous flow.

(d) Explain the values for the activation energy of viscous flow in terms of intermolecular forces.

8. The viscosities in centipoise for three liquids at various temperatures are given below:

Temperature, °C	Viscosity, cp		
	C_6H_{14}	$C_6H_5NO_2$	i-C_3H_7OH
0	4.012	28.2	45.646
20	3.258	19.8	23.602
35	–	15.5	–
40	2.708	–	13.311
60	2.288	–	–
80	–	–	5.292

(a) Explain why the viscosity of C_6H_{14} is so much lower than that of nitrobenzene.

(b) Explain why the viscosity of i-C_3H_7OH is higher than that of C_6H_{14} despite the significant difference in formula weights.

(c) Make an appropriate graph for each liquid to determine the activation energy of viscous flow.

(d) Explain the values for the activation energy of viscous flow in terms of intermolecular forces.

9. Explain why acetic acid is extensively dimerized in dilute solutions when benzene is the solvent but not when the solvent is water.

10. (a) Phenol, C_6H_5OH, hydrogen bonds to both diethyl ether, $(C_2H_5)_2O$, and diethyl sulfide, $(C_2H_5)_2S$. In one case, the OH stretching band is shifted by 280 cm^{-1}, and in the other, the shift is 250 cm^{-1}. Match the band shifts to the electron pair donors and explain your answer.

(b) The strengths of the hydrogen bonds between phenol and the two electron pair donors are 15.1 and 22.6 kJ mol^{-1}, respectively. Match the bond strengths to the donors and explain your answer.

11. Explain why the solubility of NaCl in CH_3OH is 0.237 grams per 100 grams of solvent, although in C_2H_5OH the solubility is only 0.0675 grams per 100 grams of solvent. Estimate the solubility of NaCl in i-C_3H_7OH.

12. Use structural arguments to explain why SF_4 has a boiling point of −40 °C, whereas SF_6 sublimes at −63.8 °C.

13. Molecules of both Br_2 and ICl have 70 electrons and one of the substances boils at 97.4 °C and the other boils at 58.8 °C. Which has the higher boiling point? Explain your answer.

14. Molecules of H_2S, Ar, and HCl have 18 electrons. Boiling points for these substances are −84.9, −60.7, and −185.7 °C. Match the boiling points to the compounds and explain your answer.

15. Explain why hydrazine, N_2H_4, has a viscosity of 0.97 cp at 20 °C, whereas hexane, C_6H_{14}, has a viscosity of 0.326 cp at the same temperature.

16. Although fluorobenzene and phenol have almost identical formula weights, one has a viscosity of 2.61 cp at a temperature of 60 °C, whereas the other has a viscosity of 0.389 cp at the same temperature. Assign the viscosities to the correct liquids and explain your answer.

17. The solubility of NO in water at 25 °C is more than twice that of CO. Explain this difference in terms of the difference in the structures of the molecules.

18. Explain in terms of molecular structure the following trend in solubilities of the gases in water: $C_2H_2 >> C_2H_4 > C_2H_6$. How does the trend in solubility reflect other chemical properties?

19. Make a sketch showing a water molecule approaching a molecule of Xe and how that affects the electron cloud of the Xe atom. Use this sketch to explain how the Xe$-$H$_2$O pair would interact with an additional water molecule.

20. Ammonium salts such as NH_4Cl are sometimes observed to undergo an abrupt change in heat capacity (and hence entropy) at some temperature below the melting point. Describe the processes likely responsible for these observations.

21. Use the bond moments shown in Table 6.2 and the fact that the bond angle in H_2S is 92.2° to calculate the dipole moment of the molecule.

22. Suppose the O$-$H bond dipole is that predicted by the percent ionic character. If the bond angle is 104.4°, calculate the approximate dipole moment of water.

23. The dipole moment of SiH_3Cl is 1.31 D, whereas that of CH_3Cl is 1.87 D. Explain this difference.

24. Using the electronegativities of H and Se, calculate the percent ionic character to the H$-$Se bond. If the bond lengths are 146 pm and the bond angle is 91°, what would be the dipole moment of H_2Se?

25. Using molecular structures and principles of intermolecular forces, explain why liquid BrF_3 has a viscosity that is approximately three times that of BrF_5 at a temperature of 20 °C.

26. Explain why the viscosities of m-ClC_6H_4OH and p-ClC_6H_4OH at 45 °C are approximately equal, whereas that of o-ClC_6H_4OH is about half as large.

27. Explain why the heat of vaporization of 2-pentanone is 33.4 kJ mol^{-1}, whereas that of 2-pentanol is 41.4 kJ mol^{-1}.

28. The viscosity of 1-chloropropane is only about one-seventh that of 1-propanol. Explain this difference in terms of intermolecular forces.

CHAPTER 7

Ionic Bonding and Structures of Solids

Molecular geometry and bonding in covalent molecules are covered in detail in several courses in chemistry, including inorganic chemistry. A thorough knowledge of these topics is essential for interpreting properties of inorganic compounds and predicting their reactions. However, one should not lose sight of the fact that a great many inorganic materials are solids. Some covalent solids were described briefly in Chapter 4, but many others are metals or crystals that are essentially ionic in nature. In order to deal with the chemistry of these materials, it is necessary to be familiar with the basic crystal structures and the forces that hold them together. Consequently, this chapter will present a survey of ionic bonding, describe the structures of several types of crystals, and explain the structures of metals. Crystals are never perfectly regular, so it is also essential to discuss the types of defects that occur in the structures of ionic salts and metals.

Although ionic crystals are held together by electrostatic forces, the ions become separated when the solid dissolves. Ions are strongly attracted to the ends of polar molecules that have charges opposite to those of the ions. Because the dissolution of solids is intimately related to their chemical behavior, the energy relationships for dissolving ionic solids will also be discussed. Moreover, the proton affinities of some anions can be determined by means of thermodynamic studies on the decomposition of solids, as will be illustrated in this chapter. This chapter will provide an overview of several aspects of structures and bonding in inorganic solids. Transformation of materials in the solid state is a growing area of importance, and Chapter 8 will deal with the behavior of solids from the standpoint of the rates and mechanisms of such processes.

7.1 ENERGETICS OF CRYSTAL FORMATION

When ions are formed by electron transfer, the resulting charged species interact according to Coulomb's law:

$$F = \frac{q_1 q_2}{\varepsilon r^2} \qquad (7.1)$$

In this equation, q_1 and q_2 are the charges, r is the distance separating them, and ε is the dielectric constant, which has a value of 1 for a vacuum or free space. This force law has no directional component and operates equally well in any direction. Therefore, we are concerned primarily with the energies of ionic bond formation, although the arrangement of ions in the lattice is of considerable importance.

Sodium chloride is formed from the elements in their standard states with a heat of formation of -411 kJ mol^{-1}.

$$Na(s) + \tfrac{1}{2}Cl_2(g) \rightarrow NaCl(s) \quad \Delta H_f^o = -411 \text{ kJ mol}^{-1} \qquad (7.2)$$

Inorganic Chemistry. DOI: http://dx.doi.org/10.1016/B978-0-12-385110-9.00007-8

This process can be represented as taking place in a series of steps, each of which has a well-known enthalpy associated with it. The application of Hess' law provides a useful way to obtain the enthalpy for the overall process because it is independent of the path.

The enthalpy of formation of a compound is a so-called thermodynamic state function, which means that the value depends only on the initial and final states of the system. When the formation of crystalline NaCl from the elements is considered, it is possible to consider the process as if it occurred in a series of steps that can be summarized in a thermochemical cycle known as a Born–Haber Cycle. In this cycle, the overall heat change is the same regardless of the pathway that is followed between the initial and final states. Although the *rate* of a reaction depends on the pathway, the enthalpy change is a function of initial and final states only, not the pathway between them. The Born–Haber cycle for the formation of sodium chloride is shown as follows.

In this cycle, S is the sublimation enthalpy of Na, D is the dissociation enthalpy of Cl_2, I is the ionization potential of Na, E is the energy released when an electron is added to a Cl atom, and U is the lattice energy.

Sometimes, the unknown quantity in the cycle is the lattice energy, U. From the cycle shown above, we know that the heat change is the same regardless of the pathway by which NaCl(s) is formed. Therefore, we see that

$$\Delta H_f^o = S + \frac{1}{2}D + I + E - U. \tag{7.3}$$

Solving this equation for U we obtain

$$U = S + \frac{1}{2}D + I + E - \Delta H_f^o. \tag{7.4}$$

Using the appropriate data for the formation of sodium chloride, U (kJ mol^{-1}) $= 109 + 121 + 496 - 349 - (-411) = 786$ kJ mol^{-1}. Although this is a useful approach for determining the lattice energy of a crystal, the electron affinity of the atom gaining the electron is difficult to measure experimentally. In fact, it may be impossible to measure the heat associated with an atom gaining two electrons, so the only way to obtain a value for the second electron affinity is to calculate it. As a result, the Born–Haber cycle is often used in this way, and this application of a Born–Haber cycle will be illustrated later in this chapter. In fact, electron affinities for some atoms are available only as values calculated by this procedure, and they have not been determined experimentally.

Consider the process in which one mole of Na$^+$(g) and one mole of Cl$^-$(g) are allowed to interact to produce one mole of ion pairs rather than the usual three-dimensional lattice. Energy will be released as the positive and negative charges get closer together. In this case, the energy released is approximately -439 kJ mol^{-1} of ion pairs formed when the internuclear distance is the same as it is in the crystal, 279 pm (2.79 Å). However, if one mole of Na$^+$(g) and one mole of Cl$^-$(g) are allowed to form a mole of solid crystal, the energy released is about -786 kJ. The term *lattice energy* is applied to the process of separating one mole of crystal into the gaseous ions. Therefore, if the energy *released* when a mole of crystal *forms* from the gaseous ions is -786 kJ, the energy *absorbed* when one mole of crystal is *separated* into the gaseous ions will be $+786$ kJ. This convention is exactly the same as that which applies to covalent bond energies. If we divide the lattice energy by the energy released when ion pairs form, we find that -786 kJ mol^{-1}/-439 kJ mol^{-1} = 1.78. This ratio is of particular importance in considering crystal energies, and it is known as the *Madelung constant*, which will be discussed later.

As we have seen, several atomic properties are important when considering the energies associated with crystal formation. Ionization potentials and heats of sublimation for the metals, electron affinities, and dissociation energies for the nonmetals and heats of formation of alkali halides are shown in Tables 7.1 and 7.2.

When a single ion having a $+1$ charge and one having a -1 charge are brought closer together, the electrostatic energy of the interaction can be represented by the equation

$$E = -\frac{e^2}{r} \tag{7.5}$$

where e is the charge on the electron and r is the distance of separation. If the number of ions of each type is increased to Avogadro's number, N_o, the energy released will be N_o times the energy for one ion pair.

$$E = -\frac{N_o e^2}{r} \tag{7.6}$$

It is easy to use this equation to determine what the attraction energy would be for one mole of $Na^+(g)$ interacting with one mole of $Cl^-(g)$ at a distance of 2.79 Å (279 pm). We will calculate the value first in ergs and then convert the result into kJ. Because the charge on the electron is 4.8×10^{-10} esu and 1 esu $= 1$ g$^{1/2}$ cm$^{3/2}$ s^{-1}, the attraction energy is

$$E = \frac{(6.02 \times 10^{23})(4.8 \times 10^{-10} \text{ g}^{1/2} \text{ cm}^{3/2} \text{ sec}^{-1})^2}{2.79 \times 10^{-8} \text{ cm}} = 4.97 \times 10^{12} \text{ erg} = 4.97 \times 10^5 \text{ J} = 497 \text{ kJ}$$

Table 7.1 Ionization Potentials and Heats of Sublimation of Alkali Metals and the Heats of Formation of Alkali Metal Halides

| Element | I^a, (kJ mol^{-1}) | S^b, (kJ mol^{-1}) | ΔH_f^o (kJ mol^{-1}) of Halide MX | | | |
			X = F	X = Cl	X = Br	X = I
Li	518	160	605	408	350	272
Na	496	109	572	411	360	291
K	417	90.8	563	439	394	330
Rb	401	83.3	556	439	402	338
Cs	374	79.9	550	446	408	351

[a]*Ionization potential of the metal.*
[b]*The heat of sublimation of the metal.*

Table 7.2 Electron Affinities and Dissociation Energies for Halogens

Element	Electron Affinity (kJ mol^{-1})	Dissociation Energy (kJ mol^{-1})
F_2	333	158
Cl_2	349	242
Br_2	324	193
I_2	295	151

We have already mentioned that for sodium chloride approximately 1.78 times as much energy is released when the crystal lattice forms as when ion pairs form. This value, the Madelung constant (A) for the sodium chloride lattice, could be incorporated to predict the total energy released when one mole of NaCl crystal is formed from the gaseous Na^+ and Cl^- ions. The result would be

$$E = -\frac{N_oAe^2}{r} \tag{7.7}$$

Multiplying the result obtained above for ion pairs by 1.75 indicates that the lattice energy would be 875 kJ mol^{-1}, but the actual value is 786 kJ mol^{-1}. The question to be answered is why does forming the crystal release less energy than expected?

A sodium ion has 10 electrons surrounding the nucleus, and a chloride ion has 18 electrons. Even though the sodium ion carries a positive charge and the chloride ion has a negative charge, as the ions get closer together, there is repulsion between the electron clouds of the two ions. As a result, the calculated attraction energy is larger than the lattice energy. The equation for the lattice energy should take into account the repulsion between ions that increases as the distance between them becomes smaller. This can be done by adding a term to the expression that gives the attraction energy to take into account the repulsion as a function of distance. The repulsion is expressed as

$$R = \frac{B}{r^n} \tag{7.8}$$

where B and n are constants and r is the distance separating the ion centers. The value of n depends on the number of electrons surrounding the ions, and it is usually assigned the values 5, 7, 9, 10, or 12 for ions having the electron configurations of He, Ne, Ar, Kr, or Xe, respectively. For example, if the crystal is NaF, an n value of 7 is appropriate. If the crystal is one in which the cation has the configuration of one noble gas but the anion has a different noble gas configuration, an average value for n is chosen. For example, Na^+ has the configuration of Ne ($n=7$), but Cl^- has the configuration of Ar ($n=9$). Therefore, a value of 8 can be used in calculations for NaCl. When the repulsion is included, the lattice energy U is expressed as

$$U = -\frac{N_oAe^2}{r} + \frac{B}{r^n} \tag{7.9}$$

We know the values of all the quantities shown in this equation except B.

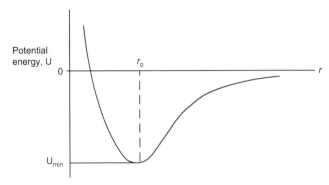

FIGURE 7.1
Variation of potential energy with distance separating a cation and an anion.

When positive and negative ions are relatively far apart, the overall electrostatic charges (which lead to attraction) dominate the interaction. If the ions are forced very close together, there is repulsion. At some distance, the energy is most favorable, which means that the total energy is a minimum at that distance as shown in Figure 7.1.

To find where the minimum energy occurs, we take the derivative dU/dr and set it equal to zero.

$$\frac{dU}{dr} = 0 = \frac{N_oAe^2}{r^2} - \frac{nB}{r^{n+1}} \tag{7.10}$$

When we solve for B, we obtain

$$B = \frac{N_oAe^2 r^{n-1}}{n} \tag{7.11}$$

We have approached this problem in terms of the lattice *forming* from the ions. However, the lattice energy is defined in terms of the energy required to separate it into the gaseous ions. Therefore, as used in Eqn (7.9), the value for U will be negative because the attraction energy is much larger than the repulsion energy at the usual internuclear distance. When we substitute the value for B into Eqn (7.9) and change the sign to show the separation of the crystal into gaseous ions, we obtain

$$U = \frac{N_oAe^2}{r}\left(1 - \frac{1}{n}\right) \tag{7.12}$$

This equation is known as the Born–Landé equation, and it is very useful for calculating lattice energies of crystals when the values are known for A, r, and n. If ions having charges other than $+1$ and -1 form the lattice, the charges on the cation and anion, Z_c and Z_a, must be included in the numerator of the factor preceding $(1 - 1/n)$ to give $(Z_cZ_aN_oAe^2/r)(1 - 1/n)$.

7.2 MADELUNG CONSTANTS

We have already defined the Madelung constant as the ratio of the energy released when a mole of crystal forms from the gaseous ions to that released when ion pairs form. In order to understand what

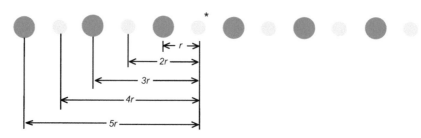

FIGURE 7.2
A "lattice" composed of a chain of alternating positive (yellow) and negative (green) ions.

this means, we will consider the following example. Suppose that a mole of Na^+ and a mole of Cl^- ions form a mole of ion pairs where the internuclear distance is r. As was shown earlier, the energy of interaction would be $-N_o e^2/r$, where the symbols have the same meaning as in Eqn (7.12). Now let us arrange a mole of Na^+ and a mole of Cl^- ions in a chain structure such as that shown in Figure 7.2.

It is now necessary to calculate the energy of interaction for the ions in this arrangement. The energy of interaction of a charge q located in an electric field of strength V that has an opposite charge is given by $-Vq$. The procedure that we will use to calculate the interaction energy for the chain of ions involves calculating the electric field strength, V, at the reference ion $+^*$ that is produced by all the other ions. Then, the total interaction energy will be given by Ve, where e is the charge on the electron. In the arrangement of ions shown above, the cation labeled as $+^*$ is located at a distance of r from two negative ions, so the contribution to the electric field potential at $+^*$ is $-2e/r$. However, two positive ions are located at a distance of $2r$ from the $+^*$ ion, which gives a contribution of $+2e/2r$ to the electric field strength. Continuing outward from $+^*$ we find two anions at a distance of $3r$, and the contribution to the field strength is $-2e/3r$. If we continue working outward from the reference ion $+^*$, we would find that the contributions to the electric field strength can be represented as a series, which can be written as

$$V = -\frac{2e}{r} + \frac{2e}{2r} - \frac{2e}{3r} + \frac{2e}{4r} - \frac{2e}{5r} + \frac{2e}{6r} - \cdots \tag{7.13}$$

If we factor out $-e/r$, the field strength V can be written as

$$V = -\frac{e}{r}\left(2 - 1 + \frac{2}{3} - \frac{2}{4} + \frac{2}{5} - \frac{2}{6} + \cdots\right) \tag{7.14}$$

The series inside the parentheses converges to a sum that is $2 \ln 2$ or 1.38629. This value is the Madelung constant for a hypothetical chain consisting of Na^+ and Cl^- ions. Thus, the total interaction energy for the chain of ions is $-1.38629 \, N_o e^2/r$, and the *chain* is more stable than *ion pairs* by a factor of 1.38629, the Madelung constant. Of course NaCl does not exist in a chain, so there must be an even more stable way of arranging the ions.

In the preceding illustration, the series happened to be one that is recognizable for which the sum is known. A more likely situation is that the series might be made up of terms that do not lead to a recognizable sum and may not converge quickly. In that case, a numerical procedure for finding the value to which the series converges is available, and it will be described by using the example above. The procedure consists of finding the sums progressively by summing the first term (only one value, 2.0000) then adding one term each time. After the sums (A) for several terms are found, the averages are obtained for adjacent sums and written in another column. Then, the average values for the previous averages are obtained and written in a different column. In each step, there is one less average value than the number of sums being taken. Eventually, only one average value results, and that gives an approximation to the summation of the series. The procedure is a numerical converging process that is illustrated below where the subscripts on A give the numbers of the terms being included in the summation. For example, $A_{1,2,3}$ indicates that the summation is over terms 1, 2, and 3 of the series.

		Averages			
A_1	$= 2.0000$				
		1.5000			
$A_{1,2}$	$= 1.0000$		1.4167		
		1.3384		1.3959	
$A_{1,2,3}$	$= 1.6667$		1.3751		1.3896
		1.4167		1.3834	
$A_{1,2,3,4}$	$= 1.1667$		1.3917		
		1.3917			
$A_{1,2,3,4,5}$	$= 1.5667$				

Note that in this case only five terms have been included in the partial summation, but the averaging procedure produces an approximate convergence to a value of 1.3896. This value differs only insignificantly from the correct sum of 1.38629, which was obtained earlier as the value 2 ln 2. Finding the Madelung constant for a crystal that has a three-dimensional lattice is by no means this easy. However, the numerical convergence technique illustrated above is a very useful technique in cases where it can be applied. For three-dimensional lattices, it becomes much more difficult to determine the distances of ions from the reference ion used as the initial point in determining the electric field strength. Although the hypothetical chain structure does not correspond to an actual crystal, it provides a convenient model to show how a Madelung constant can be obtained, but it is difficult when the lattice has three dimensions.

When a three-dimensional crystal lattice forms from ions, each ion is surrounded by several nearest neighbors, the number and geometrical distribution of which depend on the type of crystal structure. The Madelung constant takes into account the interaction of an ion with all the other ions rather than with only one ion of opposite charge. As a result, it has a numerical value that depends on the structure of the crystal. Consider the layer of ions shown in Figure 7.3 that shows the sodium chloride arrangement. Keep in mind that the layer below the one shown has ions of opposite charge directly below those in this layer, and the layer above the one shown will be arranged like the one below the layer shown. We begin by starting with the reference ion $+^*$ and work outward to determine the contributions to the electric field strength at that point.

First, there are six negative ions surrounding $+^*$ at a distance r, four in the layer shown, and one above the page and one below the page. These six negative ions generate a potential of $-6e/r$. Next, there are 12 cations at a distance of $2^{1/2}r$, four in the layer shown, four in the layer below the page, and

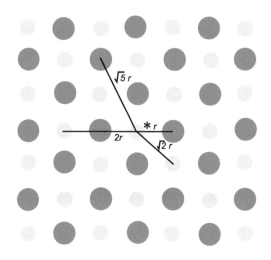

FIGURE 7.3
A layer of ions (chloride is shown in green and sodium is shown in yellow) in the sodium chloride crystal structure.

four in the layer above the page. The eight cations that are above and below the layer shown are directly below and above the four anions shown closest to $+*$. These 12 cations generate a contribution to the field that is expressed by $12e/2^{1/2}r$. Working outward from $+*$, we next encounter eight anions at a distance of $3^{1/2}r$, which give rise to a contribution to the field that is expressed as $-8e/3^{1/2}r$. There are six cations at a distance of $2r$, which give rise to a contribution of $6e/2r$. We could continue working outward and obtain many other terms in the series. The series that represents V can be written as

$$V = -\frac{6e}{r} + \frac{12e}{\sqrt{2}r} - \frac{8e}{\sqrt{3}r} + \frac{6e}{2r} + \cdots \tag{7.15}$$

from which we obtain

$$V = -\frac{e}{r}\left(6 - \frac{12}{\sqrt{2}} + \frac{8}{\sqrt{3}} - 3 + \cdots\right) \tag{7.16}$$

Table 7.3 Madelung Constants for Some Common Crystal Lattices

Crystal Type	Madelung Constant[a]
Sodium chloride	1.74756
Cesium chloride	1.76267
Zinc blende	1.63806
Wurtzite	1.64132
Rutile	2.408
Fluorite	2.51939

[a]Rutile and fluorite have twice as many anions as cations. The factor of 2 is not included in the numerical value shown.

In this series, the terms neither lead to a recognizable series nor converge very rapidly. In fact, it is a rather formidable process to determine the sum, but the value obtained is 1.74756. Note that this is approximately equal to the value given earlier for the ratio of the energy released when a crystal forms to that when only ion pairs form. As stated earlier, the Madelung constant is precisely that ratio.

Details of the calculation of Madelung constants for all the common types of crystals are beyond the scope of this book. When the arrangement of ions differs from that present in NaCl, the number of ions surrounding the ion chosen as a starting point and the distances between them may be difficult to determine. They will most certainly be much more difficult to represent as a simply factor of the basic distance between a cation and an anion. Therefore, each arrangement of ions (crystal type) will have a different value for the Madelung constants, and the values for several common types of crystals are shown in Table 7.3.

7.3 THE KAPUSTINSKII EQUATION

Although the Born—Landé equation provides a convenient way to evaluate the lattice energy of many crystals, it has some limitations. First, the crystal structure must be known so that the appropriate Madelung constant can be chosen. Second, some ions are not spherical (for example, NO_3^- is planar, SO_4^{2-} is tetrahedral, etc.), so the distance between ion centers may be different in different directions. As a result, another way to calculate lattice energies is needed. One of the most successful approaches to calculating lattice energies for a wide range of crystals is provided by the Kapustinskii equation

$$U(\text{kJ mol}^{-1}) = \frac{120,200 \, mZ_cZ_a}{r_c + r_a} \left(1 - \frac{34.5}{r_c + r_a}\right) \tag{7.17}$$

In this equation, r_c and r_a are the radii of the cation and anion in pm, Z_a and Z_c are their charges, and m is the number of ions in the formula for the compound. Note that the Madelung constant does not appear in the Kapustinskii equation and that we need only the sum of the ionic radii (distance between ion centers), not the individual radii. It is not necessary to know the crystal structure (and hence the Madelung constant) for the crystal in order to use the Kapustinskii equation. The equation gives highly reliable calculated values for crystals where the bonding is almost completely ionic (as is the case for NaCl, KI, etc.). If there is a substantial amount of covalence (as there is in cases such as AgI or CuBr), the calculated lattice energy does not agree well with the actual value. The large ions in compounds of this type are easily deformed (polarizable), so they also have a substantial attraction that results from their having charge separations because of being distorted. As was described in Chapter 6, the van der Waals forces are significant between ions of this type. In spite of these limitations, the Kapustinskii equation provides a useful way to calculate crystal energies.

There is another use of the Kapustinskii equation that is perhaps even more important. For many crystals, it is possible to determine a value for the lattice energy from other thermodynamic data or the Born—Landé equation. When that is done, it is possible to solve the Kapustinskii equation for the sum of the ionic radii, $r_a + r_c$. When the radius of one ion is known, carrying out the calculations for a series of compounds that contains that ion enables the radii of the counterions to be determined. In other words, if we know the radius of Na^+ from other measurements or calculations, it is possible to determine the radii of F^-, Cl^-, and Br^- if the lattice energies of NaF, NaCl, and NaBr are known. In fact, a radius could be determined for the NO_3^- ion if the lattice energy of $NaNO_3$ was known. Using this approach, based on

Table 7.4 Radii of Common Monatomic and Polyatomic Ions

Singly Charged		Doubly Charged		Triply Charged	
Ion	r, pm	Ion	r, pm	Ion	r, pm
Li^+	60	Be^{2+}	30	Al^{3+}	50
Na^+	98	Mg^{2+}	65	Sc^{3+}	81
K^+	133	Ca^{2+}	94	Ti^{3+}	69
Rb^+	148	Sr^{2+}	110	V^{3+}	66
Cs^+	169	Ba^{2+}	129	Cr^{3+}	64
Cu^+	96	Mn^{2+}	80	Mn^{3+}	62
Ag^+	126	Fe^{2+}	75	Fe^{3+}	62
NH_4^+	148	Co^{2+}	72	N^{3-}	171
F^-	136	Ni^{2+}	70	P^{3-}	212
Cl^-	181	Zn^{2+}	74	As^{3-}	222
Br^-	195	O^{2-}	145	Sb^{3-}	245
I^-	216	S^{2-}	190	PO_4^{3-}	238
H^-	208	Se^{2-}	202	SbO_4^{3-}	260
ClO_4^-	236	Te^{2-}	222	BiO_4^{3-}	268
BF_4^-	228	SO_4^{2-}	230		
IO_4^-	249	CrO_4^{2-}	240		
MnO_4^-	240	BeF_4^{2-}	245		
NO_3^-	189	CO_3^{2-}	185		
CN^-	182				
SCN^-	195				

thermochemical data, to determine ionic radii yields values that are known as *thermochemical radii*. For a planar ion such as NO_3^- or CO_3^{2-}, it is a sort of average or effective radius, but it is still a very useful quantity. For many of the ions shown in Table 7.4, the radii were obtained by precisely this approach.

7.4 IONIC SIZES AND CRYSTAL ENVIRONMENTS

It is apparent from the data shown in Table 7.4 that there is a great difference between the sizes of some ions. For example, the ionic radius of Li^+ is 60 pm, whereas that of Cs^+ is 169 pm. When these ions are forming a crystal with Cl^-, which has a radius of 181 pm, it is easy to understand that the geometrical arrangement of ions in the crystals may be different even though both LiCl and CsCl have equal numbers of cations and anions in the formulas.

When spherical objects are stacked to produce a three-dimensional array (crystal lattice), the relative sizes of the spheres determine what types of arrangements are possible. It is the interaction of the cations and anions by electrostatic forces that leads to stability of any ionic structure. Therefore, it is essential that each cation be surrounded by several anions and each anion be surrounded by several cations. This local arrangement is largely determined by the relative sizes of the ions. The number of ions of opposite charge surrounding a given ion in a crystal is called the *coordination number*. This is actually not a very good term because the bonds are not *coordinate* bonds (see Chapter 16). For a specific cation, there will be a limit to the number of anions that can surround the cation because the anions will be touching each other. The opposite is also true, but the problem is greater with the anions touching because most anions are larger than most cations.

Consider the arrangement of ions in which six anions surround a cation as shown in Figure 7.4. In this arrangement, the six anions have their centers at the corners of an octahedron with the cation at the center. There are four anions whose centers are in the same plane as the center of the cation in addition to one anion above and below the plane.

Calculating the minimum size for the cation that can be in contact with the six anions as the anions are just touching each other is a simple problem. The critical factor is the relative sizes of the ions, which is expressed as the radius ratio, r_c/r_a. In the development that follows, the ions are considered to be hard spheres, which is not exactly correct.

The geometry of the arrangement shown in Figure 7.4 is such that θ is 45°. With this arrangement, the four anions shown just touch the cation and just touch each other. Because $S = r_c + r_a$, the relationship between the distances is

$$\cos 45° = \frac{\sqrt{2}}{2} = \frac{r_c}{S} = \frac{r_c}{r_c + r_a} \tag{7.18}$$

The right hand side of this equation can be expanded, after which solving for r_c/r_a gives a value of 0.414. The significance of this value is that if the cation is smaller than 0.414 r_a, this arrangement will probably not be stable because the anions will be touching each other without touching the cation. A radius ratio of at least 0.414 is required for all six of the anions to touch the cation. It is possible to consider an arrangement in which four anions surround a cation in a tetrahedral arrangement. Performing a calculation similar to that just described leads to the conclusion that the four anions will touch the cation only when r_c is at least 0.225 r_a. Earlier it was shown that a radius ratio of at least 0.414 is required to fit six anions around a cation when both are considered as hard spheres. Therefore, it can be seen that when $0.225 < r_c/r_a < 0.414$, a tetrahedral arrangement of anions around each cation should result. Similar calculations can easily be carried out to determine values for r_c/r_a that lead to other stable arrangements of ions in other crystal environments. Keep in mind that this is only a guide because ions are not actually hard spheres. The results of the calculations are summarized in Table 7.5.

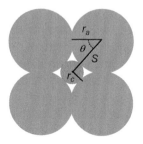

FIGURE 7.4
Anions surrounding a cation in an octahedral arrangement. Only the four ions in a plane are shown.

Table 7.5 Arrangements of Ions Predicted to be Stable from Radius Ratio Values

r_c/r_a	Cation Environment	Number of Nearest Neighbors	Example
1.000	*fcc* or *hcp*[a]	12	Ni, Ti
0.732–1.000	Cubic	8	CsCl
0.414–0.732	Octahedral	6	NaCl
0.225–0.414	Tetrahedral	4	ZnS
0.155–0.225	Trigonal	3	–
0.155	Linear	2	–

[a]*fcc and hcp are face-centered cubic and hexagonal close packing of spheres (see Section 7.8).*

Based on the ionic radii, nine of the alkali halides should not have the sodium chloride structure. However, only three, CsCl, CsBr, and CsI, do not have the sodium chloride structure. This means that the hard sphere approach to ionic arrangement is inadequate. It should be mentioned that it does predict the correct arrangement of ions in the majority of cases. It is a *guide*, not an infallible *rule*. One of the factors that is not included is related to the fact that the electron clouds of ions have some ability to be deformed. This *electronic polarizability* leads to additional forces of the types to be discussed in the next chapter. Distorting the electron cloud of an anion leads to part of its electron density being drawn toward the cations surrounding it. In essence, there is some sharing of electron density as a result. As a result, the bond has become partially covalent.

Although the structure of CsCl is quite different from that of NaCl, even CsCl can be transformed into the sodium chloride structure when heated to temperatures above 445 °C. Some of the other alkali halides that do not have the sodium chloride structure under ambient conditions are converted to that structure when subjected to high pressure. Many solid materials exhibit this type of polymorphism, which depends on the external conditions. Conversion of a material from one structure to another is known as a *phase transition*.

One interesting thing to note is that in cases where the radius ratio does not predict the correct structure, it usually fails when its value is close to one of the limiting values between two predicted structures. A value of r_c/r_a of 0.405 is very close to the limiting value for coordination numbers of 4 and 6. Subtle factors could cause the environment around the cation to be either 4 or 6, even though strictly speaking we would expect a coordination number of 4. In contrast, if r_c/r_a is 0.550, we would expect a coordination number of 6 for the cation, and, in almost every case, we would be correct. When the radius ratio is not close to the limiting values, it takes more than a *subtle* factor to cause the structure to be different from that predicted. Although the radius ratio does not always predict the correct crystal structure, it works very well except when the ratio is close to one of the limiting values.

We can see how the polarizability of ions affects lattice energies by considering the silver halides. When the lattice energies are calculated using the Kapustinskii equation, the calculated values are significantly lower than the experimental values as shown by the data presented in Table 7.6. Most of this difference is due to polarization effects, the resulting partial covalent bonding, and large London forces (see Chapter 6).

Table 7.6 Lattice Energies of Silver Halides

| Compound | Lattice Energy, kJ mol^{-1} | |
	Calculated (Eqn 7.17)	Experimental
AgF	816	912
AgCl	715	858
AgBr	690	845
AgI	653	833

Although we have illustrated the effects of polarizability of ions by considering a few cases where the effects are large, there must be *some* polarization effect for any combination of ions. However, there is an even more important consideration. It is known that the *apparent* radius of a given ion depends somewhat on the environment of the ion. For example, an ion surrounded by four nearest neighbors will appear to be slightly different in size than one that is surrounded by six ions of opposite charge. We have treated the ionic radius as if it were a fixed number that is the same in any type of crystal environment, but this is not the case. Moreover, ionic radii are determined from x-ray diffraction experiments that actually determine the distance between ion centers. For example, if the distance between ion centers is determined for NaF and the radius of the fluoride ion is known from other measurements, we can then deduce the radius of the sodium ion. Ionic radii are somewhat variable depending on the values assigned to other ions. In fact, for some ions a range of tabulated values for the radii is encountered. The ionic radii shown in Table 7.4 may not be exactly correct for certain ions in specific crystal structures.

In a crystal lattice, a cation is surrounded by a certain number of anions. Electrostatic forces exist between the oppositely charged ions. If a $+1$ ion is surrounded by six anions (most $+1$ cations are relatively large) having -1 charges, each anion is attached to other cations and a rigid lattice results. Such a lattice is characterized by a high melting point. For smaller, more highly charged cations, the coordination number is smaller and each anion is attached to a smaller number of cations. When the coordination number of a cation becomes equal to its valence, the cation and its nearest neighbors constitute a discrete neutral structure. Therefore, there will be no strong extended forces and the lattice is held together much more loosely, which results in a lower melting point. For example, the melting points of NaF, MgF_2, and SiF_4 (a molecular solid) are 1700, 2260, and $-90\,°C$, respectively.

7.5 CRYSTAL STRUCTURES

By means of the radius ratio, we have already described the type of local environment around the ions in several types of simple crystals. For example, in the sodium chloride *structure* (not restricted to NaCl itself), there are six anions surrounding each cation. The sodium chloride crystal structure is shown in Figure 7.5.

For the NaCl crystal, the radius ratio is 0.54, which is well within the range for an octahedral arrangement of anions around each cation $(0.414 - 0.732)$. However, because this is a 1:1 compound, there are equal numbers of cations and anions. This means that there must be an identical arrangement of cations around each anion. In fact, for 1:1 compounds, the environment around each type of ion

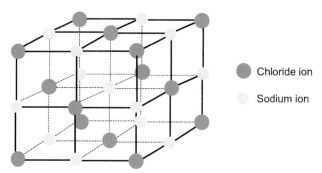

FIGURE 7.5
The rock salt or sodium chloride crystal structure.

must be identical. We can see that this is so from a very important concept known as the *electrostatic bond character*. If we predict (and find) that six Cl^- ions surround each Na^+, each "bond" between a sodium ion and a chloride ion must have a bond character of 1/6 because the sodium has a unit valence and the six "bonds" must add up to the valence of Na^+. If each "bond" has a character of 1/6, there must also be six bonds to each Cl^- because the chloride ion also has a unit valence (although it is negative in this case). Each bond has only a single magnitude regardless of which ion we are considering.

The electrostatic bond character is an extremely useful property for understanding crystal structures. Consider the structure of CaF_2 in which each Ca^{2+} is surrounded by eight F^- ions in a cubic arrangement as shown in Figure 7.6. Because calcium in CaF_2 has a valence of 2, the eight bonds to fluoride ions must total 2 electrostatic bonds so that each bond has a character of 1/4. However, because each bond between Ca^{2+} and F^- amounts to 1/4 bond, there can be only four bonds to each F^- because it has a valence of 1 (it is negative, of course, but that doesn't matter). As a result, the coordination number of Ca^{2+} is 8, whereas that of F^- is 4 in the crystal of CaF_2. Note how those

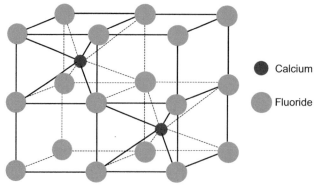

FIGURE 7.6
The calcium fluoride structure (also known as the fluorite structure).

coordination numbers correspond exactly to the fact that there are twice as many fluoride ions as there are calcium ions in the formula.

The fluorite structure is a common one for compounds that have 1:2 stoichiometry. A great many compounds have formulas that have twice as many cations as anions. Examples include compounds such as Li_2O and Na_2S. These compounds have crystal structures that are like the fluorite structure but with the roles of the cations and anions reversed. This structure is known as the *antifluorite* structure, in which there are eight cations surrounding each anion and four anions surrounding each cation. The antifluorite structure is the most common one for compounds that have formulas containing twice as many cations as anions.

In the case of the CsCl structure shown in Figure 7.7, the coordination number of the cation is eight. Because this is a 1:1 compound, the coordination number of the anion must also be eight. In terms of the electrostatic bond character approach, each bond between a cation and an anion must have a bond character of 1/8 because eight bonds must add up to the valence of Cs, which is 1. The valence of chlorine in cesium chloride is also 1, so there must be eight bonds to each Cl^-. In accord with this, the CsCl structure has 8 anions arranged at the corners of a cube around each Cs^+. Eight cubes come together at each corner, and each cube contains a Cs^+ ion, so there are eight Cs^+ ions surrounding each Cl^-.

There are two forms of zinc sulfide that have structures known as *wurtzite* and *zinc blende*. These structures are shown in Figures 7.8(a) and (b). Using the ionic radii shown in Table 7.4, we determine the radius ratio for ZnS to be 0.39, and, as expected, there are four sulfide ions surrounding each zinc ion in a tetrahedral arrangement. Zinc has a valence of 2 in zinc sulfide, so each bond must be 1/2 in character because four such bonds must satisfy the valence of 2. Because the sulfide ion also has a valence of 2, there must be four bonds to each sulfide ion. Therefore, both of the structures known for zinc sulfide have a tetrahedral arrangement of cations around each anion and a tetrahedral arrangement of anions around each cation. The difference between the structures is in the way in which the ions are arranged in layers that have different structures.

It should not be inferred that the crystal structures described so far apply only to binary compounds. Either the cation or the anion may be a polyatomic species. For example, many ammonium compounds have crystal structures that are identical to those of the corresponding rubidium or potassium compounds because the radius of NH_4^+ ion (148 pm) is similar to that of K^+ (133 pm) or

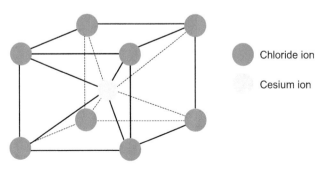

Chloride ion

Cesium ion

FIGURE 7.7
The cesium chloride structure.

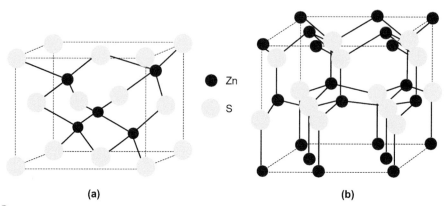

FIGURE 7.8
The zinc blende (a) and wurtzite (b) structures for zinc sulfide.

Rb^+(148 pm). Both NO_3^- and CO_3^{2-} have ionic radii (189 and 185 pm, respectively) that are very close to the ionic radius of Cl^- (181 pm), so many nitrates and carbonates have structures identical to the corresponding chloride compounds. Keep in mind that the structures shown so far are *general* types that are not necessarily restricted to binary compounds or the compounds from which they are named.

Rutile, TiO_2, which has the structure shown in Figure 7.9, is an important chemical that is used in enormous quantities as the opaque white material to provide covering ability in paints. Because the Ti^{4+} ion is quite small (56 pm), the structure of TiO_2 has only six O^{2-} ions surrounding each Ti^{4+} ion as predicted by the radius ratio of 0.39. Therefore, each Ti–O bond has an electrostatic bond character of 2/3 because the six bonds to O^{2-} ions total the valence of 4 for Ti. There can be only three bonds from Ti^{4+} to each O^{2-} ion because three such bonds would give the total valence of 2 for oxygen $(3 \times 2/3 = 2)$.

A lattice arrangement known as the ReO_3 structure (shown in Figure 7.10) provides an interesting application of the electrostatic bond character approach. In ReO_3, the valence of Re is 6, and, in the ReO_3 structure, each Re is surrounded by six oxide ions. Therefore, each Re–O bond must have a bond character of 1 because six bonds add up to a valence of 6 for Re. It follows that each oxygen can have only 2 bonds to it because each bond has a character of 1. The structure we are led to is one in which there are six oxide ions (arranged octahedrally as expected) surrounding each Re but only two Re^{6+} ions

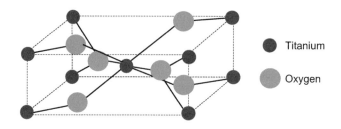

Titanium

Oxygen

FIGURE 7.9
The rutile structure.

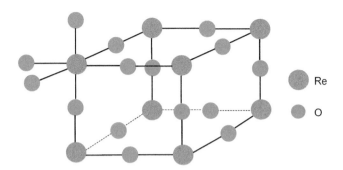

FIGURE 7.10
The ReO$_3$ structure.

surrounding each oxide ion in a linear arrangement. This is also the crystal structure of AlF$_3$ in which each Al^{3+} is surrounded by six F$^-$ ions in an octahedral arrangement and each F$^-$ ion has an Al^{3+} ion on either side.

The difference in physical properties between BeF$_2$ and BF$_3$ (melting points, 800 and -127 °C) as well as between AlF$_3$ and SiF$_4$ (melting points, 1040 °C and -96 °C) is striking. Although BF$_3$ is predominantly ionic, the B^{3+} ion is so small that no more than four F$^-$ ions can surround it (as in the BF$_4^-$ ion). In order to form an extended lattice, the B^{3+} ion would have to be surrounded by six F$^-$ ions in order for the electrostatic bond character to add up to +3 for B and -1 for F. The small size of B^{3+} prevents this, so BF$_3$ is a monomer even though it is predominantly ionic. In the case of BeF$_2$, four fluoride ions surrounding the Be^{2+} satisfy its valence, and each fluoride ion bridges between two Be^{2+} ions. Accordingly, BeF$_2$ forms an extended lattice, which is reflected by its high melting point. For SiF$_4$ to form a lattice, it would be necessary for eight F$^-$ ions to surround Si^{4+}, which is too small to allow this. The result is that SiF$_4$ exists as molecules that are predominantly ionic. Again we see the great utility of the electrostatic bond character approach of Pauling. Although not using the electrostatic bond character, R.J. Gillespie has written eloquently regarding the difference in properties of the fluorides discussed above (see *J. Chem. Educ. 75*, 923, 1998).

Aluminum oxide, which has the mineral name *corundum,* is a solid that has several important uses. Because it will withstand very high temperatures, it is a refractory material, and, because of its hardness, it is commonly used in abrasives. Corundum often contains traces of other metals, which impart a color to the crystals making them valuable as gemstones. For example, *ruby* contains a small amount of chromium oxide, which causes the crystal to have a red color. By adding a small amount of a suitable metal oxide, it is possible to produce gemstones having a range of colors.

Aluminum is produced commercially by the electrolysis of *cryolite*, Na$_3$AlF$_6$, but *bauxite*, Al$_2$O$_3$, is the usual naturally occurring source of the metal. The oxide is a widely used catalyst that has surface sites that function as a Lewis acid. A form of the oxide known as activated alumina has the ability to adsorb gases and effectively remove them. Other uses of the oxide include ceramics, catalysts, polishing compounds, abrasives, and electrical insulators.

Aluminum forms mixed oxides with other metals that have the general formula AB$_2$O$_4$, where A is an ion that has a charge of +2, whereas B is a metal ion with a +3 charge. The compound MgAl$_2$O$_4$ is the mineral known as *spinel*, which gives rise to the general name for compounds having the general formula AB_2O_4 being known as *spinels*, and this general formula can be written as $AO \cdot B_2O_3$; so $MgAl_2O_4$ can also be written as $MgO \cdot Al_2O_3$. A large number of materials are known in which Fe^{2+}, Zn^{2+}, Co^{2+}, Ni^{2+}, or other +2 ions replace Mg^{2+}. Some common minerals of this type include *ghanite* ($ZnAl_2O_4$), *hercynite* ($FeAl_2O_4$), and *galestite* ($MnAl_2O_4$).

Spinels have a crystal structure in which there is a face-centered cubic (*fcc*) arrangement of O^{2-} ions. In a structure of this type, there are two types of environments in which cations have octahedral or tetrahedral arrangements of anions surrounding them. In the *spinel* structure, it is found that the +3 ions are located in octahedral holes and the tetrahedral holes are occupied by the +2 ions. A different structure is possible for these ions. That structure has half of the +3 metal ions located in the tetrahedral holes, whereas the other half of these ions and the +2 ions are located in the octahedral holes. In order to indicate the population of the two types of lattice sites, the formula for the compound is grouped with the tetrahedral hole population indicated first (the position normally occupied by the +2 ion, A) followed by the groups populating the octahedral holes. Thus, the formula AB_2O_4 becomes $B(AB)O_4$ in order to correctly indicate the places of the ions in the lattice. Because of the reversal of the roles of the +2 and +3 ions in the lattice, the structure is called the *inverse spinel* structure. A compound that has an inverse spinel structure is lithium ferrite, $LiFeO_2$. This compound is used in fabricating electrodes in lithium batteries. The synthesis of this compound has been carried out in several ways that include solid state reactions, ion exchange methods, and reactions of compounds containing lithium and iron under the influence of sources of energy such as microwaves, heat, or mechanical stress.

Although many ternary compounds containing NH_4^+, NO_3^-, CO_3^{2-}, etc. have structures that are identical to those of binary compounds, the mineral *perovskite*, $CaTiO_3$, has a different type of structure. In fact, this is a very important structure type that is exhibited by a large number of other compounds. The *perovskite* structure is shown in Figure 7.11. By examining the structure, it is easy to see that a Ti^{4+} ion resides in the center of a cube, each corner of which is the location of a Ca^{2+} ion. The oxide ions are located at the center of the six faces of the cube. It is easy to see that the only bonds to the Ti^{4+} are those from its nearest neighbors, the six O^{2-} ions. Therefore, each Ti—O bond must have a bond character of 4/6 because six such bonds total the +4 valence of Ti.

Consider now the bonds to each O^{2-} ion in the *perovskite* structure. First, there are two bonds to Ti^{4+} ions, which have a character of 4/6, each which gives a total of 4/3. However, there are four Ca^{2+} ions

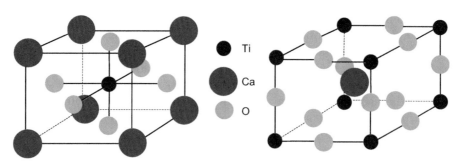

FIGURE 7.11

Two views of the structure of perovskite, $CaTiO_3$. In the structure on the left, note the coordination number of 6 for Ti that arises from six oxygen atoms surrounding it in an octahedral arrangement. In the structure on the right, note the coordination number of 12 for Ca that comes from 12 oxygen atoms surrounding it with four in each of three staggered "layers."

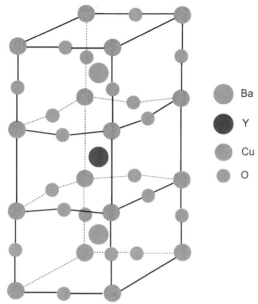

FIGURE 7.12
The structure of the unit cell of $YBa_2Cu_3O_7$.

on the corners of the face of the cube where an oxide ion resides. These four bonds must add up to a valence of 2/3 so that the total valence of 2 for oxygen is satisfied. If each Ca—O bond amounts to a bond character of 1/6, four such bonds would give the required 2/3 bond to complete the valence of oxygen. From this it follows that each Ca^{2+} must be surrounded by 12 oxide ions so that $12(1/6) = 2$, the valence of calcium. It should be apparent that the concept of electrostatic bond character is a very important tool for understanding crystal structures.

There are a large number of ternary compounds that are oxides for which the general formula is ABO_3, where A = Ca, Sr, Ba, etc. and B = Ti, Zr, Al, Fe, Cr, Hf, Sn, Cl, or I. Many have the perovskite structure, so it is an important structural type.

In recent years, a lot of study has been devoted to several classes of superconductors. One such compound has the formula $YBa_2Cu_3O_7$, and it has the structure shown in Figure 7.12. Half of the copper atoms shown in the top and bottom faces of the structure shown in Figure 7.12 are surrounded by four oxygen atoms in a square planar arrangement. The other half of the copper atoms are surrounded by five oxygen atoms arranged in a square-based pyramid. The yttrium atom is surrounded by eight oxygen atoms in a cubic arrangement. This superconductor has a critical temperature of 93 K.

7.6 SOLUBILITY OF IONIC COMPOUNDS

An enormous amount of chemistry is carried out in solutions that consist of ionic compounds that have been dissolved in a solvent. In order to separate the ions from the lattice in which they are held, there must be strong forces of interaction between the ions and the molecules of the solvent. The most common solvent for ionic compounds is water, and that solvent will be assumed for the purposes of this discussion.

When an ionic compound is dissolved in a solvent, the crystal lattice is broken apart. As the ions separate, they become strongly attached to solvent molecules by ion-dipole forces. The number of water molecules surrounding an ion is known as its *hydration number*. However, the water molecules clustered around an ion constitute a shell, which is referred to as the primary solvation sphere. The water molecules are in motion and are also attracted to the bulk solvent that surrounds the cluster. Because of this, solvent molecules move into and out of the solvation sphere, giving a hydration number that does not always have a fixed value. Therefore, it is customary to speak of the average hydration number for an ion.

From the standpoint of energy, the processes of separating the crystal lattice and solvating the ions can be related by means of a thermochemical cycle of the Born-Haber type. For an ionic compound MX, the cycle can be shown as follows:

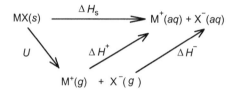

In this cycle, U is the lattice energy, ΔH^+ and ΔH^- are the heats of hydration of the gaseous cation and anion, and ΔH_s is the heat of solution. From this cycle, it is clear that

$$\Delta H_s = U + \Delta H^+ + \Delta H^- \tag{7.19}$$

As defined earlier, the lattice energy is positive, whereas the solvation of ions is strongly negative. Therefore, the overall heat of solution may be either positive or negative depending on whether it requires more energy to separate the lattice into the gaseous ions than is released when the ions are solvated. Table 7.7 shows the heats of aquation, ΔH°_{hyd}, several for ions.

The data presented in Table 7.7 show that for cations the enthalpies of aquation are dependent on the charges on the ions and their sizes. For ions having the same charge, the heat of aquation decreases as the size of the ion increases. This is reasonable because the polar solvent molecules are attracted more strongly to small compact ions where the charge is localized to a small region of space. The heat of aquation increases dramatically as the charge on the ion increases. A simple principle of electrostatics suggests that this would be the case because the negative ends of polar water molecules will be attracted more strongly to a higher positive ion as shown by Coulomb's law. The charge density, as reflected by the charge to size ratio, is one factor in determining the heats of hydration of ions.

The hydration enthalpy (H) for an ion can be expressed as

$$H = -\frac{Ze^2}{2r}\left(1 - \frac{1}{\varepsilon}\right) \tag{7.20}$$

where Z is the charge on the ion, r is its radius, and ε is the dielectric constant of the medium (78.4 for water). The hydration enthalpy increases with increasing charge on the ion, whereas it decreases with increasing size of the ion. One reason that small anions such as F^- have high hydration enthalpies is

Table 7.7 Hydration Enthalpies for Ions

Ion	r, pm	ΔH°_{hyd}, kJ mol^{-1}
H$^+$	–	−1100
Li$^+$	74	−520
Na$^+$	102	−413
K$^+$	138	−321
Rb$^+$	149	−300
Cs$^+$	170	−277
Mg^{2+}	72	−1920
Ca^{2+}	100	−1650
Sr^{2+}	113	−1480
Ba^{2+}	136	−1360
Al^{3+}	53	−4690
F$^-$	133	−506
Cl$^-$	181	−371
Br$^-$	196	−337
I$^-$	220	−296

because they are attracted to the centers of positive charge in the water molecules, which are the hydrogen atoms. As a result, a very small distance separates the negative ion from the positive centers in the water molecule.

The interactions between ions and solvent molecules are primarily electrostatic in nature, so the dipole moment of the solvent is an important consideration. However, the structure of the solvent molecules is also important. For example, nitrobenzene has a high dipole moment of 4.22 D, which is much larger than the value of 1.85 D for water. The nitrobenzene molecule has a large dipole moment, but it is a poor solvent for ionic salts such as NaCl. The high dipole moment results from the charge being separated over a long distance. Also, the molecules of nitrobenzene are too large for them to pack efficiently around small ions, so the solvation number is too small to result in strong solvation. Although the dipole moment of nitrobenzene indicates that it might be a suitable solvent for ionic compounds, this is not actually the case.

The hydration enthalpy of the Al^{3+} ion is enormous (−4690 kJ mol^{-1}), and there are some interesting effects produced as a result. When NaCl is dissolved in water and the solvent evaporated, the solid NaCl can be recovered. If AlCl$_3$ is dissolved in water, evaporation of water does not yield the solid AlCl$_3$. The Al^{3+} ion is so strongly solvated that other reactions become energetically more favorable than removing the solvent. This can be shown as follows:

$$AlCl_3(s) \xrightarrow{H_2O} Al^{3+}(aq) + 3Cl^-(aq) \tag{7.21}$$

When the solvent is evaporated, a solid is obtained that contains the aquated aluminum ion and chloride ions. This solid can be described as [Al(H$_2$O)$_6$]Cl$_3$, although the number of water molecules may depend on the conditions. When this solid is heated, water is lost until the composition

$[Al(H_2O)_3Cl_3]$ is approached. When heated to still higher temperature, this compound does lose HCl rather than water.

$$[Al(H_2O)_3Cl_3](s) \rightarrow Al(OH)_3(s) + 3\,HCl(g) \qquad (7.22)$$

When heated to a very high temperature, $Al(OH)_3$ loses H_2O to yield Al_2O_3.

$$2Al(OH)_3(s) \rightarrow Al_2O_3(s) + 3H_2O(l) \qquad (7.23)$$

The essence of this behavior is that the bonds between Al^{3+} and oxygen are so strong that reactions other than dehydration become energetically favored. When beryllium compounds are dissolved in water, the Be^{2+} ion is so strongly solvated that it also behaves in this way. The charge to size ratios for Al^{3+} and Be^{2+} are approximately equal ($+3/53$ and $+2/30$, respectively), which results in their having similar chemical behavior. This is known as a *diagonal* relationship because aluminum is one row below beryllium in the periodic table, but it is also one column farther to the right along a "diagonal."

By considering the dissolution of NaCl, it is found that the lattice energy is 786 kJ mol^{-1} and the heat of solvation of Na$^+$ is -413 kJ mol^{-1}, whereas that of Cl$^-$ is -371 kJ mol^{-1}. Using these data, the heat of solution is calculated to be only 4 kJ mol^{-1}. This indicates that there is essentially no heat absorbed or released when NaCl dissolves in water. As a result, changing the temperature has almost no effect on the solubility of NaCl in water. If a graph is made of the solubility of NaCl in water as a function of temperature, it has a slope that is almost zero. In fact, the solubility of NaCl in water at 0 °C is about 35.7 grams per 100 grams of water, and, at 100 °C, it is approximately 39.8 grams per 100 grams of water. On the other hand, for some solids, separating the crystal lattice requires much more energy than is released when the ions are hydrated. In this case, the overall process absorbs heat, so increasing the temperature favors the dissolution process, and a graph of solubility of the compound as a function of temperature yields a line that rises as the temperature is increased. If the solid is one for which more heat is released when the ions are solvated than is absorbed in separating the lattice, the heat of solution will be negative, and the compound will become less soluble as the temperature increases. Figure 7.13 shows solubility curves for these three types of behavior.

If a solvent that does not strongly solvate ions is used, the crystal will not dissolve because the lattice energy will be much larger in magnitude than the sum of the solvation energies for the ions. From this discussion, it should be clear that the solution behavior of a solid compound is also related to how strongly the crystal is held together. However, it must be kept in mind that the heats of solvation of ions

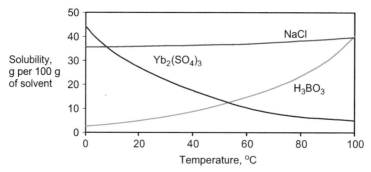

FIGURE 7.13
Variation in solubility of three inorganic compounds with temperature.

are not constant over a wide range of temperature. They are themselves variables, which means that when a wide range of temperature is considered, solubility behavior is not exactly predictable from this simple approach.

A simple approach to the effect of temperature on solubility can be illustrated by considering the cases shown in Figure 7.14. Increasing the temperature of a system at equilibrium causes a shift in the endothermic direction. In Figure 7.14(a), the endothermic direction is toward the solution phase, so increasing the temperature will cause an increase in the solubility of the solute. For the case illustrated in Figure 7.14(b), increasing the temperature will cause the system to shift in the direction to increase the amounts solute and solvent, so the solubility will decrease. In the case illustrated in Figure 7.14(c), the solubility will not change much as the temperature is changed.

Quantitatively, the effect of temperature on solubility can be explained by considering the dissolution process as represented by the population of states of unequal energy. The Boltzmann distribution law relating the populations n_1 and n_2 for states E_1 and E_2, which are of unequal energy, can be written as

$$\frac{n_2}{n_1} = e^{-\Delta E/kT} \tag{7.24}$$

where k is Boltzmann's constant, ΔE is the difference in energy between two states, and T is the temperature (K). When expressed on a molar basis where the heat of solution is ΔH_s, the expression for the variation in solubility with temperature becomes

$$\frac{n_2}{n_1} = e^{-\Delta H_s/RT} \tag{7.25}$$

Taking the logarithm on both sides of the equation yields

$$\ln n_2 - \ln n_1 = -\frac{\Delta H_s}{RT} \tag{7.26}$$

Considering the dissolution process represented as shown in Figure 7.14(a), we see that the amount of solute is immaterial as long as there is enough to give a saturated solution. Therefore, the $\ln n_1$ term can be treated as a constant, C, and the population n_2 can be replaced by the solubility, S, to give

$$\ln S = -\frac{\Delta H_s}{RT} + C \tag{7.27}$$

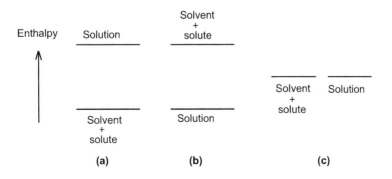

FIGURE 7.14
Thermal changes during dissolution of a solid in a liquid.

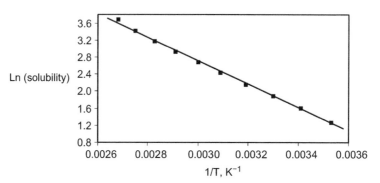

FIGURE 7.15
The linear ln (solubility) vs. 1/T relationship for boric acid in water.

From this equation, it can be seen that a plot of ln S vs. $1/T$ should yield a straight line having a slope of $-\Delta H_s/R$. Determining the solubility of a compound at several temperatures allows the heat of solution to be determined in this way. For the process illustrated in Figure 7.14(a), the heat of solution is positive, so the slope of the line will be negative. For the process illustrated in Figure 7.14(b), the heat of solution is negative, so a line having a positive slope is obtained. As shown in Figure 7.13, for NaCl the heat of solution is very close to 0, and, as a result, the solubility is almost constant in the temperature range 0–100 °C. Figure 7.15 shows the application of Eqn (7.27) to the solubility of boric acid in water. Linear regression yields a slope of -2737 K^{-1}, which is equal to $-\Delta H/R$. Therefore, from the solubility data, the value for the heat of solution of boric acid is 22.7 kJ mol^{-1}.

7.7 PROTON AND ELECTRON AFFINITIES

Some cations can be considered as neutral molecules that have accepted a hydrogen ion. For example, the ammonium ion results from the addition of H^+ to NH_3. Although the subject of acid-base chemistry will be discussed in Chapter 9, it is appropriate to discuss one related topic in this chapter because it deals with the behavior of solids. That topic is the proton affinity of a base. The proton affinity of a base is similar to the electron affinity of an atom, which was discussed in Chapter 1. Whereas the electron affinity is the energy required to remove an electron from a gaseous atom that has gained the electron, the proton affinity is the energy necessary to remove a proton from a gaseous base that has gained a proton. It is a measure of the intrinsic basicity of a species in the gas phase without the complicating effects that are often caused by solvents.

Ammonium compounds decompose in a number of ways when heated. For many of them, the heat of decomposition is either known or rather easily measured by techniques such as differential scanning calorimetry. The Kapustinskii equation can be employed to determine the lattice energy. With this information and by using the known proton affinity for NH_3 (866 kJ mol^{-1}), it is possible to determine the proton affinity of the anion in an ammonium compound. Although this has been done for numerous compounds, the procedure will be illustrated for ammonium bisulfate and ammonium sulfate.

The decomposition of NH_4HSO_4 and the appropriate thermochemical cycle for determining the proton affinity of the $HSO_4^-(g)$ ion can be shown as follows:

$$NH_4HSO_4(s) \xrightarrow{\quad \Delta H_{dec} = 169 \text{ kJ mol}^{-1} \quad} H_2SO_4(g) + NH_3(g)$$

$$\downarrow U_2 \qquad\qquad\qquad\qquad \uparrow -PA(HSO_4^-)$$

$$NH_4^+(g) + HSO_4^-(g) \xrightarrow{\quad PA(NH_3) \quad} NH_3(g) + H^+(g) + HSO_4^-(g)$$

In this cycle, ΔH_{dec} is the heat of decomposition and U_2 is the lattice energy for NH_4HSO_4, $PA(NH_3)$ is the proton affinity of $NH_3(g)$, and $PA(HSO_4^-)$ is the proton affinity of the bisulfate ion. The heat of decomposition of NH_4HSO_4 has been determined to be 169 kJ mol^{-1}, and the proton affinity of $NH_3(g)$ is 866 kJ mol^{-1}. The ionic radii for NH_4^+ and HSO_4^- are 143 pm and 206 pm, respectively, and the lattice energy calculated for NH_4HSO_4 by means of the Kapustinskii equation is 641 kJ mol^{-1}. From the cycle shown above, we find that

$$PA(HSO_4^-) = U_2 + PA(NH_3) - \Delta H_{dec} \tag{7.28}$$

and substituting for the known quantities yields a value of 1338 kJ mol^{-1} for the proton affinity of the $HSO_4^-(g)$ ion. Proton affinities for other -1 ions range from a value of 1309 kJ mol^{-1} for I$^-$ to 1695 kJ mol^{-1} for the CH_3^- ion. Therefore, the value 1338 kJ mol^{-1} found for the HSO_4^- ion is in good agreement with the values for other ions having a -1 charge.

The procedure illustrated above can also be applied to the decomposition of $(NH_4)_2SO_4$ to determine the proton affinity of the gaseous SO_4^{2-} ion. The radius of the SO_4^{2-} ion is 230 pm, so the Kapustinskii equation leads to a value of 1817 kJ mol^{-1} for the lattice energy of $(NH_4)_2SO_4$. When $(NH_4)_2SO_4(s)$ is heated, it produces $NH_4HSO_4(s)$ and NH_3, with the heat of decomposition being 195 kJ mol^{-1}. The thermochemical cycle to be used can be written as follows:

$$(NH_4)_2SO_4(s) \xrightarrow{\quad \Delta H_{dec} = 195 \text{ kJ mol}^{-1} \quad} NH_3(g) + NH_4HSO_4(s)$$

$$\downarrow U_1 \qquad\qquad\qquad\qquad\qquad\qquad \uparrow -U_2$$

$$2\,NH_4^+(g) + SO_4^{2-}(g)$$

$$\downarrow PA(NH_3)$$

$$NH_4^+(g) + NH_3(g) + H^+(g) + SO_4^{2-}(g) \xrightarrow{\quad -PA(SO_4^{2-}) \quad} NH_4^+(g) + HSO_4^-(g) + NH_3(g)$$

From this cycle, it is apparent that

$$PA(SO_4^{2-}) = U_1 + PA(NH_3) - \Delta H_{dec} - U_2 \tag{7.29}$$

where U_2 is the lattice energy for NH_4HSO_4, which was shown earlier to be 641 kJ mol^{-1}. When the values are substituted for the known quantities shown in Eqn (7.29), a value of 1847 kJ mol^{-1} is found for the proton affinity for $SO_4^{2-}(g)$. Proton affinities for most other -2 ions are somewhat higher than this value, but most of the ions are smaller in size (see Chapter 9). The SO_4^{2-} ion has extensive double bonding between the sulfur and oxygen atoms (see Chapter 4), which may decrease the availability of electron pairs on the oxygen atoms to attract H^+ ions. After all, H_2SO_4 is a very strong acid, so it loses protons readily.

By means of appropriate thermochemical cycles, it is possible to calculate proton affinities for species for which experimental values are not available. For example, using the procedure illustrated by the two examples above, the proton affinities of ions such as $HCO_3^-(g)$ (1318 kJ mol^{-1}) and $CO_3^{2-}(g)$ (2261 kJ mol^{-1}) have been evaluated. Studies of this type show that lattice energies are important in determining other chemical data and that the Kapustinskii equation is a very useful tool.

In Chapter 1 we discussed the electron affinities of atoms and how they vary with position in the periodic table. It was also mentioned that no atom accepts two electrons with a release of energy. As a result, the only value available for the energy associated with adding a second electron to O^- is one calculated by some means. One way in which the energy for this process can be estimated is by making use of a thermochemical cycle such as the one that follows showing the steps that could lead to the formation of MgO.

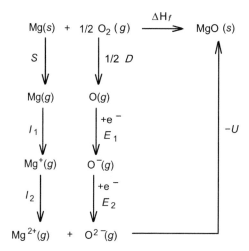

From this cycle, it is possible to calculate E_2, the electron affinity for the second electron added to an oxygen atom. The heat of formation and lattice energy for MgO(s) are -602 and 3795 kJ mol^{-1}, respectively. For Mg(g), the first two ionization potentials are 738 and 1451 kJ mol^{-1}, and one-half of the heat of dissociation of $O_2(g)$ is 249 kJ mol^{-1}. The first electron affinity for O(g) is 141 kJ mol^{-1}, so adding the electron has an enthalpy associated with it of -141 kJ mol^{-1}. Knowing values for all these quantities makes it possible to obtain a value of $+750$ kJ mol^{-1} for the second electron affinity for the oxygen atom. Energetically, this is a very unfavorable process! Even though addition of the first electron to an oxygen atom is energetically favorable, the sum of the energies for adding two electrons is $+609$ kJ mol^{-1}. It should be mentioned that this calculation is not likely to give a highly accurate value. The dominant term in the equation is the lattice energy of MgO, which is approximately 3800 kJ mol^{-1}, but that value is not known with certainty. For example, if the lattice energy is calculated by means of the

Kapustinskii equation, it must be remembered that the covalent contribution is not taken into account. Even though the uncertainty in the second electron affinity (and hence the sum of the first and second) for oxygen is rather large, there is no doubt that adding *two* electrons is energetically unfavorable. If it were not for the fact that lattices containing the doubly charged oxide ion are extremely stable, it would be very unlikely that compounds containing O^{2-} would be obtainable. This situation also exists for other ions having charges more negative than -1 (e.g. S^{2-}).

7.8 STRUCTURES OF METALS

Metals consist of spherical atoms that are arranged in three-dimensional lattices. The number of ways in which this occurs is limited, and only four types of structures (shown in Figure 7.16) are needed to show the arrangements of atoms in almost all metals. These arrangements for packing spheres are sometimes called *closest packing*. One way in which spherical atoms can be packed is with one atom on each corner of a cube. This structure is known as the simple cubic structure. When we realize that atoms in a metal are bonded to each other, we see one of the problems with the simple cubic structure. Each atom is surrounded by only six other atoms (the *coordination number*), and, even when the atoms are touching, there is a great deal of free space. When cubes are stacked, eight cubes come together at each corner and there are eight corners to each cube where an atom resides. Therefore, because eight cubes share a common corner, only one-eighth of each atom belongs in any one cube. The total occupancy of the cube is $8(1/8) = 1$, and there is one atom per cubic unit cell. Figure 7.16(a) shows the simple cubic structure.

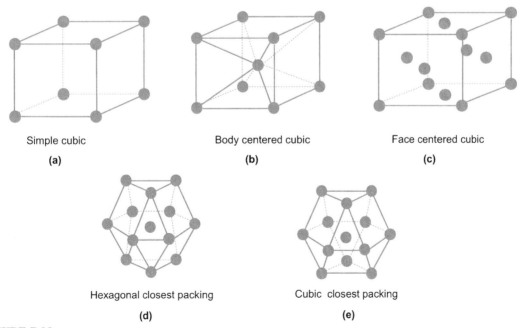

| Simple cubic | Body centered cubic | Face centered cubic |
| (a) | (b) | (c) |

Hexagonal closest packing
(d)

Cubic closest packing
(e)

FIGURE 7.16
The most common structures of metals.

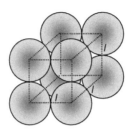

FIGURE 7.17
A space filling model of spheres arranged in the simple cubic pattern.

We can determine the amount of empty space in the simple cubic (a space filling model is shown in Figure 7.17) structure by considering it to have an edge length l, which will be twice the radius of an atom. Therefore, the radius of the atom is $l/2$, so the volume of one atom is $(4/3)\pi(l/2)^3 = 0.524l^3$, but the volume of the cube is l^3. From this we see that because the cube contains only one atom that occupies 52.4% of the volume of the cube, there is 47.6% empty space. Because of the low coordination number and the large amount of empty space, the simple cubic structure does not represent an efficient use of space and does not maximize the number of metal atoms bonded to each other. Consequently, the simple cubic structure is not a common one for metals.

The body-centered cubic (*bcc*) structure contains one atom in the center of a cubic cell that also has one atom on each corner of the cube that is touching the center atom. In this structure, shown in Figure 7.16(b), there are two atoms per unit cell. The atom in the center resides totally within the cube, and one-eighth of each atom on a corner resides in the cube. When we realize that the atoms are touching, we see that a diagonal through the cube must represent one diameter of an atom plus two times the radius of an atom. If the cube has an edge length l, the length of the diagonal will be $3^{1/2}l$, which is equal to 4 times the radius of an atom. Therefore, the radius of an atom is given as $3^{1/2}l/4$ and the volume of two atoms is $2(4/3)\pi r^3$ or $2(4\pi/3)(3^{1/2}l/4)^3 = 0.680l^3$. Thus we see that 68% of the cube is occupied by the two atoms in the cell or that 32% of the volume of the cell is empty space. Not only is this an improvement over the simple cubic structure in terms of space utilization but also, in the *bcc* structure, each atom is surrounded by eight nearest neighbors. Several metals are found to have the *bcc* structure.

In addition to the two structures already discussed, another arrangement of atoms in a cubic unit cell is possible. Atoms of a metal are identical, so the ratio of atomic sizes is 1.000, which allows a coordination number of 12. One structure that has a coordination number of 12 is known as the *fcc* structure, and it has one atom on each corner of the cube and an atom on each of the six faces of the cube. The atoms on the faces are shared by two cubes, so one-half of each atom belongs in each cube. With there being six faces, there are $6(1/2) = 3$ atoms in each cube from this source in addition to the $8(1/8) = 1$ from the atoms on the corners. The total number of atoms per unit cell is four. The *fcc* structure (also referred to as cubic closest packing) is shown in Figure 7.16(c). It can be shown that there is 26% free space in this arrangement, and the coordination number of each atom is 12. Therefore, the *fcc* arrangement is the most efficient of the three structures described so far, and many metals have this structure (see Table 7.8).

Hexagonal closest packing (*hcp*), which also involves a coordination number of 12, is shown in Figure 7.16(d). If we examine the environment around an atom, we find that there are six others arranged around it in a hexagonal pattern. It can be shown that these six atoms touch the one in the center just as they touch each other. In addition to the six atoms in the same layer, each atom is also surrounded by atoms that

Table 7.8 Closest Packing Structures of Metals. Polymorphism is Observed at Ambient Temperature for Some Metals and at Elevated Temperatures for Others

Li	Be										
bcc	hcp										
Na	Mg										
bcc	hcp										
K	Ca	Sc	Ti	V	Cr	Mn	Fe	Co	Ni	Cu	Zn
bcc	fcc[a]	hcp	hcp	bcc	bcc	hcp[b]	bcc	fcc	fcc	fcc	hcp[c]
		fcc						hcp			
Rb	Sr	Y	Zr	Nb	Mo	Tc	Ru	Rh	Pd	Ag	Cd
bcc	fcc[a]	hcp	hcp	bcc	bcc	hcp	hcp	fcc	fcc	fcc	hcp[c]
Cs	Ba	La	Hf	Ta	W	Re	Os	Ir	Pt	Au	Hg
bcc	bcc	hcp	hcp	bcc	bcc	hcp	hcp	fcc	fcc	fcc	—

[a]Calcium and strontium have other structures depending on the temperature.
[b]Manganese has a distorted hcp structure, but two other complex forms are known.
[c]Zinc and cadmium have distorted hcp structures in which the six nearest neighbors in the same plane are at one distance but atoms in the planes above and below are farther away.

are contained in the layers above and below it. Three atoms from each of those layers make contact with the atom that resides in the center of the hexagon. In *hcp*, the groups of three atoms in the layers above and below the atom being considered are aligned. The layers above and below the atom under consideration are exactly alike, although in most cases they are slightly farther away than the six atoms that are in the same plane. If we call the layers *A* and *B*, the repeating pattern in *hcp* is …*ABABAB*…. In this arrangement, the coordination number is 12, and there is 26% free space as in the *fcc* structure. In fact, the only difference between *hcp* and *fcc* is that although each has an atom surrounded by six others in the same plane, the planes above and below that plane are different. In *fcc*, the repeating pattern is …*ABCABCABC*… where the *C* layer has atoms that are not in alignment with those in the *A* layer. Figure 7.16(e) shows the layer arrangement in the *fcc* structure. Many metals have either *fcc* or *hcp* as the stable arrangement of atoms. The most common structures for a large number of metals are summarized in Table 7.8.

Having shown that the coordination numbers and the percentage of free space are equal for *fcc* and *hcp*, we would conclude that they represent arrangements of atoms that have energies that are very similar. As a result, it might be expected that a transformation of a metal from one structure to the other would be possible. Such transformations have been observed for several metals, and one example is

$$Co(hcp) \xrightarrow{417\,°C} Co(fcc) \tag{7.30}$$

It is also possible to bring about changes between other types of packing arrangements. For example, titanium undergoes a change from *hcp* to *bcc*,

$$Ti(hcp) \xrightarrow{883\,°C} Ti(bcc) \tag{7.31}$$

which means that there is a change in coordination number in this case. The ability to exist in more than one structure (*polymorphism*) is quite common in metals. As a general rule, metals that have *fcc* structures (e.g. Ag, Au, Ni, and Cu) are more malleable and ductile than are those with other structures. Metals that have the *hcp* structure (e.g. W, Mo, V, and Ti) tend to be more brittle and less ductile, which makes them harder to work into desired shapes and forms. These differences in properties are related to the ease with which the planes of metal atoms can be moved in relation to each other. Although the topic will not be discussed in more detail in this book, the structures and properties of metals are of great importance in the area of materials science.

7.9 DEFECTS IN CRYSTALS

Although several types of lattices have been described for ionic crystals and metals, it should be remembered that no crystal is perfect. The irregularities or defects in crystal structures are of two general types. The first type consists of defects that occur at specific sites in the lattice, and they are known as *point defects*. The second type of defect is a more general type that affects larger regions of the crystal. These are the *extended defects* or *dislocations*. Point defects will be discussed first.

One type of point defect that can not be entirely eliminated from a solid compound is the *substituted ion* or *impurity* defect. For example, suppose a large crystal contains 1 mole of NaCl that is 99.99 mole percent pure and that the 0.01% impurity is KBr. As a fraction, there is 0.0001 mole of both K^+ and Br^- ions, which is 6.02×10^{19} ions of each type present in the 1 mole of NaCl! Although the level of purity of the NaCl is high, there is an enormous number of impurity ions that occupy sites in the lattice. Even if the NaCl were 99.9999 mole percent pure, there would still be 6.02×10^{17} impurity cations and anions in a mole of crystal. In other words, there is a defect, known as a substituted ion or impurity defect, at each point in the crystal where some ion other than Na^+ or Cl^- resides. Because K^+ is larger than Na^+ and Br^- is larger than Cl^-, the lattice will experience some strain and distortion at the sites where the larger cations and anions reside. These strain points are frequently reactive sites in a crystal.

An analogous situation exists in crystals that are not ionic. For example, a highly pure metal might contain 99.9999 mole percent of one metal but still contain 0.0001 mole percent of another. There will be atoms of the metal impurity at specific sites in the lattice, which will constitute defects that alter the structure of the lattice slightly.

A different type of defect occurs at specific sites when an atom or ion is missing from a lattice position and is transferred to the surface of the crystal. It is also possible for pairs of ions of opposite charge to be missing in relatively close proximity, which allows the crystal to be electrically neutral in that region. Defects of the missing ion type, known as *Schottky defects*, are illustrated in Figure 7.18.

Removing an ion or atom from its lattice site leaves unbalanced forces between the atoms surrounding the site, so such a defect constitutes a high energy site. At any given temperature, the number of high energy sites is n_2 when the total number of sites is n_1. Actually, the number of occupied sites is $n_1 - n_2$, but the number of vacancies is small compared to the total number of sites in the lattice, so n_1 is essentially a constant. The Boltzmann Distribution Law gives the relationship between the numbers of sites,

$$\frac{n_2}{n_1} = e^{-\Delta E/kT} \tag{7.32}$$

where ΔE is the difference in energy between an occupied site and a vacancy, k is Boltzmann's constant, and T is the temperature (K). The defect population is increased if the temperature is increased. However, with the energy to form a vacancy being in the 0.5–1.0 eV (50 – 100 kJ mol^{-1}) range, the population of vacancies is small even at high temperature. For example, if the energy necessary to form

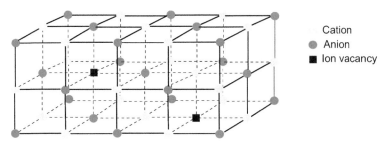

FIGURE 7.18
An illustration of Schottky defects in an ionic crystal.

a Schottky defect is 0.75 eV and the temperature is 750 K, the fraction of Schottky defects compared to total number of lattice sites, n_2/n_1, is

$$n_2/n_1 = \exp(-0.75 \text{ eV} \times 1.60 \times 10^{-12} \text{ erg/eV} / (1.38 \times 10^{-16} \text{ erg/K} \times 750 \text{ K})) = 9.2 \times 10^{-5}$$

Although this is a small fraction, for one mole of lattice sites, this amounts to 5.6×10^{18} Schottky defects. The ability of ions to move from their sites into vacancies and by so doing creating new vacancies is largely responsible for the conductivity in ionic crystals.

It is possible to create a population of Schottky defects that is much higher than the equilibrium population that is based on Eqn (7.32). If a crystal is heated to high temperature, lattice vibrations become more pronounced, and eventually ions begin to migrate from their lattice sites. If the crystal is quickly cooled, the extent of the motion of ions decreases rapidly, so that ions that have moved from their lattice sites cannot return. As a result, the crystal will contain a population of Schottky defects that is much higher than the equilibrium population at the lower temperature. If a crystal of KCl is prepared so that it contains some $CaCl_2$ as an impurity, incorporating a Ca^{2+} ion in the crystal at a K^+ site requires that another K^+ site be vacant to maintain electrical neutrality although the added Cl^- ions can occupy anion sites. However, the atomic weights of Ca and K are very similar. As a result of the vacancies, the crystal of KCl containing $CaCl_2$ has a lower density than pure KCl in which each cation site contains K^+.

A somewhat different situation is found in the type of point defect known as a *Frenkel defect.* In this case, an atom or ion is found in an interstitial position rather than in a normal lattice site as is shown in Figure 7.19. In order to place an atom or ion in an interstitial position, it must be possible for it to be close to other lattice members. This is facilitated when there is some degree of covalence in the bonding as is the case for silver halides and metals. Accordingly, Frenkel defects are the dominant type of defect in these types of solids.

When a crystal of an alkali halide has the vapor of the alkali metal passed over it, the alkali halide crystal becomes colored. The reason for this is that a type of defect that leads to absorption of light is created in the crystal. Such a defect is known as an *F*-center because the German word for color is *farbe*. It has been shown that such a defect results when an electron occupies a site normally occupied by an anion (an anion "hole"). This arises as a result of the reaction

$$K(g) \longrightarrow K^+(\text{cation site}) + e^-(\text{anion site}) \tag{7.33}$$

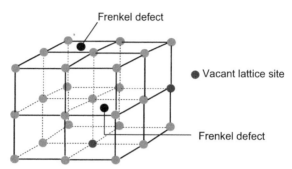

FIGURE 7.19
Frenkel defects in a crystal structure.

The potassium ions that are produced occupy cation lattice sites, but no anions are produced, so electrons occupy anion sites. In this situation, the electron behaves as a particle restricted to motion in a three-dimensional box that can absorb energy as it is promoted to an excited state. It is interesting to note that the position of the maximum in the absorption band is below 4000 Å (400 nm, 3.1 eV) for LiCl but it is at approximately 6000 Å (600 nm, 2 eV) for CsCl. One way to explain this observation is by noting that for a particle in a three-dimensional box, the difference between energy levels increases as the size of the box becomes smaller, which is the situation in LiCl. Schottky, Frenkel, and *F*-center defects are not the only types of point defects known, but they are the most common and important types.

In addition to the point defects that occur at specific lattice sites, there are types of defects, known as extended defects, that extend over a region of the crystal. The three most important types of extended defects are the *stacking fault, edge displacement,* and *screw dislocation*. A stacking fault involves an extra layer of atoms or a missing layer of atoms in the structure. For example, if the layers of atoms are represented as *A*, *B*, and *C*, the normal sequence of layers in the *fcc* structure is …*ABCABCABC*… . Stacking faults in this type of structure might be of the types …*ABCABABC*… (missing *C* layer) or …*ABCBABCABC*… (extra *B* layer). Stacking faults are normally encountered in metals in which all the atoms are identical, but the layers are distinct.

An edge dislocation occurs when an extra plane or layer of atoms extends part way into the crystal, which causes atoms in that region of the crystal to be compressed, but in the region where the extra plane does not extend they are spread apart. This type of crystal defect is shown in Figure 7.20(a). It is rather like a stacking fault that does not extend through the entire crystal. One consequence of an edge dislocation is that it is easier to produce slip or movement along a plane perpendicular to the edge dislocation. The bonds between atoms are already somewhat stretched in that region of the crystal.

A screw dislocation, shown in Figure 7.20(b), arises when the planes of atoms line up on one side of the crystal but the planes are out of register by one cell dimension on the other side of the crystal. Suppose this book were cut in such a way that the pages were cut from the outside edge half way to the binding. Then, turn the edges that are cut in such a way that the first sheet in the top half of the book lines up with the second sheet in the bottom half of the book. The second sheet in the top half lines up with the third sheet in the bottom half, and the *n*th sheet in the top half lines up with the *n*+1 sheet in the bottom half. However, at the edges of the pages that are bound, the pages are still aligned because the dislocation extends only part way through the book (crystal). This would be analogous to a screw dislocation in a crystal.

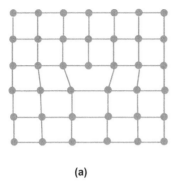 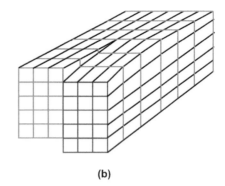

(a)   **(b)**

FIGURE 7.20
An edge dislocation (a) and screw dislocation (b) in a crystal.

7.10 PHASE TRANSITIONS IN SOLIDS

Although in this chapter the structures of many inorganic solids have been described, it has also been pointed out that some substances can be transformed from one structure to another. In fact, polymorphism is a very common occurrence in inorganic chemistry. For example, carbon exists in the form of diamond or as graphite, and, another form, C_{60}, will be described in Chapter 13. Numerous metals can be transformed from one solid structure to another. Many compounds (e.g. KSCN, $NaNO_3$, AgI, SiO_2, NH_4Cl, and $NH_4H_2PO_4$ to name a few) undergo structural changes. Such phase transitions are usually induced by changes in temperature or pressure. Because of the importance and scope of phase transitions in inorganic chemistry, a brief survey of the topic will be presented here.

Phase transitions are classified from a consideration of several factors. One type of phase transition is known as *reconstructive* based on the fact that a major rearrangement of structural units (atoms, molecules, or ions) is involved. For example, the conversion of the fused-ring layer structure of graphite into diamond, in which each carbon atom is bonded to four others in a tetrahedral arrangement, requires a radical change in structure and bonding mode. This is a reconstructive phase transition that takes place slowly under extreme conditions. The thermal stability of graphite and diamond are not greatly different (ΔH for $C_{(graphite)} \rightarrow C_{(diamond)}$ is only 2 kJ mol^{-1}), but there is no low energy pathway for the transition. This process is normally carried out at 1000−2000 °C and a pressure of up to 105 bar.

If a phase transition can occur by changing positions of structural units without rupturing bonds, it is known as a *displacive* phase transition. Because there is usually not a great change in bonding, displacive transitions are usually brought about by much milder conditions than are required to cause reconstructive transitions. For example, CsCl is converted from the CsCl structure to the NaCl structure by heating it to 479 °C, and AgI is transformed from a wurtzite structure to a *bcc* at 145 °C. Many of the phase transitions in metals can be induced by heating the metals at moderate temperatures because the bonds between metal atoms are shifted but not completely broken.

A complete discussion of the hundreds of phase transitions that are known for inorganic compounds is clearly beyond the scope of this book. However, it is possible to make some general observations regarding structural changes. From the thermodynamic principles discussed above, it is clear that when a phase transformation is induced by raising the temperature of a solid, there is an increase in volume and an increase in entropy as a result of the phase change. The phase that is stable at the higher temperature will generally have more disorder and a structure with a lower coordination number. On the other hand, if the phase transition is one that is brought about by increasing the pressure, the high pressure phase will usually be more dense and have a more ordered structure than the low temperature phase. The conversion of graphite to diamond has already been discussed in this connection. These general principles are applicable to many cases when predicting the structural changes that accompany a phase transition.

When cesium chloride is heated to a temperature of 479 °C, it changes from the CsCl structure to the NaCl structure. In this case, the coordination number changes from 8 to 6 as expected. On the other hand, when KCl is subjected to a pressure of 19.6 kbar, it is changed from the NaCl structure (coordination number 6) to the CsCl structure (coordination number 8). A very large number of other examples of this type of behavior can be given. Examples involving transformation of metals have been given earlier in this chapter. The subject of *rates* of phase transformations is one of the interesting problems in kinetics of solid state processes, and this subject will be considered in Chapter 8. Considered in its entirety, the subject of phase transformations is relevant to understanding solid state inorganic chemistry and materials science.

7.11 HEAT CAPACITY

The heat capacity of a monatomic ideal gas is the result of the molecules being able to absorb energy in three degrees of translational freedom. Each degree of freedom can absorb $\frac{1}{2}R$, which results in the heat capacity being approximately $3(1/2)R$. For a gas composed of diatomic molecules, there is also heat absorption to change the rotational energy and the vibrational energy. For a diatomic molecule, there is only one vibrational degree of freedom, which contributes R to the heat capacity. The number of degrees of vibrational freedom is given by $3N - 5$ for a linear molecule and $3N - 6$ for a nonlinear molecule, where N is the number of atoms.

Although the lattice members in a crystal do not move through space as do molecules of a gas, the lattice vibrations in a solid begin at very low temperature and are fully activated to absorb thermal energy at room temperature. For one mole of a monatomic species—for example, a metal such as Ag or Cu—the heat capacity should be given by $3R$ because there are three degrees of vibrational freedom for each particle in the lattice. For a very large number of particles, $3N - 6 \approx 3N$. Therefore, for a metal, the heat capacity, C_p, should be approximately $3R$ or 6 cal/mol deg (25 J/mol deg). The molar heat capacity is simply the specific heat multiplied by the atomic weight,

$$\text{Specific heat (cal/g K)} \times \text{Atomic weight (g/mol)} = C_p \text{ (cal/mol K)}$$

$$\text{Specific heat} \times \text{Atomic weight} \approx 6 \text{ cal/mol deg} \approx 25 \text{ J/mol K}$$

This rule was stated in 1819 by Dulong and Petit, and it indicates that the specific heat of a metal multiplied by the atomic weight is a constant. This relationship provides a way to estimate the atomic weight of a metal if its specific heat is known. How well the rule holds is indicated by the molar heat capacities of metals shown in Table 7.9.

Table 7.9 Molar Heat Capacities of Selected Metals at Room Temperature

Metal	C_p, J/mol K	Metal	C_p, J/mol K
Sb	25.1	Bi	25.6
Cd	25.8	Cu	24.5
Au	25.7	Ag	25.8
Sn	25.6	Ni	25.8
Pt	26.5	Pd	26.5

The data shown in the table indicate that the law of Dulong and Petit holds surprisingly well for metals. For a mole of NaCl, there are two moles of particles, so the heat capacity is approximately 12 cal/mol deg or 50 J/mol K. However, the heat capacity of a solid is not a constant, but rather it decreases rapidly at lower temperatures as shown in Figure 7.21 for copper. A more complete explanation of the heat capacity of a solid as outlined below was developed by Einstein.

For a harmonic oscillator, the mean energy is given in terms of the frequency, ν, by

$$E = \frac{1}{2}h\nu + \frac{\sum_n nh\nu \, e^{-nh\nu/kT}}{\sum_n h\nu \, e^{-nh\nu/kT}} \tag{7.34}$$

where h is Planck's constant and k is Boltzmann's constant. Simplifying gives

$$E = \frac{1}{2}h\nu + \frac{h\nu}{e^{h\nu/kT} - 1} \tag{7.35}$$

This can be considered as the average energy of a specific atom over time or the average energy of all the atoms at a specific time. It is useful to consider two special cases. At low temperature, $h\nu > kT$ and the

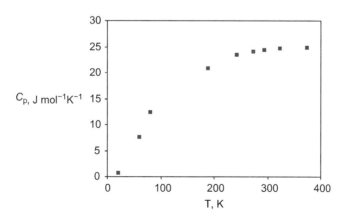

FIGURE 7.21

The variation in heat capacity of copper as a function of temperature. Note that the heat capacity increases approximately as the cube of T at low temperature and approaches the value predicted by the law of Dulong and Petit near room temperature.

average energy is approximately $\frac{1}{2}h\nu$. At high temperature, $h\nu < kT$, so $e^{h\nu/kT}$ becomes approximately equal to $1 + h\nu/kT$ so that

$$E = \frac{1}{2}h\nu + kT \approx kT \tag{7.36}$$

This is the classical limit because the energy levels expressed in terms of $h\nu$ are much smaller than the average energy of the oscillator.

When the temperature is such that $h\nu \approx kT$, neither of the limiting cases described above can be used. For many solids, the frequency of lattice vibration is of the order of 10^{13} Hz so that the temperature at which the value of the heat capacity deviates substantially from $3R$ is above 300 to 400 K. For a series of vibrational energy levels that are multiples of some fundamental frequency, the energies are 0, $h\nu$, $2h\nu$, $3h\nu$, etc. For these levels, the populations of the states (n_0, n_1, n_2, etc.) will be in the ratio $1 : e^{-h\nu/kT} : e^{-2h\nu/kT} : e^{-3h\nu/kT}$, etc. The total number of vibrational states possible for N atoms is $3N$ because there are three degrees of vibrational freedom. Therefore, from the Boltzmann distribution law we find that

$$n_1 = n_0 \, e^{-h\nu/kT}; \; n_2 = n_0 \, e^{-2h\nu/kT}; \; n_3 = n_0 \, e^{-3h\nu/kT}; \; \text{etc.} \tag{7.37}$$

Therefore, the total heat content, Q, of the crystal can be expressed as the sum of the number of particles populating each level times the energy of that level.

$$Q = n_0 \left[h\nu \, e^{-h\nu/kT} + 2h\nu \, e^{-2h\nu/kT} + 3h\nu \, e^{-3h\nu/kT} + \cdots \right] \tag{7.38}$$

The total number of atoms, N, is the sum of the populations of the states,

$$N = n_0 + n_1 + n_2 + n_3 + \ldots \tag{7.39}$$

Therefore,

$$Q = \frac{3Nh\nu}{e^{h\nu/kT} - 1} \tag{7.40}$$

The heat capacity is $\partial Q/\partial T$ so C_v can be expressed as

$$C_v = \frac{\partial Q}{\partial T} = 3Nk \left(\frac{h\nu}{kT}\right)^2 \frac{e^{h\nu/kT}}{\left(e^{h\nu/kT} - 1\right)^2} \tag{7.41}$$

If we let $x = h\nu/kT$ and define θ such that $k\theta = h\nu_{max}$, where ν_{max} is the maximum populated frequency, the total thermal energy can be written as

$$Q = \frac{9NkT^4}{\theta^3} \int_0^{x_{max}} \frac{x^3}{e^x - 1} dx \tag{7.42}$$

where $x_{max} = \theta/T$ and $x = h\nu/kT$. Also, θ is known as the Debye characteristic temperature. The expression for the heat capacity thus becomes

$$C_v = 9R \left(\frac{T}{\theta}\right)^3 \int_0^{x_{max}} \frac{e^x \, x^4}{(e^x - 1)^2} dx \tag{7.43}$$

Table 7.10 Debye Characteristic Temperatures for Metals			
Metal	**θ**	**Metal**	**θ**
Li	430	Ca	230
Na	160	Pt	225
K	99	Be	980
Au	185	Mg	330
Pb	86	Zn	240
Cr	405	Cd	165

At high temperature, the integral is approximately $\int x^2 \, dx$ and the energy is evaluated as $3RT$ so that $C_v = 3R$, the classical value of Dulong and Petit. At low temperature (where x is large), we can approximate x by letting it equal ∞ so the integral becomes equal to $\pi^4/15$. In this case, $C_v = 463.9(T/\theta)^3$ cal/mol deg, and it can be seen that C_v varies as T^3. This agrees well with the experimental variation of C_v with temperature in the region where the heat capacity curve rises steeply. Different metals have different values for the Debye characteristic temperature, and values for several metals are shown in Table 7.10.

If we assume that the complete assembly of atoms vibrates as a coupled system that vibrates as a whole, only certain energies are allowed. The energy of the system must change by some multiple of $h\nu$, and these quanta of vibrational energy involve displacements of all the atoms. The quanta of energy that correspond to changes in vibrational states are called *phonons*. As the temperature increases, the extent of atomic vibrations becomes larger, which is equivalent to saying that the number of phonons has increased. Vibrations in a solid produce longitudinal (compression) waves and transverse (perpendicular) waves. Adjacent atoms moving in the same phase give rise to so-called acoustic modes, and adjacent atoms moving with phases approximately 180° apart give rise to optical modes.

In a metal, there are excited states for electrons that lie below the ionization energy. This can be considered as an electron in a "conduction band" and a "hole" that interacts so that the combination is neutral but not of lowest energy. Such an excited state is called an *exciton*. Excitons may move by diffusion of the electron-hole pair or by transfer of a molecular exciton to another molecule. Reversion of the exciton to a lower energy state may be slow enough for the lifetime to be longer that of lattice relaxation processes.

7.12 HARDNESS OF SOLIDS

Although not one of the most frequently discussed properties of solids, hardness is an important consideration in many instances, especially in the area of mineralogy. In essence, hardness is a measure of the ability of a solid to resist deformation or scratching. It is a difficult property to measure accurately, and, for some materials, a range of values is reported. Because of the nature of hardness, it is necessary to have some sort of reference so that comparisons can be made. The hardness scale most often used is that developed by Austrian mineralogist F. Mohs in 1824. The scale is appropriately known as the Mohs scale. Table 7.11 gives the fixed points on which the scale is based.

It has long been recognized that the Mohs scale is not totally satisfactory for several reasons. One reason is that some minerals have different resistance to scratching and deformation on different

Table 7.11 Reference Materials for the Mohs Hardness Scale

Mineral	Mohs Hardness	Modified Value	Mineral	Mohs Hardness	Modified Value
Talc	1	1.0	Orthoclase	6	6.0
Graphite	2	2.0	Quartz	7	7.0
Calcite	3	3.2	Topaz	8	8.2
Fluorite	4	3.7	Corundum	9	8.9
Apatite	5	5.2	Diamond	10	10.0

surfaces of the crystals. For example, crystalline *calcite* has surfaces that differ by as much as 0.5 unit depending on the face tested. Also, some minerals do not have a fixed composition. For example, *apatite* is a calcium phosphate that contains varying amounts of chlorine and fluorine. *Fluorapatite* is $Ca_5(PO_4)_3F$ and *chloroapatite* is $Ca_5(PO_4)_3Cl$. Naturally occurring apatite is written as $Ca_5(PO_4)_3(F,Cl)$ to indicate that composition. As a result of these difficulties, a modified Mohs scale has been proposed, and Table 7.11 shows the values for the minerals that are the references on the Mohs scale. It is easy to see that the values are not significantly different.

That hardness should be related to other properties of crystals is intuitively obvious. As a general rule, the materials that have high hardness values also have high melting points and lattice energies. For ionic crystals, hardness also increases as the distance between ions decreases. This trend is illustrated by the oxides and sulfides of the Group II metals as is shown by the data presented in Table 7.12.

Also, there is a rough correlation between hardness and the effect of higher charges on the ions when the distances between ionic centers are essentially the same. As shown by the data presented in Table 7.13, this correlation is evident when several pairs of compounds are considered in which the distances between ionic centers are approximately the same.

The data show that when compounds such as LiF and MgO are considered, even though the internuclear distances are similar, there is a great difference in the hardness of the materials. Of course, there is also a great difference between the melting points and lattice energies of the compounds. A similar situation exists for NaCl compared to CaS for which the internuclear distance is approximately 280 pm in both cases, but the hardnesses are 2.5 and 4.0, respectively. Although the data allow generalizations to be made regarding hardness and other properties, they do not permit a quantitative relationship to be developed.

Table 7.12 Hardness of Group II Oxides and Sulfides

Cation	Oxides		Sulfides	
	r, pm	h	r, pm	h
Mg^{2+}	210	6.5	259	4.5
Ca^{2+}	240	4.5	284	4.0
Sr^{2+}	257	3.5	300	3.3
Ba^{2+}	277	3.3	318	3.0

Table 7.13 Hardness (Mohs Scale) of Some Ionic Crystals

	LiF	MgO	NaF	CaO	LiCl	SrO
r, pm	202	210	231	240	257	257
h	3.3	6.5	3.2	4.5	3.0	3.5
	LiCl	**MgS**	**NaCl**	**CaS**	**LiBr**	**MgS**
r, pm	257	259	281	284	275	273
h	3.0	4.5–5	2.5	4.0	2.5	3.5
	CuBr	**ZnSe**	**GaAs**	**GeGe**		
r, pm	246	245	244	243		
h	2.4	3.4	4.2	6.0		

Table 7.14 Hardness and Melting Point for Selected Metals

Metal	Hardness	Melting Point (K)	Metal	Hardness	Melting Point (K)
Cadmium	2.0	594	Palladium	4.8	1825
Zinc	2.5	693	Platinum	4.3	2045
Silver	2.5–4.0	1235	Ruthenium	6.5	2583
Manganese	5.0	1518	Iridium	6.0–6.5	2683
Iron	4.0–5.0	1808	Osmium	7.0	3325

The hardness of transition metals varies widely. Table 7.14 shows the values for several metals along with their melting points. The data show that there is a rough correlation between the hardness of many metals and their melting points. However, it should be kept in mind that the hardness scale is not a highly accurate one (the number of significant digits may be only one in some cases), so it is not possible to develop a good quantitative relationship. In spite of the limitations, it is often worthwhile to have a general understanding of the hardness of inorganic materials and how that property is related to many others.

In this chapter, a survey of the structure and properties of solids has been presented. Solid state chemistry has emerged as an important area of science, and, although it is not exclusively so, much of the work deals with inorganic substances. For more information on this important area, the references at the end of the chapter should be consulted.

References for Further Study

Anderson, J.C., Leaver, K.D., Alexander, J.M., Rawlings, R.D., 1974. *Materials Science*, 2nd ed. Wiley, New York. This book presents a great deal of information on characteristics of solids that is relevant to solid state chemistry.

Borg, R.J., Dienes, G.J., 1992. *The Physical Chemistry of Solids*. Academic Press, San Diego, CA. A good coverage of topics in solid state science.

Burdett, J.E., 1995. *Chemical Bonding in Solids*. Oxford University Press, New York. A higher level book on the chemistry and physics of solids.

Douglas, B., McDaniel, D., Alexander, J., 1994. *Concepts and Models of Inorganic Chemistry*, 3rd ed. John Wiley, New York. Chapters 5 and 6 give a good introduction to solid state chemistry.

Gillespie, R.R., 1998. A discussion of the properties of fluorides in terms of bonding. *J. Chem. Educ.* 75, 923.

Julg, A., 1978. *Crystals as Giant Molecules*. Springer Verlag, Berlin. This is Volume 9 in a lecture note series. It presents a wealth of information and novel ways of interpreting properties of solids.

Ladd, M.F.C., 1979. *Structure and Bonding in Solid State Chemistry*. John Wiley, New York. An excellent book on solid state chemistry.

Pauling, L., 1960. *The Nature of the Chemical Bond*, 3rd ed. Cornell University Press, Ithaca, New York. A classic book that presents a good description of crystal structures and bonding in solids.

Raghavan, V., Cohen, M., 1975. *Solid-state phase transformations*, chapter 2. In: Hannay, N.B. (Ed.), *Treatise on Solid State Chemistry. Changes in State*, vol. 5. Plenum Press, New York. A mathematical treatment of the subject including a good treatment of the kinetics of phase transitions.

Rao, C.N.R., 1984. *Acc. Chem. Res.* 17, 83−89. This review, *Phase Transitions and the Chemistry of Solids*, presents a general overview of phase transitions.

Rao, C.N.R., Rao, K.J., 1967. *Phase transformations in solids*, chapter 4. In: Reiss, H. (Ed.), *Progress in Solid State Chemistry*, vol. 4. Pergamon Press, New York. A good introduction to the topic of phase transitions by two of the eminent workers in the field.

Smart, L., Moore, E., 2012. *Solid State Chemistry*, 4th ed. CRC Press, Boca Raton, FL. An introductory book on solid state chemistry.

West, A.R., 1984. *Solid State Chemistry and its Applications*. Wiley, New York. One of the best introductory books on solid state chemistry. Chapter 12 in this book is devoted to a discussion of phase transitions.

QUESTIONS AND PROBLEMS

1. Consider two Na^+ Cl^- ion pairs arranged in a head-to-tail or antiparallel structure in which the distance between ionic centers is 281 pm. Calculate the energy of this arrangement.

2. (a) Explain why the hydration number for Li^+ is approximately five, whereas that for Mg^{2+} is almost twice that number.

   (b) Explain why the hydration number for Mg^{2+} is approximately 10 but that for Ca^{2+} is approximately seven.

3. Use the Kapustinskii equation to determine the lattice energies for the following:
   (a) RbCl, (b) NaI, (c) $MgCl_2$, (d) LiF.

4. Using your answers in Problem 3 and the data shown in Table 7.7, calculate the heat of solution for the compounds.

5. $RbCaF_3$ has the perovskite structure with the Ca in the center of the unit cell. What is the electrostatic bond character of each of the Ca−F bonds? How many fluoride ions must surround each Ca^{2+} ion? What is the electrostatic bond character of each Rb−F bond? How many F^- ions surround each Rb^+?

6. The nickel crystal has a cubic closest packing arrangement with an edge length of 352.4 nm. Using this information, calculate the density of nickel.

7. In PdO, each Pd is surrounded by four oxygen atoms, but *planar* sheets do not exist. Explain why they are not expected.

8. Potassium fluoride (KF) crystallizes in a sodium chloride lattice. The length of the edge of the unit cell (sometimes called the cell or lattice constant) has the value 267 pm for KF.

   (a) Calculate the attraction that exists for one mole of KF.

   (b) Using the Kapustinskii equation, calculate the lattice energy for KF. The ionic radii are 138 and 133 pm, respectively, for K^+ and F^-.

9. The H−H bond energy is 435 kJ mol^{-1}, and the heat of sublimation of Li is 160 kJ mol^{-1} and its ionization energy is 518 kJ mol^{-1}. The heat of formation of LiH is −90.4 kJ mol^{-1} and its lattice energy is 916 kJ mol^{-1}. Use this information with an appropriate thermochemical cycle to evaluate the electron affinity of hydrogen.

10. The structure of PdCl$_2$ involves linear chains that can be shown as

    Considering the structure to be ionic, explain why there are no strong forces of attraction between chains.

11. Predict the crystal type for each of the following using the radius ratio.
    (a) K$_2$S, (b) NH$_4$Br, (c) CoF$_2$, (d) TiF$_2$, (e) FeO.

12. The solubility (S) of KBrO$_3$ in water (given as grams per 100 grams of water) varies with temperature as follows.

Temp., °C	10	20	30	50	60
S, g/100 g H$_2$O	4.8	6.9	9.5	17.5	22.7

    Use these data to determine the heat of solution of KBrO$_3$ in water.

13. The length of the edge of the unit cell for Na$_2$O (which has the antifluorite structure) is 555 pm. For Na$_2$O, determine the following:
    (a) the distance between sodium ions
    (b) the distance between sodium and oxide ions
    (c) the distance between oxide ions
    (d) the density of Na$_2$O

14. Suppose that in a crystal of NaCl, contact between chloride ions occurs and contact between sodium and chloride ions occurs. Determine the percentage of free space in the NaCl crystal.

15. Although CaF$_2$ has the fluorite structure, MgF$_2$ has the rutile structure. Explain this difference.

16. The removal of two electrons from a magnesium atom is highly endothermic as is the addition of two electrons to an oxygen atom. In spite of this, MgO forms readily from the elements. Write a thermochemical cycle for the formation of MgO and explain the process from the standpoint of the energies involved.

17. Cations in aqueous solutions have an effective radius that is approximately 75 pm larger than the crystallographic radii. The value of 75 pm is approximately the radius of a water molecule. It can be shown that the heat of hydration of cations should be a linear function of Z^2/r' where r' is the effective ionic radius and Z is the charge on the ion. Using the ionic radii shown in Table 7.4 and hydration enthalpies shown in Table 7.7, test the validity of this relationship.

18. KF crystallizes in a sodium chloride lattice arrangement with a cell edge length of 267 pm.
    (a) Calculate the total attraction in one mole of KF.
    (b) Calculate the actual lattice energy by means of the Kapustinskii equation or a thermochemical cycle.

(c) Explain why the values determined in parts (a) and (b) are different.

(d) Using the results from (a) and (b), evaluate the value of n that is correct in this case according to

$$U = \frac{N_oAe^2}{r}\left(1 - \frac{1}{n}\right)$$

Note: $e = 4.8 \times 10^{-10}$ esu and 1 esu $= 1\ g^{1/2}\ cm^{3/2}\ sec^{-1}$.

19. A certain metal has a *fcc* structure with an edge length of 3.75×10^{-8} cm. If the density of the metal is $7.71\ g\ cm^{-3}$, what is the atomic weight of the metal?

CHAPTER 8

Dynamic Processes in Inorganic Solids

Although reactions in the gas phase and solutions may be better understood on the molecular level, reactions in solids are quite common and useful. Because so many inorganic compounds are solids, inorganic chemistry involves a great deal of solid-state science. However, reactions in the solid state may involve several factors that are not relevant to reactions carried out in the gas phase or in solution. Some of the reactions in inorganic solids are of economic importance, and others reveal a great deal about the behavior of inorganic materials. The study of reactions in solids is frequently given little attention in the presentation of inorganic chemistry, but a great deal is known about many of the processes. Therefore, this chapter is devoted to presenting some of the basic ideas concerning reactions in solids and discussing some of the methods used in this area of inorganic chemistry.

8.1 CHARACTERISTICS OF SOLID-STATE REACTIONS

There are several ways to induce reactions in solids. The application of heat, electromagnetic radiation, pressure, ultrasound, or some other form of energy may induce a transformation in a solid. For centuries, it has been a common practice to subject solid materials to heat to determine their thermal stability, study their physical properties, or convert one material into another. One important commercial reaction, that producing lime,

$$CaCO_3(s) \xrightarrow{\Delta} CaO(s) + CO_2(g) \tag{8.1}$$

is carried out on an enormous scale (see Chapter 13).

Reactions in solids are often vastly different from those that take place in solutions. Because many of the reactions in the solid state involve *inorganic* materials, an introduction to this important topic will be presented in this chapter to show some of the principles that are applicable to this area of inorganic chemistry. The emphasis will be on showing several types of reactions, but no attempt will be made to present comprehensive coverage of the hundreds of reactions that take place in the solid state. Although some reactions involve the reaction of two solid phases, the discussion will deal primarily with one component. An enormous number of reactions of this type involve the decomposition of a solid to produce a different solid and a volatile product as illustrated by Eqn (8.1).

Rates of reactions in solutions are usually expressed as mathematical functions of concentrations of the reacting species as variables, the *rate law*. For reactions in solids, this is not feasible because any particle of a uniform density has the same number of moles per unit volume. A different reaction variable must be chosen and that is most commonly the fraction of the reaction complete, α. At the beginning of a reaction, $\alpha = 0$; and at the completion of the reaction, $\alpha = 1$ (if the reaction goes to

243

Inorganic Chemistry. DOI: http://dx.doi.org/10.1016/B978-0-12-385110-9.00008-X

completion, which is not always the case). The fraction of the reactant remaining is given by $(1 - \alpha)$ so rate laws are generally written in terms of that quantity. Rate laws for reactions in solids are frequently determined by the geometry of the sample, formation of active sites, diffusion, or some other factors. As a result, many of the rate laws for reactions in solids are derived based on those considerations. In the majority of cases, it is not possible to interpret a rate law with the usual concepts related to bond breaking and bond making steps. Moreover, even though the rate constant for a reaction varies according to the Arrhenius equation, the calculated "activation energy" may be for some process such as diffusion rather than changes in "molecules", which often are not present. The transition state may be the motion of an ion through a potential field in a crystal rather than a molecule having stretched or bent bonds.

To test rate laws, α must be determined as a function of time using an appropriate experimental technique. If the reaction involves the loss of a volatile product as shown in Eqn (8.1), the extent of reaction can be followed by determining the mass loss either continuously or from sample weight at specific times. Other techniques are applicable to different types of reactions. After α has been determined at several reaction times, it is often instructive to make a graph of α versus time before the data are analyzed according to the rate laws. As will be shown later, one can often eliminate some rate laws from consideration because of the general shape of the α versus t curve.

Figure 8.1 shows three hypothetical α versus time curves for solid-state reactions that apply particularly to cases where a gaseous product is evolved. For some solid reactants, gases may be adsorbed on the solid before the reaction begins and may be lost quickly when the reaction begins. When the α versus t curve is examined, it is seen that there is an initial change that may be due to the loss of the volatile material that was adsorbed (Curve I, Region A in Figure 8.1). The loss of the adsorbed gas *appears* to indicate that the reaction (illustrated by Curve I) is taking place with the loss of a gaseous product so the graph shows an initial deviation from the horizontal axis. This initial response is not part of the chemical reaction. Such a condition is rather uncommon, but in a general case, it is assumed that it can be present. If no volatiles are lost, but there is an induction period (illustrated by Curve II), the value of α increases more rapidly as shown in Regions B on Curves I and II. In these regions, the rate of

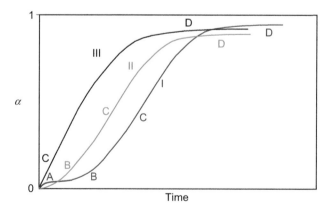

FIGURE 8.1
The variation of α versus time for a hypothetical reaction in the solid state. Curve I shows the evolution of a gas (A) followed by an induction period (B) before the maximum rate is reached, Curve II represents a reaction that shows an induction period (B) before reaching maximum rate, and Curve III represents a reaction that begins at the maximum rate.

the reaction is increasing (in what is usually a nonlinear way). This is known as the *acceleratory* period. In Region C (as shown by Curves I, II, and III), the rate of the reaction is a maximum, and after that, the rate decreases (the *deceleratory* or *decay* period, D) and approaches zero as completion or equilibrium is approached. The reaction represented by Curve III shows no acceleratory period, and it begins at the maximum rate. It must be emphasized that the majority of reactions in solids do not show all these features, and many are represented by Curve III, which does not have complicating features such as rapid loss of gas or an induction period.

For many reactions, the rate is a maximum as the reaction begins (the maximum amount of reactant is present at $t = 0$). As a solid reacts, there is frequently a coalescence of surface material as surface tension works to produce a minimum surface area. At elevated temperatures, there is increased mobility of structural units (atoms or ions) that leads to rounding of surfaces. This process, known as *sintering*, can lead to closing of pores and welding of individual particles together. As a result, it may be difficult for a volatile product to escape from the reacting solid. The situation in which a gaseous product is held by adsorption or absorption is known as *retention*. Because of retention, a reaction may never reach completion in terms of all the gaseous product being evolved.

Even though the vast majority of solid-state reactions do not exhibit all the features indicated on curve I shown in Figure 8.1, it is frequently the case that no single rate law will correlate data for the entire range of the reaction. A different function may be needed to fit data in the induction region, another in the region where the rate is maximum, and yet another in the deceleratory (decay) region. It should be remembered that subtle features of a reaction are sometimes more apparent when looking at a graph rather than looking at numerical data alone.

In addition to the complications described, other factors are important in specific reactions. If a reaction takes place on the surface of a solid, reducing the particle size (by grinding, milling, or vibration) leads to an increase in surface area. A sample of a solid treated in this way *may* react faster than an untreated sample, but in some cases, changing the particle size does not alter the rate. This has been found to be true for the dehydration of $CaC_2O_4 \cdot H_2O$, which is independent of the particle size over a wide range of α values.

In Chapter 6, it was shown how defects in solids can be produced by heating the solid to produce the defects, then quenching the crystal. On the other hand, defects can be removed by heating a crystal and allowing it to cool very slowly in an annealing process to allow the defects to be removed by rearrangement of particles. Defects represent high energy sites where a reaction may initiate. Increasing the concentration of defects will usually increase the rate at which the solid will react. These observations show that the reactivity of a particular sample of a solid may depend on prior treatment of the sample.

8.2 KINETIC MODELS FOR REACTIONS IN SOLIDS

There are significant differences between the kinetic models for reactions carried out in the solid phase and those taking place in gases or solutions. Therefore, it is appropriate to describe briefly some of the kinetic models that have been found to be particularly applicable to reactions in inorganic solids.

8.2.1 First-Order

As was discussed earlier in this chapter, the concept of a reaction order does not apply to a crystal that is not composed of molecules. However, there are numerous cases in which the rate of reaction is proportional to the amount of material present. We can show how this rate law is obtained in a simple way. If the amount of material at any time, t, is represented as W and if we let W_o be the amount of material initially present, the amount of material that has reacted at any time will be equal to $(W_o - W)$. In a first-order reaction, the rate is proportional to the amount of material. Therefore, the rate of reaction can be expressed as

$$-\frac{dW}{dt} = kW \tag{8.2}$$

When integration is performed so that the amount of reactant is W_o at $t = 0$ and is W at time t, the result is

$$\ln\frac{W_o}{W} = kt \tag{8.3}$$

We now need to transform the rate equation to one involving α, the fraction of the sample reacted, by making use of the relationship

$$\alpha = \frac{W_o - W}{W} = 1 - \frac{W}{W_o} \tag{8.4}$$

Therefore, $(1 - \alpha) = W/W_o$, which when substituted in Eqn (8.3) gives

$$-\ln(1 - \alpha) = kt \tag{8.5}$$

If the fraction reacted is α, then $(1 - \alpha)$ is the amount of sample unreacted, and we see that a plot of $-\ln(1 - \alpha)$ versus time would be linear with a slope of k. It should come as no surprise that reactions are known in which the rate law over some portion of the reaction is first-order, but in the latter states of the reaction, the reaction is diffusion controlled.

8.2.2 The Parabolic Rate Law

Reactions in which a gas or liquid reacts with the surface of a solid are rather common processes in inorganic chemistry. The product that forms as a layer on the surface of the solid may impede the other reactant from contacting the solid. There are several types of behavior that depend on how the product layer affects the mobility of reactants, but in this instance, we will assume that the rate is inversely proportional to the thickness of the product layer. When the rate law is written in terms of the thickness of the product layer, x, the result is

$$\text{Rate} = \frac{dx}{dt} \tag{8.6}$$

Because the rate decreases as x increases, the rate is proportional to $1/x$ indicating that

$$\frac{dx}{dt} = k\frac{1}{x} \tag{8.7}$$

Rearranging gives

$$x\,dx = k\,dt \tag{8.8}$$

At $t = 0$, $x = 0$ and at a later time t, the thickness of the layer is x. Integrating between these limits yields

$$\frac{x^2}{2} = kt \tag{8.9}$$

Because the rate law in this form is a quadratic equation, the rate law is known as the *parabolic rate law*. When we solve this equation for the thickness of the product layer, x, we obtain

$$x = (2\,kt)^{1/2} \tag{8.10}$$

This is the rate law that applies when the product layer is protective in nature.

If the product layer is not protective, the mobile reactant has access to the surface of the solid. In that case, it is easy to show that the rate law can be expressed as

$$x = kt \tag{8.11}$$

In another type of reaction, the penetration of the mobile reactant varies as $1/x^2$, which gives rise to the so-called cubic rate law of the form

$$x = (3\,kt)^{1/3} \tag{8.12}$$

As will be described later, a common and important type of reaction that involves the oxidation of metals during corrosion processes sometimes follows a rate law of this form.

8.2.3 Contracting Volume Rate Law

A rate law that shows some of the peculiarities of reactions in solids arises in the following way. A solid particle having a spherical shape is assumed to react only on the surface. This rate law has been found to model the shrinking of solid particles in aerosols as well as other reactions that take place on the surface of solid particles.

In this model, the rate of the reaction is determined by the surface area, $S = 4/3\pi r^3$, but the amount of the reactant is determined by the volume, $V = 4\pi r^3/3$. The amount of solid reacting is given as $-dV/dt$ and is determined by the surface area. The rate law can be written as

$$-\frac{dV}{dt} = kS = k\left(4\pi r^2\right) \tag{8.13}$$

From the expression for the volume, we can solve for r^2 to obtain $(3V/4\pi)^{2/3}$. Substituting for r^2 in Eqn (8.13) yields

$$-\frac{dV}{dt} = k(4\pi)\left(\frac{3V}{4\pi}\right)^{2/3} = k(4\pi)\left(\frac{3}{4\pi}\right)^{2/3} V^{2/3} = k'V^{2/3} \tag{8.14}$$

where $k' = k\,(4\pi)\,(3/4\pi)^{2/3}$. Therefore, we have shown that

$$-\frac{dV}{dt} = k'V^{2/3} \tag{8.15}$$

which leads to this type of process being known as a "two-thirds" order reaction, but this is inappropriate because it is not an "order" type of reaction. After integration, the rate law becomes

$$V_o^{1/3} - V^{1/3} = \frac{k't}{3} \tag{8.16}$$

To obtain the rate law in a form containing α, we recall that the fraction of the particle reacted is the change in volume divided by the original volume, $\alpha = (V_o - V)/V_o$. Rearranging gives $\alpha = 1 - (V/V_o)$, which leads to $(V/V_o) = 1 - \alpha$, the fraction remaining.

Dividing both sides of Eqn (8.16) by $V_o^{1/3}$ gives

$$\frac{V_o^{1/3} - V^{1/3}}{V_o^{1/3}} = 1 - \frac{V^{1/3}}{V_o^{1/3}} = \frac{k't}{3V_o^{1/3}} \tag{8.17}$$

Substituting the expression found above for $(1 - \alpha)$ leads to

$$1 - (1 - \alpha)^{1/3} = \frac{k't}{3V_o^{1/3}} \tag{8.18}$$

This equation can be put in the form

$$1 - (1 - \alpha)^{1/3} = k''t \tag{8.19}$$

where $k'' = k'/3V_o^{1/3}$ and $k' = k \cdot 4\pi(3/4\pi)^{2/3}$. The *observed* rate constant, k'', is associated with the geometry of the sample, but it is not related to the population of a transition state in the usual sense. If a reaction is assumed to take place on the surface of a cubic solid, the rate law turns out to be identical to that shown above, except that the observed rate law has other geometric factors subsumed in it. Although the derivation will not be shown, if a reaction involves a receding area, the rate law will contain the quantity $(1 - \alpha)^{1/2}$.

As shown by Eqn (8.15), the reaction is a "two-thirds" order, but that does not involve the concept of molecularity. As a result of the surface area being a maximum at the beginning of the reaction, the rate is maximum at that time and decreases thereafter. A rate law of this type is known as a *deceleratory* rate law. As will be shown later, there are several rate laws that show this characteristic.

8.2.4 Rate Laws for Cases Involving Nucleation

Solids are not generally equally reactive throughout the sample. Many reactions in solids begin at an active site and progress outward from that point. For example, when a solid undergoes a phase transformation, the change usually begins at active sites that may involve point defects. As the solid changes structure outward from such active sites, it may follow a rate law of the type being considered here. However, many types of reactions in solids as well as crystallization proceed from active sites so this type of rate law is a frequently occurring one. Microscopic examination and other techniques have been used to follow the spread of reactions from nuclei.

The active sites from which a reaction in a solid spreads are known as *nuclei*. It is known that nuclei may grow in one, two or three dimensions, and each case leads to a different form of the rate law. If the nuclei form in random sites in the solid (or perhaps on the surface), the rate laws are known *random nucleation* rate laws that have the form

$$[-\ln(1 - \alpha)]^{1/n} = kt \tag{8.20}$$

where n is known the *index of reaction*. As should be apparent, the concept of "order" is not applicable in these cases. This rate law is known as the Avrami−Erofeev rate law, for which the initial assumptions about the way in which the reaction spreads from the nuclei give rise to n values of 1.5, 2, 3, or 4. These rate laws are referred to as A1.5, A2, A3, and A4, respectively. Although this will be stated without proof, the case where $n = 1.5$ corresponds to a diffusion controlled process. Derivations of the Avrami–Erofeev rate laws (sometimes called simply Avrami rate laws) are somewhat tedious, and the interested reader is referred to the references listed at the end of this chapter for details (especially Young, 1966).

When testing data for α versus t, the object is to identify the appropriate value of n after which the rate constant can be calculated, and knowing the rate constants at several temperatures allows an activation energy to be determined. Taking the logarithm of both sides of Eqn (8.20) yields

$$\frac{1}{n}\ln[-\ln(1-\alpha)] = \ln(kt) = \ln k + \ln t \tag{8.21}$$

This equation shows that a graph of $\ln[-\ln(1-\alpha)]$ versus $\ln(t)$ would be linear and have a slope of n and an intercept of $n[\ln(k)]$ if the value of n correctly fits the data. For a series of (α, t) data, plots can be made using the various values of n as trial values to see which value yields a straight line. Because of the numerical idiosyncrasies, it is generally better to make the graphs using Eqn (8.20) by plotting $[\ln(1/(1-\alpha))]^{1/n}$ versus time to test the n values. In some data analysis procedures, the computations are carried out by computer in such a way that the value of n is allowed to vary from perhaps 1–4. Linear regression is carried out iteratively as n is varied by some increment until the highest correlation coefficient is obtained to the accuracy desired. Although this will find a value for n that gives the best fit of the data to an Avrami rate law, there is little *chemical* meaning or interpretation for an n value of perhaps 2.38 or 1.87.

Methods of data analysis for reactions in solids are somewhat different from those used in other types of kinetic studies. Therefore, the analysis of data for an Avrami type rate law will be illustrated by an numerical example. The data to be used are shown in Table 8.1, and they consist of (α, t) pairs that were calculated assuming the A2 rate law and $k = 0.020 \text{ min}^{-1}$.

The data shown in Table 8.1 were used to prepare Figure 8.2. The relationship shows the *sigmoidal* profile that is characteristic of autocatalysis or nucleation processes. Although the graph will not be shown, a plot of $[-\ln(1-\alpha)]^{1/2}$ versus time for these data yields a straight line as expected. When graphs are prepared of the $[-\ln(1-\alpha)]^{1/n}$ functions versus time using the various values of n, only the "correct" value of n will yield a straight line. If a trial value of n is larger than the correct value, a curve

Table 8.1 Values of α as a Function of Time for a Reaction Following an Avrami–Erofeev Rate Law with $n = 2$ and $k = 0.020 \text{ min}^{-1}$

Time, min	α	Time, min	α	Time, min	α
0	0.000	35	0.387	70	0.859
5	0.010	40	0.473	75	0.895
10	0.039	45	0.555	80	0.923
15	0.086	50	0.632	85	0.944
20	0.148	55	0.702	90	0.960
25	0.221	60	0.763	95	0.973
30	0.302	65	0.815	100	0.982

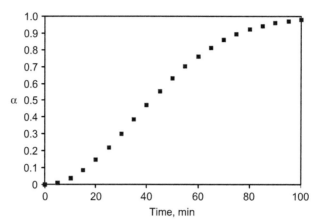

FIGURE 8.2
A graph of α versus time for an Avrami rate law with $n = 2$ and $k = 0.020\ \mathrm{min}^{-1}$.

will be obtained that is concave downward, whereas a value of n smaller than the correct value will result in a curve that is concave upward. If a sigmoidal curve is obtained for α versus time data when analyzing data for a reaction in a solid, it is generally a good indication that the rate of the reaction is controlled by a nucleation process.

A large number of inorganic compounds crystallize as hydrates. One of the most familiar examples is copper sulfate pentahydrate, $CuSO_4 \cdot 5H_2O$. Like most hydrates, when this material is heated it loses water, but because all the H_2O molecules are bound in different ways, some are lost more easily than others. Therefore, as the solid is heated the reactions observed first are

$$CuSO_4 \cdot 5\,H_2O(s) \;\rightarrow\; CuSO_4 \cdot 3\,H_2O(s) + 2\,H_2O(g) \tag{8.22}$$

$$CuSO_4 \cdot 3\,H_2O(s) \;\rightarrow\; CuSO_4 \cdot H_2O(s) + 2\,H_2O(g) \tag{8.23}$$

The first of these reactions takes place as the sample is heated in the range of 47–63 °C and the second in the range of 70.5–86 °C. When the data were analyzed to determine the rate law for the processes, it was found that both reactions followed an Avrami rate law with an index of 2 as the extent of reaction varied from $\alpha = 0.1$ to $\alpha = 0.9$ (Ng et al., 1978). Another reaction for which most data provide the best fit with an Avrami rate law is

$$[Co(NH_3)_5H_2O]Cl_3(s) \;\overset{\Delta}{\rightarrow}\; [Co(NH_3)_5Cl]Cl_2(s) + H_2O(g) \tag{8.24}$$

For this reaction, the best fit was given with the A1.5 rate law,

$$1 - (1 - \alpha)^{2/3} = kt \tag{8.25}$$

An interesting reaction of a solid coordination compound is

$$K_4[Ni(NO_2)_6] \cdot xH_2O(s) \;\rightarrow\; K_4[Ni(NO_2)_4(ONO)_2](s) + xH_2O(g) \tag{8.26}$$

in which dehydration and linkage of two of the nitrite ligands occurs. For this reaction, an Avrami rate law provided the best fit to the kinetic data, but as is often the case, the inaccuracy of the data did not allow a distinction to be made unambiguously between A.15 and A2 (House and Bunting, 1975). The essence of this discussion is to show that many types of reactions in solids follow Avrami rate laws.

The kinetic models described above are only a few of those that have been found to represent reactions in solids. Moreover, it is sometimes observed that a reaction may follow one rate law in the early stages of the reaction, but a different rate law may apply in the later stages. Because many of the rate laws that apply to reactions in solids are quite different from those encountered in the study of reactions in gases and solutions, a summary of the most common types is presented in Table 8.2. Although the rate laws shown in Table 8.2 do not encompass all those known to model solid-state reactions, they do apply to the vast majority of cases.

Fitting data for the fraction of a reaction complete as a function of time to the kinetic models is often accompanied by some difficulty. For some reactions, it may not be possible to determine the values for

Table 8.2 Some Common Rate Laws for Reactions in Solids

Type of Process		Mathematical Form of $f(\alpha) = kt$
Deceleratory α-time curves based on reaction order		
F1	First-order	$-\ln(1-\alpha)$
F2	Second-order	$1/(1-\alpha)$
F3	Third-order	$[1/(1-\alpha)]^2$
Deceleratory α-time curves based on geometrical models		
R1	One-dimensional contraction	$1-(1-\alpha)^{2/3}$
R2	Contracting area	$1-(1-\alpha)^{1/2}$
R3	Contracting volume	$1-(1-\alpha)^{1/3}$
Deceleratory α-time curves based on diffusion		
D1	One-dimensional diffusion	α^2
D2	Two-dimensional diffusion	$(1-\alpha)\ln(1-\alpha)+\alpha$
D3	Three-dimensional diffusion	$[1-(1-\alpha)^{1/3}]^2$
D4	Ginstling–Brounshtein	$[1-(2\alpha/3)]-(1-\alpha)^{2/3}$
Sigmoidal α-time curves		
A1.5	Avrami–Erofeev one-dimensional growth of nuclei	$[-\ln(1-\alpha)]^{2/3}$
A2	Avrami–Erofeev two-dimensional growth of nuclei	$[-\ln(1-\alpha)]^{1/2}$
A3	Avrami–Erofeev three-dimensional growth of nuclei	$[-\ln(1-\alpha)]^{1/3}$
A4	Avrami–Erofeev	$[-\ln(1-\alpha)]^{1/4}$
B1	Prout–Tompkins	$\ln[\alpha/(1-\alpha)]$
Acceleratory α-time curves		
	Power law	$\alpha^{1/2}$
	Exponential law	$\ln\alpha$

α with high accuracy, and examination of the rate laws shown in Table 8.2 shows that some of them do not differ much. Slight errors in the data can obscure subtle differences in how well the rate laws model a reaction. For example, a series of data for α as a function of time might give about an equally good fit to the A2 or A3 rate laws because they differ only slightly in mathematical form. In most cases, it is not possible to follow a reaction over several half-lives, and if this were done, it is likely that the early and latter stages of the reaction would not be correctly modeled by the same rate law. In an effort to reduce the difficulty in determining the correct rate law, several replicate kinetic runs can be made, and the data are analyzed as described above. In most cases, the data from a *majority* of the runs will indicate the same rate law as the best fitting one. Kinetic studies on reactions in the solid state involve a rather different protocol than is followed for studies on reactions in the gas phase or in solutions.

In general, the function that gives the best fit (as indicated by the highest correlation coefficient) to the α,t data is assumed to be the "correct" rate law. However, if several runs are made, it is usually the case that not *all* the data sets will give the best fit with the same rate law.

8.2.5 Reactions Between Two Solids

Although not particularly common, there are numerous cases in which two solids undergo a reaction. This must usually be induced by heating the mixed solids or supplying energy by some other means such as heat or pressure. It has also been found that when two solids are suspended in an inert liquid, the application of ultrasound can cause them to react. In some ways, the effect of ultrasound is similar to instantaneous application of heat and pressure because the particles are driven together as cavitation occurs. Ultrasound causes cavitation, and when the cavities implode, the suspended particles are driven together violently as a result of internal pressures that may be as much as a few thousand atmospheres (see Chapter 6). When this happens, there may be a reaction between the particles. An example of this type of process is

$$CdI_2 + Na_2S \xrightarrow[\text{dodecane}]{\text{ultrasound}} CdS + 2\,NaI \tag{8.27}$$

Although we have shown several kinetic models for reacting solids, none specifically applies to a reaction between *two* solids. A rate law that has been developed to model reacting powders is known as the Jander equation, and it is written as

$$\left[1 - \left(\frac{100 - y}{100}\right)^{1/3}\right]^2 = kt \tag{8.28}$$

where y is the percent of the reaction complete. This equation can also be written as

$$\left[1 - \left(1 - \frac{y}{100}\right)^{1/3}\right]^2 = kt \tag{8.29}$$

When y is the percent reacted, $y/100$ is the fraction reacted, which is equal to α, and the equation can be written as

$$\left[1 - (1 - \alpha)^{1/3}\right]^2 = kt \tag{8.30}$$

This equation has the same form as that for three-dimensional diffusion (see Table 8.2). The Jander equation was found to model the process shown in Eqn (8.27) quite well. The reaction between two solids begins on the surface of the particles and progresses inward. For solids in which there is no anisotropy in the structures, diffusion should take place equally in all directions so a three-dimensional diffusion model would seem to be appropriate.

8.3 THERMAL METHODS OF ANALYSIS

From the discussion of reactions in solids presented, it should be apparent that it is not practical in most cases to determine the concentration of some species during a kinetic study. In fact, it may be necessary to perform the analysis in a continuous way as the sample reacts with no separation necessary or even possible. Experimental methods that allow measurement of the progress of the reaction, especially as the temperature is increased, are particularly valuable. Two such techniques are *thermogravimetric analysis* (TGA) and *differential scanning calorimetry* (DSC). These techniques have been widely used to characterize solids, determine thermal stability, study phase changes, and so on. Because they are so versatile in studies on solids, these techniques will be described briefly.

Thermoanalytical methods comprise a series of techniques in which a property is determined at different temperatures or as the temperature changes continuously. The property measured may include the mass of the sample (TGA), the heat flow to the sample (DSC), the magnetic character of the sample or some other property, such as dimensional changes. Each of these types of measurements gives information on some change undergone by the sample, and if the change is followed over time, it is possible to derive kinetic information about the transformation.

When heated, many solids evolve a gas. For example, most carbonates release carbon dioxide when heated. Because there is a mass loss, it is possible to determine the extent of the reaction by following the mass of the sample. The technique of TGA involves heating the sample in a pan surrounded by a furnace. The sample pan is suspended from a microbalance so its mass can be monitored continuously as the temperature is raised (usually as a linear function of time). A recorder provides a graph showing the mass as a function of temperature. From the mass loss, it is often possible to establish the stoichiometry of the reaction. Because the extent of the reaction can be followed, kinetic analysis of the data can be performed. Because mass is the property measured, TGA is useful for studying processes in which volatile products are formed. In another form of thermal analysis, known as *thermodilatometry*, the volume of the sample is followed as the temperature is changed. If the sample undergoes a phase transition, there will usually be a change in density of the material so the volume of the sample will change. Other properties such as the behavior of a sample in a magnetic field can be studied as the temperature is changed.

Differential scanning calorimetry involves using sophisticated electronic circuits to compare the heat flow necessary to keep the sample and a reference at the same temperature as both are heated. If the sample undergoes an endothermic transition, it takes more heat to keep its temperature increasing at a constant rate than if no reaction occurred. Conversely, if the sample undergoes an exothermic transition, less heat is required to keep its temperature increasing at a constant rate. These situations cause the recorder to show a peak in either the endothermic or exothermic direction, respectively. The area under the peak is directly proportional to the amount of heat absorbed or liberated so if a calibration peak has been obtained, it is possible to determine ΔH for the transition from the area under the peak. The fraction of the reaction complete is obtained by comparing the area under the peak when heating to several temperatures has occurred to that corresponding to the complete reaction. Knowing the extent of the reaction as a function of time or temperature enables the rate law for the transformation to be determined. DSC can be used to study processes for which there is no mass loss as long as the transformation absorbs or liberates heat. Therefore, it can be used to study changes in crystal structure as well as chemical reactions.

The brief description of TGA and DSC presented here is intended to show the types of measurements that are possible. To see how the methods are useful for studying changes in solids, it is not necessary to discuss the operation of the instruments or to go into details of data analysis. However, it will become clear during the discussion how useful these techniques are for studying solids.

Another technique that has been employed for studying certain types of changes in solids is infrared spectroscopy, in which the sample is contained in a cell that can be heated. By monitoring the infrared spectrum at several temperatures, it is possible to follow changes in bonding modes as the sample is heated. This technique is useful for observing phase transitions and isomerizations. When used in combination, techniques such as TGA, DSC, and variable temperature spectroscopy make it possible to learn a great deal about dynamic processes in solids.

8.4 EFFECTS OF PRESSURE

Although volume changes in liquid and solids are small, the application of high pressure can be considered as doing work on the sample. Generally, a transformation from one form of a solid to another is brought about by increasing the temperature. If the two forms have different volumes, increasing the pressure will favor a shift to the form having the smaller volume. If the sample undergoes a reaction, the transition state may have smaller or larger volume than the starting material. If the volume of the transition state is smaller than that of the reactants, increasing the pressure will favor formation of the transition state and thus increase the rate of the reaction. On the other hand, if the transition state has a larger volume than the reactants, increasing the pressure will hinder the formation of the transition state, thereby decreasing the rate of the reaction. Pressure studies have provided valuable information about several types of dynamic processes in solids that include phase transitions, isomerizations, and chemical reactions.

To gain an appreciation of the effect produced by pressure, we consider the following example. Suppose that increasing the pressure on a sample by 1000 atm produces a change in volume of 10 cm^3/mol. The work done on the sample is given by $P\Delta V$, which is

$$1000 \, \text{atm} \, \times \, 0.010 \, \text{l/mol} \; = \; 10 \, \text{l atm/mol}$$

When converted to kJ/mol, the work done on the sample is found to be only 1.01 kJ/mol. To produce a significant amount of work on the sample requires enormous pressures. The effects are usually observable only when several kilobars are involved (1 bar $= 0.98692$ atm). Volume changes in the order of ± 25 cm^3/mol are typical for pressure changes of 10 kbar.

When a chemical reaction takes place, the volume change accompanying the formation of the transition state is known as the *volume of activation*, $\Delta V^{\ddagger}$. It can be expressed as

$$\Delta V^{\ddagger} = V^{\ddagger} - \Sigma V_R \tag{8.31}$$

where $V^{\ddagger}$ is the volume of the transition state and ΣV_R is the sum of the molar volumes of the reactants. The free energy of activation is given by

$$\Delta G^{\ddagger} = G^{\ddagger} - \Sigma G_R \tag{8.32}$$

where $G^{\ddagger}$ is the free energy of the transition state and ΣG_R is the sum of the molar free energies of the reactants. However, for a process at constant temperature

$$\left(\frac{\partial G}{\partial P}\right)_T = V \tag{8.33}$$

For reactants forming a transition state, it is found that

$$\left(\frac{\partial G}{\partial P}\right)_T = V^{\ddagger} - \Sigma V_R = \Delta V^{\ddagger} \tag{8.34}$$

From transition state theory, we know that the equilibrium constant for formation of the transition state ($K^{\ddagger}$) is related to the free energy by the relationship

$$\Delta G^{\ddagger} = -RT \ln K^{\ddagger} \tag{8.35}$$

The variation in rate constant (which depends on the concentration of the transition state) with pressure can be expressed as

$$\left(\frac{\partial \ln k}{\partial P}\right)_T = -\frac{\Delta V^{\ddagger}}{RT} \tag{8.36}$$

At constant temperature, the partial derivatives can be replaced, and rearrangement of this equation leads to

$$\Delta V^{\ddagger} = -RT \frac{d\ln k}{dP} \tag{8.37}$$

Solving this equation for $d\ln k$ gives

$$d\ln k = -\frac{\Delta V^{\ddagger}}{RT} dP \tag{8.38}$$

which can be integrated to give

$$\ln k = -\frac{\Delta V^{\ddagger}}{RT} P + C \tag{8.39}$$

From this equation, we see that if the rate constant is determined at a series of pressures, a plot of $\ln k$ versus P should be linear with the slope being $-\Delta V^{\ddagger}/RT$. Although this approach is valid, the graph obtained may not be exactly linear, but the interpretation of these cases does not need to be presented here. It is sufficient to note that the volume of activation can be determined by studying the effect of pressure on reaction rates.

When the value for $\Delta V^{\ddagger}$ is negative, an increase in pressure will increase the rate of reaction. This has been observed for the linkage isomerization reaction (see Chapter 20)

$$[Co(NH_3)_5ONO]^{2+} \rightarrow [Co(NH_3)_5NO_2]^{2+} \tag{8.40}$$

The negative volume of activation in this case is interpreted as indicating that the $-ONO$ group does not leave the coordination sphere of the metal but rather changes bonding mode by a sliding mechanism that leads to a transition state that has a smaller volume than the initial complex. Some complexes having a coordination number of 5 are known, which can exist in either the trigonal bipyramid or square-based pyramid structure. The transformation from one structure to another is known in some cases to be induced by high pressure.

As was discussed in Chapter 7, there are numerous solids that can exist in more than one form. It is frequently the case that high pressure is sufficient inducement for the structure to change. An example of this type of behavior is seen in KCl, which has the sodium chloride (rock salt) structure at ambient pressure but is converted to the cesium chloride structure at high pressure. Other examples illustrating the effect of pressure will be seen throughout this book (see especially Chapter 20). It should be kept in mind that studies involving pressure changes can yield information about transformations that is not easily obtainable by other means.

8.5 REACTIONS IN SOME SOLID INORGANIC COMPOUNDS

The number of inorganic solids that undergo some type of reaction in the solid state is very large. If reactions in which a solid is converted into different solid phase and a volatile product are included, the number is even larger. Even though the number of reactions known to occur in solid compounds is very large, many have not been the subject of kinetic studies. In this section, a few of the types of processes will be shown. Numerous other examples will be shown in later chapters of this book, especially in Chapters 13, 14, and 20. All the reactions shown take place at elevated temperature (some *very* elevated) so heating is understood. The temperatures required depend on the specific compound, and some of the reactions are shown as general types. Consequently, the temperatures required are not shown.

When metal carbonates are heated, they decompose to produce the metal oxide and CO_2. From an economic standpoint, the decomposition of limestone, $CaCO_3$, is perhaps the most important reaction of this type because the product, lime, is used in making mortar and concrete.

$$CaCO_3(s) \rightarrow CaO(s) + CO_2(g) \tag{8.41}$$

In another reaction of this type, metal sulfites lose SO_2 when heated,

$$MSO_3(s) \rightarrow MO(s) + SO_2(g) \tag{8.42}$$

Partial decomposition of some salts is the preparative means to others. For example, the commercial preparation of tetrasodium diphosphate involves the thermal dehydration of Na_2HPO_4.

$$2\,Na_2HPO_4 \rightarrow H_2O + Na_4P_2O_7 \tag{8.43}$$

Other compounds that can be partially decomposed include dithionates with the following being typical reactions of this type.

$$CdS_2O_6(s) \rightarrow CdSO_4(s) + SO_2(g) \tag{8.44}$$

$$SrS_2O_6(s) \rightarrow SrSO_4(s) + SO_2(g) \tag{8.45}$$

As will be discussed in Chapter 14, the thermal decomposition of solids containing the $S_2O_6^{2-}$ ion to give SO_4^{2-} and SO_2 is a general reaction of dithionates.

When most solid oxalates are heated, they are converted to the corresponding carbonate with the evolution of CO.

$$MC_2O_4(s) \rightarrow MCO_3(s) + CO(g) \tag{8.46}$$

Oxalates are frequently obtained as hydrates so the first reaction when a hydrated metal oxalate is heated is the loss of water. However, some metal oxalates decompose in a reaction that is quite different from that shown in Eqn (8.46). Several metal oxalates decompose according to the following equation:

$$Ho_2(C_2O_4)_3(s) \rightarrow Ho_2O_3(s) + 3 CO_2(g) + 3 CO(g) \tag{8.47}$$

This behavior is not unexpected given the fact that many metal carbonates lose CO_2 to produce the oxides.

In a somewhat more unusual reaction, heating an alkali metal peroxydisulfate results in the rupture of the O—O linkage in the $S_2O_8^{2-}$ ion with the loss of oxygen. For example,

$$Na_2S_2O_8(s) \rightarrow Na_2S_2O_7(s) + 1/2 O_2(g) \tag{8.48}$$

The O—O bond in most peroxides has an energy of approximately 140 kJ/mol, which is approximately the activation energy for this reaction. Although it is tempting to say that the initial step in the decomposition of the peroxydisulfate is the breaking of the O—O bond, one cannot be certain that this is so just because the activation energy is about equal to that bond energy.

Earlier in this chapter, it was pointed out that prior treatment and procedural variables can affect the kinetics of reactions of a solid substance. Although a large number of studies have been conducted to evaluate these factors, two such studies will be summarized here.

Ammonium salts usually decompose to gaseous products, and a sizeable number of such compounds have been the subject of kinetic studies (Muehling, et al. 1995). Decomposition of ammonium carbonate leads to gaseous products,

$$(NH_4)_2CO_3(s) \rightarrow 2 NH_3(g) + CO_2(g) + H_2O(g) \tag{8.49}$$

As a result, ammonium carbonate is conveniently studied by mass loss techniques, such as TGA. In one study, the decomposition of particles having different size distributions (302 ± 80, 98 ± 36, and 30 ± 10 μm, respectively) was studied by carrying out a large number of kinetic runs. It was found that decomposition of the largest particles almost always followed either a first-order or a three-dimensional diffusion rate law. The samples consisting of particles having intermediate size decomposed by a first-order rate law, and samples containing the smallest particles decomposed by a three-dimensional diffusion rate law.

A reaction that has been the subject of numerous studies is the dehydration of $CaC_2O_4 \cdot H_2O$.

$$CaC_2O_4 \cdot H_2O(s) \rightarrow CaC_2O_4(s) + H_2O(g) \tag{8.50}$$

In a highly replicated kinetic study (House and Eveland, 1993), the dehydration of samples of freshly prepared $CaC_2O_4 \cdot H_2O$ and material that had been stored in a desiccator for a year was studied. It was found that the kinetics of the dehydration reaction did not depend on the particle size, but there was a great difference between the behavior of the freshly prepared and aged samples. The R1 (one-dimensional contraction) rate law was most appropriate for the freshly prepared material, but the aged material showed considerable variation in replicate kinetic runs. The majority of the data fit the R1 rate law, but several runs gave the best fit with an A2 or A3 rate law. Moreover, dehydration of the material studied soon after its preparation had an activation energy of 60.1 ± 6.6 kJ/mol, but that for the aged material had an activation energy of 118 ± 15 kJ/mol. Although many other kinetic studies on the decomposition of inorganic solids could be described, the discussion presented shows illustrative examples of this important area.

8.6 PHASE TRANSITIONS

The transformation of a substance from the solid phase to the liquid phase (melting) is a common type of change in state. However, a transformation of a solid substance from one solid *structure* to another is also referred to as a phase transition, although no change in physical *state* occurs. In Chapter 7, the ability of metals to be converted from one structure to another was described. In some cases, there is no change in coordination number as the metal changes from one form to another, as for example, in the transition Co(*hcp*) to Co(*fcc*) where both structures have a coordination number of 12. However, in the transformation of Ti(*hcp*) to Ti(*bcc*), the coordination number of Ti changes from 12 to 8. Phase transitions are observed to occur in all types of materials. For example, sulfur exists at room temperature in the rhombic form, but at temperatures near the melting point, it is converted to the monoclinic form.

Although most phase transitions are brought about by a change in temperature, many phase transitions can also be induced by a change in pressure. As a material changes structure, there is almost always some change in volume, and as described in Section 8.4, increasing the applied pressure favors the phase that has smaller volume. When pressure is applied to KCl, it changes from the sodium chloride structure to the cesium chloride structure. The pressure required is 19.6 kbar, and the volume change is -4.11 cm^3/mol. One way to study phase transitions is by dilatometry, and instruments for measuring volume changes as a function of temperature are available.

Transformation of a solid such as KCl from the NaCl structure to the CsCl structure is a vastly different situation than that of converting graphite into diamond. In the latter transition, the bonds between carbon atoms must be broken and replaced with an entirely different bonding mode. This type of transition is called a *reconstructive* transition. Such changes are normally slow and have high activation energies associated with them. The transformation of most ionic solids from one crystal structure to another does not involve breaking *all* the ionic bonds between the ions so the process usually has a low activation energy associated because the change in structure is relatively minor. Phase transitions of this type are called *displacive* transitions. Both reconstructive and displacive transitions may involve either the groups in the primary or first bonding environment (nearest neighbors) or the secondary environment. For example, if a solid consists of tetrahedral units, breaking the primary bonding environment would require the breaking of bonds *within* the units, whereas disrupting the secondary environment would involve breaking bonds *between* the units. Examples of all these types of phase transitions are known.

An enormous number of phase transitions are known to occur in common solid compounds. For example, silver nitrate undergoes a displacive phase transition from an orthorhombic form to a hexagonal form at a temperature of approximately $162\,°C$ that has an enthalpy of $1.85\,kJ/mol$. In many cases, the nature of these transitions is known, but in other cases, there is some uncertainty. Moreover, there is frequently disagreement among the values reported for the transition temperatures and enthalpies. Even fewer phase transitions have been studied from the standpoint of kinetics, although it is known that a large number of these transformations follow an Avrami rate law. There is another complicating feature of phase transitions that we will now consider.

Suppose a solid S is transformed from Phase I (S_I) to Phase II (S_{II}). The energy profile for the transition of a solid from Phase I to Phase II resembles that for a chemical reaction. However, consider the energy diagram shown in Figure 8.3. In this case, the activation energy for the forward process ($I \rightarrow II$) is lower than that for the reverse process. If the solid is heated infinitesimally slowly (which in the limit approaches equilibrium conditions), the rate at which Phase I is converted into Phase II in the forward process ($S_I \rightarrow S_{II}$) will be equal to the rate for the reverse process ($S_I \rightarrow S_{II}$). Therefore, if a graph is made showing the fraction of the sample in the initial phase as a function of temperature, the same curve will result regardless of which direction the conversion is carried out. This situation is shown in Figure 8.4. This is true if (and usually only if) the transformation is carried out by changing the temperature at an infinitesimally low rate.

In the usual experiments, the sample may be heated (or cooled) at a rate of a fraction of a degree per minute up to several degrees per minute. Under these conditions, the rates of the forward and reverse reactions are *not* equal. The result will be that the heating and cooling curves giving α as a function of temperature will *not* coincide. The curves generate a loop that is known as *thermal hysteresis*. In the more usual case, the rate of the reverse reaction upon cooling the sample is *lower* than that of the forward reaction (as it would be for the system represented the energy profile shown in Figure 8.3). At a given temperature, the fraction of the sample converted as the sample is cooled is lower than that obtained as the sample is heated. The result is shown in Figure 8.5.

In Figure 8.5, w is the distance between the heating and cooling curves at the point where $\alpha = 0.5$ and is called the *hysteresis width*. This temperature may be quite small or it may amount to several degrees depending on the nature of the phase transition and the heating rate. Many substances exhibit this type of behavior as a result of a phase change.

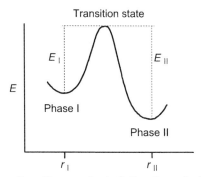

FIGURE 8.3
The energy profile for the transformation of a solid from Phase I to Phase II.

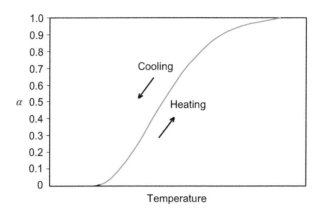

FIGURE 8.4
The fraction of the sample transformed from I → II (heating) and from II → I (cooling) as a function of temperature. In this case, there is no thermal hysteresis.

From a thermodynamic standpoint, we know that at the temperature at which the two phases are in equilibrium, the free energy, G, is the same for both phases. Therefore,

$$\Delta G = \Delta H - T\Delta S = 0 \tag{8.51}$$

As a result, there will be a continuous change in G as the transition of one phase into another takes place. However, for some phase transitions (known as *first-order transitions*), it is found that there is a discontinuity in the *first derivative* of G with respect to pressure or temperature. It can be shown that the partial derivative of G with pressure is the equal to volume, and the derivative with respect to temperature is equal to entropy. Therefore, we can express these relationships as follows:

$$\left(\frac{dG}{dP}\right) = V \tag{8.52}$$

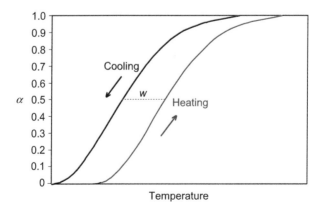

FIGURE 8.5
The thermal hysteresis that occurs when the rate of the phase change on cooling is lower than that on heating at the same temperature.

$$\left(\frac{dG}{dT}\right) = -S \tag{8.53}$$

For the first case, as the sample is heated, there will be a change in the volume of the sample that can be followed by a technique known as dilatometry. For changes in entropy, use can be made of the fact that $\Delta G = 0$ and from Eqn (8.51) we find that

$$\Delta S = \frac{\Delta H}{T} \tag{8.54}$$

Although there are other ways, one of the most convenient and rapid ways to measure ΔH is by DSC. When the temperature is reached at which a phase transition occurs, heat is *absorbed* so more heat must flow to the sample in order to keep the temperature equal to that of the reference. This produces a peak in the endothermic direction. If the transition is readily reversible, cooling the sample will result in heat being *liberated* as the sample is transformed into the original phase, and a peak in the exothermic direction will be observed. The area of the peak is proportional to the enthalpy change for transformation of the sample into the new phase. Before the sample is completely transformed into the new phase, the *fraction* transformed at a specific temperature can be determined by comparing the *partial* peak area up to that temperature to the *total* area. That fraction, α, determined as a function of temperature can be used as the variable for kinetic analysis of the transformation.

A different type of phase transition is known in which there is a discontinuity in the second derivative of free energy. Such transitions are known as *second-order transitions*. From thermodynamics, we know that the change in volume with pressure at constant temperature is the coefficient of compressibility, β, and the change in volume with temperature at constant pressure is the coefficient of thermal expansion, α. The thermodynamic relationships can be shown as follows:

$$\left(\frac{\partial^2 G}{\partial P^2}\right)_T = \left(\frac{\partial V}{\partial P}\right)_T = -\beta V \tag{8.55}$$

$$\left(\frac{\partial^2 G}{\partial P \partial T}\right) = \left(\frac{\partial V}{\partial T}\right)_P = \alpha V \tag{8.56}$$

In addition to these relationships, we know that the second derivative of free energy with temperature can be expressed by the following relationship.

$$\frac{\partial^2 G}{\partial T^2} = -\left(\frac{\partial S}{\partial T}\right)_P = -\frac{C_P}{T} \tag{8.57}$$

For certain types of phase transitions, it is possible to study the process by following changes in these variables.

As a phase transition occurs, there is a change of some type in the lattice. The units (molecules, atoms, or ions) become more mobile. If the solid is reacting in some way and the temperature is at or very near that at which the solid undergoes a phase transition, there may be a rapid increase in the rate of the reaction with a slight increase in temperature because lattice reorganization enhances the ability of the solid to react. A similar situation can occur in cases in which two solids are reacting if the temperature is at or near the temperature corresponding to a phase transition in one of the solids. This phenomenon is sometimes referred to as the *Hedvall effect*.

8.7 REACTIONS AT INTERFACES

Several types of reactions involving solids with gases or liquids occur at the interface between the two phases. The most important reaction of this type is corrosion. Efforts to control or eliminate corrosion involve research that spans the spectrum from the coatings industry to the synthesis and production of corrosion-resistant materials. The economic ramifications of corrosion are enormous. Although there are numerous types of reactions that can be represented as taking place at an interface, the oxidation of a metal will be described. Figure 8.6 represents the oxidation of a metal.

   At the interface, oxygen atoms add two electrons to become oxide ions. At the surface of the metal, metal atoms are oxidized to metal ions by losing electrons. In the process, migration of electrons occurs, but is also necessary for O^{2-} and M^{2+} ions to be combined, and that requires mobility of the ions. Even though the electrons are normally more mobile, both cations and anions diffuse. As a result of the reduction on the surface, there is a somewhat negative charge at the gas/solid interface. This results in an electric field gradient that assists the migration of metal ions to the extent that they may actually be more mobile than the oxide ions.

   When the reacting metal is iron, the process is complicated by the fact that iron has two oxidation states, $+2$ and $+3$. Therefore, possible oxidation products include the well-known oxides Fe_2O_3, Fe_3O_4 (which is $Fe_2O_3 \cdot FeO$), and FeO. Because oxygen is in excess at the surface, the product at the surface will contain the metal in its highest oxidation state, which means the product having the lowest percentage of iron (highest percentage of oxygen). That product is Fe_2O_3. Of the iron oxides mentioned above, the product having the next highest oxygen content is Fe_3O_4, Finally, farther below the surface there will be FeO, and the interior of the object will be iron. Figure 8.7 shows these features of the oxidation process.

   Because of the way in which reactions such as this take place, the phases are not normally stoichiometric compounds. The result is that there are vacancies into which the Fe^{2+} ions can move. It has been found that in the early stages of corrosion the rate varies with the partial pressure of oxygen and that the rate varies as $P(O_2)^{0.7}$ when the pressure is low ($P(O_2) < 1$ torr). This dependence of the rate on the pressure of oxygen can be shown to be consistent with chemisorption of oxygen being the rate-determining process. It can be shown that if the process were controlled by the rate of conversion of oxygen molecules into oxide ions, the rate would depend on $P(O_2)^1$. If the reaction involves equilibrium between oxygen molecules and oxide ions on the surface of the FeO, the rate at which FeO is

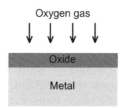

FIGURE 8.6
The oxidation of a metal object.

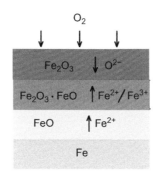

FIGURE 8.7
A representation of the phases present and the motion of ions in the corrosion of iron. *(Adapted from Borg and Dienes (1988), p. 295).*

produced would depend on $P(O_2)^{1/2}$. Neither of these mechanisms is consistent with the observed dependence of the rate on the partial pressure of oxygen.

At high temperature and a higher partial pressure of oxygen $(1 < P(O_2) < 20$ torr), the rate of growth of the FeO layer follows the parabolic rate law. The rate of formation of FeO is determined by the rate of diffusion of Fe^{2+}, but the rate of diffusion of O^{2-} determines the rate at which the thickness of Fe_2O_3 increases.

8.8 DIFFUSION IN SOLIDS

Although solids have definite shapes and the lattice members (atoms, ions, or molecules) are essentially fixed in their locations, there is still movement of units from their lattice sites. In fact, several properties of solids are determined by diffusion within a solid structure. There are two principle types of diffusion processes. *Self-diffusion* refers to diffusion of matter within a pure sample. When the diffusion process involves a second phase diffusing into another, the process is called *heterodiffusion*. Self-diffusion in metals has been extensively studied, and the activation energies for diffusion in many metals have been determined. Diffusion in a metal involves the motion of atoms through the lattice. Melting a solid requires a temperature high enough to cause the lattice members to become mobile. It has been found that there is a good linear relationship between the melting points of metals and the activation energies for self-diffusion.

It is known that if two metals having different diffusion coefficients are placed in contact (as if they are welded together), there is some diffusion at the interface. Suppose two metals, A and B, are placed in intimate contact as illustrated in Figure 8.8.

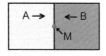

FIGURE 8.8
A marker wire (M) placed at the interface of metals A and B.

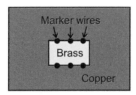

FIGURE 8.9
The experiment to show vacancy movement in diffusion of zinc.

The concentration of each metal in the other will be highest at the interface and decrease (usually exponentially) as the distance from the interface increases. If a wire (usually referred to as a marker) made of an inert material is placed the interface, the metals moving at different rates will cause wire to appear to move. Because A diffuses past the marker to a greater extent than does B, it appears that the marker has moved farther into the block of metal A. In this way, it is possible to identify the more mobile metal. If the metals diffuse at the same rate, the wire would remain stationary. An application of this principle was made in the study of diffusion of zinc in brass. The arrangement is shown in Figure 8.9.

When this system was studied over time, it was found that the marker wires move toward each other. This shows that the most extensive diffusion is zinc from the brass (an alloy of zinc and copper) outward into the copper. If the mechanism of diffusion involved an *interchange* of copper and zinc, the wires would not move. The diffusion in this case takes place by the *vacancy* mechanism described below as zinc moves from the brass into the surrounding copper. As the zinc moves *outward*, vacancies are produced in the brass and the wires move *inward* with the rate of movement of the wires being proportional to $t^{1/2}$ (the parabolic rate law shown in Eqn (8.10)). This phenomenon is known as the *Kirkendall effect*.

Displacements of lattice members are determined by energy factors and concentration gradients. To a considerable extent, diffusion in solids is related to the existence of vacancies. The "concentration" of defects (sites of higher energy) can be expressed in terms of a Boltzmann distribution as

$$N_o = N_x \, e^{-E/kT} \tag{8.58}$$

where N_x is the total number of lattice members, k is Boltzmann's constant, and E is the energy necessary to create the defect. Because the creation of a defect is somewhat similar to separating part of the lattice to give a more random structure, E is comparable to the heat of vaporization. In some cases, a lattice member in a Frenkel defect can move into a vacancy or Schottky defect to remove both defects in a recombination process.

When a crystal is heated, lattice members become more mobile. As a result, there can be removal of vacancies as they become filled by diffusion. Attractions to nearest neighbors are reestablished with the result that there is a slight increase in density and the liberation of energy. There will be a disappearance of dislocated atoms or perhaps a redistribution of locations. These events are known to involve several types of mechanisms. However, the diffusion coefficient, D, is expressed as

$$D = D_o e^{-E/RT} \tag{8.59}$$

(a)

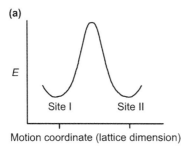

E

Site I   Site II

Motion coordinate (lattice dimension)

(b)

Interstitial position

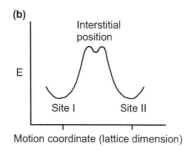

E

Site I   Site II

Motion coordinate (lattice dimension)

FIGURE 8.10

A representation of the energy change that occurs during diffusion.

where E is the energy required for diffusion, D_o is a constant, and T is the temperature (K). The similarity of this equation to the Arrhenius equation that relates the rate constant for a reaction to temperature is apparent.

One type of diffusion mechanism is known as the *interstitial* mechanism because it involves movement of a lattice member from one interstitial position to another. When diffusion involves the motion of a particle from a regular lattice site into a vacancy, the vacancy then is located where the site is vacated by the moving species. Therefore, the vacancy moves in the *opposite* direction to that of the moving lattice member. This type of diffusion is referred to as the *vacancy* mechanism. In some instances, it is possible for a lattice member to vacate a lattice site and for that site to be filled simultaneously by another unit. In effect, there is a "rotation" of two lattice members so this mechanism is referred to as the *rotation* mechanism of diffusion.

In addition to movement of lattice members within a crystal, it is also possible for there to be motion of members along the surface. Consequently, this type of diffusion is known as *surface diffusion*. Because crystals often have grain boundaries, cracks, dislocations, and pores, there can be motion of lattice members along and within these extended defects. The energy change as diffusion occurs is illustrated in Figure 8.10.

With each internal lattice site having essentially the same energy, motion of a lattice member from one regular lattice to another involves the diffusing species moving over an energy barrier, but the initial and final energies are the same as shown in Figure 8.10(a). When a lattice member moves from a regular lattice site into an interstitial position, there is an energy barrier to the motion. The interstitial position represents a higher energy than that of a regular site, and the energy profile is shown in Figure 8.10(b). However, the interstitial position represents a site of lower energy than other positions in the immediate vicinity. This gives rise to an energy relationship that is shown in Figure 8.10(b), in which there is an energy "well" at the top of the potential energy curve. Energy increases as the lattice member moves from its site, but when the member is in precisely an interstitial position, the energy is slightly lower than when it is displaced slightly from the interstitial position.

8.9 SINTERING

Sintering forms the basis for the important manufacturing process known as powder metallurgy as well as the preparation of ceramics. Objects are produced from powdered materials that include high-melting metals (such as molybdenum and tungsten), carbides, nitrides, and so on. These materials are formed to make machine parts, gears, tools, turbine blades, and many other products. To shape the objects, a mold is filled with the powdered material and pressure is applied. For a given mass of a particulate solid, the smaller the particles the larger the surface area. When heated at high temperature, the material flows, pores disappear, and a solid mass results, even though the temperature may be below the melting point of the material. Plastic flow and diffusion allow the particles to congeal to form a solid mass. By using powder metallurgy, it is possible to produce objects having high dimensional accuracy more economically than if machining were required. The nature of this important process will be described in more detail later in this section.

If a regular lattice such as the NaCl structure is considered, it will be seen that within the crystal each ion is surrounded by six others of opposite charge. However, each ion on the surface of the crystal does not have a nearest neighbor on one side so the coordination number is only 5. Along an edge of the crystal, the coordination number is 4 because there are two sides that do not have a nearest neighbor. Finally, an ion on the corner of the crystal has three sides that are not surrounded by nearest neighbors so those units have a coordination number of 3. If the crystal structure of a metal is examined, a similar difference between the coordination numbers of the internal, facial, edge, and corner atoms will be seen.

The total interaction for any lattice member with its nearest neighbors is determined by the coordination number. Consequently, lattice members in positions on faces, edges, and corners are in high energy positions with the energy of the positions increasing in that order. There is a tendency for the occupancy of high energy sites to be minimized. In a small amount of liquid (such as a droplet), that tendency is reflected by the formation of a surface of minimum area, which is spherical because a sphere gives the smallest surface area for a given volume. When a solid is heated, there is motion of individual particles as the tendency to form a minimum surface is manifested. The process is driven by "surface tension" as the solid changes structure to give a minimum surface area, which also gives the smallest number of lattice members on the surface.

Not all solids exhibit sintering, but many do. Sintering is accompanied by the removal of pores and the rounding of edges. When the solid is composed of many small particles, there will be welding of grains and a densification of the sample. For ionic compounds, both cations and anions must be relocated, which may occur at different rates. Consequently, sintering is often related to the rate of diffusion, which is in turn related to the concentration of defects. One way to increase the concentration of defects is to add a small amount of a compound that contains an ion having a different charge than that of the major component. For example, adding a small amount of Li_2O (which contains a 2:1 ratio of cations to anions) to ZnO increases the number of anion vacancies. In ZnO, anion vacancies determine the rate of diffusion and sintering. On the other hand, adding Al_2O_3 decreases the rate of sintering in ZnO because two Al^{3+} ions can replace three Zn^{2+} ions, which leads to an excess of cation vacancies.

Heating the solid in an atmosphere that removes some anions will lead to an increase in anion vacancies. For example, when ZnO is heated in an atmosphere of hydrogen there is an increase in the number of anion vacancies. Sintering of Al_2O_3 is also limited by diffusion of oxygen. Heating Al_2O_3 in a hydrogen atmosphere leads to the removal some oxide ions, which increases the rate of sintering. The rate of sintering of Al_2O_3 is dependent on particle size, and it has been found that

$$\text{Rate} \propto \left(\frac{1}{\text{Particle size}} \right)^3 \qquad (8.60)$$

For particles that measure 0.50 and 2.0 μm, the ratio of the rates is $(2.0/0.50)^3$ or 64 so the smaller particles sinter much faster than the larger ones.

If the sample being sintered is a powdered metal, the result can be a dense strong object that resembles one made from a single piece of metal. This is the basis for the manufacturing technique known as *powder metallurgy*. This is an important process in which many objects such as gears are produced by heating and compressing powdered metal in a mold of suitable shape. There is a considerable reduction in cost compared with similar objects shaped by traditional machining processes.

In powder metallurgy, the powdered material to be worked is pressed in a mold then heated to increase the rate of diffusion. The temperature required to obtain flow of the material may be significantly below the melting point. As the powder becomes more dense and less porous, the vacancies move to the surface to produce a structure that is less porous and more dense. In addition to diffusion, plastic flow and evaporation and condensation may contribute to the sintering process. As sintering of a solid occurs, it is often possible to observe microscopically the rounding of corners and edges of individual solid particles. When particles undergo coalescence, they fuse together to form a "neck" between them. Continued sintering leads to thickening of the neck regions and a corresponding reduction in the size of the pores that exist between the necks. Finally, there is a growth of particles of the solid to form a compact mass. The apparent volume of the sample is reduced as a result of surface tension causing the pores to close.

In the process of powder metallurgy, the material to be compacted may be prepared by blending the components prior to sintering. In different schemes, the components are premixed and then heated to cause annealing of the mixture or they may be prealloyed by adding the minor constituents to the major one in the liquid state. When the major constituent is powdered iron, the powder can be obtained in a variety of ways that include reducing the ore in a kiln and atomization of the metal as a liquid in a high pressure stream. For making objects of iron alloys, the mix is pressed to shape before sintering, which is carried out by heating the mixture to approximately $1100\,°C$ in a protective atmosphere. This is well below the melting point of iron ($1538\,°C$), but it is sufficient to cause diffusion. Bonding between particles occurs as grain boundaries disappear.

In making objects of bronze, the premix consists of approximately 90% copper, 10% tin, and a small amount of a lubricant. The mixture is sintered at approximately $800\,°C$ in a protective atmosphere that consists primarily of nitrogen, but it may also contain a small pressure of hydrogen, ammonia, or carbon monoxide. The properties of objects produced by powder metallurgy depend on procedural variables such as particle size distribution in the mixture, preheating treatment, sintering time, atmosphere composition, and the flow rate of the gaseous atmosphere. The results of procedural changes are not always known in advance, and much of what is known about how to carry out specific processes in powder metallurgy is determined by experience.

8.10 DRIFT AND CONDUCTIVITY

When applied to the motion of ions in a crystal, the term *drift* applies to motion of ions under the influence of an electric field. Although movement of electrons in conduction bands determines conductivity in metals, in ionic compounds it is the motion of *ions* that determines the electrical conductivity. There are no free or mobile electrons in ionic crystals. The *mobility* of an ion, μ, is defined as the velocity of the ion in an electric field of unit strength. Intuitively, it seem that the mobility of the ion in a crystal should be related to the diffusion coefficient. This is, in fact, the case and the relationship is

$$D = \frac{kT}{Z}\mu \tag{8.61}$$

where Z is the charge on the ion, k is Boltzmann's constant, and T is the temperature (K). The relationship between the *ionic conductivity*, σ, and the rate of diffusion in the crystal, D, can be expressed as

$$\sigma = \alpha \frac{Nq^2}{kT}D \tag{8.62}$$

In this equation, N is the number of ions per cubic centimeter, q is the charge on the ion, and α is a factor that varies from about $1-3$ depending on the mechanism of diffusion. Because conductivity of a crystal depends on the presence of defects, studying conductivity gives information about the presence of defects. The conductivity of alkali halides by ions has been investigated in an experiment illustrated in Figure 8.11.

As the electric current passes through this system, the cathode (negative electrode) grows in thickness as that of the anode (positive electrode) shrinks. At the cathode, M^+ ions are converted to M atoms, which results in growth of the cathode. From this observation, it is clear that the cations are primarily responsible for conductivity, and this is the result of a vacancy type of mechanism. In this case, the positive ion vacancies have higher mobility than do the vacancies that involve negative ions.

Because the number of vacancies controls the conductivity, changing the conditions so that the number of vacancies increases will increase conductivity. One way to increase the number of vacancies is to dope the crystal with an ion of different charge. For example, if a small amount of a compound containing a +2 ion is added to a compound such as sodium chloride, the +2 ions will occupy cation sites. Because one +2 cation will replace two +1 ions and still maintain overall electrical neutrality, there will be a vacant cation site for each +2 ion present. As a result, the mobility of Na^+ will be increased because of the increase in the number of vacancies. Although doping is effective at lower temperatures, it is less so at high temperature. The reason is that the number of vacancies is determined by a Boltzmann population of the higher energy states and at high temperature, the number of vacancies is already large.

In this chapter, we have described some of the types of transformations in solids that involve rate processes. This is an immensely practical area because many industrial processes involve

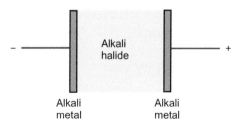

FIGURE 8.11
Arrangement of an experiment to demonstrate ionic drift.

such changes in inorganic substances, and they are an essential part of materials sciences. For a more complete discussion of these important topics, the references given below should be consulted.

References for Further Study

Borg, R.J., Dienes, G.J., 1988. An Introduction to Solid State Diffusion. Academic Press, San Diego. A thorough treatment of many processes in solids that are related to diffusion.

Gomes, W., 1961. *Nature (London)*, 192, 965. An article discussing the difficulties associated with interpreting activation energies for reactions in solids.

Hannay, N.B., 1967. *Solid-state Chemistry*. Prentice-Hall, Englewood Cliffs. An older book that gives a good introduction to solid state processes.

House, J.E., 2007. *Principles of Chemical Kinetics*, 2nd ed. Elsevier/Academic Press, San Diego. Chapter 7 is devoted to reactions in the solid state.

House, J.E., 1993. *Coord. Chem. Rev.* 128, 175−191. Mechanistic considerations for anation reactions in the solid state.

House, J.E., 1980. *Thermochim. Acta* 38, 59−66. A proposed mechanism for the thermal reactions in solid complexes.

House, J.E., 1980. *Thermochim. Acta* 38, 59. A discussion of reactions in solids and the role of free space and diffusion.

House, J.E., Bunting, R.K., 1975. *Thermochim. Acta* 11, 357−360.

House, J.E., Eveland, R.W., 1993. *J. Solid State Chem.* 105, 136−142.

Muehling, J.K., Arnold, H.R., House, J.E., 1995. *Thermochim. Acta* 255, 347−353.

Ng, W.-L., Ho, C.-C., Ng, S.-K., 1978. *J. Inorg. Nucl. Chem.* 34, 459−462.

O'Brien, P., 1983. *Polyhedron* 2, 223. An excellent review of racemization reactions of coordination compounds in the solid state.

Schmalzreid, H., 1981. *Solid State Reactions*, 2nd ed. Verlag Chemie, Weinheim. A monograph devoted to solid state reactions.

West, A.R., 1984. *Solid State Chemistry and Its Applications*. Wiley, New York. A very good introduction to the chemistry of the solid state.

Young, D.A., 1966. *Decomposition of Solids*. Pergamon Press, Oxford. An excellent book that discusses reactions of many inorganic solids and principles of kinetics of solid state reactions.

QUESTIONS AND PROBLEMS

1. A solid compound X is transformed into Y when it is heated at 75 °C. A sample of X that is quickly heated to 90 °C for a very short time (with no significant decomposition) and then quenched to room temperature is later found to be converted to Y at a rate that is 2.5 times that of a sample that has had no prior heating when both are heated at 75 °C for a long period of time. Explain these observations.

2. Suppose a solid compound A is transformed into B when it is heated at 200 °C. An untreated sample of A shows no induction period, but a sample of A that was irradiated with neutrons does show an induction period. After the induction period, the irradiated sample gave similar kinetic behavior to that of the untreated sample. Explain these observations.

3. Consider the reaction

$$A(s) \rightarrow B(s) + C(g)$$

which takes place at high temperature. Suppose that as crystals of A are transformed into B that sintering of B transforms it into rounded, glassy particles. What effect would this likely have on the latter stages of the reaction?

4. When KCN and AgCN are brought together, they react to form K[Ag(CN)$_2$]. The initial stage of the reaction can be shown as follows:

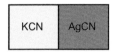

Sketch the system after some period of reaction. Discuss two possible cases for the limiting process in the reaction and how that might alter the sketch you made.

5. Describe the effects on the conductivity of KCl produced by adding a small amount of MgCl$_2$. Explain the specific origin of the change in conductivity.

6. Describe the effects on the rate of sintering of Fe$_2$O$_3$ produced by adding a small amount of MgCl$_2$. Explain the specific origin of the change in conductivity.

7. Suppose a solid is to be the subject of a kinetic study (such as decomposition). How would prior irradiating the solid with X rays or γ rays likely affect the kinetic behavior of the solid? Explain the origin of the effects.

8. The tarnishing of a metal surface follows the parabolic rate law. Discuss the units on the rate constant in comparison with those for a first-order reaction. Suppose the rate of tarnishing is studied at several temperatures and the activation energy is calculated. For reactions in gases and liquids, the activation energy is sometimes interpreted in terms of bond breaking processes. How would you interpret the activation energy determined in this case?

9. Prepare rate plots of the data shown in Table 8.1 using an Avrami rate law with n values of 2, 3, and 4.

10. If the data shown in Table 8.1 were available for only the first 30 min of reaction, explain the difficulty in deciding the value of n that applies in this reaction.

Acids, Bases, and Solvents

CHAPTER 9

Acid–Base Chemistry

In the study of chemistry, the results of observations on the transformations and properties of many materials are encountered. Schemes that provide structure to the information concerning this type of chemistry go a long way toward systematizing its study. One such approach is that of the chemistry of acids and bases. Closely related to the chemistry of acids and bases is the study of solvents other than water, the chemistry of nonaqueous solvents (see Chapter 10). In this chapter, several areas of acid–base chemistry and their application to reactions of inorganic substances will be described.

9.1 ARRHENIUS THEORY

An early attempt to provide a framework to the observations on the chemistry of substances that react in water to produce acids or bases was provided by S. A. Arrhenius. At that time, the approach was limited to aqueous solutions, and the definitions of an acid and a base were given in these terms. Of course we now know that acid–base behavior is not limited to aqueous solutions but it applies much more broadly. If we consider the reaction between gaseous HCl and water,

$$HCl(g) + H_2O(l) \rightarrow H_3O^+(aq) + Cl^-(aq) \tag{9.1}$$

we see that the solution contains H_3O^+, the *hydronium* ion or, as it is perhaps more generally known, the *oxonium* ion. In aqueous solution, HNO_3 also ionizes as illustrated in the reaction,

$$HNO_3(aq) + H_2O(l) \rightarrow H_3O^+(aq) + NO_3^-(aq) \tag{9.2}$$

In studying the properties of solutions of substances such as HCl and HNO_3, Arrhenius was led to the idea that the *acidic* properties of the compounds were due to the presence of an ion that we now write as H_3O^+ in the solutions. He therefore proposed that an *acid* is a substance whose water solution contains H_3O^+. The properties of aqueous solutions of acids are the properties of the H_3O^+ ion, a solvated proton (hydrogen ion) that is known as the *hydronium* ion in much of the older chemical literature but also referred to as the *oxonium* ion.

It is appropriate at this point to comment on the nature of the solvated hydrogen ion in aqueous solutions. We saw in Chapter 7 that there is limited number of ions that can surround one of opposite charge in a crystal. There is also a rather definite number of water molecules that can solvate an ion (the hydration number) in an aqueous solution. Hydration numbers for ions are not always fixed due to the fact that water molecules are continuously entering and leaving the solvation sphere of the ion. For many metal ions, the hydration number is approximately 6. Owing to its extremely small size, the hydrogen ion has a large charge to size ratio. Therefore, it interacts strongly with the negative ends of polar molecules, **273**

Inorganic Chemistry. DOI: http://dx.doi.org/10.1016/B978-0-12-385110-9.00009-1

such as water, and the heat of hydration of $H^+(g)$ is -1100 kJ mol^{-1}, a value that is very high for a $+1$ ion. Although the hydrogen ion in water is represented as H_3O^+ to show that it is solvated, the solvation sphere certainly includes more than one water molecule. For H^+, the hydration number is probably at least 4, particularly in dilute solutions of acids. When four water molecules solvate the proton, the species formed is $H_9O_4^+$, which has the proton in the center of a tetrahedron of water molecules. Another species that contains the solvated proton is $H_5O_2^+$, which has the proton located between two water molecules in a linear structure. This ion has been identified in a few solid compounds as one of the positive ions present. Doubtless there exist other species with different numbers of water molecules solvating a proton. Such species may be transitory in nature and have a rather large degree of fluxional character, but there is no doubt that the species in acidic solutions is not as simple as the symbol H_3O^+ would indicate. However, when we wish to show a solvated proton this symbol will be used, but it is very little more accurate at portraying the actual species than is a simple H^+, which was used for many years.

Among the substances that react with water to produce H_3O^+ are HCl, HNO_3, H_2SO_4, $HClO_4$, H_3PO_4, $HC_2H_3O_2$, and many others. Because water solutions of all of these compounds contain H_3O^+, they have similar properties. Of course there is a difference in degree because some are strong acids but others are weak. Solutions of the compounds listed conduct electric current, change the colors of indicators, neutralize bases, and dissolve some metals. In fact, these are the characteristics of the H_3O^+ ion in aqueous solutions. Thus, the compounds all behave as acids. The difference in strengths of acids is related to the extent of their ionization reactions. Acids such as HCl, HNO_3, H_2SO_4, and $HClO_4$ are strong acids because in dilute aqueous solutions they are almost completely ionized. As a result, aqueous solutions of these compounds are good conductors of electricity. In contrast, the ionization of acetic acid is only 1-3% depending on the concentration of the acid so it is a weak acid. Table 9.1 shows the dissociation constants for a large number of acids.

When $NH_3(g)$ dissolves in water, some ionization occurs as indicated by the equation

$$NH_3(g) + H_2O(l) \leftrightarrows NH_4^+(aq) + OH^-(aq) \tag{9.3}$$

In this reaction, one of the products is OH^-, which is the species of basic character in aqueous solutions. When NaOH dissolves in water, the reaction is not actually an ionization reaction because the Na^+ and OH^- ions already exist in the solid. The process is a dissolution process rather than an ionization reaction. Substances such as NaOH, KOH, $Ca(OH)_2$, NH_3, and amines (RH_2N, R_2HN, and R_3N) are all bases because their water solutions contain OH^-. The compounds dissolve in water to give solutions that conduct electric current, change the colors of indicators, and neutralize acids, which are some of the reactions that characterize aqueous solutions of bases. Substances that contain OH^- ions prior to dissolving them in water or those that react with water to give reactions in which they are extensively ionized to produce OH^- ions are strong bases. The reactions of ammonia and amines result in only a slight degree of ionization so these compounds are called weak bases.

It is interesting to note that whereas special emphasis is placed on the solvated hydrogen ion by writing it as H_3O^+, no such distinction is made for OH^- other than to write it as $OH^-(aq)$. However, this ion is also solvated strongly by several polar water molecules.

When NaOH is dissolved in water and the solution is added to a solution of HCl in water, the substances are ionized so the reaction is

$$Na^+(aq) + OH^-(aq) + H_3O^+(aq) + Cl^-(aq) \rightarrow 2\,H_2O(l) + Na^+(aq) + Cl^-(aq) \tag{9.4}$$

Table 9.1 Dissociation Constants for Some Common Acids

Acid	Conjugate Base	Dissociation Constant
$HClO_4$	ClO_4^-	Essentially complete
HI	I^-	Essentially complete
HBr	Br^-	Essentially complete
HCl	Cl^-	Essentially complete
HNO_3	NO_3^-	Essentially complete
$HSCN$	SCN^-	Essentially complete
HSO_4^-	SO_4^{2-}	2.0×10^{-2}
$H_2C_2O_4$	$HC_2O_4^-$	6.5×10^{-2}
$HC_2O_4^-$	$C_2O_4^{2-}$	6.1×10^{-5}
$HClO_2$	ClO_2^-	1.0×10^{-2}
HNO_2	NO_2^-	4.6×10^{-4}
H_3PO_4	$H_2PO_4^-$	7.5×10^{-3}
$H_2PO_4^-$	HPO_4^{2-}	6.8×10^{-8}
HPO_4^{2-}	PO_4^{3-}	2.2×10^{-13}
H_3AsO_4	$H_2AsO_4^-$	4.8×10^{-3}
H_2CO_3	HCO_3^-	4.3×10^{-7}
HCO_3^-	CO_3^{2-}	5.6×10^{-11}
$H_4P_2O_7$	$H_3P_2O_7^-$	1.4×10^{-1}
H_2Te	HTe^-	2.3×10^{-3}
HTe^-	Te^-	1.0×10^{-5}
H_2Se	HSe^-	1.7×10^{-4}
HSe^-	Se^{2-}	1.0×10^{-10}
H_2S	HS^-	9.1×10^{-8}
HS^-	S^{2-}	$\sim 10^{-19}$
HN_3	N_3^-	1.9×10^{-5}
HF	F^-	7.2×10^{-4}
HCN	CN^-	4.9×10^{-3}
$HOBr$	BrO^-	2.1×10^{-9}
$HOCl$	ClO^-	3.5×10^{-8}
HOI	IO^-	2.3×10^{-11}
$HC_2H_3O_2$	$C_2H_3O_2^-$	1.75×10^{-5}
C_6H_5OH	$C_6H_5O^-$	1.28×10^{-10}
C_6H_5COOH	$C_6H_5O_2^-$	6.46×10^{-5}
$HCOOH$	HCO_2^-	1.8×10^{-4}
H_2O	OH^-	1.1×10^{-16}

Sodium chloride is soluble in water so it is written as an ionized product. There is no change in the sodium and chloride ions so they can be omitted from both sides of the equation. As a result, the net ionic equation can be written as

$$H_3O^+(aq) + OH^-(aq) \rightarrow 2\,H_2O(l) \tag{9.5}$$

If solutions that contain other ionized acids and bases are mixed, the reaction is still one that occurs between the $H_3O^+(aq)$ and $OH^-(aq)$. Therefore, the neutralization reaction between an acid and a base is shown in Eqn (9.5) according to Arrhenius theory.

When we consider the reaction

$$NH_3(g) + HCl(g) \rightarrow NH_4Cl(s) \tag{9.6}$$

and compare it to the reaction in aqueous solution,

$$NH_4^+(aq) + OH^-(aq) + H_3O^+(aq) + Cl^-(aq) \rightarrow 2\,H_2O(l) + NH_4^+(aq) + Cl^-(aq) \tag{9.7}$$

we see that ammonium chloride is a product of both reactions. In the second reaction, we can recover solid NH_4Cl by evaporating the water. Although the second process is explained by Arrhenius acid–base chemistry, the first is not. The reactants shown in Eqn (9.6) are not dissolved in water to which the Arrhenius definitions of acid and base are applicable. In order to describe acid–base reactions in the gas phase or in solvents other than water, a different approach is needed.

9.2 BRØNSTED–LOWRY THEORY

J. N. Brønsted and T. M. Lowry independently arrived at definitions of an acid and a base that do not involve water. They recognized that the essential characteristic of an acid–base reaction was the transfer of a hydrogen ion (proton) from one species (the acid) to another (the base). According to these definitions, *an acid is a proton donor* and *a base is a proton acceptor*. The proton must be donated to some other species so *there is no acid without a base*. According to Arrhenius, HCl is an acid because its water solution contains H_3O^+, which indicates that an acid can exist independently with no base being present. When examined according to the Brønsted–Lowry theory, the reaction

$$HCl(g) + H_2O(l) \rightarrow H_3O^+(aq) + Cl^-(aq) \tag{9.8}$$

is an acid–base reaction not because the solution contains H_3O^+ but rather because a proton is transferred from HCl (the acid) to H_2O (the base).

Once a substance has functioned as a proton donor, it has the potential to accept a proton (react as a base) from another proton donor. For example, acetate ions are produced by the reaction

$$HC_2H_3O_2(aq) + NH_3(aq) \rightarrow NH_4^+(aq) + C_2H_3O_2^-(aq) \tag{9.9}$$

The acetate ion can now function as a proton acceptor from a suitable acid. For example,

$$HNO_3(aq) + C_2H_3O_2^-(aq) \rightarrow HC_2H_3O_2(aq) + NO_3^-(aq) \tag{9.10}$$

In this reaction, the acetate ion is functioning as a base. On the other hand, Cl^- has very little tendency to function as a base because it comes from HCl, which is a very strong proton donor. According to the

Brønsted–Lowry theory, the species remaining after a proton is donated is called the conjugate base of that proton donor.

$$
\begin{array}{c}
\text{A conjugate pair} \\
\overbrace{\text{Acid}_1 \quad\quad\quad\quad\quad \text{Base}_2} \\
| \quad\quad\quad\quad\quad\quad\quad | \\
HC_2H_3O_2(aq) + NH_3(aq) \longrightarrow C_2H_3O_2^-(aq) + NH_4^+(aq) \\
| \quad\quad\quad\quad\quad\quad\quad | \\
\underbrace{\text{Base}_1 \quad\quad\quad\quad\quad \text{Acid}_2} \\
\text{A conjugate pair}
\end{array}
\qquad (9.11)
$$

In this reaction, acetic acid donates a proton to produce its conjugate, the acetate ion, which is able to function as a proton acceptor. Ammonia accepts a proton to produce its conjugate, the ammonium ion, which can function as a proton donor. Two species that differ by the transfer of a proton are known as *a conjugate pair*. The conjugate acid of H_2O is H_3O^+, and the conjugate base of H_2O is OH^-.

Characteristics of the reactions described so far lead to several conclusions regarding acids and bases according to the Brønsted–Lowry theory.

1. There is no acid without a base. The proton must be donated to something else.
2. The stronger an acid is, the weaker its conjugate will be as a base. The stronger a base is, the weaker its conjugate will be as an acid.
3. An acid reacts to displace a weaker acid. A stronger base reacts to displace a weaker base.
4. The strongest acid that can exist in water is H_3O^+. If a stronger acid is placed in water, it will donate protons to water molecules to produce H_3O^+.
5. The strongest base that can exist in water is OH^-. If a stronger base is placed in water, it will accept protons from water to produce OH^-.

A reaction that is of great importance in the chemistry of aqueous solutions is related to the acidic or basic character of conjugates. For example, after acetic acid *donates* a proton, the acetate ion has the ability to *accept* a proton. Therefore, when $NaC_2H_3O_2$ is dissolved in water, the reaction that occurs is the *hydrolysis* reaction,

$$
C_2H_3O_2^-(aq) + H_2O(l) \leftrightarrows HC_2H_3O_2(aq) + OH^-(aq) \qquad (9.12)
$$

This reaction does not take place to a great degree as should be expected from the fact that both an acid $(HC_2H_3O_2)$ and a base (OH^-) are produced in the same reaction. The result is that a 0.1 M solution of sodium acetate has a pH of 8.89, which means that the solution is basic, but not strongly so. Another way to look at this reaction is to say that OH^- is a strong base but $HC_2H_3O_2$ is a weak acid. Therefore, the solution should be basic, and it is. On the other hand, a 0.1 M solution of NaCl or $NaNO_3$ has a pH of 7, because the hydrolysis represented by the equation

$$
NO_3^-(aq) + H_2O(l) \leftrightarrows HNO_3(aq) + OH^-(aq) \qquad (9.13)
$$

which would produce a strong acid and a strong base in the solution, does not take place. Accordingly, we see that the anions of strong acids (which are very weak bases) do not effectively hydrolyze in aqueous solutions. We can also see that the weaker an acid is, its stronger the conjugate will be as a base and the more extensive the hydrolysis reaction will be. Consider the series of acids $HC_2H_3O_2$, HNO_2, HOCl, and

HOI, for which the K_a values are 1.75×10^{-5}, 4.6×10^{-4}, 3.5×10^{-8}, and 2.3×10^{-11}, respectively. As a result, if the basicities of 0.1 M solutions of the sodium salts of these acids are compared, it is found that the solution of NaOI is the strongest base whereas the solution of $NaNO_2$ is the weakest base. Because carbonic acid is weak, a solution of Na_2CO_3 is strongly basic owing to the reactions

$$CO_3^{2-}(aq) + H_2O(l) \leftrightarrows HCO_3^-(aq) + OH^-(aq) \tag{9.14}$$

$$HCO_3^-(aq) + H_2O(l) \leftrightarrows H_2CO_3(aq) + OH^-(aq) \tag{9.15}$$

Hydrolysis also occurs when the conjugate acid of a weak base is placed in water. For example, when NH_4Cl is dissolved in water, the hydrolysis reaction that takes place is

$$NH_4^+(aq) + H_2O(l) \leftrightarrows NH_3(aq) + H_3O^+(aq) \tag{9.16}$$

A 0.1 M solution of NH_4Cl has a pH of approximately 5.11, so it is distinctly acidic. Strictly speaking, *lysis* means "to split" and *hydrolysis* means the splitting of a water molecule (as in the hydrolysis to produce basic solutions as shown in Eqn (9.14)). In the case of NH_4^+ reacting with water, there is proton donation and acceptance, but there is no *lysis*.

There is another type of hydrolysis reaction that leads to acidic solutions. When a compound such as aluminum chloride is dissolved in water, the cation becomes strongly solvated. The extremely energetic nature of the hydration of the Al^{3+} ion is discussed in Chapter 7. Because of the very high charge to size ratio, the Al^{3+} ion is a very high-energy ion that reacts to relieve part of the charge density. One way in which this is done is by the reaction

$$[Al(H_2O)_6]^{3+} + H_2O \leftrightarrows H_3O^+ + [AlOH(H_2O)_5]^{2+} \quad K = 1.4 \times 10^{-5} \tag{9.17}$$

As a result of this reaction, part of the charge on the aluminum ion is relieved by the loss of H^+ from one of the water molecules. The pH of a 0.10 M solution of $AlCl_3$ is 2.93, so this type of hydrolysis leads to solutions that are quite acidic. Other metal ions having large charge and small size (Fe^{3+} ($K = 4.0 \times 10^{-3}$), Be^{2+}, Cr^{3+} ($K = 1.4 \times 10^{-4}$), etc.) behave in this manner and give solutions that are acidic.

When H_2SO_4, HNO_3, $HClO_4$, and HCl are dissolved in water to produce dilute solutions, each of them reacts essentially 100% to produce H_3O^+ ions. Water is sufficiently strong as a proton acceptor and these acids are all strong enough to ionize almost totally. As a result, they appear to be equal in strength as acids. In fact, their strength in water appears to be identical to that of H_3O^+ because that ion is the conjugate of all of the acids listed, and that is the acidic species present in all cases. It is sometimes stated that the strengths of the acids are "leveled" to that of H_3O^+. This phenomenon is called the *leveling effect*, and its basis is that H_2O is sufficiently strong as a base that it accepts protons from the strong acids. If a different solvent is used which is less basic than water, the ionization reactions of even strong acids do not take place completely because the acids do not actually have the same strength. A suitable solvent is glacial acetic acid, which is not normally a base but can function as a one when in the presence of a very strong acid:

$$HC_2H_3O_2 + HClO_4 \leftrightarrows HC_2H_3O_2H^+ + ClO_4^- \tag{9.18}$$

$$HC_2H_3O_2 + HNO_3 \leftrightarrows HC_2H_3O_2H^+ + NO_3^- \tag{9.19}$$

When the extent of reactions such as these is studied, it is found that the reaction with $HClO_4$ progresses farther to the right than with any of the other acids. By this means, it is possible to rank even the strong acids that react completely with water in terms of strength as shown in Table 9.1.

When it is recalled that the ammonium ion is simply NH_3 that has gained a proton, it is clear that NH_4^+ is the conjugate acid of NH_3. Therefore, it is not unusual to expect NH_4^+ to behave as an acid, which is illustrated in Eqn (9.16). However, the NH_4^+ ion can react as an acid under other conditions. When an ammonium salt, such as NH_4Cl, is heated to its melting point, the salt becomes acidic. In fact, the reactions are similar to those of HCl. For example, metals dissolve with the release of hydrogen:

$$2\,HCl + Mg \;\rightarrow\; MgCl_2 + H_2 \tag{9.20}$$

$$2\,NH_4Cl + Mg \;\rightarrow\; MgCl_2 + H_2 + 2\,NH_3 \tag{9.21}$$

After the NH_4^+ acts as a proton donor in such reactions, the NH_3 produced escapes as gas. Carbonates react with HCl to produce CO_2, sulfites react to produce SO_2, and oxides react to produce water. Heated ammonium salts react in a similar fashion.

$$2\,NH_4Cl + CaCO_3 \;\rightarrow\; CaCl_2 + CO_2 + H_2O + 2\,NH_3 \tag{9.22}$$

$$2\,NH_4Cl + MgSO_3 \;\rightarrow\; MgCl_2 + SO_2 + H_2O + 2\,NH_3 \tag{9.23}$$

$$2\,NH_4Cl + FeO \;\rightarrow\; FeCl_2 + H_2O + 2\,NH_3 \tag{9.24}$$

Because of its ability to react with metal oxides, NH_4Cl has been used as a soldering flux for many years. Removing the oxide on the surface of the object allows a strong joint to be produced. In older nomenclature, NH_4Cl was known as *sal ammoniac*.

There is nothing unusual about the acidic behavior of ammonium salts. In fact, any protonated amine can function as a proton donor. Because of this, many amine salts have been used as acids in synthetic reactions. If the chlorides are used, the amine salts are known as amine hydrochlorides. One of the earliest amine hydrochlorides studied with regard to its behavior as an acid is pyridine hydrochloride (pyridinium chloride), $C_5H_5NH^+Cl^-$. In the molten state, this compound undergoes many reactions of the type shown above.

9.3 FACTORS AFFECTING THE STRENGTH OF ACIDS AND BASES

The range of substances that function as acids is very large. Familiar binary compounds such as hydrogen halides are included as are the oxy acids such as H_2SO_4, HNO_3, H_3PO_4, and many others. Acids range in strength from the very weak boric acid, $B(OH)_3$, to acids that are very strong such as $HClO_4$. In this section, some general guidelines will be developed for predicting and correlating the strengths of acids.

One of the considerations related to acid strength for a polyprotic acid, such as H_3PO_4, is the fact that the first proton to be removed is lost more easily than the second and third. The first proton is lost from a neutral molecule, whereas the second and third protons are removed from species that have negative charges. The stepwise dissociation of H_3PO_4 can be shown as follows:

$$H_3PO_4 + H_2O \leftrightarrows H_3O^+ + H_2PO_4^- \quad K_1 = 7.5 \times 10^{-3} \tag{9.25}$$

$$H_2PO_4^- + H_2O \leftrightarrows H_3O^+ + HPO_4^{2-} \quad K_2 = 6.8 \times 10^{-8} \tag{9.26}$$

$$HPO_4^{2-} + H_2O \leftrightarrows H_3O^+ + PO_4^{3-} \quad K_3 = 2.2 \times 10^{-13} \tag{9.27}$$

It will be noted that there is a factor of approximately 10^5 between successive dissociation constants. This relationship exists between the equilibrium constants for numerous polyprotic acids, and it is sometimes known as *Pauling's rule*. This rule is also obeyed by sulfurous acid for which $K_1 = 1.2 \times 10^{-2}$ and $K_2 = 1 \times 10^{-7}$.

Another of the important concepts in dealing with acid strength is illustrated by the dissociation constants for the chloro-substituted acetic acids. The dissociation constants are as follows: CH_3COOH, $K_a = 1.75 \times 10^{-5}$; $ClCH_2COOH$, $K_a = 1.40 \times 10^{-3}$; $Cl_2CHCOOH$, $K_a = 3.32 \times 10^{-2}$; $K_a = Cl_3CCOOH$, $K_a = 2.00 \times 10^{-1}$. The dissociation constants for the acids show clearly the effect of replacing hydrogen atoms on the methyl group with chlorine atoms. As a result of having high electronegativity, the chlorine atoms cause a migration of electron density toward the end of the molecule where they reside, which causes the electron pair in the O—H bond to be shifted farther away from the hydrogen atom. The electrons are *induced* to move under the influence of the chlorine atoms. With each substitution of Cl for H, the acid becomes stronger due to the shift of electrons, which is known as *inductive effect*. The charge separation within the Cl_3CCOOH molecule can be represented as shown in the structure below.

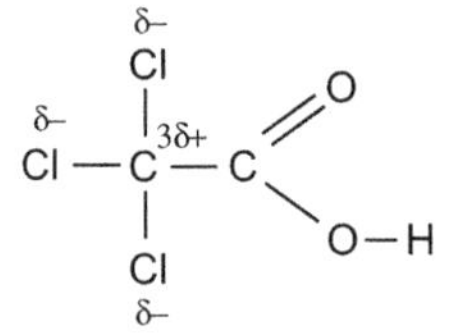

The effect of one chlorine atom in different positions can be seen from the dissociation constants for the monochlorobutyric acids. In the series of acids, a chlorine atom can be attached to the carbon atom adjacent to the COOH group or on one of the other carbon atoms. The dissociation constant for butyric acid is 1.5×10^{-5}. When a chlorine atom is attached in the three available positions, the dissociation constants are as follows:

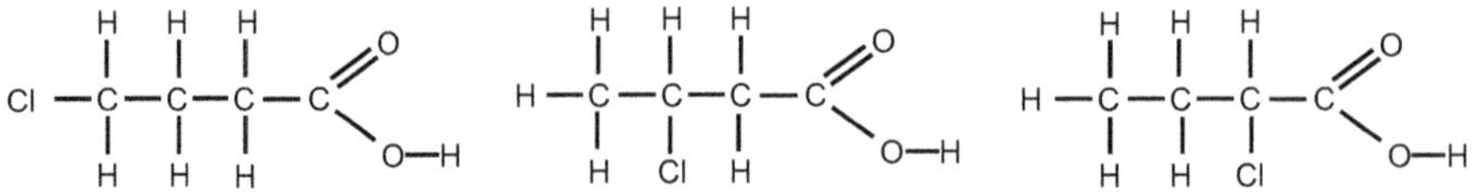

In this series, it is apparent that the effect of the chlorine atom is greatest when it is closest to the carboxyl group and least when it is farthest away.

Hydrogen ions are not normally removed from C—H bonds, but it is not impossible to do so. For example, the acidity of acetylene is well known, and many compounds exist in which the C_2^{2-} ion is present. It is known that the ease of removing H^+ from a carbon atom depends on the way in which the carbon atom is bound in a molecule. Most acidic are the −C−H groups in alkynes, whereas the least acidic are the C−H bonds in alkanes. Between these two extremes are the C−H bonds in aromatic molecules and those in alkenes with the C−H bonds in aromatics being the more acidic of the two types. Because the ease of removing H^+ depends on the charge separation in the C−H bond, we can conclude that carbon atoms behave *as if* their electronegativities vary in the following series, which also shows the hybrid orbital type used by carbon. This series also gives the acid strengths of these types of compounds.

Bond type C_{yne}−H > C_{arom}−H > C_{ene}−H > C_{ane}−H

Carbon hybridization sp sp^2 sp^2 sp^3

The inductive effect is also illustrated by the strengths of acids such as HNO_3 and HNO_2. The structures for the molecules are as follows:

In the HNO_3 molecule, there are two oxygen atoms that do not have hydrogen atoms attached. These oxygen atoms cause migration of electron density away from the H−O bond, which leaves the hydrogen ion easier to remove. In HNO_2, there is only one oxygen atom that does not have a hydrogen atom attached, so the extent of electron migration is much less than in HNO_3. The result is that nitric acid is a strong acid whereas nitrous acid is weak, although a somewhat different way of comparing the strengths of HNO_2 and HNO_3 follows.

When a proton is lost from an acid such as HNO_2, the NO_2^- ion that results is stabilized by the contributions from the resonance structures that can be shown as

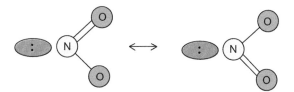

so that the −1 charge on the ion is dispersed over the structure. The result is that the NO_2^- ion does not attract H^+ as strongly as might be expected. On the other hand, the effect is even greater for nitric acid where the resulting NO_3^- ion has the −1 charge distributed over three oxygen atoms. Therefore, NO_3^- for which three resonance structures can be drawn is not as strong a base as is NO_2^-, which means that HNO_3 is a stronger acid than HNO_2.

Examples of the inductive effect abound. Although sulfuric acid is a strong acid, sulfurous acid is weak. Pauling provided a way of systematizing the inductive effect by writing the formulas for oxy acids as $(HO)_nXO_m$. Because the inductive effect is produced by the oxygen atoms that are not held in OH groups, it is the value of m in the formula that determines acid strength. There is a factor of approximately 10^5 produced in K for each unit of increase in the value of m. This principle is illustrated by the following examples.

If $m = 0$, the acid is very weak, $K_1 = 10^{-7}$ or less	$B(OH)_3$, boric acid, $K_1 = 5.8 \times 10^{-10}$
	$HOCl$, hypochlorous acid, $K = 3.5 \times 10^{-8}$
If $m = 1$, the acid is weak, $K_1 = 10^{-2}$ or less	$HClO_2$, chlorus acid, $K = 1 \times 10^{-2}$
	HNO_2, nitrous acid, $K = 4.6 \times 10^{-4}$
If $m = 2$, the acid is strong, $K_1 > 10^3$	H_2SO_4, sulfuric acid, K_1 is large
	HNO_3, nitric acid, K is large
If $m = 3$, the acid is very strong, $K_1 > 10^8$	$HClO_4$, perchloric acid, K is very large

Although the dissociation constants are very large for HCl, HBr, and HI, the range for all the hydrogen halides is from 7.2×10^{-4} for HF to 2×10^9 for HI. It is interesting to note that the dissociation constants for H_2O, H_2S, H_2Se, and H_2Te range from approximately 1×10^{-16} to 2.3×10^{-3}. As a result, there is a factor of approximately 10^{13} between the dissociation constants for the first and last member of the hydrogen compounds in Groups VI and VII.

Phenol, C_6H_5OH, is a weak acid for which $K_a = 1.3 \times 10^{-10}$ whereas aliphatic alcohols such as C_2H_5OH are not acidic under most conditions. From the standpoint of energy required for removing H^+ from the OH bonds, the energy difference for the two alcohols is not great. What is considerably different is what happens after the H^+ is removed. As shown in Figure 9.1, the phenoxide ion, $C_6H_5O^-$, has several resonance structures that contribute to its stability. These structures result in the ion residing at a lower energy than is possible for the ethoxide ion for which comparable resonance structures cannot be drawn.

The result of this resonance stabilization of the anion by delocalization of the negative charge makes the overall energy change required for ionization to be smaller for phenol than it is for ethanol. This can be represented in terms of energies as shown in Figure 9.2.

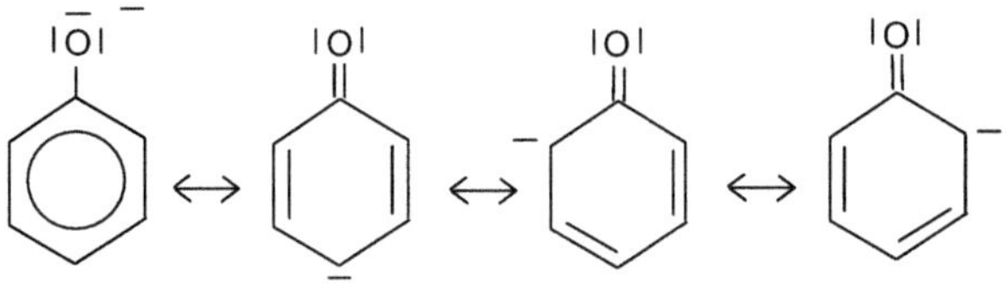

FIGURE 9.1
Resonance structures for the phenoxide ion.

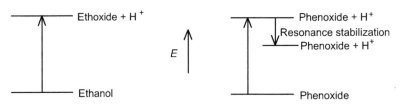

FIGURE 9.2
An energy diagram for dissociation of ethanol and phenol.

The difference is primarily in the resonance stabilization of the anion produced, not because of any great difference in the strengths of the O—H bonds.

Another example of resonance stabilization of the conjugate base after proton removal is that of acetylacetone (2,4-pentadione), which undergoes tautomerization reaction,

$$\text{(9.28)}$$

After the proton is lost, the resonance structures

stabilize the anion so that acetylacetone is slightly acidic. Many coordination compounds are known in which the anion bonds strongly to metal ions. The complexing ability of the anion is a manifestation of its behavior as a Lewis base (see Chapters 16 and 20).

Although aliphatic alcohols are not normally acidic, the OH group is sufficiently acidic to react with very active metals, such as sodium. The reaction can be written as

$$Na(s) + C_2H_5OH(l) \rightarrow H_2(g) + NaOC_2H_5(s) \qquad (9.29)$$

Except for being much less vigorous, this reaction is analogous to the reaction of sodium with water,

$$Na(s) + H_2O(l) \rightarrow H_2(g) + NaOH \qquad (9.30)$$

In both reactions, an anion that is a strong base is produced.

The factors that affect base strength are based primarily on the ability of a species to attract H^+, which is consistent with the principles that govern electrostatic interactions. The smaller and more highly charged a negative species is, the stronger the attraction for the small, positive H^+ ion will be. For example, O^{2-} has a higher negative charge than OH^-, and as a result it is a stronger base. One does not normally expect reactions that take place in aqueous solutions to produce oxides as a result of the basicity of O^{2-}. Ionic oxides normally react with water to produce hydroxides:

$$CaO + H_2O \rightarrow Ca(OH)_2 \qquad (9.31)$$

Although S^{2-} is a base, it is a weaker base than O^{2-} owing to its larger size. The small H^+ bonds better to the smaller, more "concentrated" region of negative charge on O^{2-}. In other words, the oxide ion has a higher charge density than the sulfide ion.

In the same way, the series of nitrogen species can be arranged in the order of decreasing base strength as $N^{3-} > NH^{2-} > NH_2^- > NH_3$. Likewise, NH_3 is a stronger base than PH_3 because the unshared pair of electrons on the nitrogen atom is contained in a smaller orbital, which leads to a greater attraction for the very small H^+. We will have more to say regarding such cases later in this chapter.

Another species that is a strong base is the hydride ion, H^-. Metal hydrides react with water to produce basic solutions:

$$H^- + H_2O \rightarrow H_2 + OH^- \tag{9.32}$$

This great affinity for water gives rise to the use of metal hydrides as drying agents for removing the last traces of water from organic liquids (that do not contain OH bonds). Calcium hydride, CaH_2, is commonly used for this purpose.

9.4 ACID–BASE CHARACTER OF OXIDES

In the foregoing discussion, it is clear that many acids contain a nonmetal, oxygen, and hydrogen. This suggests that one way to prepare an acid might be to carry out a reaction of the oxide of a nonmetal with water. In fact, this is exactly the case and many acids can be prepared in this way. For example,

$$SO_3 + H_2SO_4 \rightarrow H_2SO_4 \tag{9.33}$$

$$Cl_2O_7 + H_2O \rightarrow 2\, HClO_4 \tag{9.34}$$

$$CO_2 + H_2O \rightarrow H_2CO_3 \tag{9.35}$$

$$P_4O_{10} + 6\, H_2O \rightarrow 4\, H_3PO_4 \tag{9.36}$$

The general nature of this reaction indicates that oxides of nonmetals react with water to produce acidic solutions. Such oxides are sometimes referred to as *acidic anhydrides*.

Ionic metal oxides contain the oxide ion, which is a very strong base. Therefore, the addition of a metal oxide to water will result in a basic solution as a result of the reaction

$$O^{2-} + H_2O \rightarrow 2\, OH^- \tag{9.37}$$

Oxides of metals are sometimes called *basic anhydrides* because they react with water to produce basic solutions. Some examples of this type of reaction are the following:

$$MgO + H_2O \rightarrow Mg(OH)_2 \tag{9.38}$$

$$Li_2O + H_2O \rightarrow 2\, LiOH \tag{9.39}$$

As we have seen earlier in the case of proton transfer reactions such as the one that occurs between $HCl(g)$ and $NH_3(g)$, water is not necessary for the acid–base reaction to take place. This is also true of the reactions between the acidic oxides of nonmetals and the basic oxides of metals. In many cases, they react directly as illustrated in the following equations:

$$CaO + CO_2 \rightarrow CaCO_3 \tag{9.40}$$

$$CaO + SO_3 \rightarrow CaSO_4 \tag{9.41}$$

It is also true that a more acidic oxide can replace a weaker one. For example, SO_3 is a more acidic oxide than CO_2 so heating $CaCO_3$ with SO_3 liberates CO_2:

$$CaCO_3 + SO_3 \rightarrow CaSO_4 + CO_2 \tag{9.42}$$

This reaction can be interpreted as a stronger Lewis acid replacing a weaker one from its compound. If we consider CO_3^{2-} as CO_2 that has attached an O^{2-}, this reaction shows that SO_3 has a stronger affinity for O^{2-} and removes it from the CO_3^{2-} to liberate CO_2.

Not all binary oxides fall clearly into the category of acidic or basic oxides. For example, the oxides of Zn and Al have the ability to react as both acid and base depending on the other reactant. This is illustrated in the following equations.

$$ZnO + 2\,HCl \rightarrow ZnCl_2 + H_2O \tag{9.43}$$

Here, ZnO is reacting as a base, but in the reaction

$$ZnO + 2\,NaOH + H_2O \rightarrow Na_2Zn(OH)_4 \tag{9.44}$$

ZnO reacts as an acid as it forms the $[Zn(OH)_4]^{2-}$ complex ion. This reaction is formally equivalent to the reaction without water,

$$Na_2O + ZnO \rightarrow Na_2ZnO_2 \tag{9.45}$$

The oxyanion containing Zn^{2+}, ZnO_2^{2-}, is known as zincate, and it is equivalent to $Zn(OH)_4^{2-}$, which is named as the tetrahydroxozincate(II) anion (see Chapter 16). From these reactions, it is clear that ZnO can react as either an acidic or a basic oxide, and it is therefore known as an *amphoteric* oxide. In essence, there are some oxides that are clearly acidic, some that are clearly basic, and some that are in between. There is, in fact, a continuum of acid–base character for the oxides of elements that is shown in Figure 9.3.

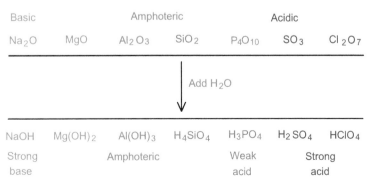

FIGURE 9.3
Acid–base character of oxides.

9.5 PROTON AFFINITIES

In Chapter 7, it was shown how the enthalpy of decomposition of an ammonium salt could be used to calculate the proton affinity of the anion. The *proton affinity* is a *gas* phase property (as is electron affinity) that gives the intrinsic basicity of a species. The reaction of H^+ with a base, B can be shown as

$$B(g) + H^+(g) \rightarrow BH^+(g) \quad \Delta H = \text{proton addition enthalpy (negative)} \quad (9.46)$$

Because most species will release energy when a proton is added, the proton addition enthalpy is *negative*. The proton affinity is the heat associated with the reverse process, the removal of the proton, which is accompanied by a *positive* enthalpy. Therefore, the proton affinity of a species B is defined as the enthalpy of the gas phase reaction,

$$BH(g)^+ \rightarrow B(g) + H^+(g) \quad (9.47)$$

The addition of the H^+ to B can be shown by means of a thermochemical cycle:

From this cycle, we can write

$$PA_B = I_H - I_B + E_{B^+ - H} \quad (9.48)$$

In this equation, I_H is the ionization potential for H (1312 kJ mol^{-1}), I_B is the ionization potential for the base B, and $E_{B^+ - H}$ is the energy of the $B^+ - H$ bond. The term I_B value is subtracted from the I_H value (the last value is of lesser importance), which leads to the conclusion that the smaller the value for I_B, the greater the proton affinity. Because H^+ reacts by removing electron density from B, the easier this process is to accomplish, the smaller the value of I_B will be. For the molecules CH_4, NH_3, H_2O, and HF, the proton affinities are 527, 841, 686, and 469 kJ mol^{-1}, respectively. These values correlate well with the ionization potentials of the *molecules*, which are in the order $NH_3 < H_2O < CH_4 < HF$.

The proton affinities for neutral molecules are generally in the range of 500–800 kJ mol^{-1}. Anions with -1 charges have proton affinities of 1400–1700 kJ mol^{-1}, whereas values for -2 ions are in the 2200–2400 kJ mol^{-1} range. Extensive tables of proton affinities for a wide range of molecules and ions are available. The proton affinity provides a useful measure of the inherent basicity of a species without the complicating effects produced by solvents. Unfortunately, although gas phase proton affinities are useful for relating absolute base strength to molecular structures, little acid–base chemistry is carried out in this way.

Because the proton is a hard Lewis acid, it would be expected to interact preferentially with hard Lewis bases. These are species that have small charge and small size. For the halide ions, a relationship between the sizes of the anions and their proton affinities should be expected. However, for a broad

Table 9.2 Proton Affinities of Some −1 and −2 Anions

Ion	r, pm	PA, kJ mol⁻¹	Ion	r, pm	PA, kJ mol⁻¹
F^-	136	1544	O^{2-}	140	2548
Cl^-	181	1393	S^{2-}	184	2300
Br^-	195	1351	Se^{2-}	198	2200
I^-	216	1314	CO_3^{2-}	185	2270
OH^-	121	1632	SH^-	181	1464

range of ions that includes polyatomic species, structural differences would come into play because only one of the atoms in the structure is the most basic site. Correlations should be attempted only for species having the same structure and charge. The proton affinities and ionic radii of the anions of Group VIA and Group VIIA atoms are shown in Table 9.2.

Figure 9.4 shows the relationship between ionic radius and proton affinity in a graphical way for monatomic ions having a −1 charge. It is clear that to a good approximation there is a correlation between the size of the anion and its proton affinity. Although this is in no way a detailed study, it is clear that the smaller (and thus *harder*) the negative ion (with the same type of structure) is, the more strongly will it bind a proton.

Although it is not surprising that anions have a rather high affinity for protons, it is also found that neutral molecules bind protons with the release of energy. Table 9.3 shows the proton affinities for some neutral molecules having simple structures.

It is interesting to note that even a "saturated" molecule, such as CH_4, has a significant attraction for a proton. This demonstrates clearly that even a pair of electrons that is shared in a bond can be a binding site for H^+. In general, the more acidic (or less basic) a compound is, the lower the value for its proton affinity will be. For example, the proton affinity for NH_3 is 866 kJ mol⁻¹, whereas for PH_3 it is 774, in keeping with the fact that PH_3 is the weaker base.

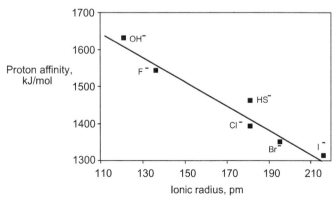

FIGURE 9.4
Variation of proton affinity with ionic radius for −1 ions.

Table 9.3 Proton Affinities (in kJ mol^{-1}) For Neutral Molecules[a]

CH$_4$	NH$_3$	H$_2$O	HF	Ne
528	866	686	548	201
SiH$_4$	PH$_3$	H$_2$S	HCl	Kr
~600	804	711	575	424
	AsH$_3$	H$_2$Se	HBr	Xe
	732	711	590	478
			HI	
			607	

[a]*Most values are from a larger table given by Porterfield (1993), p. 325.*

The data shown in Tables 9.2 and 9.3 reveal some interesting facts. We know that HI is a strong acid in aqueous solution. However, the proton affinity of $I(g)^-$ is 1314 kJ mol^{-1}, whereas that of H_2O is 686 kJ mol^{-1}. Therefore, for the reaction

$$HI(g) + H_2O(g) \rightarrow H_3O(g)^+ + I(g)^- \tag{9.49}$$

the enthalpy change would be the difference between the energy to pull a proton off $I(g)^-$, which is 1314 kJ mol^{-1} and the energy released when a proton is added to $H_2O(g)$, which is -686 kJ mol^{-1}. Consequently, the process represented in Eqn (9.49) is *endothermic* to the extent of 628 kJ mol^{-1}. However, the data shown in Table 7.7 indicate that the enthalpy of hydration of H^+ is -1100 kJ mol^{-1} and that of I^- is -296 kJ mol^{-1}. When solvation of the ions is taken into account, the situation becomes somewhat different. The fact remains that the removal of a proton from HI (a strong acid *in aqueous solutions*) and placing it on H_2O is not an energetically favorable reaction in the *gas* phase. These observations should indicate clearly why the role of a solvent (and the solvent chosen) are vitally important to how acids and bases function.

9.6 LEWIS THEORY

Up to this point, we have dealt with the subject of acid–base chemistry in terms of proton transfer. If we seek to learn what it is that makes NH_3 a base that can accept a proton, we find that it is because there is an unshared pair of electrons on the nitrogen atom where the proton can attach. Conversely, it is the fact that the hydrogen ion seeks a center of negative charge that makes it leave an acid, such as HCl, and attach to the ammonia molecule. In other words, it is the presence of an unshared pair of electrons on the base that results in proton transfer. Sometimes known as the *electronic* theory of acids and bases, it shows that the essential characteristics of acids and bases do not always depend on the transfer of a proton. This approach to acid–base chemistry was first developed by G.N. Lewis in the 1920s.

When the reaction

$$HCl(g) + NH_3(g) \rightarrow NH_4Cl(s) \tag{9.50}$$

is considered from the standpoint of electrons, we find that the proton from HCl is attracted to a center of negative charge on the base. The unshared pair of electrons on the nitrogen atom in ammonia is just that type of center. When the proton attaches to the NH_3 molecule, both of the electrons used in the bond come from the nitrogen atom. Thus, the bond is a *coordinate covalent bond* (or simply a coordinate bond) that is formed as the result of the reaction between an acid and a base.

Lewis acid–base chemistry provides one of the most useful tools ever devised for systematizing an enormous number of chemical reactions. Because the behavior of a substance as an acid or a base has nothing to do with proton transfer, many other types of reactions can be considered as acid–base reactions. For example,

$$BCl_3(g) + :NH_3(g) \rightarrow H_3N:BCl_3(s) \tag{9.51}$$

involves the BCl_3 molecule attaching to the unshared pair of electrons on the NH_3 molecule to form a coordinate bond. Therefore, this reaction is an acid–base reaction according to the Lewis theory. The product of such a reaction results from the addition of two complete molecules, and as a result, it is often referred to as an acid–base *adduct* or *complex*. The BCl_3 is an electron pair acceptor, a Lewis acid. The NH_3 is an electron pair donor, a Lewis base. According to these definitions, it is possible to predict what types of species will behave as Lewis acids and bases.

The following types of species are Lewis acids:
1. Molecules that have fewer than eight electrons on the central atom (e.g. BCl_3, $AlCl_3$, etc.)
2. Ions that have a positive charge (e.g. H^+, Fe^{3+}, Cr^{3+}, etc.).
3. Molecules in which the central atom can add additional pairs of electrons even though it already has an octet or more of electrons (e.g. $SbCl_3$, PCl_5, SF_4, etc.).

Lewis bases include the following types of species:
1. Anions that have an unshared pair of electrons (e.g. OH^-, H^-, F^-, PO_4^{3-}, etc.).
2. Neutral molecules that have unshared pairs of electrons (e.g. NH_3, H_2O, R_3N, ROH, PH_3, etc.)

We can now write equations for many reactions that involve a Lewis acid reacting with a Lewis base. The following are a few examples:

$$SbF_5 + F^- \rightarrow SbF_6^- \tag{9.52}$$

$$AlCl_3 + R_3N \rightarrow R_3N:AlCl_3 \tag{9.53}$$

$$H^+ + PH_3 \rightarrow PH_4^+ \tag{9.54}$$

$$Cr^{3+} + 6\,NH_3 \rightarrow Cr(NH_3)_6^{3+} \tag{9.55}$$

Acid–base behavior according to the Lewis theory has many of the same aspects as does acid–base theory according to the Brønsted–Lowry theory.

1. There is no acid without a base. An electron pair must be donated to one species (the acid) by another (the base).
2. An acid (or base) reacts to displace a weaker acid (or base) from a compound.
3. The interaction of a Lewis acid with a Lewis base is a type of neutralization reaction because the acidic and basic characters of the reactants are removed.

An acid–base reaction takes place readily between BF_3 and NH_3,

$$BF_3 + NH_3 \rightarrow H_3N:BF_3 \qquad (9.56)$$

However, when the product of this reaction is brought into contact with BCl_3, a reaction takes place:

$$H_3N:BF_3 + BCl_3 \rightarrow H_3N:BCl_3 + BF_3 \qquad (9.57)$$

In this reaction, a Lewis acid, BF_3, has been replaced by essentially a stronger one, BCl_3. There are two questions that arise concerning this reaction. First, why is BCl_3 a stronger Lewis acid than BF_3? Second, how does this type of reaction take place?

The strength of a Lewis acid is a measure of its ability to attract a pair of electrons from a molecule that is behaving as a Lewis base. Fluorine is more electronegative than chlorine so it appears that three fluorine atoms should withdraw electron density from the boron atom leaving it more positive. This would also happen to some extent when the peripheral atoms are chlorine, but chlorine is less electronegative than fluorine. On this basis, we would expect BF_3 to be a stronger Lewis acid. However, in the BF_3 molecule, the boron atom uses sp^2 hybrid orbitals, which leaves one empty $2p$ orbital that is perpendicular to the plane of the molecule. The fluorine atoms have filled $2p$ orbitals that can overlap with the empty $2p$ orbital on the boron atom to give some double bond character to the B–F bonds because the p orbitals of the atoms are of similar size. This can be illustrated as shown in Figure 9.5. As a result of the contribution by resonance structures having some double bond character, the boron atom in BF_3 is not as electron deficient as it is in BCl_3.

To answer the second question, we will write a general substitution reaction as

$$B:A + A' \rightarrow B:A' + A \qquad (9.58)$$

where B is a Lewis base and A and A′ are Lewis acids. Because of having unshared pairs of electrons, Lewis bases seek to interact with a center of positive charge (electron deficiency) in another species. In the reaction above, A is the electron deficient species, and that is where B has bonded. Because B seeks to interact with a *positive* site, it is known as *nucleophile* ("nucleus loving"). On the other hand, the molecules A and A′ being electron deficient seek to interact with an unshared pair of electrons on another species. Therefore, A and A′ are "electron loving" species or *electrophiles*. In the reaction above, one electrophile has replaced another so the reaction is known as *electrophilic substitution*.

The reaction

$$A:B + :B' \rightarrow A:B' + :B \qquad (9.59)$$

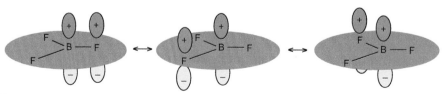

FIGURE 9.5
A representation of overlap between the $2p$ orbital on boron with the $2p$ orbitals on fluorine in BF_3. This causes the boron to be less electron deficient.

represents one nucleophile replacing another so the reaction is known as *nucleophilic substitution*. There are two limiting ways that we can envision such a reaction taking place. In the first way, the group that is leaving (B) leaves before B′ bonds to A. That bond-breaking step would be the slow or rate-determining step for the process. Following that step, the entering group, B′, enters in a bond-making step, which would be fast compared to the bond-breaking step. The process can be shown as

$$A:B \; \overset{\text{Slow}}{\rightleftharpoons} \; [A \; + \; :B]^{\ddagger} \underset{:B'}{\overset{\text{Fast}}{\longrightarrow}} A:B' \; + \; :B \tag{9.60}$$

In a kinetic study on this reaction, information is gained only about the slow step. If a system of pipes has a 1-inch pipe attached to a 6-inch pipe and a study is carried out on the rate at which water flows through the system, information will be gained only on the rate of flow of water through the 1-inch pipe. That is the *rate-determining* step. As long as there is a reasonable concentration of B′ available, the rate of the reaction shown above will be dependent only on the concentration of A:B. The rate law for the process can be written as

$$\text{Rate} = k_1[A:B] \tag{9.61}$$

(where k_1 is the rate constant) and the reaction is first-order in A:B. Changing the concentration of B′ within wide limits does not alter the rate of the reaction, which is known as an S_N1 process. The symbol S_N1 is a shorthand way to write *substitution, nucleophilic, first-order*.

   If the reaction takes place in such a way that the first (slow) step involves B′ starting to bond to A before B leaves, the reaction can be shown as follows:

$$A:B \; \overset{\text{Slow}}{\rightleftharpoons} \; [B \text{---} A \text{---} B']^{\ddagger} \underset{:B'}{\overset{\text{Fast}}{\longrightarrow}} A:B' \; + :B \tag{9.62}$$

In this case, the rate of the reaction is determined by how rapidly the transition state is formed, and that process requires both A:B and :B′. The rate law for the reaction is

$$\text{Rate} = k_2[A:B][B'] \tag{9.63}$$

According to this mechanism, there is a first-order dependence on both the concentration of [A:B] and B′, and the reaction is called S_N2 process (*substitution, nucleophilic, second-order*). Although many nucleophilic substitution reactions follow one of these simple rate laws, many others do not. More complex rate laws such as

$$\text{Rate} = k_1[A:B] + k_2[A:B][B'] \tag{9.64}$$

are observed. The substitution reactions in which one Lewis acid replaces another are known as electrophilic substitution and simple first- and second-order processes are designated as S_E1 or S_E2, respectively.

Perhaps the greatest area in which the Lewis acid–base approach is most useful is that of coordination chemistry. In the formation of coordination compounds, Lewis acids such as Cr^{3+}, Co^{3+}, Pt^{2+}, or Ag^+ bind to a certain number (usually 2, 4, or 6) groups as a result of electron pair donation and acceptance. Typical electron pair donors include H_2O, NH_3, F^-, CN^-, and many other molecules and ions. The products, known as coordination compounds or coordination complexes, have definite structures that are predictable in terms of principles of bonding. Because of the importance of this area of inorganic chemistry, Chapters 16–22 in this book are devoted to coordination chemistry.

9.7 CATALYTIC BEHAVIOR OF ACIDS AND BASES

One of the characteristics of acids and bases is that they catalyze certain reactions. Many years ago, J.N. Brønsted studied the relationship between acid strength as measured by the dissociation constant and the rate of a reaction that is catalyzed by the acid. The relationship that Brønsted recognized can be written as:

$$k = CK_a^n \tag{9.65}$$

where K_a is the dissociation constant for the acid, k is the rate constant for the reaction, and C and n are constants. Taking the logarithm of both sides of the equation yields

$$\ln k = n \ln K_a + \ln C \tag{9.66}$$

or using common logarithms the relationship becomes

$$\log k = n \log K_a + \log C \tag{9.67}$$

The pK_a of an acid is defined as $-\log K_a$ so Eqn (9.67) can be written as

$$\log k = -n\, pK_a + \log C \tag{9.68}$$

The form of this equation shows that a graph of $\log k$ vs. pK_a should be linear with a slope of $-n$ and an intercept of $\log C$. Dissociation of the acid produces the anion A^- that accepts protons from water as shown in the equation

$$A^- + H_2O \leftrightarrows HA + OH^- \tag{9.69}$$

for which the equilibrium constant, K_b, is written as

$$K_b = \frac{[HA][OH^-]}{[A^-]} \tag{9.70}$$

The equilibrium constant, K_a, is related to the free energy change for the dissociation of the acid, ΔG_a by the equation

$$\Delta G_a = -RT \ln K_a \tag{9.71}$$

Solving Eqn (9.71) for $\ln K_a$ and substituting the result in Eqn (9.66) gives

$$\ln k = -\frac{n\Delta G_a}{RT} + \ln C \tag{9.72}$$

Eqn (9.72) is known as *linear free energy relationship,* and it shows that there should be a linear relationship between the logarithm of the rate constant for a reaction and the free energy for the dissociation of the acid.

Relationships can be developed in a similar manner for reactions that are catalyzed by bases. The equation that can be obtained is

$$\log k' = n' \log K_b + \log C \qquad (9.73)$$

where the prime indicates that a base is being considered rather than an acid. The relationship between K_a and K_b is $K_a K_b = K_w.$ So substituting for K_b in Eqn (9.73) gives

$$\log k' = n \log (K_w/K_a) + \log C \qquad (9.74)$$

Although a general discussion of linear free energy relationships (LFER) is not being presented here, the approach given above can be extended to the Hammet σ and ρ parameters that are so useful in organic chemistry. Consult the references at the end of this chapter for details on the application of LFER in organic chemistry. The discussion above shows that the catalytic behavior of different acids is predictable when based on the difference in acid strength.

Although the discussion to this point has been concerned with the explanation of the behavior of Brønsted acids as catalysts, the range of reactions is enormous in which catalysis by acids and bases occurs. Many of the important types of organic reactions involve catalysis by acids or bases. In this section, several reactions will be mentioned, but the mechanistic details will not be presented in this book on inorganic chemistry. The discussion is intended to show the scope of catalysis by acids and bases.

A very important type of reaction known as enolization is catalyzed by both acids and bases. Also, the hydrolysis of an ester,

$$CH_3-\overset{\overset{\displaystyle O}{\|}}{C}-O-CH_3 + H_2O \longrightarrow CH_3COOH + CH_3OH \qquad (9.75)$$

is catalyzed by acids with the first step believed to involve the addition of a proton to an unshared pair of electrons on the carbonyl oxygen atom. In fact, several organic reactions that are catalyzed by acids involve the addition of H^+ to an unshared pair of electrons on an atom. Although an oxygen atom in a carbonyl group is a very weak base, it is certainly basic in comparison to strong acids. The enolization reaction of acetone,

$$CH_3-\overset{\overset{\displaystyle O}{\|}}{C}-CH_3 \longrightarrow CH_2=\overset{\overset{\displaystyle OH}{|}}{C}-CH_3 \qquad (9.76)$$

is catalyzed by OH^- with the first step being the removal of H^+ from a methyl group.

$$CH_3-\overset{\overset{\displaystyle O}{\|}}{C}-CH_3 \ + \ OH^- \ \longrightarrow \ CH_3-\overset{\overset{\displaystyle O}{\|}}{C}-CH_2^- \ + \ H_2O \tag{9.77}$$

Acetone is a very weak acid (K_a approximately 10^{-19}), but the presence of an oxygen atom bound to the adjacent carbon atom produces an inductive effect that makes it considerably more acidic than is a hydrocarbon molecule.

Another reaction that is catalyzed by a base is that which leads to the formation of the benzyne intermediate. In this case, the strong base is the amide ion, NH_2^-.

$$\tag{9.78}$$

The triple bond in the benzyne intermediate is very reactive toward a wide range of nucleophiles. The reaction of acetaldehyde with methanol to produce a hemiacetal is also a base catalyzed reaction. In this reaction, the methoxide ion, CH_3O^-, is the base:

$$\tag{9.79}$$

Many important reactions involve catalysis by Lewis acids or bases. One of the most important of these is the type of reaction carried out by Charles Friedel and James Crafts. These reactions, known as the Friedel–Crafts reactions, actually involve several types of important processes. One of these is alkylation, which is illustrated by the reaction of benzene with an alkyl halide in the presence of $AlCl_3$, a strong Lewis acid.

$$C_6H_6 + RCl \ \xrightarrow{\ AlCl_3\ } \ C_6H_5R + HCl \tag{9.80}$$

In this reaction, the function of $AlCl_3$ is to generate an attacking carbocation by means of the reaction

$$AlCl_3 + RCl \ \leftrightarrows \ R^+ + AlCl_4^- \tag{9.81}$$

A mixture of HF and BF_3 catalyzes the reaction

$$\tag{9.82}$$

by first generating the $(CH_3)_3C^+$ carbocation, which then attacks the ring.

$$HF + BF_3 + (CH_3)_2C=CH_2 \rightarrow (CH_3)_3C^+BF_4^- \qquad (9.83)$$

Lewis acids such as $FeCl_3$ and $ZnCl_2$ are also useful catalysts. For example, bromination of benzene by Br_2 in the presence of $FeBr_3$ can be shown as

$$\qquad (9.84)$$

The function of the catalyst is to provide a bromine with a positive charge either by separating the molecule to give Br^+ and $FeBr_4^-$ or by giving a polar Br_2 molecule that results from attaching one of the bromine atoms to $FeBr_3$ as in $Br^+- Br^- - FeBr_3$.

Another reaction that depends on the presence of an acid catalyst is nitration. In this case, the catalyst is H_2SO_4, which functions to produce the *nitronium* ion, NO_2^+:

$$HNO_3 + H_2SO_4 \leftrightarrows H_2NO_3^+ + HSO_4^- \qquad (9.85)$$

$$H_2NO_3^+ + H_2SO_4 \leftrightarrows H_3O^+ + NO_2^+ + HSO_4^- \qquad (9.86)$$

The second step is sometimes shown as

$$H_2NO_3^+ \rightarrow NO_2^+ + H_2O \qquad (9.87)$$

but in the presence of sulfuric acid the H_2O certainly would be protonated. The actual nitration process is:

$$\qquad (9.88)$$

Many reactions are catalyzed by aluminum oxide, Al_2O_3, which is also known as alumina. In the solid, there are sites on the surface where a strongly acidic aluminum ion is available to bond to an electron pair donor. One such reaction involves the dehydration of alcohols to produce alkenes. This process can be represented as follows:

$$\qquad (9.89)$$

In this reaction, an acidic site on the catalyst is shown as an aluminum atom bound to three oxygen atoms, and this is a site where an alcohol molecule can bind. If the acidic sites on the alumina are shielded by reacting it with a base, the oxide is no longer an effective acid catalyst. In other words, the acidity is removed by attaching molecules such as NH_3 to the acidic sites.

The reactions shown above are just a few of the enormous number of processes that involve catalysis by acids and bases. Some of the reactions are of economic importance and they certainly serve to illustrate that this type of chemistry is not solely the province of either the organic or inorganic chemist.

9.8 THE HARD–SOFT INTERACTION PRINCIPLE (HSIP)

As we have seen, the Lewis theory of acid–base interactions based on electron pair donation and acceptance applies to many types of species. As a result, the electronic theory of acids and bases pervades the whole of chemistry. Because the formation of metal complexes represents one type of Lewis acid–base interaction, it was in that area that evidence of the principle, *species of similar electronic character interact best*, was first noted. As early as the 1950s, Ahrland, Chatt, and Davies had classified metals as belonging to Class A if they formed more stable complexes with the first element in the periodic group or to Class B if they formed more stable complexes with the heavier elements in that group. This means that metals are classified as A or B based on the electronic character of the donor atom they prefer to bond to. The donor strength of the ligands is determined by the stability of the complexes they form with metals. This behavior is summarized in the following table.

	Donor strength
Class A metals	$N >> P > As > Sb > Bi$
	$O >> S > Se > Te$
	$F > Cl > Br > I$
Class B metals	$N << P > As > Sb > Bi$
	$O << S \approx Se \approx Te$
	$F < Cl < Br < I$

Thus, Cr^{3+} and Co^{3+} belong to Class A because they form more stable complexes with oxygen as the donor atom than when sulfur is the donor atom. On the other hand, Ag^+ and Pt^{2+} belong to Class B because they form more stable complexes with P or S as the donor atom than with N or O as the donor atom.

The examination of the hard and soft electronic character of acids and bases, as discussed here, was first put into systematic form by R. G. Pearson in the 1960s. According to the interpretation of Pearson, soft bases are those electron donors that have high polarizability, low electronegativity, empty orbitals of low energy, or those that are easily oxidizable. Hard bases have the opposite properties. Soft acids are those having low positive charge, large size, and completely filled outer orbitals. Polarizability, the ability to distort the electron cloud of a molecule, and low electronegativity depend upon these properties. Hard acids have the opposite characteristics. Based on their properties, we would expect that typical hard acids would be species such as Cr^{3+}, Co^{3+}, Be^{2+}, and H^+, for which distortion of the electron cloud would be slight. Soft acids would include species such as Ag^+, Hg^{2+}, Pt^{2+}, or uncharged

Table 9.4 Lewis Bases

Hard	Soft
OH^-, H_2O, F^-	RS^-, RSH, R_2S
SO_4^{2-}, Cl^-, PO_4^{3-}, CO_3^{2-}, NO_3^-	I^-, SCN^-, CN^-, $S_2O_3^{2-}$
ClO_4^-, RO^-, ROH, R_2O	CO, H^-, R^-
NH_3, RNH_2, N_2H_4	R_3P, R_3As, C_2H_4

Borderline bases

C_5H_5N, N_3^-, N_2, Br^-, NO_2^-, SO_3^{2-}

metal atoms, all of which are easily polarizable. Obviously, such a distinction is made in a qualitative way, and the classification of some species will not be definitive. Tables 9.4 and 9.5 show some typical acids and bases classified according to their hard or soft character. (Lists are based on those in Pearson, R.G., *J. Chem. Educ.* 1968, *45*, 581.).

The guiding principle regarding the interaction of electron pair donors and acceptors is that the *most favorable* interactions occur when the acid and base have similar electronic character. In accord with this observation, it is found that hard acids *preferentially* interact with hard bases, and soft acids interact *preferentially* with soft bases. This is related to the way in which the species interact. Hard acids interact with hard bases primarily by interactions that result from forces between ions or polar species. Interactions of these types will be favored by high charge and small size of both the acid and the base. Soft acids and soft bases interact primarily by sharing electron density, which is favored when the species have high polarizablity. Frequently, interactions between soft acids and soft bases involve bonding between neutral molecules. Orbital overlap that leads to covalent bonding is most favorable when the orbitals of the donor and acceptor atoms are of similar size and energy.

A modification of the *hard–soft acid–base* (HSAB) approach was first explained by C.K. Jørgensen in connection with the stability of a cobalt complex. Under normal circumstances, Co^{3+} is a hard Lewis acid. However, when Co^{3+} is bonded to five cyanide ions, it is found that a more stable complex results

Table 9.5 Lewis Acids

Hard	Soft
H^+, Li^+, Na^+, K^+, Be^{2+}, Mg^{2+}	Cu^+, Ag^+, Au^+, Ru^+
Ca^{2+}, Mn^{2+}, Al^{3+}, Sc^{3+}, La^{3+}, Cr^{3+}	Pd^{2+}, Cd^{2+}, Pt^{2+}, Hg^{2+}
Co^{3+}, Fe^{3+}, Si^{4+}, Ti^{4+}	$GaCl_3$, RS^+, I^+, Br^+
$Be(CH_3)_2$, BF_3, HCl, $AlCl_3$, SO_3	O, Cl, Br, I,
$B(OR)_3$, CO_2, RCO^+, R_2O, RO^-	Uncharged metals

Borderline acids

Fe^{2+}, Co^{2+}, Ni^{2+}, Zn^{2+}, Cu^{2+}, Sn^{3+},
Rh^{3+}, $(BCH_3)_3$, Sb^{3+}, SO_2, NO^+

when the sixth group is iodide rather than fluoride. Therefore, it is found that $[Co(CN)_5I]^{3-}$ is more stable than $[Co(CN)_5F]^{3-}$. If the other five ligands are NH_3, the opposite effect is observed. The apparent contradiction in the case of the cyanide complexes is the result of the "softening" of Co^{3+} that occurs when five of the ligands are soft. This causes the *aggregate* of the hard Co^{3+} and five soft ligands to behave as a soft electron pair acceptor for the I^-. However, the five CN^- ions have made the Co^{3+} in the *complex* much softer than an *isolated* Co^{3+} ion. Although Co^{3+} is normally a hard electron pair acceptor, the five CN^- ions attached cause it to behave as a soft acid. This effect is known as the *symbiotic effect*. Whether a species appears to be hard or soft depends on the groups attached and their character.

The HSAB principle is not restricted to the usual types of acid–base reactions. It is a guiding principle for all types of interactions; species of similar electronic character interact best. We have already seen some applications (such as the relative strength of HF and HI) of this principle, which we will continue to call HSAB, but we now consider a number of other types of applications.

9.8.1 Hydrogen Bonding

The HSAB principle can be applied in a qualitative way to interactions that result from hydrogen bonding. Stronger hydrogen bonds would be expected to form when the electron donor atom is a hard Lewis base in which the pair of electrons being donated are in a small, compact orbital. For example, the occurrence of hydrogen bonding in the hydrogen compounds of the first long group of elements has already been cited as being responsible for their high boiling points (see Chapter 6). Hydrogen bonding is much more extensive in NH_3, H_2O, and HF than it is in PH_3, H_2S, and HCl. Unshared pairs of electrons on second row atoms are contained in larger orbitals and do not interact as well with the very small H nucleus.

A clear effect of hydrogen bonding is afforded by considering the interaction of alcohols with acetonitrile, CH_3CN, and trimethylamine, $(CH_3)_3N$. The dipole moments of these molecules are 3.44 D and 0.7 D, respectively. However, nitriles are soft bases but amines are hard bases. In this case, when CH_3OH is hydrogen bonded to CH_3CN and $(CH_3)_3N$, the bonds have energies of about 6.3 kJ mol^{-1} and 30.5 kJ mol^{-1} respectively, in accord with the predictions of HSAB.

When phenol, C_6H_5OH, is hydrogen bonded to $(C_2H_5)_2O$, the OH stretching band in the infrared spectrum is shifted by about 280 cm^{-1} and the hydrogen bonds have energies of about 22.6 kJ mol^{-1}. When phenol hydrogen bonds to $(C_2H_5)_2S$, the corresponding values for these parameters are about 250 cm^{-1} and 15.1 kJ mol^{-1}. Ethers are considered as hard bases, whereas alkyl sulfides are soft.

If our qualitative predictions regarding softness of the base are valid, we should expect that C_6H_5OH hydrogen bonded to the bases $(C_6H_5)_3P$ and $(C_6H_5)_3As$ would also follow the same trend. In fact, when the OH of phenol is hydrogen bonded to $(C_6H_5)_3P$, the OH stretching band is shifted by 430 cm^{-1}, and when it is hydrogen bonded to $(C_6H_5)_3As$, it is shifted by 360 cm^{-1}. This is exactly as expected. When phenol hydrogen bonds to CH_3SCN, the hydrogen bond energy is 15.9 kJ mol^{-1} and the OH band is shifted by 146 cm^{-1}. When phenol is hydrogen bonded to CH_3NCS, the OH stretching band is shifted by 107 cm^{-1} and the hydrogen bond energy is 7.1 kJ mol^{-1}. The sulfur end of SCN is a soft electron donor, and the nitrogen end is significantly harder. Hydrogen bonding to these atoms reflects this difference.

9.8.2 Linkage Isomers

Ions such as SCN^- have two potential electron donor atoms. When bonding to metal ions, the bonding mode may be determined by HSAB considerations. For example, when SCN^- bonds to Pt^{2+}, it bonds through the sulfur atom. When it bonds to Cr^{3+}, it bonds through the nitrogen atom, in accord with the Class A and Class B behavior of these metal ions described earlier. Steric effects can cause a change in bonding mode in Pt^{2+} complexes. When three very large groups such as $As(C_6H_5)_3$ are bound to Pt^{2+}, the steric effects can cause SCN^- to bond through the nitrogen atom even though the electronic character would indicate the opposite. As shown in Figure 9.6, the bonding through the nitrogen atom is linear, whereas the one through the sulfur atom is not.

9.8.3 Solubility

One of the simplest applications of the HSAB principle is related to solubility. The rule, "like dissolves like", is a manifestation that solute particles interact best with solvent molecules which have similar characteristics. Small, highly charged particles or polar molecules are solvated best by solvents containing small, highly polar molecules. Large solute particles having low polarity are solvated best by solvent molecules having similar characteristics. Consequently, NaCl is soluble in water, whereas sulfur, S_8, is not. On the other hand, NaCl is insoluble in CS_2 but S_8 dissolves in CS_2.

For many years before the hard–soft interaction principle was described by Ralph G. Pearson, a rule of solubility was stated as "like dissolves like". This meant that large, nonpolar solute molecules interact best with large nonpolar solvent molecules. Water is a good solvent for most ionic solids, but carbon tetrachloride is a good solvent for solids that consist of large, nonpolar molecules. In the same way, H_2O and CH_3OH are both small, polar molecules, and the liquids are completely miscible. On the other hand, even though they contain OH groups, alcohols that have as many as five carbon atoms in the chain are not completely miscible with water. The alkyl group containing a longer carbon chain becomes the dominant part of the molecule rather than the OH group. Sodium chloride is an ionic solid that dissolves to the extent of 35.9 g in 100 g of water at 25 °C. The solubilities in CH_3OH, C_2H_5OH, and i-C_3H_7OH are 0.237, 0.0675, and 0.0041 g, respectively, in 100 g of solvent at 25 °C. In the case of the alcohols, as the chain length increases, the organic part of the molecule dominates the OH functional group so that these compounds become progressively poorer solvents for ionic solutes.

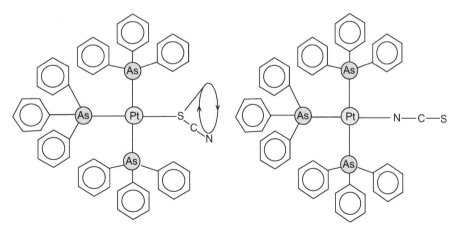

FIGURE 9.6
Influence of large ligands such as triphenylarsine on the bonding mode of SCN^-.

It is not as simple as considering the dipole moment of the solvent because nitrobenzene has a dipole moment of 4.22 D, but the large molecules cannot surround Na^+ and Cl^- effectively. Therefore, NaCl is not soluble in nitrobenzene.

In order for a solvent to remove ions from the lattice, the molecules must be polar and small enough to allow several solvent molecules to surround the ions. Of course NaCl is soluble in water because water consists of small, polar molecules. If the solvent is CH_3OH, the solubility of NaCl is much less than it is in water. The dipole moment of CH_3OH is somewhat lower than that for H_2O, but the CH_3OH molecules are large enough to prevent as many of them from surrounding Na^+ and Cl^- efficiently. When C_2H_5OH is used as a solvent for NaCl, it is found that C_2H_5OH is a considerably poorer solvent than CH_3OH. The dipole moments of the two alcohols are not very different, but the larger C_2H_5OH cannot solvate the ions as well as the smaller CH_3OH molecules. Because Na^+ and Cl^- ions are small and compact, they interact better with the smaller CH_3OH molecules than with larger molecules of another alcohol.

The HSIP allows us to correctly predict the results of many experiments. For example, suppose an aqueous solution containing Cs^+, Li^+, F^-, and I^- is evaporated. The solid products of the reaction could be CsF and LiI or CsI and LiF:

$$
\begin{array}{c}
\text{Solution containing} \\
Cs^+,\ Li^+,\ F^-,\ I^-
\end{array}
\xrightarrow[\text{Evaporate}]{}
\begin{array}{l}
\nearrow \quad CsI\ +\ LiF \\
\text{or} \\
\searrow \quad CsF\ +\ LiI
\end{array}
\tag{9.90}
$$

Based on the principles of bonding related to electronegativity, the element with *highest* electronegativity should bond best to the one with the *lowest* electronegativity, which means that CsF should be produced. However, based on the hard–soft interaction principle, the ions of similar electronic character should interact best. The small Li^+ ion should bond better to F^- and the large Cs^+ should bond better to I^-, exactly as is observed.

The conclusion that follows is that *ionic solids precipitate best from aqueous solutions when the ions are of similar size, preferably with the two ions having the same magnitude in their charges.* For the reaction

$$M^+(aq) + X^-(aq) \ \rightarrow \ MX(xtl) \tag{9.91}$$

the enthalpy of crystallization changes are ~ 0, 66.9, 58.6, and -20.9 kJ mol^{-1} for LiF, LiI, CsF, and CsI, respectively. These data can be used to show that if a solution contains Li^+, Cs^+, F^-, and I^-, the precipitation of LiF and CsI is more favorable than that of LiI and CsF. We can conclude that the small, hard Li^+ ion interacts better with F^- and the large, soft Cs^+ interacts better with the large I^-.

Another example of how the HSIP applies to precipitation can be seen in a familiar case from analytical chemistry. Because ions of similar size and magnitude of charges precipitate (interact) best, a good counter ion for precipitation of Ba^{2+} is one that is of similar size and has a -2 charge. In accord with this, Ba^{2+} is normally precipitated as the sulfate because of the favorable size and charge of the anion compared to the cation.

The application of the HSAB is of considerable importance in preparative coordination chemistry in that some complexes are stable only when they are precipitated using a counter ion conforming to the above rule. For example, $CuCl_5^{3-}$ is not stable in aqueous solution but can be isolated as $[Cr(NH_3)_6]$ $[CuCl_5]$. Attempts to isolate solid compounds containing the complex ion $Ni(CN)_5^{3-}$ as $K_3[Ni(CN)_5]$

lead to KCN and $K_2[Ni(CN)_4]$. It was found, however, that when counter ions, such as $Cr(NH_3)_6^{3+}$ or $Cr(en)_3^{3+}$, were used, solids containing the $Ni(CN)_5^{3-}$ anion were obtained.

9.8.4 Reactive Site Preference

We have already used the HSAB principle as it applies to linkage isomers in metal complexes. This application to bonding site preference can also be used to show the behavior of other systems. For example, the reactions of organic compounds also obey the principle when reacting with nucleophiles such as SCN^- or NO_2^-:

$$CH_3SCN \quad \overset{CH_3I}{\longleftarrow} \quad NCS^- \quad \overset{RCOX}{\longrightarrow} \quad RC(O)NCS \tag{9.92}$$

In this case, the acidic species RCOX is a hard acid and reacts with the nitrogen end of SCN^- to form an acyl isothiocyanate. The soft methyl group bonds to the S atom and forms methyl thiocyanate. Consider the following reactions of NO_2^-:

$$CH_3NO_2 \quad \overset{CH_3I}{\longleftarrow} \quad NO_2^- \quad \overset{t\text{-}BuCl}{\longrightarrow} \quad t\text{-}BuONO \tag{9.93}$$

Here, the $t\text{-}Bu^+$ carbocation is a hard acid, so the product is determined by its interaction with the oxygen (harder) electron donor. With CH_3I, the product is nitromethane, showing the softer character of the methyl group.

Consider the reaction of PCl_3 with AsF_3.

$$PCl_3 + AsF_3 \rightarrow PF_3 + AsCl_3 \tag{9.94}$$

Although both arsenic and phosphorus are soft, arsenic is softer. Likewise, Cl is softer than F, so we predict that a scrambling reaction would take place as shown above.

Phosphine, PH_3, is a pyramidal molecule that has an unshared pair of electrons on the phosphorus atom. Therefore, it can function as both a proton acceptor (Brønsted base) and an electron pair donor (Lewis base). For the moment, it is the ability to accept protons that will be considered. When ammonia is dissolved in water, an equilibrium

$$NH_3(aq) + H_2O(l) \leftrightharpoons NH_4^+(aq) + OH^-(aq) \tag{9.95}$$

is established for which the equilibrium constant is 1.8×10^{-5}. When phosphine is dissolved in water, the reaction

$$PH_3(aq) + H_2O(l) \leftrightharpoons PH_4^+(aq) + OH^-(aq) \tag{9.96}$$

takes place with an equilibrium constant of only $\sim 10^{-26}$. Why is PH_3 so much weaker as a base than is NH_3? The answer lies in a principle known as the *hard–soft acid–base principle* (often referred to in this book as the *hard–soft interaction principle*, HSIP, because it governs interactions that are not necessarily acid–base in character). The essential idea is that *species of similar electronic character interact best*. In the cases above, H^+ is the electron pair acceptor. It is extremely small in size, and it is nonpolarizable. Because of these traits, it is considered a hard acid. In the NH_3 molecule, it is the electron pair on the nitrogen atom that is the basic site. The orbital is small, and it readily bonds to the small H^+. In the PH_3 molecule, it is the pair of electrons on the phosphorus atom where the H^+ must attach. However, phosphorus is in the third row of the periodic table and is a considerably larger atom than the nitrogen atom. As a result, the small H^+ bonds much better to the electron pair in the more compact orbital on the nitrogen atom in NH_3 than it does to the electrons in the larger, more diffuse orbital on the phosphorus atom in PH_3. Therefore, although NH_3 is a weak base, it is a much stronger base than PH_3. In fact, PH_3 is such a weak base that only salts of strong acids with relatively large anions (to match the size of PH_4^+) are stable. Accordingly, PH_4I is stable, whereas PH_4F is not. Several important applications of the matching of sizes of cation and anion will be discussed in connection with the synthesis of coordination compounds in Chapter 20.

The HSAB principle illustrated above is one of the most useful principles in all of chemistry for predicting how many types of interactions occur. It is not restricted to acid–base interactions so it is better called the *HSIP*. It predicts that hard acids (high charge, small size, low polarizability) *prefer* to interact with hard bases (small size, high charge, low polarizability). Thus, the reaction

$$H^+(aq) + OH^-(aq) \rightarrow H_2O(l) \tag{9.97}$$

proceeds farther to the right than does the reaction

$$H^+(aq) + SH^-(aq) \rightarrow H_2S(aq) \tag{9.98}$$

In other words, OH^- is a much stronger base than SH^-. The orbital holding the unshared pair of electrons on the oxygen atom is smaller than the one holding the electron pair on the sulfur atom. Therefore, H^+ interacts more strongly with OH^- than it does with SH^-. The HSIP does not say that hard acids will *not* bond to soft bases. Rather, it says that bonding between hard acids and hard bases is *more effective* than between hard acids and soft bases. A similar statement can also be made for bonding as a result of the interaction between soft acids and soft bases.

Earlier it was described how PH_3 is a much weaker base than NH_3. That is certainly true when the interaction of these molecules with H^+ is considered. However, if the electron pair acceptor is Pt^{2+}, the situation is quite different. In this case, the Pt^{2+} ion is large and has a low charge so it is considered to be a soft (polarizable) Lewis acid. Interaction between Pt^{2+} and PH_3 provides a more stable bond that when NH_3 bonds to Pt^{2+}. In other words, the soft electron acceptor, Pt^{2+}, bonds better to the softer electron donor, PH_3, than it does to NH_3. The HSIP does not say that soft Lewis acids will not interact with hard Lewis bases. In fact, they will interact, but this is not the most favored type of interaction.

Although sometimes referred to as the hard–soft acid–base *theory*, it is actually a *principle* that relates to many types of chemical interactions. It provides a good explanation of why HF is a weak acid. If H^+ might potentially interact with either H_2O or F^-, the situation with regard to the preferred bonding mode could be shown as follows:

$$H_2O \overset{?}{\leftarrow} H^+ \overset{?}{\rightarrow} :F^-$$

In this case, the unshared pairs of electrons on F^- and H_2O to which the H^+ can bind are in orbitals of similar size. In addition, the fluoride ion has a negative charge, which attracts H^+ strongly. Therefore, H^+ bonds preferentially to F^- rather than to H_2O. What this is means is that the reaction

$$HF(aq) + H_2O(l) \rightleftharpoons H_3O^+(aq) + F^-(aq) \qquad (9.99)$$

takes place only slightly because the hydrogen ion bonds preferentially to F^-. Therefore, HF is a weak acid.

If we consider a similar situation with H^+ and Cl^- ions in water, what is the preferred bonding mode for H^+?

$$H_2O: \overset{?}{\leftarrow} \ H^+ \ \overset{?}{\rightarrow} \ :Cl^-$$

Now, the possibility exists for H^+ to bond to Cl^- or H_2O. The orbital that has an unshared pair of electrons on the oxygen is much smaller than that on the chloride ion. Therefore, H^+ preferentially bonds to H_2O, which means that the reaction

$$HCl(aq) + H_2O(l) \rightarrow H_3O^+(aq) + Cl^-(aq) \qquad (9.100)$$

goes to the right almost to completion in dilute solutions. As expected on this basis, HBr and HI are even stronger acids than HCl.

9.8.5 Formation of Lattices

The HSIP gives a qualitative explanation for the fact that equilibrium for the reaction

$$LiBr(s) + RbF(s) \rightleftharpoons LiF(s) + RbBr(s) \qquad (9.101)$$

lies far to the right. However, it is possible to verify this conclusion on a more quantitative basis. From the standpoint of thermodynamics, the lattice energies for the solids can be written in terms of equations such as

$$U = \frac{N_o Ae^2}{r}\left(1 - \frac{1}{n}\right) \qquad (9.102)$$

Each of the compounds shown in Eqn (9.101) has the same crystal structure, the sodium chloride structure, so the Madelung constant is the same for all of them. The term containing $1/n$ is considered to be a constant for the two pairs of compounds (reactants and products). Actually, an average value of n is used for each compound because the ions do not have the configuration of the same noble gas. The lattice energy is the energy required to separate one mole of crystal into the gaseous ions. In the discussion that follows, it must be kept in mind that the lattice energies of the reactants must be supplied but the lattice energies of the products will be liberated. Therefore, using an approach that is analogous to the use of bond energies, discussed in Chapter 4, the energy change will be given by

$$\Delta E = U_{LiBr} + U_{RbF} - U_{LiF} - U_{RbBr} \qquad (9.103)$$

The lattice energies of these compounds are easily calculated or found in tables. When the values are substituted in Eqn (9.103), it is found that

$$\Delta E(kJ) = 761 + 757 - 1017 - 632 = -131 \text{ kJ} \qquad (9.104)$$

which shows that the energetically favored products are LiF and RbBr.

Although the approach shown above ratifies our conclusion that was reached on the basis of HSIP, it would be convenient if a general method could be developed. The energies involved are determined by the distances between the ion centers in each compound. Whether ΔE is positive (unfavorable) or negative (favorable) depends on

$$\frac{1}{r_{LiF}} + \frac{1}{r_{RbBr}} - \frac{1}{r_{LiBr}} - \frac{1}{r_{RbF}}$$

but the distances can also be written in terms of the radii of the individual ions.

$$\frac{1}{r_{Li} + r_F} + \frac{1}{r_{Rb} + r_{Br}} - \frac{1}{r_{Li} + r_{Br}} - \frac{1}{r_{Rb} + r_F}$$

The four parameters involved are the radii of the four ions. The question to be answered is whether the sum of the first two terms is smaller in magnitude than the sum of the last two terms. If the radii are represented as a, b, c, and d, the expression

$$\frac{1}{a+c} + \frac{1}{b+d} > \frac{1}{a+d} - \frac{1}{b+c}$$

will be true if $b > a$ and $d > c$. In terms of the ions, this means that the products that give the most favorable lattice energy sum are those that have the two smallest ions combined. Of necessity, the two largest ions must also be combined in the other product. Therefore, the products will be LiF and RbBr, which is in agreement with the HSIP.

A similar approach can be used for cases in which ions having charges other than $+1$ and -1 are involved. For two cations having charges c_1 and c_2 and two anions having charges of a_1 and a_2, the energetically favored products are those where in terms of the magnitudes of the charges $c_1 > c_2$ and $a_1 > a_2$. This can be expressed in the relationship

$$(c_1 a_1 + c_2 a_2) > (c_1 a_2 + c_2 a_1)$$

The principle that embodies this relationship can be stated as *the products will be those in which the smaller ions will combine with oppositely charged ions of higher charge*. In these cases, the Madelung constants may be different, so other factors may be involved. However, the principle correctly predicts reactions such as the following:

$$2\,NaF + CaCl_2 \;\rightarrow\; CaF_2 + 2\,NaCl \tag{9.105}$$

$$2\,AgCl + HgI_2 \;\rightarrow\; 2\,AgI + HgCl_2 \tag{9.106}$$

$$2\,HCl + CaO \;\rightarrow\; CaCl_2 + H_2O \tag{9.107}$$

In the last of these reactions, Ca^{2+} has a higher charge than H^+, but it is also much larger. Although the predictions are correct for a very large number of cases, the removal of a product from the reaction zone (formation of a precipitate or a gas) can cause the system to behave differently.

$$KI + AgF \;\rightarrow\; KF + AgI(s) \tag{9.108}$$

$$H_3PO_4 + NaCl \rightarrow HCl(g) + NaH_2PO_4 \qquad (9.109)$$

The products of a large number of reactions of many types are correctly predicted by the HSIP. Some examples are the following.

$$AsF_3 + PI_3 \rightarrow AsI_3 + PF_3 \qquad (9.110)$$

$$MgS + BaO \rightarrow MgO + BaS \qquad (9.111)$$

In the first of these reactions, I is softer than F and As is softer than P. Therefore, the exchange takes place to provide a more suitable match of hard–soft properties. In the second reaction, Mg^{2+} is a small, hard ion, whereas Ba^{2+} is much larger and softer. The O^{2-} ion bonds better to Mg^{2+}, whereas S^{2-} bonds better with Ba^{2+}. The HSIP predicts correctly the direction of many reactions of diverse types.

Although many applications of the hard–soft interaction principle will be presented in later chapters, two additional applications will be illustrated here. First, consider the interaction of the Lewis acid Cr^{3+} with the Lewis base SCN^-, which could donate an electron pair from either the S or N atom:

$$Cr^{3+} + SCN^- \quad \begin{matrix} ? \nearrow & Cr^{3+} : SCN^- \\ \\ ? \searrow & Cr^{3+} : NCS^- \end{matrix}$$

Because Cr^{3+} is a small, highly charged (hard) Lewis acid, it will bond preferentially to the smaller, harder Lewis base. In the thiocyanate ion, the electron pair on the nitrogen atom is held in a smaller, more compact orbital. Therefore, when a thiocyanate complex forms with Cr^{3+}, it is bonded through the nitrogen atom, and the complex is written as $[Cr(NCS)_6]^{3-}$. If the metal ion is Pt^{2+}, the opposite result is obtained. Because Pt^{2+} is large, has a low charge, and is polarizable, it bonds preferentially to the soft sulfur atom in thiocyanate. The resulting complex is written as $[Pt(SCN)_4]^{2-}$. Cases such as these will be discussed more fully in Chapter 16.

9.9 ELECTRONIC POLARIZABILITIES

As discussed in Chapter 6, the electronic polarizability, α, of species is very useful for correlating many chemical and physical properties. Values of α are usually expressed in cm^3 per unit (atom, ion, or molecule). Because atomic dimensions are conveniently expressed in Angstroms, polarizability is also expressed as $Å^3$ so $10^{-24}\ cm^3 = 1\ Å^3$. Polarizability gives a measure of the ability of the electron cloud of a species to be distorted, so it is also related to the hard–soft character of the species in a qualitative way. Table 9.6 gives the polarizabilities for ions and molecules.

In Chapter 6, the polarizability of molecules was considered as one factor related to both London and dipole-induced dipole intermolecular forces. The data shown in Table 9.6 confirm many of the observations that can be made about physical properties. For example, in the case of F_2, Cl_2, and Br_2, the London forces that arise from the increase in polarizability result in a general increase in boiling point. It is also interesting to note that metal ions having low polarizability (Al^{3+}, Be^{2+}, etc.) are those that are acidic (as shown in Eqn (9.17)). Also, in Chapter 7 we discussed how the polarization of ions leads to

Table 9.6 Electronic Polarizabilities for Selected Ions and Molecules in $Å^{3,a}$

Atoms and ions with noble gas configurations

2 Electrons:	He	Li^+	Be^{2+}	B^{3+}	C^{4+}		
	0.201	0.029	0.008	0.003	0.0013		
8 Electrons:	Si^{4+}	Al^{3+}	Mg^{2+}	Na^+	Ne	F^-	O^{2-}
	0.0165	0.052	0.094	0.179	0.390	1.04	3.88
18 Electrons:	Ti^{4+}	Sc^{3+}	Ca^{2+}	K^+	Ar	Cl^-	S^{2-}
	0.185	0.286	0.47	0.83	1.62	3.66	10.2
36 Electrons:	Zr^{4+}	Y^{3+}	Sr^{2+}	Rb^+	Kr	Br^-	Se^{2-}
	0.37	0.55	0.86	1.40	2.46	4.77	10.5
54 Electrons:	Ce^{4+}	La^{3+}	Ba^{2+}	Cs^+	Xe	I^-	Ce^{2-}
	0.73	1.04	1.55	2.42	3.99	7.10	14.0

Molecules

	H_2	O_2	N_2	F_2	Cl_2	Br_2
	0.80	1.58	1.74	1.16	4.60	6.90
	HCl	HBr	HI	HCN	BCl_3	BBr_3
	2.64	3.62	5.45	2.58	9.47	11.87
	H_2O	H_2S	NH_3	CCl_4		
	1.45	3.80	2.33	10.53		

[a]*Values compiled from C. Kittel, Introduction to Solid State Physics, 6th ed., Wiley, New York, 1986, p. 371; L. Pauling, General Chemistry, third ed., W.H. Freeman, San Francicco, 1970, p. 397, and Lide, D.R., Handbook of Chemistry and Physics, 72nd ed., CRC Press, Boca Raton, 1991, pp. 10–197 to 10–201.*

a lattice energy that is higher than that predicted on the basis of electrostatic interactions alone. The polarizability data shown in the table makes it easy to see that certain ions are much more polarizable than others. Although we will not visit again all of the ramifications of electronic polarizability, it is a very useful and important property of molecules and ions that relates to both chemical and physical behavior.

9.10 THE DRAGO FOUR-PARAMETER EQUATION

One of the main objections to the hard–soft approach to many types of interactions is that it is qualitative in nature. That is, in fact, one of its strengths, and predictions can be made as to how many processes will take place without resorting to calculations. Although this is still largely true, attempts have been made (with varying degrees of success) to make it quantitative by producing numerical values for softness or hardness parameters. Another quantitative approach to acid–base interactions of *molecules* that behave as Lewis acids or bases was developed by R. S. Drago and his coworkers. According to this interpretation of acid–base interactions, the coordinate bond formed between a Lewis acid and a Lewis base is the sum of an electrostatic (ionic) contribution and a covalent contribution. The essence of this approach is embodied in the equation

$$-\Delta H_{AB} = E_A E_B + C_A C_B \tag{9.112}$$

In this equation, ΔH_{AB} is the enthalpy change for the formation of an acid–base adduct AB; E_A and E_B are parameters that express the electrostatic bonding capabilities of the acid and base; and C_A and C_B are parameters that relate to the covalent bonding tendencies of the acid and base. The product of the electrostatic parameters gives the enthalpy change due to ionic contributions to the bond, whereas the product of the covalent parameters gives the covalent contribution to the bond. The total bond enthalpy is the sum of the two terms that represent the contributions to the bond. One difficulty arises as a result of the fact that values for the parameters characteristic of the acid and base must be determined. This is analogous to the situation regarding electronegativity in which *differences* are determined by bond energies.

In the case of the four-parameter equation, the enthalpies of interaction of a large number of acids and bases were determined calorimetrically in an inert solvent. With these values being known, a value of 1.00 was assigned for E_A and C_A for the Lewis acid iodine. The experimental enthalpies for the interaction of iodine with several molecular Lewis bases were fitted to the data to determine E_B and C_B values for the bases. Values were thus established for the four parameters for many acids and bases so that they can be used in Eqn (9.112) to calculate the enthalpies of the interactions. The agreement of the experimental and calculated enthalpies is excellent in most cases. However, the four-parameter approach is used primarily in conjunction with interactions between *molecular* species although extensions of the approach to include interactions between charged species have been made. Table 9.7 gives the parameters for several acids and bases.

One of the problems encountered when dealing with the interaction of Lewis acids and bases in a quantitative way is in evaluating the role of the solvent. Bond energies in molecules are values based on the molecule in the gas phase. However, it is not possible to study the

Table 9.7 Acid and Base Parameters for use in the Drago Four-Parameter Equation

Acid	E_A	C_A	Base	E_B	C_B
I_2	1.00	1.00	NH_3	1.15	4.75
ICl	5.10	0.83	CH_3NH_2	1.30	5.88
SO_2	0.92	0.81	$(CH_3)_3N$	0.81	11.5
$SbCl_5$	7.38	5.13	$(C_2H_5)_3N$	0.99	11.1
$B(CH_3)_3$	6.14	1.70	C_5H_5N	1.17	6.40
BF_3	9.88	1.62	CH_3CN	0.89	1.34
C_6H_5SH	0.99	0.20	$(CH_3)_3P$	0.84	6.55
C_6H_5OH	4.33	0.44	$(C_2H_5)_2O$	0.96	3.25
$Al(C_2H_5)_3$	12.5	2.04	$(CH_3)_2SO$	1.34	2.85
$Al(CH_3)_3$	16.9	1.43	C_6H_6	0.53	0.68

interaction of many Lewis acids and bases in the gas phase because the adducts formed are not sufficiently stable to exist at the temperature necessary to convert the reactants to gases. For example, the reaction between pyridine and phenol takes place readily in solution as a result of hydrogen bonding.

$$C_6H_5OH + C_5H_5N: \rightarrow C_6H_5OH : NC_5H_5 \tag{9.113}$$

However, the adduct is not stable enough to exist in significant amounts at a temperature of 116 °C, the boiling point of pyridine. As a result, such interactions are studied in solutions although it is the strength of the bond that would be produced in the gas phase that is desired. A thermochemical cycle can be devised to show the relationship for the interaction of an acid and a base both in solution (indicated by (s)) and in the gas phase.

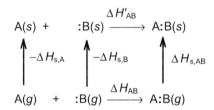

It is the interaction of the Lewis acid and base in the gas phase that gives the enthalpy of the A:B bond, ΔH_{AB}, but what is more often (and conveniently) measured is the enthalpy change when the reaction is carried out in solution (indicated by (s)), $\Delta H'_{AB}$.

In the thermochemical cycle above, the sum of the heats of solvation of A(g) and B(g) must be equal to the heat of solvation of the A:B(g) adduct if ΔH_{AB} and $\Delta H'_{AB}$ are to be equal. One way to assure that they will be approximately equal is to use a solvent that does not solvate any of the species strongly. If the heat of solution of A(g), B(g), and A:B(g) are all approximately zero, the solvent is truly an inert solvent. When such a solvent is used, the measured value for ΔH_{AB} will be approximately equal to $\Delta H'_{AB}$ and the measured enthalpy change will give the A:B bond enthalpy. Such a solvent is a hydrocarbon, such as hexane, cyclohexane, or heptane. For some systems, CCl_4 behaves as an inert solvent, but the unshared pairs of electrons on the chlorine atoms enable this solvent to interact more strongly with some solutes than do hydrocarbon molecules. Benzene is definitely not an inert solvent as a result of its ability to function as an electron donor because of the π electron system in the molecule. Calorimetric studies to determine the strengths of interactions of acids and bases in solution must be considered suspect unless the solvent is chosen carefully. There is no question but that the Drago four-parameter equation gives accurate values for the enthalpies of interaction for a very large number of molecules. When the enthalpy of interaction is calculated for the interaction of certain acid–base pairs, the calculated value is found to be higher than the experimental value. Such situations can arise when there are steric factors that cause the interaction to be less energetically favorable than expected.

For many years, there has been some controversy regarding the HSIP versus the Drago four-parameter approach. Proponents of the latter, cling to the fact that it can provide quantitative information on the interactions of Lewis acids and bases. Although this is true, the areas of applicability of that approach are somewhat restricted because it applies primarily to molecular species. On the other hand, proponents of the hard—soft interaction principle point to the fact that it applies to species of almost any type. Although it is qualitative in its approach (but there have been attempts to develop quantitative measures), it is extremely versatile for giving explanations for an enormous range of interactions. Both the four-parameter approach and the hard—soft interaction principle are valuable tools for correlating a large body of observations on Lewis acid—base interactions. Fortunately, one does not have to choose between them because in the areas where both apply, they generally support the same conclusions.

References for Further Study

Drago, R.S., Vogel, G.C., Needham, T.E., 1971. *J. Am. Chem. Soc.* 93, 6014. One of the series of papers on the four-parameter equation. Earlier papers are cited in this reference.

Drago, R.S., Wong, N., Ferris, D.C., 1991. *J. Am. Chem. Soc.* 113, 1970. One of the publications that gives extensive tables of values for E_A, E_B, C_A, and C_B. Earlier papers cited in this reference also give tables of values.

Finston, H.L., Rychtman, A.C., 1982. *A New View of Current Acid-Base Theories.* John Wiley, New York. A book that gives comprehensive coverage of all of the significant acid-base theories.

Gur'yanova, E.N., Gol'dshtein, I.P., Romm, I.P., 1975. *Donor-Acceptor Bond.* John Wiley, New York. A translation of a Russian book that contains an enormous amount of information and data on the interactions of many Lewis acids and bases.

Ho, Tse-Lok, 1977. *Hard and Soft Acid and Base Principle in Organic Chemistry.* Academic Press, New York. The applications of the hard-soft acid base principle to many organic reactions.

Lide, D.R., 2003. In: *CRC Handbook of Chemistry and Physics*, 84th ed. CRC Press, Boca Raton, FL. Extensive tables of dissociation constants for acids and bases are available in this handbook.

Luder, W.F., Zuffanti, S., 1946. *The Electronic Theory of Acids and Bases.* John Wiley, New York. A small book that is a classic in Lewis acid-base chemistry. Also available as a reprint volume from Dover.

Pearson, R.G., 1997. *Chemical Hardness.* Wiley-VCH, New York. This book is devoted to applications of the concept of hardness to many areas of chemistry.

Pearson, R.G., 1963. *J. Am. Chem. Soc.* 85, 3533. The original publication of the hard-soft acid-base approach by Pearson.

Pearson, R.G., 1966. *J. Chem. Educ.* 45, 581. A general presentation of the hard-soft interaction principle by Pearson.

Pearson, R.G., Songstad, J., 1967. *J. Am. Chem. Soc.* 89, 1827. A paper describing the application of the hard-soft interaction principle to organic chemistry.

Porterfield, W.W., 1993. *Inogranic Chemistry: A Unified Approach,* 2nd ed. Academic Press, San Diego. This book has good coverage of many topics in inorganic chemistry.

QUESTIONS AND PROBLEMS

1. Complete the following (if a reaction occurs).
   (a) $NH_4^+ + H_2O \rightarrow$
   (b) $HCO_3^- + H_2O \rightarrow$
   (c) $S^{2-} + NH_4^+ \rightarrow$
   (d) $BF_3 + H_3N{:}BCl_3 \rightarrow$
   (e) $Ca(OCl)_2 + H_2O \rightarrow$
   (f) $SOCl_2 + H_2O \rightarrow$

2. **(a)** Show by means of equations what happens when $AlCl_3$ catalyzes the reaction of an alkyl halide with benzene.

   (b) Show by means of equations what happens when H_2SO_4 catalyzes the reaction of nitric acid with toluene.

   (c) Explain the role of the catalyst in the reactions in parts (a) and (b).

3. Arrange the following in the order of decreasing acid strength:
   $H_2PO_4^-$, HNO_2, HSO_4^-, HCl, and H_2S.

4. Complete each of the following.

   (a) $HNO_3 + Ca(OH)_2 \rightarrow$

   (b) $H_2SO_4 + Al(OH)_3 \rightarrow$

   (c) $CaO + H_2O \rightarrow$

   (d) $NaNH_2 + H_2O \rightarrow$

   (e) $NaC_2H_3O_2 + H_2O \rightarrow$

   (f) $C_5H_5NH^+Cl^- + H_2O \rightarrow$

5. Complete each of the following. If NH_3 is a reactant, assume a liquid NH_3 solution.

   (a) $NH_4Cl + H_2O \rightarrow$

   (b) $Na_2NH + NH_3 \rightarrow$

   (c) $CaS + H_2O \rightarrow$

   (d) $OPCl_3 + H_2O \rightarrow$

   (e) $ClF + H_2O \rightarrow$

   (f) $K_2CO_3 + H_2O \rightarrow$

6. Complete the following equations showing reactions that occur when $NH_4Cl(s)$ is heated with the other reactants shown.

   (a) $NH_4Cl(s) + CaO(s) \rightarrow$

   (b) $NH_4Cl(s) + SrCO_3(s) \rightarrow$

   (c) $NH_4Cl(s) + Al(s) \rightarrow$

   (d) $NH_4Cl(s) + BaS(s) \rightarrow$

   (e) $NH_4Cl(s) + Na_2SO_3(s) \rightarrow$

7. Explain why PH_4I is stable whereas PH_4F and $PH_4C_2H_3O_2$ are not.

8. The Br^- and CN^- ions are approximately equal in size. Explain why NH_4Br is stable but NH_4CN is not.

9. Why is NF_3 a much weaker base than NH_3?

10. Predict the product and draw its structure for the reaction shown below.
    $Na_2SO_4 + BCl_3 \rightarrow$

11. For the reaction $(CH_3)_2O + C_2H_5F \rightarrow [(CH_3)_2OC_2H_5]^+F^-$

    (a) draw the structure for the cation

    (b) explain how the presence of BF_3 aids in this reaction.

12. Consider the reaction $(CH_3)_2S + CH_3I \rightarrow [(CH_3)_3S]^+I^-$ and explain why the cation in this case is more stable than that produced in the reaction in Problem 11.

13. Write a complete, balanced equation to show what happens (if anything) in each of the following cases.

    (a) trimethylamine is added to water

    (b) HCl is added to a mixture of acetic acid and sodium acetate

    (c) ammonium nitrate is added to water

    (d) sodium carbonate is added to water

    (e) sodium hydride is added to methanol

    (f) sulfuric acid is added to a solution of sodium acetate in water

14. Write a complete, balanced equation to show what happens (if anything) in each of the following cases.
    (a) potassium acetate is added to water
    (b) NaOH is added to a mixture of acetic acid and sodium acetate
    (c) a solution of aniline, $C_6H_5NH_2$, is added to hydrochloric acid
    (d) sodium hypochlorite is added to water
    (e) sodium nitrate is added to water
    (f) sodium amide is dissolved in water

15. When BF_3 behaves as a Lewis acid, the stability of the adducts with $(CH_3)_2O$, $(CH_3)_2S$, and $(CH_3)_2Se$ decreases in the order in which the Lewis bases are listed. However, when the Lewis acid is $B(CH_3)_3$, the most stable adduct is that formed with $(CH_3)_2S$ and the adducts with $(CH_3)_2Se$ and $(CH_3)_2O$ are about equally stable. Explain the difference in stability of the adducts.

16. Although the central atoms in $HClO_4$ and H_5IO_6 have the same oxidation state, the two compounds are greatly different in their acid strengths. Which is stronger? Explain why.

17. The molecule $(CH_3)_2NCH_2PF_2$ would bond differently to BH_3 and BF_3. Explain the difference.

18. Although HF is a weak acid in water, it is a strong acid in liquid NH_3. Write equations to show the reactions and explain the difference in acid strength.

19. Although $N(CH_3)_3$ is a stronger base than NH_3, the adduct $H_3N:B(CH_3)_3$ is more stable than $(CH_3)_3N:B(CH_3)_3$. Explain this observation.

20. In each case, pick the stronger acid and explain your answer.
    (a) H_2CO_3 or H_2SeO_4?
    (b) $Cl_2CHCOOH$ or Cl_2CHCH_2COOH?
    (c) H_2S or H_2Se?

21. Write equations to show how molten ammonium fluoride would react with each of the following.
    (a) Zn (b) FeO (c) $CaCO_3$ (d) NaOH (e) Li_2SO_3

22. Explain the trend in dissociation constants of the acids HOX, where X is a halogen.

23. Arrange the following compounds in the order of increasing pH of a 0.1 M solution of each: $NaOCl$, $NaHCO_2$, Na_2CO_3, NaN_3, and NaHS.

24. Which is more acidic, $Co(H_2O)_6^{2+}$ or $Co(H_2O)_6^{3+}$? Explain your answer.

25. What would be the products of the reaction of BF_3 with I_2? Would you expect a similar reaction between BF_3 and Cl_2 to occur? Explain your answer.

26. zThe values for K_1 and K_2 in the dissociation of $H_4P_2O_7$ are 1.4×10^{-1} and 3.2×10^{-2}. Explain in terms of molecular structure why Pauling's rule does not work as well in this case as it does for the dissociation of H_3PO_4.

27. For some very weak bases, titration in water is not feasible. Using appropriate equations, explain why this would be the case for aniline, $C_6H_5NH_2$, which has a K_b of 4.6×10^{-10}.

28. Write the equation for the reaction of aniline, $C_6H_5NH_2$, with glacial acetic acid. Explain why it would be possible to titrate aniline in this acid.

29. Arrange the following in the order of acid strength: H_2SeO_3, H_3AsO_3, $HClO_4$, H_2SO_3, H_2SO_4.

30. Explain why $H_2PO_4^-$ has two oxygen atoms without hydrogen atoms attached, but it is not a strong acid such as H_2SO_4.

31. Draw the structure for SF_4 and predict how the molecule would be attacked by a Lewis acid. How would the attack of a Lewis base occur?

32. Where is the acidic site in the SO_3 molecule? Draw structures to explain your answer.

33. Explain the reaction

$$CO_2 + OH^- \rightarrow HCO_3^-$$

as a Lewis acid–base reaction.

34. Explain why benzene is a better solvent for many Lewis acids than is CCl_4.

35. Although Br^- and CN^- have approximately the same thermochemical radii, there is a great difference in the basicity of the ions. Which is the stronger base? Explain your answer.

36. Use the data shown in Table 9.7 to calculate the enthalpy for the interactions of BF_3 and $(CH_3)_3B$ with $(CH_3)_3N$ and $(CH_3)_3P$. Explain what the values indicate.

37. In the purification of ClF_3, HF that is usually present can be separated by reaction with sodium fluoride. Explain the basis for this procedure.

38. Explain why the proton affinity of SH^- is 1464 kJ mol^{-1} but that for S^{2-} is 2300 kJ mol^{-1}.

CHAPTER 10

Chemistry in Nonaqueous Solvents

Although reactions in gases and solids are by no means rare, it is the enormous number of reactions carried out in solutions that is the subject of this chapter. However, there is no question that the vast majority of reactions are carried out in solutions where water is the solvent. It is important to note that most nonaqueous solvents present some difficulties when their use is compared to that of water as a solvent. Some of the more important nonaqueous solvents are NH_3, HF, SO_2, $SOCl_2$, N_2O_4, CH_3COOH, $POCl_3$, and H_2SO_4. Some of these compounds (NH_3, HF, and SO_2) are gases at ambient temperature and pressure. Some of them are also highly toxic. Some of the compounds are both gaseous and toxic. It is almost never as convenient to use a nonaqueous solvent as it is to use water.

In view of the difficulties that accompany the use of a nonaqueous solvent, one may certainly ask why such use is necessary. The answer includes several of the important principles of nonaqueous solvent chemistry that will be elaborated on in this chapter. First, solubilities are different. In some cases, classes of compounds are more soluble in some nonaqueous solvents than they are in water. Second, the strongest acid that can be used in an aqueous solution is H_3O^+. As was illustrated in Chapter 9, any acid that is stronger than H_3O^+ will react with water to produce H_3O^+. In some other solvents, it is possible to routinely work with acids that are stronger than H_3O^+. Third, the strongest base that can exist in aqueous solutions is OH^-. Any stronger base will react with water to produce OH^-. In some nonaqueous solvents, a base stronger than OH^- can exist, so it is possible to carry out certain reactions in such a solvent that cannot be carried out in aqueous solutions. These differences permit synthetic procedures to be carried out in nonaqueous solvents that would be impossible when water is the solvent. As a result, chemistry in nonaqueous solvents is an important area of inorganic chemistry, and this chapter is devoted to the presentation of a brief overview of this area.

10.1 SOME COMMON NONAQUEOUS SOLVENTS

Although water is used as a solvent more extensively than any other liquid, other solvents may offer some important advantages. For example, if any base stronger than OH^- is placed in water, it reacts with water to produce OH^-. If it is necessary to use a stronger base than OH^- in some reaction, the best way is to use a solvent that is more basic than water because the anion from that base will be stronger than OH^-. This situation exists in liquid ammonia in which the basic species is NH_2^- which is a stronger base than OH^-. If a reaction requires an acid that is stronger than H_3O^+, it may be necessary to carry out the reaction in a more acidic solvent than in water.

Some compounds that are important nonaqueous solvents require special conditions and apparatus for their handling. Ammonia, sulfur dioxide, hydrogen fluoride, and dinitrogen tetroxide are **313**

Inorganic Chemistry. DOI: http://dx.doi.org/10.1016/B978-0-12-385110-9.00010-8

Table 10.1 Properties of Some Common Nonaqueous Solvents

Solvent	m.p., °C	b.p., °C	Dipole Moment, D	Dielectric Constant
H_2O	0.0	100	1.85	78.5
NH_3	−77.7	−33.4	1.47	22.4
SO_2	−75.5	−10.0	1.61	15.6
HCN	−13.4	25.7	2.8	114.9
H_2SO_4	10.4	338	−	100
HF	−83	19.4	1.9	83.6
N_2H_4	2.0	113.5	1.83	51.7
N_2O_4	−11.2	21.5	−	2.42
CH_3OH	−97.8	65.0	1.68	33.6
$(CH_3)_2SO$	18	189	3.96	45
CH_3NO_2	−29	101	3.46	36
$(CH_3CO)_2O$	−71.3	136.4	2.8	20.5
H_2S	−85.5	−60.7	1.10	10.2
HSO_3F	−89	163	−	−

all gases at room temperature and atmospheric pressure. Some of the solvents such as liquid hydrogen cyanide are extremely toxic. In view of these difficulties, it may seem unusual that so much work has been carried out in nonaqueous solvents. However, in many cases, the advantages outweigh the disadvantages. The scope of reactions that can be carried out in nonaqueous solvents is very broad indeed because there is such a difference in the characteristics of the solvents. Some of the frequently used nonaqueous solvents are listed in Table 10.1 along with their relevant properties.

One of the disadvantages of using a nonaqueous solvent is that in most cases ionic solids are less soluble in a nonaqueous solvent than in water. There are exceptions to this. For example, silver chloride is insoluble in water but it is soluble in liquid ammonia. As will be illustrated later, some reactions take place in opposite directions in a nonaqueous solvent and water.

10.2 THE SOLVENT CONCEPT

It has long been known that some autoionization occurs in water and it was presumed that nonaqueous solvents behaved in a similar way. Although the reaction of sodium hydride with water,

$$NaH + H_2O \rightarrow H_2 + NaOH \tag{10.1}$$

could be presumed to take place in water that exists as H^+OH^-, it is by no means the case. In fact, even though water undergoes a *slight* degree of autoionization, the reaction takes place as shown above without assuming *prior* ionization of water. An analogous situation exists for nonaqueous solvents, and in many cases *assuming* that the solvent ionizes makes clear what parts of the molecule are reacting in what way. The fact that the solvent does not undergo autoionization is immaterial.

Knowing what species would be produced may make it easier to predict the course of reactions in a particular solvent.

One of the primary pieces of evidence that water undergoes some autoionization comes from the conductivity of pure water. The equilibrium can be written as

$$2 H_2O \leftrightarrows H_3O^+ + OH^- \tag{10.2}$$

The conductivity of liquid ammonia is sufficiently high to indicate a very slight degree of auto-ionization. In order for ions to be produced, *something* must be transferred from one molecule to another, and in solvents such as water or ammonia it is proton transfer that occurs. Accordingly, the ionization can be shown as

$$2 NH_3 \leftrightarrows NH_4^+ + NH_2^- \quad K_{am} = \sim 2 \times 10^{-29} \tag{10.3}$$

According to the Arrhenius theory of acids and bases, the acidic species in water is the solvated proton (which we write as H_3O^+). This shows that the *acidic* species is the *cation* characteristic of the solvent. In water, the *basic* species is the *anion* characteristic of the solvent, OH^-. By extending the Arrhenius definitions of an acid and a base to liquid ammonia, it becomes apparent from Eqn (10.3) that the acidic species is NH_4^+ and the basic species is NH_2^-. It is apparent that any substance that leads to an increase in the concentration of NH_4^+ is an acid in liquid ammonia. A substance that leads to an increase in the concentration of NH_2^- is a base in liquid ammonia. For other solvents, autoionization (if it occurs) leads to different ions, but in each case presumed ionization leads to a cation and an anion. Generalization of the nature of the acidic and basic species leads to the idea that in a solvent, the cation characteristic of the solvent is the acidic species and the anion characteristic of the solvent is the basic species. This is known as *the solvent concept*. Neutralization can be considered as the reaction of the cation and anion from the solvent. For example, the cation and anion react to produce unionized solvent:

$$HCl + NaOH \rightarrow NaCl + H_2O \tag{10.4}$$

$$NH_4Cl + NaNH_2 \rightarrow NaCl + 2 NH_3 \tag{10.5}$$

Note that there is no requirement that the solvent actually undergoes autoionization.

If autoionization were to take place in liquid sulfur dioxide, the process could be written as

$$2 SO_2 \leftrightarrows SO^{2+} + SO_3^{2-} \tag{10.6}$$

That this reaction does not take place to a measurable extent does not prevent us from knowing that the ions produced *would be* the acidic and basic species. Therefore, we can predict that Na_2SO_3 would be a base in liquid SO_2 because it contains SO_3^{2-} and that $SOCl_2$ would be an acid because it could formally produce SO^{2+} if it were to dissociate. Thus, we predict that Na_2SO_3 and $SOCl_2$ would react according to the equation

$$Na_2SO_3 + SOCl_2 \xrightarrow{\text{Liq. } SO_2} 2 NaCl + 2 SO_2 \tag{10.7}$$

This reaction represents a neutralization reaction in liquid sulfur dioxide. It makes no difference that the solvent does not ionize or that $SOCl_2$ is a covalent molecule. The utility of the solvent concept is *not* that it correctly predicts that solvents undergo some autoionization. The value of the solvent concept is that it allows us to correctly predict how reactions would take place *if* the solvent ionized. Note that in this case $SOCl_2$ does not ionize, but if it did it would produce SO^{2+} (the acidic species characteristic of the solvent) and Cl^-.

The concepts illustrated above can be extended to other nonaqueous solvents. For example, in liquid N_2O_4 if autoionization occurred it would produce NO^+ and NO_3^-. In this solvent, a compound that furnishes NO_3^- would be a base and a compound that would (formally rather than actually) produce NO^+ would be an acid. Therefore, the reaction of KNO_3 with $NOCl$ (actually linked as $ONCl$) is a neutralization in liquid N_2O_4.

$$NOCl + NaNO_3 \xrightarrow{\text{Liq. } N_2O_4} NaCl + N_2O_4 \tag{10.8}$$

Seemingly, another possible autoionization of liquid N_2O_4 could involve splitting the molecule in a different way to produce NO_2^+ and NO_2^-,

$$N_2O_4 \leftrightharpoons NO_2^+ + NO_2^- \tag{10.9}$$

However, in liquid N_2O_4, reactions do not occur as if NO_2^+ and NO_2^- were present and there is no evidence to indicate that they are. In most cases, autoionization of the solvent *if it occurs* is by transfer of H^+ or O^{2-}.

The conductivity of liquid HF indicates slight autoionization that can be represented as

$$3\,HF \rightarrow H_2F^+ + HF_2^- \tag{10.10}$$

In this case, both H^+ and F^- are solvated by molecules of HF, so they are not shown as the simple ions. Although we will not describe how such species are prepared now (see Chapter 15), $H_2F^+SbF_6^-$ behaves as an acid and $BrF_2^+HF_2^-$ behaves as a base in liquid HF. The neutralization reaction can be written as

$$H_2F^+SbF_6^- + BrF_2^+HF_2^- \xrightarrow{\text{Liq. HF}} 3\,HF + BrF_2^+SbF_6^- \tag{10.11}$$

10.3 AMPHOTERIC BEHAVIOR

When an aqueous solution containing Zn^{2+} or Al^{3+} has NaOH added, a precipitate of the metal hydroxide forms. Upon continued addition of base, the precipitate dissolves just as it does when an acid is added. In the first case, the aluminum hydroxide is behaving as a base, and in the second, it behaves as an acid. This behavior, the ability to react as an acid or a base, is known as *amphoterism*. The reaction of Zn^{2+} with a base and an acid can be shown as follows:

$$\begin{array}{c} Zn^{2+} + 2\,OH^- \longrightarrow Zn(OH)_2 \\ {\scriptstyle + 2\,H_3O^+} \swarrow \qquad \searrow {\scriptstyle + 2\,OH^-} \\ Zn(H_2O)_4^{2+} \qquad\qquad Zn(OH)_4^{2-} \end{array} \tag{10.12}$$

When reacting with a base, $Zn(OH)_2$ dissolves by the formation of a complex, $Zn(OH)_4^{2-}$. In the reaction with an acid, protons are transferred from H_3O^+ to the hydroxide ions, forming water molecules that remain coordinated to the Zn^{2+} ion.

The behavior of Zn^{2+} in liquid ammonia is analogous to that in water. First, a precipitate of $Zn(NH_2)_2$ forms when an amide is added, but the precipitate dissolves when a solution containing either NH_4^+ or NH_2^- is added. This behavior can be shown as follows:

$$Zn^{2+} + 2\,NH_2^- \longrightarrow Zn(NH_2)_2$$

$$+\ 2\,NH_4^+ \swarrow \qquad \searrow +\ 2\,NH_2^- \qquad (10.13)$$

$$Zn(NH_3)_4^{2+} \qquad\qquad Zn(NH_2)_4^{2-}$$

Although this behavior has been illustrated starting with the metal ion, analogous equations can be written starting with the metal oxide as well as the hydroxide. Amphoteric behavior is exhibited in other solvents, as will be illustrated later.

10.4 THE COORDINATION MODEL

For some nonaqueous solvents, the autoionization, if it occurs at all, must be to a degree so small that virtually no ions are present. If the ion product constant for a solvent is as low as 10^{-40}, the concentration of each ion would be 10^{-20} M. This can be put in perspective by considering one mole of the solvent and realizing that out of 6.02×10^{23} molecules of solvent, only about 1000 have undergone ionization. The presence of a *minute* trace of impurity (it is very difficult to obtain completely anhydrous solvents) would account for more ions by reaction than would be produced by autoionization. Therefore, some alternative must exist that removes the necessity for assuming that autoionization occurs.

When $FeCl_3$ is placed in $OPCl_3$, it can be shown by spectrophotometry that $FeCl_4^-$ is present. One way this could occur is by ionization of the solvent,

$$OPCl_3 \leftrightarrows OPCl_2^+ + Cl^- \qquad (10.14)$$

followed by the reaction of $FeCl_3$ with the Cl^- produced.

$$FeCl_3 + Cl^- \rightarrow FeCl_4^- \qquad (10.15)$$

Another plausible set of events might involve the interaction of $FeCl_3$ with $OPCl_3$ before ionization occurs. This is reasonable to expect because $FeCl_3$ is a Lewis acid and $OPCl_3$ is a molecule that contains atoms having unshared pairs of electrons. This type of interaction could be shown as follows, which shows $FeCl_3$ pulling Cl^- from $OPCl_3$ so that it is not necessary for prior autoionization to occur:

$$FeCl_3 + OPCl_3 \leftrightarrows [Cl_3Fe - ClPOCl_2] \leftrightarrows OPCl_2^+ + FeCl_4^- \qquad (10.16)$$

In this way, some chloride ions might be removed from the solvent, and by complexing with Fe^{3+}, the reaction is shifted to the right by Le Chatelier's principle. Although this scheme removes the assumption that $OPCl_3$ undergoes autoionization by having $FeCl_3$ pull the Cl^- off, it ignores the fact that toward Lewis acids, the *oxygen* atom in $OPCl_3$ is a more basic site than the chlorine atoms.

In a classic study by Devon Meek and Russell S. Drago, $FeCl_3$ was placed in a different solvent. The solvent, triethylphosphate, $OP(OC_2H_5)_3$, contained *no* chlorine, but the spectra of the solutions were similar to those of $FeCl_3$ in $OPCl_3$ and clearly showed the presence of $FeCl_4^-$! The only possible source of Cl^- that was needed to form $FeCl_4^-$ must be $FeCl_3$ itself. Drago reasoned that in order for this to happen, a molecule of the solvent must replace a chloride ion on $FeCl_3$. In other words, a *substitution* reaction occurs that involves the solvent, and the liberated Cl^- reacts with another molecule of $FeCl_3$ to produce $FeCl_4^-$. Because the complex between $FeCl_3$ and the solvent bonded by a coordinate bond (see Chapter 16), this model of solvent behavior is known as the *coordination model*. The sequence of reactions can be shown as follows in which X represents either Cl or an ethyl group:

$$FeCl_3 + OPX_3 \leftrightarrows [Cl_3FeOPX_3] \leftrightarrows [Cl_{3-x}Fe(OPX_3)_{1+x}]^{x+} + x[FeCl_4^-] \ldots$$
$$\leftrightarrows [Fe(OPX_3)_6]^{3+} + 3[FeCl_4^-] \tag{10.17}$$

As chloride ions are replaced on Fe^{3+}, the process continues sequentially until it contains solvated Fe^{3+} (as the complex $[Fe(OPX_3)_6]^{3+}$) and the chloride complex, $FeCl_4^-$.

The coordination model provides a way to explain many reactions that occur in nonaqueous solvents without having to assume that autoionization takes place. As shown in Eqn (10.17), the fact that $FeCl_4^-$ is produced can be explained by substitution rather than autoionization. However, as has been shown earlier in this chapter, it is sometimes useful to assume that the solvent concept is valid, and many reactions take place just as if the solvent has ionized to a slight degree into an acidic and a basic species.

10.5 CHEMISTRY IN LIQUID AMMONIA

In many of its properties, liquid ammonia resembles water. Both are polar and involve extensive hydrogen bonding in the liquid state. It is interesting to note that hydrogen bonding in liquid ammonia is less extensive than in water in which the oxygen atoms can form hydrogen bonds to two other water molecules. This difference is evident from the fact that the heat of vaporization of water is 40.65 kJ/mol, but that of ammonia is only 23.26 kJ/mol. Although there are differences when specific types of compounds are considered, both water and liquid ammonia dissolve many types of solids. Having a boiling point of $-33.4\,°C$ means that it is necessary to work with liquid ammonia at low temperature or high pressure. If liquid ammonia is kept in a Dewar flask, the rate of evaporation is low enough that solutions can be kept for a reasonable length of time.

We have already mentioned that silver chloride is readily soluble in liquid ammonia. Because it is slightly less polar than water and has lower cohesion energy, intermolecular forces make it possible for organic molecules to create cavities in liquid ammonia. As a result, most organic compounds are more soluble in liquid ammonia than they are in water. Physical data for liquid ammonia are summarized in Table 10.2.

Table 10.2 Physical Properties of Liquid NH_3

Melting point	$-77.7\,°C$
Boiling point	$-33.4\,°C$
Density at $-33.4\,°C$	$0.683\,g\,cm^{-3}$
Heat of fusion	$5.98\,kJ\,mol^{-1}$
Heat of vaporization	$23.26\,kJ\,mol^{-1}$
Dipole moment	$1.47\,D$
Dielectric constant	22
Specific conductance at $-35\,°C$	$2.94 \times 10^{-7}\,ohm$

Liquid ammonia is a base, so reactions with acids generally proceed to a greater degree than do the analogous reactions in water. For example, acetic acid is a weak base in water, but it ionizes completely in liquid ammonia. Even though ammonia is a base, it is possible for protons to be removed, but only when it reacts with exceedingly strong bases such as N^{3-}, O^{2-}, or H^-. Some of the important types of reactions that occur in liquid ammonia will now be illustrated.

10.5.1 Ammoniation Reactions

Most solvents have unshared pairs of electrons and they are polar. Therefore, they have the ability to attach to metal ions or to hydrogen bond to anions. As a result, when many solids crystallize from solutions, they have included a definite number of solvent molecules. When this occurs in water, we say that the crystal is a *hydrate*. An example of this is the well-known copper sulfate pentahydrate,

$$CuSO_4 + 5\ H_2O \rightarrow CuSO_4 \cdot 5H_2O \tag{10.18}$$

A similar type of behavior is observed for AgCl in liquid ammonia:

$$AgCl + 2\ NH_3 \rightarrow Ag(NH_3)_2Cl \tag{10.19}$$

Because ammonia is the solvent, the solvated species is known as an *ammoniate*. The solvent molecules are not always bound in the solid in the same way. For example, some solids may contain water of hydration, but in other cases, the water may be coordinated to the metal ion. In classifying materials as hydrates or ammoniates, the mode of attachment of the solvent is not always specified.

10.5.2 Ammonolysis Reactions

In many places in this book, the reactivity of bonds between nonmetals and halogens is stressed. Reactions such as

$$PCl_5 + 4\ H_2O \rightarrow 5\ HCl + H_3PO_4 \tag{10.20}$$

take place readily. In these reactions, molecules of water are split or *lysized*, so the reactions are known as *hydrolysis* reactions. Reactions in which ammonia molecules are split are known as *ammonolysis* reactions. Some examples of these reactions can be illustrated as follows:

$$SO_2Cl_2 + 4\,NH_3 \rightarrow SO_2(NH_2)_2 + 2\,NH_4Cl \tag{10.21}$$

$$CH_3COCl + 2\,NH_3 \rightarrow CH_3CONH_2 + NH_4Cl \tag{10.22}$$

$$BCl_3 + 6\,NH_3 \rightarrow B(NH_2)_3 + 3\,NH_4Cl \tag{10.23}$$

$$2\,CaO + 2\,NH_3 \rightarrow Ca(NH_2)_2 + Ca(OH)_2 \tag{10.24}$$

Many other reactions involving ammonolysis are known, but these illustrate the process.

10.5.3 Metathesis Reactions

In order for a metathesis reaction to occur in water, some product must be removed from the reaction. Generally, this involves the formation of a precipitate, the evolution of a gas, or the formation of an unionized product. Because solubilities are different in liquid ammonia, reactions are often unlike those in water. Although silver halides are insoluble in water, they are soluble in liquid ammonia as a result of forming stable complexes with ammonia. Therefore, the reaction

$$Ba(NO_3)_2 + 2\,AgCl \rightarrow BaCl_2 + 2\,AgNO_3 \tag{10.25}$$

takes place in liquid ammonia because $BaCl_2$ is insoluble, but in water, the reaction

$$BaCl_2 + 2\,AgNO_3 \rightarrow Ba(NO_3)_2 + 2\,AgCl \tag{10.26}$$

occurs because AgCl is insoluble.

10.5.4 Acid–Base Reactions

Although it is not necessary for autoionization to occur, the solvent concept shows that the cation characteristic of the solvent is the acidic species and the anion is the basic species. Therefore, when substances that contain these ions are mixed, neutralization occurs. For liquid ammonia, NH_4^+ is the acidic species and NH_2^- is the basic species. Such a reaction can be shown as

$$NH_4Cl + NaNH_2 \xrightarrow{\text{Liq. } NH_3} 2\,NH_3 + NaCl \tag{10.27}$$

Acid–base reactions also include those in which the solvent itself functions as an acid or a base. This can be shown as follows. Sodium hydride reacts with water to produce a basic solution,

$$H_2O + NaH \rightarrow H_2 + NaOH \tag{10.28}$$

and the analogous reaction with liquid ammonia can be shown as

$$NH_3 + NaH \rightarrow H_2 + NaNH_2 \tag{10.29}$$

Just as the oxide ion is a strong base toward water,

$$CaO + H_2O \rightarrow Ca(OH)_2 \tag{10.30}$$

the imide and nitride ions are bases toward liquid ammonia. Both are too strong to exist in liquid ammonia, so they react to produce a weaker base, NH_2^-.

$$CaNH + NH_3 \rightarrow Ca(NH_2)_2 \tag{10.31}$$

$$Mg_3N_2 + 4\,NH_3 \rightarrow 3\,Mg(NH_2)_2 \tag{10.32}$$

The oxide ion is also sufficiently strong as a base to remove protons from ammonia.

$$2\,BaO + 2\,NH_3 \rightarrow Ba(NH_2)_2 + Ba(OH)_2 \tag{10.33}$$

Because of the base strength of liquid ammonia, acids that are weak in water ionize completely in liquid ammonia.

$$HF + NH_3 \xrightarrow{\text{Liq. NH}_3} NH_4^+F^- \tag{10.34}$$

A type of deprotonation reaction that takes place in liquid ammonia but not in water occurs because it is possible to utilize the base strength of the amide ion. The reaction involves removing H^+ from a molecule of ethylenediamine, $H_2NCH_2CH_2NH_2$, (written as en) that is coordinated to Pt^{2+} in the complex $[Pt(en)_2]^{2+}$.

$$\tag{10.35}$$

When the ethylenediamine that has had a proton removed (to become en$^-$) is designated as en$-$H, the complex can be written as $[Pt(en-H)_2]$ and it is uncharged. This is a reactive species that will undergo numerous reactions in liquid ammonia. This procedure was used by Watt and coworkers to produce derivatives of the ethylenediamine ligands by means of reactions such as

$$[Pt(en-H)_2] + 2\,CH_3Cl \rightarrow [Pt(CH_3en)_2]Cl_2 \tag{10.36}$$

These interesting synthetic reactions are possible because of the basicity of NH_2^- which cannot be utilized in aqueous solutions.

10.5.5 Metal–Ammonia Solutions

If there is one radical difference between the chemistry in water and that of liquid ammonia, it is in the behavior toward Group IA metals. When placed in water, these metals liberate hydrogen in vigorous reactions such as

$$2\,Na + 2\,H_2O \rightarrow H_2 + 2\,NaOH \tag{10.37}$$

Table 10.3 Solubility of Alkali Metals in Ammonia

Metal	Temperature °C	Molality of Saturated Solution	Temperature °C	Molality of Saturated Solution
Li	0	16.31	−33.2	15.66
Na	0	10.00	−33.5	10.93
K	0	12.4	−33.2	11.86
Cs	0	−	−50.0	25.1

In contrast, these metals dissolve and undergo reaction only very slowly in liquid ammonia. Solutions containing alkali metals in liquid ammonia have been known for over 140 years, and they have properties that are extraordinary. The extent to which the metals dissolve is itself interesting. The solubilities are shown in Table 10.3.

That the dissolution does not involve a chemical change in the metal is indicated by the fact that evaporation of the solutions regenerates the metal. If the product is not placed under thermal stress, the metal can be recovered as a solvate having the formula $M(NH_3)_6$. Moreover, the solutions have densities that are lower than the solvent alone. It is clear that some expansion of the liquid occurs when the metal dissolves. In appearance, all of the solutions are blue when dilute but are bronze colored when more concentrated than approximately 1 M. The solutions exhibit conductivity that is higher than that of 1:1 electrolytes. The conductivity decreases as the concentration of the metal increases, but the conductivity of concentrated solutions is characteristic of metals. An additional anomaly is that the solutions are paramagnetic, but the magnetic susceptibility decreases for concentrated solutions. The magnitude of the magnetic susceptibility is in agreement with there being one free electron produced by each metal atom. These are the facts that must be explained by any successful model for these solutions.

The model of metal−ammonia solutions that has emerged is based on ionization of the metal atoms to produce metal ions and electrons that are both solvated. The solvated electron is believed to reside in a cavity in ammonia, and thus it may behave as a particle in a three-dimensional box with quantized energy levels. Transitions between the energy levels may give rise to absorption of light and thereby cause the solutions to be colored. The dissolution process can be represented as

$$M + (x + y)NH_3 \rightarrow M(NH_3)_x^+ + e^-(NH_3)_y \tag{10.38}$$

The form of the cavity in which an electron resides is not known, but it is reasonable to expect that the hydrogen atoms are directed around the electron in some way because of their slight positive charge. Although there may be several molecules of ammonia forming the cavity, it is reasonable to represent it as

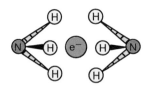

in which hydrogen atoms having a positive formal charge form a cage around the electron. The spectrum corresponding to the solvated electron in liquid ammonia exhibits a maximum at approximately 1500 nm. Although this hardly represents a particle in a box, if we assume that it does and if we assume that the band maximum represents the transition of the electron between the states $n = 1$ and $n = 2$, we can solve for the dimension of the box. In this case, the energy difference corresponds to a box having a length of about 120 nm although other estimates range from 300 to 600 nm. Even though this is not a precise calculation, the calculated length of the cavity is at least compatible with the electron residing in a small cavity of molecular dimensions and the fact that the density of a solution of this type is lower than that of the pure solvent.

Solutions of alkali metals in liquid ammonia have been studied by many techniques. These include electrical conductivity, magnetic susceptibility, NMR, volume expansion, spectroscopy (visible and infrared), and other techniques. The data obtained indicate that the metals dissolve with ionization and that the metal ion and electron are solvated. Several simultaneous equilibria have been postulated to explain the unique properties of the solutions. These are generally represented as follows:

$$M + (x+y)NH_3 \leftrightarrows M(NH_3)_x^+ + e^-(NH_3)_y \tag{10.39}$$

$$2\,M^+(am) + 2\,e^-(am) \leftrightarrows M_2(am) \tag{10.40}$$

$$2\,e^-(am) \leftrightarrows e_2^{2-}(am) \tag{10.41}$$

$$M^+(am) + e_2^{2-}(am) \leftrightarrows M^-(am) \tag{10.42}$$

$$M^+(am) + M^-(am) \leftrightarrows M_2(am) \tag{10.43}$$

The expansion of the solutions has been accounted for by considering the electron as residing in a hole in the solvent. Pairing electrons as shown in Eqn (10.41) is considered to be at least partially responsible for the decrease in paramagnetism as the concentration of the solution is increased. However, pairing may also involve free electrons being removed as shown in Eqn (10.40). Although it is not likely that all of these steps are involved, it is likely that the relative importance of these reactions depends on the particular metal dissolved and the concentration of the solution. In addition to dissolving, alkali metals also react slowly by the liberation of hydrogen,

$$2\,Na + 2\,NH_3 \rightarrow 2\,NaNH_2 + H_2 \tag{10.44}$$

This reaction is accelerated photochemically and is catalyzed by transition metal ions.

Reduction of a chemical species involves the gain of electrons by that species. Because the solutions of alkali metals in liquid ammonia contain free electrons, they are extremely strong reducing agents. This fact has been exploited in a large number of reactions. For example, oxygen can be converted to superoxide or peroxide ions.

$$O_2 \xrightarrow{e^-(am)} O_2^- \xrightarrow{e^-(am)} O_2^{2-} \tag{10.45}$$

The solutions can also convert transition metals to unusual oxidation states. For example, solutions of alkali metals in liquid ammonia are such strong reducing agents that metals bound in complexes can be reduced to the (0) oxidation state. The reaction of $[Pt(NH_3)_4]Br_2$ with potassium dissolved in liquid ammonia produces $[Pt^0(NH_3)_4]$.

$$[Pt(NH_3)_4]Br_2 + 2\,K \rightarrow [Pt^0(NH_3)_4] + 2\,KBr \tag{10.46}$$

Other cases can be shown as

$$[Ni(CN)_4]^{2-} + 2\,e^-(am) \rightarrow [Ni^0(CN)_4]^{4-} \tag{10.47}$$

$$Mn_2(CO)_{10} + 2\,e^-(am) \rightarrow 2\,[Mn(CO)_5]^- \tag{10.48}$$

Many compounds react by loss of hydrogen and the production of an anion. Some examples of this type of reaction are

$$2\,SiH_4 + 2\,e^-(am) \rightarrow 2\,SiH_3^- + H_2 \tag{10.49}$$

$$2\,CH_3OH + 2\,e^-(am) \rightarrow 2\,CH_3O^- + H_2 \tag{10.50}$$

It has also been found that alkali metals dissolve in solvents such as methylamine and ethylenediamine. These solutions have some characteristics of the solutions containing ammonia, and they undergo similar reactions.

10.6 LIQUID HYDROGEN FLUORIDE

A solvent that resembles water in many ways is liquid hydrogen fluoride. The molecule is polar, there is some autoionization, and it is a fairly good solvent for numerous ionic solids. Although the boiling point of liquid HF is rather low (19.5 °C), it has a liquid range that is comparable to that of water, partially as a result of extensive hydrogen bonding. One of the problems associated with the use of liquid HF is that it attacks glass, so containers must be made of some inert material such as Teflon®, a polytetrafluoroethylene. The data for this nonaqueous solvent are shown in Table 10.4.

As expected from the rather high heat of vaporization (that lies between the values for water and liquid ammonia), liquid HF has a liquid range that spans over 100 °C and a relatively high boiling point.

Table 10.4 Physical Properties of HF

Melting point	$-83.1\,°C$
Boiling point	$19.5\,°C$
Density at 19.5 °C	$0.991\,g\,cm^{-1}$
Heat of fusion	$4.58\,kJ\,mol^{-1}$
Heat of vaporization	$30.3\,kJ\,mol^{-1}$
Equivalent conductance	$1.4 \times 10^{-5}\,ohm^{-1}$
Dielectric constant at 0 °C	83.6
Dipole moment	$1.83\,D$

The autoionization of liquid HF can be represented by the equation

$$3\ HF \leftrightarrows H_2F^+ + HF_2^-$$ (10.51)

Water has an equivalent conductance of $6.0 \times 10^{-8}\ ohm^{-1}$ at $25\,^{\circ}C$ and that of liquid HF is $1.4 \times 10^{-5}\ ohm^{-1}$. Therefore, the ion product constant for HF ionization is approximately 8×10^{-12}, which is larger than the value of 1.0×10^{-14} for water. The HF_2^- ion has a symmetric linear structure

$$\overset{-}{F \cdots H \cdots F}$$

and it represents the strongest case of hydrogen bonding. It can be considered as a molecule of HF solvating a fluoride ion. The H_2F^+ cation has eight electrons around the central F atom, so it is isoelectronic with water, and the structure of the ion is

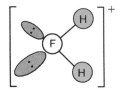

Both the dielectric constant and dipole moment are comparable to those of water, indicating that HF is a good solvent for inorganic compounds, but many organic compounds are also soluble. In general, the fluorides of +1 metals are much more soluble than those of +2 or +3 metals. At $11\,^{\circ}C$, the solubility of NaF is approximately 30 g per 100 g of liquid HF, that of MgF_2 is only 0.025 g, and that of AlF_3 is 0.002 g.

The acidity of liquid HF is high enough that it can function as an acid catalyst in many instances. The cation characteristic of the solvent, H_2F^+, is generated when HF reacts with a strong Lewis acid that is capable of forming stable fluoride complexes. The reactions with BF_3 and AsF_5 are typical:

$$BF_3 + 2\ HF \rightarrow H_2F^+ + BF_4^-$$ (10.52)

$$AsF_5 + 2\ HF \rightarrow H_2F^+ + AsF_6^-$$ (10.53)

Equations such as these can also be written in the molecular form as

$$AsF_5 + HF \rightarrow HAsF_6$$ (10.54)

The basic species in liquid HF is the fluoride ion or the solvated fluoride ion, HF_2^-. As in the case of water in which OH^- forms hydroxo complexes, fluoride ion forms complexes in liquid HF. This behavior gives rise to amphoterism with metals ions such as Zn^{2+} and Al^{3+}. In the case of Al^{3+}, AlF_3 is relatively insoluble in liquid HF, so the amphoteric behavior can be shown as follows.

$$Al^{3+} + 3\ F^- \longrightarrow AlF_3$$

$$+\ 3\ H_2F^+ \diagdown\ \diagup\ +\ 3\ F^-$$

$$Al^{3+} + 6\ HF \qquad\qquad AlF_6^{3-}$$ (10.55)

Although in this scheme the basic species is written as F^- for simplicity, the solvated species is actually HF_2^- in liquid HF.

The principal characteristics of liquid HF are its acidity and the stability of the HF_2^- anion, which represents the strongest hydrogen bond. As a result, in many reactions in liquid HF, the solvent behaves as an acid and generates the HF_2^- anion. Typical reactions in which this type of behavior is shown are as follows:

$$2\,HF + H_2O \rightarrow H_3O^+ + HF_2^- \tag{10.56}$$

$$2\,HF + R_2C{=}O \rightarrow R_2C{=}OH^+ + HF_2^- \tag{10.57}$$

$$2\,HF + C_2H_5OH \rightarrow C_2H_5OH_2^+ + HF_2^- \tag{10.58}$$

$$2\,HF + Fe(CO)_5 \rightarrow HFe(CO)_5^+ + HF_2^- \tag{10.59}$$

In addition to these types of reactions, liquid HF converts some oxides to oxyhalides. For example, the reaction of permanganate can be shown as follows:

$$5\,HF + MnO_4^- \rightarrow MnO_3F + H_3O^+ + 2\,HF_2^- \tag{10.60}$$

From the foregoing, it should be apparent that liquid HF is a versatile and useful nonaqueous solvent.

10.7 LIQUID SULFUR DIOXIDE

In addition to the hydrogen-containing solvents already discussed, chemistry has been carried out in liquid sulfur dioxide for many years. Although the SO_2 molecule has a significant dipole moment, the liquid is a good solvent for many covalent substances. The molecule is polarizable due to the π electrons, so the London forces between SO_2 and the solute molecules lead to solubility. In accordance with the hard–soft interaction principle, compounds such as $OSCl_2$, $OPCl_3$, and PCl_3 are very soluble. Although aliphatic hydrocarbons are not soluble in liquid SO_2, most aromatic hydrocarbons are appreciably soluble. This difference in solubility provides a way to separate the two types of hydrocarbons by means of solvent extraction. Ionic compounds are almost insoluble unless they consist of very large ions that contribute to greater polarizability and lower lattice energy. Physical properties of liquid sulfur dioxide are shown in Table 10.5. As will be described later, sulfur dioxide is a useful solvent for carrying out reactions utilizing superacids such as the $HOSO_2F/SbF_5$ mixture.

Sulfur dioxide can function as a very weak Lewis acid or a Lewis base and thereby form a variety of solvates:

$$SO_2 + SnCl_4 \rightarrow SnCl_4 \cdot SO_2 \tag{10.61}$$

With some very strong Lewis acids, solvolysis occurs to produce an oxyhalide:

$$SO_2 + NbCl_5 \rightarrow NbOCl_3 + OSCl_2 \tag{10.62}$$

Table 10.5 Physical Properties of SO_2

Melting point	$-75.5\,°C$
Boiling point	$-10.0\,°C$
Density	$1.46\,g\,cm^{-1}$
Heat of fusion	$8.24\,kJ\,mol^{-1}$
Heat of vaporization	$24.9\,kJ\,mol^{-1}$
Dipole moment	$1.63\,D$
Dielectric constant	15.6
Specific conductance	$3\times10^{-8}\,ohm^{-1}$

The SO_2 molecule has unshared pairs of electrons on both the sulfur and oxygen atoms. As a result, it forms numerous complexes with transition metals in which it is known to attach in several ways. These include bonding through the sulfur atom, through an oxygen atom, by both oxygen atoms, and by various bridging schemes. In most cases, the complexes involve soft metals in low oxidation states. Another important reaction of sulfur dioxide is known as the insertion reaction in which it is placed between the metal and another ligand. This type of reaction can be shown as (where X and L are other groups bonded to the metal)

$$X_nM-L + SO_2 \rightarrow X_nM-SO_2-L \tag{10.63}$$

This type of reaction will be discussed in more detail in Chapter 22.

If autoionization of the solvent shown as

$$2\,SO_2 \leftrightarrows SO^{2+} + SO_3^{2-} \tag{10.64}$$

occurs, it must be to an *infinitesimal* extent. As is the case with other nonaqueous solvents, maintaining the solvent scrupulously pure and anhydrous is essential. Such a transfer of an oxide ion would be energetically unfavorable. It is interesting to note that there is no exchange of radioactive sulfur in $OSCl_2$ with sulfur in liquid SO_2. If autoionization did occur in these liquids, both would generate SO^{2+}, so it would be expected that the radioactive sulfur would be found in both species. The fact that it is not shows that these solvents do not produce ions.

Even though liquid SO_2 does not autoionize, the acidic species *would be* SO^{2+} and the basic species *would be* SO_3^{2-}. Therefore, the neutralization reaction

$$K_2SO_3 + SOCl_2 \xrightarrow{\text{Liq. } SO_2} 2\,KCl + 2\,SO_2 \tag{10.65}$$

takes place as expected on the basis of the solvent concept. Unlike reactions in aqueous solutions, this reaction cannot be shown in ionic form as

$$SO^{2+} + SO_3^{2-} \rightarrow 2\,SO_2 \tag{10.66}$$

Even though it is now known that the solvent concept does not represent the actual reacting species in some nonaqueous solvents, it is still a useful tool.

If thionyl chloride undergoes dissociation according to the equation

$$SOCl_2 \rightarrow SOCl^+ + Cl^- \tag{10.67}$$

there would be some free Cl^- in the solution. If another chloride compound were dissolved in the solvent, there could be chloride exchange. Although based on the discussion to this point we would not expect autoionization, it is interesting to note that when other chloride compounds containing radioactive chlorine are dissolved in $SOCl_2$, there is *chloride* exchange with $SOCl_2$.

An illustration that a nonaqueous solvent such as liquid SO_2 is a versatile reaction medium is the following:

$$3\ SbF_5 + S_8 \xrightarrow{\text{Liq. } SO_2} (S_8)^{2+}(SbF_6^-)_2 + SbF_3 \tag{10.68}$$

By increasing the ratio of SbF_5 to S_8, it is also possible to generate the S_4^{2+} species. The reaction between PCl_5 and liquid SO_2 can be used to produce thionyl chloride and phosphoryl chloride.

$$PCl_5 + SO_2 \rightarrow OPCl_3 + OSCl_2 \tag{10.69}$$

The amphoteric behavior of Zn^{2+} and Al^{3+} in some nonaqueous solvents has already been described. This behavior can also be demonstrated for liquid SO_2. The aluminum compound containing the anion characteristic of the solvent forms a precipitate, which is then soluble in either the acid or the base in liquid SO_2. This can be shown as

$$2\ AlCl_3 + 3\ [(CH_3)_4N]_2SO_3 \rightarrow Al_2(SO_3)_3 + 6\ [(CH_3)_4N]Cl \tag{10.70}$$

As we have already shown, $SOCl_2$ is an acid in liquid SO_2, so the precipitate of $Al_2(SO_3)_3$ reacts as follows to give aluminum chloride and the solvent:

$$3\ SOCl_2 + Al_2(SO_3)_3 \rightarrow 2\ AlCl_3 + 6\ SO_2 \tag{10.71}$$

In liquid SO_2, the basic species is SO_3^{2-}, so $[(CH_3)_4N]_2SO_3$ gives a basic solution in which aluminum sulfite dissolves by forming a sulfite complex as shown in the equation

$$Al_2(SO_3)_3 + 3\ [(CH_3)_4N]_2SO_3 \rightarrow 2\ [(CH_3)_4N]_3Al(SO_3)_3 \tag{10.72}$$

Battery technology has developed enormously in recent years. One of the most useful types of batteries is known as the lithium battery, but there are actually several designs, only one of which is described here. In one of the types, the anode is constructed of lithium or a lithium alloy, hence the name. A graphite cathode is used, and the electrolyte is a solution of $Li[AlCl_4]$ in thionyl chloride. At the anode, lithium is oxidized,

$$Li \rightarrow Li^+ + e^- \tag{10.73}$$

At the cathode, the reduction of thionyl chloride occurs,

$$2\ SOCl_2 + 4\ e^- \rightarrow SO_2 + S + 4\ Cl^- \tag{10.74}$$

As the battery produces electricity, lithium chloride is formed on the cathode, and eventually the battery is no longer effective. Lithium batteries have long life and give high current-to-weight ratio, so they are widely used in digital cameras, watches, pacemakers, etc.

In addition to the reactions in which the solvent is also a *reactant* (such as solvolysis), there are many others in which the solvent functions only to make reactants accessible to each other. Even in those cases, the difference in solubilities can lead to different products depending on the solvent. Consider the following metathesis reactions:

$$\text{In water: } AgNO_3 + HCl \rightarrow AgCl + HNO_3 \tag{10.75}$$

$$\text{In } SO_2\text{: } 2\,AgC_2H_3O_2 + SOCl_2 \rightarrow 2\,AgCl + OS(C_2H_3O_2)_2 \tag{10.76}$$

$$\text{In } NH_3\text{: } NH_4Cl + LiNO_3 \rightarrow LiCl + NH_4NO_3 \tag{10.77}$$

$$\text{In } NH_3\text{: } AgCl + NaNO_3 \rightarrow AgNO_3 + NaCl \tag{10.78}$$

The first of these reactions takes place because AgCl is insoluble in water. The second takes place because AgCl is insoluble in sulfur dioxide. The third reaction takes place because the highly ionic LiCl is not soluble in liquid ammonia. The last of these reactions takes place because AgCl is soluble in liquid ammonia but NaCl is not. It is clear that metathesis reactions may be different in some nonaqueous solvents.

The chemistry of the specific solvents discussed in this chapter illustrates the scope and utility of nonaqueous solvents. However, as a side note, several other nonaqueous solvents should at least be mentioned. For example, oxyhalides such as $OSeCl_2$ and $OPCl_3$ (described in the discussion of the coordination model earlier in this chapter) also have received a great deal of use as nonaqueous solvents. Another solvent that has been extensively investigated is sulfuric acid which undergoes autoionization

$$2\,H_2SO_4 \leftrightarrows H_3SO_4^+ + HSO_4^- \tag{10.79}$$

For this equilibrium, the ion product constant has a value of approximately 2.7×10^{-4}. However, other species are present in sulfuric acid as a result of equilibria that can be written as

$$2\,H_2SO_4 \leftrightarrows H_3O^+ + HS_2O_7^- \tag{10.80}$$

$$2\,H_2SO_4 \leftrightarrows H_3SO_4^+ + HSO_4^- \tag{10.81}$$

$$SO_3 + H_2SO_4 \leftrightarrows H_2S_2O_7 \tag{10.82}$$

Disulfuric acid (also known as pyrosulfuric acid or oleum) also undergoes dissociation,

$$H_2SO_4 + H_2S_2O_7 \leftrightarrows H_3SO_4^+ + HS_2O_7^- \tag{10.83}$$

Many species are protonated in liquid sulfuric acid, even some that are not normally bases. For example,

$$CH_3COOH + H_2SO_4 \leftrightarrows CH_3COOH_2^+ + HSO_4^- \tag{10.84}$$

$$HNO_3 + H_2SO_4 \leftrightarrows H_2NO_3^+ + HSO_4^- \tag{10.85}$$

$$H_2NO_3^+ + H_2SO_4 \leftrightarrows NO_2^+ + H_3O^+ + HSO_4^- \tag{10.86}$$

Sulfuric acid undergoes many other reactions with metals as well as with metal oxides, carbonates, nitrates, sulfides, etc. It is a versatile nonaqueous solvent.

10.8 SUPERACIDS

In aqueous solutions, the leveling effect causes there to be an upper limit on acid strength, that of H_3O^+, and the acidity is generally defined in terms of the pH of the solution. However, the strength of H_3O^+ is not an absolute limit to the property of acidity, and it is possible to obtain acidic media which are so strong that they are generally referred to as *superacids*. A superacid can be defined as an acid that is stronger than 100% sulfuric acid. Acidity can also be defined in terms of the Hammett acidity function, H_0, which is defined in terms of the equilibrium

$$B + H^+ \leftrightarrows BH^+ \tag{10.87}$$

and the equilibrium constant for dissociation of BH^+ is

$$K_{BH^+} = \frac{[B][H^+]}{[BH^+]} \tag{10.88}$$

The Hammett function is defined in terms of the dissociation of BH^+, which leads to

$$H_0 = pK_{BH^+} - \log\frac{[BH^+]}{[B]} \tag{10.89}$$

For 100% sulfuric acid, the value of H_0 is -12 (usually written as $-H_0 = 12$), and a superacid is defined as an acid that is stronger than sulfuric acid. For example, the value of H_0 for FSO_3H is -15. Although HF is a weak acid in water, liquid HF has a value of H_0 that is -15.1 largely due to the stability of the species HF_2^-. A strong Lewis acid such as SbF_5 functions as a fluoride acceptor and can increase acidity markedly as a result of the reaction

$$SbF_5 + 2\,HF \leftrightarrows H_2F^+ + SbF_6^- \tag{10.90}$$

Superacids have H_0 values in the range of 10^{15}–10^{25}, so their acid strength can be many orders of magnitude greater than that of H_2SO_4. One way to produce a superacid is to change the peripheral atoms on a molecule of an oxyacid such as H_2SO_4 to increase the inductive effect. When one of the two $-OH$ groups on sulfuric acid is replaced by a fluorine atom, the result is $HOSO_2F$ (melting point $-89\,°C$, boiling point $163\,°C$), and it is a stronger acid than H_2SO_4 alone.

A compound that has been utilized industrially as a superacid is triflic acid, CF_3SO_3H. However, several other superacids contain other fluorocarbon groups. Three such compounds are shown in Figure 10.1.

An advantage of a superacid such as $HCF_2CF_2SO_3H$ (tetrafluoroethanesulfonic acid [TFESA]) over triflic acid is that it has lower volatility. As a result of the advantages of TFESA, it has become commercially available as the pure compound or supported on silica.

The behavior of superacids is an important aspect of nonaqueous solvent chemistry because numerous reactions can be carried out when the acids are much stronger than H_3O^+. This characteristic was recognized as early as 1927 by James Bryant Conant. One strongly acidic medium that is commonly utilized is obtained by mixing SbF_5 and HSO_3F. When the strong Lewis acid SbF_5 is added to $HOSO_2F$, it accepts a pair of electrons (probably from one of the oxygen atoms) which increases the inductive effect on the remaining O–H bond even more than that in $HOSO_2F$ alone. The result is that the $HOSO_2F/SbF_5$ mixture is a stronger acid than $HOSO_2F$, and the interaction between the two components can be represented as

$$2\ HSO_3F + SbF_5 \rightarrow H_2SO_3F^+ + F_5SbOSO_2F^- \tag{10.91}$$

One very powerful oxidizing agent that can be used in this superacid is FO_2SOOSO_2F, peroxydisulfuryl difluoride, which has the structure

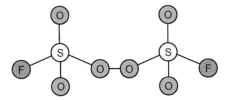

In the $HOSO_2F/SbF_5$ mixture, iodine can be oxidized by $S_2O_6F_2$ to produce polyatomic cations (see Chapter 15). The reaction in which I_3^+ is produced can be written as

$$S_2O_6F_2 + 3\ I_2 \rightarrow 2\ I_3^+ + 2\ SO_3F^- \tag{10.92}$$

This strong oxidizing agent in the superacid mixture will also oxidize sulfur to produce the S_8^{2+} and S_4^{2+} ions described earlier (there are actually some others known but these are probably the best characterized species).

$$S_2O_6F_2 + S_8 \rightarrow S_8^{2+} + 2\ SO_3F^- \tag{10.93}$$

FIGURE 10.1
Some superacids containing halogenated alkyl groups.

The $HOSO_2F/SbF_5$ mixture will dissolve substances such as paraffin with the formation of carbocations. For example, the reaction with *neo*-pentane can be shown as

$$H_2OSO_2F^+ + (CH_3)_4C \rightarrow (CH_3)_3C^+ + HOSO_2F + CH_4 \tag{10.94}$$

The $H_2SO_3F^+$ ion is such a strong acid that it will also protonate phosphorus halides as well as hydrocarbons.

$$PX_3 + H_2SO_3F^+ \rightarrow HPX_3^+ + HSO_3F \tag{10.95}$$

Another mixture that will function in a similar way is the mixture of HF and SbF_5 which results in the equilibrium

$$HF + SbF_5 \leftrightharpoons H^+ + SbF_6^- \tag{10.96}$$

which can be more correctly represented by showing H^+ as the solvated species, H_2F^+. Suitable solvents for carrying out reactions utilizing this superacid mixture include liquid HF, SO_2, SO_2FCl, and SO_2F_2.

The superacid mixture HF/SbF_5 will also protonate hydrocarbons to form carbocations, but it is a weaker acid than $H_2SO_3F^+$. In one interesting reaction, carbonic acid is converted into $C(OH)_3^+$, which is isoelectronic with boric acid, $B(OH)_3$. Superacids are such strong proton donors that they will protonate many substances that do not normally behave as bases. These include species such as hydrocarbons, ketones, organic acids, and many others. The chemistry of superacids makes possible many reactions that could not be carried out in any other way. They can function as effective catalysts, and in some cases, a smaller amount of catalyst is utilized and conditions may be milder which also results in lower energy consumption. As mentioned earlier, this is one of the aspects of nonaqueous solvents that makes that type of chemistry indispensable.

References for Further Reading

Finston, H.L., Rychtman, A.C., 1982. *A New View of Current Acid-Base Theories*. John Wiley, New York. A useful and comprehensive view of all the major acid-base theories.

Harmer, M.A., Junk, C., Rostovtsev, Carcani, L.E., Vickery, J., Schnepp, Z., 2007. *Green Chem.* 9, 30–37. An article describing the structures and properties of superacids supported on silica substrates.

Jolly, W.L., 1972. *Metal-Ammonia Solutions*. Dowden, Hutchinson & Ross, Inc, Stroudsburg, PA. A collection of research papers that serves as a valuable resource on all phases of the physical and chemical characteristics of these systems.

Jolly, W.L., Hallada, C.J., 1965. In: Waddington, T.C. (Ed.), *Non-Aqueous Solvent Systems*. Academic Press, San Diego.

Lagowski, J.J., Moczygemba, G.A., 1967. In: Lagowski, J.J. (Ed.), *The Chemistry of Non-aqueous Solvents*. Academic Press, San Diego.

Meek, D.W., Drago, R.S., 1961. *J. Amer. Chem. Soc.* 83, 432. The classic paper describing the coordination model as an alternative to the solvent concept.

Nicholls, D., 1979. *Chemistry in Liquid Ammonia: Topics in Inorganic and General Chemistry*, vol. 17. Elsevier, Amsterdam.

Olah, G.A., Surya Prakash, G.K., Sommer, J., 1985. *Superacids*. Wiley, New York.

Pearson, R.G., 1997. *Chemical Hardness*. Wiley-VCH, New York. An interesting book on several aspects of hardness including the behavior of ions in crystals.

Popovych, O., Tompkins, R.P.T., 1981. *Non-aqueous Solution Chemistry*. Wiley, New York.

Waddington, T.C., 1969. *Non-Aqueous Solvents*. Appleton-Century-Crofts, New York. An older book that presents a lot of valuable information on nonaqueous solvents.

QUESTIONS AND PROBLEMS

1. Write equations for the reactions of the following in liquid sulfuric acid.
   (a) NaF (b) $CaCO_3$ (c) KNO_3 (d) $(CH_3)_2O$ (e) CH_3OH

2. Explain why water is a better medium than liquid ammonia for a reaction that requires an acidic medium, whereas the opposite is true for a reaction that requires a basic medium.

3. (a) Suppose SO_2 forms a complex by attaching to $[Ni(PR_3)_4]$ (where $R = C_6H_5$). How would you expect it to attach? Explain your answer.
   (b) Predict the structure of the product if two SO_2 molecules were to replace H_2O in the complex $[Mn(OPR_3)_4(H_2O)_2]^{2+}$.
   (c) Suppose the hypothetical complex $[(CO)_4Fe-SO_2-Cr^{2+}(NH_3)_5]$ could be made. How might sulfur dioxide function as a bridging ligand in this complex?

4. Explain the difference in acidity of HCN in water and liquid ammonia.

5. Explain why there are no hydride complexes of the type $Co(NH_3)_5H^+$, but there are some such as $Mo(CO)_5H$.

6. Acetic anhydride, $(CH_3CO)_2O$, is assumed to undergo slight autoionization.
   (a) Write the equation to represent the autoionization process.
   (b) What is one substance that would be acidic in this solvent?
   (c) What is one substance that would be basic in this solvent?
   (d) Write the equations to show the amphoteric behavior of Zn^{2+} in liquid acetic anhydride.

7. When $AlCl_3$ is dissolved in $OSCl_2$, some $AlCl_4^-$ is produced.
   (a) Assuming that the solvent concept is applicable, explain by means of appropriate equations how the formation of $AlCl_4^-$ occurs.
   (b) Use the coordination model and write equations to show how $AlCl_4^-$ is formed.
   (c) Explain clearly the difference between how the solvent concept and the coordination model apply to the reaction of $AlCl_3$ with $OSCl_2$.

8. Explain how KNO_3 can be used as a nitrating agent in liquid HF. What species are present? How do they act?

9. Each of the following reactions takes place in an aqueous solution. Write an equation that is analogous to each when the solvent is liquid ammonia.
   (a) $KOH + HCl \rightarrow KCl + H_2O$
   (b) $CaO + H_2O \rightarrow Ca(OH)_2$
   (c) $Mg_3N_2 + 6\,H_2O \rightarrow 2\,NH_3 + 3\,Mg(OH)_2$
   (d) $Zn(OH)_2 + 2\,H_3O^+ \rightarrow Zn^{2+} + 4\,H_2O$
   (e) $BCl_3 + 3\,H_2O \rightarrow 3\,HCl + B(OH)_3$

10. Explain the role of hydrogen bonding in the autoionization of liquid HF.

11. Using equations, show how $TiCl_4$ can produce $TiCl_6^{2-}$ in $OPCl_3$ solution.

12. Using equations, show how $SbCl_5$ is an acid in liquid $OPCl_3$, in liquid HF, and in liquid BrF_3.

13. Write equations to show how each of the following would react in liquid ammonia.
    (a) LiH (b) NaH_2PO_4 (c) BaO (d) AlN (e) $SOCl_2$

14. Write the equation for the reaction of Cl_2 with water and the analogous equation for the reaction of Cl_2 with liquid ammonia.

15. Explain why $CaCl_2$ is much less soluble in liquid ammonia than is $CuCl_2$.

16. If you were trying to generate the $N_2F_3^+$ ion in a nonaqueous solvent, what would be a good solvent to use? Write the equation for the formation of $N_2F_3^+$ in that solvent.

17. Write equations to show the amphoteric behavior of Zn^{2+} in 100% acetic acid.

18. In liquid NH_3, Ir^{3+} forms a precipitate when $NaNH_2$ is added but redissolves when an excess of NH_2^- is added. Write equations to show this behavior.

19. Write equations to explain why the solution is acidic when $AlCl_3$ is dissolved in water.

20. Consider H_2O and liquid neon both of which have eight electrons in the valence shell. However, there the similarity ends. Describe the properties that are different and explain the characteristics of the molecules that give rise to those properties.

21. Write an equation to show why aluminum fluoride is an acid in liquid HF.

22. Rank the following solids in the order of decreasing solubility in liquid HF. Explain the order you give. CsF, LiF, BaF_2, CaF_2, KF.

23. Both BrF_3 and IF_5 are bases in liquid HF. Write equations to show this behavior.

24. Complete and balance the following in which the second reactant is also the solvent in each case.
    (a) $AsCl_3 + H_2O \rightarrow$
    (b) $AsCl_3 + NH_3 \rightarrow$
    (c) $SiCl_4 + H_2O \rightarrow$
    (d) $SiCl_4 + NH_3 \rightarrow$

25. The reaction of liquid ammonia with SO_2Cl_2 produces sulfamide and ammonium imidosulfide. Write the equation for the reaction and draw structures for the products.

26. A solution of sodium in liquid ammonia will dissolve zinc. Why? Write an equation as part of your answer.

4

Chemistry of the Elements

CHAPTER 11

Chemistry of Metallic Elements

Metals have characteristics that make them desirable for construction, implements, and ornamental uses, such as in jewelry. Metals have been so important that the names Chalcolithic, Bronze, and Iron Age have been applied to epochs of history. Metals have an enduring quality. They are solid, durable, and attractive. The first draft of this passage is being written with a pen made of sterling silver (an alloy consisting of 92.5% silver and 7.5% copper). The lure of metals has drawn adventurers to remote places, and they have been the spoils of war. Their intrinsic value is exemplified by the ornaments we wear and the role of metals in monetary systems.

Of the slightly more than 100 known elements, approximately three-fourths are metals. These elements have properties that vary from those of mercury (m.p. $-39\,^\circ\mathrm{C}$) to those of tungsten (m.p. $3407\,^\circ\mathrm{C}$). However, the characteristics of metals that physically distinguish them are their metallic luster, electrical conductivity, malleability, and ductility. Elements generally classified as metals exhibit wide variation in these properties. Chemically, metals are also reducing agents as a result of having comparatively low ionization potentials. Another characteristic that differs enormously is their cost. Some of the base metals sell for a few cents per pound, whereas some of the exotic metals sell for a few thousand dollars per gram.

Another issue with metals is availability. For example, cobalt is not produced in the United States, but it used extensively in a wide variety of alloys and in the production of one of the most common types of lithium batteries. The availability of cobalt is crucial to several segments of American industry. For example, batteries being developed for use in automobiles powered by alternate energy sources are currently envisioned to use a lithium ion battery that also contains cobalt. However, cobalt is not the only strategic metal, and there is concern about the availability of several metals that are vital to industries in the United States, China, and Japan. There will be competition and stockpiling of strategic metals as the reserves become less accessible.

11.1 THE METALLIC ELEMENTS

The metals found in Group IA of the periodic table are known as the *alkali metals*, and those in Group IIA are the *alkaline earths*. Metals found between Groups IIA and IIIA (the so-called *d*-block metals) are the *transition metals*. The series of elements following lanthanum ($Z = 57$, the *f*-block metals) represent those in which the 4*f* level is being filled in progressing through the group. These elements were referred to as the rare earths but are now generally referred to as the *lanthanides* because they follow lanthanum in the periodic table. The row of elements below the lanthanides is known as the *actinides* because they follow actinium. All are radioactive and the later members of the series are artificially produced. In this chapter, the discussion will not include the actinides owing to their being comparatively rare, **337**

Inorganic Chemistry. DOI: http://dx.doi.org/10.1016/B978-0-12-385110-9.00011-X

radioactive, and of generally limited use in practical chemistry. There is considerable and inevitable overlap of some topics when discussing the chemistry of metals. For example, a great deal of the chemistry of organometallic compounds of main group elements is discussed in Chapter 12 and that of transition metals is presented in Chapter 21. The chemistry of metallic elements such as tin, lead, antimony, and bismuth is discussed with that of the other members of the respective groups in Chapters 14 and 15. Some of the chemistry of the Group IA metals was presented in Chapter 10 in connection with their behavior in nonaqueous solvents, such as liquid ammonia. The basic structures of metals, their characteristics, and the phase transitions that occur between phases have been discussed in Chapter 7.

11.2 BAND THEORY

In a general way, a metal can be considered as a regular lattice of positive ions surrounded by an electron "gas" or "sea" that is mobile. This simple model accounts for many properties of metals. For example, the mobile electrons in metals enable them to be good electrical conductors. Because metal atoms can be displaced without destroying the lattice, metals can be shaped and retain their cohesiveness. Consequently, metals are *malleable* (can be hammered to change shapes) and *ductile* (can be drawn into wires). Moreover, the substitution of a different atom in the lattice is possible with the result being a solution of one metal in another, an *alloy*. The addition of other nonmetallic atoms such as carbon, nitrogen, or phosphorus increases the hardness of the metal while making it more brittle and less ductile.

Although metals are generally good conductors of electricity, there is still some resistance to electrical flow, which is known as the *resistivity* of the metal. At normal temperatures, the resistivity is the result of the flow of electrons being impeded because of motion of atoms that results from vibration about mean lattice positions. When the temperature is raised, the vibration of atoms about their mean lattice positions increases in amplitude, which further impedes the flow of electrons. Therefore, the resistivity of metals increases as the temperature increases. In a metal, electrons move throughout the structure. There are usually a small number of electrons from each atom that are considered, and because in most structures (*fcc* and *hcp*), each atom has 12 nearest neighbors, there is no possibility for the formation of the usual bonds that require two electrons for each. As a result, *individual* bonds are usually weaker than those of ionic or covalent character. Because of the large overall number of bonds, the cohesion in metals is quite high.

As metal atoms interact with nearest neighbors at relatively short distance, orbital overlap results in electron density being shared. As mentioned earlier, that electron density is delocalized in orbitals that are essentially molecular orbitals encompassing all of the atoms. The number of atoms that contribute an orbital to the molecular orbital scheme approaches the number of atoms present. As two atoms approach each other and interaction occurs, the two atomic orbitals can be considered as forming two *molecular* orbitals, one bonding and one antibonding. This is illustrated in Figure 11.1. If three atoms interact, three molecular orbitals result, etc. In many ways, this situation is similar to the Hückel molecular orbitals described in Chapter 5 where it was shown that the number of molecular orbitals is equal to the dimensionality of the secular determinant. The energy of the kth orbital in the Hückel approach can be expressed as

$$E = \alpha + 2\beta \cos\frac{k\pi}{N+1} \tag{11.1}$$

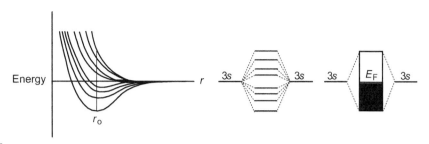

FIGURE 11.1
Interaction of atomic orbitals to produce four bonding and four antibonding molecular orbitals. As the number of atoms gets very large, the molecular orbitals form a continuum. In this case (described in the text), the atoms are assumed to be sodium.

where α is a Coulomb integral (H_{ii}), β is the resonance integral (H_{ij}), and N approaches infinity. Therefore, there are N energy levels that span an energy band that approaches 4β in overall width with the energy separating levels k and $k + 1$ approaching zero.

The *overall* difference in energy of the molecular orbitals is determined by how effectively the atomic orbitals interact, but the spacing between individual *adjacent* molecular orbitals decreases as the number of orbitals increases. When the number of atoms is very large, the spacing becomes smaller than the thermal energy, kT (where k is Boltzmann's constant). For only two orbitals, the variation in orbital energy with interatomic distance is shown in Figure 11.1. For a very large number of orbitals, there is a continuum of molecular orbitals that form a "band" as shown in Figure 11.1. This is the origin of the band theory often referred to in connection with bonding in metals. When N atoms interact to produce N molecular orbitals, each orbital can hold 2 electrons and if each atom contributes one electron, the molecular orbitals will be half-filled. This situation is also illustrated in Figure 11.1.

For a metal such as sodium, there are overlapping bands formed from the $2p$ and $3s$ atomic orbitals. In a general way, we can consider that a band is formed from each type of atomic orbitals that is occupied. However, there are energy gaps *between* the bands. If we again consider the case of sodium, there will be bands from the $1s$, $2s$, $2p$, $3s$, and $3p$ orbitals (although the $3p$ orbitals are unoccupied), which is shown in Figure 11.2. Each band can hold $2(2l + 1)N$ electrons, where N is the number of atoms.

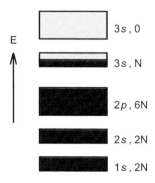

FIGURE 11.2
Bands formed by interaction of orbitals on sodium atoms and their populations. The bands shown in blue are filled, but the bands shown in yellow are empty. The band arising from the $3s$ orbitals is only half-filled because there is only one electron per atom in the $3s$ state.

Because the highest occupied band is only half-filled, electrons can move into the band and move within the band during electrical conduction. Light in the visible region of the spectrum interacts with the electrons with absorption and re-emission occurring at the surface. It is this phenomenon that gives metals their shiny appearance or *metallic luster*.

In the free electron model, the electrons are presumed to be loosely bound to the atoms making them free to move throughout the metal. The development of this model requires the use of quantum statistics that apply to particles (such as electrons) that have half-integral spin. These particles, known as Fermions, obey the Pauli exclusion principle. In a metal, the electrons are treated as if they were particles in a three-dimensional box represented by the surfaces of the metal. For such a system when considering a cubic box, the energy of a particle is given by

$$E = \frac{\hbar^2 \pi^2}{2mL^2}(n_x^2 + n_y^2 + n_z^2) \tag{11.2}$$

where n_x, n_y, and n_z are the quantum numbers for the respective coordinates. If the quantum numbers are set equal to 1 (corresponding to the ground state), Eqn (11.2) simplifies to

$$E = \frac{3\hbar^2 \pi^2}{2mL^2} \tag{11.3}$$

Making the assumption that the quantum numbers are continuous, the number of allowed energy states per unit volume that have an energy between E and $E + dE$, can be represented by the energy function

$$g(E)dE = \frac{8\sqrt{2}\pi m^{3/2}}{h^3} \cdot E^{1/2}\,dE = C^{1/2}\,dE \tag{11.4}$$

The expression $g(E)$ is frequently referred to as the *density of states*. The probability that an electron will be in a specific state having an energy E can be expressed as

$$f(E) = \frac{1}{e^{(E-E_F)/kT} + 1} \tag{11.5}$$

From this expression, known as the Fermi−Dirac distribution function, we can see that when $T = 0$, $f(E) = 1$ for $E < E_F$ and $f(E) = 0$ for $E > E_F$. The number of electrons with energy between E and $E + dE$ per unit volume is given by $f(E)\,g(E)\,dE$, which can be written as

$$N(E)dE = C\frac{E^{1/2}dE}{e^{(E-E_F)/kT} + 1} \tag{11.6}$$

When the total number of electrons per unit volume is given by

$$\chi = C\int_0^\infty \frac{E^{1/2}}{e^{(E-E_F)/kT} + 1}\,dE \tag{11.7}$$

this expression can be simplified by making use of the special conditions given earlier,

$$\chi = C \int_0^{E_F} E^{1/2} dE = \frac{2}{3} CE_F^{3/2}$$

(11.8)

After substitution for C and simplifying the result is

$$E_F = \frac{h^2}{2m} \left(\frac{3\chi}{8\pi} \right)^{3/2}$$

(11.9)

The Fermi energy is equal to the highest energy within the filled band (see Figure 11.1).

11.3 GROUPS IA AND IIA METALS

The metals in the first two groups of the periodic table are characterized as "s block" elements because of their outer shells having one and two electrons in an *s* orbital. These configurations follow the closed shell arrangements of the noble gases so the outer electrons are subject to considerable shielding from the nuclear charge. The result is that the alkali metals found in Group IA have the lowest ionization potentials of any elements. In their chemistry, they most often are found as the +1 ions, although this is not universally true. Known as the *alkaline earths*, the elements in Group IIA have an ns^2 configuration, but even though this corresponds to a filled shell, the metals are reactive partially because of shielding of the outermost electrons. Because the elements in Groups IA and IIA are so reactive, they are always found combined in nature. The reduction of the metals requires a strong reducing agent so it was not possible to produce the metals in ancient times when carbon was the reducing agent used in metallurgy. Therefore, the metals were not obtained until the early 1800s when electrochemical processes began to be used. Several compounds of the Groups IA and IIA metals were known in ancient times. These include salt, limestone ($CaCO_3$), and sodium carbonate (Na_2CO_3). In discussing the chemistry of the elements in these groups, only selected areas will be covered. It is impossible to cover the entire field in part of a chapter so the approach taken is to show representative slices of the enormous pie that this field of chemistry represents.

11.3.1 General Characteristics

The elements in Groups IA and IIA of the periodic table have the valence shell configurations of ns^1 and ns^2, respectively. Consequently, the expected pattern for the Groups IA and IIA elements is the formation of +1 and +2 ions, respectively. All naturally occurring compounds of these elements contain the atoms in those forms. Because the isotope of francium having the longest half-life is 21 min, francium is omitted from the discussion. In addition to the data shown in Table 11.1, it is frequently useful to know the charge-to-size ratio of ions. That parameter is simply the charge on the ion divided by its radius, and it gives the so-called charge density of the ion. This quantity gives a guide to how strongly the ion will be solvated as a result of ion−dipole forces. It is also a measure of the hardness of the species (see Chapter 9).

Table 11.1 Relevant Data for Group IA, Group IIA, and Aluminum

Metal	Crystal Structure	m.p., °C	b.p., °C	Density, g/cm³	$-\Delta H_{hyd}$, kJ/mol	R_{atom}, pm	R_{ion},[a] pm
Group IA							
Li	*bcc*	180.5	1342	0.534	515	152	68
Na	*bcc*	97.8	893	0.970	406	186	95
K	*bcc*	63.3	760	0.862	322	227	133
Rb	*bcc*	39	686	1.53	293	248	148
Cs	*bcc*	28	669	1.87	264	265	169
Group IIA							
Be	*hcp*	1278	2970	1.85	2487	111	31
Mg	*hcp*	649	1090	1.74	2003	160	65
Ca	*fcc*	839	1484	1.54	1657	197	99
Sr	*fcc*	769	1384	2.54	1524	215	113
Ba	*bcc*	725	1805	3.51	1360	217	135
Ra	*bcc*	700	1140	5	~1300	220	140
Group IIIA							
Al	*fcc*	660	2327	2.70	4690	143	50

[a]This is the radius of the ion having the charge equal to the group number.

A large number of compounds of the Groups IA and IIA metals occur naturally (salt, soda, limestone, etc.). These compounds have been important for thousands of years and they still are. Moreover, lime is still produced by heating limestone, and the quantity produced is enormous.

As a result of the decrease in ionization potential, the reactivity of the metals in Groups IA and IIA increases in progressing down in the group. The electronegativities of these metals range from about 0.8 to slightly over 1.0 (except for beryllium, which has an electronegativity of 1.6). All are very strong reducing agents that form binary compounds with most nonmetallic elements and even some other metals that have higher electronegativity. As has been discussed in Chapter 10, the metals in Group IA (especially) and Group IIA (to a slight extent) are somewhat soluble in liquid ammonia and amines, and the solutions have many interesting characteristics.

Because the Group IA metals are such strong reducing agents and so reactive, they are generally prepared by electrolysis reactions. For example, sodium is produced by the electrolysis of a molten mixture that contains NaCl and $CaCl_2$. Lithium is produced by the electrolysis of a mixture of LiCl and KCl. The production of potassium is carried out by using sodium as a reducing agent at 850 °C. Under these conditions, potassium is more volatile so the equilibrium

$$Na(g) + K^+ \rightleftarrows Na^+ + K(g) \tag{11.10}$$

is drawn to the right as potassium is removed. Therefore, even though potassium is normally a stronger reducing agent than sodium, removal of the potassium allows the reaction to be carried out efficiently.

11.3.2 Negative Ions

In most instances, the Group IA metals form +1 ions, but this is not always the case. Because of the low ionization potential of sodium, an unusual situation exists with regard to forming Na^+ and Na^- ion pairs. If we consider the reaction

$$Na(g) + Na(g) \rightarrow Na^+(g) + Na^-(g) \qquad (11.11)$$

we find that it requires 496 kJ/mol to remove an electron from a sodium atom (the ionization potential). Adding the electron to another sodium atom

$$Na(g) + e^- \rightarrow Na^-(g) \qquad (11.12)$$

releases approximately 53 kJ/mol (the reverse of electron affinity). Thus, the process shown in Eqn (11.11) would have an enthalpy of about +443 kJ/mol. However, if the process could be carried out in water without any other reaction occurring, the enthalpy of solvation of $Na^+(g)$ *liberating* 406 kJ/mol would make the process of electron transfer more nearly energetically favorable.

$$Na^+(g) + n\,H_2O \rightarrow Na^+(aq) \qquad (11.13)$$

The $Na^-(g)$ would also be solvated, but the heat associated with that process would be less than that for the cation. All this is hypothetical because sodium reacts vigorously with water. What is needed is a substance that strongly solvates the ions produced without otherwise reacting. One type of molecule that binds strongly to sodium ions is a polyether having the structure

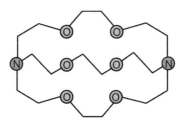

By dissolving sodium in ethylenediamine and adding this complexing agent (known as a *cryptand*) followed by evaporation of the solvent, it has been possible to recover a solid that contains Na^+(*crypt*) Na^-. This shows that although it is rather rare, it possible for Na to complete the $3s$ level. Of course, this type of behavior also occurs when H^- is formed.

11.3.3 Hydrides

Most of the Groups IA and IIA metals form compounds with hydrogen that (at least formally) contain H^-. These compounds are discussed in more detail along with the chemistry of hydrogen in Chapter 13. Because they have the characteristics of ionic compounds (white solids of low volatility), these hydrides are known as the *salt-like hydrides*. However, the hydrides of beryllium and magnesium are quite different because of their being much more covalent (these metals have electronegativities of 1.6 and 1.3, respectively). These hydrides are polymeric and consist of chains having hydrogen atoms bridging between the metal atoms. The most important characteristic of the ionic hydrides is the very strong basicity of the hydride ion. Ionic hydrides will remove protons from almost any molecule that contains an OH bond, including water, alcohols, etc. as illustrated by the following equations:

$$H^- + H_2O \rightarrow H_2 + OH^- \tag{11.14}$$

$$H^- + CH_3OH \rightarrow H_2 + CH_3O^- \tag{11.15}$$

The H^- ion is such a strong base that it will deprotonate NH_3 to generate the amide ion.

$$H^- + NH_3 \rightarrow H_2 + NH_2^- \tag{11.16}$$

Ionic hydrides such as NaH and CaH_2 can be used as drying agents because they will remove hydrogen from the traces of water present in many solvents. Lithium aluminum hydride, $LiAlH_4$, is a versatile reducing agent that is used to carry out many types of reactions in organic chemistry.

11.3.4 Oxides and Hydroxides

The Group IA metals react with oxygen, but the products are not always the "normal" oxides. Lithium reacts with oxygen in the expected way,

$$4\,Li + O_2 \rightarrow 2\,Li_2O \tag{11.17}$$

However, sodium reacts with oxygen with the product being predominantly the *peroxide*.

$$2\,Na + O_2 \rightarrow Na_2O_2 \tag{11.18}$$

The O_2^- ion is known as the *superoxide* ion, and it is produced when oxygen reacts with potassium, rubidium, and cesium.

$$K + O_2 \rightarrow KO_2 \tag{11.19}$$

Presumably, the formation of oxides of the larger atoms in Group IA is favored when they contain larger anions. It is generally more favorable when crystal lattices are formed from cations and anions of similar size and magnitude of charge (see Chapter 9). In reactions with oxygen, the lighter members of the Group IIA metals give normal oxides, but barium and radium give peroxides.

When the oxygen compounds of the Groups IA and IIA metals react with water, strongly basic solutions are produced regardless of whether an oxide, peroxide, or superoxide is involved.

$$Li_2O + H_2O \rightarrow 2\,LiOH \tag{11.20}$$

$$Na_2O_2 + 2\,H_2O \rightarrow 2\,NaOH + H_2O_2 \tag{11.21}$$

$$2\,KO_2 + 2\,H_2O \rightarrow 2\,KOH + O_2 + H_2O_2 \tag{11.22}$$

Sodium hydroxide, sometimes referred to as caustic soda or simply caustic, is produced in enormous quantities by the electrolysis of an aqueous solution of sodium chloride.

$$2\,NaCl + 2\,H_2O \xrightarrow{\text{Electrolysis}} 2\,NaOH + Cl_2 + H_2 \tag{11.23}$$

The fact that this is also the reaction used to prepare chlorine makes it especially important. However, sodium hydroxide will react with chlorine as shown in the equation

$$2\,OH^- + Cl_2 \rightarrow OCl^- + Cl^- + H_2O \tag{11.24}$$

In order to prevent this reaction, the products must be kept separated during the electrolysis. In one way, a diaphragm is used, and in the other, a mercury cell is employed. In the mercury cell, a mercury cathode reacts with sodium to form an amalgam that is continuously removed. In the diaphragm cell, the cathode and anode compartments are separated by a diaphragm made of asbestos. The solution containing sodium hydroxide is removed from the cathode compartment before diffusion through the diaphragm occurs. Billions of pounds of NaOH are produced annually, and it is used in many processes that require a strong base.

Potassium hydroxide is produced by electrolysis of aqueous KCl. Because it is more soluble than NaOH in organic solvents, KOH is widely used in certain types of processes. For example, KOH is used in the production of many types of soaps and detergents. The hydroxides of rubidium and cesium are of little importance compared to those of sodium and potassium, but they are even stronger bases.

The oxides of the Group IIA metals are ionic so they react with water to produce the hydroxides.

$$MO + H_2O \rightarrow M(OH)_2 \tag{11.25}$$

However, beryllium oxide is quite different and it exhibits amphoteric behavior.

$$Be(OH)_2 + 2\,OH^- \rightarrow Be(OH)_4^{2-} \tag{11.26}$$

$$Be(OH)_2 + 2\,H^+ \rightarrow Be^{2+} + 2\,H_2O \tag{11.27}$$

Magnesium hydroxide is a weak, almost insoluble, base that is used as a suspension known as "milk of magnesia" that is a common antacid. The utility of the hydroxides of the Group IIA metals is limited somewhat because they are only slightly soluble (only about 0.12 g of $Ca(OH)_2$ dissolves in 100 g of water, although it is a strong base). Calcium hydroxide is produced in enormous quantities by the decomposition of limestone,

$$CaCO_3 \rightarrow CaO + CO_2 \tag{11.28}$$

which is followed by the reaction of CaO (*lime*) with water.

$$CaO + H_2O \rightarrow Ca(OH)_2 \tag{11.29}$$

Calcium hydroxide is known as *hydrated lime* or *slaked lime*, and it is used extensively in some applications because it is less expensive than NaOH or KOH. It reacts with CO_2 to form $CaCO_3$, which binds particles of sand and gravel together in mortar and cement.

11.3.5 Halides

Sodium chloride is found in salt beds, salt brines, and sea water throughout the world, and it is also mined in some locations. Consequently, sodium chloride is the source of numerous other sodium compounds. A large portion of the sodium chloride utilized is consumed in the production of sodium hydroxide (Eqn (11.23)). The production of sodium metal involves the electrolysis of the molten chloride, usually in the form of a eutectic mixture with calcium chloride. Sodium carbonate is an important material that is used in many ways, such as making glass. It was formerly produced from NaCl by means of the Solvay process, in which the overall reaction is

$$NaCl(aq) + CO_2(g) + H_2O(l) + NH_3(aq) \rightarrow NaHCO_3(s) + NH_4Cl(aq) \qquad (11.30)$$

As described in Chapter 8, there are many transformations of solids that have industrial importance, and in this case, the solid product decomposes when heated to give sodium carbonate.

$$2\,NaHCO_3(s) \rightarrow Na_2CO_3(s) + H_2O(g) + CO_2(g) \qquad (11.31)$$

Although the Solvay process is still in use in some parts of the world, the chief source of sodium carbonate is the mineral *trona*, $Na_2CO_3 \cdot NaHCO_3 \cdot 2H_2O$.

11.3.6 Sulfides, Nitrides, Carbides, and Phosphides

The sulfides of the Group IIA metals generally have the sodium chloride structure, but those of the Group IA metals have the antifluorite structure as a result of the ratio of anions to cations being 2. Solutions of the sulfides are basic as a result of the hydrolysis reaction

$$S^{2-} + H_2O \rightarrow HS^- + OH^- \qquad (11.32)$$

One preparation of the Groups IA and IIA sulfides involves the reaction of H_2S with the metal hydroxides.

$$2\,MOH + H_2S \rightarrow M_2S + 2\,H_2O \qquad (11.33)$$

Because of the tendency of sulfur toward catenation, solutions containing sulfides react with sulfur to give polysulfides, which can be represented as S_n^{2-} (see Chapter 15). Sulfides of the Groups IA and IIA metals can also be produced by reducing the sulfates with carbon at high temperature.

$$BaSO_4 + 4\,C \rightarrow BaS + 4\,CO \qquad (11.34)$$

Lithopone, a commonly used pigment containing barium sulfate and zinc sulfide, is produced by the following reaction:

$$BaS + ZnSO_4 \rightarrow BaSO_4 + ZnS \qquad (11.35)$$

Because $Ca(OH)_2$ is a base and H_2S is an acid, the following reaction can be used to prepare CaS.

$$Ca(OH)_2 + H_2S \rightarrow CaS + 2\,H_2O \qquad (11.36)$$

Most nonmetallic elements will react with the Groups IA and IIA metals to give binary compounds. Heating the metals with nitrogen or phosphorus gives nitrides and phosphides of the metals.

$$12\,Na + P_4 \rightarrow 4\,Na_3P \tag{11.37}$$

$$3\,Mg + N_2 \rightarrow Mg_3N_2 \tag{11.38}$$

Although the products in these equations are written as if they were simple ionic binary compounds, this is not always the case. For example, some nonmetals form clusters containing several atoms arranged in polyhedral structures. One such species is the P_7^{3-} cluster, which has six phosphorus atoms at the vertices of a trigonal prism with the seventh occupying a position above a triangular face on one end of the prism. When the binary nitrides or phosphates react with water, basic solutions result because the anions undergo solvolysis reactions, such as those represented by the following equations:

$$Na_3P + 3\,H_2O \rightarrow 3\,NaOH + PH_3 \tag{11.39}$$

$$Mg_3N_2 + 6\,H_2O \rightarrow 3\,Mg(OH)_2 + 2\,NH_3 \tag{11.40}$$

$$Li_3N + 3\,ROH \rightarrow 3\,LiOR + NH_3 \tag{11.41}$$

Reactions similar to these provide convenient syntheses of hydrides of elements such as phosphorus, arsenic, tellurium, selenium, etc. because these elements do not react directly with hydrogen, and the hydrides are unstable.

Binary carbides are formed when the metals are heated strongly with carbon. The most important carbide of the Groups IA and IIA metals is calcium carbide, CaC_2. This carbide is actually an *acetylide* because it contains the C_2^{2-} ion and it reacts with water to produce acetylene.

$$CaC_2 + 2\,H_2O \rightarrow Ca(OH)_2 + C_2H_2 \tag{11.42}$$

The reaction between CaO and carbon (coke) at very high temperature produces CaC_2.

$$CaO + 3\,C \rightarrow CaC_2 + CO \tag{11.43}$$

Calcium acetylide is used in the manufacture of calcium cyanamide, $CaCN_2$, by the reaction of CaC_2 with N_2 at high temperature.

$$CaC_2 + N_2 \xrightarrow{1000\ °C} CaCN_2 + C \tag{11.44}$$

As in the case of the Haber process for the synthesis of ammonia, this reaction represents a way of converting elemental nitrogen into a compound (nitrogen fixation). Moreover, calcium cyanamide reacts with steam at high temperature to yield ammonia,

$$CaCN_2 + 3\,H_2O \rightarrow CaCO_3 + 2\,NH_3 \tag{11.45}$$

Calcium cyanamide is also a constituent in some fertilizers.

Another cyanamide compound that has a number of significant uses is sodium cyanamide that is prepared as shown below. Sodium amide is obtained by the reaction of ammonia with Na at 400 °C.

$$2\,Na\ +\ 2\,NH_3\ \rightarrow\ 2\,NaNH_2\ +\ H_2 \tag{11.46}$$

The reaction between sodium amide and carbon produces sodium cyanamide

$$2\,NaNH_2\ +\ C \rightarrow Na_2CN_2\ +\ 2\,H_2 \tag{11.47}$$

which reacts with carbon to produce sodium cyanide.

$$Na_2CN_2\ +\ C \rightarrow 2\,NaCN \tag{11.48}$$

The major use of sodium cyanamide is in the production of sodium cyanide, a compound that is used extensively in preparing solutions for the electroplating of metals. Another use for NaCN is in extraction processes employed to separate gold and silver from ores as a result of forming complexes with CN^-. Sodium cyanide is an extremely toxic compound that is used in the process known as *casehardening* of steel. In this process, the object to be hardened is heated and allowed to react with the cyanide to form a layer of metal carbide on the surface.

11.3.7 Carbonates, Nitrates, Sulfates, and Phosphates

Some of the important compounds containing the Groups IA and IIA metals are the carbonates, nitrates, sulfates, and phosphates. We have already mentioned the mineral *trona* as the source of sodium carbonate. Calcium carbonate is found in many forms that include chalk, *calcite*, *aragonite*, marble, etc., as well as in eggshells, coral, and seashells. In addition to its use as a building material, calcium phosphate is converted into fertilizers in enormous quantities (see Chapter 14).

Magnesium is found as the carbonate in the mineral *magnesite* and as the silicate in the mineral *olivine*. Magnesium is also found as *Epson salts*, $MgSO_4 \cdot 7H_2O$, which is used in solution for medicinal purposes. A mixed carbonate containing calcium and magnesium is *dolomite*, $CaCO_3 \cdot MgCO_3$, that is used in construction and in antacid tablets. Calcium is also found as $CaSO_4 \cdot 2H_2O$, *gypsum*. *Beryl*, which has the composition $Be_3Al_2(SiO_3)_6$, is one mineral that contains beryllium. If traces of chromium are present, the result is the green gemstone *emerald*. Among other sources, sodium and potassium are found as the nitrates, and a method for preparing nitric acid involves heating a nitrate with sulfuric acid.

$$2\,NaNO_3\ +\ H_2SO_4\ \rightarrow\ Na_2SO_4\ +\ 2\,HNO_3 \tag{11.49}$$

The carbonates, sulfates, nitrates, and phosphates of the Groups IA and IIA metals are important materials in inorganic chemistry. Some of the most important compounds of the Groups IA and IIA elements are organometallic compounds, particularly for lithium, sodium, and magnesium, and Chapter 12 will be devoted to this area of chemistry.

11.4 ZINTL PHASES

In the late 1800s, it was observed that dissolving lead in a liquid ammonia solution containing sodium gave a compound that contained the two metals. Extensive work on systems of this type was carried out in the 1930s by Eduard Zintl in Germany whose work is considered to be the foundation of this type of chemistry. As is often the case when someone makes a significant discovery, compounds of this type are now referred to as *Zintl compounds* or *Zintl phases*. They are compounds that contain a Group IA or IIA metal with heavier members of Groups IIIA, IVA, VA, or VIA. Although the usual binary compounds such as Na_2S, K_3As, and Li_3P exist, many others have more complex formulas and structures. They contain anions that consist of clusters of several or many atoms of the more electronegative element. Some of them have formulas such as Sn_4^{4-}, P_7^{3-}, Pb_9^{4-}, Sb_7^{3-}, etc. Because the cations are frequently metals from Groups IA and IIA, Zintl phases will be described in this section following the discussion of those metals.

One of the most common techniques for preparing Zintl phases is by the reaction of a solution of the alkali metal in liquid ammonia with the other element. However, many of these materials are obtained by heating the elements. For example, heating barium with arsenic leads to the reaction

$$3\,Ba \;+\; 14\,As \;\rightarrow\; Ba_3As_{14} \tag{11.50}$$

As a general definition, a Zintl phase is one that contains a Group IA or IIA metal with a metalloid from later groups in the periodic table, but some of these materials have variable composition. Therefore, many Zintl phases contain Bi, Sn, In, Pb, As, Se, or Te as the metalloid. It has been found that several compounds of this type contain 14 atoms of the metalloid in a cluster with some examples being Ba_3As_{14} and Sr_3P_{14}, all of which contain the metalloid anion in M_7^{3-} clusters. The structure of these ions is an end-capped irregular trigonal prism.

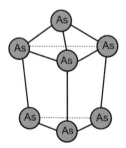

As will be seen in Chapter 14, except for dimensions and bond angles, it is the same general structure as that for P_4S_3. Because a sulfur atom has one more electron than an atom of phosphorus, a structure that contains four phosphorus atoms and three sulfur atoms is isoelectronic with the P_7^{3-} ion. This suggests, as is in fact the case, that it should be possible to prepare Zintl phases containing two different metalloids. Some examples are $Pb_2Sb_2^{2-}$, $TlSn_8^{3-}$, and $Hg_4Te_{12}^{4-}$, although many others are known.

Numerous Zintl anions are formed by selenium and tellurium with some of the more prominent species being Se_n^{2-} (where $n = 2$, 4, 5, 6, 7, 9, or 11). The species with $n = 11$ contains two rings that have five and six members that are joined by a selenium atom. Those with smaller numbers of selenium atoms generally consist of zigzag chains. Tellurium forms an extensive series of polyanions that are present in such species as $NaTe_n$ ($n = 1-4$). One tellurium anion contains the $Hg_4Te_{12}^{4-}$ ion, but other species such as $[(Hg_2Te_5)_n]^{2-}$ are also known as is the Te_{12}^{2-} anion that is present in some cases where the cation is a +1 metal.

Polyatomic species containing atoms from Group IVA are produced by reducing the elements in liquid ammonia that contains some dissolved sodium. In accord with the hard–soft interaction principle (see Chapter 9), isolation of species containing large anions is best accomplished when a large cation of similar magnitude in charge is used. When ethylenediamine is introduced in the solution, the Na^+ cation is solvated to give larger cations, such as $Na_4(en)_5^+$ and $Na_4(en)_7^+$. With these large cations of +1 charge, it is possible to isolate the Ge_9^- and Sn_9^- ions in solids. One very effective binding agent for Na^+ is a polyether known as 18-crown-6 that has the structure

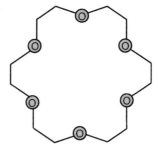

This ligand forms stable complexes known as *cryptates* (generally abbreviated as *crypt* in writing formulas) with cations of the Group IA metals. More recently, the ligand known as *crypt-222* has been used instead of ethylenediamine. This ligand has the molecular structure

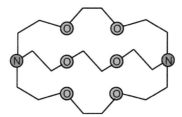

As a result of having several potential donor sites, this molecule attaches strongly to cations of the Group IA metals. Using the sequestered sodium ion, numerous Zintl phases containing some cryptands have been isolated, including $[Na(crypt)^+]_2[Sn_5]^{2-}$, $[Na(crypt)^+]_4[Sn_9]^{4-}$ and $[Na(crypt)^+]_2[Pb_5]^{2-}$. The 5-atom anions have a trigonal bipyramid structure, but the Sn_9^{4-} ion has a capped square antiprism structure.

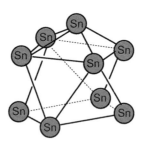

The ions having five tin or lead atoms are prepared by the reaction of a solution containing sodium and the cryptand reacting with alloys of sodium and tin or lead, respectively. It should also be mentioned that numerous derivatives of these materials have been prepared that contain alkyl and other groups.

Although numerous Zintl phases have been studied as solids and their structures determined, differences between packing in crystals and solvation in solution frequently cause the clusters to be different in the two media. One purpose for discussing Zintl phases here is to show the relationship between the alkali and alkaline earth metals to this interesting area of cluster compounds. Although these clusters also deal with atoms in other groups of the periodic table, the Group IA metals figure prominently in this chemistry because of the nature of solutions of the metals in liquid ammonia and their strength as reducing agents.

11.5 ALUMINUM AND BERYLLIUM

Because of numerous similarities in their properties and reactions, aluminum and beryllium will be described together, even though they are in different groups of the periodic table. Although not completely understood, there is some indication that the accumulation of aluminum in the brain may have some relationship to Alzheimer's disease, and beryllium compounds are extremely toxic.

The principal ore of aluminum is *bauxite*, one of the major constituents of which is $AlO(OH)$. The production of aluminum is accomplished by electrolysis of Al_2O_3 dissolved in *cryolite*, Na_3AlF_6, but the molten solution containing Al_2O_3 contains some Na_2O, and its reaction with AlF_3 generates Na_3AlF_6.

$$4\,AlF_3 + 3\,Na_2O \rightarrow 2\,Na_3AlF_6 + Al_2O_3 \tag{11.51}$$

Naturally occurring Na_3AlF_6 is not readily available in sufficient quantity for the production of aluminum, but it can be produced by the reaction

$$3\,NaAlO_2 + 6\,HF \rightarrow Na_3AlF_6 + 3\,H_2O + Al_2O_3 \tag{11.52}$$

There is a complex relationship between the oxide, hydroxide, and hydrous oxide of aluminum. Conversion between several phases is possible as a result of the relationships

$$2\,Al(OH)_3 = Al_2O_3 \cdot 3H_2O = Al_2O_3 + 3\,H_2O \tag{11.53}$$

$$2\,AlO(OH) = Al_2O_3 \cdot H_2O = Al_2O_3 + H_2O \tag{11.54}$$

The chemistry of aluminum and beryllium is strongly influenced by the high charge-to-size ratio of their usual ions. The $+2$ beryllium ion has a radius of 31 pm and that of Al^{3+} is 50 pm. Therefore, the charge-to-size ratios for the two ions are 0.065 and 0.060, respectively. Because of their relative positions along a diagonal in the periodic table, the resemblance between Be^{2+} and Al^{3+} is known as a *diagonal relationship*. One way in which the solution chemistry of beryllium and aluminum are similar can be illustrated by considering what happens when the chlorides are dissolved in water, and the water is then evaporated. As water is evaporated, the loss of HCl occurs and the final products are not the chlorides, but rather the oxides. This can be illustrated for aluminum as follows:

$$AlCl_3(aq) \xrightarrow{-nH_2O} Al(H_2O)_6Cl_3(s) \xrightarrow{-3H_2O} Al(H_2O)_3Cl_3(s) \xrightarrow{\Delta} 0.5\,Al_2O_3(s) + 3\,HCl(g) + 1.5\,H_2O(g)$$

(11.55)

Beryllium shows similar behavior as a result of forming bonds to oxygen.

$$BeCl_2(aq) \xrightarrow{-nH_2O} Be(H_2O)_4Cl_2(s) \xrightarrow{-2H_2O} Be(H_2O)_2Cl_2(s) \xrightarrow{\Delta} BeO(s) + 2\,HCl(g) + H_2O(g) \quad (11.56)$$

Compounds of beryllium and aluminum are substantially covalent as a result of the high charge-to-size ratio, which causes polarization of anions and very high heats of hydration of the ions (-2487 kJ/mol for Be^{2+} and -4690 kJ/mol for Al^{3+}).

Aluminum is amphoteric, and this has been illustrated in Chapter 10 both for water as the solvent and for some nonaqueous solvents. Because of the high charge density of Al^{3+}, solutions of aluminum salts are acidic as a result of hydrolysis.

$$Al(H_2O)_6^{3+} + H_2O \rightarrow Al(H_2O)_5OH^{2+} + H_3O^+ \tag{11.57}$$

There is apparently some aggregation of ions as a result of the formation of OH bridges to give species, such as $[(H_2O)_4Al(OH)_2Al(H_2O)_4]^{4+}$. In the vast majority of complexes, aluminum has a coordination number of 6, but the number can also be 4 as in $LiAlH_4$.

Another similarity between the behavior of aluminum and beryllium is seen in the fact that both form polymeric hydrides as a result of bridging hydrogen atoms. The structures of aluminum chloride (a dimer) and beryllium chloride (a polymer chain) (Figure 11.3) illustrate the types of bridged structures exhibited by these compounds.

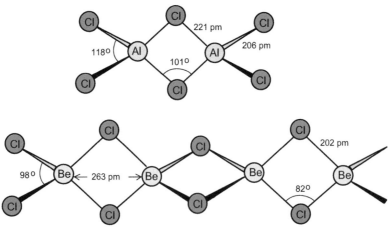

FIGURE 11.3
The structures of [AlCl$_3$]$_2$ and [BeCl$_2$]$_n$.

Although the bond angles do not equal those expected for tetrahedral coordination, the beryllium is considered to make use of sp^3 hybrid orbitals as it does in simple complexes, such as $[BeF_4]^{2-}$, $[Be(OH)_4]^{2-}$, etc. Aluminum chloride is soluble in a wide range of solvents, but the dimeric structure is retained in nonpolar solvents that do not form complexes. In solvents that have donor properties, solubility leads to complex formation to give species such as S:$AlCl_3$ (where S is a solvent molecule). Beryllium chloride is soluble in solvents such as alcohols, ether, and pyridine, but slightly soluble in benzene.

The fact that Be^{2+} and Al^{3+} have charge-to-size ratios that are so high causes them to exert a great polarizing effect on the molecules and ions to which they are bound. This results in their compounds being substantially covalent. Unlike the situation when NaCl is dissolved in water and the ions become hydrated, dissolving $BeCl_2$ or $AlCl_3$ in water results in the formation of essentially covalent complexes such as $Be(H_2O)_4^{2+}$ and $Al(H_2O)_6^{3+}$ from which the water cannot easily be removed. When heated strongly, the loss of HCl becomes the energetically favorable process rather than loss of water from compounds such as $Be(H_2O)_4Cl_2$ and $Al(H_2O)_6Cl_3$ because of the strong bonds to oxygen formed by Be^{2+} and Al^{3+}.

The system involving Al_2O_3, $AlO(OH)$, and $Al(OH)_3$ is very complex. However, some of the derivatives of these formulas are widely used compounds. For example, a compound having the formula $Al_2(OH)_5Cl$ is used in personal care products, such as deodorants. Alumina, Al_2O_3, exists in two forms that are known as α-Al_2O_3 and γ-Al_2O_3, and they differ in structure. *Corundum* is the mineral form of α-Al_2O_3. In addition to the oxides, $AlO(OH)$ is an important material that exists in two forms, α-$AlO(OH)$ or *diaspore* and γ-$AlO(OH)$ or *boehmite*. Finally, the hydroxide also exists in two forms, α-$Al(OH)_3$ (*bayerite*) and γ-$Al(OH)_3$ (*gibbsite*). Alumina is an important material because it is used as a packing medium in chromatography columns, and it is a very important catalyst for numerous reactions. The most common carbide of aluminum, Al_4C_3, is considered to be a methanide because it reacts with water to produce methane.

$$Al_4C_3 + 12\,H_2O \rightarrow 4\,Al(OH)_3 + 3\,CH_4 \qquad (11.58)$$

11.6 THE FIRST-ROW TRANSITION METALS

The transition metals include the three series of elements that are positioned between the first two groups and the last six groups in the periodic table. These series have as their general characteristic that a set of d orbitals is being filled in progressing from one element to the next. Although some aspects of the chemistry of transition metals will be discussed in this chapter, the elements have many other interesting and important aspects dealing with coordination chemistry that will be discussed in Chapters 16–22. The first series, usually referred to as the first-row transition metals, involves filling the $3d$ orbitals. The second- and third-row transition metals correspond to those in which the $4d$ and $5d$ orbitals are being filled. Because a set of d orbitals can hold a maximum of 10 electrons, there are 10 elements in each series. The groups containing the transition metals are sometimes designated as the "B" groups or as Groups 3–10. In the latter case, the groups are usually denoted as 1 through 18 in going across the periodic table.

Most of the first-row transition metals and several in the second and third groups have important uses. For example, iron is the basis of the enormous range of ferrous alloys in which the other first-row metals are often combined. The metallurgy of iron-based alloys is a vast and complex field. Among the

Table 11.2 Properties of the Transition Metals

| | Transition Metals | | | | | | | | | |
	Sc	Ti	V	Cr	Mn	Fe	Co	Ni	Cu	Zn
m.p., °C	1541	1660	1890	1900	1244	1535	1943	1453	1083	420
b.p., °C	2836	3287	3380	2672	1962	2750	2672	2732	2567	907
Crystal structure	hcp	hcp	bcc	bcc	fcc	bcc	hcp	fcc	fcc	hcp
Density, g/cm^3	2.99	4.5	6.11	7.19	7.44	7.87	8.90	8.91	8.94	7.14
Atomic radius, pm[a]	160	148	134	128	127	124	125	124	128	133
Electronegativity	1.3	1.5	1.6	1.6	1.5	1.8	1.9	1.9	1.9	1.6

[a]For coordination number = 12 because most structures are hcp or fcc.

many forms of iron are *cast iron, wrought iron,* and the myriad special steels that contain other metals as well as carbon and other main group elements. Iron and its alloys account for over 90% of all of the metals used. Nickel, manganese, and cobalt are also metals that are essential to the production of numerous types of alloys. These special alloys are used in many ways, including making of tools, engine parts, catalysis, etc. One form of nickel is an important catalyst known as *Raney nickel,* which is prepared by the reduction of NiO with hydrogen. Nickel is also used in several alloys that have many applications. For example, *Monel* is a type of alloy that contains nickel and copper in a ratio of about 2:1. Copper is used not only as a coinage metal but also in numerous types of electrical devices and conductors. Zinc is used as a constituent in brass, some types of batteries, and as a protective coating on sheet metal (*galvanizing*). Because of its low density, titanium is used in alloys that are important in the production of aircraft and aerospace components. Vanadium and aluminum are alloyed with titanium to produce alloys that are stronger than titanium alone, and one of the common alloys contains 6% aluminum and 4% vanadium. Plating metals with chromium provides protection from rust. Chromium is also used in making various types of stainless steel. Scandium is becoming increasingly important in making tools and small devices where exceptional strength is combined with extremely light weight (the density of Sc is 2.99 g/cm^3). It is impossible to imagine an industrialized society in which the transition metals would not be vital commodities. Table 11.2 contains useful data for the first-row transition metals.

Several aspects of the chemistry of the transition metals in all three series will be discussed in Chapters 19–22 in dealing with coordination compounds and organometallic chemistry. Section 11.9 will deal with several aspects of the chemistry of the first-row transition metals. Numerous correlations can be attempted between properties of elements, and one of the interesting relationships is the variation in melting point with position in the transition series. In a general way, the melting point reflects the strength of the bonding between metal atoms because melting a solid requires that the forces between atoms be overcome. Figure 11.4 shows the variation in melting points of the elements, and although it is not shown, there is a reasonably good correlation between melting point and hardness for most of the first-row metals.

As was described earlier in this chapter, bonding in metals involves electrons in energy bands that encompass the structure. Therefore, the increase in melting point for the early elements in the transition

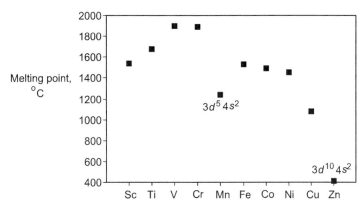

FIGURE 11.4
The melting points of the first-row transition series.

series corresponds to the increase in the number of electrons involved in populating the energy bands that are responsible for metallic bonding. After about the middle of the series, additional electrons are forced to occupy higher energy states so the bonding between atoms becomes weaker. By the time the end of the series is reached, the filled d shell is no longer a major factor in the bonding, and the atoms of zinc are held together by much weaker forces. Figure 11.4 also shows that there is a significant effect caused by the half-filled and filled shell configurations that occurs at Mn and Zn.

11.7 SECOND- AND THIRD-ROW TRANSITION METALS

Some second- and third-row transition metals are, for good reason, known as precious metals. These include silver, palladium, rhodium, iridium, osmium, gold, and platinum. As this is written, gold is over \$1500 per ounce and silver is over \$25 per ounce. Some of the other metals such as rhodium, osmium, and rhenium are also extremely expensive. Most of the second- and third-row transition metals are found as minor constituents in ores of other metals. Consequently, we will not enumerate the sources, minerals, or the processes by which these metals are obtained. Some of their most important properties are shown in Table 11.3.

It is of some interest to note how the melting points of the third-row transition metals vary with the number of valence electrons. The trend is shown graphically in Figure 11.5. There is an increase in the number of electrons occupying energy bands in progressing across the third row of the transition series. The effect of this can be seen by looking at the melting points of the metals, which show a general increase until tungsten is reached. After tungsten, the melting point decreases, and by the time mercury is reached, the element is a liquid at room temperature. A simple explanation of these observations is based on the fact that within a solid metal the number of "bonds" to nearest neighbors requires that only about six electrons are used in bonding, even though the number of nearest neighbors is 12. If the number of electrons available is greater than six, they will be forced to occupy some of the antibonding states, which decreases the net bonding effect. Consequently, the melting point will be lower than if only a total of six valence electrons are present. As shown in Figure 11.5, that is precisely the trend

Table 11.3 Properties of the Second- and Third-Row Transition Metals

	Transition Metals of the Second Row									
	Y	Zr	Nb	Mo	Tc	Ru	Rh	Pd	Ag	Cd
m.p., °C	1522	1852	2468	2617	2172	2310	1966	1552	962	321
b.p., °C	3338	4377	4742	4612	4877	3900	3727	3140	2212	765
Crystal structure	hcp	hcp	bcc	bcc	fcc	hcp	fcc	fcc	fcc	hcp
Density, g/cm³	4.47	6.51	8.57	10.2	11.5	12.4	12.4	12.0	10.5	8.69
Atomic radius, pm[a]	182	162	143	136	136	134	134	138	144	149
Electronegativity	1.2	1.4	1.6	1.8	1.9	2.2	2.2	2.2	1.9	1.7
	Transition Metals of the Third Row									
	La	Hf	Ta	W	Re	Os	Ir	Pt	Au	Hg
m.p., °C	921	2230	2996	3407	3180	3054	2410	1772	1064	−39
b.p., °C	3430	5197	5425	5657	5627	5027	4130	3827	2807	357
Crystal structure	hcp	hcp	bcc	bcc	hcp	hcp	fcc	fcc	fcc	—
Density, g/cm³	6.14	13.3	16.7	19.3	21.0	22.6	22.6	21.4	19.3	13.6
Atomic radius, pm	189	156	143	137	137	135	136	139	144	155
Electronegativity	1.0	1.3	1.5	1.7	1.9	2.2	2.2	2.2	2.4	1.9

[a]For coordination number = 12.

shown by the melting points of the third-row transition metals. Note that the effect of the half-filled shell that was so prominent in the first-row metals at manganese is essentially absent in the third-row metals.

Although much of the chemistry of the second- and third-row transition metals is similar to that of metals in the first row, there are several interesting differences. One involves a series of

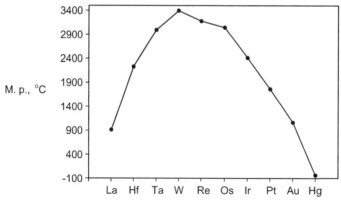

FIGURE 11.5
Melting points of the third-row transition metals.

molybdenum compounds that are ternary sulfides. A general formula can be written as MMo_6S_8 where M is a +2 metal or as $M_2Mo_6S_8$ when M is a +1 metal. Known as *Chevrel phases*, the first compound of this type was the lead compound $PbMo_6S_8$. More recently, other metals have been incorporated and others contain selenium or tellurium instead of sulfur. The $Mo_6S_8^{2-}$ ion has a structure in which the molybdenum atoms reside at the corners of an octahedron with a sulfur atom placed above each of the triangular faces. A major point of interest in Chevrel phases is that they behave as superconductors.

11.8 ALLOYS

The physical characteristics of metals that permit them to function as versatile materials for fabricating many items are their ductility, malleability, and strength. Although strength probably needs no explanation, the first two of these characteristics are related to the ability of the metal to be fabricated into a desired shape. Metals vary widely in these characteristics, and a metal or alloy that is well suited to one use may be entirely unsatisfactory for another. Addressing this branch of applied science is beyond the scope of this book, but a book on materials science provides a great deal of information that is relevant for students in inorganic chemistry.

As important as the transition metals are, the fact that they form many important alloys greatly extends the versatility of the metals. In this section, we will briefly describe some of the major factors that are relevant to the behavior of alloys. The study of alloys is a vast area of applied science so in order to illustrate some of the principles in an efficient way, we will deal primarily with the behavior of alloys of copper and iron. Alloys of some of the nontransition metals (lead, antimony, tin, etc.) will be described in subsequent chapters. Many of the principles that apply in the behavior of a specific metal are involved in the behavior of others.

Alloys are classified broadly in two categories: *single-phase* alloys and *multiple-phase* alloys. A *phase* is characterized by having a homogenous composition on a macroscopic scale, a uniform structure, and a distinct interface with any other phase present. The coexistence of ice, liquid water, and water vapor meets the criteria of composition and structure, but distinct boundaries exist between the states so there are three phases present. When liquid metals are combined, there is usually some limit to the solubility of one metal in another. An exception to this is the liquid mixture of copper and nickel, which forms a solution of any composition between pure copper and pure nickel. The molten metals are completely miscible. When the mixture is cooled, a solid results that has a random distribution of both types of atoms in an *fcc* structure. This single solid phase thus constitutes a solid solution of the two metals so it meets the criteria for a *single-phase alloy*.

Alloys of copper and zinc can be obtained by combining the molten metals. However, zinc is soluble in copper up to only about 40% (of the total). When the content of a copper/zinc alloy contains less than 40% zinc, cooling the liquid mixture results in the formation of a solid solution in which Zn and Cu atoms are uniformly distributed in an *fcc* lattice. When the mixture contains more than 40% zinc, cooling the liquid mixture results in the formation of a compound having the composition CuZn. The solid alloy consists of two phases, one of which is the compound CuZn and the other is a solid solution that contains Cu with approximately 40% Zn dissolved in it. This type of alloy is known as a *two-phase alloy*, but many alloys contain more than three phases (*multiple-phase* alloys).

The formation of solid solutions of metals is one way to change the properties (generally to increase strength) of the metals. Strengthening metals in this way is known as *solid solution strengthening*. The ability of two metals to form a solid solution can be predicted by a set of rules known as the *Hume-Rothery rules*, which can be stated as follows:

1. The atomic radii of the two kinds of atoms must be similar (within about 15%) so that lattice strain will not be excessive.
2. The crystal structure of the two metals must be identical.
3. In order to minimize the tendency for the metals to form compounds, identical valences and approximately equal electronegativities are required.

However, it must be mentioned that although these guidelines are useful, they are not always successful in predicting solubility.

When a second metal is dissolved in a host metal and the mass is cooled to produce a solid solution, the solution has increased strength compared to the host metal. This occurs because in a regular lattice consisting of identical atoms it is relatively easy to move the atoms away from each other. There is less restriction on mobility of atoms and the sharing of electrons between atoms is equal. However, when zinc or nickel is dissolved in copper, the alloy is strengthened, and the degree of strengthening is approximately a linear function of the fraction of the added metal. Adding the second metal distorts the lattice at the sites where the atoms of that metal reside, and this restricts atomic motions and strengthens the metal. However, the atomic radius of copper is 128 pm, whereas the radii of zinc and nickel are 133 and 124 pm, respectively. The atomic radius of tin is 151 pm so if tin were dissolved in copper (assuming that its limited solubility is not exceeded), there would be a greater strengthening effect because there is a greater disparity between the sizes of the host and guest metal atoms. This is exactly the effect observed, and bronze (Cu/Sn) is stronger than brass (Cu/Zn) in which the atomic radii are very similar. The atomic radius of beryllium is 114 pm so adding beryllium (within the solubility limit) causes a great strengthening of copper. In fact, the addition of atoms having a smaller atomic radius than copper causes a greater effect than adding those that are larger than copper even when the same absolute difference in size exists. In both cases, the degree of strengthening is approximately a linear function of the weight percent of the metal added to copper.

An interesting observation regarding the effect of having some dissimilar atoms in a lattice is illustrated by the alloy *Monel*. Nickel is harder than copper, but when the alloy containing the two metals is made, it is harder than nickel.

No single volume could contain the complete description of the composition, properties, and structures of ferrous alloys. Furthermore, the effect of heat treatment and other methods of changing the properties of alloys constitute an entire science unto itself. Accordingly, the description given of ferrous metallurgy will be only an overview of this enormously important area.

Steels constitute a wide range of iron-based alloys. General types include *carbon steels* (containing from 0.5 to 2.0% carbon) and only small amounts of other metals (generally less than 3–4%). Other metals in the alloys may include nickel, manganese, molybdenum, chromium, or vanadium in varying amounts. These ingredients are added to produce a steel having the desired characteristics. The properties of the steel are determined by the composition as well as the heat treatment methods employed.

If the total amount of metals added to iron exceeds about 5%, the alloy is sometimes called a *high alloy* steel. Most stainless steels are in this category because the chromium content is between 10 and

25%, and some types also contain 4−20% nickel. Stainless steels, so-called because of their resistance to corrosion, are of several types. The form of iron having the *fcc* structure is known as γ-Fe or *austenite* and one type of stainless steel (which contains nickel) is known as *austenitic stainless steel* because it has the austenite (*bcc*) structure. *Martensitic* stainless steels have a structure that contains a body-centered tetragonal arrangement that results from rapidly quenching the austenite structure. In addition to these two types, *ferritic* stainless steel has a *bcc* structure and does not contain nickel. In addition to the stainless steels, a large number of alloys known as tool steels are important. As the name implies, these are special alloys that are used to make tools for cutting, drilling, and fabricating metal. These alloys commonly include some or all of the following in various amounts: Cr, Mn, Mo, Ni, W, V, Co, C, and Si. In many cases, the alloys are engineered to have the desired properties of resistance to impact, heat, abrasion, corrosion, or thermal stress. Heat treatment of a steel having the desired composition can alter the structure of the metal so that certain properties are optimized. Thus, there are a large number of variables in the manufacture of steel. The manufacture of special steels is an important area of metallurgy that we may fail to appreciate fully when driving an automobile that has dozens of different alloys used in its construction.

Alloys that retain high strength at high temperatures (>1000 °C in some cases) are known as *superalloys*. Some of these materials are also highly resistant to corrosion (oxidation). These alloys are difficult to make, contain metals that are not readily available, and are expensive. They are used in situations where the conditions of service make them essential, for example, aircraft engines where certain designs require as much as 50% by weight of some of these special alloys.

The designation of an alloy as a superalloy is based on its strength at high temperatures. In the application of these alloys in fabricating a gas turbine, this is important because the efficiency of a turbine is greater at high temperatures. However, it is possible to study the sample for an extended period of time when it is subjected to a stress that is not sufficient to cause failure of the object. Even though the object may not break, it may elongate due to stretching of the metal. Movement of the metal under stress is called *creep*. Not only do the superalloys have high strength at elevated temperatures, they are resistant to creep, which makes them desirable for many uses.

Generally, superalloys alloys are given special names, and a few of the more common ones and their compositions are described in Table 11.4. Because several of the superalloys contain very little iron, they are closely related to some of the nonferrous alloys. Some of the second- and third-row transition metals possess many of the desirable properties of superalloys. They maintain their strength at high

Table 11.4 Composition of Some of the Superalloys	
Name	**Composition, Percent by Weight**
16-25-6	Fe 50.7; Ni 25; Cr 16; Mo 6; Mn 1.35; C 0.06
Haynes 25	Co 50; Cr 20; W 15; Ni 10; Fe 3; Mn 1.5; C 0.1
Hastelloy B	Ni 63; Mo 28; Fe 5; Co 2.5; Cr 1; C 0.05
Inconel 600	Ni 76; Cr 15.5; Fe 8.0; C 0.08
Astroloy	Ni 56.5; Cr 15; Co 15; Mo 5.5; Al 4.4; Ti 3.5; C 0.6; Fe <0.3
Udimet 500	Ni 48; Cr 19; Co 19; Mo 4; Fe 4; Ti 3; Al 3; C 0.08

temperatures, but they may be somewhat reactive with oxygen under these conditions. These metals are known as *refractory metals* and they include niobium, molybdenum, tantalum, tungsten, and rhenium.

11.9 CHEMISTRY OF TRANSITION METALS

Although much of the chemistry of transition metals is associated with coordination compounds, there are some important aspects of their behavior that are related to other types of compounds. In this section, a brief overview of the chemistry of transition metals will be given with emphasis on the first-row metals.

11.9.1 Transition Metal Oxides and Related Compounds

The reaction between a transition metal and oxygen frequently yields a product that may not be strictly stoichiometric. Part of the reason for this was shown in Chapter 8 when the dynamic nature of the reaction of a metal with a gas was considered. Moreover, it is frequently found that transition metals can exist in more than one oxidation state so mixed oxides are possible. The product that is in contact with oxygen is normally that in which the *higher* oxidation state is found.

Although scandium is becoming an increasingly important metal for fabricating objects combining light weight and strength, the oxide Sc_2O_3 is not particularly useful. In contrast, TiO_2 is an ingredient in many types of paint because it has a bright white color, is opaque, and has low toxicity. It is found as the mineral *rutile* and of course it has that crystal structure. Titanium also forms complex anions in the form of ternary oxides. Most notable of this type of compound is $CaTiO_3$, which is *perovskite*. This crystal structure was discussed in Chapter 7, and it is one of the most important structural types for ternary oxides. Another series of titanates has the formula M_2TiO_4, in which M is a +2 metal.

There are several ways in which titanium can be produced. For example, reduction of TiO_2 gives the metal,

$$TiO_2 + 2\,Mg \rightarrow 2\,MgO + Ti \tag{11.59}$$

It can also be produced from *ilmenite*, $FeTiO_3$, by the reaction

$$2\,FeTiO_3 + 6\,C + 7\,Cl_2 \rightarrow 2\,TiCl_4 + 2\,FeCl_3 + 6\,CO \tag{11.60}$$

The metal is then obtained by means of the reaction

$$TiCl_4 + 2\,Mg \rightarrow 2\,MgCl_2 + Ti \tag{11.61}$$

In accord with the +4 oxidation state of Ti and subsequent covalence, TiO_2 behaves as an acidic oxide, as illustrated by the following reaction:

$$CaO + TiO_2 \rightarrow CaTiO_3 \tag{11.62}$$

Vanadium forms a series of oxides, some of which have the formulas VO, V_2O_3, VO_2, and V_2O_5. The most important of the oxides is V_2O_5, and its most important use is as a catalyst for the oxidation of SO_2 to SO_3 in the production of sulfuric acid. However, this is not the only process in which V_2O_5 is an effective catalyst so it is used in this way in numerous other types of reactions. Although V_2O_3 reacts as a basic oxide, V_2O_5 is an acidic oxide that gives rise to numerous "vanadates" of different composition. Some of these can be shown by the following simplified equations:

$$V_2O_5 + 2\,NaOH \rightarrow 2\,NaVO_3 + H_2O \qquad (11.63)$$

$$V_2O_5 + 6\,NaOH \rightarrow 2\,Na_3VO_4 + 3\,H_2O \qquad (11.64)$$

$$V_2O_5 + 4\,NaOH \rightarrow Na_4V_2O_7 + 2\,H_2O \qquad (11.65)$$

The phosphorus(V) oxide written in simplest form is P_2O_5 so it should be expected that there would be considerable similarity between the various "phosphates" and the "vanadates". This is precisely the case, and the various forms of "vanadate" include VO_4^{3-}, $V_2O_7^{4-}$, $V_3O_9^{3-}$, HVO_4^{2-}, $H_2VO_4^{-}$, H_3VO_4, $V_{10}O_{28}^{6-}$, and others. These species illustrate the fact that vanadium is similar in some ways to phosphorus, which is also a Group V element. The numerous vanadate species can be seen to result from reactions such as the hydrolysis reaction,

$$VO_4^{3-} + H_2O \rightarrow VO_3(OH)^{2-} + OH^- \qquad (11.66)$$

The equilibria among the species are dependent on the pH of the solution in much the same way as the equilibria among the various phosphate species (see Chapter 14). Not only is the formation of polyvanadates reminiscent of phosphorus chemistry, so is the fact that vanadium forms oxyhalides, such as OVX_3 and VO_2X.

   Chromite, $Fe(CrO_2)_2$ (which can also be written as $FeO \cdot Cr_2O_3$), is the principal ore containing chromium. After obtaining the oxide, chromium is prepared by the reduction of Cr_2O_3 with aluminum or silicon.

$$Cr_2O_3 + 2\,Al \rightarrow 2\,Cr + Al_2O_3 \qquad (11.67)$$

Sodium chromate is produced as part of the process for obtaining Cr_2O_3, and it is probably the most important chromium compound. Although there are other oxides of chromium, Cr_2O_3 is very important because of its catalytic properties. One way to obtain this oxide is by the decomposition of ammonium dichromate,

$$(NH_4)_2Cr_2O_7 \rightarrow Cr_2O_3 + N_2 + 4\,H_2O \qquad (11.68)$$

or by heating a mixture of NH_4Cl and $K_2Cr_2O_7$.

$$2\,NH_4Cl + K_2Cr_2O_7 \rightarrow Cr_2O_3 + N_2 + 4\,H_2O + 2\,KCl \qquad (11.69)$$

Because of its green color, Cr_2O_3 has been used as a pigment, and it is an amphoteric oxide, as illustrated by the following equations:

$$Cr_2O_3 + 2\,NaOH \rightarrow 2\,NaCrO_2 + H_2O \qquad (11.70)$$

$$Cr_2O_3 + 6\,HCl \rightarrow 3\,H_2O + 2\,CrCl_3 \tag{11.71}$$

Another oxide of chromium is CrO_2, which owing to the similar size of Ti^{4+} and Cr^{4+} has the rutile structure. CrO_2 is a magnetic oxide that is utilized in magnetic tapes. Because it contains chromium in the $+6$ oxidation state, CrO_3 is a strong oxidizing agent that causes the ignition of some organic materials. It is produced by the reaction of $K_2Cr_2O_7$ and sulfuric acid.

$$K_2Cr_2O_7 + H_2SO_4(conc) \rightarrow K_2SO_4 + H_2O + 2\,CrO_3 \tag{11.72}$$

and, as would be expected given the $+6$ oxidation state, it is an acidic oxide,

$$CrO_3 + H_2O \rightarrow H_2CrO_4 \tag{11.73}$$

Chromium(VI) compounds include those that contain the yellow chromate (CrO_4^{2-}) and the orange dichromate $(Cr_2O_7^{2-})$ that are versatile oxidizing agents in many types of syntheses. Potassium dichromate is a primary standard in analytical chemistry, and it is a frequently employed oxidizing agent in redox titrations in which Cr^{3+} is the reduction product. In aqueous solutions, there is an equilibrium between CrO_4^{2-} and $Cr_2O_7^{2-}$ that depends on the pH of the solution. As a result of the reaction

$$2\,CrO_4^{2-} + 2\,H^+ \rightleftarrows Cr_2O_7^{2-} + H_2O \tag{11.74}$$

basic solutions are yellow, but acidic solutions are orange. In addition to being widely used oxidizing agents, chromates and dichromates are used as pigments and dyes, and in a process for tanning leather known as the *chrome process*.

When Cr^{3+} is reduced in aqueous solutions, the product is the aqua complex $[Cr(H_2O)_6]^{2+}$, which has an intense blue color. The reduction of Cr^{3+} is easily carried out by zinc and hydrochloric acid. When the solution containing Cr^{2+} is added to one containing sodium acetate, a brick red precipitate of $Cr(C_2H_3O_2)_2$ forms. This is unusual in the fact that there are few insoluble acetates. Because Al^{3+} and Cr^{3+} have similar charge-to-size ratios, there are some similarities between the chemical behaviors of the ions.

There are three common oxides of manganese, MnO, Mn_2O_3 (*hausmannite*), and MnO_2, which occurs naturally as the mineral *pyrolusite*. Reducing MnO_2 with hydrogen leads to the formation of MnO.

$$MnO_2 + H_2 \rightarrow MnO + H_2O \tag{11.75}$$

Decomposition of many hydroxides leads to the formation of oxides and such is the case with $Mn(OH)_2$.

$$Mn(OH)_2 \rightarrow MnO + H_2O \tag{11.76}$$

Although Mn_2O_3 can be prepared by the oxidation of the metal, the most important oxide of manganese is MnO_2. A common preparation of chlorine in laboratory experiments involves MnO_2 as the oxidizing agent.

$$MnO_2 + 4\,HCl \rightarrow Cl_2 + MnCl_2 + 2\,H_2O \tag{11.77}$$

The oxide Mn_2O_7 is a dangerous compound that detonates and reacts explosively with reducing agents including many organic compounds. Manganese also forms oxyanions, the most common of which is the permanganate ion, MnO_4^-. Permanganates are commonly used as oxidizing agents in many types of reactions. The manganese oxides show a transition from basic character for MnO to acidic character for oxides that contain the metal in higher oxidation states. Because of its deep purple color and the fact that the reduction product in acidic solutions, Mn^{2+}, is almost colorless, it serves as its own indicator in titrations. When MnO_4^- reacts as an oxidizing agent in acidic solutions, Mn^{2+} is the reduction product, but in basic solutions, MnO_2 is the reduction product.

For many years, one use of MnO_2 has been in the construction of batteries. In the "dry cell" battery, the anode is made of zinc and the oxidizing agent is MnO_2 with the electrolyte being a paste containing NH_4Cl and $ZnCl_2$. In the "alkaline" battery, the electrolyte is a paste of KOH. The reactions that occur in an alkaline battery are as follows:

$$Zn(s) + 2\,OH^-(aq) \rightarrow Zn(OH)_2(s) + 2\,e^- \tag{11.78}$$

$$2\,MnO_2(s) + H_2O(l) + 2\,e^- \rightarrow Mn_2O_3(s) + 2\,OH^-(aq) \tag{11.79}$$

In the older dry cell, the acidic ammonium ion slowly attacks the metal container, which can lead to leakage.

For centuries, iron oxides have been utilized in diverse ways. First, it is the oxide that is reduced to obtain the metal itself. Second, iron oxides in several forms have been used as pigments for many centuries. The three most common iron oxides are FeO, Fe_2O_3, and Fe_3O_4 (which can be described as $FeO \cdot Fe_2O_3$). Although the formula for iron(II) oxide is written as FeO, there is usually a deficiency in iron, which is not surprising because this is the oxide in the lowest oxidation state. It is the oxide residing farthest from the oxygen in terms of the phases present (see Chapter 8). In Chapter 8, it was pointed out that many metal carbonates and oxalates decompose when heated to give the oxides. Such reactions provide a way to obtain FeO.

$$FeCO_3 \rightarrow FeO + CO_2 \tag{11.80}$$

$$FeC_2O_4 \rightarrow FeO + CO + CO_2 \tag{11.81}$$

The mineral *magnetite* is a naturally occurring form of Fe_3O_4 that has an inverse spinel structure as a result of both Fe^{2+} and Fe^{3+} being present.

Although the oxide containing Fe(III) is Fe_2O_3, it also occurs as $Fe_2O_3 \cdot H_2O$, which has the same composition as FeO(OH). This oxide reacts in an amphoteric manner with both acids and basic oxides, as illustrated by the following equations:

$$Fe_2O_3 + 6\,H^+ \rightarrow 2\,Fe^{3+} + 3\,H_2O \tag{11.82}$$

$$Fe_2O_3 + CaO \rightarrow Ca(FeO_2)_2 \tag{11.83}$$

$$Fe_2O_3 + Na_2CO_3 \rightarrow 2\,NaFeO_2 + CO_2 \tag{11.84}$$

The ferric ion has a rather high charge density, which causes solutions of ferric salts to be acidic as a result of the reaction

$$Fe(H_2O)_6^{3+} + H_2O \rightarrow H_3O^+ + Fe(H_2O)_5OH^{2+} \tag{11.85}$$

Only two oxides of cobalt have been characterized, CoO and Co_3O_4 (which is actually $Co^{II}Co_2^{III}O_4$). The latter has a structure in which Co^{2+} ions are located in tetrahedral holes and Co^{3+} ions are located in octahedral holes of a spinel structure. Decomposition of either $Co(OH)_2$ or $CoCO_3$ produces CoO, and decomposition of $Co(NO_3)_2$ can be used to produce Co_3O_4.

$$3\,Co(NO_3)_2 \rightarrow Co_3O_4 + 6\,NO_2 + O_2 \tag{11.86}$$

Decomposition of either nickel hydroxide or nickel carbonate yields NiO, the only oxide of nickel of any importance. However, two oxides of copper are known, Cu_2O and CuO. Of these, Cu_2O is the more stable, and it is the product when CuO is heated to very high temperature.

$$4\,CuO \rightarrow 2\,Cu_2O + O_2 \tag{11.87}$$

A well-known test for sugars is the *Fehling test*. When a basic solution containing Cu^{2+} reacts with a carbohydrate (a reducing agent), a red precipitate of Cu_2O is produced. This oxide has also been added to glass, to which it imparts a red color. Decomposition of cupric hydroxide or carbonate yields CuO. The mineral *malachite* has the composition $CuCO_3 \cdot Cu(OH)_2$, and it decomposes at moderate temperature according to the equation

$$CuCO_3 \cdot Cu(OH)_2 \rightarrow 2\,CuO + CO_2 + H_2O \tag{11.88}$$

This oxide is used in making blue and green colored glass and in glazes for pottery, but in recent years, the preparation of superconducting materials such as $YBa_2Cu_3O_7$ has become a serious interest. Other materials containing mixed oxides have also been produced.

The only oxide of Zn is the amphoteric ZnO, which can be obtained by the reaction of the elements. This oxide reacts as a base toward H^+,

$$ZnO + 2\,H^+ \rightarrow Zn^{2+} + H_2O \tag{11.89}$$

However, toward a basic oxide, ZnO reacts as an acidic oxide to produce an oxyanion known as a *zincate*. In molecular form, the equation can be written as

$$ZnO + Na_2O \rightarrow Na_2ZnO_2 \tag{11.90}$$

When a base reacts in aqueous solution, the reaction leads to a hydroxo complex and can be shown as follows:

$$ZnO + 2\,NaOH + H_2O \rightarrow Na_2Zn(OH)_4 \tag{11.91}$$

Formally, a zincate is equivalent to a hydroxo complex that has been dehydrated as illustrated by the reaction

$$Na_2Zn(OH)_4 \xrightarrow{\Delta} 2\,H_2O + Na_2ZnO_2 \tag{11.92}$$

The reactions with acid and base that show amphoteric behavior can be summarized as follows:

$$(11.93)$$

Metallic zinc dissolves readily in both acids and bases, as illustrated in the following equations:

$$Zn + H_2SO_4 \rightarrow ZnSO_4 + H_2 \tag{11.94}$$

$$Zn + 2\,NaOH + 2\,H_2O \rightarrow Na_2Zn(OH)_4 + H_2 \tag{11.95}$$

At room temperature, zinc oxide is white, but when heated it becomes yellow. A compound that changes color on heating is said to be *thermochromic*.

11.9.2 Halides and Oxyhalides

Titanium tetrachloride is an important compound, but neither the +2 or +3 chlorides of titanium are widely used. $TiCl_4$ can be obtained by the reactions

$$TiO_2 + 2\,C + 2\,Cl_2 \rightarrow TiCl_4 + 2\,CO \tag{11.96}$$

$$TiO_2 + 2\,CCl_4 \rightarrow TiCl_4 + 2\,COCl_2 \tag{11.97}$$

In many ways, $TiCl_4$ behaves as a covalent compound of a nonmetal. It is a strong Lewis acid that forms complexes with many types of Lewis bases, and it hydrolyzes in water. It also reacts with alcohols to yield compounds having the formula $Ti(OR)_4$. However, it is the behavior of $TiCl_4$ (reacting with $[Al(C_2H_5)_3]_2$) as a catalyst in the Ziegler–Natta polymerization of ethylene that is the most important use of the compound (see Chapter 22).

Halides of vanadium are known with the metal in the +2, +3, +4, and +5 oxidation states. As might be expected, the fluoride is the only well characterized vanadium(V) halide. It can be obtained by the reaction

$$2\,V + 5\,F_2 \rightarrow 2\,VF_5 \tag{11.98}$$

The tetrahalides undergo disproportionation reactions to give the more stable +3 and +5 compounds.

$$2\,VF_4 \rightarrow VF_5 + VF_3 \tag{11.99}$$

As mentioned in the previous section, transition metals in high oxidation states exhibit behavior that is similar to that of some nonmetals. Vanadium(V) does this with the formation of oxyhalides having the formulas VOX_3 and VO_2X.

The two most common series of chromium halides have the formulas CrX_2 and CrX_3 (where $X = F$, Cl, Br, or I). However, CrF_6, is also known. Compounds having the formula CrX_3 are Lewis acids, and they also form many coordination compounds. For example, $CrCl_3$ reacts with liquid ammonia to yield $CrCl_3 \cdot 6NH_3$ (which is written in standard notation as $[Cr(NH_3)_6]Cl_3$). The reactions illustrated by the following equations can be utilized to produce $CrCl_3$.

$$Cr_2O_3 + 3C + 3Cl_2 \rightarrow 2CrCl_3 + 3CO \tag{11.100}$$

$$2Cr_2O_3 + 6S_2Cl_2 \rightarrow 4CrCl_3 + 3SO_2 + 9S \tag{11.101}$$

Because the highest oxidation state of Cr is +6, it forms oxyhalides that have the formula CrO_2X_2 that are known as *chromyl* halides. However, in the case of the fluorine compound, $CrOF_4$ is also known as well as some oxyhalides having the formula $CrOX_3$. Some reactions that yield chromyl halides are the following:

$$CrO_3 + 2HCl(g) \rightarrow CrO_2Cl_2 + H_2O \tag{11.102}$$

$$3H_2SO_4 + CaF_2 + K_2CrO_4 \rightarrow CrO_2F_2 + CaSO_4 + 2KHSO_4 + 2H_2O \tag{11.103}$$

The latter equation can be modified to produce CrO_2Cl_2 by substituting KCl or NaCl for CaF_2. As would be expected, compounds such as CrO_2F_2 and CrO_2Cl_2 react vigorously with water or alcohols.

Fluoride compounds are known that contain Re in the +7 oxidation state and Tc in the +6 oxidation state, but in the case of manganese, the highest oxidation state is found in MnF_4. It is found that higher oxidation states are more frequently encountered in the heavier members within a group in the transition series. Manganese(VII) oxyhalides have been obtained, and although MnO_3F and MnO_3Cl are known, they are not important compounds. MnF_4 can be prepared by the reaction of the elements.

In spite of its having an electron configuration that suggests a possible oxidation state of +6, iron forms no halogen compounds in which it has an oxidation state higher than +3. The halogen compounds of iron consist of the series FeX_2 and FeX_3. However, Fe^{3+} is an oxidizing agent that reacts with I^- so FeI_3 decomposes to FeI_2 and I_2. The reaction

$$Fe + 2HCl(aq) \rightarrow FeCl_2 + H_2 \tag{11.104}$$

yields the dichloride, which is obtained upon evaporation as the tetrahydrate, $FeCl_2 \cdot 4H_2O$. Anhydrous $FeCl_2$ can be obtained by reaction of the metal with gaseous HCl.

$$Fe + 2HCl(g) \rightarrow FeCl_2(s) + H_2(g) \tag{11.105}$$

The trihalides of iron can be prepared by the general reaction

$$Fe + 3X_2 \rightarrow 2FeX_3 \tag{11.106}$$

None of the oxyhalides of iron have major uses, but $FeCl_3$ is a Lewis acid that functions as a catalyst for many reactions.

Although CoF_3 is known, the other halides of Co(III) are not stable because it is a strong oxidizing agent that is capable of oxidizing halide ions. In fact, even the fluoride is so reactive that it is sometimes used as a fluorinating agent. The Lewis acidity of Co^{3+} provides the basis for the enormous number of coordination compounds that exist containing cobalt.

Nickel halides are limited to the series NiX_2 with X being F, Cl, Br, or I. Although the heavier metals of the groups have been omitted from consideration when discussing other types of compounds, it should be mentioned that tetrafluorides are known for Pd and Pt. This shows the general tendency of the heavier metals in the group to form compounds in which the metal has a higher oxidation state than is exhibited by the first member of the group. It should be mentioned that PtF_6 is such a strong oxidizing agent that it reacts with O_2 to produce the O_2^+ ion, and it also was the reactant that gave the first xenon compound.

Halides of copper include the series containing Cu(I) and Cu(II), except for the absence of CuF and CuI_2 (which is unstable because of Cu^{2+} being an oxidizing agent and I^- being a reducing agent). As a result, the reaction between Cu^{2+} and I^- takes place as shown in the following equation:

$$2\,Cu^{2+} + 4\,I^- \rightarrow 2\,CuI + I_2 \tag{11.107}$$

As expected based on the $3d^{10}\,4s^2$ configuration, zinc routinely forms +2 compounds, and all of the halides having the formula ZnX_2 are known. Anhydrous $ZnCl_2$, which has applications in the textile industry, can be prepared as shown by the following equation:

$$Zn + 2\,HCl(g) \rightarrow ZnCl_2(s) + H_2(g) \tag{11.108}$$

Evaporation of a solution produced by dissolving Zn in aqueous hydrochloric acid gives $ZnCl_2 \cdot 2H_2O$ as the solid product. Heating this compound does not result in the formation of anhydrous zinc chloride because of the reaction

$$ZnCl_2 \cdot 2H_2O(s) \rightarrow Zn(OH)Cl(s) + HCl(g) + H_2O(g) \tag{11.109}$$

This is similar to the behavior of aluminum halides discussed earlier in this chapter, and it illustrates the fact that dehydration of a hydrated solid cannot be used as a way to prepare anhydrous halide compounds in some instances.

The brief discussion of the chemistry of the first-row transition metals presented here shows only a small portion of this vast subject. However, it illustrates some of the differences between the metals and how their chemistry varies throughout the series. For additional details, the reference text by Greenwood and Earnshaw (1997) or that by Cotton, et al. (1999) should be consulted. In Chapters 16−22, many other aspects of the organometallic and coordination chemistry of these metals will be presented.

11.10 THE LANTHANIDES

Electrons generally fill orbitals in atoms as the sum $(n + l)$ increases. Therefore, after the $6s$ orbital (for which $n + l = 6$) is filled at Ba, it would be expected that the next orbitals populated would be those for which the sum $(n + l) = 7$ but having the lowest n giving that sum. The corresponding orbitals would be the 4f set. However, lanthanum ($Z = 57$) has the electron configuration (Xe) $5d^1 6s^2$, indicating that the 5d orbitals have started filling before the 4f. To further complicate the situation, element 58, cerium, does not have the configuration $5d^2 6s^2$ but rather it is $4f^1 5d^1 6s^2$. After that, the number of electrons in the 4f level increases regularly until $Z = 63$, europium, which has the configuration $4f^7 6s^2$. Gadolinium ($Z = 64$) has the configuration $4f^7 5d^1 6s^2$ as a result of the stability of the half-filled 4f shell, but at terbium, filling the 4f level resumes with the 5d shell being empty. Following terbium, the 4f level has an increasing number of electrons until the shell is filled at ytterbium ($4f^{14} 6s^2$). Lutetium has the additional electron in the 5d level to give the configuration $4f^{14} 5d^1 6s^2$. Except for minor irregularities, the 4f level is filled with 14 electrons for the elements cerium through lutetium. In referring to a lanthanide without indicating a specific atom, the generic Ln is often used. Table 11.5 shows relevant data for the lanthanide elements.

An interesting effect of the half-filled and filled 4f shell is shown when a graph is made of the melting points of the elements. Such a graph is shown in Figure 11.6. Although it is not shown, a plot of atomic radii for the metals shows a large increase in size for Eu and Yb. For example, the radii of Sm and Gd are approximately 180 pm, but Eu situated between them has a radius of 204 pm. The difference in size between Yb and the atoms before and after it also amounts to about 20 pm. Europium and ytterbium represent atoms having half-filled and filled 4f levels in addition to $6s^2$. One reason for the anomalous behavior of these metals is that other lanthanides behave as if they contribute three electrons to the conduction band and exist as $+3$ ions, but Eu and Yb behave as if they contribute only two electrons from the 6s level leaving the half-filled and filled 4f level intact with the metals having $+2$ charges. The weaker bonding between the metals is reflected by having lower melting points and larger sizes.

Table 11.5 Properties of the Lanthanides

Metal	Structure[a]	m.p. °C	Radius (+3) r, pm	$-\Delta H_{hyd}$	First Three Ionization Potential Sum, kJ/mol
Ce[a]	fcc	799	102	3370	3528
Pr	fcc	931	99.0	3413	3630
Nd	hcp	1021	98.3	3442	3692
Pm	hcp	1168	97.0	3478	3728
Sm	rhmb	1077	95.8	3515	3895
Eu	bcc	822	94.7	3547	4057
Gd	hcp	1313	93.8	3571	3766
Tb	hcp	1356	92.3	3605	3803
Dy[a]	hcp	1412	91.2	3637	3923
Ho	hcp	1474	90.1	3667	3934
Er	hcp	1529	89.0	3691	3939
Tm	hcp	1545	88.0	3717	4057
Yb[a]	fcc	824	86.8	3739	4186
Lu	hcp	1663	86.1	3760	3908

[a]Two or more forms are known.

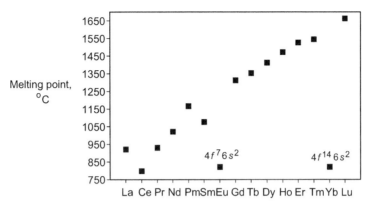

FIGURE 11.6
The melting points of the lanthanides. Note the dramatic effect that occurs for the half-filled and filled 4*f* shell.

One of the major themes in the chemistry of the lanthanides is their having a well-defined +3 oxidation state. However, with cerium having the configuration $5s^2\ 5p^6\ 4f^2\ 6s^2$, it is not surprising that the +2 and +4 oxidation states are most common for the element. Other members of the lanthanides also exhibit +2 and +4 oxidation states, although the +3 state is much more common. One of the interesting features of the lanthanide ions is the more or less regular decrease in the sizes of the +3 ions when progressing across the series, which gives rise to the term *lanthanide contraction*. This phenomenon, illustrated by the graph shown in Figure 11.7, is the result of the increasing nuclear charge coupled with ineffective screening of the 4*f* levels.

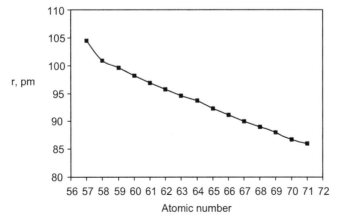

FIGURE 11.7
Radii of the +3 ions of the lanthanides as a function of atomic number. The coordination number is assumed to be 6.

Metal ions that have a high charge-to-size ratio (charge density) undergo hydrolysis to produce acidic solutions. As a result of the decrease in size of the +3 ions across the lanthanide series, there is a general increase in acidity of the aquated ions because of the reaction

$$[Ln(H_2O)_6]^{3+} + H_2O \rightleftarrows [Ln(H_2O)_5OH]^{2+} + H_3O^+ \tag{11.110}$$

One of the consequences of the lanthanide contraction is that some of the +3 lanthanide ions are very similar in size to some of the similarly charged ions of the second-row transition metals. For example, the radius of Y^{3+} is about 88 pm, which is approximately the same as the radius of Ho^{3+} or Er^{3+}. As shown in Figure 11.8, the heats of hydration of the +3 ions show clear indication of the effect of the lanthanide contraction.

The decrease in size of the +3 ions of the lanthanides is also a factor in stability of complexes of these ions, and complexes with a given ligand usually show an increase in stability in progressing through the series. However, with certain chelating agents, the stability increases up to a point and then levels off. Another facet of the coordination chemistry of the lanthanides is that coordination numbers greater than 6 are rather common. Because the +3 ions are rather large for ions having a positive charge that high (for comparison, Cr^{3+} has an ionic radius of about 64 pm and Fe^{3+} about 62 pm), the ligands frequently do not form a regular structure. However, the lanthanide ions are generally considered to be hard Lewis acids that interact preferentially with hard electron pair donors.

The lanthanides are rather reactive metals, and a good indication of their ease of oxidation and reduction can be seen by considering their reduction potentials. For comparison, the reduction potentials for magnesium and sodium are as follows:

$$Mg^{2+} + 2\,e^- \rightarrow Mg \quad E^\circ = -2.363 \text{ V} \tag{11.111}$$

$$Na^+ + e^- \rightarrow Na \quad E^\circ = -2.714 \text{ V} \tag{11.112}$$

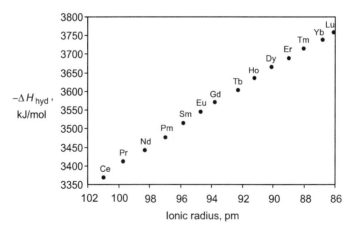

FIGURE 11.8
Heat of hydration of +3 lanthanide ions as a function of ionic radius.

For lanthanum, the reduction potential is

$$La^{3+} + 3e^- \rightarrow La \quad E° = -2.52\ V \tag{11.113}$$

and representative values for other lanthanides are Ce, -2.48; Sm, -2.40; Ho, -2.32; and Er, -2.30 V. Therefore, the lanthanides are quite reactive, and many of the reactions are readily predictable. For example, with halogens, the usual products are LnX_3. Oxidation (sometimes very vigorous!) leads to oxides having the formula Ln_2O_3, although some have complex structures. Some of the lanthanides are sufficiently reactive to replace hydrogen from water to produce the hydroxide, $Ln(OH)_3$. The hydrides are particularly interesting in that even those having the formula LnH_2 are believed to contain Ln^{3+} but with two H^- ions and a free electron that is located in a conduction band. When subjected to high pressure, additional hydrogen reacts to yield LnH_3, in which a third H^- ion is present. Sulfides, nitrides, and borides are also known for most of the lanthanides. Although details cannot be presented here, there is an enormous amount of chemistry known for the lanthanides.

References for Further Reading

Bailar, J.C., Emeleus, H.J., Nyholm, R., Trotman-Dickinson, A.F., 1973. *Comprehensive Inorganic Chemistry*, vol. 1. Pergamon Press, Oxford. The chemistry of metals is extensively covered in this five volume set.

Burdett, J.K., 1995. *Chemical Bonding in Solids*. Oxford University Press, New York. An advanced book that treats many of the aspects of structure and bonding in solids.

Cotton, F.A., Wilkinson, G., Murillo, C.A., Bochmann, M., 1999. *Advanced Inorganic Chemistry*, 6th ed. This book is the yardstick by which other books that cover the chemistry of the elements is measured. Several chapters present detailed coverage of the chemistry of metals.

Everest, D.A., 1964. *The Chemistry of Beryllium*. Elsevier Publishing Co., Amsterdam. A general survey of the chemistry, properties, and uses of beryllium.

Flinn, R.A., Trojan, P.K., 1981. *Engineering Materials and Their Applications*. Chapters 2, 5, and 6, 2nd ed. Houghton Mifflin Co., Boston. This book presents an excellent discussion of the structures of metals and the properties of alloys.

Greenwood, N.N., Earnshaw, A., 1997. *Chemistry of the Elements*. Chapters 20–29. Butterworth-Heinemann, Oxford. This book may well contain more descriptive chemistry than any other single volume, and it contains extensive coverage of transition metal chemistry.

Jolly, W.L., 1972. *Metal-Ammonia Solutions*. Dowden, Hutchinson & Ross, Inc., Stroudsburg, PA. A collection of research papers that serves as a valuable resource on all phases of the physical and chemical characteristics of these systems that involve solutions of Group IA and IIA metals.

King, R.B., 1995. *Inorganic Chemistry of the Main Group Elements*. VCH Publishers, New York. An excellent introduction to the descriptive chemistry of many elements. Chapter 10 deals with the alkali and alkaline earth metals.

Mingos, D.M.P., 1998. *Essential Trends in Inorganic Chemistry*. Oxford, New York. This books contains many correlations of data for metals.

Mueller, W.M., Blackledge, J.P., Libowitz, G.G., 1968. *Metal Hydrides*. Academic Press, New York. An advanced treatise on metal hydride chemistry and engineering.

Pauling, L., 1960. *The Nature of the Chemical Bond*, 3rd ed. Chapter 11. Cornell University Press, Ithaca, NY. This classic book contains a wealth of information about metals. Highly recommended.

Rappaport, Z., Marck, I., 2006. *The Chemistry of Organolithium Compounds*. Wiley, New York.

Wakefield, B.J., 1974. *The Chemistry of Organolithium Compounds*. Pergamon Press, Oxford. A book that provides a survey of the older literature and contains a wealth of information.

West, A.R., 1988. *Basic Solid State Chemistry*. John Wiley, New York. A very readable book that is an excellent place to start in a study of metals and other solids.

QUESTIONS AND PROBLEMS

1. Predict which of the following pairs of metals would be completely miscible and explain your line of reasoning.
   (a) Au/Ag (b) Al/Ca (c) Ni/Al (d) Ti/Al (e) Ni/Co (f) Cu/Mg

2. Cadmium and bismuth form a single liquid phase but are almost completely insoluble as solids. Explain this phenomenon.

3. Which would be more acidic, a 0.1 M solution of $Mg(NO_3)_2$ or a 0.1 M solution of $Fe(NO_3)_3$? Write equations to explain your answer.

4. On the basis of their properties, explain why separation of the lanthanides is possible but difficult.

5. On the basis of properties, explain why is yttrium frequently found with the lanthanide elements.

6. Iron(III) chloride boils at 315 °C. The density of the vapor decreases as temperature is increased above the boiling point. Explain why this observation.

7. When $CoCl_2 \cdot 6H_2O$ is heated, anhydrous $CoCl_2$ is formed. However, this is not the case when $BeCl_2 \cdot 4H_2O$ is heated. Explain the difference and write equations for the processes described.

8. The formation of the ion pair Na^+ and Na^- was described in the text. Comment on the possibility of forming the Mg^{2+} and Mg^{2-} ion pair.

9. Using the data given in Table 11.2, calculate the density of manganese.

10. Complete and balance the following assuming that heat is applied in reactions of solids.
    (a) $ZnCO_3 \cdot Zn(OH)_2(s) \rightarrow$
    (b) $Cr_2O_7^{2-} + Cl^- + H^+ \rightarrow$
    (c) $CaSO_3(s) \rightarrow$
    (d) $MnO_4^- + Fe^{2+} + H^+ \rightarrow$
    (e) $ZnC_2O_4(s) \rightarrow$

11. Which would be more acidic, a 0.2 M solution of $Ho(NO_3)_3$ or one of $Nd(NO_3)_3$?

12. Explain why the solubility of copper in aluminum is very low.

13. Complete and balance the following assuming that heat is applied.
    (a) $MgO + TiO_2 \rightarrow$
    (b) $Ba + O_2 \rightarrow$
    (c) $(NH_4)_2Cr_2O_7 \rightarrow$
    (d) $Cd(OH)_3 \rightarrow$
    (e) $NaHCO_3 \rightarrow$

14. Complete and balance the following. Heat may be required in reactions that do not involve water or ammonia.
    (a) $CrOF_4 + H_2O \rightarrow$
    (b) $VOF_3 + H_2O \rightarrow$
    (c) $Ca(OH)_2 + ZnO \rightarrow$
    (d) $CrCl_3 + NH_3(l) \rightarrow$
    (e) $V_2O_5 + H_2 \rightarrow$

15. Determine the percent of free space in *bcc* structure.

16. Draw the structures for the ortho, pyro, and meta vanadate ions and give equations showing how they can be interconverted.

17. Which of the lanthanides would likely be the most reactive? Explain the basis for your answer.

18. Would it require more energy to remove an electron from a *6s*, *5d*, or *4f* orbital from an atom of a lanthanide element? Explain your answer and describe any special cases that may be encountered.

19. When iron is heated, it undergoes a change from the *bcc* structure to *fcc* at 910 °C. In these forms, the unit cells are 293 and 363 pm, respectively. By what percent does the volume change during the phase transition? What would be the change in density?

20. Although molten $CaCl_2$ is a good conductor of electricity, molten $BeCl_2$ is not. Explain this difference. Also explain why molten $BeCl_2$ becomes a much better conductor when NaCl is added to the liquid.

21. Write complete balanced equations to show the reactions of the following with water.
    (a) CaH_2 (b) BaO_2 (c) CaO (d) Li_2S (e) Mg_3P_2

22. Complete and balance the following assuming that the reactants may be heated.
    (a) $SrO + CrO_3 \rightarrow$
    (b) $Ba + H_2O \rightarrow$
    (c) $Ba + C \rightarrow$
    (d) $CaF_2 + H_3PO_4 \rightarrow$
    (e) $CaCl_2 + Al \rightarrow$

23. Explain why $Cd(OH)_2$ is a stronger base than $Zn(OH)_2$, which is amphoteric.

24. On the basis of atomic properties, explain why copper forms solid solutions for different solid solutions that contain up to Ni 100%, Al 17%, or Cr <1%.

25. Explain why the LnF_3 compounds have higher melting points than the corresponding chlorides.

26. In the CrO_4^{2-} ion, the Cr–O bond length is 166 pm. In the $Cr_2O_7^{2-}$ ion, there are Cr–O bonds that are 163 pm in length and others that are 179 pm. Draw structures for the ions and explain these observations.

27. Write complete balanced equations to show the reactions of the following with water.
    (a) KNH_2 (b) Al_4C_3 (c) Mg_3N_2 (d) Na_2O_2 (e) RbO_2

28. How would the acidity vary for VO, V_2O_3, and V_2O_5?

29. What is the electrostatic bond character of the Ca–O bonds in perovskite?

30. When beryllium chloride is heated with $LiBH_4$ in a sealed tube, BeB_2H_8 is produced. Draw a few possible structures for this compound and speculate on their stability.

CHAPTER 12

Organometallic Compounds of the Main Group Elements

Since the 1950s, the increase in emphasis on the chemistry of organometallic compounds has been one of the dominant changes in chemistry. It is probably not appropriate to say in *inorganic* chemistry because the effect has also been great on *organic* chemistry. Organometallic compounds have been known since the discovery of Zeise's salt, $K[Pt(C_2H_4)Cl_3]$, in 1827 and the preparation of metal alkyls by Sir Edward Frankland in 1849, but with the discovery of ferrocene and Ziegler–Natta polymerization in the early 1950s, organometallic chemistry took on a different level of importance. These discoveries were followed in 1955 by the preparation of dibenzenechromium. The level of importance to which organometallic chemistry has risen can be estimated by considering that the literature dealing with this type of chemistry has grown enormously. In addition to general reference works, numerous specialized monographs have been published, and there are journals devoted specifically to publishing articles on topics that deal with organometallic chemistry.

Organometallic chemistry cuts across discipline lines (if in fact there exist any discipline lines) and has a significant part in organic, inorganic, biochemistry, materials science, and chemical engineering. Because of the enormous scope of organometallic chemistry, this chapter is included to provide background in this area as it relates to main group metals in the periodic table. Chapters 21 and 22 deal in part with organometallic chemistry of transition metals, but there is certainly some overlap of the areas. However, because of the heavy reliance of the organometallic chemistry of transition metals on coordination chemistry, that area of organometallic chemistry is discussed after coordination chemistry (Chapters 16–20). Chapters 14 and 15 deal with elements of Groups IVA–VIIA, and some of the organometallic chemistry of the heavy elements in these groups is discussed in those chapters. However, zinc, cadmium, and mercury have electron configurations nd^{10} $(n+1)s^2$ and in many cases lose the *s* electrons. In that way, they resemble the metals in Group IIA so the organometallic chemistry of these elements will also be described briefly.

Although there are numerous organometallic compounds of all of the elements in any specific group, those of one are generally more important. For example, in Group IA, the organic compounds of lithium are more numerous than those of sodium or potassium. Accordingly, most of the discussion will focus on the one or two elements that have the most extensive organometallic chemistry.

Considering the breadth of the field and the voluminous literature that has developed, it is impossible to deal with details of organometallic chemistry in a single volume much less a chapter or two in an inorganic chemistry book of a general nature. Rather, we will focus on some general aspects of the field and concentrate primarily on a relative few topics that still portray the nature of this type of **375**

Inorganic Chemistry. DOI: http://dx.doi.org/10.1016/B978-0-12-385110-9.00012-1

chemistry. We will begin by considering ways in which organometallic compounds are produced and then describe some types of reactions before considering compounds of specific elements. There will be some repetition in progressing in this way because some preparations and reactions will involve specific elements that will be subsequently described. However, this reinforcement is not a disadvantage from a pedagogical point of view.

12.1 PREPARATION OF ORGANOMETALLIC COMPOUNDS

Organometallic compounds vary widely in their properties and reactivity just as do the elements from which they are produced. These compounds may be lithium alkyls, Grignard reagents, or organotin compounds. Accordingly, there is no universal method for preparing the compounds, but we will present here some of the types of reactions that have been widely employed.

12.1.1 Reaction of Metals and Alkyl Halides

This technique is most appropriate when the metal is highly reactive. It should be kept in mind that even though a formula may be written as if the species is a monomer, several types of organometallic compounds are associated. Examples of this type of reaction are

$$2\,Li + C_4H_9Cl \rightarrow LiC_4H_9 + LiCl \tag{12.1}$$

$$4\,Al + 6\,C_2H_5Cl \rightarrow [Al(C_2H_5)_3]_2 + 2\,AlCl_3 \tag{12.2}$$

$$2\,Na + C_6H_5Cl \rightarrow NaC_6H_5 + NaCl \tag{12.3}$$

The most important reaction of this type is that in which Grignard reagents are produced.

$$RX + Mg \xrightarrow{\text{dry ether}} RMgX \tag{12.4}$$

Before its use as a fuel additive to reduce engine knock was banned in the United States, tetra-ethyllead was produced in enormous quantities. One method of producing the compound was the reaction between lead and ethyl chloride, in which the reactivity of lead was enhanced by its amalgamation with sodium.

$$4\,Pb/Na\,\text{amalgam} + 4\,C_2H_5Cl \rightarrow 4\,NaCl + Pb(C_2H_5)_4 \tag{12.5}$$

Mercury alkyls can be prepared by an analogous reaction in which an amalgam of sodium and mercury is used.

$$2\,Hg/Na\,\text{amalgam} + 2\,C_6H_5Br \rightarrow Hg(C_6H_5)_2 + 2\,NaBr \tag{12.6}$$

It is frequently found that the reactivity of a metal is higher when it is in contact with a different metal. This principle can be applied in other preparations such as the following:

$$2\,Zn/Cu + 2\,C_2H_5I \rightarrow Zn(C_2H_5)_2 + ZnI_2 + 2\,Cu \tag{12.7}$$

12.1.2 Alkyl Group Transfer Reactions

The reaction between a metal alkyl and a covalent halide of another element is sometimes made possible by the fact that transfer of the alkyl group leads to a crystalline product. An example of this type is the reaction between a sodium alkyl and a covalent halide such as $SiCl_4$.

$$4\,NaC_6H_5 + SiCl_4 \rightarrow Si(C_6H_5)_4 + 4\,NaCl \tag{12.8}$$

The formation of sodium chloride is a strong driving force in this reaction. The hard–soft interaction principle (see Chapter 9) is convenient in this case because of the favorable interaction of Na^+ with Cl^-. Other examples of this type of reaction are as follows:

$$2\,Al(C_2H_5)_3 + 3\,ZnCl_2 \rightarrow 3\,Zn(C_2H_5)_2 + 2\,AlCl_3 \tag{12.9}$$

$$2\,Al(C_2H_5)_3 + 3\,Cd(C_2H_3O_2)_2 \rightarrow 3\,Cd(C_2H_5)_2 + 2\,Al(C_2H_3O_2)_3 \tag{12.10}$$

Because mercury is easily reduced, dialkylmercury compounds are useful reagents for preparing a large number of alkyls of other metals by group transfer reactions. This is illustrated by the following equations:

$$3\,Hg(C_2H_5)_2 + 2\,Ga \rightarrow 2\,Ga(C_2H_5)_3 + 3\,Hg \tag{12.11}$$

$$Hg(CH_3)_2 + Be \rightarrow Be(CH_3)_2 + Hg \tag{12.12}$$

$$3\,HgR_2 + 2\,Al \rightarrow 2\,AlR_3 + 3\,Hg \tag{12.13}$$

$$Na(excess) + HgR_2 \rightarrow 2\,NaR + Hg \tag{12.14}$$

In the last reaction, low-boiling hydrocarbons are used as a solvent, and because the sodium alkyls are predominantly ionic, they are relatively insoluble. Alkyls of other Group IA metals (shown as M in the equation) can also be produced by this type of reaction, in which benzene is a frequently used solvent.

$$HgR_2 + 2\,M \rightarrow 2\,MR + Hg \tag{12.15}$$

In some cases, one organic group in one metal alkyl will undergo a substitution reaction to produce another metal alkyl.

$$NaC_2H_5 + C_6H_6 \rightarrow NaC_6H_5 + C_2H_6 \tag{12.16}$$

12.1.3 Reaction of a Grignard Reagent with a Metal Halide

Although Grignard reagents generally react by transferring alkyl groups and that was the general subject of the previous section, the reactions are so important that they will be described alone in this section. This is one of the most widely applicable ways in which metal alkyls are obtained. The following are typical reactions of this type:

$$3\,C_6H_5MgBr + SbCl_3 \rightarrow Sb(C_6H_5)_3 + 3\,MgBrCl \tag{12.17}$$

$$2\,CH_3MgCl + HgCl_2 \rightarrow Hg(CH_3)_2 + 2\,MgCl_2 \tag{12.18}$$

$$2\,C_2H_5MgBr + CdCl_2 \rightarrow Cd(C_2H_5)_2 + 2\,MgBrCl \tag{12.19}$$

Note that 2 MgBrCl "units" are formally equivalent to $MgBr_2 + MgCl_2$. Thus, we will continue to write the formula as MgBrCl to simplify the equations, although the products may simply be mixtures of the two magnesium halides.

The reaction

$$C_2H_5ZnI + n\text{-}C_3H_7MgBr \rightarrow n\text{-}C_3H_7ZnC_2H_5 + MgBrI \tag{12.20}$$

gives a product that contains two different alkyl groups. It is generally found that over time compounds of this type undergo a reaction to yield two products containing the same alkyl groups, ZnR_2 and ZnR'_2.

$$2\,n\text{-}C_3H_7ZnC_2H_5 \rightarrow Zn(n\text{-}C_3H_7)_2 + Zn(C_2H_5)_2 \tag{12.21}$$

The reactions of Grignard reagents constitute an enormously important type of synthetic chemistry, and they can be used in a wide variety of cases.

12.1.4 Reaction of an Olefin with Hydrogen and a Metal

In some cases, it is possible to synthesize a metal alkyl directly from the metal. An important case of this type is

$$2\,Al + 3\,H_2 + 6\,C_2H_4 \rightarrow 2\,Al(C_2H_5)_3 \tag{12.22}$$

A different type of direct synthesis involves the reaction of the metal with an alkyl halide.

$$2\,CH_3Cl + Si \xrightarrow[\;300\,°C\;]{Cu} (CH_3)_2SiCl_2 \tag{12.23}$$

12.2 ORGANOMETALLIC COMPOUNDS OF GROUP IA METALS

There is an enormous organometallic chemistry associated with the Group IA metals, particularly lithium and sodium. Lithium alkyls can be prepared by the reaction of the metal and an alkyl halide,

$$2\,Li + RX \rightarrow LiR + LiX \tag{12.24}$$

For this process, suitable solvents include hydrocarbons, benzene, and ether. Lithium alkyls can also be prepared by the reaction of the metal with a mercury alkyl,

$$2\,Li + HgR_2 \rightarrow 2\,LiR + Hg \tag{12.25}$$

Compounds of lithium with aryl groups can be prepared by the reaction of butyllithium with an aryl halide.

$$LiC_4H_9 + ArX \rightarrow LiAr + C_4H_9X \tag{12.26}$$

Lithium reacts with acetylene in liquid ammonia solution to give the mono and dilithium acetylides, $LiC\equiv CH$ and $LiC\equiv CLi$, with the evolution of hydrogen, which illustrates the slight acidity of acetylene. One commercial use of $LiC\equiv CH$ is in one step of the synthesis of vitamin A.

Lithium alkyls are used in processes such as polymerization and in transfer of alkyl groups in many types of reactions. Some examples are as follows:

$$BCl_3 + 3\,LiR \rightarrow 3\,LiCl + BR_3 \tag{12.27}$$

$$SnCl_4 + LiR \rightarrow LiCl + SnCl_3R(\text{and other products}) \tag{12.28}$$

$$3\,CO + 2\,LiR \rightarrow 2\,LiCO + R_2CO \tag{12.29}$$

One of the most interesting organometallic compounds of a transition metal is ferrocene (see Chapter 21). Butyllithium reacts with ferrocene to produce the mono- and dilithiated compounds that have the structures shown below.

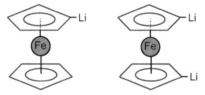

These reactive compounds are useful for preparing numerous other derivatives of ferrocene. As would be expected, lithium alkyls react with any trace of moisture.

$$LiR + H_2O \rightarrow LiOH + RH \tag{12.30}$$

These extremely reactive compounds are also spontaneously flammable in air.

A great deal of effort has been directed to determining the structures of lithium alkyls. It has been determined that in hydrocarbon solutions the dominant species is a hexamer when the alkyl groups are small. In the solid phase, the structure is body centered cubic with the $(LiCH_3)_4$ units at each lattice site. Each unit is a tetramer in which the four lithium atoms reside at the corners of a tetrahedron and the methyl groups are located above the centers of the triangular faces. The carbon atoms of the alkyl groups are bound to the three lithium atoms at the corners of the triangle. This results in a structure that can be shown (with only two methyl groups included) as

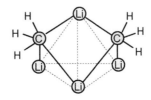

Bonding of the methyl group to three lithium atoms involves an sp^3 orbital on the methyl group simultaneously overlapping three orbitals (they may be $2s$ or hybrids of $2s$ and $2p$ orbitals) on the Li atoms. This can be represented as

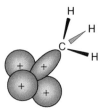

Although this picture of the bonding in methyllithium is generally adequate, it is probably an over-simplification of the true situation. Three-center two-electron bonds exist in numerous types of compounds (such as diborane), but it may be in this case that there is also some weak interaction between the lithium atoms.

12.3 ORGANOMETALLIC COMPOUNDS OF GROUP IIA METALS

Although some organometallic compounds of calcium, barium, and strontium are known, they are far less numerous and important than those of beryllium and magnesium. One of the most important discoveries in organometallic chemistry is that made in 1900 by Victor Grignard. That work is of enormous importance because it led to the chemistry of a class of compounds that are now referred to as Grignard reagents, and they are prepared by the reaction of magnesium with an alkyl halide in a solution of dry ether. The process can be represented by the equation

$$Mg + RX \xrightarrow{\text{Dry ether}} RMgX \tag{12.31}$$

Evaporation of the excess ether results in the formation of an "etherate" containing the solvent bonded to the magnesium. Generally, the formula for this product is represented as $RMgX \cdot 2R_2O$, and the structure is a distorted tetrahedron. However, even in solution, "RMgX" exists as aggregates with the dimer being the dominant species. The equilibrium

$$2\,RMgX \rightleftarrows (RMgX)_2 \tag{12.32}$$

can involve other species and the composition depends on the nature of the alkyl group, which can function as an electron pair donor; the association is hindered by complex formation between the solvent and RMgX. Moreover, the situation is complicated by the equilibrium

$$2\,RMgX \rightleftarrows MgR_2 + MgX_2 \tag{12.33}$$

which is known as the Schlenk equilibrium. The dimer, $(RMgX)_2$, is believed to have a structure that can be shown as

$$R-Mg \overset{\displaystyle X}{\underset{\displaystyle X}{\diagup\!\!\!\diagdown}} Mg-R$$

although the dimer can also be represented by the structure

There may also be some ionization that can be represented as

$$2\ RMgX \rightleftarrows RMg^+ + RMgX_2^- \tag{12.34}$$

The nature of the species present in a solution of a Grignard reagent is a complex issue with no single representation showing the true situation.

Because of the enormous importance of Grignard reagents in synthetic chemistry (organic, inorganic, and organometallic), an entire volume could be written on the subject. Shown here are only a very few of the myriad possibilities for reactions of these versatile compounds. Normally, the reactivity of Grignard reagents varies with the nature of the halogen as follows: $I > Br > Cl$. It is also found that the alkyl compounds are more reactive than the aryl compounds. One of the important types of reactions of Grignard reagents is that of lengthening a carbon chain in the reaction with a primary alcohol.

$$ROH\ +\ CH_3MgBr \rightarrow RCH_3\ +\ Mg(OH)Br \tag{12.35}$$

When a secondary alcohol is used, the reaction can be shown as

$$RR'HCOH\ +\ CH_3MgBr \rightarrow RR'HC-CH_3\ +\ Mg(OH)Br \tag{12.36}$$

Grignard reagents react with formaldehyde to give (after adding HCl to the product) primary alcohols.

$$HCHO\ +\ RMgX \rightarrow RCH_2OH\ +\ MgClX \tag{12.37}$$

The reaction of a Grignard reagent with CO_2 can be used to prepare a carboxylic acid.

$$CO_2 + RMgX\ \xrightarrow{\ H_2O\ }\ RCOOH + MgXOH \tag{12.38}$$

Reaction of RMgX with RCHO leads to a secondary alcohol, whereas the reaction with an ester, RCOOR′, gives a tertiary alcohol after acidifying the products of the initial reactions. The reaction of a Grignard reagent with sulfur is complex, but it can be represented by the equation

$$8\ RMgX\ +\ S_8 \rightarrow 4\ R_2S\ +\ 4\ MgS\ +\ 4\ MgX_2 \tag{12.39}$$

These are but a few of the ways in which Grignard reagents are used in synthesis.

In addition to magnesium, there is an extensive chemistry of organoberyllium compounds. The alkyl compounds are obtained most conveniently by the reaction of beryllium chloride with a Grignard reagent.

$$BeCl_2\ +\ 2\ CH_3MgCl \rightarrow Be(CH_3)_2\ +\ 2\ MgCl_2 \tag{12.40}$$

Because dimethylberyllium is a Lewis acid, it remains attached to ether that is the solvent in the reaction. Beryllium alkyls can also be produced by the reaction of the chloride with a lithium alkyl,

$$BeCl_2\ +\ 2\ LiCH_3 \rightarrow Be(CH_3)_3\ +\ 2\ LiCl \tag{12.41}$$

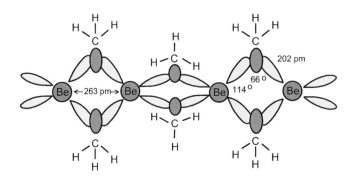

FIGURE 12.1
The structure of dimethylberyllium.

or by reaction with a dialkylmercury.

$$Be + Hg(CH_3)_2 \rightarrow Be(CH_3)_2 + Hg \tag{12.42}$$

As is the case with numerous other metal alkyls, beryllium alkyls are spontaneously flammable in air. Beryllium oxide is produced, and it has a heat of formation of -611 kJ/mol. Dimethylberyllium also reacts explosively with water, and some of its other properties resemble those of trimethylaluminum. This should not be surprising because the metals have a strong diagonal relationship that relates to their similar charge to size ratios.

The structure of dimethylberyllium is similar to that of trimethylaluminum except for the fact that the beryllium compound forms chains, whereas the aluminum compound forms dimers. Dimethylberyllium has the structure shown in Figure 12.1.

The bridges involve an sp^3 orbital on the methyl groups overlapping an orbital (probably best regarded as sp^3) on the beryllium atoms to give two-electron three-center bonds. Note, however, that the bond angle Be—C—Be is unusually small. Because dimethylberyllium is a Lewis acid, the polymeric $[Be(CH_3)_2]_n$ is separated when a Lewis base is added and adducts form. For example, with phosphine the reaction is

$$[Be(CH_3)_2]_n + 2\,nPH_3 \rightarrow n[(H_3P)_2Be(CH_3)_2] \tag{12.43}$$

An unusual type of organometallic compound of beryllium is that in which it is coordinated to a cyclopentadienyl ring.

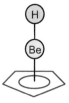

In addition to the hydrogen compound, others have been prepared that contain a halogen or a methyl group.

12.4 ORGANOMETALLIC COMPOUNDS OF GROUP IIIA METALS

The organometallic chemistry of other members of Group IIIA is relatively much less important than that of aluminum. There is an extensive organic chemistry of aluminum, and some of the compounds are commercially important. For example, triethylaluminum is used in the Ziegler–Natta process for polymerization of alkenes (see Chapter 22). As a result of extensive dimerization, aluminum alkyls have the general formula $[AlR_3]_2$. It is interesting that $B(CH_3)_3$ does not undergo molecular association and it has a boiling point of $-26\,°C$. Although it does not react with water, $B(CH_3)_3$ is spontaneously flammable in air. Trimethylgallium (b.p. $55.7\,°C$) and triethylgallium (b.p. $143\,°C$) show almost no tendency to form dimers under most conditions. In contrast to the behavior of trimethylboron and trimethylgallium, the boiling point of trimethylaluminum, which exists as $[Al(CH_3)_3]_2$, is $126\,°C$ and that of $[Al(C_2H_5)_3]$ is $186.6\,°C$, even though the aluminum compounds have lower molecular weights.

Aluminum alkyls react with many substances and are spontaneously flammable in air,

$$[Al(CH_3)_3]_2 + 12\,O_2 \rightarrow 6\,CO_2 + 9\,H_2O + Al_2O_3 \tag{12.44}$$

Aluminum alkyls can be prepared in several ways. In one process, aluminum reacts with alkyl halides to produce $R_3Al_2Cl_3$ (known as the sesquichloride),

$$2\,Al + 3\,RCl \rightarrow R_3Al_2Cl_3 \tag{12.45}$$

The product undergoes a redistribution to produce $R_4Al_2Cl_2$ and $R_2Al_2Cl_4$. The reaction of aluminum with HgR_2 results in transfer of alkyl groups,

$$6\,HgR_2 + 4\,Al \rightarrow 2\,[AlR_3]_2 + 6\,Hg \tag{12.46}$$

Mixed alkyl hydrides can be prepared by the reaction

$$2\,Al + 3\,H_2 + 4\,AlR_3 \rightarrow 6\,AlR_2H \tag{12.47}$$

The Al–H bond is sufficiently reactive that alkenes give an insertion reaction,

$$R_2AlH + C_2H_4 \rightarrow R_2AlC_2H_5 \tag{12.48}$$

As mentioned earlier, aluminum alkyls dimerize extensively. The structure of $[Al(CH_3)_2]$ is shown in Figure 12.2. In this case, an orbital on the methyl group overlaps with orbitals on two aluminum atoms to give three-center two-electron bonds. Although there are four bonds to each aluminum atom, the orientation of the bonds deviates considerably from typical tetrahedral cases. Because the distance between the aluminum atoms is relatively short, there may be some partial bonding between them. Note that the angle between the bonds to the two terminal CH_3 groups is quite close to that corresponding to sp^2 hybrid orbitals on the aluminum atoms. This would leave p orbitals available to form a σ bond between the aluminum atoms. It should also be noted that the $Al–C_t$ bonds are considerably shorter than the $Al–C_b$ bonds (b and t signify bridged and terminal, respectively).

The structure of the trimethylaluminum dimer has been discussed as if it were static, but that is not the case even at room temperature. When a solution of the compound in toluene is cooled to $-65\,°C$,

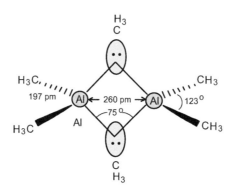

FIGURE 12.2
The structure of the trimethylaluminum dimer.

the proton nuclear magnetic resonance indicates that there are hydrogen atoms in two types of environments, bridging and terminal methyl groups. However, at room temperature, there is only one signal in the NMR indicating that the methyl groups exchange. It is believed that the rapid scrambling of methyl groups takes place by breaking one of the bridges followed by rotation of the $Al(CH_3)_3$ groups relative to each other with the bridge being reestablished to a different methyl group.

The association of aluminum alkyls provides a good opportunity to illustrate the principles discussed in Chapter 6 in regard to properties of liquids. Measurements of the molecular weights in benzene solutions indicate that the methyl, ethyl, and n-propyl compounds are completely dimerized. However, the heat of dissociation of $(AlR_3)_2$ dimers varies with nature of the alkyl group as shown by the following heats of dissociation:

$R =$	CH_3	C_2H_5	$n\text{-}C_3H_7$	$n\text{-}C_4H_9$	$i\text{-}C_4H_9$
ΔH_{diss}, kJ mol^{-1}	81.2	70.7	87.4	38	33

As was described in Chapter 6, the solubility parameter, δ, can be used as a diagnostic tool for studying the molecular association. Table 12.1 shows some of the relevant data for several aluminum

Table 12.1 Physical Data for Some Aluminum Alkyls

Compound[a]	Boiling Point, °C	ΔH_{vap}, kJ mol^{-1}	ΔS_{vap}, J mol^{-1}	δ, J$^{1/2}$ cm$^{-3/2}$
$Al(CH_3)_3$	126.0	44.92	112.6	20.82
$Al(C_2H_5)_3$	186.6	81.19	176.6	23.75
$Al(n\text{-}C_3H_7)_3$	192.8	58.86	126.0	17.02
$Al(i\text{-}C_4H_9)_3$	214.1	65.88	135.2	15.67
$Al(C_2H_5)_2Cl$	208.0	53.71	111.6	19.88
$Al(C_2H_5)Cl_2$	193.8	51.99	111.1	21.58

[a]Formula given is for the monomer, but the compound normally exist as dimers.

alkyls. The solubility parameters were calculated from vapor pressure data using the procedure described in Chapter 6.

As was described in Chapter 6, the entropy of vaporization is a worthwhile piece of evidence in studying the association that occurs in liquids and vapors. In the case of $[Al(C_2H_5)_3]_2$, the entropy of vaporization $(176.6\ J\ mol^{-1}\ K^{-1})$ is almost exactly twice the value of $88\ J\ mol^{-1}\ K^{-1}$ predicted by Trouton's rule,

$$\Delta S_{vap} = \frac{\Delta H_{vap}}{T} \approx 88\ J\ mol^{-1}\ K^{-1} \tag{12.49}$$

This indicates that in the process of vaporization 1 mol of liquid is converted into 2 mol of vapor. Therefore, we conclude that $[Al(C_2H_5)_3]_2$ dimers are present in the liquid, but the vapor consists of monomeric $Al(C_2H_5)_3$ units. Examination of the data for $[Al(CH_3)_3]_2$ shows an entropy of vaporization of $112.6\ J\ mol^{-1}\ K^{-1}$, which is greater than the value of $88\ J\ mol^{-1}\ K^{-1}$ predicted by Trouton's rule, but it is lower than twice the value. This value could be interpreted as corresponding to a liquid that is only *partially* dimerized being converted *completely* into monomer during vaporization.

However, there is a different explanation of the value for the entropy of vaporization of trimethylaluminum. If the liquid exists completely as dimers, the entropy of vaporization for $[Al(CH_3)_3]_2$ could be interpreted as indicating a liquid that is completely dimerized but which becomes only *partially* dissociated during vaporization. Because triethylaluminum fits unambiguously the case of a *completely* dimerized liquid that *completely* dissociates in the vapor, we look for other factors to consider in the trimethylaluminum case. One such property that is useful is the solubility parameter.

The solubility parameters for trimethylaluminum and triethylaluminum are 20.8 and 23.7 $J^{1/2}\ cm^{-3/2}$, respectively. These values are high enough to indicate association in the liquids, and they are quite comparable and in the expected order for compounds differing slightly only in molecular masses. From the entropy of vaporization, it is evident that triethylaluminum is associated in the liquid phase and that it is dissociated in the vapor. From the fact that the solubility parameters are so similar, it can be concluded that trimethylaluminum is also completely dimerized in the liquid phase. Therefore, because the entropy of vaporization is not *twice* the value expected on the basis of Trouton's rule, we conclude that trimethylaluminum is a completely dimerized liquid that only *partially* dissociates to monomers during vaporization. One reason for the difference between triethylaluminum and trimethylaluminum lies in the difference in their boiling points, which are 186.6 and $126.0\ °C$, respectively. This difference of $60\ °C$ is sufficient to cause complete dissociation of $[Al(C_2H_5)_3]_2$ during vaporization but only *partial* dissociation of $[Al(CH_3)_3]_2$ at its significantly lower boiling point.

Support for this conclusion is found by considering the cases of $Al(n\text{-}C_3H_7)_3$ and $Al(i\text{-}C_4H_9)_3$. For these compounds, the entropies of vaporization are 126.0 and $135.2\ J\ mol^{-1}\ K^{-1}$, respectively. These values are larger than the $88\ J\ mol^{-1}\ K^{-1}$ predicted by Trouton's rule so vaporization is accompanied by a change in the degree of molecular association. Even though the molecular weights are higher for the propyl and butyl derivatives, the solubility parameters of 17.0 and 15.7 $J^{1/2}\ cm^{-3/2}$ for $Al(n\text{-}C_3H_7)_3$ and $Al(i\text{-}C_4H_9)_3$ are considerably *smaller* than are the values for the methyl and ethyl compounds. Because the solubility parameter reflects the cohesion energy of the liquid, we conclude that $Al(n\text{-}C_3H_7)_3$ and $Al(i\text{-}C_4H_9)_3$ are only partially dimerized in the liquid phase. Thus, the high entropies of vaporization result from partially dimerized liquids being converted into vapors that are completely monomeric. Both compounds have boiling points (192.8 and $214.1\ °C$ for $Al(n\text{-}C_3H_7)_3$ and $Al(i\text{-}C_4H_9)_3$, respectively) that are higher than that of $[Al(C_2H_5)_3]_2$, which, as we have seen, dissociates completely during vaporization. Therefore, we can conclude that $[Al(n\text{-}C_3H_7)_3]_2$ and $[Al(i\text{-}C_4H_9)_3]_2$ should be completely dissociated in the *vapor* phase also, but they must be only partially associated in the *liquid* phase. The behavior of the lower aluminum alkyls can be summarized as shown below. These observations are for the pure compounds, not solutions in a solvent such as benzene where association is known to occur.

	Monomer Formula			
	$Al(CH_3)_3$	$Al(C_2H_5)_3$	$Al(n\text{-}C_3H_7)_3$	$Al(i\text{-}C_4H_9)_3$
Liquid:	Dimers	Dimers	Dimers + monomers	Dimers + monomers
Vapor:	Dimers + monomers	Monomers	Monomers	Monomers

Other types of association are found in other metal alkyls. For example, $Ga(CH_3)_3$ is a dimer that dissociates upon vaporization, but the analogous compounds containing larger alkyl groups are monomers.

The solubility parameters have been determined for several tetra-*n*-alkyl germanes for which the values are as follows (in $J^{1/2}$ cm$^{-3/2}$): $Ge(CH_3)_4$, 13.9; $Ge(C_2H_5)_4$, 17.6; $Ge(n\text{-}C_3H_7)_4$, 18.0; $Ge(n\text{-}C_4H_9)_4$, 20.3; and $Ge(n\text{-}C_5H_{11})_4$, 21.5. Note that all of these values are lower than those for $Al(CH_3)_3$ or $Al(C_2H_5)_3$. The conclusion is that the properties of the liquid tetraalkylgermanes indicate that the liquids are unassociated, a conclusion that is supported by other evidence.

We have already mentioned that the mixed alkyl halide compounds of aluminum dimerize and the solubility parameters shown in Table 12.1 for these compounds are consistent with that assessment. The solubility parameters for triethylboron and diethylzinc are 15.4 and 18.2 $J^{1/2}$ cm$^{-3/2}$, respectively. These values are indicative of liquids that are not strongly associated, which is known to be the case.

Other organometallic compounds of aluminum include the alkyl hydrides, R_2AlH. Molecular association of these compounds leads to cyclic tetramers. When the dimeric and trimeric compounds are dissolved in a basic aprotic solvent, the aggregates separate as a result of formation of bonds between Al and the unshared pair of electrons on the solvent molecule. Toward Lewis bases such as trimethylamine, aluminum alkyls are strong Lewis acids (as are aluminum halides).

$$[AlR_3]_2 \; + \; 2 \; :NR_3 \; \rightarrow \; 2 \; R_3Al{:}NR_3 \tag{12.50}$$

This is but a special case of the general principle that molecular association in solution is greatly affected by the nature of the solvent.

Aluminum alkyls figure prominently in polymerization processes and can themselves promote polymerization. In that connection, one of the important reactions of aluminum alkyls is that of adding across a double bond,

$$C_2H_4 \; + \; AlR_3 \; \rightarrow \; R_2AlCH_2CH_2R \tag{12.51}$$

Subsequent reaction in which another insertion occurs lengthens the hydrocarbon chain so polymerization occurs. In this type of process, the chains consist of a relatively few carbon atoms. When the reaction above continues in the case where $R = C_2H_5$, the products include $AlRR'R''$ in which differing numbers of carbon atoms may be present in the three alkyl groups.

The most important reaction of aluminum alkyls is in polymerization of alkenes by the Ziegler–Natta process. In that process, $Al(C_2H_5)_3$ reacts as an alkylating agent with $TiCl_4$, which then undergoes a reaction in which an alkene molecule is inserted in the bond between Ti and the ethyl group, thus lengthening the chain (see Chapter 22). Because polymers such as polyethylene and polypropylene are produced in enormous quantities and have many important uses, the Ziegler–Natta process has great industrial significance.

Aluminum alkyls undergo many reactions that are typical of covalent metal compounds. Compounds containing methyl, ethyl, and propyl groups ignite spontaneously in air. The reactions with water take place with explosive violence when the alkyl groups contain four or fewer carbon atoms.

$$Al(C_2H_5)_3 + 3\,H_2O \rightarrow Al(OH)_3 + 3\,C_2H_6 \tag{12.52}$$

The reaction of an aluminum alkyl with an alcohol is also violent, but it can be mediated by carrying out the reaction in a dilute solution of an inert solvent.

$$AlR_3 + 3\,R'OH \rightarrow Al(OR')_3 + 3\,RH \tag{12.53}$$

Aluminum alkyls also undergo reactions in which they function by transferring alkyl groups.

$$2\,Al(C_2H_5)_3 + 3\,ZnCl_2 \rightarrow Zn(C_2H_5)_2 + 2\,AlCl_3 \tag{12.54}$$

Reactions of this type can be carried out with other metal halides.

12.5 ORGANOMETALLIC COMPOUNDS OF GROUP IVA METALS

There is a well-developed organometallic chemistry of tin, and some of the compounds are used in large quantities in industrial processes. For example, compounds such as $[C_8H_{17}SnOC(O)CH{=}CHC(O)O]_n$ are used as stabilizers for PVC polymers, food packaging, protective coatings for wood, control of fungi in several types of food, and numerous other products. Tin compounds have very low toxicity so they are well suited for uses that bring them in contact with food products. Organotin compounds are important reagents for synthesizing many other compounds.

Alkyl tin compounds can be prepared by the reaction of $SnCl_4$ with a Grignard reagent,

$$SnCl_4 + 4\,RMgCl \rightarrow SnR_4 + 4\,MgCl_2 \tag{12.55}$$

Alkyl groups are also transferred from another metal alkyl, such as LiR or AlR_3.

$$4\,LiR + SnCl_4 \rightarrow SnR_4 + 4\,LiCl \tag{12.56}$$

$$4\,AlR_3 + 3\,SnCl_4 \rightarrow 3\,SnR_4 + 4\,AlCl_3 \tag{12.57}$$

The chemical behavior of tetraalkylstannanes is radically different from that of the far more reactive aluminum alkyls. The tin compounds are stable in air, they do not react with water, and do not form acid–base adducts readily. Tetramethylstannane (b.p. 26.5 °C) and tetraphenylstannane (b.p. 228 °C) are sufficiently stable and unreactive that they can be distilled without significant decomposition or oxidation.

In some cases, metallic tin reacts directly with an alkyl halide to produce mixed alkyl halide compounds.

$$Sn + RX \rightarrow R_2SnX_2 \,(\text{and other products}) \tag{12.58}$$

Both the dichlordimethyl silanes and germanes are much more reactive than the tetraalkyls and hydrolyze readily.

$$(CH_3)_2GeCl_2 + H_2O \rightarrow (CH_3)_2GeO + 2\,HCl \tag{12.59}$$

The great utility of dialkyldichlorosilane (b.p. 70 °C) lies in the fact that it is an intermediate in the production of silicon polymers.

$$x\,(CH_3)_2SiCl_2 + x\,H_2O \rightarrow [(CH_3)_2SiO]_x + 2x\,HCl \tag{12.60}$$

Alkyl tin compounds react with tin tetrahalides to produce a range of products that have the general formula R_nSnX_{4-n}. These compounds react with $LiAlH_4$ to produce tin hydrides. Other aspects of the organic chemistry of tin are present in Chapter 14.

12.6 ORGANOMETALLIC COMPOUNDS OF GROUP VA ELEMENTS

There are many organometallic compounds of arsenic, antimony, and bismuth known that constitute series having chemical properties that differ markedly. These compounds generally decrease in stability in the order As > Sb > Bi, which agrees with the increasing difference in size of the atoms and carbon atoms. Arsenic compounds include both aliphatic derivatives and heterocycles, such as arsabenzene,

The analogous bismuth compound is much less stable. The coordination chemistry of both six- and five-membered rings has been explored with one of the most unusual compounds being the complex $[Fe(AsC_4H_4)_2]$, which contains a five-membered AsC_4H_4 ring and has the structure of ferrocene. The trialkyl arsenic compounds have an unshared pair of electrons on the arsenic atoms so they function as Lewis bases toward metal ions. Consequently, there are many coordination compounds in which AsR_3 molecules are ligands. In this behavior, the decreasing tendency of the alkyls of Group VA elements to complex with metals decreases in the series As > Sb > Bi in accordance with the increasing metallic character of the elements.

Many organoarsenic compounds are prepared by reactions of $AsCl_3$ with alkyl group transfer agents such as Grignard reagents, lithium alkyls, or aluminum alkyls. Typical reactions include

$$3\,RMgBr + AsCl_3 \rightarrow AsR_3 + 3\,MgBrCl \tag{12.61}$$

$$AsCl_3 + 3\,LiR \rightarrow AsR_3 + 3\,LiCl \tag{12.62}$$

Arsenic dichloride will also react with molecules such as alcohols that have polar OH bonds.

$$AsCl_3 + 3\,ROH \rightarrow As(OR)_3 + 3\,HCl \qquad (12.63)$$

Another organic compound of arsenic that is of historical significance was discovered by P. Ehrlich in 1901. That compound is known as arsphenamine or Salvarsan, which has the structure

This compound was found to be effective in treating syphilis and African sleeping sickness, although more effective drugs soon became available. In addition to the name Salvarsan, this compound is listed in the *Handbook of Chemistry and Physics* as "606" to denote where it fell in the series of compounds tested by Ehrlich and his collaborators.

12.7 ORGANOMETALLIC COMPOUNDS OF Zn, Cd, AND Hg

Although zinc, cadmium, and mercury are not members of the so-called main group elements, their behavior is very similar because of their having complete *d* orbitals that are not normally used in bonding. By having the filled *s* orbital outside the closed *d* shell, they resemble the Group IIA elements. Zinc is an essential trace element that plays a role in the function of carboxypeptidase A and carbonic anhydrase enzymes (see Chapter 23). The first of these enzymes is a catalyst for the hydrolysis of proteins, whereas the second is a catalyst for the equilibrium involving carbon dioxide and bicarbonate,

$$H_2O + CO_2 \rightleftarrows HCO_3^- + H^+ \qquad (12.64)$$

This process is involved in the removal of CO_2 by blood as bicarbonate is formed and the release of CO_2 in the lungs. Cadmium and mercury are highly toxic. Heavy metals (those that are soft as a result of large size and low charge) are toxic with the toxicity increasing as the softness increases. On the other hand, metals that are considered hard (such as Mg^{2+}, Ca^{2+}, Fe^{3+}) are not generally considered to be toxic. In one mode of toxic action, heavy metals bond to groups such as $-SH$ that are present in proteins and enzymes.

Zinc forms numerous organometallic compounds with the dialkyls being the most important. Although these compounds do not associate to give aggregates, they are spontaneously flammable. The reaction of a zinc halide with a Grignard reagent can be used to prepare the compounds.

$$ZnBr_2 + 2\,C_2H_5MgBr \rightarrow Zn(C_2H_5)_2 + 2\,MgBr_2 \qquad (12.65)$$

Other alkyl compounds can be obtained by transfer reactions such as that with BR_3.

$$3\,Zn(CH_3)_2 + 2\,BR_3 \rightarrow 3\,ZnR_2 + 2\,B(CH_3)_3 \qquad (12.66)$$

This process is the result of the stability of trimethylboron. Alkyl group transfer also occurs in the reaction

$$Zn + HgR_2 \rightarrow ZnR_2 + Hg \qquad (12.67)$$

In terms of their chemical behavior, dialklyzinc compounds generally react to transfer alkyl groups to other metals. Dialkyls of cadmium are usually prepared by the reaction of a Grignard reagent with a cadmium halide.

$$CdX_2 + 2\,RMgX \rightarrow CdR_2 + 2\,MgX_2 \tag{12.68}$$

Mixed alkyl halide compounds result from the reaction of dialkylcadmium with cadmium halides.

$$CdR_2 + CdX_2 \rightarrow 2\,RCdX \tag{12.69}$$

In this chapter, we have briefly surveyed the chemistry of organometallic compounds of some of the main group elements. Additional aspects of organometallic chemistry will be illustrated in Chapters 14 and 15, especially for the heavier elements in these groups. Organometallic chemistry is a vast and important area, the relevance of which can be appreciated only by consulting additional references.

References for Further Study

Advances in Organometallic Chemistry, vol. 55, 2007 This series of 55 volumes has had many editors over the many years of publication.

Coates, G.E., 1960. *Organo-Metallic Compounds*. Wiley, New York. One of the classic books in the field of organometallic chemistry.

Cotton, F.A., Wilkinson, G., Murillo, C.A., Bochmann, M., 1999. *Advanced Inorganic Chemistry*, 6th ed. John Wiley, New York. This reference text contains a great deal of information on organometallic chemistry of main group elements.

Crabtree, R.H., Mingos, D.M.P. (Eds.), 2007. *Comprehensive Organometallic Chemistry III*. Elsevier, Amsterdam.

Greenwood, N.N., Earnshaw, A., 1997. *Chemistry of the Elements*, 2nd ed. Butterworth-Heinemann, Oxford Because this 1341 page book deals with chemistry of all elements, a great deal of it concerns their organometallic chemistry.

Rappaport, Z., Marck, I., 2006. *The Chemistry of Organolithium Compounds*. Wiley, New York.

Rochow, E.G., 1964. *Organometallic Chemistry*. Reinhold, New York. This small book has an elementary introduction to the field and includes a great deal of history.

Suzuki, H., Matano, Y., 2001. *Organobismuth Chemistry*. Elsevier, Amsterdam.

Thayer, J.S., 1988. *Organometallic Chemistry, An Overview*. VCH Publishers, Weinheim.

Wakefield, B.S., 1976. *The Chemistry of Organolithium Compounds*. Pergamon Press, Oxford.

QUESTIONS AND PROBLEMS

1. Complete and balance the following:
   (a) $LiC_2H_5 + PBr_3 \rightarrow$
   (b) $CH_3MgBr + SiCl_4 \rightarrow$
   (c) $NaC_6H_5 + GeCl_4 \rightarrow$
   (d) $LiC_4H_9 + CH_3COCl \rightarrow$
   (e) $Mg(C_5H_5)_2 + MnCl_2 \rightarrow$
2. Write complete equations for the following processes:
   (a) the preparation of butyllithium
   (b) the reaction of butyllithium with water

(c) dissolving of beryllium in sodium hydroxide

(d) preparation of phenyl sodium

(e) reaction between ethanol and lithium hydride

3. Complete and balance the following:

(a) $C_4H_9OH + CH_3MgCl \rightarrow$

(b) $C_3H_7Cl + Na \rightarrow$

(c) $Zn(C_2H_5)_2 + SbCl_3 \rightarrow$

(d) $NaC_6H_5 + C_2H_5Cl \rightarrow$

(e) $B(CH_3)_3 + O_2 \rightarrow$

4. Write complete equations to show the following reactions:

(a) the reaction of butyllithium with cadmium chloride

(b) the reaction of ethyl magnesium bromide with acetaldehyde

(c) the reaction of methyllithium with bromine

(d) the reaction of triethylaluminum with methanol

(e) the reaction of triethylaluminum with ethane

5. From what you know about the nature of $B(CH_3)_3$ and $Al(CH_3)_3$, what would be a reasonable value for the solubility parameter for $B(CH_3)_3$? Explain your answer.

6. Speculate on the degree of aggregation of C_2H_5MgCl in the solvents dioxane and benzene assuming that the concentration is the same in both solvents. If there would be a difference, explain why.

7. Although diethylzinc does not normally associate, C_2H_5ZnCl does to some extent in nonpolar solvents to give a tetramer. Why is there a difference between $Zn(C_2H_5)_2$ and C_2H_5ZnCl? Speculate on the structure of $[C_2H_5ZnCl]_4$.

8. Complete and balance the following:

(a) $Na_2S + RCl \rightarrow$

(b) $AsCl_3 + Na + C_6H_5Cl \rightarrow$

(c) $(C_6H_5)_2TeCl_2 + LiC_6H_5 \rightarrow$

(d) $i\text{-}C_3H_7OH + C_2H_5MgCl \rightarrow$

(e) $SiH_4 + CH_3OH \rightarrow$

9. When aluminum alkyls form complexes with $(CH_3)_2X$ (where $X = O$, S, Se, or Te), the stability of the complexes decreases as X progresses from O to Te. Explain this trend in the order of stability of the complexes.

10. Although $(CH_3)_2CCl_2$ does not react with water, $(CH_3)_2SnCl_2$ hydrolyzes readily. Explain this difference.

11. At concentrations of about 1 molal in tetrahydrofuran, C_2H_5MgBr exists almost entirely as the monomer. However, at a comparable concentration in diethyl ether, the dominant species is the dimer. Explain this difference.

12. Alkyls of Group IVA elements are essentially unreactive in air, but Group IIIA alkyls are extremely reactive. Provide an explanation for this great difference in behavior.

13. Explain the entropy of vaporization data given in Table 12.1 for $Al(C_2H_5)_2Cl$ and $Al(C_2H_5)Cl_2$ in terms of the species present in both the liquid and vapor phases.

14. At room temperature, nuclear magnetic resonance cannot used to distinguish between R_2Mg and RMgX. What does this signify? Propose a mechanism for a process to explain this observation.

CHAPTER 13

Chemistry of Nonmetallic Elements I. Hydrogen, Boron, Oxygen, and Carbon

Nonmetallic elements span the range from the extremely reactive fluorine to relatively unreactive elements such as carbon and nitrogen. One nonmetallic element, sulfur, has been known for thousands of years, but the discovery of others such as the halogens had to await the development of electrochemistry in the early 1800s. The nonmetallic elements constitute a group of approximately 20 elements found in the upper right-hand quadrant of the periodic table. A few others (such as germanium and tellurium) have properties that are typical of both metals and nonmetals. In this and the next two chapters, we will present an overview of some of the important chemistry of nonmetallic elements. Of necessity, such a summary is incomplete, so the references given at the end of the chapters should be consulted for more complete coverage.

13.1 HYDROGEN

Hydrogen is the most abundant element in the universe when considered from the standpoint of number of atoms. On Earth, hydrogen constitutes about 15.4% of all the atoms, but owing to the low mass of the atoms, it constitutes only 0.9% in terms of mass. Three isotopes are known, ^1H, ^2H or deuterium, and ^3H or tritium. Deuterium oxide, D_2O, is also known as "heavy water" and is produced by electrolysis because ordinary water is electrolyzed more easily. Many compounds undergo exchange of deuterium and hydrogen. For example, when CH_3OH is placed in D_2O, the hydrogen atom of the OH group undergoes exchange, but the three hydrogen atoms of the methyl group do not. Similarly, there is rapid exchange of the hydrogen atom in the carboxyl group when CH_3CH_2COOH is placed in D_2O. In general, hydrogen atoms that are held in polar bonds undergo exchange, but those in organic fragments (attached to carbon) do not. Isotope exchange is important in elucidating molecular structures and mechanisms of reactions.

Because the ionization potential for the hydrogen atom is over 1300 kJ/mol, there is little likelihood that compounds containing the H^+ cation would be stable. Obtaining H atoms to ionize is unfavorable because the $H-H$ bond energy is about 435 kJ/mol. Although hydrogen ions are very energetically unfavorable, there are some hydrated species such as $H_5O_2^+$ (which is actually $(H_2O)_2 \cdot H^+$) and $H_9O_4^+$ (which is the tetrahydrate $(H_2O)_4 \cdot H^+$) that are sufficiently stable that they have been identified as cations in solids. The structure of the $H_9O_4^+$ ion can be shown as

393

Inorganic Chemistry. DOI: http://dx.doi.org/10.1016/B978-0-12-385110-9.00013-3

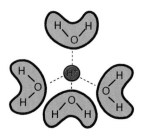

The $+1$ oxidation state is the most common for hydrogen, but it does not exist as H^+ ions in compounds.

When an electron is added to a gaseous hydrogen atom, the process is accompanied by liberation of 74 kJ/mol, which means that making H^- is energetically favorable. The result is that there are numerous stable compounds that contain the hydride ion, H^-. The reaction of hydrogen with an active metal such as sodium leads to this type of compound.

$$2\,Na + H_2 \rightarrow 2\,NaH \tag{13.1}$$

Such *hydride* compounds contain hydrogen in a negative oxidation state, which means that the other element must have a lower electronegativity than hydrogen. As we shall see, hydrides can be considered in three groups, *ionic*, *covalent*, and *interstitial*. The majority of hydrogen compounds contain hydrogen that has neither lost nor gained an electron. By that we mean that the majority of hydrogen compounds are covalent. In this section, we will present a brief overview of the simplest element.

13.1.1 Preparation of Hydrogen

Commercial preparation of hydrogen must be carried in the most economical means. Because water is abundant and inexpensive, it is the source of hydrogen in some instances. One reaction that is carried out on a large scale is the reaction of carbon with water at high temperature,

$$C + H_2O \rightarrow CO + H_2 \tag{13.2}$$

The steam reformer process involves the reaction of methane and high-temperature steam in the presence of a nickel catalyst. The reactions are

$$CH_4 + 2\,H_2O \xrightarrow{Ni} CO_2 + 4\,H_2 \tag{13.3}$$

$$CH_4 + H_2O \xrightarrow{Ni} CO + 3\,H_2 \tag{13.4}$$

The products include H_2, CO_2, CO, and steam, which are then processed in a shift converter, resulting in the reaction

$$CO + H_2O \rightarrow CO_2 + H_2 \tag{13.5}$$

Because large quantities of unsaturated organic compounds are obtained by removing hydrogen from saturated hydrocarbons, these processes are also important in production of hydrogen commercially. In one of these processes, hexane is converted to cyclohexane by the reaction

$$C_6H_{14} \xrightarrow{Catalyst} C_6H_{12} + H_2 \tag{13.6}$$

Cyclohexane is further dehydrogenated to produce benzene by the reaction

$$C_6H_{12} \xrightarrow{\text{Catalyst}} C_6H_6 + 3\,H_2 \tag{13.7}$$

Other dehydrogenation reactions carried out on a large scale are the production of 1,3-butadiene from butane and the production of styrene from ethyl benzene. These reactions can be shown as follows:

$$C_4H_{10} \xrightarrow{\text{Catalyst}} CH_2{=}CH{=}CH{=}CH_2 + 2\,H_2 \tag{13.8}$$

$$C_6H_5CH_2CH_3 \xrightarrow{\text{Catalyst}} C_6H_5CH{=}CH_2 + H_2 \tag{13.9}$$

Styrene and butadiene are the monomers utilized in producing polymers that are used in large quantities.

There are two important electrochemical processes in which hydrogen is produced. The first is the electrolysis of water,

$$2\,H_2O \xrightarrow{\text{Electrolysis}} 2\,H_2 + O_2 \tag{13.10}$$

In this method, each gas is produced in a separate compartment, so they have high purity.

In this process, deuterium oxide, D_2O, is electrolyzed more slowly, so the water becomes enriched in the heavier isotope. The other electrolytic process that produces hydrogen is the electrolysis of a solution of sodium chloride.

$$2\,Na^+ + 2\,Cl^- + 2\,H_2O \xrightarrow{\text{Electrolysis}} H_2 + Cl_2 + 2\,Na^+ + 2\,OH^- \tag{13.11}$$

All three products in this reaction are of commercial importance. The reaction is the most common method used to prepare chlorine and sodium hydroxide, so it is carried out on an enormous scale. This also means that it is an important source of hydrogen.

There are many reactions that produce hydrogen on a small scale. These include replacement reactions in which a metal (M) having a higher reduction potential than hydrogen reacts with an acid (HA).

$$2\,M + 2\,HA \rightarrow H_2 + 2\,MA \tag{13.12}$$

Another type of replacement reaction occurs when a metal such as aluminum or zinc reacts with a strong base. For example,

$$2\,Al + 2\,NaOH + 6\,H_2O \rightarrow 2\,Na[Al(OH)_4] + 3\,H_2 \tag{13.13}$$

Reactions in which the hydride ion functions as a Brønsted base also result in the production of hydrogen. Typical reactions are the following.

$$CaH_2 + 2\,H_2O \rightarrow Ca(OH)_2 + 2\,H_2 \tag{13.14}$$

$$CH_3OH + NaH \rightarrow NaOCH_3 + H_2 \tag{13.15}$$

Metal hydrides are so effective as water scavengers that they are sometimes used as drying agents to remove traces of water from solvents.

13.1.2 Hydrides

Compounds containing hydrogen and another element that has a lower electronegativity than hydrogen are properly considered to be *hydrides*. An enormous range of hydrides exist, and they are usually grouped in three categories. The first type consists of the *ionic* or *salt-like* hydrides in which the hydrogen is present as H^-. Although the electron affinity of hydrogen is 74 kJ/mol, transfer of an electron from a metal to hydrogen is unfavorable unless the ions interact to form a lattice. These are compounds that contain hydrogen and a metal that has very low electronegativity. Generally, this means that the metal resides in Group IA or IIA of the periodic table. In most cases, the compounds are formed by direct combination of the elements at high temperature. When M is a Group IA metal, the reaction is

$$2\ M + H_2 \rightarrow 2\ MH \tag{13.16}$$

Some metal nitrides react with hydrogen to produce the metal hydride.

$$Na_3N + 3\ H_2 \rightarrow 3\ NaH + NH_3 \tag{13.17}$$

The properties of the hydrides of the Group IA metals are shown in Table 13.1.

The most important characteristic of ionic hydrides is that they are strong Brønsted bases. The hydride ion will react with most molecules that contain a hydrogen atom bound to an atom of high electronegativity. Such molecules include water, alcohols, and ammonia as illustrated by the following equations:

$$H^- + H_2O \rightarrow OH^- + H_2 \tag{13.18}$$

$$H^- + ROH \rightarrow RO^- + H_2 \tag{13.19}$$

$$H^- + NH_3 \rightarrow NH_2^- + H_2 \tag{13.20}$$

The hydride ion is approximately the same size as I^-, and as a result of it having an unshared pair of electrons, it functions as a Lewis base. Many complexes are known that contain hydride ions as ligands bound to transition metals. Because the hydride ion is a soft Lewis base, most of the complexes contain metal atoms or ions of low charge, and in many cases, they are from the second or third transition series (see Chapters 16 and 22). Other complexes containing hydride ligands are the AlH_4^- and BH_4^- ions that are usually contained in compounds such as $LiAlH_4$ and $NaBH_4$. Known as lithium aluminum hydride and sodium borohydride, these compounds are versatile reducing agents that are frequently used in synthetic organic chemistry.

Table 13.1 Properties of the Hydrides of Group IA Metals

Compound	ΔH_f° (kJ mol^{-1})	U (kJ mol^{-1})	Radius (pm)	Apparent Charge on H (e units)	Density Hydride (g cm^{-3})	Density, Metal (g cm^{-3})
LiH	−89.1	916	126	−0.49	0.77	0.534
NaH	−59.6	808	146	−0.50	1.36	0.972
KH	−63.6	720	152	−0.60	1.43	0.859
RbH	−47.7	678	153	−0.63	2.59	1.525
CsH	−42.6	644	154	−0.65	3.41	1.903

When the difference in electronegativity between hydrogen and the other element is small, the hydride has covalent bonding. Although beryllium and magnesium are found in Group IIA of the periodic table, they have higher ionization potentials and higher electronegativities than other elements in this group. Because of this, the hydrides of beryllium and magnesium are polymeric covalent materials that have a chain structure in which there are hydrogen bridges between metal atoms. This type of structure can be shown as

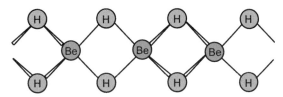

The bonding arrangement of the four H atoms around each Be is approximately tetrahedral.

The reaction of magnesium with boron produces a boride having the formula MgB_2 or Mg_3B_2.

$$3 \ Mg + 2 \ B \rightarrow Mg_3B_2 \tag{13.21}$$

As we shall see later, borides (as well as oxides, nitrides, carbides, etc.) react with water to produce a hydrogen compound of the nonmetal. Thus, the reaction of magnesium boride with water might be expected to produce BH_3, *borane*, but instead the product is B_2H_6, *diborane* (m.p. $-165.5\,°C$, b.p. $-92.5\,°C$). This interesting covalent hydride has two three-center B—H—B bonds resulting in a structure that can be shown as follows. The nature of the bonding will be described later in this chapter.

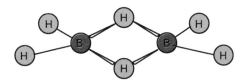

Each boron atom is surrounded by four hydrogen atoms in an arrangement that is approximately tetrahedral. Diborane can also be prepared by the reaction of BF_3 with $NaBH_4$.

$$3 \ NaBH_4 + BF_3 \rightarrow 3 \ NaF + 2 \ B_2H_6 \tag{13.22}$$

Boron and hydrogen form many compounds, and they exhibit unusual structural forms. Several of the boranes are listed in Table 13.2. Covalent hydrides are generally compounds that have low boiling points. Consequently, they are often referred to as *volatile hydrides*.

In some cases, elements having electronegativities too low to give ionic bonding with hydrogen also tend to be unreactive so that direct combination of the elements is not feasible. In such cases, the procedure described above can be used to prepare the hydride. For example, silicon hydride, SiH_4 (known as *silane*), can be produced by the reactions

$$2 \ Mg + Si \rightarrow Mg_2Si \tag{13.23}$$

$$Mg_2Si + 4 \ HCl \rightarrow SiH_4 + 2 \ MgCl_2 \tag{13.24}$$

Table 13.2 Properties of Some Covalent Hydrides

Name	Formula	Melting Point, °C	Boiling Point, °C
Diborane	B_2H_6	−165.5	−92.5
Tetraborane	B_4H_{10}	−120	18
Pentaborane-9	B_5H_9	−46.6	48
Pentaborane-11	B_5H_{11}	−123	63
Hexaborane	B_6H_{10}	−65	110
Ennaborane	B_9H_{15}	2.6	–
Decaborane	$B_{10}H_{14}$	99.7	213
Silane	SiH_4	−185	−119.9
Disilane	Si_2H_6	−132.5	−14.5
Trisilane	Si_3H_8	−117	54
Germane	GeH_4	−165	−90
Digermane	Ge_2H_6	−109	29
Trigermane	Ge_3H_8	−106	110
Phosphine	PH_3	−133	−87.7
Diphosphine	P_2H_4	−99	−51.7
Arsine	AsH_3	−116.3	−62.4
Stibine	SbH_3	−88	−18

Silane can also be produced by the reaction of $SiCl_4$ with $LiAlH_4$.

$$SiCl_4 + LiAlH_4 \rightarrow SiH_4 + LiCl + AlCl_3 \tag{13.25}$$

Silicon forms numerous hydrides that have counterparts in the hydrocarbon series. However, the silanes are named with respect to the number of silicon atoms present. For example, Si_2H_6 is known as disilane, Si_3H_8 is trisilane, etc. The situation is similar for the germanium hydrides.

When a metal phosphide undergoes hydrolysis, the product is PH_3, *phosphine*. For example,

$$Mg_3P_2 + 6\ HCl \rightarrow 2\ PH_3 + 3\ MgCl_2 \tag{13.26}$$

Phosphine is also produced when white phosphorus is heated with sodium hydroxide.

$$P_4 + 3\ NaOH + 3\ H_2O \rightarrow PH_3 + 3\ NaH_2PO_2 \tag{13.27}$$

However, PH_3 is not the only hydride of phosphorus, and it is not the only product of this reaction. The other hydride of phosphorus is *diphosphine*, P_2H_4, which is produced in the reaction above. This compound is spontaneously flammable in air, and it ignites phosphine, which is also flammable.

$$4\ PH_3 + 8\ O_2 \rightarrow 6\ H_2O + P_4O_{10} \tag{13.28}$$

The boron hydrides, silanes, phosphines, and most other covalent hydrides burn readily, some even being spontaneously flammable.

There are numerous similarities between ammonia and phosphine, but the latter is a much weaker base (see Chapter 9). In fact, phosphonium salts are stabilized by large anions that are also the conjugates of strong acids. Accordingly, the most common phosphonium salts are the iodides, bromides, tetrafluoroborates, etc. Phosphine and substituted phosphines are good Lewis bases toward soft Lewis acids, and many coordination compounds of this type are known.

Although active metals form ionic hydrides, many other metals do not. They react when placed in an atmosphere of hydrogen, but the products contain hydrogen atoms in interstitial positions in the lattice. Hydrides of this type are known as the interstitial hydrides. Because the number of interstitial positions that contain hydrogen atoms is not determined by a definite number of chemical bonds, hydrides of this type may not have simple formulas. Typical compositions are represented as $CuH_{0.96}$, $LaH_{2.78}$, $TiH_{1.21}$, $TiH_{1.7}$, or $PdH_{0.62}$. Hydrides of this type are often called *nonstoichiometric hydrides*.

Although the formation of ionic hydrides is usually exothermic, the formation of interstitial hydrides may have positive enthalpy values. Physical characteristics of interstitial hydrides are determined by the fact that hydrogen atoms in interstitial positions cause some expansion of the lattice but contribute very little mass. Consequently, the interstitial hydrides always have lower densities than the pure metals even though the crystal structure is normally the same. When interstitial positions contain hydrogen atoms, the flow of electrons in conduction bands within the metal is impeded, so the conductivity is lower than that of the pure metal. Metals are generally malleable and ductile because of the mobility of atoms within the structure. When the interstitial positions contain hydrogen atoms, the ability of the metal atoms to move in the lattice is constrained, so the metal hydride is less malleable and ductile. Typically, the hydrides are harder than the metal alone.

A hydride of a transition metal that has an electronegativity of about $1.4-1.7$ is not ionic, and bonds between the metal atoms and hydrogen atoms in interstitial positions are not really covalent. Causing the hydrogen to enter the interstitial positions requires the separation of H_2 molecules that have a bond energy of 432.6 kJ/mol and that energy is not compensated by forming either ionic or covalent bonds between the hydrogen and the metal. As a result, the heat of formation of interstitial hydrides is often positive. For some purposes, it is appropriate to consider interstitial hydrides as being solutions of atomic hydrogen in the metal. Dissolving the hydrogen usually involves separating H_2 molecules, so the metals often function as effective catalysts for hydrogenation reactions.

Because hydrogen molecules must become attached to the surface of the metal, the prior treatment of the metal has a great influence on the formation of a hydride. Cracks, pores, and other defects that aid in the adsorption of hydrogen may have been removed by heating and cooling or by the influence of radiation. Active sites may have been generated by mechanical changes in the surface. All of these factors affect the ease with which the formation of the hydride is accomplished. Because the adsorption of a gas on a solid is related to the partial pressure of the gas and the temperature, these factors influence the fraction of the available sites occupied by hydrogen. In other words, the stoichiometry of the hydride may depend on the conditions under which it is formed.

The bond angles in NH_3 are approximately $107°$, which indicates sp^3 hybridization with a small reduction in bond angle arising from the effect of the unshared pair of electrons on the nitrogen atom. In PH_3, the bond angle is only about $93°$, so the indication is that the phosphorus orbitals are *not sp^3* hybrids. Although sp^3 hybridization should reduce repulsion between electron pairs, the p orbitals on phosphorus are large enough that increasing their size by hybridization reduces the effectiveness of the overlap with hydrogen $1s$ orbitals. Bond angles in AsH_3 and SbH_3 are slightly smaller than those in PH_3, which would be expected if p orbitals on the central atoms are used in bonding. Thus, the bond angles indicate that although it is appropriate to assume sp^3 hybridization in NH_3, it is not the case for the hydrides of the heavier members of Group VA. The hydrogen compounds of Group VIA follow this same trend in bond angles.

13.2 BORON

Although boron ranks 48th among the elements in abundance, it is not found uncombined. The most common minerals containing boron are the tetraborates of sodium or calcium. *Borax*, $Na_2B_4O_7 \cdot 10H_2O$, is the most important source of boron, and large deposits of borax are found in southern California from which about three-fourths of the world demand is obtained.

Borax has been used as a soldering flux for many centuries. It was hauled from the mines in wagons pulled by teams of 20 mules which is the basis for "20-Mule Team®" borax that is used in laundry products.

13.2.1 Elemental Boron

Boron itself was first produced in 1808 by Sir Humphrey Davy, who carried out the electrolysis of molten boric acid. The reduction of boric acid by potassium was used as a preparative method by Gay-Lussac and Thenard in 1808, and reduction of B_2O_3 by magnesium was the method used by Moissan in 1895.

$$B_2O_3 + 3\ Mg \rightarrow 3\ MgO + 2\ B \tag{13.29}$$

However, the reduction reactions in which a solid reducing agent is used usually give impure products. In this case, the product contains 80%–95% boron that also contains magnesium and boron oxide as impurities. The boron produced in this way is a brownish-black form having a density of 2.37 g/cm^3.

Pure boron can be obtained in small quantities by the reduction of boron trichloride with hydrogen on a heated tungsten filament:

$$2\ BCl_3 + 3\ H_2 \rightarrow 6\ HCl + 2\ B \tag{13.30}$$

A black crystalline form of boron having a density of 2.34 g/cm^3 results. Boron exists in unit cells that have an icosahedral structure in which 20 faces that are equilateral triangles meet at 12 vertices with a boron atom at each vertex.

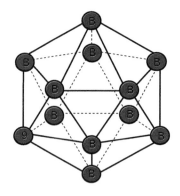

There are several ways of arranging the almost spherical B_{12} units. Three of these result in the forms of boron known as tetragonal, α-rhombohedral, and β-rhombohedral. All of the structures are very rigid, which results in boron having a hardness of 9.3 compared with the value of 10.0 for diamond (Mohs' scale).

Naturally occurring boron consists of approximately 20% of ^{10}B and 80% of ^{11}B leading to an average atomic mass of 10.8 amu. Because ^{10}B has a relatively large cross section for absorption of slow (thermal) neutrons, it is used in control rods in nuclear reactors and in protective shields. In order to obtain a material that can be fabricated into appropriate shapes, boron carbide is combined with aluminum.

It is known that ^{10}B collects in brain tumors to a greater extent than in normal tissue. Research has been conducted on the use of the isotope ^{10}B for treating brain tumors. Bombardment of the tumor with slow neutrons leads to the production of alpha particles ($^{4}He^{2+}$) and lithium nuclei that have enough energy to destroy the abnormal tissue.

$$^{10}_{5}B + ^{1}_{0}n \rightarrow ^{7}_{3}Li + ^{4}_{2}He + \gamma \tag{13.31}$$

Boron is also used in fabricating many objects. Boron produced as fibers can be added to a resin to give composites that are lighter than aluminum but which have strength comparable to that of steel. These composites are used to make fishing rods, tennis rackets, etc.

13.2.2 Bonding in Boron Compounds

With the electronic structure of the boron atom being $1s^2\, 2s^2\, 2p^1$, it might be expected that boron would lose three electrons to give compounds that contain B^{3+} ions. However, removal of three electrons requires over 6700 kJ mol^{-1}, and this is so high that it precludes compounds that are strictly ionic. Polar covalent bonds exist, and the hybridization can be considered as leading to a set of sp^2 hybrid orbitals. However, boron burns readily to produce B_2O_3, a stable oxide having a heat of formation of -1264 kJ/mol.

It should be clear that we expect boron to form three equivalent covalent bonds with 120° bond angles. As a result, boron halides have the following trigonal planar structure (D_{3h} symmetry).

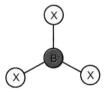

In these molecules, the boron atom has only six electrons surrounding it, so it interacts readily with species that can function as electron pair donors. For example, when F^- reacts with BF_3, the product is BF_4^- in which sp^3 hybrids are formed, so such species are tetrahedral (T_d symmetry). In most cases, molecules containing boron exhibit one of these types of bonding to boron. The boron hydrides represent a special situation that is described later.

13.2.3 Borides

Boron forms one or more borides when it reacts with most metals. For example, the reaction between magnesium and boron produces magnesium boride, Mg_3B_2.

$$3\,Mg + 2\,B \rightarrow Mg_3B_2 \tag{13.32}$$

This product reacts with acids to produce diborane, B_2H_6.

$$Mg_3B_2 + 6\ H^+ \rightarrow B_2H_6 + 3\ Mg^{2+} \tag{13.33}$$

Although BH_3 may be the expected product, it is not stable as a discrete monatomic unit. It is stable as an adduct with several Lewis bases. Some metals react with boron to form borides containing the hexaboride group, B_6^{2-}. An example of this type of compound is calcium hexaboride, CaB_6. In general, the structures of compounds of this type contain octahedral B_6^{2-} ions in a cubic lattice with metal ions. Most hexaborides are refractory materials having melting points over 2000 °C.

13.2.4 Boron Halides

As has already been mentioned, boron halides are electron-deficient molecules. As a result, they tend to act as strong Lewis acids by accepting electron pairs from many types of Lewis bases to form stable acid–base adducts. Electron donors such as ammonia, pyridine, amines, ethers, and many other types of compounds form stable adducts. In behaving as strong Lewis acids, the boron halides act as acid catalysts for several important types of organic reactions (see Chapter 9).

The molecules of all of the boron halides are planar with bond angles of 120°, which has been interpreted in terms of sp^2 hybridization of boron orbitals. However, the experimental B–X bond lengths are shorter than the values calculated using the covalent single-bond radii of the atoms. In the case of BF_3, this has been interpreted as indicating some double bond character due to π bonding that occurs when filled p orbitals on the halogen atoms donate electron density to the empty p orbital on the boron. The shorter than expected bonds may also be the result of ionic contribution to the B–X bond. The extent of this bond shortening can be estimated by means of the *Shoemaker–Stevenson equation*,

$$r_{ab} = r_a + r_b - 9.0\ [X_a - X_b] \tag{13.34}$$

where r_{ab} is the actual bond length, r_a and r_b are the covalent single-bond radii of atoms a and b, and X_a and X_b are the electronegativities of atoms a and b. The results obtained by applying this equation to the boron halides are shown in Table 13.3.

The data given in Table 13.3 show that the extent of bond shortening is greatest for B–F bonds. This is to be expected because back donation of electron density from F to B is more effective when the donor and acceptor atoms are of comparable size. The resonance structures shown below indicate how p orbitals form the multiple bonds between B and F:

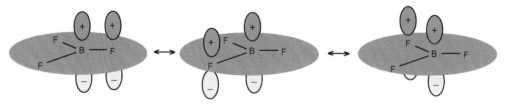

Adding electron density to the boron atom by multiple bonding would be expected to reduce its tendency to accept electron density from a Lewis base. In accordance with this, when reacting with pyridine, the strength as acceptors is $BBr_3 > BCl_3 > BF_3$. The bond shortening discussed earlier is

Table 13.3 Bond Lengths in the Boron Halide

Bond Type	Sum of Covalent Single-Bond Radii, pm	r_{ab} Calculated from Eqn 8.6, pm	Experimental r_{ab}, pm
B–F	152	134	130
B–Cl	187	179	175
B–Br	202	195	187

sometimes explained as being due to the contribution of resonance structures such as these. Because they are Lewis acids, boron halides also form complexes of the type BX_4^-. For example,

$$MCl + BCl_3 \rightarrow M^+BCl_4^- \tag{13.35}$$

The BX_4^- ions are tetrahedral because the addition of the fourth X^- means that an additional pair of electrons must be accommodated around the boron atom.

In addition to functioning as Lewis acids, boron halides undergo many other types of reactions. As is typical of most compounds containing covalent bonds between a nonmetal and a halogen, the boron halides react vigorously with water to yield boric acid and the corresponding hydrogen halide.

$$BX_3 + 3\ H_2O \rightarrow H_3BO_3 + 3\ HX \tag{13.36}$$

The BX_3 compounds will also react with other protic solvents such as alcohols to yield borate esters.

$$BX_3 + 3\ ROH \rightarrow B(OR)_3 + 3\ HX \tag{13.37}$$

An unusual type of product known as a borazine is produced by the reaction of BCl_3 with NH_4Cl.

$$3\ NH_4Cl + 3\ BCl_3 \rightarrow \underset{\text{trichloroborazine}}{B_3N_3H_3Cl_3} + 9\ HCl \tag{13.38}$$

Diboron tetrahalides, B_2X_4, are also known. These may be prepared in a variety of ways, among them the reaction of BCl_3 with mercury,

$$2\ BCl_3 + 2\ Hg \xrightarrow{\text{Hg arc}} B_2Cl_4 + Hg_2Cl_2 \tag{13.39}$$

13.2.5 Boron Hydrides

There are many compounds that contain boron and hydrogen, and they are known collectively as the *boron hydrides*. Six boron hydrides were prepared by Alfred Stock in 1910–1930 by the addition of hydrochloric acid to magnesium boride that was produced in small amounts when B_2O_3 was reduced with magnesium.

$$6\ HCl + Mg_3B_2 \rightarrow 3\ MgCl_2 + B_2H_6 \tag{13.40}$$

Because of the extreme reactivity of boron hydrides (some are spontaneously flammable in air), Stock developed procedures for handling such air-sensitive materials. The six hydrides produced by Stock were B_2H_6, B_4H_{10}, B_5H_9, B_5H_{11}, B_6H_{10}, and $B_{10}H_{14}$ with the most interesting being diborane, B_2H_6. Because there are only 12 valence shell electrons, it is impossible to make electron pair bonds between all of the atoms. It was believed that *borane*, BH_3, should be a more stable compound because a Lewis structure can be shown as

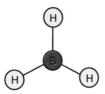

However, Stock found after numerous attempts that it was impossible to isolate borane.

The length of the B—B bond in diborane is similar to that expected for a double bond. This cannot be accounted for satisfactorily by a typical Lewis structure. It has now been determined that diborane has the structure

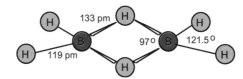

Four of the hydrogen atoms lie in the same plane as the boron atoms. The other two (forming bridges) lie above and below this plane forming an arrangement of four hydrogen atoms around each boron that is approximately tetrahedral. The bridging hydrogen atoms do not form *two* covalent bonds but rather simultaneously overlap with an sp^3 hybrid orbital from each boron atom. This produces a so-called *three-center bond*, as shown in Figure 13.1.

The combination of boron and hydrogen orbitals in the three-center bond can be shown in a molecular orbital diagram as in Figure 13.2. Using this approach to bonding, the structures of some of the other boron hydrides prepared by Stock can be described.

Since the early work of Stock, other boron hydrides have been synthesized. Some of these compounds have been used as fuel additives, and they have found some application in high-energy rocket fuels. However, because $B_2O_3(s)$ is a product of the reaction, the use of these materials in that way causes some problems. The boron hydrides will all burn readily to produce B_2O_3 and water,

$$B_2H_6 + 3\ O_2 \rightarrow B_2O_3 + 3\ H_2O \tag{13.41}$$

FIGURE 13.1

A three-center B—H—B bond in B_2H_6.

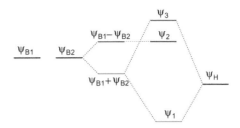

FIGURE 13.2
A molecular orbital diagram for a three-center bond in diborane. The boron group orbital $\psi_{B1} + \psi_{B2}$ has symmetry that matches that of the hydrogen orbital. It combines with the hydrogen orbital to give the bonding orbital designated as ψ_1.

The properties of some boron hydrides along with those of other volatile hydrides are shown in Table 13.2. A very interesting reaction that diborane undergoes is one in which it reacts with double bonds in hydrocarbons. The reaction can be shown as

$$B_2H_6 + 6\ RCH{=}CH_2 \rightarrow 2\ B(CH_2CH_2R)_3 \tag{13.42}$$

This type of reaction is known as *hydroboration*, and it was exploited by H. C. Brown as a route to organic derivatives of the boranes.

13.2.6 Polyhedral Boranes

In a large number of compounds containing boron and hydrogen, the boron atoms are arranged in the form of a well-defined polyhedron (octahedron, square antiprism, bicapped square antiprism, icosahedron, etc.). In these structures, the boron atoms are most often bonded to four, five, or six other atoms. The most common structure is the icosahedron exhibited by the species $B_{12}H_{12}^{2-}$. This structure consists of a B_{12} icosahedron with a hydrogen atom bonded to each boron atom and the overall structure having a -2 charge. It is prepared by the reaction of B_2H_6 with a base, which aids in the removal of hydrogen from diborane.

$$6\ B_2H_6 + 2\ (CH_3)_3N \xrightarrow{150°C} [(CH_3)_3NH]_2B_{12}H_{12} + 11\ H_2 \tag{13.43}$$

Compounds such as $Cs_2B_{12}H_{12}$ are stable up to several hundred degrees and do not react as reducing agents as does BH_4^-. A large number of derivatives of $B_{12}H_{12}^{2-}$ have been prepared in which all or part of the hydrogen atoms are replaced by other groups such as Cl, F, Br, NH_2, OH, CH_3, OCH_3, COOH, etc.

The structures of polyhedral boranes can be grouped into several classifications. If the structure contains a complete polyhedron of boron atoms, it is referred to as a *closo* borane (*closo* comes from a Greek word meaning "closed"). However, if the structure has one boron atom missing from a corner of the polyhedron, the structure is referred to as a *nido* borane (*nido* comes from a Latin word for "nest"). In this type of structure, a polyhedron having n corners has $(n-1)$ corners that are occupied by boron atoms. A borane in which there are two corners unoccupied by boron atoms is referred to as an *arachno* structure (*arachno* comes from a Greek word for "web"). Other types of boranes have structures that are classified in different ways, but they are less numerous and will not be described.

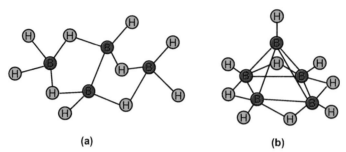

(a)   **(b)**

FIGURE 13.3
The *arachno* structure of B_4H_{10} (a) and the *nido* structure of B_5H_9 (b).

The derivative of the B_{12} icosahedron that has the formula $B_{12}H_{12}^{2-}$ has a hydrogen atom attached to a boron atom on each corner of a complete polyhedron, so it represents a *closo* type of structure. Another example of this type is $B_{10}H_{10}^{2-}$, the structure of which is described below. Because it is possible for two boron hydrides having the same formula to have different structures, these terms are often incorporated in the name as prefixes. Thus, names such as *nido*-B_5H_9, *arachno*-B_4H_{10}, etc. are frequently used.

A relatively simple boron hydride that has an *arachno* structure is B_4H_{10}, and it has the structure shown in Figure 13.3(a). Pentaborane(9), B_5H_9, has a structure in which the boron atoms form a square-based pyramid with a terminal hydrogen atom attached to each boron atom and a bridging hydrogen atom along each edge of the square base. Because this structure, shown in Figure 13.3(b), can be considered as having an octahedron of boron atoms with one vertex vacant, it represents a *nido* type of borane.

The structure of *closo* $B_6H_6^{2-}$ is relatively simple. It consists of an octahedron of boron atoms with a hydrogen atom bonded to the B atom at each apex. The *closo* $B_9H_9^{2-}$ ion has the structure shown in Figure 13.4(a) in which six boron atoms are located at the corners of a trigonal prism. The remaining three boron atoms are located above each rectangular face of the prism to form a "capped" structure.

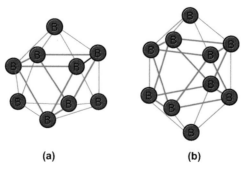

(a)   **(b)**

FIGURE 13.4
The arrangement of boron atoms in the $B_9H_9^{2-}$ (a) and $B_{10}H_{10}^{2-}$ (b) ions. Hydrogen atoms are not shown.

Each boron atom has a hydrogen atom bonded to it. The *closo* structure of $B_{10}H_{10}^{2-}$ has eight boron atoms located on the corners of a square antiprism (Archimedes antiprism). The remaining two boron atoms are located above and below the horizontal faces of the antiprism, and a hydrogen atom is bonded to each boron atom. Figure 13.4(b) shows the structure of the $B_{10}H_{10}^{2-}$ ion. An enormous number of derivatives of these basic polyhedral units (and also larger ones) exist, and a great deal of boron chemistry is concerned with their structures and reactions.

In order to describe derivatives of B_{12} or $B_{12}H_{12}^{2-}$ having icosahedral structures, it is necessary to have a way to designate positions of atoms or substituent groups. In order to do that, the positions are identified by a numbering system that is illustrated in the structure below.

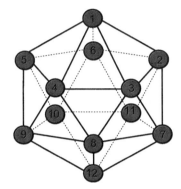

One well-known derivative of the $B_{12}H_{12}^{2-}$ ion is the *carborane*, $B_{10}C_2H_{12}$. Note that this species is neutral because each carbon atom has one more electron than does a boron atom. Because the carbon atoms can be located in any two positions in an icosahedron, there are three isomers of $B_{10}C_2H_{12}$ that differ in the location of the two carbon atoms in the structure. These isomers have the structures shown in Figure 13.5.

One derivative of the $B_{10}C_2H_{12}$ species is $B_9C_2H_{11}^{2-}$ in which a B—H group is missing from one vertex position in the icosahedron. The result is an ion that has a vacant bonding position where a metal ion

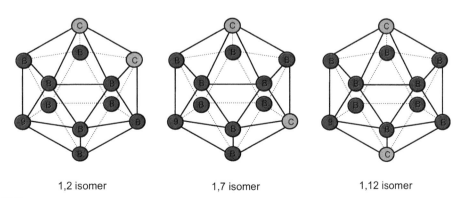

| 1,2 isomer | 1,7 isomer | 1,12 isomer |

FIGURE 13.5
The three isomers of $B_{10}C_2H_{12}$. The hydrogen atoms have been omitted to simplify the structures.

can attach. As a result, it forms a large number of *metallocarboranes* such as $C_5H_5FeB_9C_2H_{11}$ in which Fe^{2+} is also bonded to C_5H_5 giving the structure that can be shown as follows:

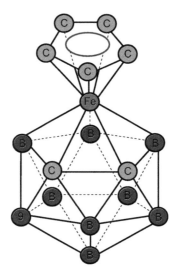

Other complexes such as $[Co(B_9C_2H_{11})_2]^-$ in which the cobalt is present as Co^{3+} are also known. This interesting complex has a structure that can be shown as

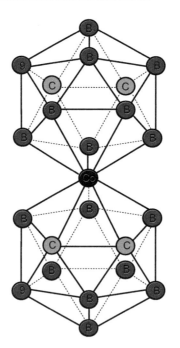

in which there is one hydrogen atom attached to each B and C atom, all of which are not shown.

The area of chemistry involving the polyhedral boranes and carboranes has seen enormous growth in recent years. Accordingly, the brief survey here does not include many interesting facets of the chemistry of these interesting compounds. For more details on the polyhedral boranes, the references at the end of this chapter should be consulted.

13.2.7 Boron Nitrides

Because boron has three valence shell electrons and nitrogen has five, the molecule BN is isoelectronic with the C_2 molecule. Also, some of the allotropic forms that exist for carbon (graphite and diamond) also exist for materials that have the formula $(BN)_x$. The form of $(BN)_x$ having the graphite structure is very similar to graphite in many ways. Its structure consists of layers of hexagonal rings containing alternating boron and nitrogen atoms. Unlike graphite, the layers of boron nitride fall directly in line with one another rather than being staggered. The structure of this form of boron nitride is shown in Figure 13.6.

In $(BN)_x$, the van der Waals forces holding the sheets in line with each other are stronger so that boron nitride is not as good a lubricant as graphite. However, the use of boron nitride as a high temperature lubricant has been investigated because of its chemical stability.

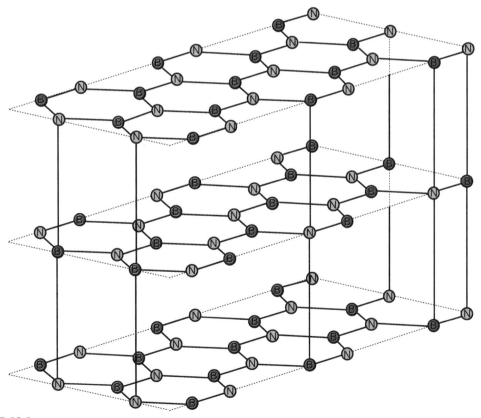

FIGURE 13.6
The layered structure of boron nitride. Compare this structure to that of graphite shown in Figure 13.7.

Under high pressure and temperature, boron nitride can be converted to a cubic form. The cubic form of $(BN)_x$ is known as *borazon*, and it has a structure similar to that of diamond. Its hardness is similar to that of diamond, and it is stable to higher temperatures. The extreme hardness results from the fact that the B—N bonds possess not only the covalent strength comparable to C—C bonds but also some ionic stabilization due to the difference in electronegativity between B and N.

Boron also forms many other compounds with nitrogen. One of the most interesting of these is borazine, $B_3N_3H_6$ (m.p. $-58\,^\circ$C, b.p. $54.5\,^\circ$C). Shown below is the structure of borazine, which is similar to the structure of benzene. In fact, borazine has sometimes been referred to as "inorganic benzene" and it has the structure shown as

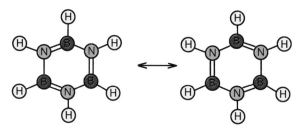

When subjected to high temperature under vacuum, borazine loses hydrogen and polymerizes to yield products known as *biborazonyl* and *naphthazine* whose structures are similar to biphenyl and naphthalene, respectively.

The borazine molecule has D_{3h} symmetry, whereas benzene has D_{6h} symmetry. The B—N bond length in H_3N—BF_3 is 160 pm, but in borazine, it is 144 pm. Although borazine resembles the aromatic benzene molecule in some respects, the electronic structure is considerable different. Theoretical studies have shown that although there is some delocalization in borazine, it is not as complete as in benzene. One reason is that because the nitrogen atom has a higher electronegativity than boron, the electron density is higher at the positions occupied by nitrogen atoms. The electron density is determined by both the π and σ bonds, both of which have polarity, but in opposite directions.

Borazine was first prepared in 1926 by the reaction of B_2H_6 with NH_3, which can be shown as

$$B_2H_6 + 2\,NH_3 \rightarrow 2\,H_3N:BH_3 \tag{13.44}$$

$$3\,H_3N:BH_3 \xrightarrow{200\,^\circ C} B_3N_3H_6 + 6\,H_2 \tag{13.45}$$

However, it can also be prepared by the following reactions:

$$3\,NH_4Cl + 3\,BCl_3 \xrightarrow[140-150\,^\circ C]{C_6H_5Cl} B_3N_3H_3Cl_3 + 9\,HCl \tag{13.46}$$

$$6\,NaBH_4 + 2\,B_3N_3H_3Cl_3 \rightarrow 2\,B_3N_3H_6 + 6\,NaCl + 3\,B_2H_6 \tag{13.47}$$

On a larger scale, the reaction of $(NH_2)_2CO$ with $B(OH)_3$ at high temperature in an atmosphere of ammonia is used to produce borazine. Trichloroborazine, $B_3N_3Cl_3H_3$, has the structure

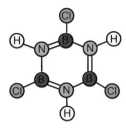

The chlorine atoms are bonded to boron atoms as expected on the basis of the difference in electronegativity.

   Although boron forms a large number of unusual compounds, many of the better-known ones are important and widely used. For example, the oxide B_2O_3 is used extensively in the making of glass. Borosilicate glass, glass wool, and fiberglass are widely used because they are chemically unreactive and can stand great changes in temperature without breaking. About 30−35% of the boron consumed is used in making types of glass. Borax has long been used for cleaning purposes in laundry products (detergents, water softeners, soaps, etc.). Boric acid, H_3BO_3 (or more accurately, $B(OH)_3$), is a very weak acid that has been used as an eye wash. It is also used in flame retardants. Boron fiber composites are used in fabricating many items such as tennis rackets, aircraft parts, and bicycle frames. Borides of such metals as titanium, zirconium, and chromium are used in the fabrication of turbine blades and rocket nozzles. Although it is a somewhat scarce element, boron and some of its compounds are quite important and are used in large quantities.

13.3 OXYGEN

Oxygen constitutes approximately 21% of the atmosphere and water contains 89% oxygen. An enormous range of minerals including silicates, phosphates, nitrates, carbonates, and sulfates also contain oxygen. Added to this scope of oxygen chemistry is the fact that all forms of animal life require oxygen for survival. Sulfuric acid, the chemical produced in largest quantity, and lime are two important materials in the chemical industry, and both are compounds that contain oxygen. Consequently, the chemistry of oxygen and its compounds constitutes a broad area that encompasses many principles, reactions, and structures.

13.3.1 Elemental Oxygen

The atmosphere contains approximately 21% oxygen (b.p. $-183\,^\circ$C) and 78% nitrogen (b.p. $-196\,^\circ$C). Both gases are colorless, odorless, and tasteless, and both are slightly soluble in water. The solubility of oxygen is sufficient to give a mole fraction of 2.29×10^{-5} at $25\,^\circ$C, whereas that of nitrogen is 1.18×10^{-5} at the same temperature. Although the solubility is low, it is sufficient for the existence of aquatic life. As with other gases, the solubility of both oxygen and nitrogen increases as pressure is increased and decreases with increasing temperature. Oxygen is appreciably soluble in many organic solvents. Oxygen is produced by plants in the photosynthesis process.

   There are three stable isotopes of oxygen, ^{16}O, ^{17}O, and ^{18}O, which have abundances of 99.762%, 0.038%, and 0.200%, respectively. The atomic masses of these isotopes are 15.994915,

16.999134, and 17.999160 amu. Both gases and water enriched in the heavier isotopes are available, and they are useful in kinetic studies and determination of molecular structure (see Chapter 21). In addition to the many other important aspects of the chemistry of oxygen is the fact that it forms complexes with metals. In that connection, we need to mention here only the vital interaction between oxygen and iron in hemoglobin, which results in the transport of oxygen in living systems. However, formation of oxygen complexes is by no means of so limited scope as will be discussed in later chapters.

The electronic ground state for the oxygen atom is $^3P_2$, which is consistent with two unpaired electrons in the $2p^4$ configuration. Unlike most paramagnetic atoms, the O_2 molecule is also paramagnetic. Although the structure has been shown as

$$\bar{\underline{O}}{=}\bar{\underline{O}}$$

this structure does not account for the paramagnetism of the molecule. One of the successes of the molecular orbital approach to bonding was the fact that the orbital diagram shown in Figure 3.8 correctly predicted both a double bond and the paramagnetic character of O_2.

The molecular orbital energy-level diagram allows the bond order to be calculated as B.O. $= (N_b - N_a)/2 = (8 - 6)/2 = 2$, the molecular orbital equivalent of a double bond. The orbital scheme shown in Figure 3.8 is also useful in predicting some of the chemical behavior of O_2. For example, it is clear that there are two vacancies in the π^*_{2px} and π^*_{2py} orbitals where one or two additional electrons could be placed to generate the O_2^- (superoxide) and O_2^{2-} (peroxide) ions. Both species are well known, and the predicted bond orders are 1.5 and 1, respectively. It is also apparent that if an electron is removed from one of the π^* orbitals, the O_2^+ (dioxygenyl ion) would result and that the bond order would increase to 2.5. Because the ionization potential of the O_2 molecule is about 12.06 eV (1163 kJ mol^{-1}), O_2^+ is a chemically viable species. Liquid oxygen has a light blue color that is the result of an electronic transition that results when the triplet ground state is excited to the singlet state having a higher energy. Characteristics of the bonds in several dioxygen species are summarized in Table 13.4.

13.3.2 Ozone

In 1785, it was noted by Van Marum that when an electric spark is passed through oxygen, an unusual odor results. That odor is caused by ozone, O_3, which has a characteristic pungent smell. In fact, ozone can be detected at very low levels by its intense odor, and the Greek word *ozein* meaning to smell is

Table 13.4 Characteristics of Dioxygen Species

Property	O_2^+	O_2	O_2^-	O_2^{2-} [a]
Bond length, pm	112	121	128	149
Bond energy, kJ mol^{-1}	623	494	–	213[a]
Force constant, mdyne Å^{-1}	16.0	11.4	5.6	4.0
Bond order	2.5	2	1.5	1

[a]In H_2O_2.

origin of the name ozone. Van Marun also noted that the gas reacted with mercury, and in 1840, Schönbein noted that the gas reacted with potassium iodide solution to liberate iodine. Ozone (m.p. $-193\ °C$ and b.p. $-112\ °C$) has a heat of formation of $+143$ kJ/mol, so it is unstable with respect to O_2. Mixtures of ozone and oxygen are explosive.

The O_3 molecule is isoelectronic with SO_2, NO_2^-, and other triatomic species containing 18 valence shell electrons. As a result, the structure can be represented by the resonance structures (having C_{2v} symmetry)

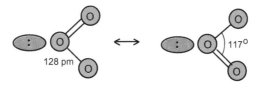

The bent O_3 molecule has infrared absorptions at $1103\ \text{cm}^{-1}$ (symmetric stretch), $701\ \text{cm}^{-1}$ (bending), and $1042\ \text{cm}^{-1}$ (asymmetric stretch).

When the bonding is considered in the molecular orbital approach, it can be seen that each oxygen atom contributes one p orbital to yield three molecular orbitals. The orbital overlap combinations lead to the wave functions

$$\psi_a = \frac{1}{2}(\psi_1 - \sqrt{2}\ \psi_2 + \psi_3)$$

$$\psi_n = \frac{\sqrt{2}}{2}(\psi_1 - \psi_3)$$

$$\psi_b = \frac{1}{2}(\psi_1 + \sqrt{2}\psi_2 + \psi_3)$$

Only the bonding (ψ_b) and nonbonding (ψ_n) orbitals are doubly occupied. The orbital overlap can be shown as

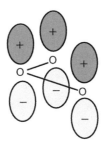

The ozone molecule has a dipole moment of 0.534 D, which is largely due to the unshared pair of electrons on the central atom. With its heat of formation being $+143$ kJ/mol, ozone is unstable with respect to oxygen. It can decompose explosively in the presence of catalysts or ultraviolet radiation. Several substances catalyze the decomposition including Na_2O, K_2O, MgO, Al_2O_3, and Cl_2. The real value of ozone lies in the fact that in the upper atmosphere, it absorbs ultraviolet radiation. The maximum in the absorption band is at 255 nm. It is believed that chlorofluorocarbons used as a refrigerating gas and as a propellant for aerosols are contributing to a reduction in the ozone layer. This has been the subject of much research and discussion and has led to discontinuing the widespread use of chlorofluorocarbons.

For laboratory use, ozone is normally produced at the time and location of use by means of an ozonizer. This device bombards oxygen with low-frequency electrical oscillation (so-called silent electric discharge) in a flow system. The effluent gas contains several percent O_3, which is usually passed directly into the reaction vessel. The reaction can be shown as

$$3\ O_2 \xrightarrow[\text{discharge}]{\text{electric}} 2\ O_3 \tag{13.48}$$

The ozone molecule can accept an electron to produce an *ozonide*, O_3^-. One of the most common stable ozonides is KO_3, which is a powerful and useful oxidizing agent. It is stable at temperatures below 60 °C, but it reacts with water,

$$4\ KO_3 + 2\ H_2O \rightarrow 4\ KOH + 5\ O_2 \tag{13.49}$$

Ozone itself is a useful and strong oxidizing agent that is comparable in strength to fluorine and atomic oxygen. It is particularly useful in situations where a "clean" oxidizing agent is needed because the reduction product is only oxygen. Water purification is such a situation, and ozone is an effective germicide. It will also oxidize some toxic materials to less harmful forms. For example, it reacts with cyanides, cyanates, and other anions. The reaction with cyanate can be shown as

$$2\ OCN^- + H_2O + 3\ O_3 \rightarrow 2\ HCO_3^- + 3\ O_2 + N_2 \tag{13.50}$$

Ozone can also be used to transform metals into their highest oxidation states, which is often desirable when separation is desired.

Ozone is a useful oxidizing agent in organic reactions especially when double bonds are involved. The product of such a reaction is known as an *ozonide*, and the general reaction can be shown as

(13.51)

These reactive intermediates are useful precursors in preparing other types of compounds.

13.3.3 Preparation of Oxygen

The atmosphere is the most common source of oxygen, and oxygen ranks third in terms of production of individual chemicals. When liquid air is distilled, nitrogen having the lower boiling point (−196 °C) is separated from oxygen (−183 °C). Electrolysis of water produces both hydrogen and oxygen, and the decomposition of 30% hydrogen peroxide is also a suitable preparative technique.

When it is desired to produce oxygen on a laboratory scale, the decomposition of some oxygen containing compound is the usual procedure. Historically, oxygen was prepared by the decomposition of a metal oxide such as HgO.

$$2 \text{ HgO} \xrightarrow{\text{heat}} 2 \text{ Hg} + O_2 \qquad (13.52)$$

When heated, peroxides decompose to produce oxides and oxygen. For example,

$$2 \text{ BaO}_2 \xrightarrow{\text{heat}} 2 \text{ BaO} + O_2 \qquad (13.53)$$

In a classic experiment that was conducted for decades in general chemistry laboratories, oxygen can be prepared by the decomposition of $KClO_3$ in the presence of MnO_2.

$$2 \text{ KClO}_3 \xrightarrow[\text{Heat}]{\text{MnO}_2} 2 \text{ KCl} + 3 O_2 \qquad (13.54)$$

The equation is deceptively simple in appearance because the reaction is very complex. It has been found that some of the MnO_2 is converted into potassium permanganate, which eventually decomposes according to the equation

$$2 \text{ KMnO}_4 \rightarrow \text{K}_2\text{MnO}_4 + \text{MnO}_2 + O_2 \qquad (13.55)$$

The glassware used to hold the $KClO_3$ must be scrupulously clean from any material that is combustible. Molten $KClO_3$ is an exceedingly strong oxidizing agent and many materials will react explosively with it. As shown in Eqn (13.55), the MnO_2 is eventually regenerated even though it is a participant in the reaction. Although Eqn (13.54) gives the appearance that the reaction is relatively simple, it is actually quite complex.

When solutions containing hydroxides are electrolyzed, the hydroxide ion becomes discharged at the anode as the oxidation process.

$$4 \text{ OH}^- \xrightarrow{\text{electricity}} 2 \text{ H}_2\text{O} + O_2 + 4 \text{ e}^- \qquad (13.56)$$

13.3.4 Binary Compounds of Oxygen

When considering the formation of ionic oxides, it should be remembered that the addition of two electrons is a very unfavorable process. Adding the first electron *liberates* about 142 kJ mol^{-1}, but adding the second to produce O^{2-} *absorbs* 703 kJ mol^{-1}. Therefore, the process

$$O + 2 \text{ e}^- \rightarrow O^{2-} \qquad (13.57)$$

is not energetically favorable. It is the interaction of the ions to produce a crystal lattice that makes the existence of ionic oxides possible. Although oxygen compounds exist for almost all other elements, the oxides of soft metals are not very stable. Oxides of metals such as silver and mercury are easily decomposed. On the other hand, oxides that contain hard cations such as Mg^{2+}, Fe^{3+}, Al^{3+}, Be^{2+}, or Cr^{3+} are some of the most stable compounds known. As expected, the lattice energies of such compounds are very high.

Although most metals react with oxygen to form oxides, the reactions of the Group IA metals do not always give the expected products as shown by the following equations:

$$4 \, Li + O_2 \rightarrow 2 \, Li_2O \text{ (a normal oxide)} \tag{13.58}$$

$$2 \, Na + O_2 \rightarrow Na_2O_2 \text{ (a peroxide)} \tag{13.59}$$

$$K + O_2 \rightarrow KO_2 \text{ (a superoxide)} \tag{13.60}$$

Both rubidium and cesium react with oxygen to give to superoxides. Group II metals follow a similar pattern with Be, Mg, Ca, and Sr giving normal oxides and Ba giving the peroxide. Radium may give either a peroxide or superoxide depending on the reaction conditions. It is important to remember that as oxygen reacts with a metal, there is a higher concentration of oxygen at the surface. Therefore, the metal will be found in an oxide in which it is in the highest oxidation state at the surface. Below the surface, the oxidation state of the metal is generally lower, and many oxides are not stoichiometric in composition.

With air in contact with the earth's surface, it is not surprising that several metals are found in ores that contain the metal oxide. Metal replacement reactions are possible due to differences in lattice energies and reduction potentials. One interesting reaction of this type is

$$Fe_2O_3 + 2 \, Al \rightarrow Al_2O_3 + 2 \, Fe \tag{13.61}$$

Known as the *thermite reaction*, this process is so strongly exothermic that the iron is produced in the molten state. In this case, the replacement Fe^{3+} by Al^{3+} is very favorable because Al^{3+} is a smaller, harder, less polarizable ion, so this reaction is in agreement with the hard-soft interaction principle (see Chapter 9).

Because of its high negative charge density, the oxide ion is a very strong Brønsted base. Therefore, when an ionic oxide is placed in water, there is proton transfer to produce hydroxide ions.

$$Na_2O + H_2O \rightarrow 2 \, NaOH \tag{13.62}$$
$$CaO + H_2O \rightarrow Ca(OH)_2 \tag{13.63}$$

Many metal hydroxides are insoluble. For example, when MgO is made into a slurry, the white suspension is known as milk of magnesia. Calcium oxide (*lime*) is produced by heating limestone (*calcium carbonate*) at high temperature.

$$CaCO_3 \xrightarrow{\text{heat}} CaO + CO_2 \tag{13.64}$$

When the oxide is added to water, the product is known as *hydrated* lime or *slaked* lime. Lime is one ingredient in mortar and cement, so it is produced on an enormous scale. When $Ca(OH)_2$ reacts with CO_2 from the atmosphere, it forms calcium carbonate, which binds particles of concrete together.

When water is added to a metal oxide, it may react to produce the hydroxide, but the reaction may not be complete. Thus, if the metal has a +3 charge, the product may consist of a mixture of M_2O_3, $M(OH)_3$, $M(OH)_x^{z-}$, and $M_2O_3 \cdot xH_2O$. The first of these is an oxide, the second and third are hydroxides, and the last is a hydrated oxide (also known as a *hydrous* oxide). In many cases, there is a complex equilibrium involving all of these species, so the exact nature of the products when a metal oxide reacts with water may be variable in composition.

Several elements form polyanions of several types. This type of behavior depends on the concentration and pH of the solution. The species can be considered as arising when additional metal oxide is added to a solution of the parent acid (as when SO_3 is dissolved in H_2SO_4 to give $H_2S_2O_7$, which contains the $S_2O_7^{2-}$ "condensed" sulfate). One type of polyanion is referred to as being isopoly because it contains only one element other than oxygen. Heteropolyanions result when oxides of different metals are condensed. The polyanions of several metals have been investigated in detail, especially those of tungsten and vanadium. Some of the species that have been identified are $V_2O_7^{2-}$, $W_2O_8^{4-}$, $W_4O_{16}^{8-}$, $W_{10}O_{32}^{4-}$, etc. In many cases, the structures of the larger ions involve octahedra of MO_6 species that are joined on edges.

13.3.5 Covalent Oxides

When oxygen combines with nonmetals, covalent oxides are produced. Also, there are numerous polyatomic ions that contain oxygen covalently bound to metals. These include species such as MnO_4^-, CrO_4^{2-}, and VO_4^{3-} as well as numerous cationic species such as VO^{3+}, UO_2^{2+}, CrO^{3+}, etc. In many cases, nonmetals react directly with oxygen to produce covalent oxides. Generally, if the reaction takes place in air (an excess of oxygen), the product will contain the nonmetal in its highest oxidation state. For example,

$$P_4 + 5\ O_2 \rightarrow P_4O_{10} \tag{13.65}$$

$$C + O_2 \rightarrow CO_2 \tag{13.66}$$

One notable exception is the reaction with sulfur, which produces the dioxide not SO_3.

$$S + O_2 \rightarrow SO_2 \tag{13.67}$$

When the nonmetal is in excess, the product contains the nonmetal in a lower oxidation state as illustrated by the reactions

$$2\ C + O_2 \rightarrow 2\ CO \tag{13.68}$$

$$P_4 + 3\ O_2 \rightarrow P_4O_6 \tag{13.69}$$

Structures of several nonmetal oxides were discussed in Chapter 4, and others will be discussed with the chemistry of the central atom.

As was discussed in Chapter 9, covalent oxides react with water to produce acids. Some examples are the following:

$$SO_3 + H_2O \rightarrow H_2SO_4 \tag{13.70}$$

$$P_4O_{10} + 6\ H_2O \rightarrow 4\ H_3PO_4 \tag{13.71}$$

$$CO_2 + H_2O \rightleftharpoons H^+ + HCO_3^- \tag{13.72}$$

Acidic and basic oxides frequently react directly to produce salts because they are the anhydrides of acids and bases.

$$CaO + SO_3 \rightarrow CaSO_4 \tag{13.73}$$

$$BaO + CO_2 \rightarrow BaCO_3 \tag{13.74}$$

13.3.6 Amphoteric Oxides

Although oxides of metals and nonmetals have been generally considered to give bases and acids, respectively, when they react with water, there are other oxides that can behave in both ways. These are the amphoteric oxides, and they include the oxides of zinc and aluminum. For example, ZnO undergoes the following reactions.

$$ZnO + 2\,HCl \rightarrow ZnCl_2 + H_2O \tag{13.75}$$

$$ZnO + 2\,NaOH + H_2O \rightarrow Na_2Zn(OH)_4 \tag{13.76}$$

The first reaction is typical of a basic metal oxide, whereas in the second, Zn^{2+} is reacting as an acid. When written starting with the hydroxide, the equivalent equations are

$$Zn(OH)_2 + 2\,H^+ \rightarrow Zn^{2+} + 2\,H_2O \tag{13.77}$$

$$Zn(OH)_2 + 2\,OH^- \rightarrow Zn(OH)_4^{2-} \tag{13.78}$$

As was shown in Chapter 9 (see Figure 9.3), the oxides of elements in the second period span the range from those that are the anhydrides of strong bases to those that are the anhydrides of strong acids. In the middle, there are a few that give solutions that are neither acidic nor basic, and these are the amphoteric cases. It is also interesting to note the trend in acid–base behavior when progressing down in a given group. For example, CO_2 is a weakly acidic oxide, whereas PbO_2 is a weakly basic oxide. This is in agreement with the fact that the elements become more metallic when progressing down the group.

Although they are oxides of nonmetals, CO and N_2O do not give acidic solutions when added to water. However, they are formally the anhydrides of formic and hyponitrous acid, respectively.

$$CO + H_2O \rightarrow H_2CO_2 \; (HCOOH, Formic\ acid) \tag{13.79}$$

$$N_2O + H_2O \rightarrow H_2N_2O_2 \; (Hyponitrous\ acid) \tag{13.80}$$

13.3.7 Peroxides

As shown earlier, the alkali metals do not react with oxygen to produce normal oxides when there is an excess of oxygen available. In some cases, a peroxide is obtained, which will react with water to produce hydrogen peroxide.

$$Na_2O_2 + 2\,H_2O \rightarrow H_2O_2 + 2\,NaOH \tag{13.81}$$

The most common form of H_2O_2 available in retail stores is a 3% solution, but this solution is an effective disinfectant. It is possible to concentrate hydrogen peroxide by distillation up to a concentration of approximately 30%. When that concentration is reached, the rate of decomposition is high enough that no further concentration occurs. Decomposition of hydrogen peroxide is catalyzed by minute traces of transition metal ions, which is easily seen when a drop of 3% H_2O_2 is placed on a cut where blood is exposed. When handling concentrated H_2O_2, the containers must be clean and free of metal compounds. Solutions that contain about 90% H_2O_2 are very strong oxidizing agents that have been used as the oxidizing agent in rockets. The decomposition of hydrogen peroxide can be shown as

$$2 \ H_2O_2 \rightarrow 2 \ H_2O + O_2 \qquad \Delta H = -100 \ kJ \ mol^{-1} \qquad (13.82)$$

The hydrogen peroxide molecule has an interesting structure (shown below) that has been described as an "open book":

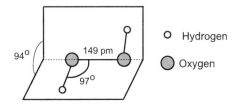

One preparation of H_2O_2 (m.p. $-0.43 \ °C$ and b.p. $150.2 \ °C$) involves converting sulfuric acid to peroxydisulfuric acid, $H_2S_2O_8$, which is done by electrolysis of cold concentrated sulfuric acid. H_2O_2 is obtained by hydrolysis of peroxydisulfuric acid,

$$H_2S_2O_8 + 2 \ H_2O \rightarrow H_2O_2 + 2 \ H_2SO_4 \qquad (13.83)$$

and it is then concentrated by distillation. Currently, H_2O_2 is produced by a process involving the reduction of 2-ethylanthroquinone dissolved in a mixture of an ester and a hydrocarbon using Raney nickel as a catalyst. After reduction of the quinone to the quinol, the catalyst is removed, and the quinol is oxidized back to the quinone. Separation of H_2O_2 is carried out by extraction. Hydrogen peroxide has many uses that vary from bleaching paper pulp to production of polymers.

The reaction of oxygen with ethers yields organic peroxides, $R-O-O-R$.

$$2 \ R-O-R + O_2 \rightarrow 2 \ R-O-O-R \qquad (13.84)$$

Organic peroxides are sensitive explosives that are so powerful that a few milligrams can cause serious injury. Some accidents have happened as a container of an ether was being opened and the small amount of peroxide that formed around the cap detonated.

13.3.8 Positive Oxygen

In addition to the considerable array of oxygen compounds discussed thus far, it is possible for oxygen to assume a positive oxidation state. This would be expected in covalent compounds only in cases where the other element is fluorine, the only atom having a higher electronegativity. Several compounds containing oxygen and fluorine have been studied with the best known being OF_2, oxygen difluoride. It is a pale yellow poisonous gas that has a b.p. of $-145\,°C$. Oxygen difluoride can be prepared by the reaction of fluorine with a dilute solution of sodium hydroxide,

$$2\,F_2 + 2\,NaOH \rightarrow OF_2 + 2\,NaF + H_2O \tag{13.85}$$

As would be expected, OF_2 is a very strong oxidizing agent that is also a fluorinating agent capable of producing both oxides and fluorides of several elements. Other known compounds that contain oxygen and fluorine include O_2F_2 and O_4F_2.

In addition to fluorine compounds that contain oxygen in a positive oxidation state, there are also compounds that contain oxygen in a positive ion. The ionization potential of the oxygen atom is 13.6 eV ($1312\,kJ\,mol^{-1}$) but that of the O_2 molecule is only 12.06 eV ($1163\,kJ\,mol^{-1}$). The NO molecule has one electron in a π^* orbital, and it has an ionization potential of 9.23 eV ($891\,kJ\,mol^{-1}$). Numerous compounds contain the NO^+ species, so it is reasonable to assume that under the right conditions, an oxygen molecule could lose an electron. What is needed is a very strong oxidizing agent, and such a compound is PtF_6. The reaction can be shown as

$$PtF_6 + O_2 \rightarrow O_2PtF_6 \tag{13.86}$$

The product contains the O_2^+ (dioxygenyl) cation, which is close enough to the size of K^+ that O_2PtF_6 and $KPtF_6$ are isomorphous. Other compounds are known that contain the O_2^+ ion including O_2AsF_6 and O_2BF_4, which can be prepared by the reaction

$$O_2 + BF_3 + 1/2\,F_2 \xrightarrow[-78\,°C]{h\nu} O_2BF_4 \tag{13.87}$$

It is interesting to note that the ionization potential of xenon is 12.127 eV ($1170\,kJ\,mol^{-1}$), which is very close to that of the O_2 molecule. Noting this, it was speculated many years ago that it should be possible to prepare compounds containing xenon. This was proved correct when in the early 1960s, Neil Bartlett carried out such a reaction and obtained a product containing Xe.

13.4 CARBON

In addition to the organic chemistry of carbon compounds, the element is also important in inorganic chemistry. In recent years, an extensive chemistry of the fullerenes, C_{60} and its derivatives, has become one of the most active new areas of inorganic and organic chemistry. There is no clear separation of the two fields even though they were believed to be separate for many years. In 1828, Friedrich Wöhler converted ammonium cyanate into urea,

$$NH_4OCN \rightarrow (H_2N)_2CO \tag{13.88}$$

This reaction showed that it was not necessary for organic compounds to be have been produced from living species. With the tremendous growth in organometallic chemistry, there is even less distinction between inorganic and organic chemistry.

13.4.1 The Element

The most extensively occurring compounds of carbon are the organic materials found in coal, petroleum, natural gas, and living plants and animals. The amount of carbon found as graphite and diamond is relatively small in comparison, but these are important materials. When all of the sources of carbon are considered, the element ranks fourteenth in terms of abundance. Elements in Group IVA of the periodic table show clearly the trend of increasing metallic character in going downward in the group. Carbon is a nonmetal, silicon and germanium are metalloids, whereas tin and lead are metallic in character.

The two isotopes of carbon that occur naturally are ^{12}C (98.89%) and ^{13}C (1.11%). Cosmic rays produce neutrons that interact with ^{14}N in the upper atmosphere to produce ^{14}C and protons,

$$^{14}N + n \rightarrow {}^{14}C + p \tag{13.89}$$

The half-life of ^{14}C is 5570 years, and living organisms contain a relatively constant amount of ^{14}C as more is ingested to replace that which decays. When the organism dies, ^{14}C is not taken in, so it is possible later to determine when the intake of ^{14}C ended by determining the amount remaining. This type of radiocarbon dating provides a convenient way to estimate the age of nonliving materials.

Carbon atoms have the ability to bond to themselves to a greater extent than those of any other element. Known as *catenation*, this ability gives rise to the several allotropic forms of the element. The most common form of elemental carbon is graphite, which has the layered structure shown in Figure 13.7.

Graphite is often used as a dry lubricant in locks because the layers that are held together only by van der Waals forces can slide easily. Graphite can have various groups between the layers to form what are known as *intercalation* compounds. This can occur in two ways. In the first, the layers move slightly farther apart but remain planar. In the second, the layers become distorted or buckled and part of the system of π bonding is interrupted. Although the reaction of graphite with fluorine at high temperature produces CF_4, the reaction carried out at low temperature produces a compound having the stoichiometry $(CF)_n$. The structure of $(CF)_n$ has layers that are not planar owing to the disruption of the π bonding system and the formation of bonds to the fluorine atoms. Each carbon atom is bonded to a fluorine atom with alternating C−F bonds being above and below each layer. As a result, the layers are much farther apart than in graphite (about 800 pm), but the layers are able to slide along each other, so the material is a lubricant.

As a result of the mobility of the electrons in π orbitals, graphite is a conductor of electricity. It is also the form of carbon used as the thermodynamic standard state. On the other hand, diamond contains carbon atoms that are bonded to four others, so all of the electrons are used in localized bonding, and it is a nonconductor that has the structure shown in Figure 13.8.

The density of diamond is 3.51 g cm^{-3}, whereas that of graphite is 2.22 g cm^{-3}. Although for the reaction

$$C(graphite) \rightarrow C(diamond) \tag{13.90}$$

$\Delta H = 2.9$ kJ mol^{-1} at 300 K and 1 atm, there is no low-energy pathway for the transformation, so the process is difficult to carry out. However, synthetic diamonds are produced on a large scale at high temperature and pressure (3000 K and 125 kbar). The conversion of graphite to diamonds is catalyzed

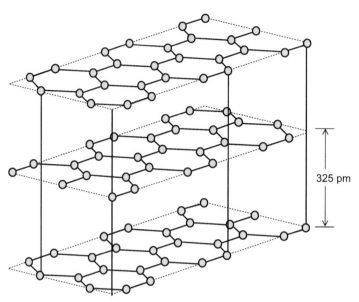

325 pm

FIGURE 13.7
The layered structure of graphite.

by several metals (i.e. chromium, iron, and platinum) that are in the liquid state. It is believed that the molten metal dissolves a small amount of graphite that crystallizes as diamond because it is less soluble in the liquid metal. Diamond is extremely hard, so diamond-tipped tools are important in manufacturing processes.

In 1985, a form of carbon having the formula C_{60} was identified by Smally, Kroto, and coworkers. Because the structure of C_{60} resembles that of the famous geodesic dome designed by R. Buckminster Fuller, the molecule has become known as *fullerene* or *buckminsterfullerene*, and the enormous number of derivatives that have been produced are known collectively as the *fullerenes*. When a high-density electric current is passed between graphite rods in a helium atmosphere, soot is produced some of which dissolves in toluene. It was found that this soluble soot contained C_{60}, which was identified by a peak at 720 amu in the

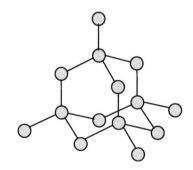

FIGURE 13.8
The structure of diamond showing four covalent bonds to each carbon atom.

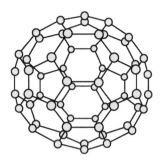

FIGURE 13.9
The structure of buckminsterfullerene, C_{60}. The structure consists of five- and six-membered rings.

mass spectrum. It was also determined that there were other aggregates of carbon atoms, which included C_{70} and others in minor amounts. The structure of C_{60} (shown in Figure 13.9) has a surface in which the atoms are arranged in pentagons and hexagons as does the geodesic dome.

When graphite is impacted by a high-powered laser, a large number of fragments are produced that have from 44 to 90 carbon atoms. Many smaller aggregates containing both linear and cyclic structures have also been identified, some from "amorphous" forms of carbon.

Like graphite, C_{60} can be transformed into diamond, but the process requires less stringent conditions. It has also been found that C_{60} becomes a superconductor at low temperature. Another interesting characteristic of C_{60} is that when it is prepared in the presence of certain metals, the C_{60} cage can enclose a metal atom. In some cases, other materials can be enclosed within the C_{60} cage in a "shrink-wrapped" manner to form "complexes" that are described as *endohedral*. It has also been possible to prepare metal complexes of C_{60} that contain metal–carbon bonds. A compound of this type is $((C_6H_5P)_2PtC_{60})$.

In addition to graphite, diamond, and C_{60}, carbon exists in several "amorphous" forms. These include charcoal, soot, lampblack, and coke, some of which have important industrial uses.

$$Coal \rightarrow C(coke) + Volatiles \qquad (13.91)$$

$$Wood \rightarrow C(charcoal) + Volatiles \qquad (13.92)$$

These forms of carbon are also known to have some order, so they are not completely amorphous. When appropriately prepared (so-called *activated* charcoal), charcoal has an enormous surface area, so it is capable of adsorbing many substances from both gases and solutions. As was described in Chapter 11, coke is used on an enormous scale as a reducing agent in the production of metals. The "amorphous" forms of carbon can be transformed into graphite by means of the *Acheson process* in which an electric current heats a rod of the "amorphous" form.

13.4.2 Carbon in Industry
Composite materials are made up of two or more materials that have different properties. These materials are combined (in many cases involving chemical bonding) to produce a new material that has properties superior to either material alone. An example of this type of composite is fiberglass, in which glass fibers are held together by a polymeric resin.

The term *advanced composite* is usually applied to a matrix of resin material that is reinforced by fibers of carbon, boron, glass, or other material that have high-tensile strength. The materials can be layered to achieve the desired results. A common type of advanced composite is an epoxy resin that is reinforced with carbon fibers, alone or in combination with glass fibers, in a multilayered pattern. Such composites are rigid, light weight, and have excellent resistance to weakening (fatigue resistance). In fact, their fatigue resistance may be better than that of steel or aluminum. Such properties make these materials suitable for aircraft and automobile parts, tennis racquets, golf club shafts, skis, bicycle parts, fishing rods, etc.

Fibers in reinforced composites may be of different lengths. In one type of composite, the fibers are long, continuous, and parallel. In another type, the fibers may be chopped discontinuous fibers that are arranged more or less randomly in the resin matrix. The properties of composites are dependent on their construction. Composites of carbon fiber and resins have high strength to weight ratio and great stiffness. Their chemical resistance is also high, and they are unreactive toward bases. It is possible to prepare such materials that are stronger and stiffer than steel objects of the same thickness yet weigh 50–60% less.

Because carbon fibers of different diameters are available and because construction parameters can be varied, it is possible to engineer composites having desired characteristics. By varying the orientation, concentration, and type of fiber, materials can be developed for specific applications. The fibers can be layered at different angles to minimize directional differences in properties. Also, layers of fibers can be impregnated with epoxy resin to form sheets that can be shaped prior to polymerization of the resin.

As a result of their stiffness, strength, and light weight, carbon fiber composites are used in many applications in aircraft and aerospace fabrication, e.g., panels, cargo doors, etc. Also, because of their high-temperature stability and lubricating properties, they are also used in bearings, pumps, etc. One limitation to the use of carbon fibers is the high cost of producing them, which can be as high as hundreds of dollars per pound. Military and aerospace applications have been the major uses of composite materials containing carbon fibers.

In addition to the use of carbon fibers in composites, there is extensive use of carbon in other manufacturing processes. For example, a mixture of coke and graphite powder can be prepared and then bonded with carbon. Typically, the carbon is added to the mixture in the form of a binder such as coal tar, pitch or a resin. The mixture is formed into the desired shape by compression molding or extrusion. Firing the object at high temperature (up to 1300 °C) in the absence of oxygen causes the binder to be converted to carbon that binds the mass together. The wear resistance and lubricating properties of the finished part can be controlled by the composition and characteristics of the coke, graphite, and binder. However, the process results in an object that may have pores. Impregnation with metals, resins, fused salts, or glasses is sometimes carried out to fill the pores. By so doing, the properties of the object can be controlled to some degree. It is possible to machine objects made of these materials to close tolerances. Materials produced in this way are also good conductors of heat and electricity. Because of the graphite present, they are also self-lubricating, and they are relatively inert toward most solvents, acids, and bases. At high temperatures, oxygen slowly attacks these materials, and they react slowly with oxidizing agents such as concentrated nitric acid. Although the carbon/graphite materials are brittle, they are stronger at high temperatures (2500–3000 °C) than they are at room temperature.

The properties described above make manufactured carbon a very useful material. It is used in applications such as bearings, valve seats, seals, dies, tools, molds, fixtures, etc. Specific uses of the final object may require materials that have been prepared to optimize certain properties. The fact that manufactured carbon can be prepared in the form of rods, rings, plates, tubes, and other configurations makes it possible to machine parts of many types. Manufactured carbon represents a range of materials that have many important industrial uses.

13.4.3 Chemical Behavior of Carbon

Perhaps the most important use of carbon is as a reducing agent because it is the least expensive reducing agent used on a large scale. Two of the major uses of carbon as a reducing agent are the production of iron,

$$Fe_2O_3 + 3\ C \rightarrow 2\ Fe + 3\ CO \tag{13.93}$$

and the production of phosphorus,

$$2\ Ca_3(PO_4)_2 + 6\ SiO_2 + 10\ C \rightarrow P_4 + 10\ CO + 6\ CaSiO_3 \tag{13.94}$$

Reduction of metals from their ores using carbon (charcoal) has been utilized for many centuries. One problem associated with using carbon as a reducing agent is that carbon is a solid, and when an excess is used (which is necessary for a reaction to go to completion in a reasonable time), the product contains some carbon. It is not a "clean" reducing agent like hydrogen where the reducing agent itself and its oxidation product are gases. However, when cost is a factor, as it is in all large-scale industrial processes, carbon may be the reducing agent of choice. Another use of carbon is as an absorbent because in the form of activated charcoal, it absorbs many substances. Carbon forms a large number of binary compounds that vary widely in properties. They range from gases such as CO and CO_2 to tungsten carbide, a hard refractory abrasive.

13.4.4 Carbides

Compounds containing carbon in a negative oxidation state are properly called *carbides* and many such compounds are known. In a manner analogous to the behavior of hydrogen and boron, carbon forms three types of binary compounds, which are usually called *ionic, covalent,* and *interstitial carbides*.

If carbon is bonded to metals having low electronegativities, the bonds are considered to be ionic with carbon having the negative charge. Such metals include Groups IA and IIA, Al, Cu, Zn, Th, V, etc. Because carbon is in a negative oxidation state, reactions of these compounds with water produce a hydrocarbon. In some cases, the hydrocarbon is methane, but in others, acetylene is produced. Therefore, the carbides are called *methanides* and *acetylides*, respectively. One of the most useful acetylides is calcium acetylide, CaC_2, which is also referred to as calcium carbide. It has the structure shown in Figure 13.10.

The acetylide ion, C_2^{2-}, is isoelectronic with N_2, CO, and CN^-. The reaction producing acetylene from calcium carbide can be shown as

$$CaC_2 + 2\ H_2O \rightarrow Ca(OH)_2 + C_2H_2 \tag{13.95}$$

When water is allowed to drip on CaC_2, acetylene is produced, and it is flammable.

$$2\ C_2H_2 + 5\ O_2 \rightarrow 4\ CO_2 + 2\ H_2O \tag{13.96}$$

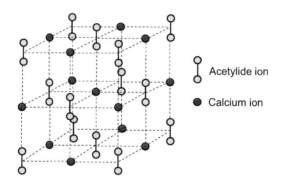

FIGURE 13.10

The structure of CaC_2. Note that the cubic arrangement is similar to that of NaCl except for the diatomic C_2^{2-} ion.

One type of portable light source consists of two chambers, the upper for water and the lower for CaC_2, that are connected by a tube having a valve to control the rate of dripping of the water. The acetylene produced is allowed to escape through an orifice in the middle of a reflector. When the acetylene burns, the light is directed by the reflector. This type of light is referred to as "carbide light" or "miners lamp" that was in common use in the past by coal miners.

Calcium carbide can be produced by the direct reaction of carbon with the metal or metal oxide:

$$CaO + 3\ C \rightarrow CaC_2 + CO \tag{13.97}$$

Other carbides, e.g., Be_2C and Al_4C_3, behave as if they contain C^{4-} and react with water to produce CH_4.

$$Al_4C_3 + 12\ H_2O \rightarrow 4\ Al(OH)_3 + 3\ CH_4 \tag{13.98}$$

$$Be_2C + 4\ H_2O \rightarrow 2\ Be(OH)_2 + CH_4 \tag{13.99}$$

When carbon forms compounds with other elements (Si, B, etc.) that have an electronegativity close to that of carbon, the bonds are considered to be covalent. The compounds, especially SiC, have the characteristics of being hard, unreactive refractory materials. Silicon carbide has a structure similar to diamond, and it is widely used as an abrasive. It is prepared by the reaction of SiO_2 with carbon.

$$SiO_2 + 3\ C \rightarrow SiC + 2\ CO \tag{13.100}$$

When most transition metals are heated with carbon, the lattice expands and carbon atoms occupy interstitial positions. The metals become harder, have higher melting points, and more brittle as a result. For example, when an iron object is heated and placed in a source of carbon atoms (charcoal or oils have been used historically), some iron carbide, Fe_3C, forms on the surface. If the object is cooled quickly (quenched), the carbide remains primarily on the surface, and a hard durable layer results. This process, known as *casehardening*, was very important prior to the development of modern heat treatment processes for steels. Other carbides, ZrC, TiC, MoC, and WC, are sometimes used in making tools for cutting, drilling, and grinding.

13.4.5 Carbon Monoxide

The reaction of carbon with oxygen can lead to different products. When there is a deficiency of oxygen, the product is carbon monoxide.

$$2\,C + O_2 \rightarrow 2\,CO \qquad (13.101)$$

Carbon monoxide can also be obtained by the reaction of carbon with CO_2.

$$C + CO_2 \rightarrow 2\,CO \qquad (13.102)$$

Because the oxides of nonmetals are acid anhydrides, CO is formally the anhydride of formic acid.

$$CO + H_2O \rightarrow HCOOH \qquad (13.103)$$

Although this reaction does not take place readily, the dehydration of formic acid does produce CO.

$$HCOOH \xrightarrow{H_2SO_4} CO + H_2O \qquad (13.104)$$

Because CO is a slightly acidic oxide, it reacts with bases to produce formates.

$$CO + OH^- \rightarrow HCOO^- \qquad (13.105)$$

As we have already mentioned, carbon is a reducing agent of great importance, and CO is often the oxidation product (see Eqns (13.93) and (13.94)). The reaction of carbon with steam at high temperature also produces CO.

$$C + H_2O \rightarrow CO + H_2 \qquad (13.106)$$

This reaction is the basis for the *water gas process* that was discussed earlier in this chapter in connection with the preparation of hydrogen.

The CO molecule is isoelectronic with N_2, CN^-, and C_2^{2-} and has a structure that can be shown as

$$|\overset{\ominus}{C} \equiv \overset{\oplus}{O}|$$

and the carbon end of the molecule carries a negative formal charge. This is the electron-rich end of the molecule and when it forms complexes with metals (metal carbonyls), it is the carbon end that binds to the metal (see Chapter 22). In some cases, CO forms bridges between two metals atoms. A few metal carbonyls can be formed by direct combination of the metal and CO.

$$Ni + 4\,CO \rightarrow Ni(CO)_4 \qquad (13.107)$$

$$Fe + 5\,CO \xrightarrow{T,\,P} Fe(CO)_5 \qquad (13.108)$$

The formulas for the metal carbonyls are determined by the number of pairs of electrons needed by the metal to reach the number of electrons in the next noble gas atom. Thus, the stable carbonyl with nickel contains four CO molecules, that with iron contains five, and that with chromium contains six. The bonding in these complexes will be discussed in more detail in Chapter 16.

Carbon monoxide is a very toxic gas that forms a stable complex by binding to iron in the heme structure in the blood. Because CO bonds to iron more strongly than does O_2, it prevents O_2 molecules from binding. Therefore, CO destroys the oxygen-carrying capacity of the blood. Moreover, CO is a cumulative poison because it requires a long time for the bound CO to be lost. When someone has been exposed to CO, treatment usually involves placing the subject in an atmosphere that is richer in oxygen to improve O_2 binding by Le Chatelier's principle.

Carbon monoxide is a reducing agent, and it burns readily,

$$2\ CO + O_2 \rightarrow 2\ CO_2 \quad \Delta H = -283\ kJ\ mol^{-1} \tag{13.109}$$

In addition to being a reducing agent in the production of metals, CO is used to produce methanol by the reaction

$$CO + 2\ H_2 \xrightarrow[\text{catalyst}]{250°,\ 50\ atm} CH_3OH \tag{13.110}$$

A commonly used catalyst in this process consists of ZnO and Cu. Because methanol is a widely used solvent and fuel, the reaction is economically significant.

13.4.6 Carbon Dioxide and Carbonates

The most familiar oxide of carbon is CO_2. Solid CO_2 sublimes at $-78.5\ °C$ and because no liquid phase is present when this occurs, solid CO_2 is called "dry" ice, and it is widely used in cooling operations. It is obtained by burning carbon in an excess of oxygen,

$$C + O_2 \rightarrow CO_2 \tag{13.111}$$

or by the reaction of a carbonate with an acid,

$$CO_3^{2-} + 2\ H^+ \rightarrow H_2O + CO_2 \tag{13.112}$$

A reaction of carbon dioxide that is essential for life forms is that of photosynthesis in which CO_2 is converted by plants into glucose. The process can be summarized as

$$6\ CO_2 + 6\ H_2O \xrightarrow{h\nu} C_6H_{12}O_6 + 6\ O_2 \tag{13.113}$$

However, the process is very complex and involves several intermediates such as chlorophyll (of which there is more than one type). Chlorophylls are porphyrins that contain magnesium. The growth of plants is responsible for the production of oxygen in the atmosphere as well as food and natural fibers.

The carbon dioxide molecule has a linear structure that can be shown as

$$\bar{\underline{O}}{=}c{=}\bar{\underline{O}}$$

and it is nonpolar. It is the anhydride of carbonic acid, H_2CO_3, so solutions of CO_2 are slightly acidic as a result of the reaction.

$$2\ H_2O + CO_2 \rightleftarrows H_3O^+ + HCO_3^- \tag{13.114}$$

Also, CO_2 will react with metal oxides to produce carbonates,

$$CO_2 + O^{2-} \rightarrow CO_3^{2-} \tag{13.115}$$

Carbon dioxide in the atmosphere has some influence on the composition of ores. The decay of organic matter can produce carbon dioxide, CO_2, and CO_2 reacts with many metal oxides to produce metal carbonates. For example,

$$CaO + CO_2 \rightarrow CaCO_3 \tag{13.116}$$

$$CuO + CO_2 \rightarrow CuCO_3 \tag{13.117}$$

Because the carbonate ion can react as a base, it can react with water to produce hydroxide ions, and H^+ and CO_3^{2-} form bicarbonate ions.

$$CO_3^{2-} + H_2O \rightarrow HCO_3^- + OH^- \tag{13.118}$$

Therefore, as an oxide mineral "weathers", a bicarbonate may result from the reactions of metal oxides with water and CO_2. Eventually, the presence of OH^- can cause part of the metal-containing mineral to be converted into a metal hydroxide, and most metal oxides react with water to produce hydroxides. For example,

$$CaO + H_2O \rightarrow Ca(OH)_2 \tag{13.119}$$

As a result of these processes, a metal oxide may be converted into a metal carbonate, a metal hydroxide, or a so-called *basic* metal carbonate (a compound containing both carbonate and hydroxide ions). A case of this type is $CuCO_3 \cdot Cu(OH)_2$ or $Cu_2CO_3(OH)_2$, which is the mineral *malachite*. The mineral *azurite* is quite similar and has the composition $2CuCO_3 \cdot Cu(OH)_2$ or $Cu_3(CO_3)_2(OH)_2$. *Azurite* and *malachite* are usually found together because both are secondary minerals produced by weathering processes on CuO. The weathering process clarifies why metals are sometimes found in compounds having unusual formulas.

In addition to being a base, the carbonate ion is also an electron pair donor that forms many complexes with metals (see Chapter 16). It has the ability to bond at either one or two sites around the metal or it can form bridges between metal ions.

Many carbonates have important uses. One of the most important carbonates is calcium carbonate, and it is found in several mineral forms such as *calcite*. The most widely occurring form of calcium carbonate is *limestone*. It is found in many places, and it has been used as a building material for thousands of years. When heated to high temperature, most carbonates lose carbon dioxide and are converted to the oxides. The most important reaction of this type involves heating $CaCO_3$ strongly (known as *calcining*) to convert it into another useful material, lime (CaO). The process of producing lime is sometimes called *lime burning*.

$$CaCO_3 \rightarrow CaO + CO_2 \tag{13.120}$$

Lime has been produced in this way for thousands of years. The loss of CO_2 corresponds to a 44% mass loss, but in ancient times, the material was considered ready for use if the mass loss was as much as one third of the original. Lime is used in huge quantities in making mortar, glass, etc. and to make calcium hydroxide (*hydrated lime*). Although $Ca(OH)_2$ is only slightly soluble in water, it is a strong base and is cheaper than NaOH, so it is widely used as a strong base.

Mortar, a mixture of lime, sand, and water, has been used in construction for thousands of years. The Appian Way, many early Roman and Greek buildings, and the Great Wall of China were constructed using mortar containing lime. In the Western Hemisphere, the Incas and Mayans used lime in mortar. The composition of mortar can vary rather widely, but the usual composition is about one-fourth lime, three-fourths sand and a small amount of water to make the mixture into a paste. Essential ingredients are a solid such as sand and lime that is converted to $Ca(OH)_2$ by reaction with water.

$$CaO + H_2O \rightarrow Ca(OH)_2 \qquad (13.121)$$

Calcium hydroxide reacts with carbon dioxide to produce calcium carbonate, $CaCO_3$,

$$Ca(OH)_2 + CO_2 \rightarrow CaCO_3 + H_2O \qquad (13.122)$$

so the particles of solid are held together by $CaCO_3$ to form a hard durable mass. Essentially, artificial limestone is reformed by the reaction of $Ca(OH)_2$ with CO_2.

Concrete is the man-made material used in the largest quantity. It is made from inexpensive materials that are found in most places throughout the world. The raw materials are often assembled from large stockpiles at the construction site. The ingredients are *aggregate* (sand, gravel, crushed rock, etc.) and *cement*, the binding agent. Lime is used in the manufacture of cement, the most common type being Portland cement, which also contains sand (SiO_2) and other oxides (aluminosilicates). Particles of aggregate bond more effectively when they have rough surfaces, so good adhesion occurs in all directions around each particle. A good mix should also have aggregate composed of a range of particle sizes rather than particles of a uniform size so that smaller particles can fill the spaces between larger ones. Basalt rock, crushed limestone, and quartzite are common materials used as aggregate.

Portland cement is made by heating calcium carbonate, sand, aluminosilicates, and iron oxide to about 870 °C. Kaolin, clay, and powdered shale are the sources of aluminosilicates. Heated strongly, this mixture loses water and carbon dioxide to form a solid mass. A small amount of calcium sulfate is added to this solid material after it is pulverized. Mortar used today consists of sand, lime, water, and cement. Concrete usually consists of aggregate, sand, and cement, which binds the materials together by reacting with water to form a mixture known as *tobermorite gel*. This material consists of layers of crystalline material that has water interspersed between them. To develop strength in the product, the correct amount of water must be used. Using too little water results in air being trapped in the mass giving a porous structure. If too much water is used, its evaporation and escape from the solid can give a structure that is porous. In either case, the concrete is not sufficiently strong. For some applications, concrete is reinforced by allowing it to harden around metal rods or wire. The complex chemical reactions involved in the hardening of concrete will not be discussed here, but it is worthwhile to emphasize the importance of the reactions of calcium hydroxide with CO_2 that take place in concrete, mortar, and related materials.

Another indispensable carbonate is sodium carbonate, which is also known as *soda ash*. Centuries ago, impure sodium carbonate was obtained from the places where brine solutions had evaporated and from dry lake beds. The major source of soda ash today is once again from a natural source, but prior to 1985, it was synthesized in large quantities. The synthetic process most often used was the *Solvay process* represented by the equations

$$NH_3 + CO_2 + H_2O \rightarrow NH_4HCO_3 \tag{13.123}$$

$$NH_4HCO_3 + NaCl \rightarrow NaHCO_3 + NH_4Cl \tag{13.124}$$

$$2\,NaHCO_3 \rightarrow Na_2CO_3 + H_2O + CO_2 \tag{13.125}$$

An enormous amount of Na_2CO_3 is produced annually because of its use in glass, laundry products, water softeners, paper, baking soda, and in making sodium hydroxide by the reaction

$$Na_2CO_3 + Ca(OH)_2 \rightarrow 2\,NaOH + CaCO_3 \tag{13.126}$$

In the United States, the source of sodium carbonate is the mineral *trona*, which has the formula $Na_2CO_3 \cdot NaHCO_3 \cdot 2H_2O$. The fact that sodium bicarbonate ($NaHCO_3$) is present is no surprise because the carbonate reacts with water and carbon dioxide to produce sodium bicarbonate.

$$Na_2CO_3 + H_2O + CO_2 \rightarrow 2\,NaHCO_3 \tag{13.127}$$

As shown in Eqn (13.125), heating the bicarbonate converts it to the carbonate with the loss of water and CO_2. The world's largest deposits of trona are found in Wyoming, but trona is also found in Mexico, Kenya, and Russia. The trona deposits in Wyoming are estimated to be 100 billion tons and they account for 90% of the U.S. production. The trona mines in Wyoming alone account for 30% of the world production.

Trona is processed to obtain sodium carbonate by crushing it to produce small particles and then heating it in a rotary kiln. This dehydration process produces impure sodium carbonate.

$$2\,Na_2CO_3 \cdot NaHCO_3 \cdot 2\,H_2O \rightarrow 3\,Na_2CO_3 + 5\,H_2O + CO_2 \tag{13.128}$$

For most uses, the Na_2CO_3 must be purified. This is done by dissolving it in water and separating insoluble rocky material by filtration. Organic impurities are removed by adsorption using activated charcoal. The hydrated crystals of $Na_2CO_3 \cdot H_2O$ are obtained by boiling off the excess water to concentrate the solution. The hydrated crystals are heated in a rotary kiln to obtain anhydrous sodium carbonate.

$$Na_2CO_3 \cdot H_2O \rightarrow Na_2CO_3 + H_2O \tag{13.129}$$

13.4.7 Tricarbon Dioxide

This oxide (m.p. -111.3 and b.p. $7\,^{\circ}C$), also known as carbon suboxide, C_3O_2, contains carbon in the formal oxidation state of $+4/3$. Because this is lower than its oxidation state in either CO or CO_2, the oxide is called carbon *sub*oxide. The molecule is linear and has the structure

$$\bar{\underline{o}} = c = c = c = \bar{\underline{o}}$$

The molecule contains three carbon atoms bonded together, so this suggests a method for its preparation by dehydrating an organic acid containing three carbon atoms. Because C_3O_2 is formally the anhydride of malonic acid, $HOOC-CH_2-COOH$, one way to prepare C_3O_2 is by the dehydration of that acid using a strong dehydrating agent such as P_4O_{10}.

$$3\ C_3H_4O_4 + P_4O_{10} \rightarrow 3\ C_3O_2 + 4\ H_3PO_4 \tag{13.130}$$

The reaction of C_3O_2 with water produces malonic acid, and the gas also reacts with NH_3,

$$C_3O_2 + 2\ NH_3 \rightarrow H_2C(CONH_2)_2 \tag{13.131}$$

The reaction of C_3O_2 with HCl to gives a diacyl chloride.

$$C_3O_2 + 2\ HCl \rightarrow H_2C(COCl)_2 \tag{13.132}$$

Although it is stable at low temperatures, carbon suboxide will readily burn, and it polymerizes when heated. Pentacarbon dioxide, C_5O_2, has been prepared, but like C_3O_2, it has no important uses.

13.4.8 Carbon Halides

Of the halogen compounds of carbon, the most important is CCl_4 (b.p. 77 °C) that is widely used as a solvent. However, the fully halogenated compounds are usually considered as derivatives of methane, so they are usually considered as "organic" in origin. Carbon tetrachloride is produced by the reaction

$$CS_2 + 3\ Cl_2 \rightarrow CCl_4 + S_2Cl_2 \tag{13.133}$$

The S_2Cl_2 obtained has many uses, including the vulcanization of rubber. The reaction of methane with chlorine also produces CCl_4.

$$CH_4 + 4\ Cl_2 \rightarrow CCl_4 + 4\ HCl \tag{13.134}$$

Unlike most covalent bonds between halogens and nonmetals, the C–Cl bonds do not hydrolyze in water. Even though it is still a useful solvent, CCl_4 is not used in dry cleaning as it was. The oxyhalides, $X_2C{=}O$, undergo many reactions owing to the very reactive C–X bonds. Although it was used as a war gas in WWI, phosgene, $COCl_2$, is a versatile chlorinating agent that has industrial uses. Metal bromides are produced when $COBr_2$ reacts with metal oxides in a sealed tube.

13.4.9 Carbon Nitrides

The most common compound of carbon and nitrogen is cyanogen, $(CN)_2$. The cyanide ion, CN^-, is a *pseudohalide* ion, which means that it resembles a halide ion because it forms an insoluble silver compound and it can be oxidized to the X_2 species. Cyanogen was first obtained by Gay-Lussac in 1815 by heating heavy metal cyanides.

$$2\ AgCN \rightarrow 2\ Ag + (CN)_2 \tag{13.135}$$

$$Hg(CN)_2 \rightarrow Hg + (CN)_2 \tag{13.136}$$

It can also be prepared from carbon and nitrogen by electric discharge between carbon electrodes in a nitrogen atmosphere. Numerous derivatives of cyanogens are known including cyanogen halides, XCN. These compounds form trimers known as the cyanuric halides, which have the cyclic structure

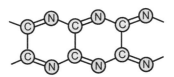

The $(CN)_2$ molecule has the linear $D_{\infty h}$ structure

116 pm   137 pm

$$|N{\equiv}C - C{\equiv}N|$$

It is a highly toxic colorless gas that burns with a violet flame to produce CO_2 and N_2.

$$(CN)_2 + 2\,O_2 \rightarrow 2\,CO_2 + N_2 \tag{13.137}$$

Polymerization of cyanogen produces *paracyanogen*.

$$n/2\,(CN)_2 \xrightarrow{\text{400-500°C}} (CN)_n \tag{13.138}$$

which has the structure

Although relatively few compounds contain only carbon and nitrogen, derivatives such as the cyanides have commercially important uses. Also, calcium cyanamide, $CaCN_2$, can be prepared by the reaction

$$CaC_2 + N_2 \rightarrow CaCN_2 + C \tag{13.139}$$

This process is important because it represents an efficient way to obtain nitrogen compounds directly using the gas, which is ordinarily unreactive. $CaCN_2$ has been used as a fertilizer because it reacts with water to produce ammonia.

$$CaCN_2 + 3\,H_2O \rightarrow 2\,NH_3 + CaCO_3 \tag{13.140}$$

This use of $CaCN_2$ is not as extensive as it formerly was. The CN_2^{2-} ion has 16 electrons so the structure can be shown as

$$\bar{N}{=}C{=}\bar{N}$$

When this ion reacts with two protons to form the parent compound, the product is $H_2N-C\equiv N$ (*cyanamide*). When this compound trimerizes, the result is cyanuric amide, which is also known as *melamine*:

Cyanamides are converted to cyanides by reaction with carbon,

$$CaCN_2 + C \rightarrow Ca(CN)_2 \tag{13.141}$$

$$CaCN_2 + C + Na_2CO_3 \rightarrow CaCO_3 + 2\,NaCN \tag{13.142}$$

Cyanides are extremely toxic, and acidifying a solution containing CN^- produces HCN.

$$CN^- + H^+ \rightarrow HCN \tag{13.143}$$

Hydrogen cyanide (b.p. 26 °C) is a very toxic gas. It is a weak acid ($K_a = 7.2 \times 10^{-10}$), so solutions of ionic cyanides are basic due to hydrolysis.

$$CN^- + H_2O \rightleftarrows HCN + OH^- \tag{13.144}$$

The CN^- ion is a good coordinating group, and it forms many stable complexes with metals (see Chapter 16). Although the carbon end of the ion is the usual coordinating site, the unshared pair of electrons on the nitrogen end makes possible complexes in which the CN^- ion is a bridging group. The $-CN$ (nitrile) group is an important one in the chemistry of organic compounds.

Cyanates (OCN^-) can be prepared from cyanides by oxidation reactions. For example,

$$KCN + PbO \rightarrow KOCN + Pb \tag{13.145}$$

Having the same atomic composition as cyanates but drastically different properties are the fulminates, which contain the CNO^- ion. Many organic compounds having the formula $R-N=C=O$ are known (the isocyanates). Cyanides undergo an addition reaction with sulfur to produce thiocyanates.

$$KCN + S \rightarrow KSCN \tag{13.146}$$

Unlike HCN, HSCN is a strong acid (comparable to HCl), and it can readily form amine hydro-thiocyanates, $R_3NH^+SCN^-$, that are analogous to amine hydrochlorides. In the molten state, these acidic salts react with oxides, carbonates, etc., and in some cases produce thiocyanate complexes of the metals. Like the cyanide ion, SCN^- is a good coordinating ion and it bonds to soft metals (e.g., Pt^{2+} or Ag^+) through the sulfur atom and to hard metals (e.g., Cr^{3+} or Co^{3+}) through the nitrogen atom (see Chapter 9).

13.4.10 Carbon Sulfides

The most common compound of carbon and sulfur is CS_2, carbon disulfide (b.p. 46.3 °C). It can be prepared by the reaction of carbon and sulfur in an electric furnace or by passing sulfur vapor over hot carbon.

$$4\,C + S_8 \rightarrow 4\,CS_2 \tag{13.147}$$

The reaction of sulfur with methane at high temperature with a suitable catalyst (SiO_2 or Al_2O_3) also produces CS_2.

$$S_8 + 2\,CH_4 \rightarrow 2\,CS_2 + 4\,H_2S \tag{13.148}$$

CS_2 is a good solvent for many substances including sulfur, phosphorus, and iodine. The compound has a high density (1.3 g/ml), and it is slightly soluble in water although it is completely miscible with alcohol, ether, and benzene. It is also quite toxic, highly flammable, and forms explosive mixtures with air. It is used in the preparation of CCl_4 as described earlier in this chapter. One interesting reaction of CS_2 is analogous to that of CO_2 in reacting with metal oxides. For example,

$$BaO + CO_2 \rightarrow BaCO_3 \tag{13.149}$$

$$BaS + CS_2 \rightarrow BaCS_3 \tag{13.150}$$

The CS_3^{2-} ion is the known as the thiocarbonate ion, and it has a trigonal planar structure.

Two other compounds containing carbon and sulfur should be mentioned. The first of these is carbon monosulfide, CS. This compound has been reported to be produced by the reaction of CS_2 with ozone. The second compound is COS or, more correctly, OCS (m.p. −138.2 °C and b.p. −50.2 °C). It is prepared by the reaction

$$CS_2 + 3\,SO_3 \rightarrow OCS + 4\,SO_2 \tag{13.151}$$

Unlike CS_2, neither CS nor OCS have any large-scale industrial uses.

References for Further Reading

Bailar Jr., J.C., Emeleus, H.J., Nyholm, R., Trotman-Dickinson, A.F., 1973. *Comprehensive Inorganic Chemistry*. Pergamon Press, Oxford. This is a five volume reference work in inorganic chemistry.

Billups, W.E., Ciufolini, M.A., 1993. *Buckminsterfullerenes*. VCH Publishers, New York. A useful survey of the early literature.

Cotton, F.A., Wilkinson, G., Murillo, C.A., Bochmann, M., 1999. *Advanced Inorganic Chemistry*, 6th ed. Chapter 5. John Wiley, New York. A 1300 page book that covers an incredible amount of inorganic chemistry. Several chapters are devoted to the elements described in this chapter.

Garrett, D.E., 1998. *Borates*. Academic Press, San Diego, CA. An extensive reference book on the recovery and utilization of boron compounds.

Greenwood, N.N., Earnshaw, A., 1997. *Chemistry of the Elements*, 2nd ed. Butterworth-Heinemann, Oxford. Probably the most comprehensive single volume on the chemistry of the elements.

Hammond, G.S., Kuck, V.J. (Eds.), 1992. *Fullerenes*. American Chemical Society, Washington, D.C. This is ACS Symposium Series No. 481, and it presents a collection of symposium papers on fullerene chemistry.

King, R.B., 1995. *Inorganic Chemistry of the Main Group Elements*. VCH Publishers, New York. An introduction to the descriptive chemistry of many elements.

Kroto, H.W., Fisher, J.E., Cox, D.E., 1993. *The Fullerenes*. Pergamon Press, New York. A reprint collection with articles on most phases of fullerene chemistry.

Liebman, J.F., Greenberg, A., Williams, R.E., 1988. *Advances in Boron and the Boranes*. VCH Publishers, New York. A collection of advanced topics on all phases of boron chemistry.

Muetterties, E.F. (Ed.), 1975. *Boron Hydride Chemistry*. Academic Press, New York. One of the early standard references on boron chemistry.

Muetterties, E.F., Knoth, W.H., 1968. *Polyhedral Boranes*. Marcel Dekker, New York. An excellent introduction to the chemistry of boranes.

Muetterties, E.F. (Ed.), 1967. *The Chemistry of Boron and Its Compounds*. John Wiley, New York. A collection of chapters on different topics in boron chemistry.

Niedenzu, K., Dawson, J.W., 1965. *Boron-Nitrogen Compounds*. Academic Press, New York. An early introduction to boron-nitrogen compounds that contains a wealth of relevant information.

Razumovskii, S.D., Zaikov, G.E., 1984. *Ozone and Its Reactions with Organic Compounds*. Elsevier, New York. Volume 15 in a series, Studies in Organic Chemistry. A good source of information on the uses of ozone in organic chemistry.

Zingaro, R.A., Cooper, W.C. (Eds.), 1974. *Selenium*. Van Nostrand Reinhold, New York. An extensive treatment of selenium chemistry.

QUESTIONS AND PROBLEMS

1. If the O_3 contains only oxygen atoms, would it be polar? Explain your answer.
2. Write complete balanced equations for the following processes.
   (a) The reaction of magnesium boride with water
   (b) The combustion of B_2H_6
   (c) The reaction of BCl_3 with C_2H_5OH
   (d) The preparation of borazine
   (e) The reaction of H_3BO_3 with CH_3COCl
3. Explain why the electron affinity of the O_2 molecule is 0.451 eV, whereas that of C_2 is 3.269 eV.
4. Describe the bonding in the BO molecule.
5. Would $FB(OH)_2$ be a stronger or weaker acid than boric acid? Explain your answer.
6. On the basis of its structure, explain why boric acid is a weak acid that functions by complexing with OH^-. Draw the structure for the product.
7. If BF_3 and $B(CH_3)_3$ were to react with the molecule

   What would you expect the products to be? Why?
8. Pyridine cannot be used as the solvent when producing the adduct of ether with $B(CH_3)_3$, but when producing the pyridine adduct, ether can be used as the solvent. Explain this difference.
9. Write equations to show the amphoteric behavior of Zn^{2+} in aqueous solutions.
10. Explain why a solution containing ferric chloride in water is acidic.
11. In this chapter, the reaction between O_2 and PtF_6 was described. Would you expect N_2 to react in a similar way? Why or why not?
12. The ionization potential for O_2 is 12.06 eV and that for O_3 is 12.3 eV. If you were trying to obtain the O_3^+ species, what strategy would you follow?
13. Determine the spectroscopic state for each of the following.
    (a) B_2; (b) O_2^+; (c) C_2

14. Consider the BF_3 and BH_3 molecules even though the latter is not stable alone. One of these forms a stronger bond to $(C_2H_5)_2S$ and the other to $(C_2H_5)_2O$. Explain how the molecules would preferentially bond.

15. Draw structures for the following. List all symmetry elements and determine the point group for each species.
    (a) ONF; (b) NCN^{2-}; (c) OCN^-; (d) C_3O_2

16. Under certain conditions (such as in interstellar space), the OH radical has been observed. Construct the molecular orbital diagram for this species. Determine the bond order and determine what type of orbital contains the unpaired electron.

17. Construct a molecular orbital energy-level diagram for the SO molecule and speculate on the nature of the bond and other characteristics of the molecule.

18. In BF_3, the B—F bond length is 130 pm, but in BF_4^-, it is approximately 145 pm. Explain this difference in the B—F bond lengths.

19. Would $(C_2H_5)_3N$ or $(C_2H_5)_3P$ react more energetically with BCl_3 if the reactants are dissolved in an inert solvent? Why?

20. Complete and balance the following.
    (a) $C_2H_5OH + CaH_2 \rightarrow$
    (b) $Al + NaOH$ (in water) $\rightarrow$
    (c) $SiCl_4 + LiAlH_4 \rightarrow$
    (d) $BCl_3 + C_2H_5MgBr \rightarrow$
    (e) $Fe_2O_3 + Al \rightarrow$

21. Describe the major industrial use of oxygen, including equations if necessary.

22. The process of adding two electrons to an oxygen atom,

$$O(g) + 2\ e^- \rightarrow O^{2-}(g)$$

absorbs 652 kJ mol^{-1}. Why are there so many ionic oxides?

23. Write balanced equations to show the difference between methanides and acetylides in reacting with water.

24. When CO bonds to BH_3, how does it bond?

25. Explain why BF_3 is a weaker acid than BCl_3.

26. Write balanced equations for each of the following processes:
    (a) preparation of $B_3N_3H_6$
    (b) preparation of $(C_2H_5)BH_2$
    (c) preparation of $NaBH_4$
    (d) combustion of diborane
    (e) preparation of $(C_6H_5)_3B$

27. Describe the process by which H_2O_2 is prepared.

28. Heating many solid carbonates leads to decomposition. What products are obtained? Suppose $CaCS_3$ were heated strongly. What would happen?

29. Consider a -1 ion that contains one atom each of C, S, and P. Draw the correct structure and explain why any other arrangements of atoms are unlikely.

30. Write complete balanced equations to show the following processes.
    (a) the reaction of BaO with SO_3
    (b) the preparation of calcium cyanamide
    (c) the oxidation of potassium cyanide with H_2O_2
    (d) the preparation of hydrogen selenide

31. The H_3^+ ion has a trigonal planar structure. Rationalize this structure in terms of electron density within the structure. If one, two, or three H_2 molecules add to the H_3^+ ion, where and how would they bond? Sketch these structures.

CHAPTER 14

Chemistry of Nonmetallic Elements II. Groups IVA and VA

As has been previously described, the chemistry of the first member of each group in the periodic table is quite different from that of the others. Having surveyed the chemistry of carbon in the previous chapter, we now address the chemistry of the other members of Group IVA with emphasis on silicon and tin. The chemistry of nitrogen is extensive so it is discussed separately before moving on to the chemistry of the remaining members of Group VA where the major emphasis is on the chemistry of phosphorus. The discussion of the Group VIA elements (with emphasis on sulfur), the halogens, and the noble gases in Chapter 15 rounds out the survey of the chemistry of the nonmetals.

14.1 THE GROUP IVA ELEMENTS

Unlike silicon, which was discovered by Berzelius in 1824, and germanium, which was discovered in 1886 by Winkler, tin and lead have been used since ancient times. The Bronze Age spans the period from about 2500 BC to 1500 BC. Bronze is an alloy of copper and tin that was a prominent material used in that period preceding the Iron Age. One of the most important minerals containing tin is *cassiterite*, SnO_2, which was reduced by heating it with carbon to obtain the metal. Producing metals in early times could be accomplished only by the simple techniques available at the time. Technology limited the temperatures that could be employed, and the available reducing agents limited which metals could be produced. In the case of tin and lead (which occurs as *galena*, PbS), reduction at relatively low temperature could be accomplished with charcoal.

Although silicon *compounds* have been used since the earliest times, the element was not obtainable until advances in technology came about. Minerals containing silicon include sand and silicates that are widely distributed (silicon constitutes approximately 23% of the earth's crust), and they have been used in making glass, pottery, and mortar for many centuries. In addition to these uses, silicon is now highly purified for use in integrated circuits (chips) and as an alloy known as *Duriron* that has many uses. Silicon has the diamond structure with a density of 2.3 g/cm^3.

Although germanium was unknown, Mendeleev predicted the properties of the missing element (which he referred to as *ekasilicon*) based on the properties of other elements. While analyzing the mineral *argyrodite*, Winkler found that it contained about 7% of some element that had not been identified, and that element turned out to be germanium. The element is now obtained as a byproduct in residues from the production of zinc. It is obtained by treating the residue with concentrated HCl, which converts germanium to $GeCl_4$. Like most other covalent halides, this compound hydrolyzes.

439

Inorganic Chemistry. DOI: http://dx.doi.org/10.1016/B978-0-12-385110-9.00014-5

$$GeCl_4 + 2 H_2O \rightarrow GeO_2 + 4 HCl \tag{14.1}$$

Reduction of the oxide is then accomplished with hydrogen.

$$GeO_2 + 2 H_2 \rightarrow Ge + 2 H_2O \tag{14.2}$$

Germanium is one of the materials that is used in producing semiconductors. When combined with phosphorus, arsenic, or antimony (which have five valence electrons), an *n*-type semiconductor results, and when combined with gallium (which has three valence electrons), the product is known as a *p*-type semiconductor.

Although tin is normally a soft silvery metal, the form of metallic tin (m.p. 232 °C) depends on the temperature. *White* tin (density 7.28 g/cm^3), which is the stable form above 13.2 °C, is metallic in appearance and properties. Below 13.2 °C, the stable form of tin is known as *gray* tin, and it crumbles to give a gray powder. The phase transition from white to gray form takes place slowly even at temperatures below 13.2 °C. The conversion of tin from the highly versatile white form to the powdery gray form is referred to as "tin disease" or "tin pest". A third form of tin is known as *brittle* tin (density 6.52 g/cm^3), and it is produced when white tin is heated above 161 °C. As its name implies, this form shows nonmetallic character by fracturing when hammered.

Many of the uses of tin are also those of lead because the metals form useful alloys. When lead is alloyed with a few percent of tin, it becomes harder and more durable. Although other compositions are produced, common solder consists of about an equal mixture of tin and lead. An alloy known as type metal contains about 82% Pb, 15% Sb, and 3% Sn, and *pewter* contains approximately 90% tin that is alloyed with copper and antimony. *Babbitt,* an alloy used in making bearings, contains 90% Sn, 7% Sb, and 3% Cu. Tin is also used to coat other metal objects to retard corrosion, and a tin-niobium alloy is used in superconducting magnets.

Lead (m.p. 328 °C, density 11.4 g/cm^3) has been used for thousands of years, and its chemical symbol is taken from its Latin name, *plumbum.* Some lead is found uncombined, but most is found as the sulfide, *galena,* from which it is obtained by roasting the ore to produce the oxide and then reducing the oxide with carbon.

$$2 PbS + 3 O_2 \rightarrow 2 PbO + 2 SO_2 \tag{14.3}$$

$$PbO + C \rightarrow Pb + CO \tag{14.4}$$

The metal was used widely in ancient Rome as plumbing, roofing, and containers for food and water. Lead compounds have also been extensively used in pigments because of their bright colors. In some parts of the world, lead compounds are still used in paints and glazes for pottery. In most cases, titanium dioxide has replaced lead oxide as the white pigment in paints. Although it is no longer in use, tetraethyl lead was a fuel additive for many years. As much as 40% of the lead used today is recovered from scrap. Lead is used in automobile batteries in which the plates are made of an alloy containing about 88–93% Pb and 7–12% Sb.

14.1.1 Hydrides of the Group IVA Elements

Except for silicon, both the +2 and +4 oxidation states are rather common for the elements in Group IVA. Compounds containing Sn^{2+} include the polymeric solids SnF_2 and $SnCl_2$, and germanium +2 compounds include GeO, GeS, and GeI_2. If the Group IVA elements are represented as E, the important hydrides are covalent or volatile compounds that can be represented as EH_4. The stability of the compounds decreases in descending order for the series of elements Si, Ge, Sn, and Pb (which, like PbH_2, is unstable). The names of the EH_4 compounds are *silane, germane, stannane,* and *plumbane.* Higher hydrides such as Si_2H_6 are named as *disilane,* etc.

As was discussed in Chapter 13, hydrogen does not react directly with some elements so the hydrides must be prepared in a different way. Alfred Stock prepared silicon hydrides by first making the magnesium compound and then reacting it with water.

$$2\,Mg + Si \rightarrow Mg_2Si \tag{14.5}$$

$$Mg_2Si + H_2O \rightarrow Mg(OH)_2 + SiH_4, Si_2H_6, \text{etc.} \tag{14.6}$$

Other reactions that can be used to prepare silane include the following.

$$SiO_2 + LiAlH_4 \xrightarrow{150-175\,°C} SiH_4 + LiAlO_2(Li_2O + Al_2O_3) \tag{14.7}$$

$$SiCl_4 + LiAlH_4 \xrightarrow{\text{ether}} SiH_4 + LiCl + AlCl_3(LiAlCl_4) \tag{14.8}$$

Silicon hydrides containing more than two silicon atoms are unstable and decompose to produce SiH_2, Si_2H_6, and H_2. These compounds are spontaneously flammable in air.

$$SiH_4 + 2O_2 \rightarrow SiO_2 + 2\,H_2O \tag{14.9}$$

The heat of formation of SiO_2 is -828 kJ/mol so the reactions are extremely exothermic. Silane and disilane do not react readily with water, but the reaction in basic solutions is

$$SiH_4 + 4\,H_2O \rightarrow 4\,H_2 + Si(OH)_4(SiO_2 \cdot 2H_2O) \tag{14.10}$$

Germane is prepared from the oxide by reaction with $LiAlH_4$,

$$GeO_2 + LiAlH_4 \rightarrow GeH_4 + LiAlO_2 \tag{14.11}$$

and stannane is obtained from the chloride in a similar reaction,

$$SnCl_4 + LiAlH_4 \xrightarrow{-30\,°C} SnH_4 + LiCl + AlCl_3 \tag{14.12}$$

14.1.2 Oxides of the Group IVA Elements

Although SiO (bond energy 765 kJ/mol) can be considered to be analogous to carbon monoxide (bond energy 1070 kJ/mol), it is not an important compound. It is formed when silicon reacts in a deficiency of oxygen or when SiO_2 is reduced with carbon.

$$SiO_2 + C \rightarrow SiO + CO \tag{14.13}$$

Because the dioxide is so stable, the monoxide disproportionates.

$$2\,SiO \rightarrow Si + SiO_2 \tag{14.14}$$

In SiO_2, the main structural feature is the tetrahedral bonding of each silicon atom to four oxygen atoms.

The monoxides and dioxides of germanium, tin, and lead are all known. Especially when obtained as precipitates, the oxides contain water and consist of a mixture of the oxide, hydroxide, and hydrous oxide. For example, GeO, $GeO \cdot xH_2O$, and $Ge(OH)_2$ (which could be written as $GeO \cdot H_2O$) all exist in equilibria or in mixtures. GeO is obtained from the reaction

$$GeCl_2 + H_2O \rightarrow GeO + 2\,HCl \qquad (14.15)$$

At high temperature, it disproportionates to produce the dioxide.

$$2\,GeO \rightarrow Ge + GeO_2 \qquad (14.16)$$

The hydrolysis of $SnCl_2$ produces $Sn(OH)_2$,

$$SnCl_2 + 2\,H_2O \rightarrow Sn(OH)_2 + 2\,HCl \qquad (14.17)$$

which can be dehydrated thermally to produce SnO:

$$Sn(OH)_2 \rightarrow SnO + H_2O \qquad (14.18)$$

The oxyanions of tin(IV), SnO_3^- and SnO_4^{4-} (known as *stannates*), result from disproportionation of SnO in basic solutions.

$$2\,SnO + 2\,KOH \rightarrow K_2SnO_3 + Sn + H_2O \qquad (14.19)$$
$$2\,SnO + 4\,KOH \rightarrow Sn + K_4SnO_4 + 2\,H_2O \qquad (14.20)$$

Litharge (red) and *massicot* (yellow) are pigments that are forms of PbO produced by reacting lead with oxygen.

$$2\,Pb + O_2 \xrightarrow{\Delta} 2\,PbO(yellow) \xrightarrow{\Delta} 2\,PbO(red) \qquad (14.21)$$

The +4 oxides of the Group IVA elements are generally acidic (as is CO_2) or amphoteric in character. As shown here for CO_2, acidic oxides form oxyanions,

$$CaO + CO_2 \rightarrow CaCO_3 \qquad (14.22)$$

and the +4 oxides of Group IVA elements give rise to numerous silicates, stannates, etc.

Polymorphism in SiO_2 leads to approximately 20 different forms of the compound. Among others, it occurs in *quartz*, *tridymite*, and *cristobalite* each of which exists in α and β forms. The structure for CO_2,

$$\overline{\underline{o}} = c = \overline{\underline{o}}$$

involves double bonds (806 kJ mol^{-1}), which are *more* than twice as strong as C—O single bonds (360 kJ mol^{-1}). In contrast, the Si=O bond (642 kJ/mol) is *not* as strong as *two* Si—O bonds

$(2 \times 460 = 920 \text{ kJ mol}^{-1})$ so it is energetically more favorable for Si to form four *single* bonds rather than two *double* bonds. This results in Si being attached by single bonds to O in SiO_4 tetrahedra that can be arranged in numerous ways to give several forms of SiO_2, and numerous phase changes can be brought about by high temperature.

Common forms of SiO_2 include sand, flint, agate, and quartz. Although SiO_2 melts at 1710 °C, some of the O—S—O bonds are broken at lower temperature causing the material to soften and become a glass. When quartz is stressed, it produces an electric current by the *piezoelectric effect*. Applying voltage across the crystal causes it to undergo vibrations with a frequency that resonates with the current frequency. This behavior of quartz is why it can be used in watches and crystals used in electronic devices. Quartz glass is obtained by fusing quartz and allowing it to cool. It is used in optical devices because it does not absorb electromagnetic radiation over a wide range of wavelengths.

There are several ways to prepare GeO_2; some of which are as follows:

$$Ge + O_2 \rightarrow GeO_2 \tag{14.23}$$

$$3\,Ge + 4\,HNO_3 \rightarrow 3\,GeO_2 + 4\,NO + 2\,H_2O \tag{14.24}$$

$$GeCl_4 + 4\,NaOH \rightarrow GeO_2 + 2\,H_2O + 4\,NaCl \tag{14.25}$$

GeO_2 is a slightly acidic oxide that reacts with water to give acidic solutions.

$$GeO_2 + H_2O \rightarrow H_2GeO_3 \tag{14.26}$$

$$GeO_2 + 2\,H_2O \rightarrow H_4GeO_4 \tag{14.27}$$

The naturally occurring form of SnO_2 (*cassiterite*) has the rutile structure (see Chapter 7), and it is an amphoteric oxide as illustrated by the following equations:

$$SnO_2 + 2\,H_2O \rightarrow H_4SnO_4 (\text{or } Sn(OH)_4) \tag{14.28}$$

$$H_4SnO_4 + 4\,NaOH \rightarrow Na_4SnO_4 + 4\,H_2O \tag{14.29}$$

$$Sn(OH)_4 + 2\,H_2SO_4 \rightarrow Sn(SO_4)_2 + 4\,H_2O \tag{14.30}$$

The oxyanions SnO_4^{4-} and SnO_3^{2-} are both considered to be *stannates* because they contain Sn(IV). In a way that is analogous to that used with the phosphorus oxyanions, the stannates are known as the *orthostannate* and *metastannate* ions.

Like SnO_2, PbO_2 has the rutile structure, but unlike SnO_2, it is a strong oxidizing agent that can be produced by the reaction

$$PbO + NaOCl \rightarrow PbO_2 + NaCl \tag{14.31}$$

PbO_2 is the oxidizing agent and lead is the reducing agent in the lead storage battery. The chemistry of the battery can be summarized by the following equation:

$$Pb + PbO_2 + 2\,H_2SO_4 \underset{\text{Discharging}}{\overset{\text{Charging}}{\rightleftharpoons}} 2\,PbSO_4 + 2\,H_2O \tag{14.32}$$

A cell in this type of battery has electrodes made of lead and spongy lead impregnated with PbO_2. Sulfuric acid is the electrolyte, and the cell voltage is approximately 2.0 V. Depending on the number of cells linked in series, the overall voltage can be 6 or 12 V (the more common type of battery).

Another oxide of lead has the formula Pb_2O_3 (also sometimes written as $PbO \cdot PbO_2$). This compound is a lead *plumbate*, $PbPbO_3$, which contains Pb(II) and Pb(IV). The lead oxide used in making lead crystal is Pb_3O_4, and it is also used as a red pigment. This oxide, for which the formula can be written more correctly as $2PbO \cdot PbO_2$ or Pb_2PbO_4, is obtained by heating PbO to 400 °C in air.

14.1.3 Glass

It is believed that glass has been made for at least 5000 years. There is no way to tell exactly how the process of making glass was discovered but was made by heating sand, sodium carbonate, and limestone. A typical composition of a common type of glass known as soda-lime is about 62.5% SiO_2, 25% Na_2CO_3, and 12.5% CaO. When cool, the glass is transparent and rigid, but it can be shaped by rolling, blowing, or molding when hot. This type of glass, sometimes referred to as flat glass, is used in windows and bottles. After an object is shaped, the hot glass is put into a furnace and cooled slowly to anneal the piece. If glass is cooled too slowly, a large fraction of the $-Si-O-Si-O-$ linkages reform making the glass brittle. Flat sheets of glass are produced by placing the molten glass on the surface of molten tin in a large shallow container. When the mass cools, sheets having flat surfaces are produced.

Glass contains silicon atoms surrounded by four oxygen atoms in a tetrahedral arrangement. These tetrahedra are joined by having oxygen atoms bonded to two Si atoms. Some of the linkages are broken when the glass is softened by heating it, and the linkages are reformed as the glass cools. *Borosilicate* glass, which does not break when subjected to rapid changes in temperature, is prepared by adding B_2O_3 to the mixture. *Lead crystal* and *flint glass* are very dense and highly refractive glasses that are prepared by replacing lime with PbO and Pb_3O_4. Colored glasses are prepared by adding other materials to the mixture. For example, CoO gives a blue glass, FeO imparts a green color, and CaF_2 produces an opaque white (milk) glass.

Because glass is a stable versatile material that is produced from inexpensive ingredients, it is used in many ways. It can be shaped or made into sheets, it is durable, and specialty glasses can be made to have specific properties. Hardened or tempered glass is a versatile construction material, and as metals become more scarce and expensive, buildings are constructed of glass supported by a metal framework. It is not surprising that millions of tons of glass are produced each year.

14.1.4 Silicates

We have already mentioned the polymorphism that exists for SiO_2. The widespread distribution of SiO_2 and unusual environmental conditions under which it has interacted with other materials (particularly oxides) have resulted in there being a large number of naturally occurring silicates. The acidic nature of SiO_2 causes it to undergo reactions that can be represented as

$$SiO_2 + 2\,O^{2-} \rightarrow SiO_4^{2-} \tag{14.33}$$

Literature that presents information on minerals often shows the composition in terms of the constituent oxides. For example, *benitoite* is described as 36.3% BaO, 20.2% TiO_2, and 43.5% SiO_2. Therefore, many silicates can formally be regarded as combinations of oxides as shown in Table 14.1.

Table 14.1 Composition of Some Silicate Minerals

Oxides Combined	Mineral Equivalent
$CaO + TiO_2 + SiO_2$	$CaTiSiO_5$, *titanite*
$\frac{1}{2} K_2O + \frac{1}{2} Al_2O_3 + 3 SiO_2$	$KAlSi_3O_8$, *orthoclase*
$2 MgO + SiO_2$	Mg_2SiO_4, *forsterite*
$BaO + TiO_2 + 3 SiO_2$	$BaTiSi_3O_9$, *benitoite*

Because this book presents information on many areas of inorganic chemistry, a detailed discussion of silicates cannot be presented. Although silicate chemistry is a very complex field, there are some pervasive principles. It should first be mentioned that an oxide is easily converted to a hydroxide by reaction with water,

$$O^{2-} + H_2O \rightarrow 2\,OH^- \tag{14.34}$$

and that an oxide (a base) reacts with carbon dioxide (an acid) to produce a carbonate.

$$O^{2-} + CO_2 \rightarrow CO_3^{2-} \tag{14.35}$$

Furthermore, a carbonate will react with water to produce a bicarbonate,

$$CO_3^{2-} + H_2O \rightarrow HCO_3^- + OH^- \tag{14.36}$$

These processes can take place naturally, and they are involved in the process of changing one mineral into another (known as *weathering*).

The tetrahedral SiO_4^{4-} ion is known as the *orthosilicate* ion. It can be regarded as the fundamental unit in the structures of most complex silicates. Several minerals including *phenacite*, Be_2SiO_4, and *willemite*, Zn_2SiO_4, contain this ion. Both of these minerals have tetrahedral coordination of SiO_4^{4-} units around the metal ion that can be illustrated by the structure

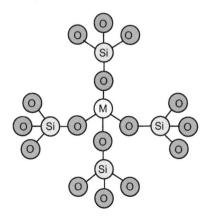

Not all cases that involve the SiO_4^{4-} units have four of them surrounding the metal ion. In Mg_2SiO_4 and Fe_2SiO_4 (forms of *olivene*), the coordination number of the metals is six, and *zircon*, $ZrSiO_4$, has a structure in which the coordination number of Zr is eight.

Tetrahedral structures can be joined at a corner or along an edge. The number of tetrahedra and how they are joined determine the overall structure. An efficient way to represent silicate structures is illustrated in Figure 14.1, and the SiO_4^{4-} ion is shown as in Figure 14.1(a). The structures are interpreted as if seen from above the tetrahedral unit so the large open circle represents an oxygen atom projected upward out of the page. The filled circle represents a silicon atom directly below the oxygen atom, and

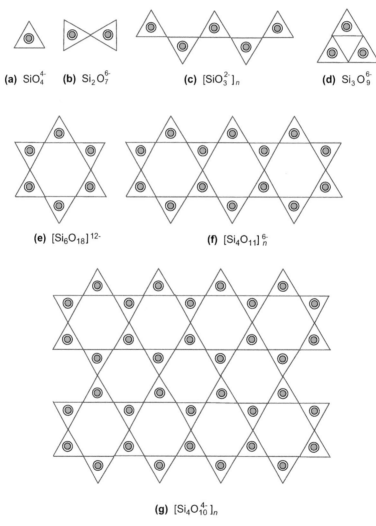

FIGURE 14.1
Representations of silicate structures.

each vertex in the triangle represents an oxygen atom in the base of the tetrahedron. The various silicate structures are then made up by sharing one or more corners or edges of the tetrahedra where oxygen atoms form bridges.

The structure formed by two tetrahedral units joined at a corner is known as the *pyrosilicate* or *disilicate* ion (see Figure 14.1(b)).

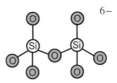

It should be noted that this ion, which occurs in the minerals *thortveitite*, $Sc_2Si_2O_7$, and *hemimorphite*, $Zn_4(OH)_2Si_2O_7$, is isoelectronic with $P_2O_7^{4-}$, $S_2O_7^{2-}$, and Cl_2O_7. Except for the fact that the SiO_4 unit has a -4 charge, whereas the PO_4 unit has a -3 charge, there is a great deal of similarity between the "polyphosphates" and the "polysilicates." If three SiO_3^{2-} (metasilicate) ions are linked in a ring, the cyclic $Si_3O_9^{6-}$ ion is formed, in which each Si is surrounded by four oxygen atoms forming a tetrahedron (Figure 14.1(d)).

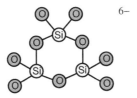

This structure is analogous to that of $P_3O_9^{3-}$ and S_3O_9 (a trimer of SO_3), and it occurs in the mineral *benitoite*, $BaTiSi_3O_9$. When six SiO_3^{2-} ions are linked in a 12-membered ring containing alternating Si and O atoms, the result is the $Si_6O_{18}^{12-}$ ion that occurs in *beryl*, $Be_3Al_2Si_6O_{18}$ (see Figure 14.1(e)).

Pyroxenes are structures that consist of long chains in which the repeating unit is SiO_3^{2-}. The general structure for these materials is shown below (see Figure 14.1(c)).

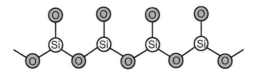

The cations located between the chains hold them together. Some pyroxenes are *diopside* ($CaMgSi_2O_6$), *hendenbergite* ($Ca(Fe,Mg)Si_2O_6$), and *spodumene* ($LiAlSi_2O_6$).

A different type of chain structure is present in the *amphiboles* (Figure 14.1(f)). Metal ions located between the chains bond them together, but half of the silicon atoms share *two* oxygen atoms and the other half share three oxygen atoms. *Tremolite*, $Ca_2Mg_5Si_8O_{22}(OH)_2$, and *hornblende*, $CaNa(Mg,Fe)_4(Al,Fe,Ti)_2Si_6(O,OH)_2$, are examples of amphiboles. When the chains are bound together by sharing three oxygen atoms, a sheet structure known as *mica* is produced (see Figure 14.1(g)). *Muscovite*, $KAl_3Si_3O_{10}(OH)_2$, and *biotite*, $K(Mg,Fe)_3AlSi_3O_{10}(OH)_2$, have this type of structure. In formulas such as those just shown, there can be substitution of one metal ion for another. For example, if a formula would require two Mg^{2+} ions, it possible that some of the lattice sites could be occupied by Fe^{2+} ions. This is indicated by showing (Mg,Fe) in the formula indicating that there are two metal ions, but they may be either Mg^{2+} or Fe^{2+} so the composition is variable. In a similar way, a +3 ion such as Al^{3+} may be replaced by a +2 and a +1 ion (such as Mg^{2+} and Na^+) so a formula could appear as Al[X] or (Mg,K)[X] and electrical neutrality still be satisfied.

The aluminosilicates are a class of silicates in which some of the Si^{4+} ions have been replaced by Al^{3+} ions. One important mineral of this type is *kaolin*, $Al_2Si_2O_5(OH)_4$, (which is described as 39.5% Al_2O_3, 46.5% SiO_2, and 14.0% H_2O). Kaolin is a very useful material for making ceramics and as clay in making china. *Orthoclase*, $KAlSi_3O_8$, is also used in making ceramics and some types of glass. *Leucite*, $KAlSi_2O_6$, has been used as a source of potassium in fertilizer.

14.1.5 Zeolites

Hydrous minerals known as *zeolites* (a word which comes from the Greek *zeo* meaning to boil and *lithos* meaning rock) are secondary minerals that form in igneous rocks. Zeolites are aluminosilicates that have anions that are porous as a result of the channels through them. Although approximately 30 naturally occurring zeolites have been found, several times that number have been prepared. The general formula for a zeolite is $M_{a/z}[(AlO_2)_a(SiO_2)_b] \cdot xH_2O$, where M is a cation having a charge +z. Because H_2O and SiO_2 are neutral, the charges on the AlO_2^- ions must be balanced by the charge on M. When M has a +1 charge, the number of AlO_2^- and M^+ will be equal. When M has a +2 charge, there will be twice as many AlO_2^- ions as M^{2+}. The ratio $a/z:a$ gives the number of M ions to the number of AlO_2^- ions. Some of the most common naturally occurring zeolites are the following:

Analcime	$NaAlSi_2O_6 \cdot H_2O$
Edingtonite	$BaAl_2Si_3O_{10} \cdot 4H_2O$
Cordierite	$(Mg,Fe)_2Mg_2Al_4Si_5O_{18}$
Stilbite	$(Ca,Na)_3Al_5(Al,Si)Si_{14}O_{40} \cdot 15H_2O$
Chabazite	$(Ca,Na,K)_7Al_{12}(Al,Si)_2Si_{26}O_{80} \cdot 40H_2O$
Sodalite	$Na_2Al_2Si_3O_{10} \cdot 2H_2O$

Some zeolites have the ability to exchange sodium for calcium and thereby function as water softeners by removing Ca^{2+}. After the zeolite has become saturated with Ca^{2+}, it can be renewed by washing it in a concentrated NaCl solution to restore the Na^+ ions. Zeolites are also used to prepare ion exchange resins, as molecular sieves and as catalysts.

As are other silicates, zeolites are made up of tetrahedra that are SiO_4 or AlO_4 units. Because of the difference in charges on Si and Al, the presence of Al^{3+} necessitates a +1 cation as well. The $Si_6O_{18}^{12-}$ ion is a basic unit in many zeolites. The six Si (and/or Al) ions define a hexagon, and these hexagons can be combined to give a structure like that shown in Figure 14.2(a). This unit is known as the *sodalite* structure, which is also called a β-cage. Eight sodalite units can be joined to give a cubic structure like that shown in Figure 14.2(b). In this structure, the sodalite units are joined by faces that have four

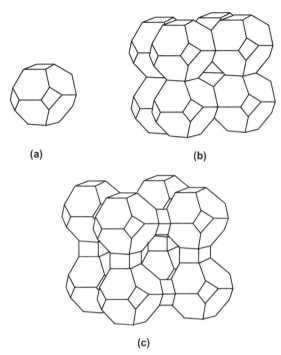

FIGURE 14.2
The structure of zeolites derived from the sodalite structure shown in (a).

members in the ring. This type of structure contains channels that allow it to function as a molecular sieve. In a closely related structure, shown in Figure 14.2(c), the eight sodalite units are joined by bridging oxygen atoms to give a structure known as *zeolite-A*. This structure contains equal numbers of Al^{3+} and Si^{4+} ions so the formula can be written as $Na_{12}(AlO_2)_{12}(SiO_2)_{12} \cdot 27H_2O$, which can also be written as $Na_{12}Al_{12}Si_{12}O_{48} \cdot 27H_2O$.

Some zeolites are useful catalysts as a result of their having very large surface area and some sites where oxygen atoms have been converted to $-OH$ groups. These sites are potential proton donors so they are referred to as *Brønsted sites*. Certain zeolites are particularly effective catalysts for cracking hydrocarbons. The chemistry of silicates truly ranges from mineralogy to organic reactions.

14.1.6 Halides of the Group IVA Elements

Each of the elements in Group IVA forms dihalides and tetrahalides. There are also a few silicon compounds of the type E_2X_6 as a result of silicon having a greater tendency for catenation. The halogen compounds of the +2 elements have greater ionic character to the bonds and thus have higher melting and boiling points than the halides containing the +4 elements. Table 14.2 presents a summary of melting points and boiling points for these compounds.

In the gas phase, the dihalides have an angular structure, but in the solid state, most have complex structures held together by bridging halogen atoms. The gaseous EX_4 molecules are tetrahedral, nonpolar, and more soluble in organic solvents than the EX_2 compounds.

Table 14.2 Melting and Boiling Points for Group IV Halides

Compound		EX₂		EX₄	
		Melting Point (°C)	Boiling Point (°C)	Melting Point (°C)	Boiling Point (°C)
Si	X = F	dec	–	−90.2	−86 subl
	X = Cl	dec	–	−68.8	57.6
	X = Br	–	–	5.4	153
	X = I	–	–	120.5	287.5
Ge	X = F	111	dec	−37 subl	–
	X = Cl	dec	–	−49.5	84
	X = Br	122 dec	–	−26.1	186.5
	X = I	dec	–	144	440 dec
Sn	X = F	704 subl	–	705	–
	X = Cl	246.8	652	−33	114.1
	X = Br	215.5	620	31	202
	X = I	320	717	144	364.5
Pb	X = F	855	1290	–	–
	X = Cl	501	950	−15	105 expl
	X = Br	373	916	–	–
	X = I	402	872	–	–

The EX₄ halides are Lewis acids that form complexes with halide ions,

$$SnCl_4 + 2\,Cl^- \rightarrow SnCl_6^{2-} \tag{14.37}$$

$$GeF_4 + 2\,HF(aq) \rightarrow GeF_6^{2-} + 2\,H^+(aq) \tag{14.38}$$

although this tendency is much lower for the +2 halides, which are less acidic. Dihalides of silicon and germanium are polymeric solids that contain halide bridges. Tin and lead dihalides have more ionic character, although they also exist in extended structures. Dihalides of silicon are unstable with respect to disproportionation.

$$2\,SiX_2 \rightarrow Si + SiX_4 \tag{14.39}$$

The dihalides of tin have a chain structure, which can be shown as

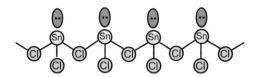

The Sn atoms in SnX_2 molecules have only three pairs of electrons surrounding them so they behave as Lewis acids and react with halide ions to give complexes.

$$SnX_2 + X^- \rightarrow SnX_3^- \qquad (14.40)$$

These complexes have pyramidal structures that can be shown as

Large cations give a favorable match of cation and anion characteristics, so in accord with the hard-soft interaction principle, the salts that have been isolated contain ions such as R_4P^+. Because of having an unshared pair of electrons, the SnX_3^- complexes can function as Lewis bases.

$$SnCl_3^- + BCl_3 \rightarrow Cl_3B\!:\!SnCl_3^- \qquad (14.41)$$

Although direct combination of the elements is feasible, germanium dihalides are usually prepared by the reaction

$$Ge + GeX_4 \rightarrow 2\,GeX_2 \qquad (14.42)$$

Germanium dibromide can be prepared in the following ways:

$$Ge + 2\,HBr \rightarrow GeBr_2 + H_2 \qquad (14.43)$$
$$GeBr_4 + Zn \rightarrow GeBr_2 + ZnBr_2 \qquad (14.44)$$

It is an unstable compound with respect to disproportionation, which occurs when it is heated.

$$2\,GeBr_2 \rightarrow GeBr_4 + Ge \qquad (14.45)$$

The dihalides of tin and lead behave as ionic metal compounds, and they will not be discussed further.

There is a vast difference in the behavior of the tetrahalides of the Group IVA elements. Although the silicon compounds are stable, lead with a +4 oxidation state is a strong oxidizing agent that oxidizes Br^- and I^- with the result that $PbBr_4$ and PbI_4 are extremely unstable. $PbCl_4$ is not a stable compound, and it explodes when heated. Silicon reacts with the halogens to give the tetrahalides. The tetrafluoride will react with fluoride ions to give complexes,

$$SiF_4 + 2\,F^- \rightarrow SiF_6^{2-} \qquad (14.46)$$

but the chloride, bromide, and iodide do not behave in this way, probably because of repulsion of the large anions around the silicon. The heavier elements in Group IVA form hexahalo complexes. The tetrahalides are Lewis acids that form complexes with many electron pair donors.

The reaction of germanium with bromine at elevated temperatures produces the tetrabromide.

$$Ge + 2\,Br_2(g) \xrightarrow{220\,^\circ C} GeBr_4 \qquad (14.47)$$

Hydrolysis of the Group IVA tetrahalides produces oxides, hydroxides, or hydrous oxides or a mixture of these products (EO_2, $E(OH)_4$, and $EO_2 \cdot 2H_2O$). The following are typical reactions:

$$GeF_4 + 4\,H_2O \rightarrow GeO_2 \cdot 2H_2O + 4\,HF \qquad (14.48)$$

$$SnCl_4 + 4\,H_2O \rightarrow Sn(OH)_4 + 4\,HCl \qquad (14.49)$$

In the case of silicon, the tendency for catenation allows halides having the formula Si_2X_6 to be prepared. The chloride can be prepared by the reaction

$$3\,SiCl_4 + Si \rightarrow 2\,Si_2Cl_6 \qquad (14.50)$$

14.1.7 Organic Compounds

There is an extensive organic chemistry of the elements in Group IVA, and entire volumes have been devoted to the subject. In the survey presented here, only some of the more common types of structures and reactions can be shown. In general, the reactions shown are meant to illustrate the *types* of behavior that are exhibited by other organic compounds and other elements in the group.

One of the most common types of reactions carried out to prepare organic derivatives of the Group IVA elements is that of transferring an alkyl group by means of a Grignard reagent. Alkylation of $SnCl_4$ can be illustrated as follows:

$$SnCl_4 + RMgX \rightarrow RSnCl_3 + MgClX \qquad (14.51)$$

By using the appropriate ratio of Grignard reagent, the stepwise alkylation can continue until the product is R_4Sn. Alkylation can also be accomplished by using a lithium alkyl.

$$GeCl_4 + 4\,LiR \rightarrow R_4Ge + 4\,LiCl \qquad (14.52)$$

The reaction of a halogen with a tetraalkyl tin gives mixed alkyl halides.

$$R_4Sn + X_2 \rightarrow R_3SnX + RX \qquad (14.53)$$

Thionyl chloride is a good chlorinating agent that reacts with R_4Sn in the following way:

$$R_4Sn + 2\,SO_2Cl_2 \rightarrow R_2SnCl_2 + 2\,SO_2 + 2\,RCl \qquad (14.54)$$

Scrambling reactions occur between R_4Sn and SnX_4 to give mixed alkyl halides.

$$R_4Sn + SnX_4 \rightarrow RSnX_3,\ R_2SnX_2,\ R_3SnX \qquad (14.55)$$

As it does in many organic reactions, sodium produces a coupling reaction with R_3SnX that can be shown as

$$2\,R_3SnX + 2\,Na \rightarrow R_3Sn{-}SnR_3 + 2\,NaX \tag{14.56}$$

At high temperature, silicon reacts with HCl to produce a useful intermediate, $HSiCl_3$.

$$Si + 3\,HCl \rightarrow HSiCl_3 + H_2 \tag{14.57}$$

Both $HSiCl_3$ and $HGeCl_3$ will add across double bonds in a process known as the *Speier reaction*. This reaction can be shown as follows:

$$HSiCl_3 + CH_3CH{=}CH_2 \rightarrow CH_3CH_2CH_2SiCl_3 \tag{14.58}$$

$$HGeCl_3 + CH_2{=}CH_2 \rightarrow CH_3CH_2GeCl_3 \tag{14.59}$$

Derivatives that are not fully alkylated undergo hydrolysis and react with numerous organic compounds. These products are extremely useful for carrying out the synthesis of a large number of derivatives. The following reactions illustrate this type of chemistry:

$$R_2SnCl_2 + 2\,H_2O \rightarrow R_2Sn(OH)_2 + 2\,HCl \tag{14.60}$$

$$R_2Sn(OH)_2 + 2\,R'COOH \rightarrow R_2Sn(OOCR')_2 + 2\,H_2O \tag{14.61}$$

$$R_2Sn(OH)_2 + 2\,R'OH \rightarrow R_2Sn(OR')_2 + 2\,H_2O \tag{14.62}$$

Tin alkyl hydrides can be prepared from the halides by the reaction with lithium aluminum hydride.

$$4\,RSnX_3 + 3\,LiAlH_4 \rightarrow 4\,RSnH_3 + 3\,LiX + 3\,AlX_3 \tag{14.63}$$

By employing sequences of reactions of these types, the organic chemistry of the Group IVA elements has become a vast area at the interface of inorganic and organic chemistry.

One additional aspect of the chemistry of tin should be mentioned. That aspect concerns the preparation of unusual tin compounds that contain rings in which there are Sn—Sn bonds. These interesting compounds include $(R_2Sn)_3$, in which there is a three-membered ring, that is prepared by the reaction

$$6\,RLi + 3\,SnCl_2 \xrightarrow{\text{THF}} (R_2Sn)_3 + 6\,LiCl \tag{14.64}$$

This area of tin chemistry has expanded greatly as many new compounds have been synthesized and characterized. One of the most unusual compounds of this type is Sn_6H_{10}, which has the structure shown below.

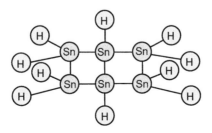

Even though the organic chemistry of lead has not been mentioned, it has been commercially important over the years. Although tetraethyl lead is no longer used as a fuel additive, that use amounted to approximately 250,000 tons/yr at one time.

The organic chemistry of lead will not be described in detail, but the following reactions can be used to prepare organic compounds of lead:

$$4\,R_3Al + 6\,PbX_2 \rightarrow 3\,R_4Pb + 3\,Pb + 4\,AlX_3 \tag{14.65}$$

$$Pb + 4\,Li + 4\,C_6H_5Br \rightarrow (C_6H_5)_4Pb + 4\,LiBr \tag{14.66}$$

$$2\,PbCl_2 + 2\,(C_2H_5)_2Zn \rightarrow (C_2H_5)_4Pb + 2\,ZnCl_2 + Pb \tag{14.67}$$

$$2\,RLi + PbX_2 \rightarrow PbR_2 + 2\,LiX \tag{14.68}$$

As in the case with the other members of Group IVA, the mixed alkyl and aryl halides of lead are also known, and their reactions with water, alcohols, amines, etc. can be used to prepare a large number of other derivatives.

14.1.8 Miscellaneous Compounds

In addition to the types of compounds discussed so far, the Group IVA elements also form several other interesting compounds. Silicon has enough nonmetallic character that it reacts with many metals to form binary *silicides*. Some of these compounds can be considered as alloys of silicon and the metal that result in formulas such as Mo_3Si and $TiSi_2$. The presence of Si_2^{2-} ions is indicated by a Si–Si distance that is virtually identical to that found in the element, which has the diamond structure. Calcium carbide contains the C_2^{2-} so it is an acetylide that is analogous to the silicon compounds.

Silicon carbide, SiC or *carborundum*, has the diamond structure, and it is widely used as an abrasive in grinding wheels. These are made by crushing the SiC, adding clay, and then heating the material in molds. Silicon carbide is prepared by the reaction

$$3\,C + SiO_2 \xrightarrow{1950\,°C} SiC + 2\,CO \tag{14.69}$$

Heating silicon and nitrogen in an electric furnace produces a silicon nitride, Si_3N_4.

$$3\,Si(g) + 2\,N_2(g) \rightarrow Si_3N_4 \tag{14.70}$$

Germanium nitride, Ge_3N_2, is produced from the imine GeNH that forms when germanium iodide reacts with ammonia.

$$GeI_2 + 3\,NH_3 \rightarrow GeNH + 2\,NH_4I \tag{14.71}$$

$$3\,GeNH \xrightarrow[\text{vacuum}]{300\,°C} NH_3 + Ge_3N_2 \tag{14.72}$$

Lead is found as the sulfide, but the other members of the group also form compounds with sulfur. Although PbS has the sodium chloride crystal structure, a silicon sulfide having the empirical formula SiS_2 is known that has a chain structure in which there are four bonds to each silicon atom. The structure can be shown as

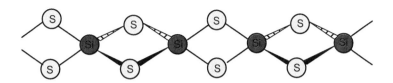

Two sulfides of germanium are known, and they are prepared by the reactions

$$GeCl_4 + 2\,H_2S \rightarrow GeS_2 + 4\,HCl \tag{14.73}$$

$$GeS_2 + Ge \rightarrow 2\,GeS \tag{14.74}$$

The formation of stannates can be accomplished by first preparing hydroxo complexes in basic solution in a reaction that can be shown as follows:

$$SnO + OH^- + H_2O \rightarrow Sn(OH)_3^- \tag{14.75}$$

The $Sn(OH)_3^-$ ion is sufficiently stable to permit forming solids containing the ion. In the $+4$ state, $Sn(OH)_6^{2-}$ results, and a solid compound containing this ion loses water to yield the stannate.

$$K_2Sn(OH)_6 \rightarrow K_2SnO_3 + 3\,H_2O \tag{14.76}$$

Silicon reacts with alkyl halides at high temperature to yield dialkyl dichloro silanes.

$$Si + 2\,CH_3Cl \xrightarrow{300\,°C} (CH_3)_2SiCl_2 \tag{14.77}$$

The Si–Cl bonds will react in ways that are typical of covalent bonds between halogens and nonmetals. One such reaction is

$$n\,H_2O + n\,(CH_3)_2SiCl_2 \rightarrow [(CH_3)_2SiO-]_n + 2n\,HCl \tag{14.78}$$

If $n = 3$, the product is a cyclic trimer that has the structure

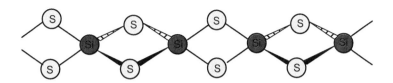

When the reaction shown in Eqn (14.78) is carried out in dilute sulfuric acid, linear polymers having oxygen bridges are produced that have the structure

When there is oxygen bridging between chains, the polymers are known as silicones, and they are synthetic oils that have many uses.

14.2 NITROGEN

Nitrogen was discovered in 1772, and it is such a versatile element that entire volumes have been devoted to its chemistry. It comprises 78% of the atmosphere, but it also occurs in many other naturally occurring materials including nitrates, amino acids, etc.

14.2.1 Elemental Nitrogen

Elemental nitrogen occurs as a very stable diatomic molecule. Only two naturally occurring isotopes are found, ^{14}N (99.635%) and ^{15}N. The occurrence mainly as ^{14}N is somewhat unusual in view of the fact that so few odd−odd nuclei are stable. Nitrogen molecules are small having a bond length of only 110 pm (1.10 Å) owing to the strong triple bond holding the atoms together (945 kJ mol^{-1}). This very strong and stable bond (the force constant is 22.4 mdyne Å^{-1}) is partially responsible for the explosive nature of some nitrogen compounds. The molecular orbital diagram for N_2 shown in Figure 3.8 indicates that the bond order is 3 in this extremely stable molecule.

The fact that the atmosphere contains 78% of elemental nitrogen suggests that the element is not very reactive. This may be surprising given the fact that nitrogen is a nonmetal having an electronegativity of 3.0, but as mentioned earlier, the strong N≡N bond implies that there are no low energy pathways for most reactions of the element.

Air can be liquefied and liquid nitrogen boils at $-195.8\,°C$, but liquid oxygen boils at $-183\,°C$ so nitrogen can be vaporized first. One laboratory preparation of nitrogen is the decomposition of sodium azide, NaN_3.

$$3\,NaN_3 \xrightarrow{\text{heat}} Na_3N + 4\,N_2 \tag{14.79}$$

Nitrogen can also be obtained by the *careful* decomposition of ammonium *nitrite*,

$$NH_4NO_2 \xrightarrow{\text{heat}} N_2 + 2\,H_2O \tag{14.80}$$

If phosphorus is burned in a closed container, it combines with oxygen leaving impure nitrogen. The oxygen can also be removed from air using pyrogallol to leave nitrogen.

14.2.2 Nitrides

At high temperature, metals will react in an atmosphere of nitrogen to produce metal nitrides. An example of this behavior is

$$3\,Mg + N_2 \rightarrow Mg_3N_2 \tag{14.81}$$

Depending on the metallic element, the nitrides may designated as ionic, covalent, or interstitial as is the case with hydrides, carbides, and borides. Boron nitrides are known that have the diamond or graphite structure (these materials were discussed in Chapter 13). Binary compounds containing nitrogen and most other elements are known, but because N_2 is not very reactive, most are not prepared by combination of the elements. If the difference between the electronegativity of nitrogen and that of the other element is about 1.6 or greater, the compounds will be essentially ionic in nature, and they can be prepared by direct combination of the elements as shown above.

The decomposition of hydroxides produces oxides, and in an analogous way, heating amides results in the production of nitrides as illustrated by the equation

$$3\,Ba(NH_2)_2 \xrightarrow{\text{heat}} Ba_3N_2 + 4\,NH_3 \tag{14.82}$$

Compounds containing the N^{3-} ion are exceedingly strong bases and react with almost any proton donor to produce NH_3. For example,

$$Mg_3N_2 + 6\,H_2O \rightarrow 3\,Mg(OH)_2 + 2\,NH_3 \tag{14.83}$$

The "nitrides" of most nonmetals exist. By calling the compounds nitrides, it is indicated that the other element has an electronegativity that is lower than that of nitrogen. Therefore, NO_2, NF_3, N_2F_2, etc. would not be considered "nitrides" because the other element is the more electronegative. This leaves quite a number of compounds, such as HN_3, S_4N_4, $(CN)_2$, that are covalent nitrides. Chemically, these compounds are quite different, and as will be shown later, methods for synthesizing them vary enormously.

The compounds formed by transition metals and nitrogen result when the metal is heated with N_2 or NH_3 at high temperature. Nitrogen atoms occupy interstitial position in the metals so the "compounds" frequently deviate from exact stoichiometry. Rather, the composition depends on the temperature and pressure used in the reaction. As in the case of interstitial hydrides, placing nitrogen atoms in interstitial positions gives predictable changes in the properties of the metal. For example, the materials are hard, brittle, high melting metallic-appearing solids. Some metal nitrides of this type are important in making cutting tools and drills.

14.2.3 Ammonia and Aquo Compounds

There are many similarities between ammonia and water (see Chapter 10). Both can function as either a proton acceptor or a proton donor. Substituting alkyl groups for one or more of the hydrogen atoms leads to a large number of organic compounds containing $-OH$ and $-NH_2$ groups. Many of the derivatives of ammonia and water are shown in Table 14.3.

14.2.4 Hydrogen Compounds

Ammonia is produced in huge quantities, and it is by far the most common and important compound of nitrogen and hydrogen. Approximately 30 billion pounds of NH_3 are used annually with a large portion being used as fertilizer or in the production of nitric acid. Ammonia is produced by the *Haber Process*, which can be shown as

$$N_2 + 3\,H_2 \xrightarrow[\text{Catalyst}]{300\text{ atm, }450\,^\circ\text{C}} 2\,NH_3 \tag{14.84}$$

Table 14.3 The Ammono and Aquo Series of Compounds

Ammono Species	Aquo Species
NH_4^+	H_3O^+
NH_3	H_2O
NH_2^-	OH^-
NH^{2-}	O^{2-}
N^{3-}	–
H_2N-NH_2	$HO-OH$
RNH_2	ROH
$RNHR$	ROR
R_3N	–
$HN=NH$	–
NH_2OH	–

Although the reaction proceeds faster at high temperature, NH_3 has a heat of formation of -46 kJ/mol so it becomes less stable. The decomposition of organic compounds that contain nitrogen during the heating of coal to produce coke also produces ammonia.

For the preparation of small amounts of ammonia, a convenient reaction involves heating an ammonium salt with a strong base.

$$(NH_4)_2SO_4 + 2\,NaOH \rightarrow Na_2SO_4 + 2\,H_2O + 2\,NH_3 \tag{14.85}$$

As shown in the previous chapter, hydrides of several nonmetals can be obtained by adding water to a binary metal compound of the nonmetal. For producing ammonia, Na_3N is an appropriate starting compound because it reacts with water:

$$Na_3N + 3\,H_2O \rightarrow 3\,NaOH + NH_3 \tag{14.86}$$

Ammonia is a colorless gas (m.p. $-77.8\,°C$ and b.p. $-33.35\,°C$) with a characteristic odor. As a result of the polarity of the $N-H$ bonds, there is extensive hydrogen bonding in the liquid and solid states. Although it is often convenient to use the formula NH_4OH, it does not represent a stable molecule. Ammonia is extremely soluble in water, but it ionizes only slightly ($K_b = 1.8 \times 10^{-5}$),

$$NH_3 + H_2O \leftrightarrows NH_4^+ + OH^- \tag{14.87}$$

Ammonia can also function as a proton donor toward extremely strong Brønsted bases such as the oxide ion.

$$NH_3 + O^{2-} \rightarrow NH_2^- + OH^- \tag{14.88}$$

There is an extensive chemistry associated with the use of liquid ammonia as a nonaqueous solvent (see Chapter 10). As a result of it having a dielectric constant of 22 and a dipole moment of 1.46 D, ammonia dissolves many ionic and polar substances. However, reactions are frequently different than in water because of differences in solubility. For example, in water, the following reaction takes place because of the insolubility of AgCl:

$$BaCl_2 + 2\,AgNO_3 \rightarrow 2\,AgCl(s) + Ba(NO_3)_2 \tag{14.89}$$

In liquid ammonia, AgCl is soluble, and the reaction

$$Ba(NO_3)_2 + 2\,AgCl \rightarrow BaCl_2(s) + 2\,AgNO_3 \tag{14.90}$$

takes place due to the insolubility of $BaCl_2$.

One difference between water and liquid ammonia involves the reactivity of Group IA metals. For example, potassium reacts very vigorously with water,

$$2\,K + 2\,H_2O \rightarrow H_2 + 2\,KOH \tag{14.91}$$

but the analogous reaction with ammonia,

$$2\,K + 2\,NH_3 \rightarrow 2\,KNH_2 + H_2 \tag{14.92}$$

takes place very slowly. Because the amide ion is a stronger base than OH^-, some reactions requiring a strongly basic solution take place better in liquid ammonia than in water (see Chapter 10).

Ammonium salts are important and useful compounds. Many have structures that are identical to those of the corresponding potassium or rubidium compounds because the ions are of similar size: $K^+ = 133$ pm, $Rb^+ = 148$ pm, and $NH_4^+ = 148$ pm. Ammonium nitrate is widely used as a fertilizer, and it contains both an oxidizing agent (NO_3^-) and a reducing agent (NH_4^+) so it is also a powerful explosive. Careful heating at temperatures below about 200 °C leads to decomposition giving N_2O.

$$NH_4NO_3 \xrightarrow{170-200\,^{\circ}C} N_2O + 2\,H_2O \tag{14.93}$$

Ammonia is also the starting material for the production of nitric acid, and the first step is oxidization of ammonia by the *Ostwald Process*.

$$4\,NH_3 + 5\,O_2 \xrightarrow{\quad Pt \quad} 4\,NO + 6\,H_2O \tag{14.94}$$

Hydrazine has the structure H_2N-NH_2, and it is the nitrogen analog of hydrogen peroxide, $HO-OH$. Although hydrazine is unstable (the heat of formation is +50 kJ mol^{-1}), it reacts as a weak diprotic base having $K_{b1} = 8.5 \times 10^{-7}$ and $K_{b2} = 8.9 \times 10^{-16}$. Hydrazine is easily oxidized and reacts vigorously with strong oxidizing agents. For example,

$$N_2H_4 + O_2 \rightarrow N_2 + 2\,H_2O \tag{14.95}$$

Hydrazine, methyl hydrazine, and unsymmetrical dimethylhydrazine, $(CH_3)_2N-NH_2$, have been used as rocket fuels that are oxidized by liquid N_2O_4. The structure of the hydrazine molecule ($\mu = 1.75$ D) can be shown as

The $N_2H_5^+$ cation has a shorter $N-N$ bond length because attaching H^+ to one of the unshared pairs of electrons reduces the repulsion between it and the other unshared pair.

Hydrazine is prepared by the *Raschig process*, the first step of which involves the production of chloramine, NH_2Cl. The process can be summarized by the equations

$$NH_3 + NaOCl \xrightarrow{\text{gelatin}} NaOH + NH_2Cl \tag{14.96}$$

$$NH_2Cl + NH_3 + NaOH \rightarrow N_2H_4 + NaCl + H_2O \tag{14.97}$$

However, another reaction competes in the process:

$$2\,NH_2Cl + N_2H_4 \rightarrow 2\,NH_4Cl + N_2 \tag{14.98}$$

This reaction is catalyzed by traces of metal ions, and gelatin is added to minimize the effect. Hydrazine is sometimes used as the reducing agent for silver in making mirrors.

Another compound of nitrogen and hydrogen is diimine, N_2H_2, that has the structure $HN{=}NH$. The compound decomposes to give N_2 and H_2, but it is believed to be a transient species in some reactions. The reactions that lead to formation of diimine can be summarized as follows:

$$H_2NCl + OH^- \rightarrow HNCl^- + H_2O \tag{14.99}$$

$$HNCl^- + H_2NCl \rightarrow Cl^- + HCl + HN{=}NH \tag{14.100}$$

Hydrogen azide (or *hydrazoic acid*) is a volatile compound (m.p. $-80\,°C$, b.p. $37\,°C$), and it is a weak acid having $K_a = 1.8 \times 10^{-5}$. It is dangerously explosive (it is 98% nitrogen!), and it is also highly toxic. In general, covalent azides or those that are substantially covalent are also explosive. Azides such as $Pb(N_3)_2$ and AgN_3 are also sensitive to shock so they have been used as primary explosives (detonators). In contrast, *ionic* azides are stable and decompose only slowly upon heating.

$$2\,NaN_3 \xrightarrow{300\,°C} 2\,Na + 3\,N_2 \tag{14.101}$$

The difference in stability of ionic and covalent azides is sometimes explained in terms of resonance structures. The azide ion, N_3^-, can be represented by the three resonance structures.

Structure I is the most important of the three. A covalent azide such as HN_3 (dipole moment $= 1.70$ D) can be represented by the resonance structures

$$\overset{\ominus}{N}-\overset{\oplus}{N}\equiv\overset{0}{N}I \;\longleftrightarrow\; \overset{0}{N}=\overset{\oplus}{N}=\overset{\ominus}{\underline{N}} \;\longleftrightarrow\; \overset{\oplus}{N}\equiv\overset{\oplus}{N}-\overset{-2}{\underline{N}}I$$

$$\text{I} \qquad\qquad \text{II} \qquad\qquad \text{III}$$

(with H attached to the left N in each structure)

Structure III makes virtually no contribution because of the positive formal charges on adjacent atoms and the overall higher formal charges. In HN_3, the bond lengths are

$$\underset{109°}{\overset{124\text{ pm} \quad 113\text{ pm}}{N_A\!\!-\!\!N_B\!\!-\!\!N_C}}$$

(with H attached to N_A)

Bonds between nitrogen atoms have lengths that are normally as follows: N—N, 145 pm; N=N, 125 pm; N≡N, 110 pm. Because structure I for HN_3 shows a single bond between N_A and N_B and Structure II shows a double bond between these atoms, the observed bond length should be somewhere between the values expected for those bonds (125 and 145 pm). However, it is almost the same as that for a double bond indicating that Structures I and II do not contribute equally and that Structure II is the more significant. Because the dipole moment of HN_3 is 1.70 D and the H—N bond moment is only about 1.33 D, it appears that Structure II, which places a negative formal charge on N_C, is dominant. Therefore, it appears that only one resonance structure makes a significant contribution to the actual structure of HN_3, whereas in N_3^- three resonance structures contribute to the true structure. It is generally true that the greater the number of contributing resonance structures, the more stable the species (the lower the energy of the species), and ionic azides are much more stable than covalent azides. This analysis is overly simplistic because the distance between N_B and N_C is close to that expected for a triple bond as shown in Structure I.

A few azides are useful, and sodium azide can be prepared by the following reactions:

$$3\,NaNH_2 + NaNO_3 \xrightarrow{175\,°C} NaN_3 + 3\,NaOH + NH_3 \tag{14.102}$$

$$2\,NaNH_2 + N_2O \rightarrow NaN_3 + NaOH + NH_3 \tag{14.103}$$

An aqueous solution containing hydrazoic acid results when a solution of nitrous acid reacts with a hydrazinium salt.

$$N_2H_5^+ + HNO_2 \rightarrow HN_3 + H^+ + 2\,H_2O \tag{14.104}$$

There is an extensive chemistry associated with coordination compounds containing azide ions as ligands. Like CN^-, the azide ion is a pseudohalide ion, which means that it forms an insoluble silver salt, exists as the acid H—X, X—X is volatile, and it can combine with other pseudohalogens to give X—X'. Although the pseudohalogen $(CN)_2$ results from the oxidation of the CN^- ion,

$$2\,CN^- \rightarrow (CN)_2 + 2\,e^- \tag{14.105}$$

the diazide N_3-N_3 is unknown. Because of the extreme stability of the N_2 molecule, the $(N_3)_2$ allotrope of nitrogen would not be expected to be stable. However, some halogen azides $(ClN_3, BrN_3, $ and $IN_3)$ are known. Chlorazide (ClN_3) is a toxic explosive compound obtained by the reaction of OCl^- and N_3^-. Other azides that contain the $-SO_2-$ group are also known. Two of them are FSO_2N_3 and $O_2S(N_3)_2$, the latter of which has the structure

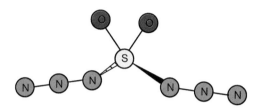

As would be expected, this compound is dangerously explosive.

14.2.5 Nitrogen Halides

Nitrogen halides having the general formula NX_3 are known, but unlike the case of phosphorus, the pentavalent compounds, NX_5, are not stable. Some mixed halides such as NF_2Cl have been studied. Nitrogen also forms N_2F_4 and N_2F_2 (which will be discussed later) that are fluorine analogs of hydrazine and diimine, respectively. Except for NF_3 (b.p. $-129\,°C$), the compounds are explosive and are little used.

The heat of formation of NF_3 is -109 kJ mol^{-1}. It is a stable compound that does not hydrolyze as do most other covalent halides of nonmetals including nitrogen trichloride.

$$NCl_3 + 3\,H_2O \rightarrow NH_3 + 3\,HOCl \qquad (14.106)$$

In this reaction, the trichloride compound behaves as though nitrogen is in the negative oxidation state. It could be argued that with an electronegativity difference of about 1.0 units that NF_3 is not typical of covalent halides. Nitrogen and chlorine have almost the same electronegativity, and NCl_3 reacts as if it were covalent. It has a heat of formation of $+232$ kJ/mol and is a violent explosive.

Nitrogen trifluoride is a compound that has interesting chemical properties. Because of the-considerable stability of the HF_2^- ion, ammonium fluoride reacts with HF to produce $NH_4F\cdot HF$, which is an ionic compound, $NH_4^+HF_2^-$. When this compound is electrolyzed in the molten state, NF_3 is produced although fluorine reacts with nitrogen to produce several products that include NF_3, N_2F_4, N_2F_2, and NHF_2. It is interesting to note that NH_3 ($\mu = 1.47$ D) is more polar that NF_3 ($\mu = 0.24$ D). In looking at the structures of the molecules, it is seen that the sum of the bond moments is in the *opposite* direction to the unshared pair in NF_3 but in the *same* direction in NH_3.

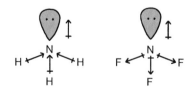

Thus, the effect of the unshared pair *reinforces* the bond moments in NH_3 but partially *cancels* them in NF_3. As a result of the fluorine atoms pulling the unshared pair of electrons inward toward the nitrogen atom, NF_3 has almost no tendency to react as a base.

Dinitrogen tetrafluoride, N_2F_4 (b.p. $-73\,°C$), can be prepared by the reaction of NF_3 with copper. The N_2F_4 molecule undergoes dissociation in the gas phase to produce $\cdot NF_2$ radicals

$$N_2F_4 \rightarrow 2 \cdot NF_2 \quad \Delta H = 84 \text{ kJ mol}^{-1} \tag{14.107}$$

There are two structures possible for dinitrogen difluoride, which is also known as difluorodiazine. These structures can be shown as follows:

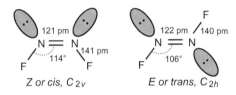

The Z (*cis*) form of N_2F_2 is more reactive than the E (*trans*) form, and the Z form slowly reacts with glass to produce SiF_4. Because it is so reactive, many compounds can be fluorinated by Z-N_2F_2 as shown in the following equations:

$$N_2F_2 + SO_2 \rightarrow SO_2F_2 + N_2 \tag{14.108}$$

$$N_2F_2 + SF_4 \rightarrow SF_6 + N_2 \tag{14.109}$$

When reacting with a strong Lewis acid that has an affinity for fluoride ions, N_2F_2 loses a fluoride ion, thereby producing N_2F^+.

$$N_2F_2 + SbF_5 \rightarrow N_2F^+SbF_6^- \tag{14.110}$$

It is also possible to prepare compounds having the general formula NH_nX_{3-n}, but only $ClNH_2$, chloramine, and HNF_2 have been well characterized.

14.2.6 Oxyhalides

A number of gaseous oxyhalides of nitrogen are known including the types XNO (*nitrosonium* or *nitrosyl* halides) with X = F, Cl, or Br, and XNO_2 (*nitryl* halides) with X = F or Cl. Nitrosonium halides are prepared by the reactions of halogens and NO.

$$2\,NO + X_2 \rightarrow 2\,XNO \tag{14.111}$$

Nitryl chloride can be prepared by the reaction

$$ClSO_3H + \text{anh. } HNO_3 \rightarrow ClNO_2 + H_2SO_4 \tag{14.112}$$

14.2.7 Nitrogen Oxides

The oxides of nitrogen that have been well characterized are described in Table 14.4. Nitrous oxide (m.p. $-91\,°C$, b.p. $-88\,°C$) is a 16-electron triatomic molecule having a linear structure. Three resonance structures can be drawn for this molecule as follows.

$$\underset{I}{\underline{\overline{N}}=N=\underline{\overline{O}}} \longleftrightarrow \underset{II}{|N\equiv N-\underline{\overline{O}}|} \longleftrightarrow \underset{III}{|\underline{\overline{N}}-N\equiv O|}$$

In Section 4.2, an analysis based on bond lengths and dipole moments was presented showing that Structures I and II contribute about equally to the actual structure. Nitrous oxide functions as an oxidizing agent that can react explosively with H_2,

$$N_2O + H_2 \rightarrow N_2 + H_2O \tag{14.113}$$

and it will oxidize metals. For example, magnesium will burn in N_2O.

$$Mg + N_2O \rightarrow N_2 + MgO \tag{14.114}$$

The mixture of acetylene and N_2O is flammable so it is used in welding.

$$5\,N_2O + C_2H_2 \rightarrow 2\,CO_2 + 5\,N_2 + H_2O \tag{14.115}$$

Nitrous oxide is relatively soluble in water, and it is used as a propellant gas in canned whipped cream. It is also used as an anesthetic (laughing gas).

Nitric oxide (m.p. $-163\,°C$, b.p. $-152\,°C$) is an important compound primarily because it is a precursor of nitric acid that is prepared in the *Ostwald process*.

$$4\,NH_3 + 5\,O_2 \xrightarrow{Pt} 4\,NO + 6\,H_2O \tag{14.116}$$

In the laboratory, reactions that can be used to produce NO include the following:

$$3\,Cu + 8\,HNO_3(dil) \rightarrow 3\,Cu(NO_3)_2 + 2\,NO + 4\,H_2O \tag{14.117}$$

Table 14.4 Oxides of Nitrogen

Formula	Name	Characteristics
N_2O	Nitrous oxide	Colorless gas, weak ox. agent
NO	Nitric oxide	Colorless gas, paramagnetic
N_2O_3	Dinitrogen trioxide	Blue solid, dissociates in gas
NO_2	Nitrogen dioxide	Brown gas, equilibrium mixture
N_2O_4	Dinitrogen tetroxide	Colorless gas
N_2O_5	Dinitrogen pentoxide	Solid is $NO_2^+NO_3^-$, gas unstab.

$$6\,NaNO_2 + 3\,H_2SO_4 \rightarrow 4\,NO + 2\,H_2O + 2\,HNO_3 + 3\,Na_2SO_4 \qquad (14.118)$$

The bonding in the NO molecule provides a basis for interpreting several properties of the molecule. From the molecular orbital diagram (see Figure 3.9), it can be seen that there is one electron in a π^* orbital. Molecules that have an odd number of electrons usually dimerize, but unlike NO_2, NO does not dimerize in the gas phase. From the molecular orbital diagram, it is seen that the bond order in NO is 2.5, and if an electron is removed, it comes from the π^* orbital leaving NO^+ that has a bond order of 3. The ionization potential for NO is 9.2 eV (888 kJ mol^{-1}) so it loses an electron rather easily to generate NO^+ (known as the *nitrosonium* or *nitrosyl* ion) that is isoelectronic with N_2, CN^-, and CO. The NO^+ ion is a good coordinating group, and many complexes are known that contain this ligand (see Chapters 16 and 21). It functions as a three-electron donor with one electron being lost to the metal and two others forming a coordinate bond.

Halogens react with NO to produce XNO, nitrosonium halides.

$$2\,NO + X_2 \rightarrow 2\,XNO \qquad (14.119)$$

and NO is easily oxidized as shown by the following equation:

$$2\,NO + O_2 \rightarrow 2\,NO_2 \qquad (14.120)$$

This reaction is one of the steps in preparing HNO_3 from NH_3. As mentioned earlier, the fact that NO does not dimerize is unusual. From the molecular orbital diagram for NO, it can be seen that the bond order is 2.5. If dimers were formed, the structure could be represented as

This structure contains a total of five bonds, which is an average of 2.5 bonds per NO unit. Therefore, there is no net increase in the number of bonds in the dimer compared with two separate molecules. The result is that there is not much energy advantage if dimers form. The melting point of NO is $-164\,°C$, and the boiling point is $-152\,°C$. The low boiling point and small liquid range, about $12\,°C$, is indicative of only very weak intermolecular forces. The Lewis structure of the molecule can be shown as

$$\underset{\cdot}{\overset{\cdot\cdot}{N}} = \overset{\cdot\cdot}{\overset{\cdot\cdot}{O}}$$

This structure shows that the molecule is essentially nonpolar ($\mu = 0.159$ D) and because it has a low molecular weight, its small liquid range is very close to that of N_2 (m.p. $-210\,°C$, b.p. $-196\,°C$).

In some ways, N_2O_3 (m.p. $-101\,°C$) behaves as a 1:1 mixture of NO and NO_2. In the gas phase, it dissociates to give NO and NO_2.

$$N_2O_3 \rightarrow NO + NO_2 \qquad (14.121)$$

At -20 to $-30\,°C$, the reverse reaction can be used to prepare N_2O_3. Two forms of N_2O_3 are represented by the structures shown below.

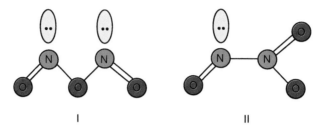

The dominant form is II, and this oxide reacts with water to yield a nitrous acid solution.

$$H_2O + N_2O_3 \rightarrow 2\,HNO_2 \tag{14.122}$$

Compounds containing NO^+ result when a strong acid such as $HClO_4$ reacts with N_2O_3.

$$N_2O_3 + 3\,HClO_4 \rightarrow 2\,NO^+ClO_4^- + H_3O^+ + ClO_4^- \tag{14.123}$$

The NO_2 molecule has a bond angle of $134°$ and unlike NO, it extensively dimerizes.

$$\underset{\text{brown}}{2\,NO_2} \leftrightarrows \underset{\text{lt. yel.}}{N_2O_4} \tag{14.124}$$

At $135\,°C$, the mixture contains only about 1% N_2O_4, but at $25\,°C$, it contains about 80% N_2O_4. The structure of N_2O_4 can be shown as

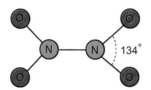

The structures $NO^+NO_3^-$, $ONONO_2$, etc. have also been identified. The structure shown has a very long $N-N$ bond (~ 175 pm), but the $N-N$ bond in N_2H_4 is only 147 pm.

Nitrogen dioxide is a toxic gas that can be prepared by the oxidation of NO.

$$2\,NO + O_2 \rightarrow 2\,NO_2 \tag{14.125}$$

When heated, some heavy metal nitrates such as $Pb(NO_3)_2$ decompose to produce NO_2.

$$2\,Pb(NO_3)_2 \xrightarrow{\text{heat}} 2\,PbO + 4\,NO_2 + O_2 \tag{14.126}$$

The major importance of the compound is that it reacts with water as one step in the preparation of nitric acid.

$$2\,NO_2 + H_2O \rightarrow HNO_3 + HNO_2 \tag{14.127}$$

The nitronium ion, NO_2^+, derived from NO_2, is the attacking species in nitration reactions (see Chapter 9). Also, liquid N_2O_4 has been rather extensively studied as a nonaqueous solvent, and autoionization, to the extent it occurs, appears to be as follows:

$$N_2O_4 \leftrightarrows NO^+ + NO_3^- \tag{14.128}$$

Accordingly, compounds such as NOCl (actually ONCl in structure) are acids and nitrates are bases in liquid N_2O_4. Therefore, the reaction

$$NaNO_3 + NOCl \rightarrow NaCl + N_2O_4 \tag{14.129}$$

is a neutralization reaction in liquid N_2O_4. Metals react with N_2O_4 to give the nitrates,

$$M + N_2O_4 \rightarrow MNO_3 + NO \tag{14.130}$$

and liquid N_2O_4 is an oxidizing agent for rocket fuels such as hydrazine.

Dinitrogen pentoxide, N_2O_5, is the anhydride of nitric acid, from which it can be prepared by dehydration at low temperatures.

$$4\,HNO_3 + P_4O_{10} \rightarrow 2\,N_2O_5 + 4\,HPO_3 \tag{14.131}$$

and it can also be prepared by the oxidation of NO_2 with ozone.

$$O_3 + 2\,NO_2 \rightarrow N_2O_5 + O_2 \tag{14.132}$$

Because solid N_2O_5 is an ionic solid, $NO_2^+\,NO_3^-$, it is suggested that a molecule formally containing NO_2^+, and a nitrate could produce N_2O_5. Such as reaction is

$$AgNO_3 + NO_2Cl \rightarrow N_2O_5 + AgCl \tag{14.133}$$

N_2O_5 is a white ionic material, $NO_2^+\,NO_3^-$, which sublimes at 32 °C. The N−O bond length in the linear NO_2^+ ion is 115 pm. The structure of the N_2O_5 molecule is

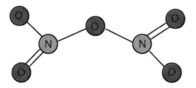

In accord with the oxidation state of nitrogen in this oxide being +5, it is also a good oxidizing agent. There is some evidence that NO_3 exists in mixtures of N_2O_5 and ozone.

14.2.8 Oxyacids of Nitrogen

Hyponitrous acid, $H_2N_2O_2$ is produced in the following reactions:

$$2\,NH_2OH + 2\,HgO \rightarrow H_2N_2O_2 + 2\,Hg + 2\,H_2O \tag{14.134}$$

$$NH_2OH + HNO_2 \rightarrow H_2N_2O_2 + H_2O \qquad (14.135)$$

When a solution of $Ag_2N_2O_2$ is treated with a solution of HCl in ether, a solution containing hyponitrous acid $H_2N_2O_2$ is obtained.

$$Ag_2N_2O_2 + 2\,HCl \xrightarrow{\text{ether}} 2\,AgCl + H_2N_2O_2 \qquad (14.136)$$

Evaporation of the ether produces the solid acid $H_2N_2O_2$, which an explosive compound. Even in an aqueous solution, the acid decomposes.

$$H_2N_2O_2 \rightarrow H_2O + N_2O \qquad (14.137)$$

Although N_2O is formally the anhydride of $H_2N_2O_2$, the acid does not result from the reaction of N_2O and water. The acid oxidizes in air to produce nitric and nitrous acids.

$$2\,H_2N_2O_2 + 3\,O_2 \rightarrow 2\,HNO_3 + 2\,HNO_2 \qquad (14.138)$$

Reduction of nitrates or nitrites by sodium amalgam in the presence of water has been used to prepare salts of hyponitrous acid.

$$2\,NaNO_3 + 8(H) \xrightarrow{\text{Na/Hg}} Na_2N_2O_2 + 4\,H_2O \qquad (14.139)$$

The $N_2O_2^{2-}$ ion exists in two forms having structures that can be shown as follows:

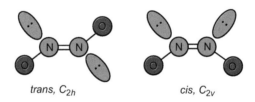

trans, C_{2h} cis, C_{2v}

The *trans* form is more stable, and it is the product of the reactions shown above. The anhydride of nitrous acid is N_2O_3, but the unstable acid is produced by acidifying a solution of a nitrite. For example, a convenient reaction is

$$Ba(NO_2)_2 + H_2SO_4 \rightarrow BaSO_4 + 2\,HNO_2 \qquad (14.140)$$

Because it is insoluble, $BaSO_4$ can be easily removed. Heating nitrates of Group IA metals results in the loss of oxygen to produce nitrites.

$$2\,KNO_3 \xrightarrow{\text{heat}} 2\,KNO_2 + O_2 \qquad (14.141)$$

At low temperature, slow decomposition of nitrous acid gives nitric acid and NO,

$$3\,HNO_2 \rightarrow HNO_3 + 2\,NO + H_2O \qquad (14.142)$$

but at high temperature, disproportionation yields NO and NO_2.

$$2\,HNO_2 \rightarrow NO + NO_2 + H_2O \tag{14.143}$$

The structure of nitrous acid can be shown in several ways, but the structure is usually considered to be

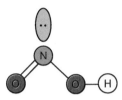

Nitrous acid is a weak acid with $K_a = 4.5 \times 10^{-4}$. Because it contains nitrogen in an intermediate oxidation state, nitrous acid can function as both an oxidizing agent and a reducing agent as shown in the following equations:

$$2\,MnO_4^- + 5\,HNO_2 + H^+ \rightarrow 5\,NO_3^- + 2\,Mn^{2+} + 3\,H_2O \tag{14.144}$$

$$HNO_2 + Br_2 + H_2O \rightarrow HNO_3 + 2\,HBr \tag{14.145}$$

$$2\,HI + 2\,HNO_2 \rightarrow 2\,H_2O + 2\,NO + I_2 \tag{14.146}$$

The acid reacts with ammonia to produce N_2.

$$NH_3 + HNO_2 \rightarrow [NH_4NO_2] \rightarrow N_2 + 2\,H_2O \tag{14.147}$$

As shown in the following structures, NO_2^- ion has unshared pairs of electrons on both the oxygen and nitrogen atoms.

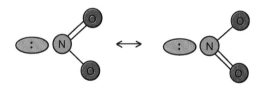

Electron pair donation can take place through either atom to produce $M-ONO$ or $M-NO_2$ linkages or it can bridge between two metal centers (to give $M-ONO-M$ linkages). The first known case of linkage isomerization involved the ions $[Co(NH_3)_5NO_2]^{2+}$ and $[Co(NH_3)_5ONO]^{2+}$, isomers that were studied by S.M. Jørgensen in the 1890s.

Nitric acid is by far the most important of the acids containing nitrogen, and it is used in enormous quantities (~30 billion lbs/yr). It is used in the manufacture of explosives, propellants, fertilizers, organic nitro compounds, dyes, plastics, etc. It is a strong acid because $b = 2$ in the formula $(HO)_aXO_b$, and it is also a strong oxidizing agent. Although it was believed to have been produced earlier, nitric acid was produced as early as 1650 by J.R. Glauber who prepared it using the reaction.

$$KNO_3 + H_2SO_4 \xrightarrow{\text{heat}} HNO_3 + KHSO_4 \tag{14.148}$$

Pure nitric acid (m.p. $-41.6\,°C$, and b.p. $82.6\,°C$) has a density of $1.503\ g/cm^3$. It forms a constant boiling mixture (having b.p. $120.5\,°C$) with water that contains 68% HNO_3 and has a density of $1.41\ g/cm^3$. Several well-defined hydrates are known such as $HNO_3 \cdot H_2O$ and $HNO_3 \cdot 2H_2O$. Concentrated nitric acid may have a light yellow-brown color because of NO_2 that results from slight decomposition,

$$4\,HNO_3 \rightarrow 2\,H_2O + 4\,NO_2 + O_2 \qquad (14.149)$$

For many years, nitric acid was prepared by heating nitrates with sulfuric acid.

$$NaNO_3 + H_2SO_4 \xrightarrow{\text{heat}} NaHSO_4 + HNO_3 \qquad (14.150)$$

In the early 1900s, it was discovered that ammonia could be oxidized in the presence of a platinum catalyst (the *Ostwald process*).

$$4\,NH_3 + 5\,O_2 \xrightarrow{\text{Pt}} 4\,NO + 6\,H_2O \qquad (14.151)$$

This is the first step in the production of nitric acid. Nitric oxide is a reactive gas that can easily be oxidized,

$$2\,NO + O_2 \rightarrow 2\,NO_2 \qquad (14.152)$$

and NO_2 disproportionates in water yielding nitric acid.

$$2\,NO_2 + H_2O \rightarrow HNO_3 + HNO_2 \qquad (14.153)$$

The nitric acid can be concentrated to about 68% by weight by distillation. The HNO_2 produced in the disproportionation decomposes to produce N_2O_3

$$2\,HNO_2 \rightarrow H_2O + N_2O_3 \qquad (14.154)$$
$$N_2O_3 \rightarrow NO + NO_2 \qquad (14.155)$$

and the nitrogen oxides are recycled.

Although the nitrate ion is planar (D_{3h}), the HNO_3 molecule has the structure

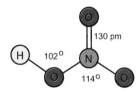

As a result of being both a strong acid and a strong oxidizing agent, nitric acid will dissolve most metals. However, aluminum forms an oxide layer on the surface, and it becomes passive to further action. Sulfur is oxidized to sulfuric acid with the reduction product being nitric oxide.

$$S_8 + 16\,HNO_3 \rightarrow 8\,H_2SO_4 + 16\,NO \tag{14.156}$$

Sulfides are oxidized to sulfates by concentrated nitric acid.

$$3\,ZnS + 8\,HNO_3 \rightarrow 3\,ZnSO_4 + 4\,H_2O + 8\,NO \tag{14.157}$$

The mixture containing one volume of concentrated HNO_3 and three volumes of concentrated HCl is known as *aqua regia*, and it will dissolve even gold and platinum.

Nitric acid and nitrates are important chemicals. For example, black powder (also known as "gunpowder") has been used for centuries, and it is a mixture containing 75% KNO_3, 15% C, and 10% S. The mixture is made into flakes while wet and then dried. Except for large guns on naval vessels, it has been replaced by nitrocellulose (*smokeless*) powder that also contains small amounts of certain additives.

Nitration of toluene by a mixture of nitric and sulfuric acids produces an explosive material known as TNT (trinitrotoluene). The overall reaction can be shown as

$$\tag{14.158}$$

This explosive is remarkably stable to shock so it requires a powerful detonator to initiate the explosion. Ammonium nitrate will explode if the explosion is initiated by another primary explosive, and mixtures of NH_4NO_3 and TNT (2,4,6-trinitrotoluene) are known as the military explosive *AMATOL*.

In the explosives industry, precise control of temperature, mixing time, concentrations, heating and cooling rates, etc. is maintained. Making these materials safely requires sophisticated equipment and technology to carry out the process even though the chemistry may appear simple. Under less than optimal conditions, some 2,3,5-trinitrotoluene, 3,5,6-trinitrotoluene, and 2,4,5-trinitrotoluene are produced, and they are much less stable than 2,4,6-TNT. A mixture of explosives is only as stable as its *least* stable component! Work with explosive materials requires specialized equipment and procedures that are available only in facilities designed for that purpose. Without sophisticated equipment and the knowledge that comes from specialized experience, no one outside the industry should ever experiment with these materials.

As mentioned earlier, there are many nitrogen compounds that are either explosives or propellants. That is one reason why nitrates and nitric acid have been so important throughout several centuries of history.

14.3 PHOSPHORUS, ARSENIC, ANTIMONY, AND BISMUTH

The elements in Group VA have a broad range of chemical properties, and there is an increase in metallic character progressing downward in the group. Each of the elements has possible oxidation states from -3 to $+5$, although not all of the range applies equally to all of the elements. The organic chemistry of phosphorus and arsenic is also extensive as is the organometallic chemistry of antimony.

All of the elements are important, and they are found in many common compounds. Some phosphorus compounds are among the most useful and essential of any element. As a result, there is a great deal more extensive chemistry of phosphorus, and more space will be devoted to it in this chapter. Much of the chemistry of the other elements can be inferred by comparison to the analogous phosphorus compounds but realizing the greater metallic character of As and Sb.

14.3.1 Occurrence

Phosphorus compounds occur widely in nature with some of the most common forms being phosphate rocks and minerals, bones, teeth, etc. Phosphate minerals include calcium phosphate, $Ca_3(PO_4)_2$; *apatite*, $Ca_5(PO_4)_3OH$; *fluorapatite*, $Ca_5(PO_4)_3F$; and *chloroapatite*, $Ca_5(PO_4)_3Cl$. Elemental phosphorus was first obtained by H. Brand, and its name is derived from two Greek words meaning "light" and "I bear" because of the phosphorescence of white phosphorus due to slow oxidation.

Several minerals contain arsenic, but most important are the sulfides *orpiment*, As_2S_3; *realgar*, As_4S_4; and *aresenopyrite*, FeAsS, and the oxide *arsenolite*, As_4O_6. Antimony is also found as the sulfide, *stibnite*, Sb_2S_3, and the sulfide has been used in pigments, production of special types of glass, and in pyrotechnics. Other antimony-containing minerals are *ullmannite*, NiSbS; *tetrahedrite*, Cu_3SbS_3; and a number of other complex sulfides. Bismuth is rather unreactive so it is sometimes found free. It is also found as *bismite*, Bi_2O_3, and *bismuth glance*, Bi_2S_3.

14.3.2 Preparation and Properties of the Elements

Phosphorus is prepared on an industrial scale by reduction of naturally occurring phosphates. Crushed phosphate rock, carbon, and silica (SiO_2) are heated to 1200–1400 °C in an electric furnace from which phosphorus is removed by distillation.

$$2\,Ca_3(PO_4)_2 + 6\,SiO_2 + 10\,C \xrightarrow{1200-1400\ ^\circ C} 6\,CaSiO_3 + 10\,CO + P_4 \tag{14.159}$$

Below 800 °C, phosphorus exists primarily as tetrahedral P_4 molecules,

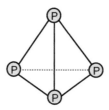

but at high temperature, some of the P_4 molecules dissociate to P_2.

Several allotropic forms of phosphorus are known, the most common of which are the white, red, and black forms. Heating the white form at 400 °C for several hours produces red phosphorus, which is known to include several forms. A red form that is amorphous can be prepared by subjecting white phosphorus to ultraviolet radiation. In the thermal process, several substances (I_2, S_8, and Na) are known to catalyze the conversion of phosphorus to other forms. Black phosphorus consists of four identifiable forms that result when white phosphorus is subjected to heat and pressure. Phosphorus is used in large quantities in the production of phosphoric acid and other chemicals. White phosphorus has been used extensively in making incendiary devices, and red phosphorus is used in making matches.

Roasting arsenic sulfide in air produces the oxide, and the element is obtained from the oxide by reduction with carbon.

$$As_4O_6 + 6\,C \rightarrow As_4 + 6\,CO \tag{14.160}$$

The stable form of arsenic is the gray or metallic form, although other forms are known. Cooling the vapor rapidly produces yellow arsenic, and an orthorhombic form is obtained if the vapor is condensed in the presence of mercury. Arsenic compounds are used in insecticides, herbicides, medicines and pigments, and arsenic is used in alloys with copper and lead. A small amount of arsenic increases the surface tension of lead, which allows droplets of molten lead to assume a spherical shape, and this fact is utilized in the production of lead shot.

Antimony is obtained by reduction of the sulfide with iron

$$Sb_2S_3 + 3\,Fe \rightarrow 2\,Sb + 3\,FeS, \tag{14.161}$$

or by heating Sb_2S_3 in air to produce the oxide, which can be reduced with carbon. The stable form of antimony has a rhombohedral structure, although high pressure converts it to cubic and hexagonal closest packing structures. There are also several amorphous forms of antimony. A small amount of antimony added to lead produces a stronger harder alloy, and the lead plates used in automobile batteries are made from such an alloy. These alloys expand on cooling and give a sharp casting without shrinking away from the mold. For this reason, antimony has long been used in type metal.

14.3.3 Hydrides

The elements in Group VA of the periodic table form binary compounds with hydrogen, some of which are analogous to the hydrogen compounds of nitrogen (NH_3, N_2H_4, and N_2H_2). However, the hydrides of the heavier elements are much less basic and less stable than NH_3. In nitrogen compounds, the unshared pair of electrons function as a hard base (see Chapter 9), and they are good proton acceptors. In PH_3, PR_3, AsH_3, AsR_3, etc., the unshared pair of electrons resides in a larger orbital so they are soft bases that have lower basicity toward proton donors. Accordingly, PH_3 ($K_b = {\sim}1 \times 10^{-28}$) is a much weaker Brønsted base than is NH_3 ($K_b = 1.8 \times 10^{-5}$). Formation of stable phosphonium salts requires that the acid be strong and the anion be large so there is a close match in the size of anion and cation. These conditions are met in the reaction with HI.

$$PH_3 + HI \rightarrow PH_4I \tag{14.162}$$

However, with soft electron pair acceptors such as Pt^{2+}, Ag^+, Ir^+, etc., phosphines are stronger Lewis bases than are NH_3 and amines so phosphines and arsines interact better with Class B metals than do amines. Generally, phosphines and arsines form stable complexes with second and third row transition metals in low oxidation states.

Arsenic, antimony, and bismuth do not normally form stable compounds as the protonated species AsH_4^+, SbH_4^+, and BiH_4^+, although a few compounds such as $[AsH_4]^+[SbF_6]^-$ and $[SbH_4]^+[SbF_6]^-$ are known. These compounds are obtained by reactions of the hydrides with the HF/SbF_5 superacid (see Section 10.8).

Phosphine is a less stable compound than NH_3 because orbital overlap between hydrogen and phosphorus is less effective than between hydrogen and nitrogen. The thermal stability of the hydrogen compounds of P, As, Sb, and Bi (named as *phosphine*, *arsine*, *stibine*, and *bismuthine*) decreases in that order, and SbH_3 and BiH_3 are unstable at room temperature. Similar trends apply to the hydrogen compounds of the Group IVA, VIA, and VIIA elements. The physical properties of the hydrogen compounds of the heavier Group VA elements are shown in Table 14.5 with the properties of NH_3 included for comparison. The effects of hydrogen bonding in NH_3 are apparent from the data.

The bond angles in PH_3, AsH_3, and SbH_3 are interesting in view of the fact that in NH_3, the bond angle is about equal to that expected for sp^3 hybridization on the nitrogen atom with a small decrease in bond angle caused by the unshared pair of electrons. The bond angles in the hydrides of other Group VA elements are close enough to $90°$ that it appears that almost pure p orbitals are used by the central atom. This may be because the larger sp^3 hybrid orbitals on the heavier atoms do not overlap as effectively with hydrogen $1s$ orbitals as in the case of nitrogen. A similar trend is seen in the hydrogen compounds of the Group VIA elements. However, another explanation is based on the fact that in NH_3, the *shared* pairs of electrons reside closer to the nitrogen atom so the effect of the unshared pair on bond angles is less. In PH_3, the bonding electrons are shared about equally (which puts them closer to the hydrogen atoms than in NH_3) so the effect of the unshared pair is greater because the electron density is less near the phosphorus atom.

The hydrazine analogs P_2H_4, As_2H_4, and Sb_2H_4 are toxic and unstable as indicated by the fact that P_2H_4 is spontaneously flammable in air, and phosphine burns readily.

$$4\,PH_3 + 8\,O_2 \rightarrow P_4O_{10} + 6\,H_2O \tag{14.163}$$

The extremely toxic trihydrides of the heavier atoms in Group VA are not generally prepared by reaction of the elements. They are usually prepared by making a metal compound of the Group VA element and then hydrolyzing it. For example,

$$6\,Ca + P_4 \rightarrow 2\,Ca_3P_2 \tag{14.164}$$

$$Ca_3P_2 + 6\,H_2O \rightarrow 3\,Ca(OH)_2 + 2\,PH_3 \tag{14.165}$$

Table 14.5 Properties of the EH_3 Compounds of the Group VA Elements

	NH_3	PH_3	AsH_3	SbH_3	BiH_3
m.p., °C	−77.7	−133.8	−117	−88	−
b.p., °C	−33.4	−87.8	−62.5	−18.4	17
ΔH_f°, kJ mol^{-1}	−46.11	−9.58	66.44	145.1	277.8
H−E−H angle, °	107.1	93.7	91.8	91.3	−
μ, D	1.46	0.55	0.22	0.12	−

Phosphine also results from the reaction of phosphorus with a hot strongly basic solution.

$$P_4 + 3\,NaOH + 3\,H_2O \rightarrow PH_3 + 3\,NaH_2PO_2 \tag{14.166}$$

Sodium arsenide can be prepared by the reaction of the elements, and it reacts with water to produce arsine according to the following equation:

$$Na_3As + 3\,H_2O \rightarrow 3\,NaOH + AsH_3 \tag{14.167}$$

Arsine is also produced in the reaction of As_2O_3 with $NaBH_4$.

$$2\,As_2O_3 + 3\,NaBH_4 \rightarrow 4\,AsH_3 + 3\,NaBO_2 \tag{14.168}$$

The hydrides of arsenic, antimony, and bismuth are unstable at elevated temperature. The Marsh test for arsenic depends on this instability when an arsenic mirror forms as arsine is passed through a heated tube:

$$2\,AsH_3 \rightarrow 2\,As + 3\,H_2 \tag{14.169}$$

14.3.4 Oxides

Although other less important oxides of the Group VA elements are known, only those in which the oxidation states are $+3$ and $+5$ are important. When the amount of oxygen is controlled, phosphorus can be oxidized to yield phosphorus(III) oxide.

$$P_4 + 3\,O_2 \rightarrow P_4O_6 \tag{14.170}$$

A tetrahedral arrangement of phosphorus atoms is retained in the P_4O_6 molecule giving a structure like that shown in Figure 14.3(a). Although dotted lines show the arrangement of phosphorus atoms, they do not represent bonds between the atoms.

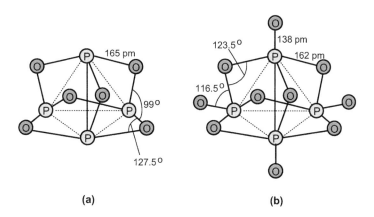

(a)   (b)

FIGURE 14.3
The structures of P_4O_6 and P_4O_{10}. The dotted lines show the original tetrahedral structure of P_4.

This oxide (m.p. 23.9 °C, b.p. 175.4 °C) reacts with cold water to give phosphorous acid, H_3PO_3, in which the arrangement of atoms is $HP(O)(OH)_2$. Disproportionation occurs when the oxide reacts with hot water, and phosphine, phosphorus, and phosphoric acid are produced. When heated above its boiling point, P_4O_6 decomposes into phosphorous and complex oxides that are described by the general formula P_nO_{2n}.

The oxides As_4O_6 and Sb_4O_6, which have structures like that of P_4O_6, are obtained by burning the elements in air. The +5 oxides of As and Sb can be obtained, but bismuth forms only Bi_2O_3. Solid As_4O_6 and Sb_4O_6 exist in several forms having oxygen atoms in bridging positions.

Although P_4O_{10} is an important material, the +5 oxides of arsenic, antimony, and bismuth are much less important. The molecular formula for the +5 oxide of phosphorus is P_4O_{10}, not P_2O_5, although it is frequently convenient to use the empirical formula. The structure of the P_4O_{10}, shown in Figure 14.3(b), is derived from the tetrahedral structure of the P_4 molecule, and it has six bridging oxygen atoms and the other four in terminal positions. At least three crystalline forms of P_4O_{10} are known. The hexagonal or *H*-form is the most common, and the others are orthorhombic forms designated as the *O*- and *O'*-forms, both of which are less reactive than the *H*-form. Heating the *H*-form at 400 °C for 2 h transforms it into the *O*-form, and heating the *H*-form at 450 °C for 24 h produces the *O'*-form.

Phosphorus(V) oxide or tetraphosphorus decaoxide, P_4O_{10}, is the anhydride of the series of phosphoric acids. It is produced in the first step of the manufacture of H_3PO_4 by burning phosphorus,

$$P_4 + 5\,O_2 \;\rightarrow\; P_4O_{10} \quad \Delta H = -2980 \text{ kJ mol}^{-1} \tag{14.171}$$

and it is a powerful dehydrating agent that can be used to prepare other oxides by dehydrating the appropriate acid. For example, dehydrating $HClO_4$ produces Cl_2O_7.

$$12\,HClO_4 + P_4O_{10} \;\rightarrow\; 6\,Cl_2O_7 + 4\,H_3PO_4 \tag{14.172}$$

When used as a desiccant, the oxide removes water by the reaction

$$P_4O_{10} + 6\,H_2O \;\rightarrow\; 4\,H_3PO_4 \tag{14.173}$$

Organic phosphates are prepared by the reactions of P_4O_{10} with alcohols.

$$P_4O_{10} + 12\,ROH \;\rightarrow\; 4\,(RO)_3PO + 6\,H_2O \tag{14.174}$$

but some monoalkyl and dialkyl derivatives are also produced. Phosphate esters also result when $OPCl_3$ reacts with alcohols. Alkyl phosphates have many uses as catalysts, lubricants, and in producing flame-proofing compounds.

Arsenic(V) oxide and antimony(V) oxide are produced when the elements react with concentrated nitric acid,

$$4\,As + 20\,HNO_3 \;\rightarrow\; As_4O_{10} + 20\,NO_2 + 10\,H_2O \tag{14.175}$$

Some compounds of antimony appear to involve the +4 oxidation state. For example, an oxide of antimony is known that has the formula Sb_2O_4, but this oxide actually contains equal numbers of Sb(III) and Sb(V) atoms. The chloride complex $Sb_2Cl_{10}^{2-}$ also contains Sb(III) and Sb(V) rather than Sb(IV).

14.3.5 Sulfides

At elevated temperatures, phosphorus, arsenic, and antimony react with sulfur to produce several binary compounds. Of the phosphorus compounds, P_4S_{10}, P_4S_7, P_4S_5, and P_4S_3 have been the subject of considerable study. Except for sulfur replacing oxygen, the structure of P_4S_{10} is like that of P_4O_{10}. The structures of the other phosphorus sulfides contain P—P bonds and P—S—P bridges. As shown in Figure 14.4, these structures can be considered as being derived from the P_4 tetrahedron by insertion of bridging sulfur atoms, but there are no bonds between P atoms.

Tetraphosphorus trisulfide (P_4S_3), which is also called phosphorus sesquisulfide, can be obtained by heating a stoichiometric mixture of phosphorus and sulfur at $180\,^{\circ}C$ in an inert atmosphere. The compound (m.p. $174\,^{\circ}C$) is soluble in toluene, carbon disulfide, and benzene, and it is used with potassium chlorate, sulfur, and lead dioxide in matches.

Tetraphosphorus pentasulfide, P_4S_5, is obtained when a solution of sulfur in CS_2 reacts with P_4S_3 in the presence of I_2. Tetraphosphorus heptasulfide, P_4S_7, is one of the products obtained when phosphorus and sulfur are heated in a sealed tube. Neither P_4S_5 nor P_4S_7 is commercially important. Tetraphosphorus decasulfur, P_4S_{10}, is prepared by reaction of a stoichiometric mixture of the elements.

$$4\,P_4 + 5\,S_8 \rightarrow 4\,P_4S_{10} \tag{14.176}$$

The reaction of this sulfide with water can as shown as

$$P_4S_{10} + 16\,H_2O \rightarrow 4\,H_3PO_4 + 10\,H_2S \tag{14.177}$$

A mixed sulfide oxide, $P_4O_6S_4$, has a structure similar to that of P_4O_{10} except sulfur atoms replace the four oxygen atoms in terminal positions.

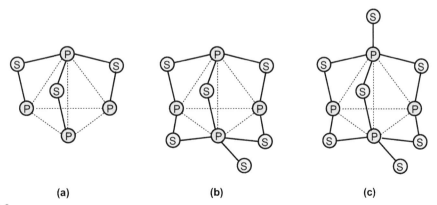

(a)   (b)   (c)

FIGURE 14.4
The most important compounds containing phosphorus and sulfur are (a) P_4S_3, (b) P_4S_5, and (c) P_4S_7. Note the outline (dotted lines) of the tetrahedron of P_4 atoms.

Arsenic, antimony, and bismuth form several sulfides that include As_4S_4 (*realgar*), As_2S_3, Sb_2S_3, and Bi_2S_3, which are the common minerals containing these elements. Some of the sulfides can also be precipitated from aqueous solutions because the sulfides of As(III), As(V), Sb(III), and Bi(III) are insoluble. As in the case of phosphorus, arsenic forms a sesquisulfide, As_4S_3, which has a structure like that of P_4S_3 shown in Figure 14.4(a). The structures of As_4S_4 and As_4S_6 are shown in Figure 14.5. Sulfides of arsenic and antimony are brightly colored, which has given rise to their use as pigments. Some of the sulfides, selenides, and tellurides of arsenic and antimony also function as semiconductors.

14.3.6 Halides

Both the +3 and +5 halogen compounds of the Group VA elements contain reactive nonmetal-halogen bonds. As a result, they can be used as starting materials for preparing many other compounds. Halogen compounds of the Group VA elements having the formula E_2X_4 are also known. They are relatively unimportant, and only the phosphorus fluoride will be described, although the chloride and iodide are also known. P_2F_4 is prepared by the coupling reaction of PF_2I brought about by mercury.

$$2\,PF_2I + 2\,Hg \;\rightarrow\; Hg_2I_2 + P_2F_4 \tag{14.178}$$

Electric discharge in a mixture of PCl_3 and H_2 produces P_2Cl_4, and white phosphorus dissolved in carbon disulfide reacts with I_2 to produce P_2I_4. All of the trihalides of the Group VA elements are known, and they can be prepared by reaction of the elements, although there are other preparative methods. The fluorides are prepared as follows:

$$PCl_3 + AsF_3 \;\rightarrow\; PF_3 + AsCl_3 \tag{14.179}$$

$$As_4O_6 + 6\,CaF_2 \;\rightarrow\; 6\,CaO + 4\,AsF_3 \tag{14.180}$$

$$Sb_2O_3 + 6\,HF \;\rightarrow\; 3\,H_2O + 2\,SbF_3 \tag{14.181}$$

Adding an excess of F^- to a solution containing Bi^{3+} gives a precipitate of BiF_3.

Phosphorus trichloride is obtained by the reaction of excess phosphorus with chlorine.

$$P_4 + 6\,Cl_2 \;\rightarrow\; 4\,PCl_3 \tag{14.182}$$

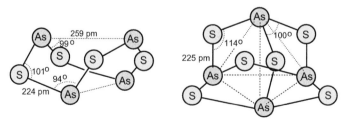

FIGURE 14.5
The structures of As_4S_4 and As_4S_6.

Oxides and sulfides of As, Sb, and Bi are converted to the chlorides by concentrated HCl.

$$As_4O_6 + 12\,HCl \rightarrow 6\,H_2O + 4\,AsCl_3 \tag{14.183}$$

$$Sb_2S_3 + 6\,HCl \rightarrow 2\,SbCl_3 + 3\,H_2S \tag{14.184}$$

$$Bi_2O_3 + 6\,HCl \rightarrow 2\,BiCl_3 + 3\,H_2O \tag{14.185}$$

Table 14.6 shows some of the properties of the trihalides of the Group VA elements.

Some of the compounds listed in Table 14.6 have low melting and boiling points indicating that they consist of discrete covalent molecules. However, others have melting points that indicate a lattice in the solid state. In general, the bond angles are intermediate between the values expected for pure p and sp^3 orbitals on the central atom. Because the molecules have an unshared pair of electrons on the central atom, the pyramidal (C_{3v}) trihalides behave as Lewis bases. However, the central atoms are not from the first long row so the trihalides are soft electron pair donors that complex better with soft Lewis acids such as second and third row transition metals in low oxidation states. Like CO, which binds to iron in blood, PF_3 also forms similar complexes, which makes it extremely toxic. Some trihalides containing two types of halogens are known, and exchange of halogen atoms takes place when two different trihalides of the same element are mixed.

$$PBr_3 + PCl_3 \rightarrow PBr_2Cl + PCl_2Br \tag{14.186}$$

$$2\,AsF_3 + AsCl_3 \rightarrow 3\,AsF_2Cl \tag{14.187}$$

Table 14.6 Physical Properties of the Trihalides of Group VA Elements

Compound	Melting Point, °C	Boiling Point, °C	μ, D	X–E–X Angle
PF_3	−151.5	−101.8	1.03	104
PCl_3	−93.6	76.1	0.56	101
PBr_3	−41.5	173.2	–	101
PI_3	61.2	dec.	–	102
AsF_3	−6.0	62.8	2.67	96.0
$AsCl_3$	−16.2	103.2	1.99	98.4
$AsBr_3$	31.2	221	1.67	99.7
AsI_3	140.4	370d	0.96	100.2
SbF_3	292	–	–	88
$SbCl_3$	73	223	3.78	99.5
$SbBr_3$	97	288	3.30	97
SbI_3	171	401	1.58	99.1
BiF_3	725	–	–	–
$BiCl_3$	233.5	441	–	100
$BiBr_3$	219	462	–	100
BiI_3	409	–	–	–

Because of the reactive covalent bonds to halogen atoms, all of the trihalides of the Group VA elements hydrolyze in water. It is found that the rates decrease in the order P > As > Sb > Bi, which agrees with the decrease in covalent bond character that results from the increase in metallic character of the central atoms. Not all of the trihalides react in the same way. The phosphorus trihalides react according to the equation

$$PX_3 + 3\,H_2O \rightarrow H_3PO_3 + 3\,HX \tag{14.188}$$

When the phosphorus halide is PI_3, this reaction is a convenient way to produce HI. Arsenic trihalides hydrolyze in an analogous way, but the trihalides of antimony and bismuth react to produce oxyhalides.

$$SbX_3 + H_2O \rightarrow SbOX + 2\,HX \tag{14.189}$$

Antimonyl chloride (sometimes called antimony oxychloride) is known as a "basic chloride." It is insoluble in water, but aqueous solutions of the trihalides can be made if enough HX is present to prevent hydrolysis. Adding water to reduce the concentration of acid causes the oxychloride to precipitate.

From the standpoint of use, PCl_3 is the most important of the trihalides. In addition to hydrolysis, PCl_3 reacts with oxygen and sulfur to produce $OPCl_3$ and $SPCl_3$, respectively.

$$2\,PCl_3 + O_2 \rightarrow 2\,OPCl_3 \tag{14.190}$$

Phosphoryl chloride (bp. 105 °C) has been used extensively as a nonaqueous solvent (see Chapter 10). Reactions of PCl_3 with other halogens give mixed pentahalides.

$$PCl_3 + F_2 \rightarrow PCl_3F_2 \tag{14.191}$$

This type of reaction involves both *oxidation* and *addition* of groups so it is known as an *oxad* reaction. Alkyl derivatives of PCl_3 can be prepared by the reactions with Grignard reagents and metal alkyls illustrated in the following equations:

$$PCl_3 + RMgX \rightarrow RPCl_2 + MgXCl \tag{14.192}$$

$$PCl_3 + LiR \rightarrow RPCl_2 + LiCl \tag{14.193}$$

If a higher ratio of alkylating agent to PX_3 is used, the dialkyl and trialkyl compounds can also be obtained. At high temperature, the reaction of benzene and PCl_3 produces phenyldichlorophosphine, $C_6H_5PCl_2$, an intermediate in the preparation of parathion.

$$C_6H_6 + PCl_3 \rightarrow HCl + C_6H_5PCl_2 \tag{14.194}$$

The organic chemistry of phosphorus includes many compounds, but the phosphite esters are especially useful. These compounds can be prepared by the reaction of PCl_3 with an alcohol.

$$PCl_3 + 3\,ROH \rightarrow (RO)_2HPO + 2\,HCl + RCl \tag{14.195}$$

Because HCl is a product, the presence of a base facilitates the reaction. If the base is an amine, an amine hydrochloride is formed, and the reaction can be shown as

$$PCl_3 + 3\,ROH + 3\,RNH_2 \rightarrow (RO)_3P + 3\,RNH_3^+Cl^- \qquad (14.196)$$

The presence of an amine makes the formation of HCl more energetically favorable because of the formation of the ionic amine hydrochloride.

There are sixteen possible pentahalides that could result from combinations of P, As, Sb, and Bi with F, Cl, Br, and I. Although all of the pentafluorides can be prepared, none of the pentaiodides is stable. The pentachloride and pentabromide are known for phosphorus, and antimony forms a pentachloride. In the solid phase, PCl_5 exists as $[PCl_4^+][PCl_6^-]$, whereas the bromide exists as $[PBr_4^+]\,[Br^-]$. In solid $PCl_4^+PCl_6^-$, the P—Cl bond lengths are 198 pm in the cation and 206 pm in the anion. Solid $SbCl_5$ exists as $SbCl_4^+Cl^-$. The PCl_5 *molecule* has the D_{3h} structure

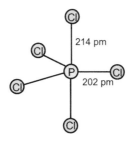

The structure of PF_5 is also a trigonal bipyramid with axial bonds of 158 pm and equatorial bonds of 152 pm. When studied by ^{19}F NMR, a single peak split into a doublet by coupling with ^{31}P is observed. Therefore, it appears that all five F atoms are equivalent indicating that there is rapid exchange between the axial and equatorial positions. An explanation of this phenomenon was provided by R.S. Berry. Known as the Berry pseudorotation, the mechanism involves the trigonal bipyramid (D_{3h}) passing through a square based pyramid (C_{4v} if all the atoms in the base are identical) as shown in Figure 14.6.

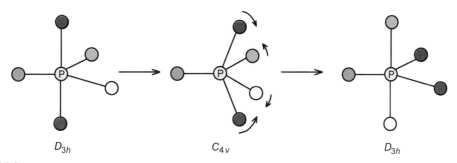

FIGURE 14.6
The Berry pseudorotation that leads to interchange of axial and equatorial groups. The colors indicate only *positions* because all five peripheral atoms are presumed to be identical.

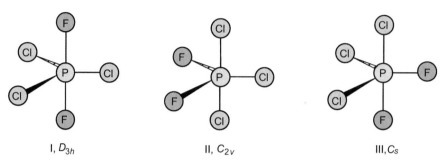

FIGURE 14.7
Structures for the isomers of PCl_3F_2.

This behavior is similar to the inversion of the ammonia molecule (C_{3v}) as it passes through a planar (D_{3h}) structure.

Phosphorus pentachloride can be prepared by chlorinating P_4 or PCl_3.

$$P_4 + 10\,Cl_2 \rightarrow 4\,PCl_5 \tag{14.197}$$

$$PCl_3 + Cl_2 \rightarrow PCl_5 \tag{14.198}$$

Phosphorus pentabromide can be prepared by analogous reactions. As mentioned earlier, several mixed halide compounds having formulas such as PCl_3F_2, PF_3Cl_2, PF_3Br_2, etc. are known. These compounds are produced when a halogen undergoes an oxad reaction with a trihalide containing a different halogen (Eqn (14.191)). The structures of these molecules provide verification of the nonequivalence of positions in a trigonal bipyramid. As a result of the difference between the axial and equatorial positions, there are three possible isomers for PCl_3F_2, which have the structures shown in Figure 14.7.

We have seen that unshared pairs of electrons occupy *equatorial* positions in structures based on a trigonal bipyramid. Therefore, the larger chlorine atoms should occupy *equatorial* positions as shown in Structure I above, and PCl_3F_2 has D_{3h} symmetry as expected. By similar reasoning, we expect the structure of PF_3Cl_2 to have C_{2v} symmetry as shown below rather than D_{3h}.

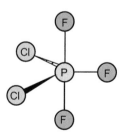

The pentahalides of Group VA elements are strong Lewis acids that react readily with electron pair donors such as halide ions to form complexes.

$$PF_5 + F^- \rightarrow PF_6^- \tag{14.199}$$

This tendency is so great that it provides a way to produce interhalogen cations such as ClF_2^+ as shown in the following reaction:

$$ClF_3 + SbF_5 \xrightarrow{\text{Liq. ClF}_3} ClF_2^+ SbF_6^- \tag{14.200}$$

Because they are strong Lewis acids, PCl_5, PBr_5, $SbCl_5$, and SbF_5 are effective catalysts for reactions such as the Friedel–Crafts reaction (see Chapter 9). The pentahalides also function as halogen transfer reagents to a variety of inorganic and organic substrates. Useful intermediates such as oxyhalides can be produced in this way. For example, thionyl chloride and phosphoryl chloride are produced by the reaction

$$SO_2 + PCl_5 \rightarrow SOCl_2 + OPCl_3 \tag{14.201}$$

Partial hydrolysis of pentahalides also leads to the formation of oxychlorides that are useful intermediates that contain reactive P–Cl bonds.

$$PCl_5 + H_2O \rightarrow OPCl_3 + 2\,HCl \tag{14.202}$$

With excess water, complete hydrolysis of PCl_5 yields H_3PO_4 and HCl. Phosphoryl chloride can also be prepared by oxidizing PCl_3 or by the reaction of P_4O_{10} with PCl_5.

$$2\,PCl_3 + O_2 \rightarrow 2\,OPCl_3 \tag{14.203}$$

$$6\,PCl_5 + P_4O_{10} \rightarrow 10\,OPCl_3 \tag{14.204}$$

The C_{3v} structure of $OPCl_3$ can be shown as

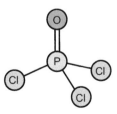

and the phosphorus-halogen bonds undergo reactions with water and alcohols.

$$OPCl_3 + 3\,H_2O \rightarrow H_3PO_4 + 3\,HCl \tag{14.205}$$

$$OPCl_3 + 3\,ROH \rightarrow (RO)_3PO + 3\,HCl \tag{14.206}$$

Organic phosphates, $(RO)_3PO$, have a wide range of industrial uses. Both $OPCl_3$ and $OP(OR)_3$ are Lewis bases that usually coordinate to Lewis acids through the terminal oxygen atom. The sulfur analog of phosphoryl chloride, $SPCl_3$, and several mixed oxyhalides, such as $OPCl_2F$ and $OPCl_2Br$, have been prepared. Oxyhalides of arsenic, antimony, and bismuth are known, but they are less useful than those of phosphorus.

The heavier members of Group VA also have a strong tendency toward complex formation in keeping with their more metallic character. For example, in HCl solutions, $SbCl_5$ adds Cl^- to form a hexachloro complex.

$$SbCl_5 + Cl^- \rightarrow SbCl_6^- \tag{14.207}$$

There is evidence that when $SbCl_5$ and $SbCl_3$ are mixed they interact to form a complex,

$$SbCl_5 + SbCl_3 \leftrightarrows Sb_2Cl_8 \tag{14.208}$$

In aqueous HCl solutions, $SbCl_5$ and $SbCl_3$ are present as the chloride complexes $SbCl_6^-$ and $SbCl_4^-$, respectively. The equilibrium between complexes can be shown as

$$SbCl_6^- + SbCl_4^- \leftrightarrows Sb_2Cl_{10}^{2-} \tag{14.209}$$

The structure of $Sb_2Cl_{10}^{2-}$ probably has bridging chloride ions.

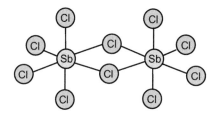

Numerous complexes such as AsF_6^-, SbF_6^-, and PF_6^- have been characterized.

14.3.7 Phosphazine Compounds

Although there are many compounds that contain phosphorus and nitrogen, the *phosphazines* compounds are unique. Because both N and P have five electrons in the valence shell, it is possible to have a six-membered heterocyclic ring in which the phosphorus atoms are bonded to two other atoms. Phosphazines having both linear and cyclic structures have been prepared. Historically, chlorides having the general formula $(PNCl_2)_n$ were the first compounds of this type prepared, and the original preparation of $(PNCl_2)_n$ was carried out in 1832 by J. von Liebig using the reaction of NH_4Cl and PCl_5.

$$n\,NH_4Cl + n\,PCl_5 \rightarrow (NPCl_2)_n + 4n\,HCl \tag{14.210}$$

This reaction can be carried out in a sealed tube or in a solvent such as $C_2H_2Cl_4$, C_6H_5Cl, or $OPCl_3$. The cyclic trimer shown below has been extensively studied.

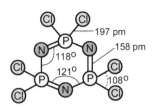

There is significant multiple bonding in the ring as indicated by the P—N bond length of about 158 pm, which is much shorter than the usual P—N single bond length (175 pm). As a result of resonance, there is only one P—N bond length. The tetramer, $(NPCl_2)_4$, has a structure in which the ring is puckered, whereas the pentamer, $(NPCl_2)_5$, has a planar ring structure. By utilizing the reactivity of the P—Cl bonds, many derivatives of $(NPCl_2)_3$ can be prepared. For example, hydrolysis produces P—OH bonds that undergo subsequent reactions to produce many derivatives.

$$(NPCl_2)_3 + 6\,H_2O \rightarrow (NP(OH)_2)_3 + 6\,HCl \tag{14.211}$$

If a reaction is carried out so that two chlorine atoms are replaced, the groups may be bound to the same phosphorus atom or to different phosphorus atoms. When the entering groups are attached to the same phosphorus atom, the product has a *geminal* structure. If the groups are on different atoms, they may be on the same side of the ring (a *cis* product) or on opposite sides of the ring (a *trans* product). The structures of these products are shown in Figure 14.8.

By means of substitution reactions, derivatives containing alkoxide, alkyl, amine, and other groups can be prepared as illustrated by the following equations:

$$(NPCl_2)_3 + 6\,NaOR \rightarrow [NP(OR)_2]_3 + 6\,NaCl \tag{14.212}$$

$$(NPCl_2)_3 + 12\,RNH_2 \rightarrow [NP(NHR)_2]_3 + 6\,RNH_3Cl \tag{14.213}$$

$$(NPCl_2)_3 + 6\,LiR \rightarrow (NPR_2)_3 + 6\,LiCl \tag{14.214}$$

$$(NPCl_2)_3 + 6\,ROH \rightarrow [NP(OR)_2]_3 + 6\,HCl \tag{14.215}$$

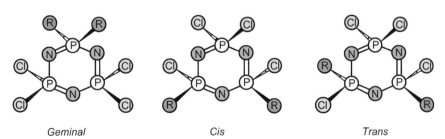

Geminal   Cis   Trans

FIGURE 14.8
Possible products of disubstitution in $(NPCl_2)_3$ to give $N_3P_3Cl_4R_2$.

In addition to $P_3N_3X_6$ molecules that contain X as a halogen or organic moiety, the azide derivative having the structure

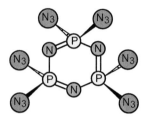

is also known.

Several products are possible when reactions are carried out with difunctional molecules such as ethylene glycol (ethanediol). For example, rings can form in which one phosphorus atom is contained in the ring or in which a larger ring contains a P—N—P unit. These structures can be illustrated as

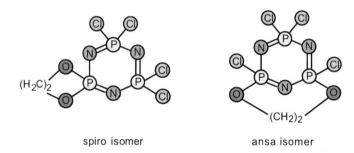

spiro isomer   ansa isomer

The reactions are carried out in the presence of a base, and the distribution depends on the nature of the solvent and the base utilized. In recent years, other interesting phosphazine derivatives have been reported that are produced when $P_3N_3Cl_6$ reacts with diols, $HO(CH_2)_nOH$, where $n = 2 - 10$. In addition to products similar to those shown above, singly, doubly, and triply bridged structures have been obtained. The structure of the doubly bridged molecules can be represented as

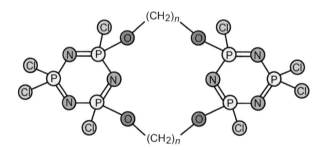

By starting with $P_3N_3Cl_6$ and carrying out a reaction with $F(C_6H_4)CH_2NH(CH_2)_nNHR$ (where $n = 3$ or 4), cyclic derivatives can be prepared that have secondary ring structures. One such compound is that which can be shown as

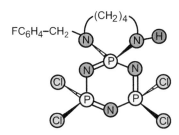

These compounds indicate the great versatility of the $P_3N_3Cl_6$ molecule when reacting with numerous organic compounds.

When $(NPCl_2)_3$ is heated to 250 °C, polymerization occurs giving polymers that contain up to 15,000 monomer units.

$$n/3\,(NPCl_2)_3 \xrightarrow{250\ °C} (NPCl_2)_n \qquad (14.216)$$

By carrying out reactions of the types shown here, a very large number of phosphazines have been prepared. In addition to synthetic work, a great deal of theoretical work has been done to elucidate the bonding in phosphazines.

14.3.8 Acids and Salts

When considering the acids of the Group VA elements, the first acid that comes to mind is probably phosphoric acid, H_3PO_4, and it should be. Phosphoric acid is one of the chemicals produced in enormous quantities, and it is used in many industrial processes. However, there are other acids that contain the Group VA elements even though none is very important in comparison to phosphoric acid. In many ways, arsenic acid, H_3AsO_4, is similar to phosphoric acid, but the similarities are all but nonexistent when bismuth is considered. The acid that bismuth forms is H_3BiO_3, which can also be written as $Bi(OH)_3$ showing that it is a very weak acid. Phosphorus(III) is also contained in an acid, H_3PO_3, but the structure of the molecule is $OP(H)(OH)_2$. The discussion of the acids of the Group VA elements will deal primarily with those containing phosphorus.

Phosphorous acid, H_3PO_3, is produced when P_4O_6 reacts with water.

$$P_4O_6 + 6\,H_2O \rightarrow 4\,H_3PO_3 \qquad (14.217)$$

It is also produced when PCl_3 is hydrolyzed.

$$PCl_3 + 3\,H_2O \rightarrow H_3PO_3 + 3\,HCl \qquad (14.218)$$

Although the formula is sometimes written as H_3PO_3, it is a *diprotic* acid that has the structure

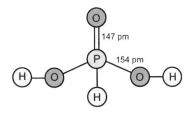

As shown by the dissociations constants $K_{a1} = 5.1 \times 10^{-2}$ and $K_{a2} = 1.8 \times 10^{-7}$, it is a weak acid. Because there is a factor of about 10^{-5} difference in the dissociation constants, it is possible to remove one proton to produce salts such as NaH_2PO_3, a so-called acid salt.

Organic phosphites are used in a variety of ways as solvents and as intermediates in synthesis. Two series of compounds having the formulas $(RO)_2P(O)H$ and $(RO)_3P$ are known, and their structures can be shown as follows.

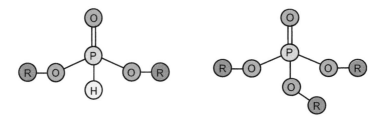

Chlorine will replace the hydrogen atom bound to phosphorus in the dialkyl phosphites yielding a dialkylchloro phosphite.

$$(RO)_2P(O)H + Cl_2 \rightarrow (RO)_2P(O)Cl + HCl \tag{14.219}$$

Partial hydrolysis reactions lead to a monoalkyl phosphite, but complete hydrolysis gives phosphorous acid. The hydrogen atom bound to phosphorus can also be replaced by reaction with sodium showing that the hydrogen atom has slight acidity.

$$2\,(RO)_2P(O)H + 2\,Na \rightarrow 2\,(RO)_2PO^-Na^+ + H_2 \tag{14.220}$$

The bonds in PCl_3 are reactive toward many compounds that contain OH groups. Consequently, alcohols react with PCl_3 to produce trialkyl phosphite esters.

$$PCl_3 + 3\,ROH \rightarrow (RO)_3P + 3\,HCl \tag{14.221}$$

Dialkyl phosphates can also be obtained if the ratio of alcohol to PCl_3 is properly maintained. Halogens will react with alkyl phosphites to give products that have a halogen bound to the phosphorus atom

$$(RO)_3P + X_2 \rightarrow (RO)_2P(O)X + RX \tag{14.222}$$

When a trialkyl phosphite reacts with PX_3, there is halogen transfer as illustrated in the following equation:

$$(RO)_3P + PX_3 \rightarrow (RO)_2PX + ROPX_2 \tag{14.223}$$

As in the case phosphorus trihalides, the phosphorus atom in trialkyl phosphites will undergo addition reactions in which oxygen, sulfur, or selenium is added. The latter two react as elements, but a suitable source of oxygen is hydrogen peroxide.

$$(RO)_3P + H_2O_2 \rightarrow (RO)_3PO + H_2O \qquad (14.224)$$

In spite of their toxicity, alkyl phosphites have been used extensively as lubricant additives, corrosion inhibitors, and antioxidants. In addition to their use as intermediates in synthesis, organophosphorus compounds are useful for separating heavy metals by solvent extraction. Several insecticides that were formerly in widespread use are derivatives of organic phosphates. Two such compounds are *malathion* and *parathion*.

As a result of the hazards in using compounds such as these, the use of parathion has been banned in the United States since 1991. Another of the toxic organophosphorus compounds is *sarin* (see structure below), a nerve gas that was produced for military use.

Although these organic phosphate derivatives are highly toxic and have been used in ways to exploit that property, many organophosphorus compounds have significant utility in the chemical industry.

14.3.9 Phosphoric Acids and Phosphates

Phosphorus(V) forms several acids that can be considered as partial dehydration products of the hypothetical compound $P(OH)_5$. For example, if that compound lost one molecule of water, the product would be H_3PO_4,

$$P(OH)_5 \rightarrow H_2O + H_3PO_4 \qquad (14.225)$$

The structure of H_3PO_4, known as *ortho*phosphoric acid, can be shown as

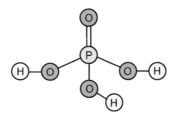

Removal of another molecule of water would result in HPO_3, which is *meta*phosphoric acid. Viewed in a different way, the phosphoric acids can be considered in terms of the ratio of water to P_2O_5, the empirical formula for P_4O_{10}. If the ratio of H_2O to P_2O_5 is 5:1, the reaction can be shown as

$$5\,H_2O + P_2O_5 \rightarrow 2\,P(OH)_5 \tag{14.226}$$

The ratios 3:1 and 1:1 would indicate, respectively, the reactions

$$3\,H_2O + P_2O_5 \rightarrow 2\,H_3PO_4 \tag{14.227}$$

$$H_2O + P_2O_5 \rightarrow 2\,HPO_3 \tag{14.228}$$

From the foregoing examples, it can be seen that although the ratio H_2O/P_2O_5 could be as great as 5, the acid resulting when the ratio is 3 (H_3PO_4) represents the upper practical limit. When water is removed from H_3PO_4, oxygen bridges are formed between phosphate groups to produce the so-called condensed phosphates. Removal of a water molecule from one molecule of H_3PO_4 produces the acid HPO_3, which is known as metaphosphoric acid, and it is a strong acid. We can view the formation of $H_4P_2O_7$ (known as diphosphoric acid or pyrophosphoric acid) as arising from the addition of P_2O_5 to H_3PO_4 or the removal of a molecule of water from two molecules of H_3PO_4. The first process can be shown as

$$4\,H_3PO_4 + P_2O_5 \rightarrow 3\,H_4P_2O_7 \tag{14.229}$$

The condensation reaction is shown as follows.

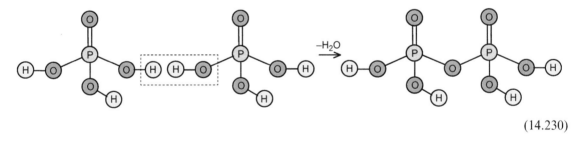

$$\tag{14.230}$$

Pyrophosphoric acid represents the H_2O/P_2O_5 ratio of 2:1, and it has four dissociation constants: $K_{a1} = 1.4 \times 10^{-1}$; $K_{a2} = 1.1 \times 10^{-2}$; $K_{a3} = 2.9 \times 10^{-7}$; $K_{a4} = 4.1 \times 10^{-10}$. The first two dissociation constants are rather close together, and they are much larger than the last two. As a result, two protons are replaced more easily than the others resulting in acid salts that have the formula $M_2H_2P_2O_7$ (where M is a univalent ion). One common example is $Na_2H_2P_2O_7$, which is used as a solid acid that reacts with $NaHCO_3$ in baking powder. However, not all of the salts are produced by neutralization reactions. For example, $Na_4P_2O_7$ is prepared by the thermal dehydration of Na_2HPO_4.

$$2\,Na_2HPO_4 \xrightarrow{\Delta} H_2O + Na_4P_2O_7 \tag{14.231}$$

If the concentration of base is high enough, the last two protons in $H_4P_2O_7$ can be removed to give salts having the formula $M_4P_2O_7$. One such compound, $Na_4P_2O_7$, is used in detergents, an emulsifier in making cheese, a dispersant in paints, and in water softening. The potassium salt is used in liquid detergents and shampoos, as a pigment dispersant in paints, and in the manufacture of synthetic rubber. The calcium salt, $Ca_2P_2O_7$, is used in toothpaste as a mild abrasive. Salts having the formulas $MH_3P_2O_7$ and $M_3HP_2O_7$ are less numerous and less useful. One compound of this type, NaH_2PO_4, is used to prepare $Na_2H_2P_2O_7$, by dehydration.

$$2\,NaH_2PO_4 \xrightarrow{\Delta} Na_2H_2P_2O_7 + H_2O \qquad (14.232)$$

Further condensation of phosphate groups leads to the polyphosphoric acids, which contain several phosphate units. The acids themselves are not particularly useful, but salts of some of these acids are used extensively. *Triphosphoric* acid, also known as *tripolyphosphoric* acid, $H_5P_3O_{10}$, can be considered as a product of the reaction

$$10\,H_2O + 3\,P_4O_{10} \rightarrow 4\,H_5P_3O_{10} \qquad (14.233)$$

but it could also be formed by loss of H_2O from $H_4P_2O_7$ and H_3PO_4. It has the structure

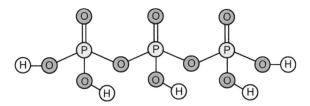

It is a rather strong acid so the first step of its dissociation in water is extensive. Other condensed *polyphosphoric* acids have the general formula $H_{n+2}P_nO_{3n+1}$, and they are formally produced by the elimination of water between the $(n-1)$ acid and a molecule of H_3PO_4. The general structure of the polyphosphoric acids can be represented as

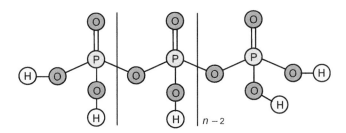

Sodium tripolyphosphate, $Na_5P_3O_{10}$, is the most important salt of the polyphosphoric acids. It is used in sulfonate detergents in which it functions to complex with Ca^{2+} and Mg^{2+} (known as *sequestering*) to prevent the formation of insoluble soaps in hard water.

Another "phosphoric acid," *trimetaphosphoric* acid, is a trimer of HPO_3 representing the H_2O/P_2O_5 ratio of 1. The structure of $H_3P_3O_9$ can be shown as

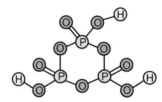

Note that the $P_3O_9^{3-}$ anion is isoelectronic with $(SO_3)_3$ and $Si_3O_9^{6-}$, which also have ring structures. When the formula for $H_3P_3O_9$ is written as $[(HO)PO_2]_3$, it can be seen that it is a strong acid. Some salts of trimetaphosphoric acid are used in foods and toiletries. The acid can be viewed as being produced by the partial dehydration of H_3PO_4,

$$3 H_3PO_4 \rightarrow (HPO_3)_3 + 3 H_2O \tag{14.234}$$

or a hydration product of P_2O_5,

$$3 P_2O_5 + 3 H_2O \rightarrow 2 (HPO_3)_3 \tag{14.235}$$

Although a cyclic tetramer is also known, it is not an important compound.

The most important acid containing phosphorus is orthophosphoric acid, H_3PO_4. This is the acid usually indicated when the name phosphoric acid is used in most contexts. Approximately 30 billion pounds of this acid are produced annually. The commercial form of the acid is usually a solution containing 85% acid. The process used to produce phosphoric acid is related to the intended use of the acid. If the acid is intended for use in foods, it is prepared by burning phosphorus and dissolving the product in water.

$$P_4 + 5 O_2 \rightarrow P_4O_{10} \tag{14.236}$$

$$P_4O_{10} + 6 H_2O \rightarrow 4 H_3PO_4 \tag{14.237}$$

A "fertilizer grade" product is obtained by treating phosphate rock with sulfuric acid.

$$3 H_2SO_4 + Ca_3(PO_4)_2 \rightarrow 3 CaSO_4 + 2 H_3PO_4 \tag{14.238}$$

Liquid H_3PO_4 has a high viscosity and it is extensively associated by hydrogen bonding.

Adding SO_3 to sulfuric acid produces $H_2S_2O_7$, which can be diluted to produce 100% sulfuric acid. However, when P_4O_{10} is added to 85% phosphoric acid, condensation occurs to produce a complex mixture that contains H_3PO_4, $H_4P_2O_7$, higher acids, and a trace of water. As a result, if enough P_4O_{10} is added to theoretically give 100% H_3PO_4, approximately 10% of the P_4O_{10} is contained in other species, especially $H_4P_2O_7$.

Phosphoric acid, $(HO)_3PO$, is a weak acid that has the following dissociation constants: $K_{a1} = 7.5 \times 10^{-3}$; $K_{a2} = 6.0 \times 10^{-8}$; and $K_{a3} = 5 \times 10^{-13}$. Salts having formulas M_3PO_4, M_2HPO_4, and

MH_2PO_4 are known where M is a univalent ion. These salts dissolve in water to give basic solutions as a result of hydrolysis reactions such as

$$PO_4^{3-} + H_2O \rightarrow HPO_4^{2-} + OH^- \tag{14.239}$$

$$HPO_4^{2-} + H_2O \rightarrow H_2PO_4^- + OH^- \tag{14.240}$$

Phosphoric acid is used in many ways that include foods, beverages, etc. It is also used in cleaning metal surfaces, electroplating, fertilizer production, preparing flame-proofing compounds, and other processes in the chemical industry making it one of the most important of the chemicals of commerce.

14.3.10 Fertilizer Production

Feeding a world population of over six billion requires the use of tools of all types. Not only is the *machinery* of agriculture important but so are the *chemicals* of agriculture. The use of effective fertilizers is essential to increase food production from a tillable landmass that (at least in the United States) is shrinking. Phosphates are an important constituent in many types of fertilizers, and their production involves primarily inorganic chemistry.

Naturally occurring calcium phosphate is the primary resource utilized in producing fertilizer. It is found in many places, and it is available in almost unlimited quantity. Converting this insoluble material into a soluble form involves changing the phosphate to some other compound that has greater solubility. In $Ca_3(PO_4)_2$, the ions have charges of +2 and −3 so the lattice energy is high, and such compounds are generally not soluble in water. Because the phosphate ion is basic, $Ca_3(PO_4)_2$ will react with an acid to produce $H_2PO_4^-$ (as the process is carried out). With the charges being lower, the compound is more soluble than calcium phosphate. For this conversion, the strong acid having the lowest cost is utilized, and that acid is sulfuric acid. The reaction can be shown as

$$Ca_3(PO_4)_2 + 2\,H_2SO_4 + 4\,H_2O \rightarrow Ca(H_2PO_4)_2 + 2\,CaSO_4 \cdot 2H_2O \tag{14.241}$$

The product, $Ca(H_2PO_4)_2$, is more soluble than the phosphate. Sulfuric acid is produced in the largest quantity of any compound with production that approaches 100 billion pounds annually. Approximately two-thirds of this amount is used in the production of fertilizers. The mixture containing of calcium dihydrogen phosphate and calcium sulfate (*gypsum*) is known as *superphosphate of lime*, and it contains a higher percent of phosphorus than does calcium phosphate.

Fluorapatite, $Ca_5(PO_4)_3F$, is sometimes found with $Ca_3(PO_4)_2$, and it also reacts with sulfuric acid,

$$2\,Ca_5(PO_4)_3F + 7\,H_2SO_4 + 10\,H_2O \rightarrow 3\,Ca(H_2PO_4)_2 \cdot H_2O + 7\,CaSO_4 \cdot H_2O + 2\,HF \tag{14.242}$$

This is one source of hydrogen fluoride, which is also produced by the action of sulfuric acid on *fluorite*, CaF_2.

Another type of fertilizer is obtained by utilizing phosphoric acid instead of H_2SO_4. The reactions are as follows:

$$Ca_3(PO_4)_2 + 4\,H_3PO_4 \rightarrow 3\,Ca(H_2PO_4)_2 \tag{14.243}$$

$$Ca_5(PO_4)_3F + 5\,H_2O + 7\,H_3PO_4 \rightarrow 5\,Ca(H_2PO_4)_2 \cdot H_2O + HF \tag{14.244}$$

The product, $Ca(H_2PO_4)_2$, contains an even higher percent of calcium, and it is often referred to as *triple superphosphate*.

Plants require nutrients that also contain nitrogen. As discussed earlier in this chapter, ammonium nitrate is an important fertilizer, and it can be produced by the reaction

$$HNO_3 + NH_3 \rightarrow NH_4NO_3 \tag{14.245}$$

An enormous quantity of ammonium nitrate is produced annually primarily for use as a fertilizer and also as an explosive. Ammonium sulfate, ammonium phosphate, and urea are also used as nitrogen-containing fertilizers. They are produce by the reactions

$$2\,NH_3 + H_2SO_4 \rightarrow (NH_4)_2SO_4 \tag{14.246}$$

$$3\,NH_3 + H_3PO_4 \rightarrow (NH_4)_3PO_4 \tag{14.247}$$

$$2\,NH_3 + CO_2 \rightarrow (NH_2)_2CO + H_2O \tag{14.248}$$

A fertilizer that provides both phosphorus and nitrogen is prepared by the reaction of calcium phosphate and nitric acid.

$$Ca_3(PO_4)_2 + 4\,HNO_3 \rightarrow Ca(H_2PO_4)_2 + 2\,Ca(NO_3)_2 \tag{14.249}$$

Inorganic chemistry is vital for production of food. The materials described are produced on an enormous scale, and they are vital to our way of life and well-being. It is safe to assume that given a projected world population of perhaps 12 billion by the year 2030, there will not be a decrease in this importance.

References for Further Reading

Allcock, H.R., 1972. *Phosphorus-Nitrogen Compounds*. Academic Press, New York. A useful treatment of linear, cyclic, and polymeric phosphorus-nitrogen compounds.

Bailar, J.C., Emeleus, H.J., Nyholm, R., Trotman-Dickinson, A.F., 1973. *Comprehensive Inorganic Chemistry*, vol. 3. Pergamon Press, Oxford. This is one volume in the five volume reference work in inorganic chemistry.

Carbridge, D.E.C., 1974. *The Structural Chemistry of Phosphorus*. Elsevier, New York. An advanced treatise on an enormous range of topics in phosphorus chemistry.

Cotton, F.A., Wilkinson, G., Murillo, C.A., Bochmann, M., 1999. *Advanced Inorganic Chemistry*, sixth ed. John Wiley, New York. A 1300 page book that has chapters dealing with all main group elements.

Glockling, F., 1969. *The Chemistry of Germanium*. Academic Press, New York. An excellent introduction to the inorganic and organic chemistry of germanium.

Goldwhite, H., 1981. *Introduction to Phosphorus Chemistry*. Cambridge University Press, Cambridge. A small book that contains a lot of information and organic phosphorus chemistry.

Gonzales-Moraga, G., 1993. *Cluster Chemistry*. Springer-Verlag, New York. A comprehensive survey of the chemistry of clusters containing transition metals as well as cages composed of main group elements such as phosphorus, sulfur, and carbon.

King, R.B., 1995. *Inorganic Chemistry of the Main Group Elements*. VCH Publishers, New York. An excellent introduction to the reaction chemistry of many elements.

Liebau, F., 1985. *Structural Chemistry of Silicates*. Springer-Verlag, New York. Thorough discussion of this important topic.

Mark, J.E., Allcock, H.R., West, R., 1992. *Inorganic Polymers*. Prentice Hall, Englewood Cliffs, NJ. A modern treatment of polymeric inorganic materials.

Rochow, E.G., 1946. *An Introduction to the Chemistry of the Silicones*. John Wiley, New York. An introduction to the fundamentals of silicon chemistry.

Toy, A.D.F., 1975. *The Chemistry of Phosphorus*. Harper & Row, Menlo Park, CA. One of the standard works on phosphorus chemistry.

Van Wazer, J.R., 1958. *Phosphorus and Its Compounds*, vol. 1. Interscience, New York. This is the classic book on all phases of phosphorus chemistry. Highly recommended.

Van Wazer, J.R., 1961. *Phosphorus and Its Compounds*, vol. 2. Interscience, New York. This volume is aimed at the technology and application of phosphorus-containing compounds.

Walsh, E.N., Griffith, E.J., Parry, R.W., Quin, L.D., 1992. *Phosphorus Chemistry*, Developments in American Science. American Chemical Society, Washington, D.C. This is ACS Symposium Series No. 486, a symposium volume that contains 20 chapters dealing with many aspects of phosphorus chemistry.

QUESTIONS AND PROBLEMS

1. Write balanced equations for the reactions of AsF_5 with each of the following compounds:
   (a) H_2O; (b) H_2SO_4; (c) CH_3COOH; (d) CH_3OH

2. Complete and balance the following. Some of the reactions may require elevated temperatures.
   (a) $As_2O_3 + HCl \rightarrow$
   (b) $As_2O_3 + Zn + HCl \rightarrow$
   (c) $Sb_2S_3 + O_2 \rightarrow$
   (d) $AsCl_3 + H_2O \rightarrow$
   (e) $Sb_2O_3 + C \rightarrow$

3. Complete and balance the following:
   (a) $AsCl_3 + LiC_4H_9 \rightarrow$
   (b) $PCl_5 + P_4O_{10} \rightarrow$
   (c) $OPCl_3 + C_2H_5OH \rightarrow$
   (d) $(NPCl_2)_3 + LiCH_3 \rightarrow$
   (e) $P(OCH_3)_3 + S_8 \rightarrow$

4. Starting with elemental phosphorus, show a series of equations to synthesize the following:
   (a) $P(OCH_3)_3$; (b) $OP(OC_2H_5)_3$; (c) $SP(OC_2H_5)_3$

5. Explain why N_2O is almost 40 times as soluble as N_2 in water at 25 °C and 1 atm pressure.

6. Trifluoramine oxide, F_3NO, has a dipole moment of 0.039 D, whereas that of NF_3 is 0.235 D. Explain this difference.

7. Although the difference in electronegativity between N and O is less than that between C and O, the NO molecule has a dipole moment of 0.159 D, whereas CO has a dipole moment of 0.110 D. Explain this difference.

8. Describe the bonding that results in the P−O distance in OPR_3 (R = alkyl group) to be somewhat shorter than the usual single bond length.

9. Suppose the −1 ion containing one atom each of nitrogen, phosphorus, and carbon could be prepared. Describe the structure of this ion in some detail with regard to the type of bonding and resonance.

10. Would phosphonium acetate likely be stable? Explain your answer.

11. A compound known as nitrosyl azide, N_4O, has been reported. Draw the structure of this unusual compound and discuss the bonding. What would be a different arrangement of atoms for a compound having this formula? Comment on its stability.

12. The oxidation of NO to NO_2 may involve an intermediate having the formula N_2O_2. Describe the structure of this compound. It is believed that there is a less stable isomer that has lower symmetry. Draw the structure for that isomer.

13. When NO reacts with $ClNO_2$, the products are $ClNO$ and NO_2. There are two ways in which this reaction could occur. Write equations showing the possible mechanisms. Explain how the use of a radioactive isotope could clarify the mechanism.

14. Complete and balance the following equations for reactions in liquid ammonia:
    (a) $NH_4Cl + NaNH_2 \rightarrow$
    (b) $Li_2O + NH_3 \rightarrow$
    (c) $CaNH + NH_4Cl \rightarrow$

15. There are two different $N-O$ distances, 143 pm and 118 pm, in the nitrous acid molecule. Draw the structure of the molecule and explain the bond lengths.

16. Phosphorous acid disproportionates at high temperature to produce phosphine. What is the other product? Write the equation for this reaction.

17. Where would the fluorine atom be located in the PCl_4F molecule? Explain how you know.

18. Coupling reactions in which a metal such as mercury removes a halogen are rather common. What would be produced by a reaction of this type when mercury reacts with PF_2I?

19. Sketch the structure of the P_4O_{10} molecule. There are two different $P-O$ bond lengths in the molecule. After deciding which is shorter and longer, explain the difference in terms of bonding.

20. When burning magnesium is placed in a bottle containing N_2O, it continues to burn. Write the equation for the reaction, and explain how it is consistent with the structure of N_2O being what it is.

21. Write balanced equations for each of the following processes:
    (a) the preparation of triethylphosphate
    (b) the reaction of PCl_5 with NH_4Cl
    (c) the preparation of arsine
    (d) the reaction of P_4O_{10} with $i\text{-}C_3H_7OH$

22. What types of reactions would the nitrogen atoms in a phosphazine ring undergo?

23. At high temperature, arsenic dissolves in molten sodium hydroxide with the liberation of hydrogen. Write the equation for the reaction.

24. Explain how the presence of SbF_5 increases the acidity of a liquid HF solvent system.

25. In the H_3PO_4 molecule there are $P-O$ bonds having lengths of 152 and 157 pm. Draw the structure of the molecule, assign the bond lengths, and discuss the differences in bonding.

26. Although phosphorous acid has the formula H_3PO_3, the titration with sodium hydroxide gives Na_2HPO_3. Why?

27. Although H_3PO_4 is a stable acid containing phosphorus, there is no corresponding H_3NO_4. Explain why this is true.

28. Even though H_3NO_4 is not stable, the reaction between Na_2O and $NaNO_3$ at high temperature produces Na_3NO_4. What would be the structure of the NO_4^{3-} ion?

29. The $N-O$ single bond is typically about 146 pm. In the NO_4^{3-} ion, the bond length is 139 pm. Explain the difference. Keep in mind that the $N-O$ bond length in the NO_3^- ion is about 124 pm.

30. The bond energy of the $P\equiv P$ bond is 493 kJ/mol and the $P-P$ bond energy is 209 kJ/mol. Use these values to show that the form of elemental phosphorus is expected to be different from that of elemental nitrogen.

31. Write balanced equations for each of the following processes:
    (a) the preparation of $(CH_3)_3PO$
    (b) the preparation of POF_3 from P_4
    (c) the reaction of P_4 with NaOH
    (d) the preparation of superphosphate fertilizer

32. Complete and balance the following:
  (a) $NaCl + SbCl_3 \rightarrow$
  (b) $Bi_2S_3 + O_2 \rightarrow$
  (c) $Na_3Sb + H_2O \rightarrow$
  (d) $BiBr_3 + H_2O \rightarrow$
  (e) $Bi_2O_3 + C \rightarrow$
33. Explain why NH_3 is more polar than NCl_3.
34. Both hyponitrous acid and nitroamide (also known as nitramide) have the formula $H_2N_2O_2$. Draw the structures for these molecules and explain any difference in acid-base properties.
35. By making use of the following bond energies (in kJ/mol), explain why CO_2 exists as discrete molecules whereas SiO_2 does not. Estimate the strength of the $Si{=}O$ bond.
  $C{-}O$, 335; $C{=}O$, 707; $Si{-}O$, 464
36. Explain why the boiling point of $SnCl_2$ is 652 °C, whereas that of $SnCl_4$ is 114 °C.
37. Write complete, balanced equations for the processes indicated.
  (a) the combustion of silane
  (b) the preparation of PbO_2
  (c) the reaction of CaO with SiO_2 at high temperature
  (d) the preparation of $GeCl_2$
  (e) the reaction of $SnCl_4$ with water
  (f) the preparation of tetraethyl lead

CHAPTER 15

Chemistry of Nonmetallic Elements III. Groups VIA–VIIIA

In the study of inorganic chemistry, it becomes apparent that the chemistry of elements having greatly different properties must be considered. Although phosphorus and xenon are both nonmetals and are distinctly different, some compounds of these elements react in similar ways. Even within a specific group of elements, there are vast differences. That is why oxygen was discussed separately from the remaining members of the *chalcogens*, Group VIA, which are considered in this chapter. In the case of the halogens, the chemistry of fluorine is more like that of the heavier members of the group so the entire group is considered together in this chapter. Consequently, this chapter presents the chemistry of Group VIA (except for oxygen), Group VIIA, and the noble gases. Entire volumes have been written on each of the elements so only a glimpse of the whole view of these elements can be presented.

15.1 SULFUR, SELENIUM, AND TELLURIUM

The chemistry of sulfur is a broad area that includes chemicals such as sulfuric acid (the compound prepared in the largest quantity) as well as unusual compounds containing nitrogen, phosphorus, and halogens. Although there is an extensive chemistry of selenium and tellurium, much of it follows logically from the chemistry of sulfur if allowance is made for the more metallic character of the heavier elements. All isotopes of polonium are radioactive, and compounds of the element are not items of commerce or great use. Therefore, the chemistry of sulfur will be presented in more detail.

15.1.1 Occurrence of the Elements

Sulfur has been known and used for thousands of years. The reason for this is that the element is found uncombined in many areas of the earth where volcanic activity is present. Such areas include the region of the Mediterranean Sea, Mexico, Chile, and Japan, among others. The recorded history of man in these areas indicates that sulfur was utilized in several ways. There are many minerals that contain sulfur, among them are *galena* (PbS), *zinc blende* (ZnS), *cinnabar* (HgS), *iron pyrites* (FeS_2), *gypsum* ($CaSO_4$), and *chalcopyrite* (CuS_2). Black gunpowder, a mixture of sulfur, charcoal, and potassium nitrate, has been known for about a thousand years, and its use in firearms influenced the course of history.

Tellurium was discovered in 1782 by Baron F.J. Müller von Reichenstein, and selenium was discovered in 1817 by Berzelius. The last member of the group was discovered in 1898 by Madame Curie, and the name polonium comes from her native country, Poland. The name tellurium comes from the Latin word, *tellus*, meaning "earth," and selenium is named from a Greek word, *selene*, **499**

Inorganic Chemistry. DOI: http://dx.doi.org/10.1016/B978-0-12-385110-9.00015-7

meaning "moon". Compounds containing sulfur also contain small amounts of selenium and tellurium, and this is the major source of these elements. In the electrolytic refining of copper, the anode sludge contains some selenium and tellurium. Sulfur is obtained from vast deposits of the element that are found below the surface in Louisiana and Texas by the *Frasch process*, which makes use of three concentric pipes. In this process, sulfur (which melts at 119 °C) is melted by superheated water forced underground in one pipe while compressed air is forced down a second pipe. The molten sulfur is forced to the surface through the third pipe.

15.1.2 Elemental Sulfur, Selenium, and Tellurium

The usual form of sulfur is known as the rhombic form, which is stable at temperatures up to about 105 °C. Above that temperature, the monoclinic structure is stable. A plastic form of sulfur can be obtained by pouring liquid sulfur into water to cool it quickly, but on standing, it is converted to the rhombic form. On a molecular level, the element exists as S_8 rings that have the structure

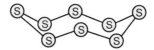

Sulfur vapor consists of a mixture of species that include S_8, S_6, S_4, and S_2 (which like O_2 is paramagnetic). Because the S_8 molecule is nonpolar, it is soluble in liquids such as CS_2 and C_6H_6. Selenium also consists of cyclic molecules that contain eight atoms, and tellurium is essentially metallic in character. In their vapors, several species are found that contain 2, 6, or 8 atoms. Both are useful as semiconductors, and selenium has been used in rectifiers. Because the electrical conductivity of selenium increases as the intensity of illumination increases, it has been used to operate electrical switches that open or close as a light beam is broken. Selenium was also used in light meters, but other types of meters are now available that are more sensitive. Table 15.1 gives a summary of the properties of the Group VIA elements.

Liquid sulfur has the unusual characteristic of the viscosity increasing with increasing temperature, and it goes through a maximum at about 170–180 °C. At higher temperature, the viscosity decreases as the temperature is increased. When S_8 molecules rupture, the long chains interact to give larger aggregates that do not flow as readily as do the molecules having a ring structure. At higher temperatures, the chains dissociate and the viscosity becomes lower. Selenium does not exhibit this unusual behavior in viscosity.

Table 15.1 Some Properties of the Group VIA Elements

	O	S	Se	Te	Po
Melting point, °C	−218.9	118.9[a]	217.4	449.8	—
Boiling point, °C	−182.96	444.6	648.8 (subl)	1390	—
Atomic radius, pm	74	104	117	137	152
Ionic radius (X^{4+}), pm	—	51	64	111	122
Ionic radius (X^{2-}), pm	126	170	184	207	—
Electronegativity	3.5	2.5	2.4	2.1	—
Ionization potential, eV	13.62	10.36	6.54	9.01	8.42

[a]*The monoclinic form.*

Although many reactions and uses of sulfur will be described, about 85% of the sulfur produced is used in making sulfuric acid, and about two thirds of the acid is used in the production of fertilizer (see Chapter 14). Sulfur is rather reactive so it reacts with most other elements. It produces a blue flame when it burns in air,

$$S_8 + 8\,O_2 \rightarrow 8\,SO_2 \tag{15.1}$$

and it produces a dark coating when it tarnishes silver.

$$16\,Ag + S_8 \rightarrow 8\,Ag_2S \tag{15.2}$$

Sulfur also forms binary compounds by reacting with phosphorus and halogens.

$$4\,P_4 + 5\,S_8 \rightarrow 4\,P_4S_{10} \tag{15.3}$$

$$24\,F_2 + S_8 \rightarrow 8\,SF_6 \tag{15.4}$$

Sulfur also reacts by undergoing addition reactions with several types of species. For example, thiosulfates are produced by the reaction of sulfur with sulfites.

$$8\,SO_3^{2-} + S_8 \rightarrow 8\,S_2O_3^{2-} \tag{15.5}$$

Thiocyanates are prepared by addition of sulfur to cyanides.

$$S_8 + 8\,CN^- \rightarrow 8\,SCN^- \tag{15.6}$$

It will also add to phosphorus atoms in a variety of molecules as shown below:

$$8\,(C_6H_5)_2PCl + S_8 \rightarrow 8\,(C_6H_5)_2PSCl \tag{15.7}$$

$$8\,PCl_3 + S_8 \rightarrow 8\,SPCl_3 \tag{15.8}$$

As a result of the tendency toward catenation, sulfur reacts with solutions of sulfides to produce polysulfides.

$$S^{2-} + (x/8)\,S_8 \rightarrow S-S_{x-1}-S^{2-} \tag{15.9}$$

In this behavior, sulfur resembles iodine, which reacts in an analogous way to form polyiodides. Sulfur will also remove hydrogen from saturated hydrocarbons to produce H_2S with the formation of carbon−carbon double bonds. Sulfur dissolves in hot concentrated nitric acid as a result of being oxidized, as shown in the following reaction:

$$S_8 + 32\,HNO_3 \rightarrow 8\,SO_2 + 32\,NO_2 + 16\,H_2O \tag{15.10}$$

In the following equations, selenium and tellurium exhibit behavior that is analogous to that of sulfur.

$$Se_8 + 8\,O_2 \rightarrow 8\,SeO_2 \tag{15.11}$$

$$3\,Se_8 + 16\,Al \rightarrow 8\,Al_2Se_3 \tag{15.12}$$

$$Se_8 + 24\ F_2 \rightarrow 8\ SeF_6 \tag{15.13}$$

$$Te + 2\ Cl_2 \rightarrow TeCl_4 \tag{15.14}$$

However, selenium and tellurium do not react with hydrogen so the hydrogen compounds are prepared by reacting the elements with a metal and then treating it with an acid. Selenium and tellurium undergo addition reactions with cyanides to yield selenocyanates and tellurocyanates.

$$8\ KCN + Se_8 \rightarrow 8\ KSeCN \tag{15.15}$$

15.1.3 Hydrogen Compounds

By far the most common compounds containing the Group VIA elements and hydrogen are those having the formula H_2E, and the properties of these compounds are summarized in Table 15.2, with H_2O included for comparison. The H_2E compounds are extremely toxic.

As a result of the bonding between sulfur atoms, several hydrogen compounds exist in which there are several sulfur atoms in a chain with a hydrogen atom bonded to the terminal sulfur atoms. These compounds include H_2S_2 (m.p. $-88\ °C$, b.p. $74.5\ °C$) and H_2S_6. Compounds of this type are known as *sulfanes*.

The data shown in Table 15.2 reveal several interesting characteristics of the H_2E molecules. The fact that they have much lower boiling points than H_2O reflects the strong hydrogen bonding in the latter. As evidenced by the heats of formation, the stability of the compounds decreases markedly in progressing down the group. This is expected on the basis of the increase in size of the atoms resulting in less effective overlap with $1s$ orbitals on hydrogen atoms. It is also apparent that there is an increase in acidity of the compounds in going downward in the group. The bond angles show that hybridization results in sp^3 orbitals on the oxygen atom in water, but the bond angles in the remaining compounds are not far from $90°$. This probably indicates that hybridization is not extensive in these cases where the orbitals on the central atom are larger, and it parallels the trend shown by the analogous Group VA compounds.

There are numerous ways in which hydrogen sulfide can be obtained. One of the simplest is to treat a metal sulfide with an acid as illustrated by the reaction of *galena*.

$$PbS + 2\ H^+ \rightarrow Pb^{2+} + H_2S \tag{15.16}$$

Table 15.2 Properties of Hydrogen Compounds of Group VI Elements

Property	H_2O	H_2S	H_2Se	H_2Te
Melting point, °C	0.00	−85.5	−65.7	−51
Boiling point, °C	100.0	−60.7	−41.3	−2.3
ΔH_f° (gas), kJ mol^{-1}	−242	−20	+66.1	+146
Dipole moment, Debye	1.85	0.97	0.62	0.2
Acid K_{a1}	1.07×10^{-16}	1.0×10^{-9}	1.7×10^{-4}	2.3×10^{-3}
Acid K_{a2}	–	1.2×10^{-15}	1.0×10^{-10}	1.6×10^{-11}
Bond angle, degrees	104.5	92.3	91.0	89.5
H–X bond energy, kJ mol^{-1}	464	347	305	268

Hydrogen sulfide can be prepared from the elements, but the unfavorable heat of formation of H_2Se and H_2Te indicates that producing them by direct combination is not an efficient route. Instead, they are first obtained as selenides and tellurides, which then react with an acid.

$$Mg + Te \rightarrow MgTe \tag{15.17}$$

$$MgTe + 2\,H^+ \rightarrow Mg^{2+} + H_2Te \tag{15.18}$$

All of the H_2E compounds dissociate slightly in water, but the second step is very slight. Because the acids are weak, solutions containing the E^{2-} anions are basic due to hydrolysis.

$$S^{2-} + H_2O \rightleftarrows HS^- + OH^- \tag{15.19}$$

A series of bisulfide salts exist, and they are also basic in solution because even the first step in the dissociation of the acids is slight.

$$HS^- + H_2O \rightleftarrows H_2S + OH^- \tag{15.20}$$

15.1.4 Polyatomic Species

Compounds having the formula H_2S_n (where n is typically 2–8) are known as *sulfanes*, with the number of sulfur atoms indicated as the prefix to the name sulfane (H_2S_3 is *trisulfane*, H_2S_6 is *hexasulfane*, etc.). Sulfanes containing longer chains of sulfur atoms can be prepared by the reaction of H_2S_2 with disulfur dichloride, S_2Cl_2 (also known as sulfur monochloride).

$$3\,H_2S_2 + S_2Cl_2 \rightarrow 2\,H_2S_4 + 2\,HCl \tag{15.21}$$

The length of the sulfur chain can be increased by subsequent reactions with S_2Cl_2.

$$2\,H_2S_2 + S_nCl_2 \rightarrow H_2S_{n+4} + 2\,HCl \tag{15.22}$$

Sulfanes decompose to produce H_2S, and sulfur in a thermodynamically favored reaction.

$$H_2S_n \rightarrow H_2S + (n-1)\,S \tag{15.23}$$

Although many compounds of the Group VIA elements are covalent or contain the elements as anions, cations that contain sulfur, selenium, and tellurium have also been studied. Sulfide ions in solution react with sulfur to produce polyatomic anions known as the *polysulfides*. When these solutions are acidified, a series of compounds are formed that contain chains of sulfur atoms. The reaction can be shown as follows:

$$S_n^{2-} + 2\,H^+ \rightarrow H_2S_n \tag{15.24}$$

Polysulfides of several metals can be prepared by reaction of the metals with excess sulfur in liquid NH_3 (Group IA metals) or by heating sulfur with the molten metal sulfide. The polysulfide ion binds to metals to form coordination compounds in which it is attached to the metal by both sulfur atoms (as a so-called bidentate ligand). One example is an unusual titanium complex containing the S_5^{2-} ion that is produced by the following reaction (the use of η to denote the bonding mode of the cyclopentadienyl ion is explained in Chapter 16).

$$(\eta^5\text{-}C_5H_5)_2TiCl_2 + (NH_4)_2S_5 \rightarrow (\eta^5\text{-}C_5H_5)_2TiS_5 + 2\ NH_4Cl \qquad (15.25)$$

Although H_2S is normally a weak acid, it functions as a base in a superacid such as HF/SbF_5 in liquid HF. The H_3S^+ ion is generated, and although solid $H_3S^+SbF_5^-$ has been obtained, there is very limited chemistry associated with this type of compound.

Cations containing only S, Se, or Te atoms have been known for many years. One of the most common ions of this type is S_4^{2+}, which has the square planar structure as shown below. This is also the structure of Se_4^{2+}, and Te_4^{2+}, which have the dimensions shown.

S, 198 (204) pm
Se, 228 (234) pm
Te, 266 (284) pm

Values in parentheses
are single bond lengths

Polyatomic cations of S, Se, and Te are produced as shown in the following reactions:

$$S_8 + 6\ AsF_5 \xrightarrow{\text{Liq. SO}_2} 2\ S_4^{2+}(AsF_6^-)_2 + 2\ AsF_3 \qquad (15.26)$$

$$S_8 + 3\ AsF_5 \xrightarrow{\text{Liq. SO}_2} S_8^{2+}(AsF_6^-)_2 + AsF_3 \qquad (15.27)$$

$$TeCl_4 + 7\ Te + 4\ AlCl_3 \xrightarrow{\text{AlCl}_3} 2\ Te_4^{2+}(AlCl_4^-)_2 \qquad (15.28)$$

$$Se_8 + 3\ AsF_5 \rightarrow Se_8^{2+}(AsF_6^-)_2 + AsF_3 \qquad (15.29)$$

These reactions are essentially equivalent to those of superacids in which there is a strong Lewis acid (SbF_5, AsF_5, etc.) involved (see Section 10.8). Although S_4^{2+}, Se_4^{2+}, and Te_4^{2+} ions have the square planar structures, the ions S_8^{2+}, Se_8^{2+}, Te_8^{2+} exist as puckered rings.

2+

The ions containing two and eight atoms are the most common, but they are by no means the only cations known for the Group VIA elements. Tellurium forms the Te_6^{4+} ion, which is a slightly distorted trigonal prism. The Se_{10}^{2+} cation has been studied, and sulfur forms the S_{19}^{2+} ion having five-atom chain, with a seven-atom ring attached on each end. The use of nonaqueous solvents and superacids makes it possible to generate many unusual and interesting species.

15.1.5 Oxides of Sulfur, Selenium, and Tellurium

Sulfur is known to form several oxides that include S_2O, S_6O, S_8O, S_7O_2, and SO that have no significant use. The oxides that contain several sulfur atoms generally have rings containing sulfur atoms with an oxygen atom appended to one of them. There do not appear to be as many oxides of selenium and tellurium, but the dioxides of all the elements have an extensive chemistry associated with them. Table 15.3 shows the properties of the Group VIA dioxides.

Sulfur dioxide is an economically important gas that is used as a refrigerant, disinfectant, and reducing atmosphere for preserving food. Although it is also used in the manufacture of many other sulfur compounds, the most important use of SO_2 is as a precursor in producing sulfuric acid. It can be obtained by burning sulfur, but it is also produced in numerous other reactions. Sulfites react with acids by liberating SO_2.

$$2\,H^+ + SO_3^{2-} \rightarrow SO_2 + H_2O \tag{15.30}$$

When sulfide ores are roasted to prepare the oxide that is subsequently reduced to obtain the metal, SO_2 is produced.

$$4\,FeS_2 + 11\,O_2 \xrightarrow{\text{heat}} 2\,Fe_2O_3 + 8\,SO_2 \tag{15.31}$$

Sulfur dioxide is released during the burning of sulfur-containing coal. Much of that gas is now removed from the effluent and used in the production of sulfuric acid. The dioxides of Se and Te are recovered from the residue after the elements have reacted with concentrated HNO_3.

Although the boiling point of SO_2 is $-75.5\,^\circ$C, those of SeO_2 and TeO_2 are 340 and 733 $^\circ$C, respectively. It hardly needs to be stated that these data suggest a drastic difference in bonding in the molecules. In the gas phase, the SO_2 molecule has a structure that can be represented by two resonance structures,

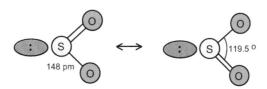

Unlike SO_2, SeO_2 and TeO_2 exist as solids having structures that are extended networks. Liquid SO_2 ($\mu = 1.63$ D) is a versatile nonaqueous solvent (see Chapter 10) that is a good solvent for many types of organic compounds, and aromatic hydrocarbons are much more soluble than aliphatic. There is little if

Table 15.3 Properties of Dioxides of the Group VIA Elements

Property	SO_2	SeO_2	TeO_2
Melting point, °C	−75.5	340	733
Boiling point, °C	−10.0	subl.	—
ΔH_f°, kJ mol^{-1}	−296.9	−230.0	−325.3
ΔG_f°, kJ mol^{-1}	−300.4	−171.5	−269.9

any ionization of SO_2 as described by the solvent concept, and it reacts primarily by forming complexes. It can behave as both a Lewis acid and a Lewis base. It also forms complexes with metals, particularly the second and third transition metals in low oxidation states, in which it can bond through the sulfur atom, an oxygen atom, or form bridges between metal ions. Sulfur dioxide reacts with PCl_5 to produce $OPCl_3$ and thionyl chloride, $SOCl_2$. Sulfuryl chloride is produced by the reaction of SO_2 with chlorine.

$$SO_2 + Cl_2 \ \rightarrow \ SO_2Cl_2 \tag{15.32}$$

The catalytic oxidation of SO_2 produces SO_3, which reacts with water to produce sulfuric acid, and it dissolves in sulfuric acid to produce *disulfuric acid* or *oleum*, $H_2S_2O_7$. Solid SO_3 exists as the trimer $(SO_3)_3$, but there are different crystalline forms known.

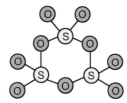

This structure is isoelectronic with $((SiO_3)_3)^{6-}$ and $(PO_3)_3^{3-}$. Gaseous SO_3 has a trigonal planar structure that has several contributing resonance structures. When the structure is drawn with only one double bond, the sulfur atom has a $+2$ formal charge that is relieved by structures having two double bonds. Therefore, multiple bonding between sulfur and oxygen is extensive.

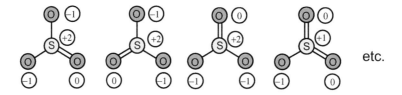

etc.

Selenium trioxide is unstable with respect to the dioxide.

$$SeO_3(s) \ \rightarrow \ SeO_2(s) + 1/2 \ O_2(g) \quad \Delta H = -54 \text{ kJ mol}^{-1} \tag{15.33}$$

It can be prepared by dehydrating H_2SeO_4 or the reaction of that acid with SO_3. Selenium trioxide is soluble in organic solvents. Dehydration of $Te(OH)_6$ gives TeO_3, which exists in two forms in the solids state. Neither SeO_3 nor TeO_3 has large-scale commercial use.

15.1.6 Halogen Compounds

There are a considerable number of well-known halogen compounds of the Group VIA elements, but the majority of them are fluorides and chlorides. Table 15.4 summarizes data for most of the Group VIA halides.

The structure of S_2F_2 resembles that of H_2O_2, with a dihedral angle of about $88°$, and SF_4 is an irregular tetrahedron. As expected, the SF_6 molecule is a regular octahedron. The structures of all these sulfur fluorides are shown in Figure 15.1.

Table 15.4 Halogen Compounds of the Group VI Elements

Compound	m.p., °C	b.p., °C
S_2F_2	−133	15
SF_4	−121	−38
SF_6	−51 (2 atm)	−63.8 (subl)
S_2F_{10}	−52.7	30
SCl_2	−122	59.6
S_2Cl_2	−82	137.1
S_3Cl_2	−45	−
SCl_4	−31 (dec)	−
S_2Br_2	−46	90 (dec)
SeF_4	−9.5	106
SeF_6	−34.6 (1500 torr)	−34.8 (945 torr)
Se_2Cl_2	−85	127 (733 torr)
$SeCl_4$	191 (subl)	−
Se_2Br_2	−146	225 d
TeF_4	129.6	194 d
TeF_6	−37.8	−38.9 (subl)
Te_2F_{10}	−33.7	59
$TeCl_2$	208	328
$TeCl_4$	224	390
$TeBr_4$	380	414 (dec)

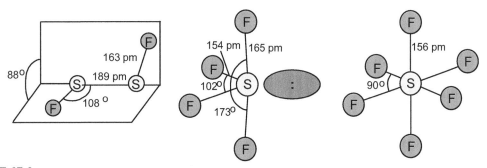

FIGURE 15.1
Structures of the most important sulfur fluorides.

In addition to these well-known fluorides, there are others that include a form of S_2F_2 that is actually thiothionyl fluoride, SSF_2, SF_2, and S_2F_{10}, in which two SF_5 groups are linked by an S–S bond.

The preparations of the sulfur fluorides include a wide variety of reactions. The hexafluoride can be prepared by fluorinating sulfur with ClF_3, BrF_5, or F_2. When fluorine is used, some SF_4 and S_2F_{10} are also produced. The hexafluoride is virtually unique among covalent halides of nonmetals because of its inertness. It is so inert that it undergoes very few reactions under ordinary conditions, and it is used as a gaseous insulator. The reason is that there is no low energy pathway because of the stability of the S–F bonds and the shielding of the sulfur atom by six fluorine atoms. In contrast, the other sulfur fluorides react with water.

$$SF_4 + 3\,H_2O \;\rightarrow\; 4\,HF + H_2O + SO_2 \tag{15.34}$$

Sulfur tetrafluoride is a fluorinating agent that undergoes many useful reactions with some examples being the following.

$$P_4O_{10} + 6\,SF_4 \;\rightarrow\; 4\,POF_3 + 6\,SOF_2 \tag{15.35}$$

$$P_4S_{10} + 5\,SF_4 \;\rightarrow\; 4\,PF_5 + 15\,S \tag{15.36}$$

$$UO_3 + 3\,SF_4 \;\rightarrow\; UF_6 + 3\,SOF_2 \tag{15.37}$$

$$CH_3COCH_3 + SF_4 \;\rightarrow\; CH_3CF_2CH_3 + SOF_2 \tag{15.38}$$

Sulfur tetrafluoride is produced in several reactions with fluorinating agents, such as IF_7.

$$7\,S_8 + 32\,IF_7 \xrightarrow{100-200\ ^\circ C} 56\,SF_4 + 16\,I_2 \tag{15.39}$$

A coupling reaction like the following can be used to prepare S_2F_{10}.

$$H_2 + 2\,SF_5Cl \xrightarrow{h\nu} S_2F_{10} + 2\,HCl \tag{15.40}$$

This fluoride is more reactive than SF_6, and at a temperature of 150 °C, it dissociates into SF_4 and SF_6.

The hexafluorides of SeF_6 and TeF_6 are similar to SF_6, but they react more readily. This is likely the result of the bonds being more polar and the fact that the larger Se and Te atoms are less shielded by six fluorine atoms than are smaller S atoms. Both SeF_6 and TeF_6 react slowly in water.

$$SeF_6 + 4\,H_2O \;\rightarrow\; H_2SeO_4 + 6\,HF \tag{15.41}$$

$$TeF_6 + 4\,H_2O \;\rightarrow\; H_2TeO_4 + 6\,HF \tag{15.42}$$

SF_4 can also react as a Lewis base, but in some cases, the fluorine atom is the electron pair donor. With the strong Lewis acids, such as BF_3 and SbF_5, the reactions can lead to the formation of the SF_3^+ cation. It is also possible for SF_4 to react with Lewis bases and thereby expand the number of electrons in the valence shell. When reacting with F^-, the SF_5^- species results.

The tetrafluorides of selenium and tellurium are similar to SF_4 both structurally (in the gaseous molecules) and in terms of reactivity. However, the bonding in condensed phases is different and leads to several types of structures. Selenium tetrafluoride can be prepared by combining the elements, but it can also be obtained by the reaction of SeO_2 with SF_4 at high temperature. The tellurium compound can be prepared by the reaction of SeF_4 and TeO_2 at 80 °C.

The most important binary compounds of sulfur and chlorine are S_2Cl_2 and SCl_2. Both of these compounds are used in large quantities in many ways, but one of the important uses for S_2Cl_2 is in the vulcanization of rubber. This compound is prepared by chlorinating sulfur.

$$S_8 + 4\,Cl_2 \;\rightarrow 4\,S_2Cl_2 \tag{15.43}$$

The structure of S_2Cl_2 is similar to that shown in Figure 15.1 for S_2F_2. Most other chlorides of sulfur have the sulfur atoms in a chain with the chlorine atoms in terminal positions, and they are produced by the reaction of S_2Cl_2 with sulfur.

$$S_2Cl_2 + n/8\ S_8 \rightarrow S_{n+2}Cl_2 \tag{15.44}$$

The reaction of S_2Cl_2 with sulfanes produces the chlorosulfanes.

$$2\ S_2Cl_2 + H_2S_x \rightarrow S_{x+4}Cl_2 + 2\ HCl \tag{15.45}$$

Although SCl_4 is unstable, $SeCl_4$ and $TeCl_4$ are stable probably as a result of the larger sizes of the central atoms and the greater polarity of the bonds. In the gaseous state, $SeCl_4$ dissociates, but $TeCl_4$ is stable up to approximately 500 °C. The structures of both Se and Te tetrahalides are similar to that of SF_4. When heated to melting, $TeCl_4$ becomes a good electrical conductor, possibly due to the ionization reaction.

$$TeCl_4 \rightarrow TeCl_3^+ + Cl^- \tag{15.46}$$

There are numerous other halogen compounds of the Group VIA elements as well as ionic species derived from them. Space does not permit a full review of the halogen compounds of Group VIA elements, but the discussion shows much of their behavior.

15.1.7 Oxyhalides of Sulfur and Selenium

The oxyhalides of the chalcogens constitute a reactive and useful group of compounds. As expected, the sulfur compounds have greater use, but $SeOF_2$ and $SeOCl_2$ are good solvents for many materials, which has led to their use as nonaqueous solvents (see Chapter 10). The two most important types of oxyhalides are the EOX_2 and EO_2X_2 series, and Table 15.5 shows a summary of relevant data for these types of compounds.

Thionyl chloride and sulfuryl chloride have structures in which the oxygen and halogen atoms are attached to the sulfur. Even though the formulas are often shown as $SOCl_2$ and SO_2Cl_2, they are

Table 15.5 Properties of Oxyhalides of Sulfur and Selenium			
Compound	**m.p., °C**	**b.p., °C**	**μ, D**
SOF_2	−110.5	−43.8	−
$SOCl_2$	−106	78.8	1.45
$SOBr_2$	−52	183	−
$SOClF$	−139.5	12.1	−
$SeOF_2$	4.6	124	−
$SeOCl_2$	8.6	176.4	−
$SeOBr_2$	41.6	dec.	−
SO_2F_2	−136.7	−55.4	1.12
SO_2Cl_2	−54.1	69.1	1.81
SO_2ClF	−124.7	7.1	1.81

correctly shown as $OSCl_2$ and O_2SCl_2. These molecules have C_s and C_{2v} symmetry, respectively. Thionyl chloride can be prepared by chlorinating SO_2,

$$PCl_5 + SO_2 \ \rightarrow \ POCl_3 + SOCl_2 \tag{15.47}$$

or by the reaction of SO_3 with S_2Cl_2 or SCl_2,

$$SO_3 + SCl_2 \ \rightarrow \ SOCl_2 + SO_2 \tag{15.48}$$

Mixed halogen compounds such as SOClF can be prepared by reactions such as

$$SOCl_2 + NaF \ \rightarrow \ SOClF + NaCl \tag{15.49}$$

An exchange reaction between SeO_2 and $SeCl_4$ produces selenyl chloride.

$$SeO_2 + SeCl_4 \ \rightarrow \ 2\,SeOCl_2 \tag{15.50}$$

Although such reactions must take place to only a very slight extent, autoionization of thionyl-chloride and selenyl chloride is presumed to produce $EOCl^+$ and $EOCl_3^-$ ions. The molecules are pyramidal in the gas phase, but there is extensive bridging between molecules in solid $SeOCl_2$.

Thionyl chloride reacts as both a Lewis acid and a Lewis base, and both the S and Se compounds are very reactive toward many other materials. The hydrolysis reactions take place readily, and $SOCl_2$ has such an affinity for water that is used as a dehydrating agent.

$$SOCl_2 + H_2O \ \rightarrow \ SO_2 + 2\,HCl \tag{15.51}$$

When dehydrating a compound such as a metal chloride, the gaseous products of the reaction make it easy to separate the anhydrous compound.

$$CrCl_3 \cdot 6H_2O + 6\,SOCl_2 \ \rightarrow \ CrCl_3 + 6\,SO_2 + 12\,HCl \tag{15.52}$$

Thionyl chloride can also be used to convert metal oxides or hydroxides to the chlorides, and it will also react with many organic compounds. Alcohols react as follows:

$$2\,ROH + SOCl_2 \ \rightarrow \ (RO)_2SO + 2\,HCl \tag{15.53}$$

Chlorine is a strong oxidizing agent that converts SO_2 into sulfuryl chloride, SO_2Cl_2. In the reaction, sulfur is *oxidized* as chlorine is *added* so the process is an *oxad* reaction.

$$SO_2 + Cl_2 \ \rightarrow \ SO_2Cl_2 \tag{15.54}$$

The SO_2Cl_2 molecule can be considered as sulfuric acid in which the OH groups have been replaced by Cl to produce an *acid chloride*. As expected, it undergoes solvolysis reactions as illustrated by the following cases:

$$SO_2Cl_2 + 2\,H_2O \ \rightarrow \ H_2SO_4 + 2\,HCl \tag{15.55}$$

$$SO_2Cl_2 + 2\,NH_3 \ \rightarrow \ SO_2(NH_2)_2 + 2\,HCl \tag{15.56}$$

Sulfuryl fluoride is produced by the reaction of SO_2 or SO_2Cl_2 with fluorine.

$$SO_2 + F_2 \rightarrow SO_2F_2 \tag{15.57}$$

$$SO_2Cl_2 + F_2 \rightarrow SO_2F_2 + Cl_2 \tag{15.58}$$

An exchange reaction between SF_6 and SO_3 also produces sulfuryl fluoride.

$$SF_6 + 2\,SO_3 \rightarrow 3\,SO_2F_2 \tag{15.59}$$

Even though sulfuryl chloride and sulfuryl fluoride can be considered as the disubstituted acid halides of sulfuric acid, the monosubstituted derivatives are also useful compounds that undergo many reactions as a result of the reactive −OH group. The monosubstituted compounds can be prepared by the following reactions:

$$HCl + SO_3 \rightarrow ClSO_3H \tag{15.60}$$

$$PCl_5 + H_2SO_4 \rightarrow POCl_3 + ClSO_3H + HCl \tag{15.61}$$

$$SO_3 + HF \rightarrow FSO_3H \tag{15.62}$$

$$ClSO_3H + KF \rightarrow KCl + FSO_3H \tag{15.63}$$

Both $ClSO_3H$ and FSO_3H react rapidly with water and alcohols.

$$FSO_3H + H_2O \rightarrow HF + H_2SO_4 \tag{15.64}$$

Fluorosulfonic acid can be used in fluorination reactions, and it functions as a catalyst in reactions such as alkylation and polymerization. One of the most important reactions of FSO_3H and $ClSO_3H$ is as sulfonating agents to introduce the −SO_3H group into various organic materials.

15.1.8 Nitrogen Compounds

Because nitrogen compounds of Se and Te are much less important than those of sulfur, this section will be devoted to the sulfur compounds. The binary compounds containing sulfur and nitrogen have several unusual structures and properties that make them an interesting series. Probably the most studied compound of this type is S_4N_4, tetrasulfur tetranitride, which is prepared by the following reactions:

$$6\,S_2Cl_2 + 4\,NH_4Cl \rightarrow S_4N_4 + S_8 + 16\,HCl \tag{15.65}$$

$$6\,S_2Cl_2 + 16\,NH_3 \rightarrow S_4N_4 + S_8 + 12\,NH_4Cl \tag{15.66}$$

Compounds that undergo a color change when heated are known as *thermochromic* compounds, and tetrasulfur tetranitride exhibits this characteristic. It changes from almost colorless at very low temperature to orange at 25 °C to dark red at 100 °C. The compound is soluble in organic solvents, but insoluble in water. Although S_4N_4 is stable under some conditions, it has a heat of formation of +460 kJ/mole partially because of the great stability of the N_2 molecule. As a result, it will detonate if subjected to shock. The structure of the S_4N_4 molecule is a quite unusual ring in which the four nitrogen atoms line in the same plane.

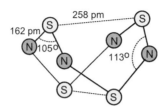

The distance between the sulfur atoms is significantly shorter than the sum of the van der Waals radii (approximately 330 pm) so some interaction between the sulfur atoms is indicated. However, the typical S–S single bond distance is approximately 210 pm so the interaction is weak. The structure involves several resonance structures, some of which are shown as follows (the unshared pairs of electrons on N and S are omitted):

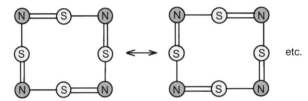

Molecular orbital calculations have also been carried out to describe the bonding by considering electron delocalization over the entire structure.

An enormous number of reactions have been carried out with tetrasulfur tetranitride to produce many ring structures, including some that have bridges across the rings. It is interesting to note that when S_4N_4 is hydrolyzed in basic solution, nitrogen ends up in a negative oxidation state in NH_3 because it has the higher electronegativity.

$$S_4N_4 + 6\, OH^- + 3\, H_2O \; \rightarrow \; S_2O_3^{2-} + 2\, SO_3^{2-} + 4\, NH_3 \qquad (15.67)$$

Reactions of S_4N_4 include some in which the ring is preserved when groups are added to S or N as well as some in which the ring is changed into one having a different number of atoms. Other reactions lead to complete disruption of the ring so there is a very broad chemistry of this interesting molecule. In halogenation reactions with Cl_2 and Br_2, the product, $S_4N_4Cl_4$, retains the ring structure of S_4N_4, and the chlorine atoms are bonded to the sulfur atoms.

$$S_4N_4 + 2\, Cl_2 \; \xrightarrow{CS_2} \; S_4N_4Cl_4 \qquad (15.68)$$

As a result of having unshared pairs of electrons, S_4N_4 undergoes addition reactions in which it functions as a Lewis base. For example, with BCl_3, the reaction can be shown as

$$S_4N_4 + BCl_3 \; \rightarrow \; S_4N_4{:}BCl_3 \qquad (15.69)$$

As shown in the following equation, reduction takes place when S_4N_4 is treated with $SnCl_2$ in ethanol.

$$S_4N_4 \; \xrightarrow[C_2H_5OH]{SnCl_2} \; S_4(NH)_4 \qquad (15.70)$$

Fluorine atoms can be added to the S_4N_4 ring to give $(F-SN)_4$ (the fluorine atoms are attached to the sulfur) by a reaction with AgF_2 in CCl_4.

When S_4N_4 reacts with $SOCl_2$, one of the products is $(S_4N_3)^+Cl^-$.

$$S_4N_4 \xrightarrow{SOCl_2} (S_4N_3)^+Cl^- \tag{15.71}$$

A reaction between S_4N_4 and S_2Cl_2 also produces the $S_4N_3^+$ ring.

$$3\ S_4N_4 + 2\ S_2Cl_2 \rightarrow 4\ (S_4N_3)^+Cl^- \tag{15.72}$$

The *thiotrithiazyl*, $S_4N_3^+$, ring that has a planar structure that can be shown as

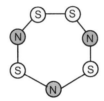

The S—N unit is the *thiazyl* group and $S_4N_3^+$ contains three of these units plus one sulfur atom (designated as *thio*) hence the name. When S_4N_4 reacts with NOCl, a five-membered ring is produced.

$$S_4N_4 + 2\ NOCl + S_2Cl_2 \rightarrow 2\ (S_3N_2Cl)Cl + 2\ NO \tag{15.73}$$

In the $S_3N_2Cl^+$ cation, the Cl is bonded to a sulfur atom giving the structure

Another binary compound of sulfur and nitrogen is the explosive compound known as disulfur dinitride (S_2N_2), which can be obtained by passing S_4N_4 vapor through silver wool at 300 °C. The structure of S_2N_2 can be shown as

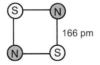

166 pm

with bond angles that are very close to 90°. Several resonance structures having double bonds are possible. Although it is a Lewis base, the main interest in the chemistry of S_2N_2 centers on the fact that it will polymerize to give $(S_2N_2)_n$, a bronze material known as *polythiazyl* that has metallic properties. The polymerization is accomplished by allowing the material to stand at room temperature or below for several days. These materials have been known for almost a century.

In addition to the compounds of sulfur and nitrogen already discussed, there are numerous others that have been studied. Tetrasulfur tetranitride is the starting point for preparing the compounds, and a wide range of reactions have been employed. Binary compounds containing selenium or tellurium with nitrogen have not been as fully characterized as those of sulfur, but Se_4N_4 (an explosive orange compound) and a tellurium nitride have been prepared. There are also cations known that contain selenium and nitrogen. Although these will not be described, the examples discussed show the general nature of this field.

15.1.9 The Oxyacids of Sulfur

It would be possible to write an entire book on the oxyacids of sulfur alone. In fact an entire book devoted to *sulfuric acid* alone has been published! By comparison, the oxyacids of selenium and tellurium are relatively insignificant. Thus, most of this section will be devoted to the acids containing sulfur. As mentioned elsewhere, sulfuric acid is produced in the largest quantity of any compound and is used so extensively in industrial chemistry that the quantity used has been referred to the "barometer" of the chemical industry. Although sulfuric acid is by far the most common and important of the oxyacids of sulfur, there are several others, most of which are shown in Table 15.6.

15.1.10 Sulfurous Acid and Sulfites

The anhydride of sulfurous acid is sulfur dioxide, which is very soluble in water. Although most of the gas is physically dissolved, the ionization takes place slightly as indicated by the equations

$$2\,H_2O + SO_2 \rightleftarrows H_3O^+ + HSO_3^- \tag{15.74}$$

$$HSO_3^- + H_2O \rightleftarrows H_3O^+ + SO_3^{2-} \tag{15.75}$$

In the solution, there appears to be a very slight tendency of SO_2 to add to SO_3^{2-} to give a small amount of $S_2O_5^{2-}$, which can also result from the dehydration reaction

$$2\,HSO_3^- \rightarrow H_2O + S_2O_5^{2-} \tag{15.76}$$

Although sulfurous acid is not an important compound, many sulfites are widely used. Sulfites react with sulfur to produce thiosulfates (see Eqn (15.5)), and sulfites are reducing agents that are used in large quantities, particularly in the sulfite process for making paper pulp. Sulfites are oxidized to sulfates by oxidizing agents, such as MnO_4^-, Cl_2, I_2, and Fe^{3+}. Most bisulfite salts are those in which the cation is a +1 ion of large size.

Sulfites can be prepared by reactions such as the following:

$$NaOH + SO_2 \rightarrow NaHSO_3 \tag{15.77}$$

$$2\,NaHSO_3 + Na_2CO_3 \rightarrow 2\,Na_2SO_3 + H_2O + CO_2 \tag{15.78}$$

The acid produced when a molecule of water is removed from two molecules of H_2SO_3 is disulfurous (*pyrosulfurous*) acid, $H_2S_2O_5$. Even though the acid is not stable, a few salts are known that contain the $S_2O_5^{2-}$ ion, which has the structure

Table 15.6 The Major Oxyacids of Sulfur[a]

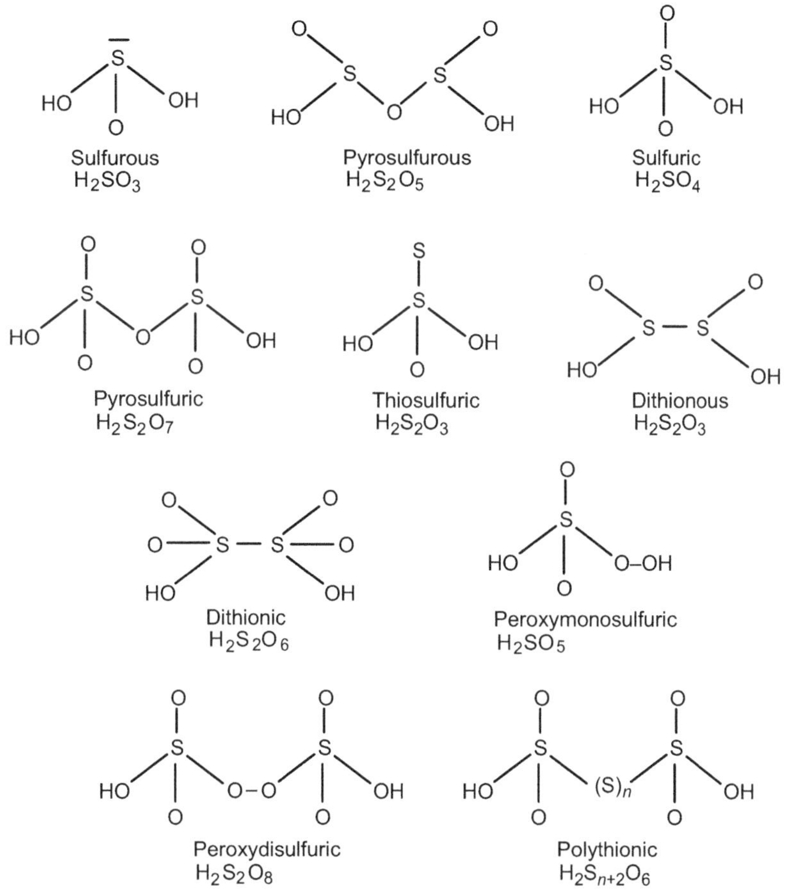

Sulfurous
H_2SO_3

Pyrosulfurous
$H_2S_2O_5$

Sulfuric
H_2SO_4

Pyrosulfuric
$H_2S_2O_7$

Thiosulfuric
$H_2S_2O_3$

Dithionous
$H_2S_2O_3$

Dithionic
$H_2S_2O_6$

Peroxymonosulfuric
H_2SO_5

Peroxydisulfuric
$H_2S_2O_8$

Polythionic
$H_2S_{n+2}O_6$

[a]Only salts exist for sulfurous, dithionic, and dithionous acids.

When SO_2 is added to a solution containing a sulfite, the $S_2O_5^{2-}$ ion is produced so the solution contains a salt of disulfurous (pyrosulfurous) acid. In aqueous solutions, these salts decompose when an acid is added.

$$S_2O_5^{2-} + H^+ \rightarrow HSO_3^- + SO_2 \tag{15.79}$$

15.1.11 Dithionous Acid and Dithionites

One of the unstable oxyacids of sulfur is dithionous acid, $H_2S_2O_4$, which contains sulfur in the $+3$ oxidation state. Reduction of sulfur $+4$ to produce the dithionite ion is accomplished as shown in the following equation:

$$2\,HSO_3^- + SO_2 + Zn \rightarrow ZnSO_3 + S_2O_4^{2-} + H_2O \tag{15.80}$$

The interesting eclipsed or "sawhorse" structure (C_{2v}) of the $S_2O_4^{2-}$ ion can be shown as

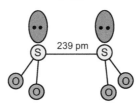

in which the O—S—O bond angle is 108°. The ion is reactive as indicated by the fact there is isotope exchange when labeled SO_2 is added to a solution of a dithionite. A typical S—S bond length is approximately 210 pm so the bond length of 239 pm indicates a weak reactive bond, which is illustrated by the following equation:

$$2\,S_2O_4^{2-} + H_2O \;\rightarrow\; S_2O_3^{2-} + 2\,HSO_3^- \tag{15.81}$$

It is believed that $S_2O_4^{2-}$ establishes an equilibrium that can be represented as

$$S_2O_4^{2-} \;\rightleftarrows\; 2\cdot SO_2^- \tag{15.82}$$

Sodium dithionite, $Na_2S_2O_4\cdot 2H_2O$, is a widely used reducing agent in several processes.

15.1.12 Dithionic Acid and Dithionates

This is another of the unstable oxyacids of sulfur that appears to contain sulfur +5. The structure of the dithionate anion is

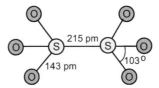

Writing the formula as $(HO)S(O)_2$—$S(O)_2OH$ shows that it is a strong acid. The S—S bond is slightly longer than a typical single bond, but the S—O bond length is close to that in sulfuric acid. Sulfites can be oxidized to give dithionates.

$$2\,SO_3^{2-} + MnO_2 + 4\,H^+ \;\rightarrow\; Mn^{2+} + S_2O_6^{2-} + 2\,H_2O \tag{15.83}$$

When heated, metal dithionates generally disproportionate to give a sulfate and SO_2.

$$MS_2O_6 \;\rightarrow\; MSO_4 + SO_2 \tag{15.84}$$

The dithionate ion is a good coordinating ion that forms stable chelates. In addition to dithionic acid, some polythionic acids having the formula $H_2S_nO_6$ are known in which the sulfur chain contains more than two sulfur atoms.

15.1.13 Peroxydisulfuric and Peroxymonosulfuric Acids

Peroxydisulfuric acid, $H_2S_2O_8$, is a solid that melts at 65 °C. The acid and its salts are such strong oxidizing agents that combustible materials can be ignited by them. The usual salts that are widely used as oxidizing agents are the sodium, potassium, and ammonium salts. Electrolytic oxidation of bisulfate is the usual method of producing the acid.

$$2 \, HSO_4^- \rightarrow H_2S_2O_8 + 2 \, e^- \tag{15.85}$$

When $H_2S_2O_8$ is hydrolyzed, it produces H_2O_2 and sulfuric acid.

$$H_2S_2O_8 + 2 \, H_2O \rightarrow 2 \, H_2SO_4 + H_2O_2 \tag{15.86}$$

Equations (15.85) and (15.86) constitute the basis for producing hydrogen peroxide. When heated, solid peroxydisulfates decompose with the liberation of oxygen.

$$K_2S_2O_8(s) \xrightarrow{\text{heat}} K_2S_2O_7(s) + 1/2 \, O_2(g) \tag{15.87}$$

Peroxymonosulfuric acid (historically known as Caro's acid), H_2SO_5, results when peroxydisulfuric acid reacts with a limited amount of water,

$$H_2S_2O_8 + H_2O \rightarrow H_2SO_4 + HOOSO_2OH \tag{15.88}$$

or by the reaction of H_2O_2 with chlorosulfonic acid.

$$H_2O_2 + HOSO_2Cl \rightarrow HCl + HOOSO_2OH \tag{15.89}$$

15.1.14 Oxyacids of Selenium and Tellurium

Oxyacids containing selenium and tellurium are known, but they and their salts are generally less important than the sulfur compounds. When selenium and tellurium tetrahalides hydrolyze, the solutions contain selenous and tellurous acids.

$$TeCl_4 + 3 \, H_2O \rightarrow 4 \, HCl + H_2TeO_3 \tag{15.90}$$

In the case of selenium, both the acid and its salts are stable. Salts can be prepared by reactions such as the following:

$$CaO + SeO_2 \rightarrow CaSeO_3 \tag{15.91}$$

$$2 \, NaOH + SeO_2 \rightarrow Na_2SeO_3 + H_2O \tag{15.92}$$

Several polytellurites ($Te_4O_9^{2-}$, $Te_6O_{13}^{2-}$, etc.) are stable as solid salts.

Although selenic and telluric acids contain the central atom in the +6 oxidation state, they are very different. The properties of selenic acid, H_2SeO_4, are very similar to those of H_2SO_4, and many of their salts are similar. The oxyacid that contains Te in the +6 state is H_6TeO_6, which can also be written as $Te(OH)_6$. This acid can be prepared from Te or TeO_2 by suitable oxidation reactions, and it can be also obtained as a solid hydrate. As expected from the formula, telluric acid is a weak acid, although some salts can be obtained in which a variable number of protons are replaced.

15.1.15 Sulfuric Acid

Sulfuric acid is a vital commodity and that is underscored by the fact that over 90 billion pounds of the acid are used annually. Sulfuric acid is utilized on a such a large scale because it is used in so many ways, and it is relatively inexpensive. Some of the important chemistry of H_2SO_4 will be illustrated in this section.

It is believed that H_2SO_4 was discovered in about the tenth century. In the 1800s, most of the H_2SO_4 was produced by the *lead chamber process*, although it was also produced by pyrolysis of $FeSO_4 \cdot xH_2O$. Today, sulfuric acid is produced by a method known as the *contact process* in which SO_2 is oxidized to SO_3, which then reacts with water to give the acid. Oxidation of SO_2 requires a suitable catalyst, such as spongy platinum or sodium vanadate. In many cases, SO_3 is dissolved in 98% sulfuric acid to produce disulfuric acid (*oleum*), which can be shipped, diluted, and still give 100% sulfuric acid. The concentration of SO_3 varies from 10 to 70% in commercial oleum.

$$SO_3 + H_2SO_4 \rightarrow H_2S_2O_7 \tag{15.93}$$

$$H_2S_2O_7 + H_2O \rightarrow 2\,H_2SO_4 \tag{15.94}$$

Concentrated sulfuric acid contains about 98% of the acid, which is equivalent to an 18 M solution. There is extensive hydrogen bonding as indicated by the high viscosity and boiling point of the liquid. The reaction of sulfuric acid with water is exothermic so dilution should always be carried out by adding the acid to water while stirring the solution. Because of its affinity for water, sulfuric acid is a strong dehydrating agent. Several hydrates of sulfuric acid have been identified, some of which are $H_2SO_4 \cdot H_2O$ (m.p. 8.5 °C), $H_2SO_4 \cdot 2H_2O$ (m.p. −39.5 °C), and $H_2SO_4 \cdot 4H_2O$ (m.p. −28.2 °C). Some of the properties of sulfuric acid are as follows:

Melting point	10.4 °C
Boiling point	290 °C (with decomp.)
Dielectric constant	100
Density (25 °C)	1.85 g cm^{-3}
Viscosity (25 °C)	24.54 cp

The H_2SO_4 molecule has the structure shown as follows:

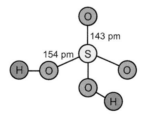

The fact that the S–O bonds to oxygen atoms that do not have hydrogen atoms attached are significantly shorter than the others is a manifestation of π bonding. If the structure contained only single bonds (as shown above), the sulfur would have a +2 formal charge, and the two types of S–O bonds would not differ much in length. To reduce the positive formal charge, multiple bonding occurs to two oxygen atoms, which reduces the bond length. That this is correct is shown by the structure of the HSO_4^- ion that remains after the H_2SO_4 molecule functions as a proton donor.

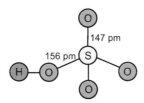

Note that the bond between sulfur and the OH group is not much different from that in the H_2SO_4 molecule, but the other three S—O bonds are lengthened compared with those in H_2SO_4 the molecule. This is the result of the π bonding being distributed to three oxygen atoms rather than just two.

The chemistry of H_2SO_4 is primarily related to its acidity. The first dissociation

$$H_2SO_4 + H_2O \rightarrow H_3O^+ + HSO_4^- \tag{15.95}$$

is considered to be complete in dilute solution and for the second step shown as

$$HSO_4^- + H_2O \rightleftarrows H_3O^+ + SO_4^{2-} \tag{15.96}$$

the dissociation constant is $K_{a2} = 1.29 \times 10^{-2}$ at $18\,°C$. In addition to the acid itself, the number of sulfates and bisulfates that have important uses is large. Sulfuric acid is used extensively as a nonaqueous solvent, and as a result of having a high dielectric constant (approximately 100) and considerable polarity, it dissolves many substances. Some of the uses of sulfuric acid will be described later in this section.

Salts of H_2SO_4 include both the *normal* sulfates containing SO_4^{2-} and *acid* sulfates or bisulfates that contain HSO_4^-. When bisulfates dissolve in water, the solutions are somewhat acidic because of the dissociation of the HSO_4^- ion as shown in Eqn (15.96). Bisulfates are produced by reactions such as the following:

$$NaOH + H_2SO_4 \rightarrow NaHSO_4 + H_2O \tag{15.97}$$

$$H_2SO_4 + NaCl \rightarrow HCl + NaHSO_4 \tag{15.98}$$

Some of the types of reactions that can be used to produce sulfates are as follows:

$$Ca(OH)_2 + H_2SO_4 \rightarrow CaSO_4 + 2\,H_2O \tag{15.99}$$

$$Zn + H_2SO_4 \rightarrow ZnSO_4 + H_2 \tag{15.100}$$

$$CaCl_2 + H_2SO_4 \rightarrow CaSO_4 + 2\,HCl \tag{15.101}$$

Sulfides, sulfites, and other sulfur-containing compounds can be oxidized to sulfates by using the suitable oxidizing agents under appropriate conditions. At high temperature, concentrated H_2SO_4 reacts as an oxidizing agent that dissolves copper.

$$2\,H_2SO_4\,(hot\ conc) + Cu \rightarrow CuSO_4 + SO_2 + 2\,H_2O \tag{15.102}$$

Bisulfate salts, such as $NaHSO_4$, can be dehydrated to produce *disulfates* (*pyrosulfates*).

$$2\,NaHSO_4 \xrightarrow{heat} Na_2S_2O_7 + H_2O \tag{15.103}$$

Sulfuric acid has been used as a nonaqueous solvent, and some proton transfer may take place as a result of autoionization in 100% H_2SO_4.

$$2 \, H_2SO_4 \;\rightarrow\; HSO_4^- + H_3SO_4^+ \tag{15.104}$$

The presence of ionic species is demonstrated by the conductivity of the solutions. It is a strongly acidic solvent that protonates alcohols, ethers, and acetic acid. These substances are not normally bases, but they have an unshared pair of electrons that can function as proton acceptors.

$$(C_2H_5)_2O + H_2SO_4 \;\rightleftarrows\; HSO_4^- + (C_2H_5)_2OH^+ \tag{15.105}$$

In the reaction with nitric acid, the *nitronium* ion, NO_2^+, is generated,

$$HNO_3 + 2 \, H_2SO_4 \;\rightleftarrows\; H_3O^+ + NO_2^+ + 2 \, HSO_4^- \tag{15.106}$$

which is the attacking species in nitration reactions. Sulfuric acid increases the concentration of the positive attacking species, which is the essence of an acid catalyst (see Chapter 9). There are many derivatives of sulfuric acid that are very useful compounds. These include salts of alkyl sulfuric acid that are detergents, such as $CH_3(CH_2)_nC_6H_4SO_3^-Na^+$, chlorosulfonic acid, $ClSO_3H$, and sulfuryl chloride. The latter compounds are prepared by the reactions

$$HCl + SO_3 \;\rightarrow\; ClSO_3H \tag{15.107}$$

$$Cl_2 + SO_2 \;\rightarrow\; SO_2Cl_2 \tag{15.108}$$

Sulfuric acid is manufactured on an enormous scale with an annual output of around 90 billion pounds. During the mid-1900s (when the production of sulfuric acid was less than half what it is now), about a third of the sulfuric acid produced was used in the production of fertilizer, but that use rose to about two thirds in the later 1900s. During that time, the world population grew from perhaps 3×10^9 to about 6×10^9.

The manufacture of fertilizers was discussed in Chapter 14. Phosphate rock is digested with sulfuric acid to convert $CaCO_3$ into a more soluble form that contains a higher percentage of phosphorus. Sulfuric acid is used as a catalyst in alkylation reactions, petroleum refining, manufacture of detergents, paints, dyes, fibers, etc. It is also used as the electrolyte in the lead-acid battery that is the usual battery in automobiles. Sulfuric acid is an enormously important chemical commodity that it would be hard to do without.

15.2 THE HALOGENS

The Group VIIA elements are known as the *halogens*, a word that has its origin in the Greek words *halos* meaning "salt" and *genes* meaning "born" or "formed". In other words, the halogens are the "salt formers", which at least partially describes their chemical behavior. The halogens are reactive elements with the result that they are always found combined. In fact, many halogen compounds are quite stable so they are not easy to convert to the elements. As a result, the free halogens were not obtained until comparatively modern times.

Chlorine was prepared in 1774 by Scheele by the reaction of HCl with MnO_2, but it was Sir Humphrey Davy who suggested the name based on the Greek *chloros* meaning greenish-yellow. Bromine was discovered in 1826 by Balard, and the name of the element comes from the Greek word *bromos* meaning "stench". Iodine was obtained from kelp in 1812 by Courtois, and its name is also from the Greek word *iodides*, which means violet. Although elemental fluorine was not available centuries ago, a common mineral that contains fluorine is *fluorspar* or *fluorite*, CaF_2. When heated with sulfuric acid, hydrogen fluoride is produced, and it was known long ago that the gas would etch glass. Fluorine is such a strong oxidizing agent that no practical means was available to obtain the element from its compounds until electrochemical processes were developed, although it has been prepared by chemical means in recent years. As a result, it was 1886 before Moissan succeeded in preparing fluorine. All isotopes of astatine, which was obtained in 1940, are radioactive so there is little chemistry to consider for the element.

15.2.1 Occurrence
There are numerous minerals that contain halogens. Minerals that contain fluorine include fluorite (CaF_2), *cryolite* (Na_3AlF_6), and *fluorapatite* ($Ca_5(PO_4)_3F$). As was discussed in Chapter 14, fluorapatite is found with calcium phosphate, which is very important in the production of fertilizers. Fluorite is found in Southeastern Illinois and Northwestern Kentucky, and cyrolite is found in Greenland, although it is produced synthetically because of its use in the electrochemical production of aluminum.

There are many natural sources of chlorine compounds, which is not surprising considering that it is the 20th most abundant element. Salt and salt water are widely available, and the Great Salt Lake contains 23% salt and the Dead Sea contains about 30%. Because salt is so abundant, most minerals that contain chlorine are not important sources for economic reasons. Bromine is found in some salt brines and in the sea as are some iodine compounds.

15.2.2 The Elements
Although the most common source of fluorine is CaF_2, the element is not prepared directly from fluorite. When HF (obtained by treating fluorite with sulfuric acid) is added to a solution containing potassium fluoride, hydrogen bonding of F^- to HF produces the HF_2^- ion. Other salts such as $KF \cdot 2HF$ and $KF \cdot 3HF$ can also be obtained if the excess of HF is sufficient. The potassium salt, KHF_2, has a melting point of 240 °C in contrast to that of CaF_2, which melts at 1430 °C. The mixture containing the solvates with HF has an even lower melting point, so the electrolysis of a KHF_2/HF mixture makes it feasible to carry out electrolysis of the molten salt, which would be difficult with CaF_2.

Although producing fluorine by chemical means had not been accomplished for many years, it was accomplished in 1986 by making use of the reaction

$$K_2MnF_6 + 2\,SbF_5 \rightarrow 2\,KSbF_6 + MnF_3 + {}^1/_2\,F_2 \tag{15.109}$$

The reason why this reaction produces fluorine is that MnF_4 is thermodynamically unstable. Therefore, if the very strong Lewis acid SbF_5 removes two fluoride ions from MnF_6^{2-}, the result is MnF_4, and it decomposes to produce fluorine and MnF_3.

The production of chlorine is based on the electrolysis reaction

$$2\,Na^+ + 2\,Cl^- + 2\,H_2O \xrightarrow{\text{Elect.}} 2\,Na^+ + 2\,OH^- + Cl_2 + H_2 \tag{15.110}$$

This process not only produces chlorine but also a way in which enormous quantities of sodium hydroxide are produced with hydrogen being the other product. Two types of cells are in use. The first and by far the most important employs a diaphragm to separate the anode and cathode compartments. A second type of cell utilizes a mercury cathode with which the sodium forms an amalgam.

$$Na^+ + 2\,Cl^- \xrightarrow{Hg} Na/Hg + Cl_2 \tag{15.111}$$

Another but less common industrial reaction is based on the oxidation of HCl in which oxides of nitrogen function as a catalyst.

$$4\,HCl + O_2 \rightarrow 2\,Cl_2 + 2\,H_2O \tag{15.112}$$

Although more important as a means of producing sodium, the electrolysis of molten NaCl is carried out on an industrial scale.

$$2\,NaCl(l) \xrightarrow{Elect.} 2\,Na(l) + Cl_2(g) \tag{15.113}$$

Chlorine can be prepared in a laboratory by the oxidation of Cl^- by means of a suitable oxidizing agent. A procedure used by Scheele in 1774 involves heating a solution of HCl with MnO_2.

$$MnO_2 + 4\,HCl \rightarrow 2\,H_2O + MnCl_2 + Cl_2 \tag{15.114}$$

Because chlorine will oxidize Br^- to give Br_2, seawater is treated with chlorine to produce bromine, which is then removed by blowing air through the mixture. Numerous reactions in which a bromide is oxidized can be carried on a laboratory scale. Iodine is obtained primarily from seawater in a process similar to that for producing bromine. However, because I^- is easier to oxidize that Br^-, there is considerable latitude in the choice of oxidizing agents. Astatine is produced by making use of a radiochemical technique in which ^{209}Bi is the target that is subjected to α particles.

$$^{209}Bi + {}^4He^{2+} \rightarrow 2\,n + {}^{211}At \tag{15.115}$$

Table 15.7 shows a summary of some of the important properties of the halogens.

Fluorine occurs only as ^{19}F, but chlorine occurs as 75% ^{35}Cl and 25% ^{37}Cl. Bromine occurs about equally as ^{79}Br and ^{81}Br, but iodine occurs only as ^{127}I. The oxidizing strength of the halogens decreases in the series $F_2 > Cl_2 > Br_2 > I_2$. As expected, the halogens will react with many substances, and they

Table 15.7 Properties of the Halogens

	F_2	Cl_2	Br_2	I_2
Melting point, °C	−219.6	−101	−7.25	113.6
Boiling point, °C	−188	−34.1	59.4	185
X–X bond energy, kJ mol^{-1}	153	239	190	149
X–X distance, pm	142	198	227	272
Electronegativity (Pauling)	4.0	3.0	2.8	2.5
Electron affinity, kJ mol^{-1}	339	355	331	302
Single bond radius, pm	71.0	99.0	114	133
Anion (X$^-$) radius, pm	119	170	187	212

produce a very large number of covalent and ionic compounds. The reaction of fluorine with water liberates oxygen.

$$2\,F_2 + 2\,H_2O \;\rightarrow\; O_2 + 4\,H^+ + 4\,F^- \tag{15.116}$$

The other halogens react with water in a disproportionation reaction that yields the halide ion and the hypohalous acid.

$$X_2 + H_2O \;\rightarrow\; H^+ + X^- + HOX \tag{15.117}$$

15.2.3 Interhalogen Molecules and Ions

One of the interesting attributes of the halogens is the fact that atoms of different types readily form bonds. When chlorine and fluorine are combined and heated, they react to form ClF, a compound known as an *interhalogen*.

$$Cl_2 + F_2 \;\xrightarrow{\;250\,°C\;}\; 2\,ClF \tag{15.118}$$

Because the halogen atoms have seven valence shell electrons, the outer shell is filled when two atoms form a bond, but when an unshared pair is separated, two additional bonding sites are available. As a result, if ClF were to react with additional fluorine, the expected result would be ClF_3. If another unshared pair of electrons were uncoupled, there would be five bonding sites and the product would be ClF_5. A general formula for compounds containing two different halogens is XX'_n, where X' is the lighter halogen and n is an odd number. The maximum value for n is 7, but the only interhalogen having that formula is IF_7. When $n = 5$, the known compounds are ClF_5, BrF_5, and IF_5, and none of the penta-chlorides are stable. This is undoubtedly due to the effect of the larger size of the chlorine atom and the fact that it is a weaker oxidizing agent than fluorine. Bonds between other halogens and fluorine are also more polar so there is some stability imparted by partial charges on the atoms. As a result of these factors, the only XX'_7 compound is IF_7. Table 15.8 shows a summary of the properties of the interhalogens.

The simplest interhalogen molecules are diatomic molecules, XX'. All of the molecules having X' = F are known, and they are generally prepared by combination of the elements. The preparation of ClF is shown above, and the preparation of BrF is also from the elements at 10 °C when the elements are diluted with nitrogen.

$$Br_2 + F_2 \;\xrightarrow{\;10\,°C\;}\; 2\,BrF \tag{15.119}$$

Disproportionation of BrF occurs with the formation of Br_2 and BrF_3 or BrF_5. Little is known about IF because it so unstable and disproportionates to I_2 and IF_5. Because of the favorable hard–soft interaction of Ag^+ and I^-, IF can be obtained by the reaction

$$AgF + I_2 \;\rightarrow\; IF + AgI \tag{15.120}$$

The halogen chlorides are of course limited to those of Br and I, but although BrCl exists in equilibrium with Cl_2 and Br_2, it is so unstable that it is not obtained in a pure form. Iodine monochloride is a much more stable compound that exists in two forms. The α form consists of ruby red needles, and the β form is a reddish-brown solid. The forms differ in the way that ICl molecules are linked by intermolecular forces in the solid state.

Table 15.8 Properties of the Interhalogens

	Formula	m.p., °C	b.p., °C	μ, D
Type XX'	ClF	−156	−100	0.88
	BrF	−33	20	1.29
	IF	—	—	—
	BrCl	−66	−5	0.57
	ICl	27	97	0.65
	IBr	36	116	1.21
Type XX'$_3$	ClF$_3$	−83	12	0.56
	BrF$_3$	8	127	1.19
	IF$_3$	—	—	—
	ICl$_3$	101 (16 atm)	—	—
Type XX'$_5$	ClF$_5$	−103	−14	—
	BrF$_5$	−60	41	1.51
	IF$_5$	10	101	2.18
Type XX'$_7$	IF$_7$	6.45 (triple pt.)	—	0

Reactions of the XX' interhalogens are numerous, but they generally constitute reactions in which halogenation occurs, those in which the interhalogen is a nonaqueous solvent, or Lewis acid–base reactions. In keeping with their being more numerous and important, the discussion will deal primarily with the fluorides. In contact with water, the XX' molecules react according to the reaction

$$XX' + H_2O \rightarrow H^+ + X'^- + HOX \tag{15.121}$$

The halogen having lower electronegativity is found as the hypohalous acid because it is the atom that has +1 oxidation state. The monofluorides are strong fluorinating agents and undergo many reactions typical of the halogens. For example, sulfuryl chloride results from the reaction of chlorine with SO_2, and a mixed halogen compound is obtained by the reaction with ClF.

$$ClF + SO_2 \rightarrow ClSO_2F \tag{15.122}$$

Oxidation-addition (oxad) reactions of this type also occur with other compounds as shown in the following examples:

$$ClF + CO \rightarrow ClCOF \tag{15.123}$$

$$ClF + SF_4 \rightarrow SF_5Cl \tag{15.124}$$

In the reaction with a strong Lewis acid such as AsF_5, ClF behaves as a Lewis base with the fluorine atom being the electron pair donor resulting in transfer of the F^- ion.

$$2\,ClF + AsF_5 \rightarrow Cl_2F^+[AsF_6]^- \tag{15.125}$$

ClF can also react as a Lewis acid by accepting a fluoride ion as illustrated by the reaction

$$ClF + KF \rightarrow K^+[ClF_2]^- \tag{15.126}$$

Only three of the possible XX_3' interhalogens (ClF_3, BrF_3, and IF_3, which is stable only at low temperature) are important. The structure of BrF_3 can be shown as

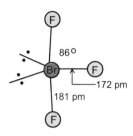

The structure of ClF_3 is similar except for the bond angle being 87.5°, and the equatorial and axial bond lengths being 160 and 170 pm, respectively. ClF_3 is produced in large quantities for use as a fluorinating agent, even though it reacts violently in many cases.

$$Cl_2 + 3\ F_2 \xrightarrow{250\ °C} 2\ ClF_3 \tag{15.127}$$

One use of ClF_3 is in the preparation of UF_6 by the reaction

$$3\ ClF_3 + U(s) \rightarrow UF_6(l) + 3\ ClF(g) \tag{15.128}$$

Many organic compounds sometimes react explosively with ClF_3, and it is such a strong fluorinating agent that it can be used to prepare other interhalogens as shown.

$$2\ ClF_3 + Br_2 \rightarrow 2\ BrF_3 + Cl_2 \tag{15.129}$$

$$10\ ClF_3 + 3\ I_2 \rightarrow 6\ IF_5 + 5\ Cl_2 \tag{15.130}$$

$$Br_2 + 3\ F_2 \xrightarrow{200\ °C} 2\ BrF_3 \tag{15.131}$$

As in the case of ClF, the trifluoride can behave as both a fluoride ion donor and acceptor, as illustrated by the reactions carried out in liquid ClF_3.

$$ClF_3 + AsF_5 \rightarrow ClF_2^+[AsF_6]^- \tag{15.132}$$

$$ClF_3 + NOF \rightarrow NO^+[ClF_4]^- \tag{15.133}$$

Bromine trifluoride has also been used extensively as a nonaqueous solvent that behaves in some reactions as if it ionized slightly according to the equation

$$2\ BrF_3 \rightleftarrows BrF_2^+ + BrF_4^- \tag{15.134}$$

It is interesting to note that the electrical conductivity of liquid BrF_3 is six orders of magnitude higher than that of liquid ClF_3, in which autoionization is assumed not to occur. In liquid BrF_3, SbF_5 is an acid because it generates the BrF_2^+ ion.

$$SbF_5 + BrF_3 \rightleftarrows BrF_2^+ + SbF_6^- \tag{15.135}$$

Because $KBrF_4$ contains the BrF_4^- ion, it is a base. Therefore, the reaction

$$BrF_2^+SbF_6^- + KBrF_4 \rightarrow 2\ BrF_3 + KSbF_6 \qquad (15.136)$$

is a neutralization reaction in liquid BrF_3. Because ClF_3 reacts so violently with many substances, BrF_3, which is a slightly milder fluorinating agent, has been widely used to prepare metal fluorides from the oxides or the metals.

Of the compounds having the formula XX_5', only the fluorides of Cl, Br, and I have been thoroughly studied. The chlorine compound is prepared by heating chlorine with an excess of fluorine at high temperature and pressure.

$$Cl_2 + 5\ F_2 \rightarrow 2\ ClF_5 \qquad (15.137)$$

A similar reaction but with less stringent conditions is used to produce BrF_5. IF_5 is prepared from the elements at room temperature. The only heptafluoride is IF_7, which is prepared from the elements using an excess of fluorine and at a temperature of 250–300 °C. All of these penta- and heptafluorides are extremely strong fluorinating agents. There are 12 electrons surrounding the central atom, which results in a square base pyramid structure. The structure of the pentafluoro compounds is represented by IF_5, which can be shown as

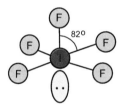

in which the iodine resides below the base of the pyramid as a result of the repulsion of the unshared pair on the bonding pairs. The structure of IF_7 is a pentagonal bipyramid.

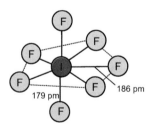

All three of the pentafluorides react violently if not explosively with many substances. In the case of ClF_5, the reaction with water is

$$ClF_5 + 2\ H_2O \rightarrow 4\ HF + FClO_2 \qquad (15.138)$$

BrF_5 reacts with water to produce HF and $HBrO_3$ if the water is diluted with a less reactive solvent such as acetonitrile. IF_5 reacts with water to produce HF and HIO_3.

$$IF_5 + 3\ H_2O \rightarrow 5\ HF + HIO_3 \qquad (15.139)$$

The low conductivity of liquid IF_5 is believed to be as a result of slight ionization,

$$2\,IF_5 \;\rightleftarrows\; IF_4^+ + IF_6^- \tag{15.140}$$

and compounds that contain each of the product ions have been isolated. IF_5 boils at approximately $101\,°C$, and at that temperature, it reacts with KF to give IF_6^-.

$$KF + IF_5 \;\rightarrow\; KIF_6 \tag{15.141}$$

Although IF_5 is quite stable, when heated to $500\,°C$, it disproportionates to give I_2 and IF_7.

Because of the size of the iodine atom and the fact that it is easier to oxidize to the $+7$ state, IF_7 is the only XX'_7 interhalogen. It is prepared by the reaction of IF_5 and F_2 at elevated temperatures, and like other halogen fluorides, it is a strong fluorinating agent. When it reacts with water, HF and HIO_4 are produced.

$$IF_7 + 4\,H_2O \;\rightarrow\; HIO_4 + 7\,HF \tag{15.142}$$

Interhalogens such as ClF_3, BrF_3, and BrF_5 will make fluorides out of almost anything. Metals, nonmetals, and many compounds are converted to the fluorides, and to make the reactions less vigorous, diluents are frequently added. If the reaction is being carried out in the gas phase, a relatively unreactive gas such as nitrogen or a noble gas is added. In most cases, the fluorination of a nonmetal produces the fluoride of the nonmetal in its highest oxidation state. For example, phosphorus is converted to PF_5 and sulfur is converted to SF_6, although it is sometimes possible to limit the fluorination somewhat by controlling the ratio of the reactants. When ClF_3 or BrF_3 is the fluorinating agent, the halogen monofluoride is frequently a product. For example,

$$B_{12} + 18\,ClF_3 \;\rightarrow\; 12\,BF_3 + 18\,ClF \tag{15.143}$$

Halogen fluorides convert oxides to fluorides or oxyfluorides. The following equations illustrate reactions of this type:

$$P_4O_{10} + 4\,ClF_3 \;\rightarrow\; 4\,POF_3 + 2\,Cl_2 + 3\,O_2 \tag{15.144}$$

$$SiO_2 + 2\,IF_7 \;\rightarrow\; 2\,IOF_5 + SiF_4 \tag{15.145}$$

$$6\,NiO + 4\,ClF_3 \;\rightarrow\; 6\,NiF_2 + 2\,Cl_2 + 3\,O_2 \tag{15.146}$$

The reaction of ClF with NOF (which is not a typical "oxide") can be represented as

$$NOF + ClF \;\rightarrow\; NO^+ClF_2^- \tag{15.147}$$

Numerous cases have been described elsewhere in which NO^+ is produced, but the ClF_2^- represents an *anion* of an interhalogen. Polyhalide anions are not rare because when iodine is dissolved in a solution containing I^-, the I_3^- ion is formed.

$$I^- + I_2 \;\rightarrow\; I_3^- \tag{15.148}$$

Anions of the X_3^- and $X'X_2^-$ types are linear with three unshared pairs of electrons around the central atom. In the case of ClF_2^-, the chlorine atom is the central atom.

There are numerous examples of cations of the interhalogens, and a great deal is known about the behavior of such species. The species that have been more fully studied involve only one type of halogen, such as I_3^+, Br_3^+, and Cl_3^+. In general, the production of these species requires rather stringent conditions that may include nonaqueous solvent systems. For example, a reaction that takes place in anhydrous sulfuric acid can be used to produce I_3^+.

$$8\ H_2SO_4 + 7\ I_2 + HIO_3 \ \rightarrow \ 5\ I_3^+ + 3\ H_3O^+ + 8\ HSO_4^- \tag{15.149}$$

The superacid system $HSO_3F/SbF_5/SO_3$ is employed when generating the Br_3^+ ion by the reaction

$$3\ Br_2 + S_2O_6F_2 \ \rightarrow \ 2\ Br_3^+ + 2\ SO_3F^- \tag{15.150}$$

but Br_3^+ has also been generated by the reaction

$$2\ O_2^+AsF_6^- + 3\ Br_2 \ \rightarrow \ 2\ Br_3^+AsF_6^- + 2\ O_2 \tag{15.151}$$

A reaction that also involves the strong Lewis acid AsF_5 generates Cl_3^+.

$$Cl_2 + ClF + AsF_5 \ \rightarrow \ Cl_3^+AsF_6^- \tag{15.152}$$

The cations discussed so far have contained only one type of halogen, but several cations are known that contain different halogens. These ions are generally of the XY_2^+ type, but others of the XYZ^+ type are possible. Although there are many possible species, the most thoroughly studied are ClF_2^+, BrF_2^+, IF_2^+, and ICl_2^+, all of which have a bent (C_{2v}) structure. Dissociation of the trihalides as nonaqueous solvents (although slight) leads to cations of this type.

$$2\ BrF_3 \ \rightleftarrows \ BrF_2^+ + BrF_4^- \tag{15.153}$$

The cation produced is the acidic species in the solvent, so it would be expected that to generate such an acidic species it would be appropriate to use a *very* strong Lewis acid. As illustrated in the following equations, this approach is successful.

$$ClF_3 + SbF_5 \ \rightarrow \ ClF_2^+SbF_6^- \tag{15.154}$$

$$BrF_3 + SbF_5 \ \rightarrow \ BrF_2^+SbF_6^- \tag{15.155}$$

$$IF_3 + AsF_5 \ \rightarrow \ IF_2^+AsF_6^- \tag{15.156}$$

$$ICl_3 + AlCl_3 \ \rightarrow \ ICl_2^+AlCl_4^- \tag{15.157}$$

The Cl_2F^+ cation is known to exist. When AsF_5 reacts with ClF, an adduct is produced having the formula $2ClF \cdot AsF_5$ that has a cation containing chlorine and fluorine atoms. The cation is believed to be $ClClF^+$ rather than $ClFCl^+$. In $XX_3'^+$ species, the bond angle is known to depend somewhat on the nature of the anion. For example, in $ClF_2^+SbF_6^-$ and $ClF_2^+AsF_6^-$, the F–Cl–F bond angles are 95.9° and 103.2°, respectively, and there are bridging fluorine atoms between the cation and the SbF_6^- and AsF_6^- anions. This behavior has also been found for other $XX_2'^+$ ions.

A slight degree of autoionization in liquid IF_5 is indicated by the conductivity.

$$2\ IF_5 \ \rightleftarrows \ IF_4^+ + IF_6^- \tag{15.158}$$

Because the cation characteristic of the solvent is IF_4^+, it is reasonable to assume that the ion may have some existence in other situations. As with the other polyatomic ions shown so far, the reaction selected involves a strong Lewis acid such as SbF_5 to remove F^-, and the same approach is taken for producing ClF_4^+ and BrF_4^+ in the interhalogen solvents.

$$ClF_5 + SbF_5 \rightarrow ClF_4^+ SbF_6^- \tag{15.159}$$

$$BrF_5 + SbF_5 \rightarrow BrF_4^+ SbF_6^- \tag{15.160}$$

$$IF_5 + SbF_5 \rightarrow IF_4^+ SbF_6^- \tag{15.161}$$

When IF_7 reacts with SbF_5 or AsF_5, the IF_6^+ cation is produced.

The formation of I_3^- by dissolving I_2 in a solution containing I^- was mentioned earlier. However, addition of I_2 continues with the formation of more complex species when the concentration of I_2 is increased.

$$I^- \overset{I_2}{\rightarrow} I_3^- \overset{I_2}{\rightarrow} I_5^- \overset{I_2}{\rightarrow} I_7^- \tag{15.162}$$

Similar behavior is exhibited by other halogens to a smaller degree. For example,

$$Cl^- + Cl_2 \rightarrow Cl_3^- \tag{15.163}$$

Some anions composed of two halogens result from similar processes.

$$Br^- + Cl_2 \rightarrow BrCl_2^- \tag{15.164}$$

$$BrCl_2^- + Cl_2 \rightarrow BrCl_4^- \tag{15.165}$$

The $BrCl_2^-$, I_3^-, and Cl_3^- species are linear ($D_{\infty h}$), but ICl_4^- and $BrCl_4^-$ are square planar (D_{4h}) with two unshared pairs of electrons on the central atom. In these structures, the halogen having the lower electronegativity is the central atom. Solids have been obtained that contain these anions, but almost all contain large cations having $+1$ charge.

It is interesting to note that for the reaction

$$Br^- + Br_2 \rightleftarrows Br_3^- \tag{15.166}$$

the equilibrium constant depends on the solvent. For example, in H_2O, $K = 16.3$; in a 50–50 weight percent mixture of CH_3OH and H_2O, $K = 58$; and in 100% CH_3OH, $K = 176$. Solvents consisting of small polar molecules solvate ions more strongly, and as a result, the Br^- ion is solvated more strongly in water, which hinders the formation of Br_3^-.

Halogens can behave as Lewis acids, but interhalogens also exhibit this type of behavior as is illustrated by the reaction of ICl and IBr with pyridine.

$$C_5H_5N + ICl \rightarrow C_5H_5NICl \rightarrow C_5H_5NI^+ + ICl_2^- \tag{15.167}$$

In reactions in which charged species are generated, the solvent often has a great effect. Because the reaction produces ions, solvation of the ionic species is energetically favorable only in polar solvents. Therefore, reactions of this type take place only in solvents that can strongly solvate the ions of the product. This is also true when the reaction is a substitution reaction such as

$$CH_3I + Cl^- \rightarrow CH_3Cl + I^- \tag{15.168}$$

The rate of this reaction varies by a factor of 10^6 depending on the solvent with the reaction being slower in methanol and much faster in N,N-dimethylformamide. Strong solvation of Cl^- impedes the reaction by hindering the formation of the transition state. Although the product shown in Eqn (15.167) is ionic, when the solvent is $CHCl_3$, the second step does not occur. In the reaction of pyridine with IBr, it appears that the ionization involves the production of IBr_2^-.

$$2\ C_5H_5NIBr \rightleftarrows (C_5H_5N)_2I^+ + IBr_2^- \tag{15.169}$$

When the reaction is carried out in aqueous HBr, it is found that the nature of the products (identified spectrophotometrically) depends on the concentration of the acid. When the acid is very dilute, the product is C_5H_5NIBr, but when the acid concentration is above 1 M, the product is $C_5H_5NH^+IBr_2^-$.

15.2.4 Hydrogen Halides

Among the most useful compounds of the halogens are the hydrogen halides. Hydrogen iodide is not very stable, and it can not be efficiently prepared by direct combination of the elements. Rather, as is the case with Te, P, and some other elements, it is easier to prepare a compound containing iodine and then carry out a hydrolysis reaction. For example,

$$P_4 + 6\ I_2 \rightarrow 4\ PI_3 \tag{15.170}$$

$$PI_3 + 3\ H_2O \rightarrow 3\ HI + H_3PO_3 \tag{15.171}$$

The properties of HF reflect the strong hydrogen bonding that persists even in the vapor state. As a result of its high polarity and dielectric constant, liquid HF dissolves many ionic compounds. Some of the chemistry of HF as a nonaqueous solvent has been presented in Chapter 10. Properties of the hydrogen halides are summarized in Table 15.9.

All of the hydrogen halides are very soluble in water, and acidic solutions result. Although HF is a weak acid, the others are strong and are almost completely dissociated in dilute solutions. HCl, HBr, and HI form constant boiling mixtures with water that contain 20.2, 47.6, and 53% of the acid, respectively.

Table 15.9 Properties of Hydrogen Halides

	HF	HCl	HBr	HI
Melting point, °C	−83	−112	−88.5	−50.4
Boiling point, °C	19.5	−83.7	−6.7	−35.4
Bond length, pm	91.7	127.4	141.4	160.9
Dipole moment, Debye	1.74	1.07	0.788	0.382
Bond energy, kJ mol^{-1}	574	428	362	295
ΔH_f°, kJ mol^{-1}	−273	−92.5	−36	+26
ΔG_f°, kJ mol^{-1}	−271	−95.4	−53.6	+1.6

There are many ways in which the hydrogen halides can be prepared. Heating a salt containing the halide ion with a nonvolatile acid is the usual way in which HF, HCl, and HBr are obtained in laboratory experiments.

$$2\,NaCl + H_2SO_4 \;\rightarrow\; 2\,HCl + Na_2SO_4 \tag{15.172}$$

$$KBr + H_3PO_4 \;\rightarrow\; HBr + KH_2PO_4 \tag{15.173}$$

$$CaF_2 + H_2SO_4 \;\rightarrow\; 2\,HF + CaSO_4 \tag{15.174}$$

Because HI is relatively unstable, this method is not appropriate for producing the compound. Instead, as mentioned above, HI is better prepared by hydrolysis reactions. If a halide salt is treated with an acid that is an oxidizing agent, a redox reaction occurs in which I_2 is produced.

$$8\,HI + H_2SO_4 \;\rightarrow\; H_2S + 4\,H_2O + 4\,I_2 \tag{15.175}$$

$$2\,HI + HNO_3 \;\rightarrow\; HNO_2 + I_2 + H_2O \tag{15.176}$$

Other than the iodine compound, the hydrogen halides can be obtained by reactions of the elements, but the reactions can be explosive. A mixture of H_2 and Cl_2 will explode if the reaction is initiated with a burst of light, which separates some Cl_2 molecules to produce $Cl\cdot$. The reaction between H_2 and Cl_2 is a classic case of a free radical reaction. A vast number of hydrolysis reactions yield hydrogen halides.

$$SOBr_2 + 2\,H_2O \;\rightarrow\; 2\,HBr + H_2SO_3 \tag{15.177}$$

$$SiCl_4 + 3\,H_2O \;\rightarrow\; 4\,HCl + H_2SiO_3 \tag{15.178}$$

$$PI_3 + 3\,H_2O \;\rightarrow\; 3\,HI + H_3PO_3 \tag{15.179}$$

On an industrial scale, HCl is also obtained in the chlorination of hydrocarbons.

15.2.5 Halogen Oxides

Compounds containing fluorine and oxygen are actually fluorides, and although O_2F_2 and OF_2 will be described, neither has significant uses. The compounds of chlorine and oxygen include some that are commercially important, but they are explosive in nature. Consequently, the discussion of halogen oxides will deal more heavily with the compounds containing chlorine.

Oxygen difluoride, OF_2 (m.p. $-223.8\,^{\circ}C$, b.p. $-145\,^{\circ}C$) is a pale yellow poisonous gas. The molecule has a bent structure (C_{2v}), and the bond angle is 103.2°. OF_2 can be prepared by the reaction of fluorine with dilute NaOH or the electrolysis of aqueous solutions containing HF and KF. The reaction of OF_2 with water can be shown as

$$H_2O + OF_2 \;\rightarrow\; O_2 + 2\,HF \tag{15.180}$$

The reaction of OF_2 with HCl produces HF and liberates chlorine.

$$4\,HCl + OF_2 \;\rightarrow\; 2\,HF + H_2O + 2\,Cl_2 \tag{15.181}$$

When heated to a temperature above $250\,^{\circ}C$, OF_2 decomposes into O_2 and F_2.

Dioxygen difluoride, O_2F_2, is a yellow-orange solid with a melting point of $-163\ °C$. The structure of the molecule is

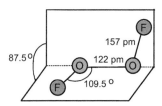

The compound is produced as a result of glow discharge through a mixture of O_2 and F_2 at -180 to $-190\ °C$. As should be expected, O_2F_2 is an extremely reactive fluorinating agent. Under the conditions that produce OF_2, small amounts of O_3F_2 and O_4F_2 are produced, but these unstable compounds decompose at liquid nitrogen temperatures.

Oxides of chlorine are both more numerous and more useful than those of fluorine. These oxides are the anhydrides of several important acids, and the oxyanions of those acids constitute the hypochlorites, chlorites, chlorates, and perchlorates. The first oxide of chlorine, Cl_2O (m.p. $-20\ °C$, b.p. $+2\ °C$), contains chlorine in the $+1$ oxidation state. It can be prepared by the reaction

$$2\ HgO + 2\ Cl_2 \ \rightarrow \ HgCl_2 \cdot HgO + Cl_2O \tag{15.182}$$

This oxide is soluble in water with which it reacts to produce hypochlorous acid.

$$H_2O + Cl_2O \ \rightarrow \ 2\ HOCl \tag{15.183}$$

The bond angle in the bent (C_{2v}) molecule is $110.8°$. One interesting reaction of Cl_2O is that with N_2O_5, which can be shown as

$$Cl_2O + N_2O_5 \ \rightarrow \ 2\ ClNO_3 \tag{15.184}$$

but other products that depend on the reaction conditions are also produced. Cl_2O converts SbF_5 into an oxychloride.

$$Cl_2O + SbCl_5 \ \rightarrow \ SbOCl_3 + 2\ Cl_2 \tag{15.185}$$

Chlorine dioxide, ClO_2 (m.p. $-60\ °C$, b.p. $11\ °C$), is an explosive gas $(\Delta H_f^o = +105\ kJ/mol)$ that has been used industrially as a gaseous bleach for wood pulp, textiles, and flour, and as a bactericide in water treatment. For most industrial applications, ClO_2 is produced on site. The molecule has a bent structure with a bond angle of $118°$. Among the reactions that can be used to prepare ClO_2, one is the following:

$$2\ HClO_3 + H_2C_2O_4 \ \rightarrow \ 2\ ClO_2 + 2\ CO_2 + 2\ H_2O \tag{15.186}$$

SO_2 acts as a reducing agent toward ClO_3^- with the reduction product being ClO_2.

$$2\ NaClO_3 + H_2SO_4 + SO_2 \ \rightarrow \ 2\ ClO_2 + 2\ NaHSO_4 \tag{15.187}$$

In basic solution, ClO_2 disproportionates to produce chlorite and chlorate ions.

$$2\ ClO_2 + 2\ OH^- \ \rightarrow \ ClO_2^- + ClO_3^- + H_2O \tag{15.188}$$

In industrial settings, ClO_2 is prepared from a solution of $NaClO_3$.

$$2\,ClO_3^- + 4\,HCl \rightarrow 2\,ClO_2 + Cl_2 + 2\,H_2O + 2\,Cl^- \qquad (15.189)$$

There is a competing reaction that can be shown as

$$ClO_3^- + 6\,HCl \rightarrow 3\,Cl_2 + 3\,H_2O + Cl^- \qquad (15.190)$$

and as a result, the overall reaction is sometimes shown as

$$8\,ClO_3^- + 24\,HCl \rightarrow 6\,ClO_2 + 9\,Cl_2 + 12\,H_2O + 8\,Cl^- \qquad (15.191)$$

In Eqn (15.187), the preparation of ClO_2 is by the reduction of ClO_3^- with SO_2. On an industrial scale, the reduction is carried out using gaseous methanol as the reducing agent.

$$2\,NaClO_3 + 2\,H_2SO_4 + CH_3OH \rightarrow 2\,ClO_2 + 2\,NaHSO_4 + HCHO + 2\,H_2O \qquad (15.192)$$

When solid $NaClO_3$ reacts with gaseous chlorine, ClO_2 is produced.

$$4\,NaClO_3 + 3\,Cl_2 \rightarrow 6\,ClO_2 + 4\,NaCl \qquad (15.193)$$

Photolysis of ClO_2 at low temperature produces several products, one of which is Cl_2O_3, which has the structure

This compound is a dangerous explosive. When chlorosulfonic acid reacts with a perchlorate, the product is Cl_2O_4, which is more accurately written as $ClOClO_3$, chlorine perchlorate.

$$CsClO_4 + ClOSO_2F \xrightarrow{-45\,°C} CsSO_3F + ClOClO_3 \qquad (15.194)$$

Although technically a chlorine oxide, chlorine perchlorate is of little importance. When ozone reacts with chlorine dioxide, the reaction produces Cl_2O_6, dichlorine hexoxide.

$$2\,O_3 + 2\,ClO_2 \rightarrow Cl_2O_6 + 2\,O_2 \qquad (15.195)$$

This compound is sometimes represented as $[ClO_2^+][ClO_4^-]$ on the basis of its reactions. One additional oxide of chlorine needs to be described. That compound is dichlorine heptoxide, Cl_2O_7 (m.p. $-91.5\,°C$, b.p. $82\,°C$), which is produced when $HClO_4$ is dehydrated with P_4O_{10}.

$$4\,HClO_4 + P_4O_{10} \rightarrow 2\,Cl_2O_7 + 4\,HPO_3 \qquad (15.196)$$

The oxides of bromine and iodine are not numerous, and they are not particularly important. Bromine monoxide, Br_2O, is obtained by the reaction

$$2\,HgO + 2\,Br_2 \rightarrow HgO \cdot HgBr_2 + Br_2O \qquad (15.197)$$

and this oxide can be oxidized by ozone to produce BrO_2.

$$4\ O_3 + 2\ Br_2O \xrightarrow{-78\ °C} 3\ O_2 + 4\ BrO_2 \tag{15.198}$$

Of the oxides of iodine that have been reported, the most common is I_2O_5, which is prepared by the dehydration of HIO_3.

$$2\ HIO_3 \xrightarrow{>170\ °C} I_2O_5 + H_2O \tag{15.199}$$

The I_2O_5 molecule has the structure

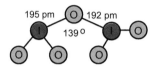

I_2O_5 is relatively stable but decomposes at 300 °C. It is a strong oxidizing agent that is used in quantitative determination of CO, which it oxidizes to CO_2.

$$I_2O_5 + 5\ CO \rightarrow I_2 + 5\ CO_2 \tag{15.200}$$

The reaction of I_2O_5 with water produces iodic acid, HIO_3. Although I_2O_4, I_4O_9, and I_2O_7 have been reported, they are of minor importance.

15.2.6 Oxyhalides

Oxyhalides of sulfur, nitrogen, and phosphorus are important compounds that have been described in other sections of this book. A considerable number of compounds of this type are known for the Group VIIA elements with most of them being fluorides. The known compounds include $FClO_2$, F_3ClO, $FClO_3$, F_3ClO_2, ClO_3OF, $FBrO_2$, $FBrO_3$, FIO_2, F_3IO, FIO_3, F_3IO_2, and F_5IO. Most of these compounds react as strong oxidizing or fluorinating agents. They also undergo hydrolysis reactions such as

$$FClO_2 + H_2O \rightarrow HClO_3 + HF \tag{15.201}$$

As we have seen in other instances, strong Lewis acids can remove F^- from many types of molecules to generate cations. That is also the case with oxyfluorides.

$$FClO_2 + AsF_5 \rightarrow ClO_2^+AsF_6^- \tag{15.202}$$

Perchloryl fluoride, $FClO_3$, is stable at temperatures up to about 500 °C, and it hydrolyzes only slowly. The molecule has the pyramidal (C_{3v}) structure

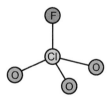

15.2.7 Hypohalous Acids and Hypohalites

When halogens dissolve in water, a reaction takes place to a slight extent that can be shown as

$$H_2O + X_2 \rightleftharpoons H^+ + X^- + HOX \tag{15.203}$$

The weak acids having the formula HOX are the hypohalous acids (they have K_a values of 3×10^{-8}, 2×10^{-9}, and 1×10^{-11} for HOCl, HOBr, and HOI, respectively). They are oxidizing agents as are the salts of the acids. HOF (m.p. $-117\,°C$) has been prepared, but it is unstable and has no serious uses. It does react with fluorine to produce OF_2 and HF.

$$HOF + F_2 \xrightarrow{H_2O} HF + OF_2 \tag{15.204}$$

Although solutions containing the hypohalous acids are produced by dissolving the halogens in water, the process must carried out at low temperature. Hypohalites are also produced when halogens react in basic solution.

$$X_2 + OH^- \rightarrow HOX + X^- \tag{15.205}$$

However, if the solutions are hot, the hypohalites undergo disproportionation to produce XO_3^- and X^-.

$$3\,HOX \rightarrow HXO_3 + 2\,HX \tag{15.206}$$

As a result, the products of reactions of halogens in basic solutions depend on pH, temperature, and the concentrations of the reactants. The acids are unstable compounds that decompose in two ways:

$$2\,HOX \rightarrow O_2 + 2\,HX \tag{15.207}$$

$$3\,HOX \rightarrow HXO_3 + 2\,HX \tag{15.208}$$

Although solutions of sodium hypochlorite are useful oxidizing agents, the solid is not very stable. Calcium hypochlorite is used in bleaches, swimming pool treatment, and so on. The decomposition of OCl^- is catalyzed by compounds containing transition metals.

$$2\,OCl^- \rightarrow 2\,Cl^- + O_2 \tag{15.209}$$

The reaction has been studied kinetically, and it is found that when a solid catalyst (CoO and/or Co_2O_3) formed by adding $Co(NO_3)_2$ to a solution of NaOCl is present, the rate of the reaction is determined by the surface area of the catalyst. Hypobromites and hypoiodites are good oxidizing agents, though less commonly used than hypochlorites. One use of NaOBr is in analysis where it is used to oxidize urea and NH_4^+ to produce N_2.

$$2\,NH_4Cl + 3\,NaOBr + 2\,NaOH \rightarrow 3\,NaBr + 2\,NaCl + 5\,H_2O + N_2 \tag{15.210}$$

$$CO(NH_2)_2 + 3\,NaOBr + 2\,NaOH \rightarrow 3\,NaBr + Na_2CO_3 + 3\,H_2O + N_2 \tag{15.211}$$

15.2.8 Halous Acids and Halites

Of the acids having the formula HXO_2, only the chlorine compound is stable enough to be of much use. Sodium chlorite is a useful oxidizing agent that results when chlorine dioxide reacts with sodium hydroxide.

$$2\,NaOH + 2\,ClO_2 \rightarrow NaClO_2 + NaClO_3 + H_2O \tag{15.212}$$

Acidifying a solution containing a chlorite gives a solution of $HClO_2$. Halite salts are used primarily as bleaching agents.

15.2.9 Halic Acids and Halates

The *halic* acids may not be industrially important, but their salts certainly are. Sodium chlorate is produced in enormous quantities and used in processes in which its oxidizing strength makes it a versatile bleach. One such use is in making paper, and potassium chlorate is used as the oxidizing agent in matches. The decomposition of potassium chlorate was discussed in Chapter 14 in connection with the laboratory preparation of oxygen.

The halic acids are strong as would be expected on the basis of the formula $(HO)_aXO_b$ with $b = 2$, but they are not stable as pure compounds. The anions are isoelectronic with SO_3^{2-}, and they have the C_{3v} structure

In the anions, the observed bond angles are $106°$ in ClO_3^- and BrO_3^- and $98–100°$ in IO_3^- depending on the nature of the cation.

Numerous reactions are useful for producing the acids and salts containing the halogens in the $+5$ oxidation state. As was shown earlier, the disproportionation reaction

$$3\,HOX \;\rightarrow\; HXO_3 + 2\,HX \tag{15.213}$$

is one such reaction. Another important type of disproportionation reaction of the halogens is illustrated by the general equation

$$3\,X_2 + 6\,OH^- \;\rightarrow\; XO_3^- + 5\,X^- + 3\,H_2O \tag{15.214}$$

For producing bromates, oxidation with hypochlorite as shown here is a useful reaction.

$$Br^- + 3\,OCl^- \;\rightarrow\; BrO_3^- + 3\,Cl^- \tag{15.215}$$

The acids are frequently prepared only as aqueous solutions by acidifying a solution of a metal halate. The following reaction is useful because the other product, $BaSO_4$, is insoluble.

$$Ba(ClO_3)_2 + H_2SO_4 \;\rightarrow\; BaSO_4 + 2\,HClO_3 \tag{15.216}$$

The industrial production of sodium chlorate is by electrolysis of an aqueous solution of NaCl, which can be shown as

$$NaCl + 3\,H_2O \xrightarrow{\text{Elect.}} NaClO_3 + 3\,H_2 \tag{15.217}$$

but this equation is an oversimplification because the process involves several steps.

At temperatures near or above the melting points, chlorates disproportionate.

$$4 \, ClO_3^- \; \rightarrow \; 3 \, ClO_4^- + Cl^- \tag{15.218}$$

However, decomposition of the acids takes place as shown in the equations

$$4 \, HBrO_3 \; \rightarrow \; 2 \, H_2O + 5 \, O_2 + 2 \, Br_2 \tag{15.219}$$

$$8 \, HClO_3 \; \rightarrow \; 4 \, HClO_4 + 2 \, H_2O + 3 \, O_2 + 2 \, Cl_2 \tag{15.220}$$

Chlorates, bromates, and iodates are strong oxidizing agents that are useful in many industrial and synthetic processes.

15.2.10 Perhalic Acids and Perhalates

The oxyacids containing chlorine and bromine in the $+7$ oxidation state are very strong acids. This can be seen when the formulas are written as $(HO)XO_3$, which corresponds to $b = 3$ in the general formula $(HO)_aXO_b$. The acids have as their anhydrides the oxides X_2O_7, but the situation with iodine is different from that of the other halogens. The reaction of I_2O_7 with excess water produces an acid having the formula H_5IO_6, and the (C_{4v}) structure of the molecule can be shown as

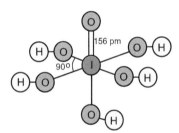

However, when I_2O_7 reacts with water, a series of reactions take place that can be shown as follows:

$$I_2O_7 \xrightarrow{H_2O} HIO_4 \xrightarrow{H_2O} H_4I_2O_9 \xrightarrow{H_2O} H_3IO_5 \xrightarrow{H_2O} H_5IO_6 \tag{15.221}$$

Even this set of reactions is an oversimplification because of reactions in which the acids are deprotonated and dehydrated in a manner somewhat similar to that leading to polyphosphates (see Chapter 14). The "periodic acid" that results is a function of the ratio of I_2O_7 to water. One of the reasons for this behavior is the relatively large size of the iodine atom. The acids HIO_4 and H_5IO_6 are sometimes referred to as metaperiodic acid and orthoperiodic acid, respectively, and solid compounds containing both IO_4^- and IO_6^{5-} are known. Periodate salts normally refer to those that contain the IO_4^- ion, such as potassium periodate. The periodates are very strong oxidizing agents that are widely used as such in both inorganic and organic chemistry.

Perchloric and perbromic acids are strong acids. Although perchloric acid can be obtained as a pure compound, it is normally available as a solution that contains 70% of the acid. This acid and solid perchlorates are such a strong oxidizing agents that they oxidize many combustible materials explosively. Even if the reaction is not immediate it may be initiated later and then become explosive. Perchloric acid is obtained from sodium perchlorate, which is prepared from the chlorate by electrolytic oxidation. If $HClO_4$ is dehydrated (as with P_4O_{10}), Cl_2O_7 results, which has the structure

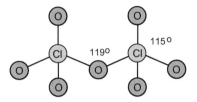

Perbromic acid was first prepared in 1969 in a very unusual way that made use of the conversion of ^{83}Se into ^{83}Br as a result of β-decay. The ^{83}Se was in the form of the sulfate, $^{83}SeO_4^{2-}$, so when decay occurred, the product was $^{83}BrO_4^-$. Satisfactory synthetic routes to the perbromates now include oxidation of BrO_3^- with XeF_2 or F_2.

There is significant multiple bonding in ClO_4^-. If the structure is drawn with four single Cl–O bonds, the chlorine atom has a formal charge of $+3$.

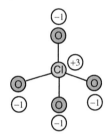

Including structures in which there are multiple bonds results in resonance structures such as

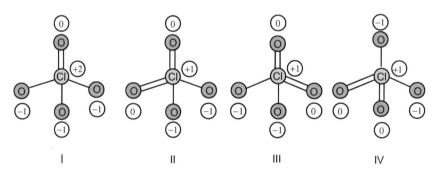

which give a lower formal charge in the chlorine atom. As a result, the observed bond length is somewhat shorter than that expected for a single Cl–O bond, so these structures make significant contributions to the structure of ClO_4^-.

15.3 THE NOBLE GASES

The noble gases are not very reactive and because they have eight electrons in the valence shell, we have come to consider the octet of electrons as being the circumstance that almost defines the essence of bonding in many molecules. The noble gases are, in fact, comparatively unreactive, but some (especially xenon) are not "inert," even though that label was once applied to the noble gases. The gases have high ionization potentials, but except for helium and neon not so extremely high that reactions are impossible. Prior to 1962, materials that held noble gases in a solid matrix such as graphite were known, but they are not compounds (they are known as *clathrates*) in the traditional sense. However, in 1962, a reaction was carried out by Neil Bartlett and D. H. Lohmann that resulted in the first "real" compound of a noble gas.

The reaction was carried out with the knowledge that oxygen reacts with the very strong oxidizing agent, PtF_6.

$$O_2(g) + PtF_6(g) \rightarrow O_2^+[PtF_6^-](s) \tag{15.222}$$

One reason that this reaction is so interesting is that the ionization potential for the oxygen *molecule* is 1177 kJ/mol (compared with that of a hydrogen *atom*, which is 1312 kJ/mol) and that for Xe is 1170 kJ/mol. It is evident that if PtF_6 will remove an electron from O_2, it should also do so from the Xe atom. When the reaction was attempted, a yellow-orange solid was obtained, and it was thought (later shown to be slightly incorrect) that the product was $Xe^+[PtF_6]^-$. It was later shown that the product was $XeF^+[Pt_2F_{11}]^-$, but much more xenon chemistry was on the way. Other fluorides including XeF_4, XF_2, and XeF_6 were prepared, and structural studies began to reveal details of these interesting molecules. In the intervening years, many other compounds containing xenon have been prepared, and the chemistry of the noble gases has also been expanded to include some krypton compounds. The major emphasis is on xenon chemistry in the discussion that follows:

15.3.1 The Elements

Helium has long been related to nuclear chemistry because of the formation of alpha particles $(\alpha = {}^4He^{2+})$ during the decay of heavy nuclei an example of which is

$$^{238}U \rightarrow {}^{234}Th + {}^4He^{2+} \tag{15.223}$$

An α particle can abstract two electrons from some other atom or molecule (and given the extremely high ionization potential of helium, the highest of any atom, it would be difficult to prevent it) to become a helium atom. Helium also is a constituent in stars as a result of the fusion reaction

$$4\,{}^1H \rightarrow {}^4He + 2\,e^+ + 2\,\nu \text{ (neutrinos)} \tag{15.224}$$

and it was known as early as 1868 that solar spectral lines indicated a new element. The element was named helium from the Greek word *helios* for sun. Several years later, the spectrum of gas from a volcano was shown to contain the same line. Argon was discovered in 1885 by Sir William Ramsay as a constituent in the residual gas after oxygen and nitrogen were removed from air. The name comes from the Greek word *argos*, which means inactive. Argon is generated by the electron capture decay of ^{40}K.

$$^{40}K \xrightarrow{\text{E.C.}} {}^{40}Ar \tag{15.225}$$

In 1898, neon, krypton, and xenon were isolated in the residual gas after nitrogen and oxygen were removed from liquid air. Their names come from the Greek words *xeneos* (stranger), *kryptos* (hidden), *neos* (new), respectively. Radon is produced in the three naturally occurring decay series of ^{235}U, ^{238}U, and ^{232}Th. Each of these series consists of numerous steps before a stable nuclide results, but the final product is radon in each case.

$$\text{From } ^{238}U: \quad ^{223}Ra \rightarrow \, ^{219}Rn + {}^4He^{2+} \quad (t_{1/2} = 11.4 \text{ days}) \quad \quad (15.226)$$

$$\text{From } ^{232}Th: \quad ^{224}Ra \rightarrow \, ^{220}Rn + {}^4He^{2+} \quad (t_{1/2} = 3.63 \text{ days}) \quad \quad (15.227)$$

$$\text{From } ^{235}U: \quad ^{226}Ra \rightarrow \, ^{222}Rn + {}^4He^{2+} \quad (t_{1/2} = 1600 \text{ yrs}) \quad \quad (15.228)$$

All of the isotopes of radon are radioactive and decay by α-emission to produce isotopes of polonium by transformations shown in the following equations:

$$^{219}Rn \rightarrow \, ^{215}Po + {}^4He^{2+} \quad (t_{1/2} = 3.96 \text{ sec}) \quad \quad (15.229)$$

$$^{220}Rn \rightarrow \, ^{216}Po + {}^4He^{2+} \quad (t_{1/2} = 55.6 \text{ sec}) \quad \quad (15.230)$$

$$^{222}Rn \rightarrow \, ^{218}Po + {}^4He^{2+} \quad (t_{1/2} = 3.825 \text{ days}) \quad \quad (15.231)$$

Radon constitutes a serious problem because being a heavy gas it collects in such places as basements and mine shafts. When inhaled, radon decays in areas where little penetration is require to cause tissue damage. Radiation from α- and β-decay is not of a highly penetrating type, but inside the lungs, it does not have to be in order to still cause damage. Table 15.10 shows several properties of the noble gases.

Compounds that consist of molecules having only very weak dispersion forces of attraction between them typically have a very narrow liquid range. The noble gases illustrate this dramatically because the largest liquid range for any of the gases is only about 9 degrees. The polarizability of the noble gases increases going downward in the group, which leads to increased solubility in water as a result of dipole-induced dipole forces of attraction (see Chapter 6).

Some of the noble gases have substantial uses. In Chapter 1, the stability of nuclei having particular numbers of nucleons was discussed. As a result of having the first shell filled with both protons and

Table 15.10 Some Properties of the Noble Gases

Property	He	Ne	Ar	Kr	Xe	Rn
Melting point, K	0.95	24.5	83.78	116.6	161.3	202
Boiling point, K	4.22	27.1	87.29	120.8	166.1	211
ΔH_{fusion}, kJ mol^{-1}	0.021	0.324	1.21	1.64	3.10	2.7
ΔH_{vap}, kJ mol^{-1}	0.082	1.74	6.53	9.70	12.7	18.1
Ioniz. Pot., kJ mol^{-1}	2372	2081	1520	1351	1170	1037
Atomic radius^a, pm	122	160	191	198	218	~220
Density^b, g L^{-1}	0.1785	0.900	1.784	3.73	5.88	9.73
Sol. in H$_2$O, $10^3 \times X^c$	7.12	8.73	30.2	57.0	105	230

^{a}*van der Waals radii.*
^{b}*At 0 °C and 1 atmosphere.*
^{c}*Where X is the mole fraction of the gas in the solution.*

neutrons, helium has a stable arrangement as does ^{20}Ne, which has an equal number of protons and neutrons. Partially as a result of this stability, both helium and neon are reasonably abundant on a cosmic scale. Having a density of only 0.18 g/L, helium is used in lighter-than-air aircraft, and it is also used as a coolant, especially for superconductors. Neon is used in neon signs, and argon is used as an inert atmosphere in some types of welding.

15.3.2 The Xenon Fluorides

Although all of the noble gases have many interesting properties, it is their *chemistry* that we consider at this point. Some compounds of krypton will be mentioned but is the chemistry of xenon that is most significant. In 1933, Linus Pauling speculated that it should be possible to produce XeF$_6$, and this has since been shown to be the case. The reaction of oxygen with PF$_6$ mentioned earlier suggested that a similar reaction with xenon might be successful in light of the similarity in ionization potentials. It should be apparent that if xenon is to react it should be with an extremely strong oxidizing agent, and F$_2$ is a suitable candidate. By this means, the difluoride and tetrafluoride of xenon are prepared as a mixture of the elements is heated or subjected to electromagnetic radiation.

$$Xe(g) + F_2(g) \rightarrow XeF_2(s) \tag{15.232}$$

When preparing XeF$_4$, a mixture containing a 5:1 ratio of fluorine to xenon was heated and subjected to several atmospheres pressure.

$$Xe(g) + 2\,F_2(g) \rightarrow XeF_4(s) \tag{15.233}$$

When a higher ratio of fluorine to xenon is used, XeF$_6$ can be obtained.

   The difference in the ionization potentials of xenon and krypton (1170 versus 1351 kJ/mol) indicates that krypton should be the less reactive of the two. Some indication of the difference can be seen from the bond energies, which are 133 kJ/mol for the Xe–F bond and only 50 kJ/mol for the Kr–F bond. As a result, XeF$_2$ is the considerably more stable of the difluorides, and KrF$_2$ is much more reactive. Krypton difluoride has been prepared from the elements, but only at low temperature using electric discharge. When irradiated with ultraviolet light, a mixture of liquid krypton and fluorine reacts to produce KrF$_2$. As expected, radon difluoride can be obtained, but because all isotopes of radon undergo rapid decay, there is not much interest in the compound. In this survey of noble gas chemistry, the main focus will be on the behavior of xenon compounds, which generally contain F, Cl, O, or N as the other atoms.

   There are 10 electrons around the central atom in XeF$_2$ so the linear structure is derived from a trigonal bipyramid with the unshared pairs of electrons in equatorial positions ($D_{\infty h}$ symmetry).

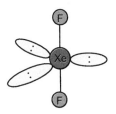

However, a different view of the bonding in XeF_2 is provided by a molecular orbital approach in which a p orbital on Xe combines with a p orbital from each F to form a three-center four-electron linear bond. Actually, the three atomic orbitals form three molecular orbitals, but only the bonding and nonbonding orbitals are populated. This population of the orbitals places nonbonding electron density on the F atoms to give polar Xe–F bonds.

The XeF_4 molecule is isoelectronic with IF_4^- (12 electrons around the central atom), and it has the same square planar (D_{4h}) structure.

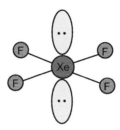

In the XeF_6 molecule, there are 14 electrons around the central atom, which gives the required six bonds and one unshared pair of electrons. Although the IF_7 molecule also has 14 electrons around the central atom, they are all bonding pairs, and the molecule has a pentagonal bipyramid structure. As a result of the unshared pair, the XeF_6 molecule has an irregular nonrigid structure that has C_{3v} symmetry, which can be shown as

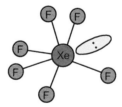

Xenon hexafluoride is known to exist in condensed phases as an equilibrium mixture that can be shown as

$$4\,XeF_6 \rightleftharpoons [XeF_5^+F^-]_4 \qquad\qquad (15.234)$$

The structure of the tetramer is complex and involves fluoride bridges as do the hexamers that exist in solid XeF_6.

In general, intermolecular forces are related to the structures of the molecules (see Chapter 6). Both XeF_2 and XeF_4 are nonpolar, so it is interesting that the melting points of XeF_2, XeF_4, and XeF_6 are 129, 117, and 50 °C, respectively, but the solids also readily sublime. The solubility parameter (discussed in Chapter 6) is useful when considering intermolecular forces and that is the case here. From the vapor pressures of the di- and tetrahalides, the calculated solubility parameters for XeF_2 and XeF_4 are 33.3 and 30.9 $J^{1/2}$ cm$^{-3/2}$, respectively. These values are high when compared to other nonpolar covalent molecules, and it is noteworthy that the *heavier* molecule has the *lower* solubility parameter. To provide a base of comparison, consider the solubility parameters for $SnCl_4$

$(260.5 \text{ g mol}^{-1})$ is $17.8 \text{ J}^{1/2} \text{ cm}^{-3/2}$ and $SiBr_4$ $(347.9 \text{ g mol}^{-1})$ is $18.0 \text{ J}^{1/2} \text{ cm}^{-3/2}$. These values follow the expected trend for molecules that interact as a result of London dispersion forces. It would be expected that London forces would be stronger between XeF_4 molecules than those between XeF_2 molecules. However, the bonds in xenon fluorides are polar. In XeF_2, the residual charges on Xe and F are likely higher than those on the atoms in the XeF_4 molecule, in which electron density is unequally shared in four directions. Therefore, although the *molecules* are not polar, the *bonds* within them are, and the bond polarity would be expected to be greater in XeF_2. This results in greater intermolecular attraction, which is reflected by the fact that the solubility parameter (and melting point) for XeF_2 is greater than that for XeF_4.

15.3.3 Reactions of Xenon Fluorides and Oxyfluorides

Many of the reactions that xenon fluorides undergo are similar in some ways to those of interhalogens. However, the xenon halides differ markedly in terms of their reactivity with XeF_2 being much less reactive than either XeF_4 or XeF_6. The difluoride reacts only slowly with water,

$$2 \, XeF_2 + 2 \, H_2O \; \rightarrow \; 2 \, Xe + O_2 + 4 \, HF \tag{15.235}$$

but in basic solution a different reaction takes place rapidly, which can be shown as

$$2 \, XeF_2 + 4 \, OH^- \; \rightarrow \; 2 \, Xe + O_2 + 2 \, H_2O + 4 \, F^- \tag{15.236}$$

Xenon tetrafluoride reacts rapidly with water by undergoing disproportionation

$$6 \, XeF_4 + 12 \, H_2O \; \rightarrow \; 2 \, XeO_3 + 4 \, Xe + 3 \, O_2 + 24 \, HF \tag{15.237}$$

as xenon +4 is converted into xenon +6 and Xe. The oxide is an explosive compound that is also produced by the hydrolysis of XeF_6.

$$XeF_6 + 3 \, H_2O \; \rightarrow \; XeO_3 + 6 \, HF \tag{15.238}$$

The hydrolysis may take place by the formation of $XeOF_4$ as an intermediate.

$$XeF_6 + H_2O \; \rightarrow \; XeOF_4 + 2 \, HF \tag{15.239}$$

$$XeOF_4 + 2 \, H_2O \; \rightarrow \; XeO_3 + 4 \, HF \tag{15.240}$$

As has been mentioned in other places, strong Lewis acids such as SbF_5 or AsF_5 have the ability to remove F^- from a variety of covalent fluorine compounds to generate polyatomic cations. Similar reactions occur between xenon fluorides and strong Lewis acids with the formation of products such as $XeF^+Sb_2F_{11}$, $XeF^+SbF_6^-$, and $Xe_2F_3^+SbF_6^-$. The structures of the cations that contain two xenon atoms have fluoride ion bridges between them. For example, the structure of $Xe_2F_3^+$ is

The XeF^+ ion bonds to the SbF_6^- anion to give a complex structure that can be shown as

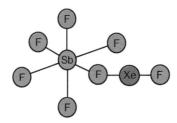

Xenon tetrafluoride undergoes reactions that are similar to those of the interhalogens. For example, the XeF_3^+ cation is generated when XeF_4 reacts a very strong Lewis acid such as BiF_5.

$$XeF_4 + BiF_5 \rightarrow XeF_3^+ BiF_6^- \tag{15.241}$$

In the solid state, PCl_5 and PBr_4 exist as ionic compounds. In a similar way, solid XeF_6 contains XeF_5^+ ions with fluoride ions bridging between them. The XeF_5^+ cation is produced by the reaction of XeF_6 with RuF_5.

$$XeF_6 + RuF_5 \rightarrow XeF_5^+ RuF_6^- \tag{15.242}$$

A cation containing two xenon atoms, $Xe_2F_{11}^+$, is also known, and it has a structure that can be shown as $F_5Xe^+ \cdots F^- \cdots XeF_5^+$. Polyatomic anions containing xenon are produced because XeF_6 is also a Lewis acid. An example of this type of reaction can be shown as

$$XeF_6 + MF \rightarrow MXeF_7 \tag{15.243}$$

where M is a +1 cation. In recent years, the ability of xenon halides to react as versatile fluorinating agents has been exploited. The higher fluorides XeF_4 and XeF_6 are very vigorous fluorinating agents, but the difluoride is less reactive, although it will fluorinate organic compounds, such as olefins.

$$\diagdown C = C \diagup + XeF_2 \longrightarrow \ \diagdown \overset{F}{\underset{}{C}} - \overset{F}{\underset{}{C}} \diagup + Xe \tag{15.244}$$

One interesting application that involves XeF_2 as a fluorinating agent involves the fluorination of uracil, which has the structure

The reaction of XeF_2 with uracil produces the 5-fluoro derivative, which can be applied topically for treating some types of skin diseases, including some types of skin cancer. The reaction producing 5-fluorouracil can be shown as

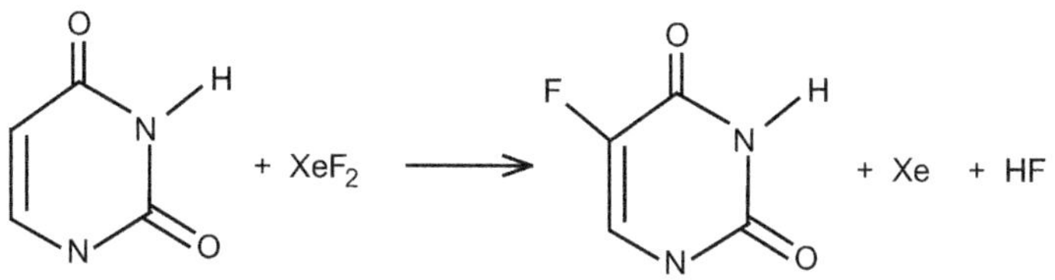

$$\text{(15.245)}$$

One of the attractive features of using XeF_2 as a fluorinating agent is that the other products, Xe and HF, are frequently more easily removed than those produced when other fluorinating agents are used.

Some oxides of xenon are known, and like most other compounds of xenon, they are usually obtained from the fluorides. Two reactions that yield XeO_3 have already been shown in Eqns (15.237) and (15.238). The heat of formation of XeO_3 is approximately $+400$ kJ/mol so it should be no surprise that the compound is a very sensitive explosive material. The structure of XeO_3 (which is isoelectronic with SO_3^{2-} and ClO_3^-) can be shown as follows:

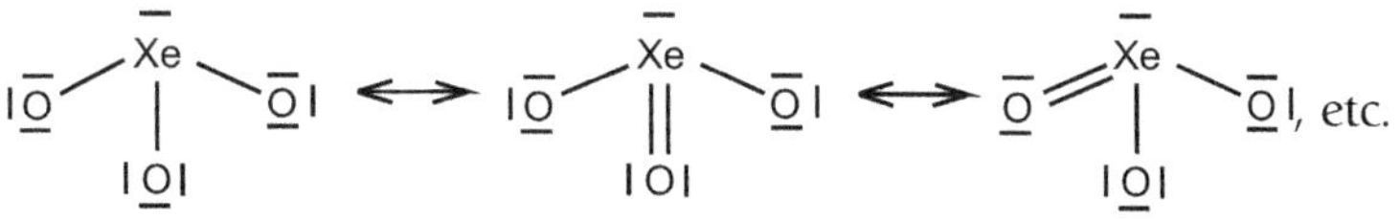

In the first structure showing only single bonds, the formal charge on Xe is $+3$ so contributions from structures showing double bonding are significant.

In basic solution, a reaction between OH^- and XeO_3 occurs, which can be shown as

$$XeO_3 + OH^- \;\rightarrow\; HXeO_4^- \qquad\qquad (15.246)$$

One of the reactions that $HClO_3$ undergoes is the disproportionation that produces perchoric acid and chlorine.

$$8\,HClO_3 \;\rightarrow\; 4\,HClO_4 + 2\,H_2O + 3\,O_2 + 2\,Cl_2 \qquad\qquad (15.247)$$

A reaction of $HXeO_4^-$ in basic solution is very similar, and it can be shown as

$$2\,HXeO_4^- + 2\,OH^- \;\rightarrow\; XeO_6^{4-} + Xe + O_2 + 2\,H_2O \qquad\qquad (15.248)$$

which results in the production of the perxenate ion, XeO_6^{4-}. This is just one of many ways in which some compounds of xenon resemble those of the halogens. Several solids containing the perxenate ion have been isolated, and the ion is the conjugate base of a weak acid, H_4XeO_6. Therefore, the salts hydrolyze to product basic solutions.

$$XeO_6^{4-} + H_2O \;\rightleftarrows\; HXeO_6^{3-} + OH^- \qquad\qquad (15.249)$$

$$HXeO_6^{3-} + H_2O \;\rightleftarrows\; H_2XeO_6^{2-} + OH^- \qquad\qquad (15.250)$$

With the oxidation state of Xe in perxenates being $+8$, they are, as expected, very strong oxidizing agents.

In Chapter 14, it was shown that oxyhalides of phosphorus can be obtained by the reaction of oxides with halides. One such reaction is

$$6\,PCl_5 + P_4O_{10} \rightleftarrows 10\,OPCl_3 \qquad (15.251)$$

In an analogous reaction, xenon oxyfluorides are produced by the reactions

$$XeF_6 + 2\,XeO_3 \rightarrow 3\,XeO_2F_2 \qquad (15.252)$$

$$2\,XeF_6 + XeO_3 \rightarrow 3\,XeOF_4 \qquad (15.253)$$

Although most of the early chemistry of Xe involved compounds containing fluorine and oxygen, a much more extensive chemistry of the element is known. One of the first known compounds that contained an organic group was C_6H_5XeF. However, the reaction

$$2\,C_6H_5XeF + Cd(C_6H_5)_2 \rightarrow 2\,Xe(C_6H_5)_2 + CdF_2 \qquad (15.254)$$

yields the diaryl compound. When C_6H_5XeF reacts with a molecule that has a very strong affinity for fluorine, one F is lost and compounds result that contain the $C_6H_5Xe^+$ ion. Another interesting compound containing a bond between xenon and carbon is produced when C_6H_5XeF reacts with $(CH_3)_3SiCN$.

$$C_6H_5XeF + (CH_3)_3SiCN \rightarrow C_6H_5XeCN + (CH_3)_3SiF \qquad (15.255)$$

Compounds of noble gases that contain H–Xe bonds have been known since 1995. Under extreme conditions such as ultraviolet photolysis of mixtures of Xe with HCl, HF, or HCN at very low temperature in a matrix of Xe, compounds such as HXeCN, HXeF, and HXeCl have been identified. Under these conditions, compounds such as HXeH and HXeCXeH have been identified and with acetylene one product is HXeCCH. These unstable products can be identified by means of infrared spectroscopy.

As mentioned earlier, krypton is known to form several compounds, but they are fewer and less well characterized than the compounds of xenon. The difluoride has been obtained by electric discharge through a mixture of Kr and F_2 at low temperature. As in the case of xenon difluoride, a cation is produced in the reaction with a strong Lewis acid, such as SbF_5.

$$KrF_2 + SbF_5 \rightarrow KrF^+SbF_6^- \qquad (15.256)$$

There is some association between the cation and another molecule of KrF_2 to produce $Kr_2F_3^+$.

As might be expected, KrF_2 is less stable than XeF_2 so it is an even stronger fluorinating agent as illustrated by the following reactions:

$$3\,KrF_2 + Xe \rightarrow XeF_6 + 3\,Kr \qquad (15.257)$$

$$I_2 + 7\,KrF_2 \rightarrow 2\,IF_7 + 7\,Kr \qquad (15.258)$$

Numerous other reactions of krypton difluoride are known, but they will not be reviewed here. The chemistry of krypton is well established, but it is still much less extensive than that of xenon. Although a rather extensive chemistry of the noble gases has developed, the vast majority of the studies have dealt with the xenon compounds.

References for Further Reading

Bailar, J.C., Emeleus, H.J., Nyholm, R., Trotman-Dickinson, A.F., 1973. *Comprehensive Inorganic Chemistry*, Vol. 3. Pergamon Press, Oxford. This is one volume in the five volume reference work in inorganic chemistry.

Bartlett, N., 1971. *The Chemistry of the Noble Gases*. Elsevier, New York. A good survey of the field by its originator.

Claassen, H.H., 1966. *The Noble Gases*. D. C. Heath, Boston. A very useful introduction to the chemistry of noble gas compounds and structure determination.

Cotton, F.A., Wilkinson, G., Murillo, C.A., Bochmann, M., 1999. *Advanced Inorganic Chemistry*, 6th ed. John Wiley, New York. Chapter 14, A 1300 page book that covers an incredible amount of inorganic chemistry. Chapter 14 deals with the noble gases.

Greenwood, N.N., Earnshaw, A., 1997. *Chemistry of the Elements*, 2nd ed. Butterworth-Heineman, Burlington, MA. Chapter 18 of this comprehensive book is devoted to the noble gases.

Holloway, J.H., 1968. *Noble-Gas Chemistry*. Methuen, London. A thorough discussion of some of the early work on noble gas chemistry. A good introductory reference.

King, R.B., 1995. *Inorganic Chemistry of the Main Group Elements*. VCH Publishers, New York. An excellent introduction to the descriptive chemistry of many elements. Chapter 7 deals with the chemistry of the noble gases.

Khriachtchev, L., Tanskanen, H., Lundell, J., Pettersson, M., Kiljunen, H., Rasanen, M., 2003. *J. Am. Chem. Soc.* 125, 4696−4697. A report on xenon compounds formed by photolysis at low temperature.

QUESTIONS AND PROBLEMS

1. Draw structures for S_2N_2 showing all valence shell electrons.
2. What are the characteristics of XeF_2 that make it a desirable fluorinating agent for organic compounds?
3. Which would you expect to be more stable, $NaIF_4$ or $CsIF_4$? Explain your answer.
4. Complete and balance the following:
   (a) $Cl_2 + H_2O \rightarrow$
   (b) $ICl_3 + H_2O \rightarrow$
   (c) $RbF + IF_3 \rightarrow$
   (d) $SiO_2 + ClF_3 \rightarrow$
   (e) $AsBr_3 + H_2O \rightarrow$
5. Write complete balanced equations for the following:
   (a) $XeF_4 + PF_3 \rightarrow$
   (b) $XeOF_4 + H_2O \rightarrow$
   (e) $XeF_4 + SOF_2 \rightarrow$
   (d) $XeF_2 + S_8 \rightarrow$
   (e) $XeF_4 + SF_2 \rightarrow$
6. If the Kr—F bond enthalpy is 50 kJ mol^{-1} and the F—F bond enthalpy is 159 kJ mol^{-1}, what would be the heat of formation of gaseous KrF_2 from the gaseous elements?
7. Some compounds have positive heats of formation so let us assume that the heat of formation of $ArF_2(g)$ is +100 kJ/mol. What would the energy of the Ar—F bond have to be to make it possible to produce $ArF_2(g)$ if the F—F bond enthalpy is 159 kJ mol^{-1} and the heat of formation could be no more positive than +100 kJ/mol?
8. Explain why $(SCl)_4N_4$ has the structure it has but that of $S_4(NH)_4$ is different.
9. Draw the structure for $S_2O_6F_2$. Predict some of the reactions for the compound.

10. One cation containing a ring of sulfur and nitrogen atoms is $S_3N_2Cl^+$. Draw the structure of this ion and describe the importance of resonance.

11. Supply the formula for a missing reactant and/or product that would complete the each of the equations indicated.
    (a) $SO_2 +$ _____ $\rightarrow SO_2Cl_2 +$ _____
    (b) $HCl +$ _____ $\rightarrow HF + H_2O +$ _____
    (c) $SO_2Cl_2 +$ _____ $\rightarrow SO_2(NH_2)_2 +$ _____
    (d) $NiS +$ _____ $\rightarrow NiO +$ _____
    (e) $HOSO_2Cl + H_2O_2 \rightarrow$ _____ $+$ _____

12. Discuss the reasons why the reaction $KF + ClSO_3H \rightarrow KCl + FSO_3H$ takes place.

13. Draw structures for the following species:
    (a) XeO_6^{4-}; (b) XeO_4; (c) XeF_3^+; (d) XeF_8^{2-}; (e) XeF_5^+

14. When SO_2 reacts with BF_3 it does so in a different way than when it reacts with $N(CH_3)_3$. Write the equations for the reactions and describe the products. What is the essential difference between the two cases?

15. Draw the structure for each of the following molecules:
    (a) $SeCl_4$
    (b) peroxydisulfuric acid
    (c) dithionic acid
    (d) SF_4
    (e) SO_3

16. Write an equation for the reaction of each compound listed in Question 15 with water.

17. Calculate the heats of formation for gaseous XeF_2 and XeF_4 if the Xe–F bond enthalpy is 133 kJ mol^{-1} and the F–F bond enthalpy is 159 kJ mol^{-1}.

18. Write equations for the processes listed.
    (a) The reaction of sulfuryl chloride with methanol.
    (b) The reaction of sulfur tetrafluoride with boron.
    (c) The reaction of sulfur with phosphorus trichloride
    (d) The reaction of sulfur with arsenic pentafluoride in liquid sulfur dioxide.
    (e) Roasting arsenic(III) sulfide.
    (f) The reaction of hydrogen chloride with sulfur trioxide.

19. Predict some of the ways in which the $S_2O_5^{2-}$ ion could conceivably bond to a metal ion when forming complexes.

20. If the compound S_4P_4 could be made, how would its structure likely differ from that of S_4N_4? Speculate on the products that would be formed if S_4P_4 were to undergo hydrolysis in basic solution.

21. What should be the enthalpy change for the reaction

$$Xe(g) + 2\ F_2(g) \rightarrow XeF_4(s)$$

    if the lattice energy of $XeF_4(s)$ is 62 kJ mol^{-1}?

22. Write balanced equations for each of the following processes:
    (a) Reaction of iodine pentachloride with water
    (b) Preparation of NaOCl
    (c) The reaction of ClF_3 with NOF
    (d) Disproportionation of OCl^-
    (e) Electrolysis of dilute sodium chloride solution

23. Explain why the reaction $2\ ClF_3 + Br_2 \rightarrow 2\ BrF_3 + Cl_2$ takes place.

24. Draw the structure of each of the following molecules:
   (a) Thiosulfuric acid
   (b) Thionyl fluoride
   (c) S_2Cl_2
   (d) H_2S_4
   (e) Difluoro disulfuric acid

25. By looking at Table 15.6, tell which of the oxyacids of sulfur are strong acids.

26. Explain how $SOCl_2$ could behave as a Lewis acid in two different ways.

27. Complete and balance the following:
   (a) $SO_2Cl_2 + C_2H_5OH \rightarrow$
   (b) $KBF_4 + SeO_3 \rightarrow$
   (c) $H_2S_2O_7 + H_2O \rightarrow$
   (d) $HOSO_2Cl + H_2O_2 \rightarrow$
   (e) $Sb_2S_3 + HCl \rightarrow$

28. Draw the molecular orbital energy level diagram for SO. What properties do you predict for this molecule?

29. Complete and balance the following:
   (a) $HNO_3 + S_8 \rightarrow$
   (b) $S_2Cl_2 + NH_3 \rightarrow$
   (c) $S_4N_4 + Cl_2 \rightarrow$
   (d) $SO_2 + PCl_5 \rightarrow$
   (e) $CaS_2O_6 \xrightarrow{\Delta}$

30. Complete and balance the following:
   (a) $ONF + ClF \rightarrow$
   (b) $Sb_2S_3 + HCl \rightarrow$
   (c) $Na_2CO_3 + S \rightarrow$
   (d) $HClO_3 + P_4O_{10} \rightarrow$
   (e) $KMnO_4 + HCl \rightarrow$

31. Write the equation for the reaction of chlorine with water and the analogous equation for the reaction of chlorine with liquid ammonia.

Chemistry of Coordination Compounds

CHAPTER 16

Introduction to Coordination Chemistry

The chemistry of coordination compounds comprises an area of chemistry that spans the entire spectrum from theoretical work on bonding to the synthesis of organometallic compounds. The essential feature of coordination compounds is that they involve *coordinate* bonds between Lewis acids and bases. Metal atoms or ions function as the Lewis acids, and the range of Lewis bases (electron pair donors) can include almost any species that has one or more unshared pairs of electrons. Electron pair donors include neutral molecules such as H_2O, NH_3, CO, phosphines, pyridine, N_2, O_2, H_2, and ethylenediamine ($H_2NCH_2CH_2NH_2$). Most anions, such as OH^-, Cl^-, $C_2O_4^{2-}$, and H^-, contain unshared pairs of electrons that can be donated to Lewis acids to form coordinate bonds. The scope of coordination chemistry is indeed very broad and interdisciplinary.

Some of the important types of coordination compounds occur in biological systems (e.g. heme and chlorophyll). There are also significant applications of coordination compounds that involve their use as catalysts. The formation of coordination compounds provides the basis for several techniques in analytical chemistry. Because of the relevance of this area, an understanding of the basic theories and principles of coordination chemistry is essential for work in many related fields of chemistry. In the next few chapters, an introduction to the basic principles of the chemistry of coordination compounds will be described.

16.1 STRUCTURES OF COORDINATION COMPOUNDS

Coordination compounds are also known as coordination complexes, complex compounds, or simply complexes. The essential feature of coordination compounds is that coordinate bonds form between electron pair donors, known as the *ligands*, and electron pair acceptors, the metal atoms or ions. The number of electron pairs donated to the metal is known as its *coordination number*. Although many complexes exist in which the coordination numbers are 3, 5, 7, or 8, the majority of complexes exhibit coordination numbers of 2, 4, or 6.

In order for a pair of electrons to be donated from a ligand to a metal ion, there must be an *empty* orbital on the metal ion to accept the pair of electrons. This situation is quite different from that where covalent bonds are being formed because in that case, one electron in a bonding pair comes from each of the atoms held by the bond. One of the first factors to be described in connection with the formation of coordinate bonds is that of seeing what type(s) of orbitals are available on the metal. If the metal ion is Zn^{2+}, the electron configuration is $3d^{10}$. Therefore, the 4s and 4p orbitals are empty and can be hybridized to give a set of four empty sp^3 hybrid orbitals. This set of hybrid orbitals could accommodate four pairs of electrons donated by ligands with the bonds pointing toward the corners **553**

Inorganic Chemistry. DOI: http://dx.doi.org/10.1016/B978-0-12-385110-9.00016-9

of a tetrahedron. Accordingly, it should be expected that $[Zn(NH_3)_4]^{2+}$ would be tetrahedral and that is correct.

In 1893, Alfred Werner proposed a theory to explain the existence of complexes such as $CoCl_3 \cdot 6NH_3$, $CoCl_3 \cdot 4NH_3$, $PtCl_2 \cdot 2NH_3$, and $Fe(CN)_3 \cdot 3KCN$. He began by assuming that a metal ion has two kinds of valence. The first, the *primary* valence, is satisfied by negative groups that balance the charge on the metal. For example, in $CoCl_3 \cdot 6NH_3$, the $+3$ valence of cobalt is satisfied by three Cl^- ions. A *secondary* valence is used to bind other groups that are usually specific in number. In the case above, the six NH_3 molecules satisfy the secondary valence of cobalt, so the coordination number of cobalt is 6. As a result of the NH_3 molecules being bonded directly to the cobalt ion, the formula is now written as $[Co(NH_3)_6]Cl_3$, where the square brackets are used to identify the actual complex that contains metal and the ligands bound directly to it.

In a complex such as $CoCl_3 \cdot 4NH_3$, the coordination number of 6 for cobalt is met by the four NH_3 molecules and two Cl^- ions being bonded directly to the cobalt. That leaves one Cl^- ion that satisfies part of the primary valence, but it is not bonded directly to the cobalt ion by means of a secondary valence. The formula for this coordination compound is written as $[Co(NH_3)_4Cl_2]Cl$. Support for these ideas is provided by dissolving the compounds in water and adding a solution containing Ag^+. In the case of $CoCl_3 \cdot 6NH_3$ or $[Co(NH_3)_6]Cl_3$, all of the chloride is immediately precipitated as AgCl. For $CoCl_3 \cdot 4NH_3$ or $[Co(NH_3)_4Cl_2]Cl$, only one third of the chloride precipitates as AgCl because two chloride ions are bound to the cobalt ion by coordinate bonds, the secondary valence. Two Cl^- ions satisfy both primary and secondary valences of cobalt, but one Cl^- ion satisfies only a primary valence. For the compound $[Co(NH_3)_3Cl_3]$, none of the chloride ions precipitate when Ag^+ is added because all of them are bound to the cobalt ion by coordinate bonds.

The electrical conductivity of solutions containing the complexes described above provides additional confirmation of the correctness of this view. For example, when $[Co(NH_3)_3Cl_3]$ is dissolved in water, the compound behaves as a nonelectrolyte. There are no ions present because the Cl^- ions are part of the coordinate structure of the complex. On the other hand, $[Co(NH_3)_6]Cl_3$, $[Co(NH_3)_5Cl]Cl_2$, and $[Co(NH_3)_4Cl_2]Cl$ behave as 1:3, 1:2, and 1:1 electrolytes, respectively. A similar situation exists for $[Pt(NH_3)_4Cl_2]Cl_2$, which behaves as a 1:2 electrolyte, and has only half of the chloride ions precipitated when Ag^+ is added to a solution of the complex. A large number of compounds having formulas such as $FeCl_3 \cdot 3KCl$ are known, and they are often called *double salts* because they consist of two complete formulas linked together. Many of the compounds are actually coordination compounds with formulas such as $K_3[FeCl_6]$.

As will be shown later, there are many complexes that have coordination numbers of 2 (linear complexes such as $[Ag(NH_3)_2]^+$), 4 (tetrahedral complexes such as $[CoCl_4]^{2-}$ or square planar complexes such as $[Pt(NH_3)_4]^{2+}$), or 6 (octahedral complexes such as $[Co(NH_3)_6]^{3+}$). A considerably smaller number of complexes having coordination numbers of 3 (trigonal planar), 5 (trigonal bipyramid or square-based pyramid), 7 (pentagonal bipyramid or capped trigonal prism), or 8 (cubic or anticubic, which is also known as Archimedes antiprism) are also known. These structures are shown in Figure 16.1.

In a tetrahedral structure, all the positions around the central atom are equivalent, so there is no possibility for geometrical or *cis/trans* isomerism. If all four groups bonded to the metal are different, there can be optical isomers. A compound such as $[Pd(NH_3)_2Cl_2]$ would exist in only one isomer if the

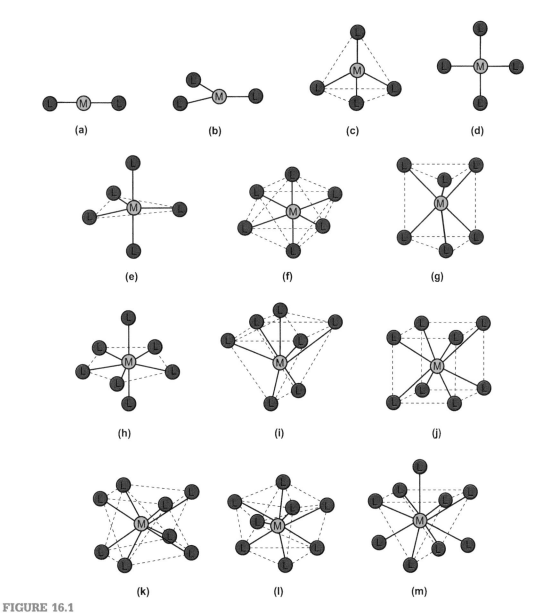

FIGURE 16.1

Some of the most common structures for coordination compounds: (a) linear, (b) trigonal planar, (c) tetrahedron, (d) square plane, (e) trigonal bipyramid, (f) octahedron, (g) trigonal prism, (h) pentagonal bipyramid, (i) single-capped trigonal prism, (j) cubic, (k) Archimedes (square) antiprism, (l) dodecahedron, and (m) triple-capped trigonal prism.

compound were tetrahedral. However, as shown below, two isomers of the compound exist because the bonding around the metal is square planar.

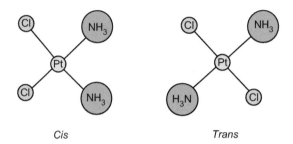

Cis   Trans

Alfred Werner isolated these two isomers, which showed conclusively that the complex is square planar rather than tetrahedral. The *cis* isomer is now known by the trade name *cisplatinol* or *cisplatin*, and it is used in treating certain forms of cancer.

If a complex has a coordination number of 6, there are several ways to place the ligands around the metal. A completely random arrangement is not expected because chemical bonds do not ordinarily form that way. Three regular types of geometry are possible for a complex containing six ligands. The six groups could be arranged in a planar hexagon (analogous to benzene), or the ligands could be arranged around the metal in a trigonal prism or in an octahedral structure. For a complex having the formula MX_4Y_2, these arrangements lead to different numbers of isomers as shown in Figure 16.2.

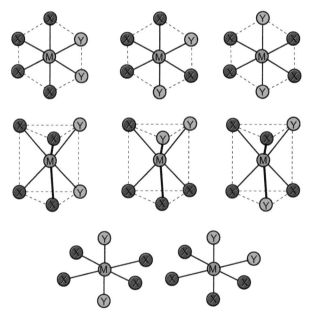

FIGURE 16.2
Geometrical isomers for MX_4Y_2 complexes having planar hexagonal, trigonal prism, and octahedral structures.

With the availability of modern experimental techniques, the structure of a compound can be established unequivocally. However, over 100 years ago, only chemical means were available for structure elucidation. The fact that only two isomers could be isolated was a good *indication* that a complex having the formula MX_4Y_2 has an octahedral structure. The argument could be made that the synthetic chemist was not able to prepare a third possible isomer or that it exists but was so unstable that it could not be isolated.

Although the majority of complexes have structures that are linear, tetrahedral, square planar, or octahedral, a few compounds have a trigonal bipyramid structure. Most notable of these are $Fe(CO)_5$, $[Ni(CN)_5]^{3-}$, and $[Co(CN)_5]^{3-}$. Some complexes having a coordination number of 5 have the square base pyramid structure including $[Ni(CN)_5]^{3-}$. Although it is not particularly common, the coordination number 8 is found in the complex $[Mn(CN)_8]^{4-}$, which has a cubic structure with CN^- ions on the corners.

Because the coordinate bonds are the result of Lewis acid–base interactions, the number of species that can form complexes with metal ions is large. Lewis bases such as H_2O, NH_3, F^-, Cl^-, Br^-, I^-, CN^-, SCN^-, and NO_2^- all form a wide range of coordination compounds. Added to these are compounds such as amines, arsines, phosphines, and carboxylic acids that are all potential ligands. The ethylenediamine molecule, $H_2NCH_2CH_2NH_2$, is a ligand that forms many very stable complexes because each nitrogen atom has an unshared pair of electrons that can be donated to a metal ion. This results in the ethylenediamine molecule being attached to the metal ion at two sites giving a ring with the metal ion being one of the members. A ring of this type is known as a *chelate* (pronounced "key-late") ring (from the Greek word *chelos* meaning "claw"). A complex that contains one or more chelate rings is called a *chelate complex* or simply a *chelate*. There are many other groups, known as *chelating agents*, that can bind to two sites. Table 16.1 shows several common ligands including several that form chelates.

16.2 METAL–LIGAND BONDS

Although the subject of stability of complexes will be discussed in greater detail in Chapter 19, it is appropriate to note here some of the general characteristics of the metal–ligand bond. One of the most relevant principles in this consideration is the hard–soft interaction principle. Metal–ligand bonds are acid–base interactions in the Lewis sense, so the principles discussed in Sections 9.6 and 9.8 apply to these interactions. Soft electron donors in which the donor atom is sulfur or phosphorus form more stable complexes with soft metal ions, such as Pt^{2+} and Ag^+, or with metal atoms. Hard electron donors such as H_2O, NH_3, or F^- generally form stable complexes with hard metal ions such as Cr^{3+} or Co^{3+}.

When the structures for many ligands (e.g. H_2O, NH_3, CO_3^{2-}, and $C_2O_4^{2-}$) are drawn, there is no question as to which atom is the electron pair donor. Ligands such as CO and CN^- normally bond to metals by donation of an electron pair from the carbon atom. It is easy to see why this is so when the structures are drawn for these species and the formal charges are shown.

$$|C \equiv O| \qquad |C \equiv N|$$

In each case, it is the carbon atom that has the negative formal charge that makes it the "electron-rich" end of the structure. Moreover, there are antibonding orbitals in both CO and CN^- that can accept electron density transferred from the nonbonding *d* orbitals on the metal (as described in

Table 16.1 Some of the Most Common Ligands

Group	Formula	Name
Water	H_2O	Aqua
Ammonia	NH_3	Ammine
Chloride	Cl^-	Chloro
Cyanide	CN^-	Cyano
Hydroxide	OH^-	Hydroxo
Thiocyanate	SCN^-	Thiocyanato
Carbonate	CO_3^{2-}	Carbonato
Nitrite	NO_2^-	Nitrito
Oxalate	$C_2O_4^{2-}$	Oxalato
Carbon monoxide	CO	Carbonyl
Nitric oxide	NO	Nitrosyl
Ethylenediamine	$H_2\ddot{N}CH_2CH_2\ddot{N}H_2$	Ethylenediamine
Acetylacetonate	$\overset{:O:}{\underset{}{CH_3-\overset{\|}{C}-CH=\overset{:\ddot{O}:^-}{\underset{\|}{C}}-CH_3}}$	Acetylacetonato
2,2'-dipyridyl		2,2'-dipyridyl
1,10-phenanthroline		1,10-phenanthroline

Section 16.10). Because of the nature of these ligands, it is not uncommon for them to function also as bridging groups in which they are bonded to two metal atoms or ions simultaneously.

Metal ions having high charge and small size are considered to be hard Lewis acids that bond best to hard electron donors. Complexes such as $[Co(NH_3)_6]^{3+}$ and $[Cr(H_2O)_6]^{3+}$ result from these hard–hard interactions. On the other hand, CO is a soft ligand, so it would not be expected that the complex $[Cr(CO)_6]^{3+}$ would be stable because Cr^{3+} is a hard Lewis acid. Complexes containing ligands such as CO would be expected for metals having low or zero charge because they are soft. As a general rule (with many known exceptions!), stable complexes are formed between metal ions having high charge and small size when they are bonded to hard Lewis bases. Stable complexes are generally formed between metals having low charge and large size when the ligands are soft. Keep in mind that when describing electronic character, the terms "hard" and "soft" are relative.

Table 16.2 Hard–Soft Classification of Metals (Lewis acids) and Ligands (Lewis bases)

Lewis Acids		Lewis Bases	
Hard	**Soft**	**Hard**	**Soft**
Al^{3+}, Be^{2+}	Ag^+, Cu^+	NH_3, OH^-	CO, I^-
Cr^{3+}, H^+	Pd^{2+}, Cd^{2+}	F^-, Cl^-	SCN^-, CN^-
Co^{3+}, Mg^{2+}	Hg^{2+}, Pt^{2+}	H_2O, SO_4^{2-}	RSH, R_2S
Fe^{3+}, Ti^{4+}	Au^+, Ir^+	ROH, PO_4^{3-}	C_2H_4, R_3P
Li^+, BF_3	Cr^0, Fe^0	RNH_2, en	R_3As, $SeCN^-$
	Borderline acids: Fe^{2+}, Co^{2+}, Ni^{2+}, Zn^{2+}, Cu^{2+}		
	Borderline bases: C_5H_5N, N_3^-, Br^-, NO_2^-		

Of the ligands shown in Table 16.1, the thiocyanate ion

$$\bar{S} = C = \bar{N}$$

is one that presents an obvious choice of bonding modes in which the electron pair donor atoms have different hard–soft character. As was mentioned in Chapter 9, SCN^- normally bonds to first-row transition metals through the nitrogen atom but to second- and third-row transition metals through the sulfur. However, the apparent hard character of a first-row metal ion can be altered by the other ligands present. For example, in $[Co(NH_3)_5NCS]^{2+}$, the thiocyanate bonds through the harder nitrogen atom, whereas in $[Co(CN)_5SCN]^{3-}$, the thiocyanate is bonded by the softer sulfur atom. In the first case, the Co^{3+} is hard and the presence of five hard NH_3 molecules does not change that character. In the second case, the hard Co^{3+}, when bonded to five CN^- ligands that are soft, has become soft as a result of the *symbiotic effect*. Ligands with the ability to bond by using different donor atoms are known as ambidentate ligands.

Because of the importance of being able to predict which complexes will be stable, it is necessary to know which ligands and metal species are soft and which are hard in terms of their electronic character. Table 16.2 gives a list of several species of each type grouped according to their hard–soft character. Some species are listed as borderline because they do not fit neatly into the hard or soft category.

16.3 NAMING COORDINATION COMPOUNDS

Because of the enormous number of coordination compounds that are known, it is essential to have a systematic way to deal with nomenclature. Many years ago, some complexes were named for their discoverers. For example, $K[C_2H_4PtCl_3]$ was known as Zeise's salt, $NH_4[Cr(NCS)_4(NH_3)_2]$ was known as Reinecke's salt, $[Pt(NH_3)_4][PtCl_4]$ was known as Magnus' green salt, and $[Pt(NH_3)_3Cl]_2[PtCl_4]$ was known as Magnus' pink salt. This procedure is not satisfactory if a large number of compounds must be named. As in other areas of chemistry, an elaborate system of nomenclature for inorganic compounds has been developed by the International Union of Pure and Applied Chemistry (IUPAC). The system for naming coordination compounds takes into account many types of compounds that are not encountered frequently enough in a beginning study of inorganic chemistry to warrant a thorough treatment of the formal rules. Presented here is a brief set of rules that will suffice for most purposes.

The number of rules that need to be followed in naming coordination compounds is small, but they are sufficient to permit naming the majority of complexes. The rules will be stated and then illustrated by working through several examples.

1. In naming a coordination compound, the cation is named first followed by the name of the anion. One or both may be complexes.
2. In naming a complex ion, the ligands are named in alphabetical order. A prefix that is used to indicate the number of ligands is not considered as part of the name of the ligand. For example, *trichloro* is named in the order indicated by the name *chloro*. However, if the ligand is diethylamine, $(C_2H_5)_2NH$, the prefix "di" is part of the ligand name, and it is used in determining alphabetical order.
   a. The names of any coordinated anions end in *o*. For example, Cl^- is chloro, CN^- is cyano, SCN^- is thiocyanato, etc.
   b. Neutral ligands are named by using their usual chemical names. For example, $H_2NCH_2CH_2NH_2$ is ethylenediamine, C_5H_5N is pyridine, etc. Four exceptions to this rule are H_2O, which is named as aqua, NH_3, which is named as ammine, CO, which is named as carbonyl, and NO, which is named as nitrosyl. Table 16.1 shows the names of several common ligands.
   c. Any coordinated cations end in *ium*. Few cases of this type are encountered, but the situation could arise with a ligand such as hydrazine, N_2H_4 (the structure of which is NH_2NH_2). One end could accept a proton, whereas the other end could coordinate to a metal. In that case, $NH_2NH_3^+$ would be named as hydrazinium.
3. To indicate the number of ligands of a particular type, the prefixes *di*, *tri*, *tetra*, etc. are used. If the name of the ligand contains one of these prefixes, the number of ligands is indicated by using the prefixes *bis*, *tris*, *tetrakis*, etc. For example, two ethylenediamine ligands would be indicated as *bis*(ethylenediamine) rather than diethylenediamine.
4. After naming the ligands, the name of the metal is given next with its oxidation state indicated by Roman numerals in parentheses with no spaces between the name of the metal and the parentheses.
5. If the complex ion containing the metal is an anion, the name of the metal ends in *ate*.

For certain types of complexes (particularly organometallic compounds), other rules are needed, but the majority of complexes can be correctly named using the short list of rules above. In the complex $[Co(NH_3)_6]Cl_3$, the cation is $[Co(NH_3)_6]^{3+}$, and it is named first. The coordinated ammonia molecules are named as ammine with the number of them being indicated by the prefix *hexa*. Therefore, the name for the compound is hexaamminecobalt(III) chloride. There are no spaces in the name of the cation. $[Co(NH_3)_5Cl]Cl_2$ has five NH_3 molecules and one Cl^- coordinated to Co^{3+}. Following the rules above leads to the name pentaamminechlorocobalt(III) chloride. Potassium hexacyanoferrate(III) is $K_3[Fe(CN)_6]$. Reinecke's salt, $NH_4[Cr(NCS)_4(NH_3)_2]$, would be named as ammonium diamminetetrathiocyanatochromate(III). In Magnus' green salt, $[Pt(NH_3)_4][PtCl_4]$, both cation and anion are complexes. The name of the complex is tetraammineplatinum(II) tetrachloroplatinate(II). The compound $[Co(en)_3](NO_3)_3$ is named as tris(ethylenediamine)cobalt(III) nitrate.

Some ligands contain more than one atom that can function as an electron pair donor. For example, SCN^- is known to bond to some metal ions through the nitrogen atom but to others through the sulfur atom. In some instances, this situation is indicated in the name as thiocyanato-*N*- and thiocyanato-*S*-. In some publications, the mode of attachment is indicated before the name, *N*-thiocyanato and *S*-thiocyanato.

Because some ligands contain more than one pair of electrons that can be donated to metal ions, it is possible for the same ligand to be bonded simultaneously to two metal centers. In other words, the ligands function as bridging groups. A bridging group is indicated by μ before the name of the ligand and separating the name of that group from the remainder of the complex by hyphens. $[(NH_3)_3Pt(SCN)Pt(NH_3)_3]Cl_3$ is named as hexaammine-μ-thiocyanatodiplatinum(II) chloride.

An additional aspect of nomenclature is the procedure of identifying the charge on a complex cation or anion by a number in parentheses after the name. The numbers indicating the charges are known as Ewens–Bassett numbers. Some examples showing both procedures for specifying charges and oxidation states are as follows:

$[Fe(CN)_6]^{3-}$	hexacyanoferrate(3−) or hexacyanoferrate(III)
$[Co(NH_3)_6]^{3+}$	hexaamminecobalt(3+) or hexamminecobalt(III)
$[Cr(H_2O)_6][Co(CN)_6]$	hexaaquachromium(3+) hexacyanocobaltate(3−) or hexaaquachromium(III) hexacyanocobaltate(III)

A more specialized set of rules for nomenclature is available as an appendix in the book by Huheey et al. (1993).

16.4 ISOMERISM

One of the interesting aspects of the chemistry of coordination compounds is the possibility of the existence of isomers. Isomers of a compound contain the same numbers and types of atoms, but they have different structures. Several types of isomerism have been demonstrated, but only a few of the most important types will be described here.

16.4.1 Geometrical Isomerism

The most common type of geometrical isomerism involves *cis* and *trans* isomers in square planar and octahedral complexes. If the complex MX_2Y_2 is tetrahedral, only one isomer exists because all the positions in a tetrahedron are equivalent. If the complex MX_2Y_2 is square planar, *cis* and *trans* isomers are possible as is shown below.

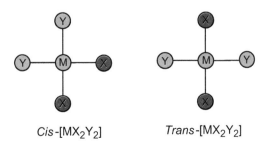

Cis-[MX$_2$Y$_2$]   *Trans*-[MX$_2$Y$_2$]

For an octahedral complex, all six positions are equivalent, so only one compound having the formula MX_5Y exists. For an octahedral complex having the formula MX_4Y_2, there will be two isomers. For $[Co(NH_3)_4Cl_2]^+$, the two possible isomers can be shown as

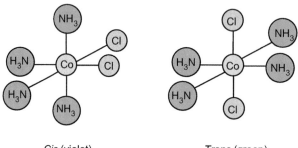

Cis (violet)   *Trans* (green)

If an octahedral complex has the formula MX_3Y_3, there are two possible isomers. In an octahedron, the positions are assigned numbers, so the locations of ligands in the structure can be identified. The usual numbering system for ligands in an octahedral complex is shown below.

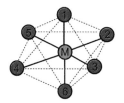

The two isomers of $[Co(NH_3)_3Cl_3]$ have structures that can be shown as follows.

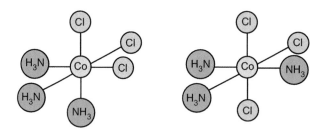

1,2,3- or *facial (fac)*   1,2,6- or *meridional (mer)*

In the *fac* isomer, the three chloride ions are located on the corners of one of the triangular faces of the octahedron. In the *mer* isomer, the three chloride ions are located around an edge (meridian) of the octahedron. The IUPAC system of nomenclature does not use this approach. A summary of the IUPAC procedures is presented in the book by Huheey et al. (1993).

Geometrical isomers are possible for complexes having a square-based pyramid structure. For example, the structures shown below illustrate *cis* and *trans* isomers for a complex having the formula MLX_2Y_2. These structures show that *cis* and *trans* arrangements are possible for the ligands in the base.

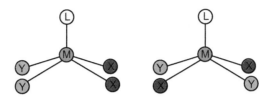

16.4.2 Optical Isomerism

For structures that do not possess a plane of symmetry, the mirror images are not superimposable. Known as *chiral* structures, such molecules rotate a beam of polarized light. If the beam is rotated to the right (when looking along the beam in the direction of propagation), the substance is said to be *dextrorotatory* (or simply *dextro*) and indicated by (+). Those substances that rotate the plane of polarized light to the left are said to be *levorotatory* or *levo* and indicated as (−). A mixture of equal amounts of the two forms is called a racemic mixture, and it produces no net rotation of the polarized light.

The dichlorobis(ethylenediamine)cobalt(II) ion can exist in two geometrical isomers. For the *trans* isomer, there is a plane of symmetry that bisects the cobalt ion and the ethylenediamine ligands leaving one Cl on either side of the plane. However, the *cis* isomer has no plane of symmetry, so two optical isomers exist. This is also the case for $[Co(en)_3]^{3+}$ as is illustrated in Figure 16.3.

Light behaves as though it consists of waves vibrating in all directions around the direction of wave propagation. For polarized light, the propagation can be regarded as a vector, which can be resolved into two circular vectors. If there is no rotation of the plane, it is expected that motion along each vector is equivalent so that each vector traverses an equal distance around the circle as shown in Figure 16.4.

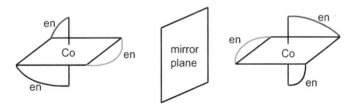

Nonsuperimposable mirror images of $[Co(en)_3]^{3+}$

Λ (lambda) isomer   Δ (delta) isomer

FIGURE 16.3
The optically active isomers of $[Co(en)_3]^{3+}$ (top).

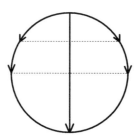

FIGURE 16.4
Polarized light represented as a vector with no rotation.

If polarized light passes through a medium that exhibits optical rotation, the motion along one of the circular vectors is slower than that of the other. The resultant vector is thus displaced from the original vector by some angle, ϕ. Figure 16.5 shows the vector model in which the phase difference is ϕ and α is one half of the phase difference.

The index of refraction of a medium, n, is the ratio of the velocity of light in a vacuum, c, to the velocity in the medium, v.

$$n = \frac{c}{v} \tag{16.1}$$

When a material exhibits different indices of refraction for the right and left-hand components of the circular vectors, the velocities in these directions are different and the plane of polarized light undergoes rotation. For the left- and right-hand vectors, the indices of refraction are

$$n_l = \frac{c}{v_l} \text{ and } n_r = \frac{c}{v_r} \tag{16.2}$$

which leads to the relationship

$$\frac{v_l}{v_r} = \frac{n_r}{n_l} \tag{16.3}$$

If we represent the length of the path of the polarized light in the sample as d and let ϕ be the phase difference in the two directions, we find that

$$\phi = \frac{2\pi d v}{v_r} - \frac{2\pi d v}{v_l} \tag{16.4}$$

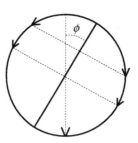

FIGURE 16.5
The rotation of polarized light represented as a rotated vector.

For light, the velocity and frequency are related by

$$\lambda \nu = c \tag{16.5}$$

Therefore, after simplification, we obtain

$$\phi = \frac{2\pi d}{\lambda_o}(n_r - n_l) \tag{16.6}$$

where λ_o is the wavelength of the light undergoing rotation. When expressed in terms of α, the relationship is

$$\alpha = \frac{\phi}{2} = \frac{\pi d}{\lambda_o}(n_r - n_l) \tag{16.7}$$

Solving for the difference in index of refraction, we obtain

$$(n_r - n_l) = \frac{\alpha \lambda_o}{\pi d} \tag{16.8}$$

When a solution is studied, the specific rotation of light having a wavelength of λ at a temperature t is written as $[\alpha]_\lambda^t$, and it is defined by the relationship

$$[\alpha]_\lambda^t = \frac{\alpha}{ds} = \frac{\alpha}{d\sigma\rho} \tag{16.9}$$

In this relationship, α is the observed rotation, d is the path length, s is the concentration of the solution in g solute/ml of solution, σ is the concentration in g solute/g of solution, and ρ is the density of the solution. The most commonly used light source is the sodium lamp, which has $\lambda = 589$ nm. Therefore, designations of a specific rotation are indicated by the use of symbols, such as $(+)_{589}-[Co(en)_3]^{3+}$, etc.

The discussion above makes use of the relationship between optical rotation and the index of refraction. However, the optical rotation varies with the wavelength of the light as does the index of refraction. When the variation of optical rotation with wavelength is studied, it is found that the curve undergoes a change in slope in the region of the maximum in an absorption band arising from an electronic transition (see Chapter 18). The change in rotation as a function of wavelength is known as *optical rotatory dispersion* (ORD). Figure 16.6 shows a schematic diagram that illustrates the change in rotation that occurs in the region of an absorption band. The rapid change in rotation at the wavelength where absorption by the complex occurs is known as the *Cotton effect* because of it having been discovered in 1895 by A. Cotton.

The diagram shown in Figure 16.6 illustrates a positive Cotton effect in which the change in rotation is from negative to positive as the wavelength is changed where the absorption occurs. A negative Cotton effect results when the rotation shifts from positive to negative at the wavelength range represented by the absorption band.

Compounds having the same optical configuration show similar Cotton effects. If the absolute configuration is known (e.g. from x-ray diffraction) for one optically active compound, a similar Cotton effect exhibited by another compound indicates that it has the same optical configuration as the known.

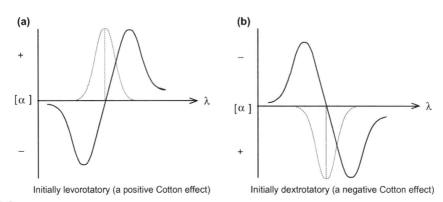

FIGURE 16.6

Variation of optical rotation with wavelength in the region of a hypothetical absorption band (shown as the dotted line).

In other words, if two compounds give electronic transitions that show Cotton effects that are the same (either both positive or both negative), the compounds have the same chirality or optical configuration. Although other methods for studying the absolute configuration of complexes exist, the methods described here have been widely used and are historically important. Consult the references at the end of this chapter for more details on the Cotton effect and ORD.

16.4.3 Linkage Isomerism

Linkage isomers result when a ligand can bond to metal ions in more than one way. Ligands having this ability are known as *ambidentate ligands* and include electron pair donors such as NO_2^-, CN^-, and SCN^- that have unshared pairs of electrons at two sites. The nitrite ion can bind to metal ions through both the nitrogen and oxygen atoms. The first case involving linkage isomers was studied in 1890s by S.M. Jørgensen, and the complexes were $[Co(NH_3)_5NO_2]^{2+}$ and $[Co(NH_3)_5ONO]^{2+}$. The second of these complexes (containing Co−ONO) is less stable, and it is converted into the −NO$_2$ isomer both in solution and the solid state either by heating or by exposure to ultraviolet light:

$$[Co(NH_3)_5ONO]^{2+} \rightarrow [Co(NH_3)_5NO_2]^{2+} \tag{16.10}$$
$$\text{red, nitrito} \qquad\qquad \text{yellow, nitro}$$

Many studies have been carried out on this reaction. Some of the unusual features of this reaction will be described in Chapter 20. Note that in writing the formulas for linkage isomers, it is customary to write the ligand with the atom that functions as the electron pair donor closest to the metal ion. Special consideration will be given in Chapter 20 to the behavior of cyanide complexes because CN^- is also an ambidentate ligand.

In Chapter 9, the hard−soft interaction principle was described as a guide to how electron donors and acceptors form bonds. This principle is particularly useful for some of the cases involving linkage isomerism. For example, SCN^- can bind to metals through either the sulfur or nitrogen atom. When thiocyanate complexes are prepared with metal ions such as Cr^{3+} or Fe^{3+} (hard Lewis acids), the bonding is through the nitrogen atom (harder electron donor). This arrangement (N-bonded SCN^-) is sometimes known as isothiocyanate bonding. However, thiocyanate bonds to Pd^{2+} and Pt^{2+} (soft Lewis acids) through the sulfur atom (softer electron donor). The potential exists for a thiocyanate ion that is coordinated in one way to change bonding mode under some conditions. However, this is sometimes dependent on the other groups coordinated to the metals as a result of either steric or electronic factors. The electronic factors will be discussed later in this chapter.

When the Lewis structures are drawn for SCN^-, the principal resonance structure is

$$\bar{\underline{S}}=C=\bar{\underline{N}}$$

Coordination compounds containing SCN^- appear to involve the structure

$$|\underline{S}-C\equiv N|$$

that leads to angular bonds when sulfur donates electrons and linear bonds when nitrogen is the electron pair donor. The result is that SCN^- forms bonds to metal ions that can be shown as

Rotation around the M−L bonds requires no spatial considerations for the linear N-bonded SCN^- but structure II shows a different situation for S-bonded SCN^-. In that case, which can be shown as

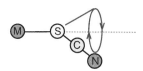

a volume represented as a cone (sometimes referred to as a *cone of revolution*) is swept out as the rotation occurs. The presence of bulky ligands can inhibit this rotation and lead to a change in bonding mode. Such a case involves the platinum complex $[Pt((C_6H_5)_3As)_3SCN]^+$, which contains large triphenylarsine groups, and it is easily converted to $[Pt((C_6H_5)_3As)_3NCS]^+$.

16.4.4 Ionization Isomerism

Although the compounds $[Pt(en)_2Cl_2]Br_2$ and $[Pt(en)_2Br_2]Cl_2$ have the same empirical formulas, they are quite different compounds. For example, the first gives Br^- when dissolved in water, whereas the second gives Cl^-. This occurs because in the first case, the Cl^- ions are coordinated to the Pt^{4+}, whereas in the second case, the Br^- ions are coordinated to the metal ion. The isomerism in cases such as these is known as *ionization isomerism*. It is easy to see that many pairs of compounds could be considered as ionization isomers among them are the following examples:

$[Cr(NH_3)_4ClBr]NO_2$	and	$[Cr(NH_3)_4ClNO_2]Br$
$[Co(NH_3)_4Br_2]Cl$	and	$[Co(NH_3)_4ClBr]Br$
$[Cr(NH_3)_5Cl]NO_2$	and	$[Co(NH_3)_5NO_2]Cl$

16.4.5 Coordination Isomerism

Coordination isomerism refers to cases where there are different ways to arrange several ligands around two metal centers. For example, there are several ways to arrange six CN^- ions and six NH_3 molecules around two metal ions that total +6 in oxidation number. One way is $[Co(NH_3)_6][Co(CN)_6]$, but $[Co(NH_3)_5CN][Co(NH_3)(CN)_5]$ and $[Co(NH_3)_4(CN)_2][Co(NH_3)_2(CN)_4]$ also have the same composition. Other examples of coordination isomerism are the following:

$[Co(NH_3)_6][Cr(CN)_6]$	and	$[Cr(NH_3)_6][Co(CN)_6]$
$[Cr(NH_3)_5CN][Co(NH_3)(CN)_5]$	and	$[Co(NH_3)_5CN][Cr(NH_3)(CN)_5]$

16.4.6 Hydrate Isomerism

A great many metal complexes are prepared by reactions carried out in aqueous solutions. Consequently, solid complexes are frequently obtained as hydrates. Water is also a potential ligand, so various possibilities exist for compounds to be prepared with water held in both ways. For example, $[Cr(H_2O)_4Cl_2]Cl \cdot 2H_2O$ and $[Cr(H_2O)_5Cl]Cl_2 \cdot H_2O$ have the same formula, but they are obviously different compounds. In the first case, two chloride ions are coordinated and one is present as an anion, whereas in the second case, the numbers are reversed. Many other examples of hydrate isomerism are known.

16.4.7 Polymerization Isomerism

Polymers are high-molecular weight materials that consist of smaller units (monomers) that become bonded together. In coordination chemistry, it is possible to have two or more compounds that have the same empirical formula but different molecular weights. For example, $[Co(NH_3)_3Cl_3]$ consists of one cobalt ion, three ammonia molecules, and three chloride ions. This is the same ratio of these species as is found in $[Co(NH_3)_6][CoCl_6]$, which has a molecular weight that is twice that of $[Co(NH_3)_3Cl_3]$. Other compounds that have the same empirical formula are $[Co(NH_3)_5Cl][Co(NH_3)Cl_5]$ and $[Co(NH_3)_4Cl_2][Co(NH_3)_2Cl_4]$. These compounds are called polymerization isomers of $[Co(NH_3)_6][CoCl_6]$, but there is no similarity to the case of polymerization of monomer units that results in a material of higher molecular weight. Polymerization isomerism is not a very accurate way to describe such materials although the term has been in use for many years.

16.5 A SIMPLE VALENCE BOND DESCRIPTION OF COORDINATE BONDS

One of the goals in describing any structure on a molecular level is to interpret how the bonding arises by making use of atomic orbitals. In the area of coordination chemistry, a simple valence bond approach is successful in explaining some of the characteristics of the complexes. In this approach, the empty orbitals on the metal ions are viewed as hybrid orbitals in sufficient number to accommodate the number of electron pairs donated by the ligands. It is generally true that first-row

Table 16.3 Hybrid Orbital Types in Coordination Compounds

Atomic Orbitals	Hybrid Type	Number of Orbitals	Structure
s,p	sp	2	Linear
s,d	sd	2	Linear
s,p,p	sp^2	3	Trigonal planar
s,p,p,p	sp^3	4	Tetrahedral
s,d,d,d	sd^3	4	Tetrahedral
d,s,p,p	dsp^2	4	Square planar
d,s,p,p,p	dsp^3	5	Trigonal bipyramid
s,p,p,d,d	sp^2d^2	5	Square base pyramid
d,d,s,p,p,p	d^2sp^3	6	Octahedral
s,p,p,p,d,d	sp^3d^2	6	Octahedral
s,p,d,d,d,d	spd^4	6	Trigonal prism
s,p,p,p,d,d,d	sp^3d^3	7	Pentagonal bipyramid
s,p,p,p,d,d,d,d	sp^3d^4	8	Dodecahedron
s,p,p,p,d,d,d,d	sp^3d^4	8	Archimedes antiprism
s,p,p,p,d,d,d,d,d	sp^3d^5	9	Capped trigonal prism

In dsp^2, the $d_{x^2-y^2}$ orbital is used. In sp^2d^2 and d^2sp^3, the $d_{x^2-y^2}$ and d_{z^2} orbitals are used.

transition metals form complexes that in many cases contain six coordinate bonds. This is especially true if the charge on the metal ion is +3. The reason for this is that a +3 metal ion has a high charge to size ratio that exerts a strong attraction for pairs of electrons. If the metal ion has a +2 charge, it is sometimes found that only four ligands will bind to the metal ion, but it also depends on the availability of empty orbitals on the metal ion. In this case, the charge to size ratio is considerably smaller (the charge is only 67% as high and the size is usually larger), so the metal ion has a much smaller affinity for electron pairs. This is the case for +2 ions such as Cu^{2+} and Zn^{2+}, which form many complexes in which the coordination number is 4 although as we shall see, there are other factors involved.

The simplified problem in describing the structure of complexes becomes one of deciding what hybrid orbital type will accommodate the number of electron pairs donated by the ligands and also correspond to the known structure of the complex. Some of the structures associated with specific types of hybrid orbitals were presented in Figure 4.1, but there are other cases that apply to complexes of metal ions. In addition to regular geometrical structures, there are numerous complexes that have irregular structures. In some cases, Jahn–Teller distortion causes the structure to deviate from ideal geometry (see Chapter 17). In spite of the difficulties, it is useful to know the hybrid orbital types associated with complexes having various structures. Table 16.3 gives a summary of the major hybrid orbital types that are applicable to complexes.

When Zn^{2+} forms a complex such as $[Zn(NH_3)_4]^{2+}$, it is easy to rationalize the bonding in terms of hybrid orbitals. Zn^{2+} has a d^{10} configuration, but the $4s$ and $4p$ orbitals are empty. Therefore, a set of four empty sp^3 hybrids can result. As shown below, four orbitals can accept four pairs of electrons that are donated by the ligands.

$$
\begin{array}{ccc}
& 3d & 4s & 4p \\
& \uparrow\downarrow \; \uparrow\downarrow \; \uparrow\downarrow \; \uparrow\downarrow \; \uparrow\downarrow & & \\
Zn^{2+} & \underline{\quad}\;\underline{\quad}\;\underline{\quad}\;\underline{\quad}\;\underline{\quad} & \underline{\quad} & \underline{\quad}\;\underline{\quad}\;\underline{\quad}
\end{array}
$$

$$
\begin{array}{cc}
& \uparrow\downarrow \; \uparrow\downarrow \; \uparrow\downarrow \; \uparrow\downarrow \\
sp^3 & \underline{\quad}\;\underline{\quad}\;\underline{\quad}\;\underline{\quad} \quad [Zn(NH_3)_4]^{2+}
\end{array}
$$

Likewise, it is easy to rationalize how Ag^+ (a d^{10} ion) can form a linear complex with two ligands such as NH_3.

$$
\begin{array}{ccc}
& 4d & 5s & 5p \\
& \uparrow\downarrow \; \uparrow\downarrow \; \uparrow\downarrow \; \uparrow\downarrow \; \uparrow\downarrow & & \\
Ag^+ & \underline{\quad}\;\underline{\quad}\;\underline{\quad}\;\underline{\quad}\;\underline{\quad} & \underline{\quad} & \underline{\quad}\;\underline{\quad}\;\underline{\quad}
\end{array}
$$

$$
\begin{array}{cc}
& \uparrow\downarrow \; \uparrow\downarrow \\
sp & \underline{\quad}\;\underline{\quad} \quad [Ag(NH_3)_2]^{+}
\end{array}
$$

When the scheme above is examined, it is easy to see that if two of the $5p$ orbitals are used, a set of sp^2 hybrid orbitals could be produced that would accommodate pairs of electrons from three ligands. It is not surprising that only a few complexes are known in which Ag^+ exhibits a coordination number of 3. If all the $5p$ orbitals were used in the hybrids, the resulting sp^3 hybrids could accommodate four pairs of electrons and a tetrahedral complex could result. Although the complexes $[Ag(CN)_3]^{2-}$ and $[Ag(CN)_4]^{3-}$ have been identified in solutions in which the concentration of CN^- is high, such complexes are unusual and of low stability. The vast majority of complexes of Ag^+ have a coordination number of 2 and linear structures. Part of the reason for this is that the silver ion is relatively large and has a low charge, which results in a low charge density. Such an ion does not assimilate the electron density from added ligands as readily as do those for which the charge density is high.

For metal ions having configurations d^0, d^1, d^2, or d^3, there will always be two of the d orbitals empty to form a set of d^2sp^3 hybrids. Therefore, we expect complexes of these metal ions to be octahedral in which the hybrid orbital type is d^2sp^3. If we consider Cr^{3+} as an example, the formation of a complex can be shown as follows:

$$
\begin{array}{ccc}
& 3d & 4s & 4p \\
& \uparrow\;\;\uparrow\;\;\uparrow & & \\
Cr^{3+} & \underline{\quad}\;\underline{\quad}\;\underline{\quad}\;\underline{\quad}\;\underline{\quad} & \underline{\quad} & \underline{\quad}\;\underline{\quad}\;\underline{\quad}
\end{array}
$$

$$
\begin{array}{cc}
& \uparrow\downarrow \; \uparrow\downarrow \; \uparrow\downarrow \; \uparrow\downarrow \; \uparrow\downarrow \; \uparrow\downarrow \\
d^2sp^3 & \underline{\quad}\;\underline{\quad}\;\underline{\quad}\;\underline{\quad}\;\underline{\quad}\;\underline{\quad} \quad [Cr(H_2O)_6]^{3+}
\end{array}
$$

In an octahedral complex, the coordinate system is set up so that the ligands lie on the axes. The d orbitals that have lobes lying along the axes are the d_{z^2} and the $d_{x^2-y^2}$. The d_{xy}, d_{yz}, and d_{xz} orbitals directed between the axes are considered to be nonbonding. It should not be inferred that only octahedral complexes are produced by Cr^{3+}. Although tetrahedral complexes would not be expected in solutions that contain Cr^{3+} because of its high charge density, solid $[PCl_4][CrCl_4]$ contains the tetrahedral $CrCl_4^-$ ion. With the $4s$ and $4p$ orbitals being empty, it is easy to see how sp^3 hybrids are obtained.

When the number of electrons in the d orbitals is four, as in the case of Mn^{3+}, there exists more than one possible type of hybrid orbital. For example, if the electrons remain unpaired in the d orbitals, there is only one orbital in the set that is empty. As a result, if an octahedral complex is formed, making use of two d orbitals requires that the $4d$ orbitals be used so that sp^3d^2 hybrid orbitals would be used by the metal. This case can be shown as follows:

$$
\begin{array}{ccccc}
 & 3d & 4s & 4p & 4d \\
 & \uparrow\ \uparrow\ \uparrow\ \uparrow & & & \\
\text{Mn}^{3+} & \underline{\ }\ \underline{\ }\ \underline{\ }\ \underline{\ }\ \underline{\ } & \underline{\ } & \underline{\ }\ \underline{\ }\ \underline{\ } & \underline{\ }\ \underline{\ }
\end{array}
$$

$$
\begin{array}{c}
\uparrow\downarrow\ \uparrow\downarrow\ \uparrow\downarrow\ \uparrow\downarrow\ \uparrow\downarrow\ \uparrow\downarrow \\
sp^3d^2\ \underline{\ }\ \underline{\ }\ \underline{\ }\ \underline{\ }\ \underline{\ }\ \underline{\ } \qquad [\text{Mn}(\text{H}_2\text{O})_6]^{3+}
\end{array}
$$

This bonding arrangement results in four unpaired electrons in the $3d$ orbitals, but with some ligands, the situation is different. In those cases, two of the $3d$ orbitals are made available to form a set of hybrid orbitals by means of electron pairing. When this occurs, the bonding arrangement can be shown as follows:

$$
\begin{array}{ccccc}
 & 3d & 4s & 4p & 4d \\
 & \uparrow\downarrow\ \uparrow\ \uparrow & & & \\
\text{Mn}^{3+} & \underline{\ }\ \underline{\ }\ \underline{\ }\ \underline{\ }\ \underline{\ } & \underline{\ } & \underline{\ }\ \underline{\ }\ \underline{\ } & \underline{\ }\ \underline{\ }
\end{array}
$$

$$
\begin{array}{c}
\uparrow\downarrow\ \uparrow\downarrow\ \uparrow\downarrow\ \uparrow\downarrow\ \uparrow\downarrow\ \uparrow\downarrow \\
d^2sp^3\ \underline{\ }\ \underline{\ }\ \underline{\ }\ \underline{\ }\ \underline{\ }\ \underline{\ } \qquad [\text{Mn}(\text{NO}_2)_6]^{3-}
\end{array}
$$

Only two unpaired electrons are present in this complex, and as will be shown in the next section, the two types of manganese complexes can be distinguished by means of their magnetic character. We have not explained *why* electron pairing occurs with some ligands but not others. The discussion of this aspect of bonding in complexes will be treated fully in Chapter 17.

In $[\text{Mn}(\text{H}_2\text{O})_6]^{3+}$, the $4d$ orbitals used to form hybrids are those outside the usual valence shell that consists of $3d$, $4s$, and $4p$ orbitals. Consequently, such a complex is often referred to as an *outer orbital* complex. To identify the hybrid orbitals, the symbol sp^3d^2 is used to indicate that the d orbitals are part of the shell with $n = 4$, and they follow the s and p orbitals in filling. In $[\text{Mn}(\text{NO}_2)_6]^{3-}$, the d orbitals are those in the valence shell, so the complex is called an *inner orbital* complex, and the hybrid orbital type is designated as d^2sp^3 to show that the principle quantum number of the d orbitals is lower than that of the s and p orbitals. Another way in which the two types of complexes are distinguished is by the terms *high spin* and *low spin*. In $[\text{Mn}(\text{H}_2\text{O})_6]^{3+}$, there are four unpaired electron spins, whereas in $[\text{Mn}(\text{NO}_2)_6]^{3-}$, there are only two unpaired electrons. Therefore, the former is referred to as a high-spin complex, and the latter is designated as a low-spin complex.

For a d^5 ion such as Fe^{3+}, the five electrons may be unpaired in the set of five d orbitals, or they may be present as two orbitals occupied by pairs of electrons and one orbital having single occupancy. A complex such as $[\text{Fe}(\text{H}_2\text{O})_6]^{3+}$ is typical of the first bonding mode (five unpaired electrons, high spin, outer orbital), whereas $[\text{Fe}(\text{CN})_6]^{3-}$ is typical of the second (one unpaired electron, low spin, inner orbital). Three orbitals can accommodate six electrons, so a d^6 ion such as Co^{3+} should form two series of complexes in which the six electrons occupy all five orbitals with two in one orbital and one in each of the others. This results in four unpaired electrons being found in the outer orbital, high-spin complex $[\text{CoF}_6]^{3-}$. On the other hand, a low-spin inner orbital complex such as $[\text{Co}(\text{NH}_3)_6]^{3+}$ has no unpaired electrons because all six of the $3d$ electrons are paired in three of the orbitals.

When the d^7 and d^8 electron configurations are considered, there are additional possibilities for forming bonding orbitals. With seven or eight electrons in the five orbitals, there is no possibility of having two of the $3d$ orbitals vacant, so if an octahedral complex is formed, it must be of the high spin outer orbital type. Both Co^{2+} (d^7) and Ni^{2+} (d^8) form many complexes of this type. With all the $3d$ orbitals being populated (not necessarily *filled*), the possibility exists for using the $4s$ and $4p$ orbitals in forming tetrahedral complexes and $[CoCl_4]^{2-}$ and $[Ni(NH_3)_4]^{2+}$ have tetrahedral structures. Although it is not possible to have two of the $3d$ orbitals vacant, it is certainly possible to have one empty so that the possibility of the dsp^3 hybrid orbital type exists. With regard to the orbitals used, the Co^{2+} ion behaves as follows:

$$\begin{array}{ccccc} & 3d & 4s & 4p & 4d \\ Co^{2+} & \uparrow\downarrow\ \uparrow\downarrow\ \uparrow\downarrow\ \uparrow & & & \end{array}$$

$$dsp^3 \quad \uparrow\downarrow\ \uparrow\downarrow\ \uparrow\downarrow\ \uparrow\downarrow\ \uparrow\downarrow \qquad [Co(CN)_5]^{3-}$$

For the d^8 Ni^{2+} ion, the orbitals and bonding can be shown as follows:

$$\begin{array}{ccccc} & 3d & 4s & 4p & 4d \\ Ni^{2+} & \uparrow\downarrow\ \uparrow\downarrow\ \uparrow\downarrow\ \uparrow\downarrow & & & \end{array}$$

$$dsp^3 \quad \uparrow\downarrow\ \uparrow\downarrow\ \uparrow\downarrow\ \uparrow\downarrow\ \uparrow\downarrow \qquad [Ni(CN)_5]^{3-}$$

If a complex having a coordination number of 4 is produced, only two of the $4p$ orbitals are used and the hybrid orbital type is dsp^2, which is characteristic of a square planar complex. For Ni^{2+}, this is illustrated by the following scheme:

$$\begin{array}{ccccc} & 3d & 4s & 4p & 4d \\ Ni^{2+} & \uparrow\downarrow\ \uparrow\downarrow\ \uparrow\downarrow\ \uparrow\downarrow & & & \end{array}$$

$$dsp^2 \quad \uparrow\downarrow\ \uparrow\downarrow\ \uparrow\downarrow\ \uparrow\downarrow \qquad [Ni(CN)_4]^{2-}$$

Both Co^{2+} and Ni^{2+} also have the ability to make use of $4s$, $4p$, and $4d$ orbitals in another way by forming sp^2d^2 hybrids in the formation of complexes having a square-based pyramid structure. In fact, trigonal bipyramid and square-based pyramid structures are observed for $[Ni(CN)_5]^{2-}$.

Although the simple valence bond approach to the bonding in coordination compounds has many deficiencies, it is still useful as a first attempt to explain the structure of many complexes. The reasons why certain ligands force electron pairing will be explored in Chapter 17, but it is clear that high and low-spin complexes have different magnetic character, and the interpretation of the results of this technique will now be explored.

16.6 MAGNETISM

In the previous sections, we have shown how a complex of a d^4 metal ion such as Mn^{3+} can form complexes utilizing sp^3d^2 hybrid orbitals or in other cases make use of d^2sp^3 hybrid orbitals. In

Chapter 17, it will be described why the nature of the bonding depends on the type of ligands present. For the time being, we will concentrate on the difference in the number of unpaired electrons that will be found in each type of complex. In a complex of Mn^{3+} where the bonding is sp^3d^2, the 3d orbitals are not used, so all four of the electrons will remain unpaired in the set of five orbitals. If the complex makes use of d^2sp^3 orbitals, two of the set of five 3d orbitals will be available only if two of the four electrons occupy one orbital and the other two electrons are found singly in two other orbitals. As a result, there will be two unpaired electrons if the bonding is d^2sp^3, but there will be four unpaired electrons if the orbitals used are sp^3d^2 hybrids. We must now develop the relationships needed to be able to use the magnetic character of a complex to determine the number of unpaired electrons it has.

To determine the magnetic character of a sample of a compound, it is weighed using a balance that allows the mass of the sample to be determined with a magnetic field either on or off. If there are unpaired electrons in the sample, it will interact with the magnetic field as a result of the magnetic field generated by the spinning electrons. If the sample is attracted to the magnetic field, it is said to be *paramagnetic*. A sample that has no unpaired electrons will be weakly repelled by the magnetic field and it is said to be *diamagnetic*. When any substance is placed in a magnetic field, there is an induced magnetization that is opposite to the field that produced it. This results in a weak repulsion between the sample and the magnetic field that is corrected for by subtracting the so-called diamagnetic contribution. The resulting magnetism of the sample is measured by the molar magnetic susceptibility, χ_M, which is related to the magnetic moment, μ, by the relationship

$$\mu = (3k/N_o)\chi_M T \tag{16.11}$$

In this equation, N_o is Avogadro's number, k is Boltzmann's constant, and T is the temperature (K). Magnetism is measured in terms of a unit known as the Bohr magneton, μ_o, (BM) that is given as

$$\mu_o = eh/4\pi mc \tag{16.12}$$

where e and m are the charge and mass of an electron, h is Planck's constant, and c is the velocity of light. An electron has an intrinsic spin of 1/2, which generates a magnetic moment that is expressed as

$$\mu_s = g[s(s+1)]^{1/2} = 2[1/2(1/2+1)]^{1/2} \tag{16.13}$$

In this equation, μ_s is the magnetic moment resulting from a spinning electron and g is the Landé "g" factor or the gyromagnetic ratio that gives the ratio of the spin magnetic moment to the spin angular momentum vector, $[s(s+1)]^{1/2}$. For a free electron, g is 2.00023, which is normally taken to be 2.00, so for one electron, μ_s has the value given by

$$\mu_s = 2[1/2(1/2+1)]^{1/2} = 3^{1/2} = 1.73 \tag{16.14}$$

As a result of orbital motion, an additional magnetic effect is produced, and taking it into account leads to the total magnetic moment for one electron,

$$\mu_{s+l} = [4s(s+1) + l(l+1)]^{1/2} \tag{16.15}$$

Table 16.4 Magnetic Moments Expected for Spin-Only Contributions

Unpaired Electrons	S	Calculated μ_s, BM	Example Complex	Experimental μ, BM
1	1/2	1.73	$(NH_4)_3TiF_6$	1.78
2	1	2.83	$K_3[VF_6]$	2.79
3	3/2	3.87	$[Cr(NH_3)_6]Br_3$	3.77
4	2	4.90	$[Cr(H_2O)_6]SO_4$	4.80
5	5/2	5.92	$[Fe(H_2O)_6]Cl_3$	5.90

where s is the spin of the electron and l is its angular momentum quantum number. When the species has more than one electron, the total magnetic moment makes use of the sum of spins, S, and orbital contributions, L, as in Russell–Saunders or L–S coupling to yield

$$\mu_{S+L} = [4S(S + 1) + L(L + 1)]^{1/2} \tag{16.16}$$

Although this equation gives the maximum magnetism, the observed value is usually somewhat smaller than the predicted value. For many complexes, the orbital contribution to the magnetism is small and can be ignored to give the so-called *spin-only* magnetic moment, which is adequate for analyzing the magnetism when the object is to obtain the number of unpaired electrons in the complex. When the orbital contributions are omitted and the sum of spins is represented as $S = n/2$, where n is the number of unpaired electrons, the spin-only magnetic moment, μ_S, is given by

$$\mu_S = \mu_0[n(n + 2)]^{1/2} \tag{16.17}$$

From this equation, the magnetic moments expected for complexes can be listed in terms of the number of unpaired electrons as shown in Table 16.4.

In many cases, there is a considerable difference between the spin-only magnetic moment and the experimental values. This is especially true for complexes that contain ligands such as cyanide that are capable of substantial π-bonding. The changes in the orbitals are such that the actual magnetic moments are usually lower than those expected on the basis of spin-only contributions.

16.7 A SURVEY OF COMPLEXES OF FIRST-ROW METALS

When examining the enormous range of complexes that are known, it sometimes seems as if any metal can form any type of complex. This is not necessarily the case, and there are some general trends that are followed. It must be emphasized that these are only rough guidelines and there are numerous exceptions. Complexes form as a result of the attraction of a metal ion for electron pairs on the ligands (electron pair donors or nucleophiles). One of the most important considerations is what types of orbitals are available on the metal ion to accept electron pairs. If a metal ion has at least two empty d orbitals, it is possible to form a set of d^2sp^3 hybrid orbitals so that six pairs of electrons can be accommodated. Therefore, for ions that have the d^0, d^1, d^2, and d^3 configurations, complexes having a coordination number of 6 and octahedral geometry are the norm.

A second important consideration in the formation of complexes is the charge to size ratio of the metal ion. Ions of high charge and small size (high charge density) have the greatest affinity for electron pairs and can generally accommodate the greater build up of negative charge that results from the acceptance of electron pairs. As a general rule, metal ions of high charge tend to form complexes in which the coordination numbers are higher especially if they are larger ions from the second and third transition series. In Chapter 8, it was shown that ions having high charge and small size are hard Lewis acids, so interactions with hard bases will be more favorable.

In view of the number and types of hybrid orbitals that are exhibited by ions of the transition metals when complexes are formed, it possible to arrive at a summary of the most important types of complexes that each metal ion should form. Although there are many exceptions, the summary presented in Table 16.5 is a useful starting point for considering the types of complexes that are formed by metals in the first transition series.

Table 16.5 A Summary of Types of Complexes Formed by First-Row Metal Ions

Number of d Electrons	Most Common Ion	Usual Geometry[b]	Hybrid Orbital	Example
0	Sc^{3+}	Octahedral	d^2sp^3	$Sc(H_2O)_6^{3+}$
1	Ti^{3+}	Octahedral	d^2sp^3	$Ti(H_2O)_6^{3+}$
2	V^{3+}	Octahedral	d^2sp^3	VF_6^{3-}
3	Cr^{3+}	Octahedral	d^2sp^3	$Cr(NH_3)_6^{3+}$
4[a]	Mn^{3+}	Octahedral (h.s.)	sp^3d^2	$Mn(H_2O)_6^{3+}$
4[a]	Mn^{3+}	Octahedral (l.s.)	d^2sp^3	$Mn(CN)_6^{3-}$
5	Fe^{3+}	Octahedral (h.s.)	sp^3d^2	$Fe(H_2O)_6^{3+}$
5	Fe^{3+}	Octahedral (l.s.)	d^2sp^3	$Fe(CN)_6^{3-}$
5	Fe^{3+}	Tetrahedral	sp^3	$FeCl_4^{-}$
6	Co^{3+}	Octahedral (h.s.)	sp^3d^2	CoF_6^{3-}
6	Co^{3+}	Octahedral (l.s.)	d^2sp^3	$Co(H_2O)_6^{3+}$
7	Co^{2+}	Octahedral (h.s.)	sp^3d^2	$Co(H_2O)_6^{2+}$
7	Co^{2+}	Trigonal bipyramid	dsp^3	$Co(CN)_5^{3-}$
7	Co^{2+}	Tetrahedral	sp^3	$CoCl_4^{2-}$
7	Co^{2+}	Square planar	dsp^2	$Co(CN)_4^{2-}$
8	Ni^{2+}	Tetrahedral	sp^3	$Ni(NH_3)_4^{2+}$
8	Ni^{2+}	Octahedral	sp^3d^2	$Ni(NH_3)_6^{2+}$
8	Ni^{2+}	Trigonal bipyramid	dsp^3	$Ni(CN)_5^{3-}$
8	Ni^{2+}	Square base pyramid	sp^2d^2	$Ni(CN)_5^{3-}$
8	Ni^{2+}	Square planar	dsp^2	$Ni(CN)_4^{2-}$
9[a]	Cu^{2+}	Octahedral	sp^3d^2	$CuCl_6^{4-}$
9[a]	Cu^{2+}	Tetrahedral	sp^3	$Cu(NH_3)_4^{2+}$
10	Zn^{2+}	Octahedral	sp^3d^2	$Zn(H_2O)_6^{2+}$
10	Zn^{2+}	Tetrahedral	sp^3	$Zn(NH_3)_4^{2+}$
10	Ag^{+}	Linear	sp	$Ag(NH_3)_2^{+}$

[a]Irregular structure due to Jahn–Teller distortion.
[b]h.s. and l.s. indicate high and low spin, respectively.

16.8 COMPLEXES OF SECOND- AND THIRD-ROW METALS

Much of what has been said so far in this chapter applies equally well to complexes of second- and third-row transition metals. However, there are some general differences that result from the fact that atoms and ions of the second- and third-row metals are larger in size than those of the first-row metals. For example, because of their larger size (when in the same oxidation state as a first-row ion), ions of metals in the second and third rows form many more complexes in which they have a coordination number >6. Whereas chromium usually has a coordination number of 6, molybdenum forms $[Mo(CN)_8]^{4-}$ and other complexes in which the coordination number is 8. Other complexes of second- and third-row metals exhibit coordination numbers of 7 and 9.

Because of their having larger sizes and more filled shells of electrons between the outer shell and the nucleus, the ionization energies of second- and third-row metals are lower than those of the first-row metals. Consequently, it is easier for the heavier metals to achieve higher oxidation states, which also favors higher coordination numbers. In general, there is also a greater tendency of the heavier metals to form metal–metal bonds. The result is that there are numerous complexes that contain clusters of metal atoms or ions when the metals are from the second and third rows. This will be shown in more detail in later chapters.

Another significant difference between complexes of first-row metals and those of the second and third rows involves the pairing of electrons. Earlier in this chapter, it was shown that for the d^4 ion Mn^{3+}, there are two series of complexes. One series has all four of the electrons unpaired, and the bonding was interpreted as involving hybrid orbitals of the sp^3d^2 type. In the other series of compounds, there are only two unpaired electrons and the hybrid orbital type is considered to be d^2sp^3. The reasons for this difference are explained in the next chapter. However, because the second- and third-row metals are larger and their orbitals are larger, it requires less energy to pair the electrons than it does for a first-row metal ion. As a result, ligands that do not force the pairing of electrons in complexes of a first-row metal may do so if the metal is from the second or third row. For example, the complex $[CoF_6]^{3-}$ is high spin, but $[RhF_6]^{3-}$ is a low-spin complex. As a general rule, there are far fewer high-spin complexes of second- and third-row metals than there are of first-row metals. In the case of Ni^{2+}, which is a d^8 ion, both tetrahedral (sp^3 hybrid orbitals) and square planar (dsp^2 hybrid orbitals) complexes are known. Complexes of the ions Pd^{2+} and Pt^{2+} (both of which are also d^8) are square planar because it is easier for the electrons to pair, which gives one empty d orbital making it possible for the dsp^2 hybrid orbitals to be used. This difference in behavior will be interpreted in terms of the effects of the ligands on the d orbitals in the next chapter when the topic of ligand field theory is discussed.

16.9 THE 18-ELECTRON RULE

The noble gas Kr has a total of 36 electrons, and that number represents a stable electron configuration. When complexes of the metal atoms of the first transition series are considered, it is found that in many cases, the number of electrons donated to the metal atom is the number required to bring the total number of electrons to 36. That number represents filled outer shells (s, p, and d) capable of holding 18 electrons. For example, when Ni bonds to carbonyl ligands (CO), the stable complex is $Ni(CO)_4$. The nickel atom has a total of 28 electrons, so the addition of eight electrons (four pairs donated by four CO ligands) brings the total number of electrons on the nickel atom to 36. The iron atom has 26 electrons, so 10 additional electrons would complete the configuration around the atom and bring it to that of krypton. Consequently, when Fe forms a complex with CO as the ligands, the most stable product is $Fe(CO)_5$. A chromium atom contains 24 electrons, so the addition of 12 electrons donated from six CO ligands gives the chromium a total of 36, and the stable product is $Cr(CO)_6$. Of course, the ligands need not be CO in all cases.

When the manganese atom ($Z = 25$) is considered, we see that the addition of five CO molecules would bring the total number of electrons to 35, whereas six CO ligands would bring the total to 37. In neither case is the 18-electron rule obeyed. In accord with these observations, neither $Mn(CO)_5$ nor $Mn(CO)_6$ is a stable complex. What is stable is the complex $[Mn(CO)_5]_2$ (sometimes written as $Mn_2(CO)_{10}$) that has the structure

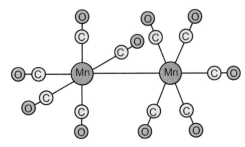

in which there is a metal–metal bond between the manganese atoms, which allows the 18-electron rule to be obeyed.

Although the 18-electron rule is not always rigorously obeyed, there is a strong *tendency* especially for complexes of uncharged metals to bind to ligands in such numbers that the metal gains the electron configuration of the next noble gas. Complexes of transition metal *ions* are not so regular in this regard. For example, Fe^{3+} has a d^5 configuration (a total of 23 electrons), but it forms many stable complexes having the FeX_6 formula that do not obey the 18-electron rule. On the other hand, Co^{3+} has a d^6 configuration (a total of 24 electrons), so the addition of six pairs of electrons from six ligands brings the total to 36. Therefore, many low-spin complexes of Co^{3+} having the formulas CoX_6 do obey the 18-electron rule. It should be emphasized that the 18-electron rule is of greatest utility when considering complexes composed of uncharged metals and soft ligands (e.g. CO, alkenes, etc.).

The 18-electron rule is especially useful when considering complexes containing ligands such as cycloheptatriene, C_7H_8, abbreviated as *cht*. This ligand, which has the following structure, can bond to metals in more than one way because each double bond can function by donating two electrons.

When a complex having the formula $[Ni(CO)_3cht]$ is considered, it is the 18-electron rule that provides the basis for determining how the *cht* ligand is bonded to the nickel. Because three CO ligands donate six electrons to the nickel, there must be only two electrons donated from the *cht* because the nickel atom needs only eight to bring its total up to 36. Therefore, the *cht* ligand is functioning as a two-electron donor and only one of the double bonds is attached to the nickel resulting in the structure

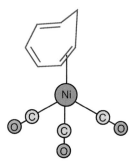

On the other hand, in the complex [Ni(CO)$_2$cht], the *cht* ligand is a four-electron donor, so two double bonds are functioning as electron pair donors and the structure is

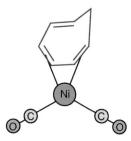

In the complex [Cr(CO)$_3$cht], the *cht* ligand functions as a six-electron donor because three CO ligands donate six electrons and 12 are needed by the chromium atom to reach a total of 36. Therefore, this complex has the structure

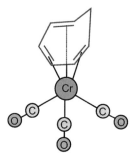

The number of complexes for which the 18-electron rule is an overriding bonding condition is very large. It is an essential tool for interpreting the bonding in organometallic compounds, and it will be considered many times in the chapters to follow.

Because some molecules such as *cht* have the ability to function as ligands with different numbers of electron pairs being donated, the bonding mode of such ligands must be indicated. The bonding mode is designated by the term *hapticity* (derived from the Greek word *haptein*, which means "fasten") and it is abbreviated as *h* or η. A superscript is added to the upper right of the symbol to indicate the number of atoms (in many cases carbon) that are attached to the metal. When a double bond functions as the electron pair donator, both carbon atoms connected by the double bond are considered as bound to the metal, so the hapticity is 2 (indicated as η^2). Thus, in the structure of [Ni(CO)$_3$cht] above, cht is an η^2 ligand, whereas in [Ni(CO)$_2$cht], the cht is bonded in an η^4 manner. In the chromium complex, [Cr(CO)$_3$cht], three of the double bonds are used, so the cht is η^6.

Naming of coordination compounds that have ligands with variable hapticity includes designating the bonding mode. For example, [Ni(CO)$_3$cht] is named as tricarbonyl(η^2-cycloheptatriene)nickel and [Ni(CO)$_2$cht] is dicarbonyl(η^4-cycloheptatriene)nickel. Tricarbonyl(η^6-cycloheptatriene)chromium is the name for [Cr(CO)$_3$cht]. In all these structures, the 18-electron rule is obeyed.

Another example of a complex that obeys the 18-electron rule is ferrocene or bis(cyclopentadienyl) iron. The cyclopentadienyl anion is generated by the reaction of cyclopentadiene with sodium, and ferrocene is obtained by the subsequent reaction with ferrous chloride.

$$2 \, Na^+C_5H_5^- + FeCl_2 \; \rightarrow \; Fe(C_5H_5)_2 + 2 \, NaCl \tag{16.18}$$

In ferrocene, written as Fe(cp)$_2$, the iron is formally in the +2 oxidation state, and the cyclopentadienyl ligands have a -1 charge and function as six-electron donors. With Fe^{2+} being a d^6 ion, the additional 12 electrons from the two ligands bring the total number of electrons on Fe to 36, so the 18-electron rule is obeyed. Because the entire five-membered ring of the cyclopentadienyl is part of the π system, the ligands are designated as η^5 with respect to bonding to the iron. Therefore, the name of this compound is bis(η^5-cyclopentadienyl)iron. Although the bonding of η^5-cyclopentadienyl has been described above, it can also bond in other ways. When bound by a σ bond to only one carbon atom, the ligand is named as η^1-cyclopentadiene.

The NO molecule is an interesting ligand that provides an illustration of some unique features of bonding in complexes. Describing the bonding in NO in terms of molecular orbitals gives (omitting the electrons from 1s orbitals) $\sigma_g^2 \, \sigma_u^2 \, \sigma_g^2 \, \pi_u^2 \, \pi_u^2 \, \pi_g^1$, which shows that one electron resides in an antibonding π molecular orbital (designated as π_g). The bond order for the molecule is 2.5. When the electron in the antibonding orbital is lost, NO$^+$ results, which has a bond order of 3 and is isoelectronic with CO and CN$^-$. With the ionization potential for NO being only 9.25 eV, complexes containing NO as a ligand involve transfer of the electron from the π_g orbital on NO to the metal (which results in an increase in bond order) concurrent with the donation of a pair of electrons from NO$^+$ to the metal. The process can be represented as

$$NO + L_5M \; \rightarrow \; [NO^+ + L_5M^-] \; \rightarrow \; L_5MNO \tag{16.19}$$

or if ligand replacement is involved,

$$NO + L_5MX \; \rightarrow \; [NO^+ + L_5M^- + X] \; \rightarrow \; L_5MNO + X \tag{16.20}$$

where L and X are other ligands bound to the metal. In this process, the NO molecule has functioned as a three-electron donor. One electron has been transferred from the π^* (also designated as π_g) orbital to the metal, and two others have been donated in the formation of a σ coordinate bond.

With its unusual coordination mode, NO forms complexes with a wide variety of metals, especially in cases where the metal can accept the transfer of an electron from the π_g orbital. With cobalt having 27 electrons, it is evident that the addition of no integral number of ligands that function as two-electron donors can bring the total to 36. However, when one ligand is an NO molecule, the cobalt has a total of

30 electrons, so three CO ligands can raise the total to 36. Therefore, the stable complex that obeys the 18-electron rule is [Co(CO)$_3$NO]. It should be apparent that complexes such as Mn(CO)$_4$(NO), Fe(CO)$_2$(NO)$_2$, and Mn(CO)(NO)$_3$ obey the 18-electron rule.

16.10 BACK DONATION

Ligands are Lewis bases that form complexes with metals by donating electron pairs to metal atoms or ions. In determining the formal charge on an atom, each bonding pair of electrons is divided equally between the two bonded atoms. When six pairs of electrons are donated to a metal ion, the metal ion has, in effect, gained six electrons. If the metal initially had a +3 charge, it will have a −3 formal charge after receiving six donated pairs of electrons. If the metal initially had a +2 charge, the formal charge on the metal will be −4 in a complex where the metal has a coordination number of 6.

Many ligands have empty orbitals in addition to the filled orbital that holds the pair of electrons that is donated to the metal. For example, the isoelectronic ligands CN$^-$, NO$^+$ and CO have a molecular orbital arrangement like that shown in Figure 16.7. The filled molecular orbitals beyond the σ_g and σ_u that arise from combining the 2s atomic functions are $(\sigma_g)^2 \, (\pi_u)^2 \, (\pi_g)^2$. However, there is an empty antibonding orbital (designated as π^* or π_g) on CN$^-$ that can be shown as

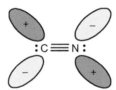

The π_u molecular orbitals have the correct symmetry to overlap with the d_{xy}, d_{yz}, or d_{xz} orbitals on the metal. Except for a metal ion having a d^0 configuration, the nonbonding d orbitals on the metal contain some electrons. With the metal having a *negative* formal charge as a result of accepting electron pairs from ligands, part of that unfavorable charge can be alleviated by donating electron density from the nonbonding d orbitals on the metal to the antibonding orbitals on a CN$^-$ ligand. This situation, known as *back donation*, can be illustrated by the structure

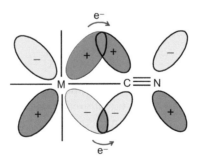

In this case, the donation is sometimes indicated as $d \rightarrow \pi$ to show the direction of the flow of electron density.

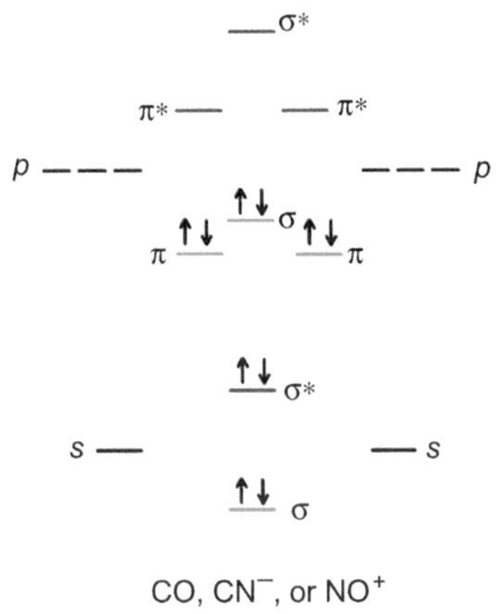

$$\sigma^*$$

$$\pi^* \quad\quad \pi^*$$

$$p \quad\quad\quad\quad\quad p$$

$$\pi \quad \sigma \quad \pi$$

$$\sigma^*$$

$$s \quad\quad\quad s$$

$$\sigma$$

CO, CN$^-$, or NO$^+$

FIGURE 16.7
Molecular orbital diagram for some common diatomic ligands.

Because the electron density is flowing from the metal onto the ligands, this donation is known as *back* donation. It is in the reverse direction to that in the normal donation of electrons in forming coordinate bonds. The term *back bonding* is sometimes used instead of back donation, but it is not as descriptive because the ligands are functioning as acceptors of electron density from the metal. The essential feature of electron *donation* is that there must be an *acceptor*, which in this case is the ligand. The ligands are referred to as π acceptors because of their receiving electron density donated from the metal to π_g orbitals. Back donation results in increasing the bond order between the metal and ligand, so it results in *additional* bonding.

There are several consequences of back donation of electron density from the metal to the ligands in a complex. First, the bonds between the metal and ligands involve more shared electron density, so the bond order of the M–L bond increases. As a result, the bonds between the metal and ligands are stronger and shorter than they would be if no back donation occurred. The degree of bond strengthening and shortening varies with the extent of back donation. Back donation also causes the infrared band attributed to stretching the metal–ligand bond to be shifted to higher wavenumbers.

Adding electron density to *antibonding* orbitals on the CN$^-$ ligand reduces the bond order of the C–N bond. The bond order is given by $(N_b - N_a)/2$, so increasing the value of N_a reduces the bond order to a value <3 even if only fractionally. With the reduction in bond order of the C–N bond comes an increase in the bond length and a shift of the absorption band for the C–N stretching vibration to lower wavenumbers. In fact, the position of the C–N stretching vibration in the infrared spectrum is one of the best indicators of the extent to which back donation has occurred.

It is interesting to note the difference between the positions of the C–N stretching vibration in $[Fe(CN)_6]^{3-}$ and $[Fe(CN)_6]^{4-}$. In the first of these complexes, the iron is $+3$, so its formal charge in the complex is -3. In the second complex, the iron is $+2$, so its formal charge is -4. Therefore, the extent of back donation is greater in the second complex. This difference is reflected by the positions of the C–N stretching bands, which are at 2135 and 2098 cm^{-1}, respectively. The situation is not actually this simple because there are three bands observed that are attributable to C—N stretching vibrations. Only the band corresponding to the highest energy is listed here.

As the extent of back donation increases, the metal—carbon bond becomes stronger and shorter. For example, in the series of complexes having the formula $[M(CN)_6]^{3-}$ where M is Cr^{3+} (d^3), Mn^{4+} (d^4), Fe^{3+} (d^5), or Co^{3+} (d^6), the M—C bond lengths are 208, 200, 195, and 189 pm, respectively. As the bond lengths decrease, there is a corresponding increase in M—C stretching frequency with the values being 348, 375, 392, and 400 cm^{-1} for the complexes in the order listed. The extent of back donation is greater when there are more electrons in the d orbitals that can be donated. Obviously, there would be no back donation from Sc^{3+} because this ion has the configuration d^0.

Even though the effects are easy to observe in such cases, back donation is by no means limited to cyanide complexes. It is required that the ligands have *empty* orbitals of appropriate symmetry to overlap with the nonbonding d orbitals on the metal. Certainly, CO is another example of a ligand that can accept back donation. Moreover, CO normally forms complexes with metals in the 0 or low oxidation states, which results in metals having a substantial negative formal charge, so back donation is significant in such cases. The C—O stretching vibration in gaseous CO is observed at 2143 cm^{-1}, but in many metal carbonyls, it is observed at lower wavenumbers with the frequency dependent on the extent of back donation.

Comparison of the C—O stretching frequencies for a series of metal carbonyl complexes can reveal interesting trends. The complexes listed below all obey the 18-electron rule, but with different numbers of CO ligands attached, the metal atoms do not have the same increase in electron density on them because the coordination numbers are different.

$Ni(CO)_4$	2057 cm^{-1}
$Fe(CO)_5$	2034 cm^{-1}
$Cr(CO)_6$	1981 cm^{-1}

For these complexes, the extent of back donation increases as the number of CO ligands increases, which causes the stretching frequencies to be found at lower wavenumbers. A similar trend is seen for the following complexes (all of which obey the 18-electron rule) showing the effect of the charge on the metal ion:

$V(CO)_6^-$	1859 cm^{-1}
$Cr(CO)_6$	1981 cm^{-1}
$Mn(CO)_6^+$	2090 cm^{-1}

In the first of these complexes, the vanadium has a -1 charge and with the addition of six CO ligands, there is a tendency to relieve part of the negative charge by extensive back donation. The tendency is less in $Cr(CO)_6$ and even less when the metal has a positive charge as in $Mn(CO)_6^+$.

Ligands such as NH_3, H_2O, en, or F^- have no empty orbitals of suitable energy or symmetry for accepting electron density back donated from d orbitals on a metal ion. This results in an interesting situation when the complexes shown in Figure 16.8 are considered.

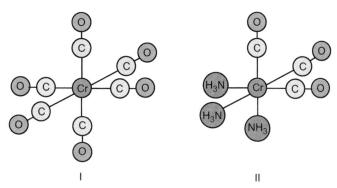

FIGURE 16.8
The structures of [Cr(CO)$_6$] and [Cr(NH$_3$)$_3$(CO)$_3$].

In both complexes, the Cr has a −3 formal charge and back donation occurs. However, because NH$_3$ has no orbitals available for accepting electron density from the metal, all the back donation in complex II goes to the three CO ligands. In Complex I, the back-donated electron density is spread among six CO ligands. Because of this difference, the CO stretching vibration in Cr(CO)$_6$ is found at 2100 cm^{-1}, whereas in Cr(CO)$_3$(NH$_3$)$_3$, it is observed at 1900 cm^{-1}.

In the case of ligands such as CN$^-$ and CO, the electron density is transferred from the metal to π^* antibonding orbitals on the ligands. Other ligands that have orbitals that are suitable for receiving electron density include those that have empty d orbitals. Ligands that have phosphorus, arsenic, or sulfur as the donor atom have empty d orbitals, so multiple bonding by back donation is possible for ligands such as PR$_3$, AsR$_3$, R$_2$S, and many others. These ligands are soft Lewis bases, which are normally found bonded to soft metal ions that have large size and low charges. As a result, the electrons on the metal ion are polarizable and can be shifted toward the ligands relatively easily.

When spectral studies are conducted to see how different ligands function as π acceptors, it is found that the ability to accept electron density varies in the order

$$NO^+ > CO > PF_3 > AsCl_3 > PCl_3 > As(OR)_3 > P(OR)_3 > PR_3.$$

It is not surprising that NO$^+$ is so effective in accepting electron density because it not only has π^* orbitals of correct symmetry but also it has a positive charge, which makes it have a great attraction for electrons in the nonbonding metal d orbitals.

There is extensive back donation in the anion of Zeise's salt, [Pt(C$_2$H$_4$)Cl$_3$]$^-$, because the π^* orbitals on C$_2$H$_4$ have correct symmetry. Figure 16.9 shows the interaction of the metal and ligand orbitals. As a result of the increased electron population in the π^* orbitals, the stretching frequency of the C=C bond decreases from 1623 cm^{-1} for gaseous C$_2$H$_4$ to 1526 cm^{-1} in Zeise's salt.

In the absence of steric effects, the effect of π bonding that results from back donation is partially responsible for ambidentate ligands such as SCN$^-$ bonding in different ways. For example, SCN$^-$ is

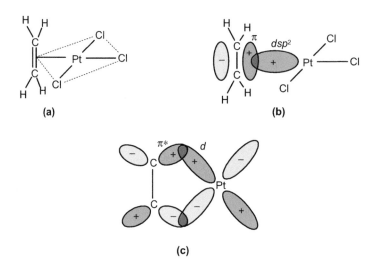

FIGURE 16.9
(a) The structure and bonding in the anion of Zeise's salt. (b) A σ bond results from the overlap of a dsp^2 hybrid orbital on the metal and the π orbital on ethylene. (c) Back donation from a d orbital on the metal to the π^* orbital on ethylene gives some π bonding.

bonded differently in the stable complexes cis-$[Pt(NH_3)_2(SCN)_2]$ and cis-$[Pt(PR_3)_2(NCS)_2]$. In the first of these, the bonding is shown as

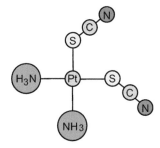

whereas in the second, the arrangement is

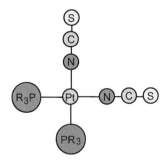

Because NH_3 has no low energy orbitals of suitable symmetry, π bonds do not form between Pt and NH_3. Sulfur has empty d orbitals available to accept electron density from the Pt^{2+}. Therefore, to the extent that π bonding occurs, it is to the sulfur end of SCN^-. When it is bonded to Pt^{2+} opposite NH_3, SCN^- bonds to Pt^{2+} through the sulfur atom.

In the second case, the phosphorus atom in the phosphine also has empty d orbitals that can accept electron density donated from the Pt^{2+}. In fact, it is more effective in this regard than is the sulfur atom in SCN^-. This results in the more stable bonding to SCN^- being to the nitrogen atom when PR_3 is in the *trans* position. In essence, the presence of π bonding ligands in *trans* positions that compete for back donation leads to a complex of lower stability. As will be discussed in Chapter 20, this phenomenon (known as the *trans effect*) has a profound effect on the rates of substitution reactions in such complexes.

16.11 COMPLEXES OF DINITROGEN, DIOXYGEN, AND DIHYDROGEN

In the survey of coordination compounds presented so far, several types of ligands have been used as illustrations. As we progress further into the topic of coordination chemistry, a great many other types of complexes will be described, especially in Chapter 21 where organometallic compounds of transition metals will be discussed. To round out the introduction to coordination compounds in this chapter, there are other types of ligands that should be included. One such ligand is the oxygen molecule, O_2 (technically known as dioxygen), which is interesting for several reasons. One reason is that it is a complex of oxygen with iron that transports oxygen in living animals. When the molecular orbital diagram for oxygen is examined, it is seen that the π^* orbitals are half filled (see Figure 16.10(a)). As a result, the oxygen molecule has orbitals that can overlap effectively with d orbitals on a metal with the result that a complex may be formed. When O_2 forms a complex, the bond gets slightly longer, and the stretching frequency is shifted to lower wavenumber.

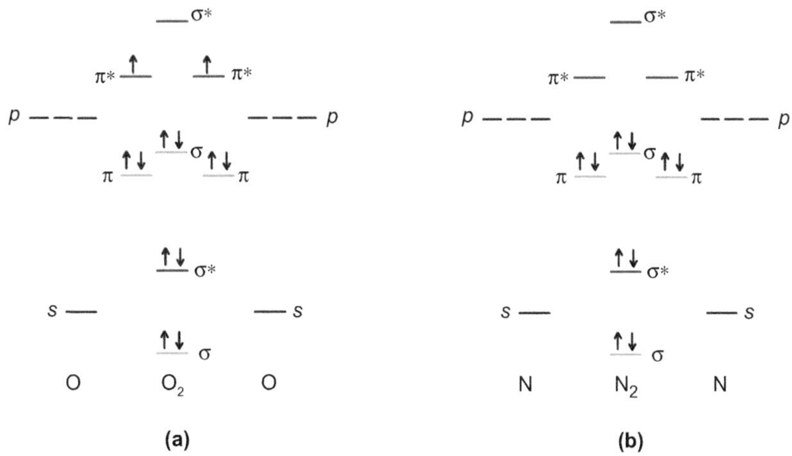

FIGURE 16.10
The molecular orbital energy-level diagrams for O_2 and N_2.

The first preparation of a complex containing oxygen was carried out by Vaska (1963) utilizing the reaction

$$Ir(CO)Cl(P\phi_3)_2 + O_2 \rightarrow Ir(O_2)(CO)Cl(P\phi_3)_2 \tag{16.21}$$

where ϕ represents the phenyl group, $-C_6H_5$, and $P\phi_3$ is triphenylphosphine. The product of the reaction shown in Eqn (16.21) can be shown as

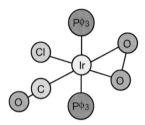

When oxygen bonds to a metal, it can do so in more than one way, and the most common ways are shown below. If the oxygen molecule is attached at only one end, the bond involving π^* orbitals on oxygen will not be linear. A classification of bond types has been used to correlate properties for compounds containing oxygen in various bonding modes. The types of bonds are categorized and illustrated as shown in Figure 16.11.

Although X-ray diffraction is the definitive way to establish structure and determine bond lengths, other techniques are also of use. As in the case of complexes of CO and ethylene, spectral studies have yielded information about the bonding in complexes containing O_2. In general, the O—O stretching band is found in similar regions for both Type Ia and Type Ib, that being roughly 1100–1200 cm^{-1}. In complexes of Type IIa and Type IIb, the position of the O—O stretching vibration is at roughly 800–900 cm^{-1}. The linkages in both Type I complexes are considered as being approximately a superoxide linkage, whereas that in Type II complexes is considered to be characteristic of a peroxide linkage. For comparison, the stretching frequency in the superoxide ion, O_2^-, is approximately 1100 cm^{-1}, whereas that for the peroxide ion, O_2^{2-}, is ~825 cm^{-1} with both depending somewhat on the nature of the cation. Because the O_2 molecule has two electrons in the π^* orbitals, it is not likely that the filled σ^* orbital could be used to form a σ bond to the metal by electron pair donation. Therefore, this type of bonding, which is significant in complexes containing CO and CN$^-$, is not thought to be important in oxygen complexes. Part of the problem in interpreting the bonding mode of O_2 to metals is that from the standpoint of charges, $M^{2+}-O_2^0$ is equivalent to $M^{3+}-O_2^-$ and to $M^{4+}-O_2^{2-}$. Since the initial report of the first dioxygen complex in 1963, a very large number of others have been prepared and characterized.

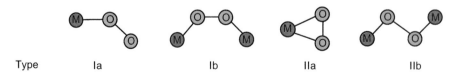

Type   Ia   Ib   IIa   IIb

FIGURE 16.11
Types of bonding exhibited by O_2 in complexes.

There is a great deal of interest in nitrogen complexes because nitrogenase is an enzyme that apparently assists in the process of nitrogen fixation in plants as a result of nitrogen forming a complex with iron or molybdenum. Although the N_2 molecule (dinitrogen) is isoelectronic with CO and CN^- (see Figure 16.10(b)), it is much less reactive than those ligands when it comes to forming coordinate bonds by electron pair donation. But, in 1965, A.D. Allen and C.V. Senoff reported the first coordination compound containing dinitrogen. It was formed by the reaction of ruthenium chloride with hydrazine, but the ruthenium complex has also been prepared by the reaction

$$[Ru(NH_3)_5H_2O]^{2+} + N_2 \ \rightarrow \ [Ru(NH_3)_5N_2]^{2+} + H_2O \tag{16.22}$$

Since that first compound was reported, hundreds of others have been prepared, and a molecule that was once thought to be nearly "inert" is known to form numerous complexes. Although several bonding modes have been proposed for N_2 in complexes, the two most common modes are with the N_2 molecule functioning as a σ donor or as a bridging ligand. These arrangements can be shown as $M-N\equiv N$ and $M-N\equiv N-M$. Thus, the $M-N-N$ arrangement is either linear or very nearly so. In complexes of dinitrogen, the N–N distance is normally in the range 110–113 pm, which is about the same as in the gaseous N_2 molecule, but the stretching band is shifted by about 100–300 cm^{-1}. Because the π^* orbitals are not populated in the N_2 molecule, it functions primarily as a σ donor. This is quite different from the coordination of O_2 in which the electrons donated to the metal are from the π^* orbitals. However, the $N\equiv N$ stretching band corresponds to 2331 cm^{-1}, but in $[Ru(NH_3)_5N_2]^{2+}$, it is at 2140 cm^{-1} indicating some population of the π^* orbitals in N_2. In the bridged $[(NH_3)_5Ru-N\equiv N-Ru(NH_3)_5]^{4+}$ complex, the stretching band is at approximately 2100 cm^{-1}.

Another milestone in coordination chemistry dates from 1980 when the first complex containing dihydrogen was reported by G.J. Kubas. It was found that hydrogen reacts with $[W(CO)_3(Pchx_3)_2]$ to give $[W(H_2)(CO)_3(Pchx_3)_2]$ (where chx = cyclohexyl and $Pchx_3$ is tricyclohexylphosphine). The complex has the structure

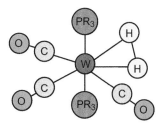

Bonding in this type of complex is believed to involve electron donation from the σ-bonding orbital. In some ways, this is similar to the formation of H_3^+ from H_2 and H^+ (see Chapter 5). However, it is also believed that there may be some back donation from the metal d orbitals to the σ^* orbital on H_2. These bonding characteristics are illustrated in Figure 16.12. To the extent that back donation occurs, it places electron density in the σ^* orbital, which weakens the H–H bond. As a result of bond weakening, the H–H bond is easier to dissociate. This could be a factor in the ability of certain metal complexes to function as catalysts for hydrogenation reactions (see Chapter 22). The formation of some hydrogen compounds can also be considered as oxidation of the metal with subsequent coordination of H^- ions.

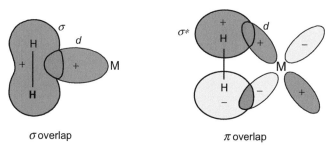

FIGURE 16.12
The bonding of dihydrogen to a metal in a complex.

In this chapter, a survey of the chemistry of coordination compounds has been presented with emphasis on elementary aspects of bonding, nomenclature, and isomerism. The discussion of complexes containing dioxygen, dinitrogen, and dihydrogen not only shows some of the characteristics of these interesting compounds but also it shows the dynamic nature of the chemistry of coordination compounds. It is a field that has come a long way since the preparation of Zeise's salt in the early 1800s and the pioneering work of Alfred Werner over a century ago. Coordination compounds are important in life processes as well as numerous industrial processes. For that reason, the following chapters will be devoted to bonding, spectra, stability, and reactions of coordination compounds. Collectively, these chapters will provide the background necessary for more advanced study.

References for Further Study

Bailar, J.C., 1956. *The Chemistry of Coordination Compounds*. Reinhold Publishing Co., New York. A classic in the field by the late Professor Bailar (arguably the most influential figure in coordination chemistry in the last half of the 20th century) and his former students.

Cotton, F.A., Wilkinson, G., Murillo, C.A., Bochmann, M., 1999. *Advanced Inorganic Chemistry*, 6th ed. John Wiley, New York. One of the great books in inorganic chemistry. In almost 1400 pages, an incredible amount of inorganic chemistry is presented.

DeKock, R.L., Gray, H.B., 1980. *Chemical Structure and Bonding*. Benjamin/Cummings, Menlo Park, CA. Chapter 6 presents a good introduction to bonding in coordination compounds.

Greenwood, N.N., Earnshaw, A., 1997. *Chemistry of the Elements*, 2nd ed. Butterworth-Heinemann, Oxford. A monumental reference work that contains a wealth of information about many types of coordination compounds.

Huheey, J.E., Keiter, E.A., Keiter, R.L., 1993. *Inorganic Chemistry: Principles of Structure and Reactivity*, 4th ed. HarperCollins College Publishers, New York. Appendix I consists of 33 pages devoted to nomenclature in inorganic chemistry.

Kettle, S.F.A., 1969. *Coordination Chemistry*, Appleton, Century. Crofts, New York. The early chapters of this book give a good survey of coordination chemistry.

Kettle, S.F.A., 1998. *Physical Inorganic Chemistry: A Coordination Approach*. Oxford University Press, New York. An excellent book on the chemistry of coordination compounds.

Porterfield, W.W., 1993. *Inorganic Chemistry-A Unified Approach*, 2nd ed. Academic Press, San Diego, CA. Chapters 10-12 give a good introduction to coordination chemistry.

Szafran, Z., Pike, R.M., Singh, M.M., 1991. *Microscale Inorganic Chemistry: A Comprehensive Laboratory Experience*. John Wiley, New York. Chapter 8 provides synthetic procedures for numerous transition metal complexes. This book also provides a useful discussion of many instrumental techniques.

Vaska, L., 1963. *Science*, 140, 809. A report on an oxygen complex with iridium.

QUESTIONS AND PROBLEMS

1. Draw structures for the following ions and indicate the overall charge:
   (a) *cis*-dichlorobis(ethylenediamine)cobalt(III)
   (b) *cis*-difluordioxalatochromate(III)
   (c) *trans*-dinitritotetranitronickelate(II)
   (d) *cis*-diamminetetrathiocyanatochromate(III)
   (e) trichloro(ethylene)platinate(II)

2. Name the following compounds:
   (a) $[Co(en)_2Cl_2]Cl$
   (b) $K_3[Cr(CN)_6]$
   (c) $Na[Cr(en)C_2O_4Br_2]$
   (d) $[Co(en)_3][Cr(C_2O_4)_3]$
   (e) $[Mn(NH_3)_4(H_2O)_2][Co(NO_2)_2Br_4]$

3. Give names for the following compounds. Do not try to decide whether the complex has a *cis* or *trans* structure if both isomers are possible.
   (a) $K_4[Co(NCS)_6]$
   (b) $[Pt(en)Cl_2]$
   (c) $[Cr(NH_3)_3(NCS)_2Cl]$
   (d) $[Pd(py)_2I_2]$
   (e) $[Pt(en)_2Cl_2]Cl_2$
   (f) $K_4[Ni(NO_2)_4(ONO)_2]$

4. For each of the following, tell whether it represents a stable compound and why or why not.
   (a) $Ni(CO)_3NO$
   (b) $Cr(CO)_3(NO)_2$
   (c) $Fe(CO)_3(\eta^2\text{-}cht)$.

5. Draw the structure for $S_2O_3^{2-}$. How would it bond to Pt^{2+}? How would it bond to Cr^{3+}? Explain how you determine how $S_2O_3^{2-}$ is bound by means of infrared spectroscopy.

6. Sketch all the geometrical isomers possible for trigonal bipyramid complexes having the formulas (where AA is a bidentate ligand)
   (a) MX_4Y; (b) MX_3Y_2; (c) $MX_3(AA)$.

7. Only one compound having the formula $[Zn(py)_2Cl_2]$ (where py = pyridine) exists but two different compounds are known having the composition $[Pt(py)_2Cl_2]$. Explain these observations and describe the bonding in each complex.

8. Two types of complexes are known that contain SO_4^{2-} as ligands. Explain how sulfate bonds to the metal ions in two different ways.

9. For the series of silver complexes with cyanide, the C—N stretching vibrations are observed as follows. The species containing three or four cyanide ions are observed only in solutions containing high concentrations of CN^-.

Species	AgCN solid	$Ag(CN)_2^-$	$Ag(CN)_3^{2-}$	$Ag(CN)_4^{3-}$
ν_{CN}, cm^{-1}	2170	2135	2105	2092

Explain the differences in position of the C—N stretching bands.

10. Suppose a cubic complex has the formula MX_5Y_3 where X and Y are different ligands. Sketch the structures of all the geometric isomers possible.

11. A nickel complex has the formula NiL_4^{2+}
    (a) What are the possible structures for the complex?
    (b) Explain how the magnetic moment could be used to determine the structure of the complex.

12. The complex $[NiX_2(PR_3)_2]$ (where X is a halide ion and R is an alkyl group) has a dipole moment that is close to zero. When R is replaced by an aryl group, the dipole moment is much larger. Explain the difference.

13. Draw the structure for $(CO)_2(NO)Co(cht)$. What is the hapticity of *cht* in this complex?

14. Draw the structure for $(C_6H_6)Cr(CO)_4$. What is the hapticity of C_6H_6 in this complex?

15. Sketch structures for all the isomers of $[Co(en)_2ClNCS]^+$ assuming that SCN^- can be ambidentate.

16. The compound $[Co(NH_3)_6]Cl_3$ is yellow-orange in color, but $[Co(H_2O)_3F_3]$ is blue. Discuss why they are so different in terms of the bonding in the complexes.

17. Describe a simple chemical test to distinguish between $[Co(NH_3)_5Br]SO_4$ and $[Co(NH_3)_5SO_4]Br$.

18. Write the formula or draw the structure requested:
    (a) the formula for a polymerization isomer of $[Pd(NH_3)_2Cl_2]$
    (b) the structure of *cis*-bis(oxalato)dichlorochromium(III) ion
    (c) the formula for a coordination isomer of $[Zn(NH_3)_4][Pd(NO_2)_4]$
    (d) the structure of diamminedithiocyanatoplatinum(II).

19. For each of the following complexes, give the hybrid orbital type and the number of unpaired electrons.
    (a) $[Co(H_2O)_6]^{2+}$; (b) $[FeCl_6]^{3-}$; (c) $[PdCl_4]^{2-}$; (d) $[Cr(H_2O)_6]^{2+}$

20. For each of the following, give the hybrid bond type and the expected magnetic moment.
    (a) $[Co(en)_3]^{3+}$; (b) $[Ni(H_2O)_6]^{2+}$; (c) $[FeBr_6]^{3-}$; (d) $Ni(NH_3)_4]^{2+}$; (e) $Ni(CN)_4]^{2-}$; (f) $[MnCl_6]^{3-}$

21. The magnetic moment of $[Co(py)_2Cl_2]$ is 5.15 B.M. What is the structure of this compound?

22. Sketch structures for all the isomers possible for $[Co(en)_2NO_2Cl]^+$.

23. Discuss how the ligand *cht* (cycloheptatriene) would be bonded in each of the following:
    (a) $[Ni(CO)_3cht]$; (b) $[Fe(CO)_3cht]$; (c) $[Co(CO)_3cht]$

24. Assume that the complex MLX_2Y_2 has a square base pyramid structure with all ligands able to bond in all positions in the coordination sphere. How many isomers are possible?

25. Draw the structure for $[Nb(\eta^5\text{-}C_5H_5)(CO)_3(\eta^2\text{-}H_2)]$.

26. Does the compound $[Co(CO)_2(\eta^2\text{-}H_2)(NO)]$ obey the 18-electron rule? Explain your answer.

27. In a recent study, it was shown that a compound has been isolated that has formula $[Ni(\eta^6\text{-}1,3,5\text{-}(CH_3)_3C_6H_3)NO]^+PF_6^-$. Draw the structure for the cation and comment on the applicability of the 18-electron rule in this case.

CHAPTER 17

Ligand Fields and Molecular Orbitals

In the previous chapter, elementary ideas about the nature of bonding in coordination compounds were presented. Because of the importance of this area of chemistry, it is essential that a more complete description of the bonding be presented. As we have seen in other cases, it is the interaction of electromagnetic radiation with matter that provides the experimental basis for this information. Consequently, the next chapter will have as its major objective the interpretation of spectra as it relates to the study of complexes. It is from studying the spectra of complexes that we obtain information about the energy states that exist and how the presence of ligands affects the orbitals of the metal ions. Magnetic studies give information about the presence of unpaired electrons, but the valence bond approach described in the last chapter does not adequately explain *why* the electrons pair in some cases but not in others. In this chapter, several of these problems will be addressed more fully in terms of ligand fields and molecular orbitals.

17.1 SPLITTING OF *d* ORBITAL ENERGIES IN OCTAHEDRAL FIELDS

There are several characteristics of coordination compounds that are not satisfactorily explained by a simple valence bond description of the bonding. For example, the magnetic moment of $[CoF_3]^{3-}$ indicates that there are four unpaired electrons in the complex, whereas that of $[Co(NH_3)_6]^{3+}$ indicates that this complex has no unpaired electrons although in each case Co^{3+} is a d^6 ion. In the last chapter, we interpreted the bonding types in these complexes as involving sp^3d^2 and d^2sp^3 hybrid orbitals, respectively, but that does not provide an explanation as to why the two cases are different. Another area that is inadequately explained by a simple valence bond approach is the number of absorption bands seen in the spectra of complexes. One of the most successful approaches to explaining these characteristics is known as crystal field theory.

When a metal ion is surrounded by anions in a crystal, there is an electrostatic field produced by the anions that alters the energies of the *d* orbitals of the metal ion. The field generated in this way is known as a *crystal field*. Crystal field theory was developed in 1929 by Hans Bethe in an attempt to explain the spectral characteristics of metal ions in crystals. It soon became obvious that anions surrounding a metal in a crystal gave a situation that is very similar to the ligands (many of which are also anions) surrounding a metal ion in a coordination compound. In cases where the ligands are not anions, they may be polar molecules, and the negative ends of the dipoles are directed toward the metal ion generating an electrostatic field. Strictly speaking, the *crystal field* approach is a purely electrostatic one based on the interactions between point charges, which is never exactly the case for complexes of transition metal ions. In view of the fact that coordinate bonds result from electron pair donation and have some covalency, the term *ligand field* is used to describe the effects of the field produced by the **591**

Inorganic Chemistry. DOI: http://dx.doi.org/10.1016/B978-0-12-385110-9.00017-0

ligands in a complex. In the 1930s, J.H. Van Vleck developed *ligand* field theory by adapting the crystal field approach to include some covalent nature of the interactions between the metal ion and the ligands. Before we can show the effects of the field around a metal ion produced by the ligands, it is essential to have a clear picture of the orientation of the d orbitals of the metal ion. Figure 17.1 shows a set of five d orbitals, and for a gaseous ion, the five orbitals are degenerate.

If the metal ion were surrounded by a spherical electrostatic field, the energy of the d orbitals would be raised, but all would be raised by the same amount. As shown in the structure below, an octahedral complex can be considered as a metal ion surrounded by six ligands that are located on the axes.

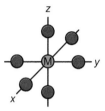

When six ligands surround the metal ion, the degeneracy of the d orbitals is removed because three of the orbitals, the d_{xy}, d_{yz}, and d_{xz} orbitals, are directed *between* the axes, but the others, the $d_{x^2-y^2}$ and the d_{z^2}, are directed *along* the axes pointing at the ligands. Therefore, there is greater repulsion between the electrons in orbitals on the ligands and the $d_{x^2-y^2}$ and d_{z^2} orbitals than there is toward the d_{xy}, d_{yz}, and d_{xz} orbitals. Because of the electrostatic field generated by the ligands, *all* of the d orbitals are raised in energy, but two of them are raised more than the other three. As a result, the d orbitals have energies that can be represented as shown in Figure 17.2.

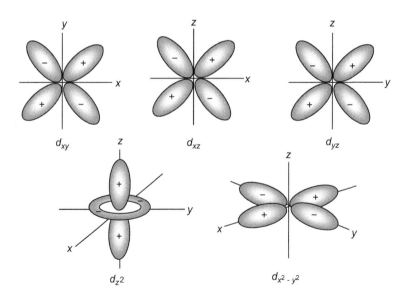

FIGURE 17.1
The spatial orientations of the set of five d orbitals for a transition metal.

The two orbitals of higher energy are designated as the e_g orbitals, and the three orbitals of lower energy make up the t_{2g} orbitals. These designations will be described in greater detail later, but the "g" subscript refers to being symmetrical with respect to a center of symmetry that is present in a structure that has O_h symmetry. The "t" refers to a triply degenerate set of orbitals whereas "e" refers to a set that is doubly degenerate. The energy separating the two groups of orbitals is called the *crystal* or *ligand field splitting*, Δ_o. Splitting of the energies of the d orbitals as indicated in Figure 17.2 occurs in such a way that the overall energy remains unchanged and the "center of energy" is maintained. The e_g orbitals are raised 1.5 times as much as the t_{2g} orbitals are lowered from the center of gravity. Although the splitting of the d orbitals in an octahedral field is represented as Δ_o, it is also sometimes designated as 10 Dq where Dq is an energy unit for a particular complex. The two orbitals making up the e_g pair are raised by $3/5\Delta_o$ but the t_{2g} orbitals are lowered by $2/5\Delta_o$ relative to the center of energy. In terms of Dq units, the e_g orbitals are raised by 6 Dq whereas the three t_{2g} orbitals are 4 Dq lower than the center of energy.

The effect of crystal field splitting is easily seen by studying the absorption spectrum of $[Ti(H_2O)_6]^{3+}$ because the Ti^{3+} ion has a single electron in the $3d$ orbitals. In the octahedral field produced by the six water molecules, the $3d$ orbitals are split in energy as shown in Figure 17.2. The only transition possible is promotion of the electron from an orbital in the t_{2g} set to one in the e_g set. This transition gives rise to a single absorption band, the maximum of which corresponds directly to the energy represented as Δ_o. As expected, the spectrum shows a single, broad band that is centered at $1/\lambda = 20,300\ cm^{-1}$, which corresponds directly to Δ_o. The energy associated with this band is calculated as follows:

$$E = h\nu = hc/\lambda = 6.63 \times 10^{-27}\,\text{erg sec} \times 3.00 \times 10^{10}\ cm/sec \times 20,300\ cm^{-1}$$
$$E = 4.04 \times 10^{-12}\ \text{erg}$$

We can convert this energy per molecule into kJ mol^{-1} by the following conversion.

$$4.04 \times 10^{-12}\ \text{erg/molecule} \times 6.02 \times 10^{23}\ \text{molecule/mole} \times 10^{-7}\ \text{J/erg} \times 10^{-3}\ \text{kJ/J}$$
$$= 243\ \text{kJ/mol}$$

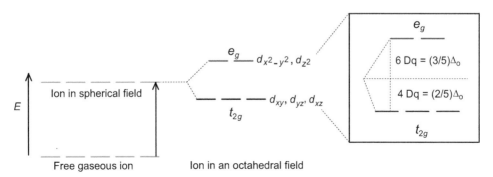

FIGURE 17.2
Splitting of the d orbitals in a crystal field of octahedral symmetry.

This energy ($243\ kJ\ mol^{-1}$) is large enough to give rise to other effects when a metal ion is surrounded by six ligands. However, only for a d^1 ion is the interpretation of the spectrum this simple. When more than one electron is present in the d orbitals, the electrons interact by spin-orbit coupling. Any transition of an electron from the t_{2g} to the e_g orbitals is accompanied by changes in the coupling scheme when more than one electron is present. As we shall see in Chapter 18, the interpretation of spectra to determine the ligand field splitting in such cases is considerably more complicated that in the d^1 case.

The ordering of the energy levels for a metal ion in an octahedral field makes it easy to visualize how high and low spin complexes arise when different ligands are present. If there are three or fewer electrons in the $3d$ orbitals of the metal ion, they can occupy the t_{2g} orbitals with one electron in each orbital. If the metal ion has a d^4 configuration (e.g. Mn^{3+}), the electrons can occupy the t_{2g} orbitals only if pairing occurs, which requires that Δ_o be larger in magnitude than the energy necessary to force electron pairing. The result is a low spin complex in which there are two unpaired electrons. If Δ_o is smaller than the pairing energy, the fourth electron will be in one of the e_g orbitals, which results in a high spin complex having four unpaired electrons. These cases are illustrated by the following diagrams.

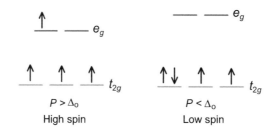

Of course, we have not yet fully addressed the factors that are responsible for the magnitude of the ligand field splitting. The splitting of the d orbitals by the ligands depends on the nature of the metal ion and the ligands as well as the extent of back donation and π bonding to the ligands. These topics will be discussed more fully in Section 17.3 and Chapter 18.

17.2 SPLITTING OF d ORBITAL ENERGIES IN FIELDS OF OTHER SYMMETRY

Although the effect on the d orbitals produced by a field of octahedral symmetry has been described, we must remember that not all complexes are octahedral or even have six ligands bonded to the metal ion. For example, many complexes have tetrahedral symmetry so we need to determine the effect of a tetrahedral field on the d orbitals. Figure 17.3 shows a tetrahedral complex that is circumscribed in a cube. Also shown are lobes of the d_{z^2} orbital and two lobes (those lying along the x-axis) of the $d_{x^2-y^2}$ orbital.

Note that in this case none of the d orbitals will point directly at the ligands. However, the orbitals that have lobes lying along the axes ($d_{x^2-y^2}$ and d_{z^2}) are directed toward a point that is midway along a diagonal of a face of the cube. That point lies at ($2^{1/2}/2$)l from each of the ligands. The orbitals that have lobes projecting between the axes (d_{xy}, d_{yz}, and d_{xz}) are directed toward the midpoint of an edge

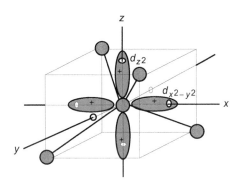

FIGURE 17.3
A tetrahedral complex shown with the coordinate system. Two lobes of the d_{z^2} orbital are shown along the *z*-axis and two lobes of the $d_{x^2-y^2}$ orbital are shown along the *x*-axis.

that is only $l/2$ from sites occupied by ligands. The result is that the d_{xy}, d_{yz}, and d_{xz} orbitals are higher in energy than are the $d_{x^2-y^2}$ and d_{z^2} orbitals because of the difference in how close they are to the ligands. In other words, the splitting pattern produced by an octahedral field is inverted in a tetrahedral field. The magnitude of the splitting in a tetrahedral field is designated as Δ_t and the energy relationships for the orbitals are shown in Figure 17.4.

There are several differences between the splitting in octahedral and tetrahedral fields. Not only are the two sets of orbitals inverted in energy, but also the splitting in the tetrahedral field is much smaller than that produced by an octahedral field. First, there are only four ligands producing the field rather than the six ligands present in the octahedral complex. Second, none of the *d* orbitals point directly at the ligands in the tetrahedral field. In an octahedral complex, two of the orbitals point directly toward the ligands and three point between them. As a result, there is a maximum energy splitting effect on the *d* orbitals in an octahedral field. In fact, it can be shown that if identical ligands are present in the complexes and the metal to ligand distances are identical, $\Delta_t = (4/9)\Delta_o$. The result is that *there are no low spin tetrahedral complexes* because the splitting of the *d* orbitals is not large enough to force electron pairing. Third, because there are only four ligands surrounding the metal ion in a tetrahedral field, the energy of all of the *d* orbitals is raised less than they are in an octahedral complex. The subscripts "*g*" do not appear on the subsets of orbitals because there is no center of symmetry in a tetrahedral structure.

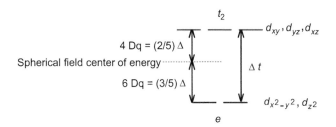

FIGURE 17.4
The orbital splitting pattern in a tetrahedral field that is produced by four ligands.

Suppose we start with an octahedral complex and place the ligands lying on the z-axis farther away from the metal ion. As a result, the d_{z^2} orbital will experience less repulsion, and its energy will decrease. However, not only do the five d orbitals obey a "center of gravity" rule for the complete set, but also each subset has a center of energy that would correspond to spherical symmetry for that subset. Therefore, if the d_{z^2} orbital is reduced in energy, the $d_{x^2-y^2}$ orbital must increase in energy to correspond to an overall energy change of zero for the e_g subset. The d_{xz} and d_{yz} orbitals have a z component to their direction. They project between the axes in such a way that moving ligands on the z-axis farther from the metal ion reduces repulsion of these orbitals. As a result, the d_{xz} and d_{yz} orbitals have lower energy, which means that the d_{xy} orbital has higher energy in order to preserve the center of energy (2) for the t_{2g} orbitals. The result is a set of d orbitals that are arranged as shown in Figure 17.5(a). With the metal to ligand bond lengths being greater in the z direction, the field is now known as a *tetragonal* field with z elongation. If the ligands on the z-axis are forced closer to the metal ion to produce a tetragonal field with z compression, the two sets of orbitals are inverted, and Figure 17.5(b) shows the d orbitals in this type of field.

From the survey of coordination chemistry presented in Chapter 16, it was seen that there are numerous complexes in which the ligands lie in a square plane around the metal ion. A square planar complex can be considered as a tetragonal complex in which the ligands along the z-axis have been drawn to an infinite distance from the metal ion. The arrangement of the d orbitals in such a complex is like that shown for the z elongation, except the splitting is much more pronounced with the d_{xy} lying above the d_{z^2}. The energy level diagram for the d orbitals in a square planar field is shown in Figure 17.6. It can be shown that the energy separating the d_{xy} and $d_{x^2-y^2}$ orbitals is exactly Δ_o, the splitting between the t_{2g} and e_g orbitals in an octahedral field.

In Chapter 16, we saw that d^8 ions such as Ni^{2+}, Pd^{2+}, and Pt^{2+} form square planar complexes that are diamagnetic. From the orbital energy diagram shown in Figure 17.6, it is easy to see why. Eight electrons can pair in the four orbitals of lowest energy leaving the $d_{x^2-y^2}$ available to form a set of dsp^2 hybrid orbitals. If the difference in energy between the d_{xy} and the $d_{x^2-y^2}$ is not sufficient to force electron pairing, none of the d orbitals is unoccupied, and a complex having four bonds would be expected to utilize sp^3 hybrid orbitals, which would result in a tetrahedral structure.

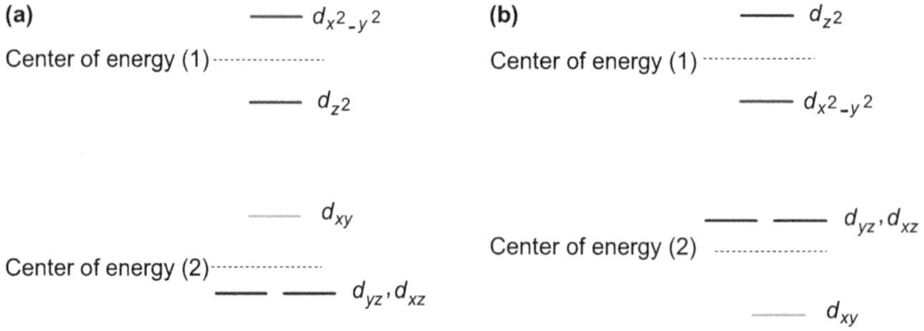

(a)

$\overline{\quad}\ d_{x^2-y^2}$

Center of energy (1) ··················

$\overline{\quad}\ d_{z^2}$

$\overline{\quad}\ d_{xy}$

Center of energy (2)··················

$\overline{\quad}\ \overline{\quad}\ d_{yz}, d_{xz}$

(b)

$\overline{\quad}\ d_{z^2}$

Center of energy (1) ··················

$\overline{\quad}\ d_{x^2-y^2}$

$\overline{\quad}\ \overline{\quad}\ d_{yz}, d_{xz}$

Center of energy (2) ··················

$\overline{\quad}\ d_{xy}$

FIGURE 17.5

The arrangement of the d orbitals according to energy in fields with elongation by moving the ligands on the z-axis farther from the metal ion (a) and with compression along the z-axis (b).

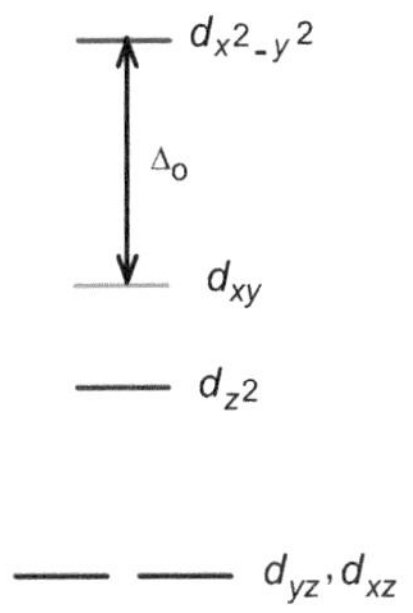

FIGURE 17.6
Energies of *d* orbitals in a square planar field produced by four ligands.

It is interesting to show how the energies of the *d* orbitals are evaluated by making use of quantum mechanics. A brief description of the approach used will be described here. Transition metal ions do not have the spherical symmetry that results for ions having all filled shells (e.g. Na^+, F^-, Ca^{2+}, O^{2-}, etc.). The angular or directional character of the *d* orbitals renders them nonequivalent in their interaction with ligands located around the metal ion in a specific geometric pattern. Ligands arranged in different ways around a metal ion will produce different effects on the *d* orbitals that result in their having different energies. The effect is the manifestation of electrostatic repulsion of an electron in a *d* orbital on the metal by the charges on the ligands (which are considered as points in space). For an electron in the d_{z^2} orbital of the metal ion, the energy can be expressed in terms of a quantum mechanical integral following the same procedure for calculating the average or expectation value for a dynamical variable. In this case, the calculation involves the wave function for the orbital and the operator for potential energy, V,

$$E = \int_{\text{all space}} \Psi^*(d_{z^2}) \; V \; \Psi(d_{z^2}) \; d\tau \tag{17.1}$$

Evaluation of the integral yields a result that can be expressed in terms of factors that are written as

$$35qe^2/4R^5 \quad \text{and} \quad (2/105)\langle a^4 \rangle$$
$$D \qquad \quad \text{and} \qquad q$$

which is usually written as the result when D and q are multiplied, an energy unit known as "Dq" results.

$$Dq = \frac{qe^2 \langle a^4 \rangle}{6R^5} \tag{17.2}$$

When expressed in terms of the crystal field splitting Δ_o, the result is

$$\Delta_o = 10 \, Dq = \frac{5qe^2 \langle a^4 \rangle}{3R^5} \tag{17.3}$$

The orbital energy is obtained by multiplying the value of Dq by a factor that takes into account the angular dependence of each orbital. Depending on which *d* orbital is being considered, the angular

Table 17.1 Energies of d-Orbitals in Ligand Fields in Dq Units

Coord. No.	Symmetry	d_{z^2}	$d_{x^2-y^2}$	d_{xy}	d_{xz}	d_{yz}
2	Linear[a]	10.28	−6.28	−6.28	1.14	1.14
3	Trigonal[b]	−3.21	5.46	5.46	−3.86	−3.86
4	Tetrahedral	−2.67	−2.67	1.78	1.78	1.78
4	Sq. plane[b]	−4.28	12.28	2.28	−5.14	−5.14
5	Trigonal bipy	7.07	−0.82	−0.82	−2.72	−2.72
5	Sq. pyramid	0.86	9.14	−0.86	−4.57	−4.57
6	Octahedral	6.00	6.00	−4.00	−4.00	−4.00
6	Trigonal prism	0.96	−5.84	−5.84	5.36	5.36
8	Cubic	−5.34	−5.34	3.56	3.56	3.56

[a]The ligands are on the z-axis, the axis of highest symmetry.
[b]The ligands are in the xy plane.

factor is +6 and −4 for d orbitals in an octahedral field. Therefore, the energies of the d orbitals are +6 Dq (for the d_{z^2} and $d_{x^2-y^2}$ orbitals) or −4 Dq (for the d_{xy}, d_{yz}, and d_{xz} orbitals). If the ligands are polar molecules, the charge, q, is replaced by the dipole moment, μ.

For a metal ion in a tetrahedral field, there are only four ligands and none of them lie on the axes. Because $4/6 = 2/3$, a factor of $2/3$ is applied to the potential energy operator V for the octahedral case. Also, evaluation of the integral giving the angular portion of the wave function gives a factor of $−2/3$ for the d_{z^2} and $d_{x^2-y^2}$ orbitals and $+2/3$ for the d_{xy}, d_{yz}, and d_{xz} orbitals. Multiplied together, a factor of $−4/9$ is obtained so the energies of the orbitals in the tetrahedral field are $−4/9(+6 \text{ Dq}) = −2.67$ Dq (for the d_{z^2} and $d_{x^2-y^2}$) orbitals and $−4/9(−4 \text{ Dq}) = 1.78$ Dq (for the d_{xy}, d_{yz}, and d_{xz} orbitals). These results and those shown above for an octahedral complex agree with the results given earlier. When similar calculations are carried out for fields having other symmetry, the orbital energies shown in Table 17.1 are obtained. The energies are based on the splitting pattern of the orbitals and their energies relative to the center of energy in a spherical field.

The orbital energies are useful when comparing the stability of complexes in different structures. As usual, electrons are placed in the orbitals starting with the orbitals of lowest energy. We will have opportunity to make use of the orbital energies when we consider reactions of complexes in which the transition state has a different structure than that of the starting complex (see Chapter 20).

17.3 FACTORS AFFECTING Δ

Because the crystal field splitting arises from the interaction of ligands with metal orbitals, it should be expected that the magnitude of the splitting would depend on the nature of the metal ion and the ligand. Earlier in this chapter, the discussion of $[Ti(H_2O)_6]^{3+}$ illustrated how the maximum in the single absorption band in the spectrum corresponds directly to the magnitude of Δ_o. If several complexes of Ti^{3+} are prepared using different ligands, the positions of the absorption bands will be shifted to higher or lower wave numbers depending on the nature of the ligand. In this way, it is possible to arrange ligands according to their ability to cause a ligand field splitting. The series of ligands arranged this way is known as the *spectrochemical series,* and the order for numerous ligands is as follows:

$$CO > CN^- > NO_2^- > en > NH_3 \approx py > NCS^- > H_2O > ox > OH^- > F^- > Cl^- > SCN^- > Br^- > I^-$$
Strong field   Weak field

Differences between the splittings produced by adjacent members in the series are small, and the order is approximate in some cases, especially when different metal ions or metal ions in different rows of the periodic table are considered. For example, the order of the halide ions is changed if the metal ion is from the second transition series in accord with the hard-soft interaction principle. The spectrochemical series is a very useful guide because reversals between rather widely separated members such as NO_2^- and NH_3 do not occur. Reversals in the order of the magnitude of the ligand field from that shown above are sometimes observed for *closely spaced* members of the series. Ligands that force electron pairing are known as *strong field* ligands, and these ligands can be expected to give low spin octahedral complexes with first row metal ions. *Weak field* ligands such as F^- and OH^- will normally give low spin complexes only with second and third row metals.

In general, the aqua complexes of first row transition metals have ligand field splittings of approximately 8,000–10,000 cm^{-1} for the +2 complexes, $[M(H_2O)_6]^{2+}$, and approximately 14,000–21,000 cm^{-1} for the complexes of the +3 ions, $[M(H_2O)_6]^{3+}$. In most cases, there is a 50–100% increase in Δ_o for the +3 ion compared to the +2 ion of the same metal. For example, the value of Δ_o is 22,870 cm^{-1} for $[Co(NH_3)_6]^{3+}$ but it is only 10,200 cm^{-1} for $[Co(NH_3)_6]^{2+}$. Table 17.2 shows representative values for Δ_o in complexes containing ions of first row metals with H_2O, NH_3, F^-, and CN^- as ligands. It is interesting to note that for $[CoF_6]^{3-}$, the value of Δ_o is only about 13,000 cm^{-1}, but the energy required to force pairing of electrons in Co^{3+} is approximately 20,000 cm^{-1}. Therefore, $[CoF_6]^{3-}$ is a high spin complex. The crystal field splitting in $[Co(NH_3)_6]^{3+}$ is 22,900 cm^{-1}, which is sufficient to force pairing so a low spin complex results.

There is an increase of approximately 30–50% in Δ_o on going from a first row transition metal to a second row metal and another 30–50% increase on going from a second row to a third row metal when the same d^n configuration and oxidation state are involved. Data for several complexes that illustrate this trend are shown in Table 17.3. In some cases, the splitting is approximately doubled in going down one row in the transition series. A result of the much larger ligand field splitting in second

Table 17.2 Values of Δ_0 for Octahedral Complexes of First Row Transition Metal Ions

Metal Ion	Splittings for Several Ligands in cm^{-1}			
	F^-	H_2O	NH_3	CN^-
Ti^{3+}	17,500	20,300		23,400
V^{3+}	16,100	18,500		26,600
Cr^{3+}	15,100	17,900	21,600	35,000
Fe^{3+}	14,000	14,000		34,800
Co^{3+}	13,000	20,800	22,900	33,800
Fe^{2+}		10,400	11,000	32,200
Co^{2+}		9,300	10,200	
Ni^{2+}		8,500	10,800	

Table 17.3 Variation in Δ_0 with Row in the Periodic Table. Ligand Field Splitting (in cm^{-1}) is Shown Below the Formula for the Complexes Shown

$[Fe(H_2O)_6]^{3+}$	$[Fe(ox)_3]^{3-}$	$[Co(H_2O)_6]^{3+}$	$[Co(NH_3)_6]^{3+}$
14,000	14,140	20,800	22,900
$[Ru(H_2O)_6]^{3+}$	$[Ru(ox)_3]^{3+}$	$[Rh(H_2O)_6]^{3+}$	$[Ru(NH_3)_6]^{3+}$
28,600	28,700	27,200	34,000
			$[Ir(NH_3)_6]^{3+}$
			41,200

and third row transition metal ions is that almost all complexes of these metals are low spin. When a substitution reaction occurs in such a complex, the ligand being replace almost always leaves before the entering ligand attaches, and the substitution occurs with the product having the same configuration as the starting complex.

From a more complete set of data similar to those shown in Table 17.3, it is possible to rank metal ions in terms of the splitting of the d orbitals produced by a given ligand. The series for many common metal ions can be given as follows:

$$Pt^{4+} > Ir^{3+} > Pd^{4+} > Ru^{3+} > Rh^{3+} > Mo^{3+} > Mn^{3+} > Co^{3+} > Fe^{3+} > V^{2+} > Fe^{2+} > Co^{2+} > Ni^{2+}$$

This series illustrates clearly the effects of charge and position in the periodic table that were described above.

As described earlier, the splitting in tetrahedral fields is usually only about 4/9 what it is for octahedral fields. For example, the tetrahedral complex $[Co(NH_3)_4]^{2+}$ has $\Delta_t = 5,900\ cm^{-1}$, whereas the octahedral complex $[Co(NH_3)_6]^{2+}$ has $\Delta_o = 10,200\ cm^{-1}$. When complexed with the Co^{2+} ion, the values of Δ_t for the Cl^-, Br^-, I^-, and NCS^- ions are 3,300, 2,900, 2,700, and 4,700 cm^{-1}, respectively. In general, the energy required to force pairing of electrons in a first row transition metal ion is in the range of $250-300\ kJ\ mol^{-1}$ (approximately $20,000-25,000\ cm^{-1}$). The result is that the splitting caused by ligands in a tetrahedral field is not sufficient to cause pairing of electrons so there are no low spin tetrahedral complexes of first row metal ions.

17.4 CONSEQUENCES OF CRYSTAL FIELD SPLITTING

In addition to the changes produced in the spectra that arise from $d-d$ transitions, there are other effects produced as a result of the splitting of the d orbital energies. Suppose a gaseous metal ion is placed in water and the ion becomes solvated with six water molecules. If the $+2$ first row transition metal ions are considered, there is a general increase in the heat of hydration in moving across the series as a result of the decreasing ionic radius brought on by the increase in nuclear charge as illustrated by the ionic radii shown in the following series.

Ion	Ca^{2+}	Ti^{2+}	V^{2+}	Cr^{2+}	Mn^{2+}	Fe^{2+}	Co^{2+}	Ni^{2+}	Cu^{2+}	Zn^{2+}
Radius, pm	99	90	88	84	80	76	74	69	72	74

The hydration process for an ion can be shown as

$$M^{z+}(g) + 6 \; H_2O(l) \rightarrow \left[M(H_2O)_6\right]^{z+}(aq) \tag{17.4}$$

The heat of hydration of an ion is related to its size and charge (see Chapter 7). However, in this case the aqua complex that is formed causes the d orbitals to be split in energy, and if the metal ion has electrons in the d orbitals, they will populate the t_{2g} orbitals, which have lower energy. This results in a release of energy over and above that produced by the hydration of an ion having a specific size and charge. High spin (weak field) aqua complexes normally result from the hydration of first row transition metal ions. The actual amount of energy released will be dependent on the number of electrons in the d orbitals. For a d^1 ion, the heat of hydration will be augmented by 4 Dq (see Figure 17.2). If the electron configuration is d^2, there will be 8 Dq released in addition to the heat of hydration for an ion of that size and charge. The process of increasing the number of electrons that are present in the d orbitals would produce the results shown in Table 17.4.

Hydration of the metal ions produces an enthalpy change that is commensurate with the size and charge of the ion with the addition of the number of Dq units shown in the weak field column in Table 17.4. For d^0, d^5, and d^{10} there is no additional stabilization of the aqua complex because in these cases there is no ligand field stabilization. Figure 17.7 shows a graph of the heats of hydration for the first row $+2$ metal ions. The graph shows what has become known as the "double humped" appearance that reflects the fact that the ligand field stabilization energy for the aqua complexes begins at 0, increases to 12 Dq, then drops to 0 on going from d^0 to d^5 and repeats the trend on going from d^6 to d^{10} (see Table 17.4).

As gaseous ions form a crystal lattice as represented by Eqn (17.5), the cations become surrounded by anions and vice versa in the solid crystal.

$$M^{2+}(g) + 2 \; X^-(g) \rightarrow MX_2(s) \tag{17.5}$$

Table 17.4 Ligand Field Stabilization Energies in Dq Units

Number of electrons	Weak Field (Dq)	Strong Field (Dq)
0	0	0
1	4	4
2	8	8
3	12	12
4	6	16
5	0	20
6	4	24
7	8	18
8	12	12
9	6	6
10	0	0

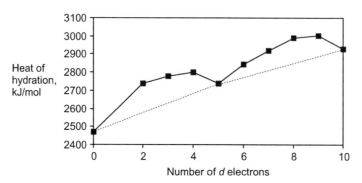

FIGURE 17.7
Heat of hydration of +2 metal ions of the first transition series.

If the cations are surrounded by six anions in an octahedral arrangement, the d orbitals will become split in energy as illustrated earlier. When the orbitals are populated with one or more electrons, the energy released when the lattice forms will be greater by an amount that reflects the ligand field stabilization energy corresponding to the number of electrons in the d orbitals. A graph of the lattice energy for the chloride compounds of the first row transition metals in the +2 oxidation state is shown in Figure 17.8. This figure shows the same general shape as the graph representing the hydration enthalpies of the metal ions. This is because the ligand field around the metal ion stabilizes both the hydrates and the crystal lattices in a way that depends on the number of electrons in the d orbitals.

Although the heat of hydration (or lattice energy) can be predicted for an ion having a specific charge and size, measuring the actual heat of hydration (or lattice energy) is not a good way to determine by difference the ligand field stabilization energy. It is large enough to produce demonstrable effects, but the ligand stabilization energy is small compared to either the heat of

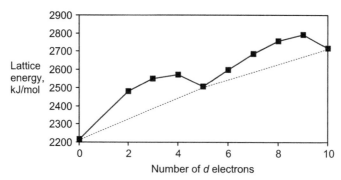

FIGURE 17.8
Lattice energies of the chlorides of the +2 metal ions of the first transition series.

hydration of a doubly charged metal ion or the lattice energy of a solid ionic compound. As a result, taking a small difference between large numbers is not a good way to determine Dq. In Chapter 18, the use of spectroscopic techniques to determine ligand field parameters will be described.

17.5 JAHN–TELLER DISTORTION

Although the application of elementary ligand field theory is adequate to explain many properties of complexes, there are other factors that come into play in some cases. One of those cases involves complexes that have structures that are distorted from regular symmetry. Complexes of copper(II) are among the most common ones that exhibit such a distortion.

Consider the Cu^{2+} ion, which has a d^9 configuration. If six ligands are arranged in a regular octahedral structure, the electrons would be arranged as shown in Figure 17.9(a). There are three electrons populating the two degenerate e_g orbitals. If the ligands on the z-axis were moved farther from the Cu^{2+}, the d_{z^2} orbital would have a lower energy, but the $d_{x^2-y^2}$ orbital would be raised in energy an equivalent amount. There would be a splitting of the d_{xy}, d_{yz}, and d_{xz} orbitals, but because they are filled there is no net energy change based on that splitting. However, two of the electrons in the $d_{x^2-y^2}$ and d_{z^2} orbitals would occupy the d_{z^2} orbital, whereas only one would be in the higher energy $d_{x^2-y^2}$ orbital. As a result, the energy of this arrangement would be lower than that for the complex having O_h symmetry. The splitting of the two sets of orbitals (e_g and t_{2g}) is not equal. The magnitude by which the $d_{x^2-y^2}$ and d_{z^2} orbitals have changed in energy is δ, but the d_{xz} and d_{yz} are lowered by an amount represented as δ' as the d_{xy} orbital is raised by $2\delta'$ as shown in Figure 17.9(b).

Keep in mind that in this diagram the splittings are not to scale! As a result of the distortion by elongation of the complex along the z-axis, the overall energy is lower by an amount δ in accord with the Jahn–Teller Theorem. That principle states that *if a system has unequally populated degenerate orbitals, the system will distort to remove the degeneracy.* When the degeneracy is removed, the state of lower energy

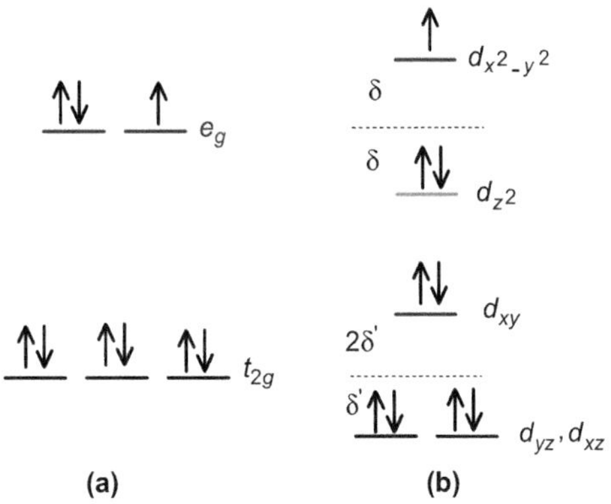

FIGURE 17.9
Energy of the *d* orbitals of a d^9 ion in an octahedral field (a) and with z elongation produced by Jahn-Teller distortion (b).

will be more fully populated. The resulting splitting pattern for the d orbitals is as shown in Figure 17.9(b).

One experimental verification of this phenomenon comes in the form of bond lengths. In $CuCl_2$, the Cu^{2+} is surrounded by six Cl^- ions, and the equatorial $Cu-Cl$ bonds are 230 pm in length, whereas the axial bonds measure 295 pm. Bond distances in many other copper(II) compounds could also be cited to illustrate Jahn–Teller distortion.

The d^9 configuration is not the only one for which distortion leads to a lower energy. For example, the d^4 high spin configuration would place one electron in the four lowest lying orbitals with an energy that is lower as a result of splitting the $d_{x^2-y^2}$ and d_{z^2} orbitals. The Cr^{2+} ion has a d^4 configuration so we should expect some distortion to occur for complexes of this ion and this is indeed the case. Moreover, a distortion would be predicted for a d^1 or d^2 ion, for which the electrons would be found in the two orbitals of lowest energy. However, the magnitude of δ' is much smaller than that of δ because the orbitals involved are directed between the axes, not at the ligands. These are nonbonding orbitals so the effect of moving the ligands on the z-axis farther from the metal ion is much smaller. The result is that any distortion involving metals with these configurations is very slight.

17.6 SPECTRAL BANDS

When a sample absorbs light in the visible region of the spectrum, the sample appears colored. The majority of complexes of transition metal ions are colored. The reason that these complexes absorb visible light is that there are electronic transitions of appropriate energy difference possible between the d orbitals. These orbitals are split according to the patterns that we have just described. The absorptions are not limited to the visible region and some are seen in the ultraviolet and infrared regions of the spectrum. However, there are striking differences between the absorption of light by $[Ti(H_2O)_6]^{3+}$ and by a hydrogen atom. As we saw in Chapter 1, the radiation emitted during electronic transitions in a hydrogen atom appear as *lines* in the spectrum. The absorption of light by $[Ti(H_2O)_6]^{3+}$ appears as a broad *band*, which has a maximum at 20,300 cm^{-1}.

In simple crystal field theory, the electronic transitions are considered to be occurring between the two groups of d orbitals of different energy. We have already alluded to the fact that when more than one electron is present in the d orbitals it is necessary to take into account the spin-orbit coupling of the electrons. In ligand field theory, these effects are taken into account as are the parameters that represent interelectronic repulsion. In fact, the next chapter will deal extensively with these factors.

A very simplistic view of the reason that absorption bands are broad arises from considering what happens during the vibration of a complex having the formula ML_6. The crystal field undergoes instantaneous changes during a vibration, and for simplicity, we will consider the symmetric stretching vibration that preserves the O_h symmetry of the complex. As a result of the ligands moving in and out from the metal ion, there will be a slight difference produced in Δ_o as the ligands are moving the small distance that accompanies the vibration. Because the electronic transition takes place on a time scale that is much shorter than that corresponding to vibration, the transition actually occurs between ligand

$$\int \psi_1 r \psi_2 d\tau$$

field states that are changing slightly during the vibration. As a result, a *range* of energies is absorbed, which results in a *band* rather than a *single line* as is the case for a gaseous atom or ion.

Transitions of the *d–d* type are known as *electric dipole* transitions. The transition between states of different multiplicity is forbidden but under certain circumstances still may be seen if only weakly. For example, Fe^{3+} has a ^6S ground state and all of the excited spectroscopic states have a different multiplicity. As a result, solutions containing Fe^{3+} are almost colorless because the transitions are spin forbidden. In order for an electric dipole transition to occur, it is necessary for the integral where ψ_1 and ψ_2 are the wave functions for the orbitals between which the transition is occurring to be nonzero. In an octahedral field, all of the *d* orbitals are "*g*", which would require the integral to be zero. The selection rule known as the *Laporte Rule* requires that *allowed transitions be between states of different symmetry*. Therefore, the *d–d* transitions in a field of *g* symmetry are *Laporte forbidden*. Forbidden transitions usually have low intensities so other factors must be operating in these cases.

For a particular bond (as in a diatomic molecule) in the lowest vibrational state within the lowest electronic level, there is an equilibrium internuclear distance, R_0. Suppose that by absorbing electromagnetic radiation there is a change to a higher electronic state. Excited states of bonds (or molecules) do not have equilibrium internuclear distances that are identical to those in the ground state. Electronic transitions occur on a time scale that is so short that nuclei can not reposition to a different equilibrium distance. Therefore, as the electronic state is changed from the ground state to an excited state the molecule is produced with internuclear distances that are not those characteristic of the excited state. If the energy of a particular bond is represented as a function of internuclear distance, the result is a potential energy curve like that shown in Figure 17.10.

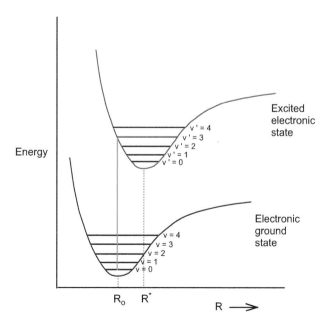

FIGURE 17.10
An illustration of the Franck–Condon principle. In this case, the transition is from $v = 0$ in the electronic ground state to the state with $v' = 3$ in the excited electronic state.

Several vibrational states are shown within the electronic ground state. A potential energy curve such as this also exists for the excited state except that it is displaced to a slightly larger equilibrium internuclear distance than is found for the ground state. Because the electronic transition to the excited state does not allow for the internuclear distance to adjust, the transition takes place with a constant value of R, a so-called *vertical transition*. Therefore, the transition with the highest probability is between the lowest vibrational level in the electronic ground state and a higher vibrational level in the excited electronic state. This phenomenon, known as the *Franck-Condon principle*, is illustrated in Figure 17.10. In fact, the overlap integrals for several vibrational levels have non-zero values, which results in the absorption of several energies that are closely spaced.

The intensity of the electronic transition described above is primarily due to changes in vibrational energy. Because both vibrational and electronic changes are involved, this type of transition is known as a *vibronic* transition. Although we will not go into details of the theory here, it can be shown that if changes in the complex occur that temporarily change its symmetry, electric dipole transitions become possible. One such change is *vibronic coupling*, which is the result of combining vibrational and electronic wave functions. The products of the wave functions must take into account *all* vibrational modes some of which change the symmetry of the complex by removing the center of symmetry (the subscript *g* no longer applies). The Laporte selection rule no longer applies so some transitions are no longer forbidden. Another factor that is associated with vibronic coupling involves an allowed transition that is of similar energy to an excited spectroscopic state that is forbidden. Vibrations in the excited state can result in a symmetry that does allow the wave functions to combine so that the transition integral is no longer zero.

As we have seen, spectral bands for absorption by transition metal complexes are broad. For absorptions to cover a range of frequencies, the energy levels must be "smeared out" rather than discrete. Earlier in this chapter we saw that one means by which the symmetry of a complex is changed is by Jahn–Teller distortion. Such a distortion has the effect of removing the degeneracy of the two sets of *d* orbitals as they exist in the ligand field. Another factor that causes broad absorption bands is that as the electronic energy is changed, so are vibrational energies, which causes a *range* of energy to be absorbed rather than a discrete line. Even though the bands are broad, they provide the basis for analysis of the spectrum of a transition metal complex to extract information about the ligand field and electronic interactions. Therefore, this important topic will be discussed in the next chapter.

17.7 MOLECULAR ORBITALS IN COMPLEXES

Up to this point in the discussion of bonding in complexes, the emphasis has been on ligand field theory and its application to interpretation of structures and magnetic properties of complexes. There are substantial differences in the effects produced by different ligands with the softer, more polarizable ligands causing the greatest effect. Ligands of that type attach to metal ions by means of bonds that are more covalent in character and have a greater tendency to form π bonds. The initial development of crystal field theory dealt with the effects caused by the electrostatic interaction of point charges in fields of specific geometry. Although the ligands were considered as *charges*, the metal was described in terms of its *orbitals* as interpreted from the point of view of quantum mechanics. Therefore, in its simplest approach crystal field theory does not deal with complexes that have substantial covalency in the bonds. In this section, the molecular orbital approach will be used to describe bonding in complexes by considering the ligands to be only σ donors and then making some additional comments about including π bonding. However, there will be no attempt to develop bonding theory to a level that makes interpretations and applications obscure. Bonding theory for the sake of theory is more appropriate in textbooks in other fields of chemistry.

When describing a complex in terms of molecular orbitals, we need to establish a model by which we can identify the orbitals utilized by both the metal and the ligands. We will first consider an octahedral complex with the positions of the ligands identified on the coordinate system shown in Figure 17.11, and the orbitals will be designated by the numbers assigned to the ligands in the positions indicated.

The basic approach to combining wave functions to represent molecular orbitals was described in Chapter 5, and the procedures used here represent an extension of those ideas. In dealing with the bonding in complexes of first row transition metal atoms and ions, it is generally acknowledged that the metal orbital energies are $3d < 4s < 4p$. Although there are nine valence orbitals having these designations, those of the $3d$ set that are used in bonding are the d_{z^2} and the $d_{x^2-y^2}$. In the case of the p and d orbitals, the orbitals utilized are those indicated that have lobes lying along the axes. The d_{xy}, d_{yz}, and d_{xz} orbitals have lobes lying *between* the axes. As a result, they do not interact with ligands lying *on* the axes, which have lobes that are unidirectional. If the ligands form π bonds to the metal ion, then these orbitals become part of the bonding orbital scheme. When only σ bonding is involved, the d_{xy}, d_{yz}, and d_{xz} are considered to be nonbonding in character.

If we consider six ligands such as H_2O or NH_3 bound to a metal ion, the first task in applying the molecular orbital approach is to deduce the form of the molecular wave functions. Ligands such as H_2O and NH_3 do not form π bonds so this type of complex has fewer subtle factors to take into account. Each ligand contributes a pair of electrons that are contained in an orbital, and we will assume that it is a σ orbital. At this point, the actual type of the ligand orbital (sp^3, p, etc.) is not important as long as it is one that does not form π bonds. For a first row transition metal, the valence shell orbitals used in the bonding will be the $3d_{z^2}$, $3d_{x^2-y^2}$, $4s$ and three $4p$ orbitals. As shown in Figure 17.12(a), the s orbital requires the wave functions for ligand orbitals to have positive signs.

The $4s$ orbital on the metal has a positive sign that is invariant in all directions. The combination of ligand orbitals, which are referred to as *symmetry adjusted linear combinations* (SALC) or *ligand group*

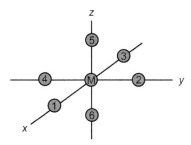

FIGURE 17.11
The coordinate system to designate orbitals used in constructing molecular orbitals for an octahedral complex.

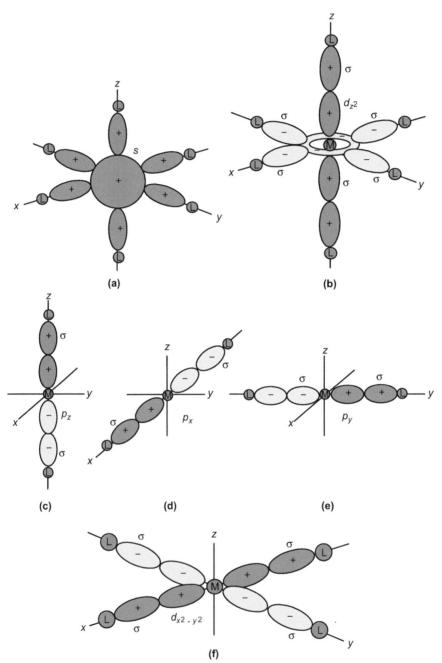

FIGURE 17.12
Combinations of ligand orbitals with the s, p_x, p_y, p_z, d_{z^2}, and $d_{x^2-y^2}$ orbitals on the metal ion.

orbitals (LGO), matching the symmetry of the *s* orbital will be the *sum* of the six ligand wave functions, ϕ_i which is written as

$$\frac{1}{\sqrt{6}}(\phi_1 + \phi_2 + \phi_3 + \phi_4 + \phi_5 + \phi_6)$$

Thus, the combination of wave functions making up the first molecular orbital will be

$$\psi(\sigma_{4s}) = a_1\ \phi_{4s} + a_2\left[\frac{1}{\sqrt{6}}(\phi_1 + \phi_2 + \phi_3 + \phi_4 + \phi_5 + \phi_6)\right] \tag{17.6}$$

where a_1 and a_2 are weighting coefficients, ϕ_{4s} is the wave function for the 4s orbital on the metal, and ϕ_i is a wave function for a ligand orbital. When combining ligand group orbitals with the *p* orbitals on the metal, it is necessary to observe the symmetry of the metal orbitals and combine the orbitals from the ligands to match that symmetry. Figure 17.12(c–e) provides an illustration of this requirement.

Because the p_x orbital is positive in the direction toward ligand 1 and negative in the direction toward ligand 3, the appropriate combination of ligand orbitals will be

$$\frac{1}{\sqrt{2}}(\phi_1 - \phi_2)$$

Consequently, the molecular wave function utilizing the p_x orbital can be expressed as

$$\psi(\sigma_{p_x}) = a_3\phi_{p_x} + a_4\left[\frac{1}{\sqrt{2}}(\phi_1 - \phi_2)\right] \tag{17.7}$$

In a similar way, we obtain the wave functions for combinations of ligand group orbitals with the p_y and p_z orbitals from the metal ion. They can be written as

$$\psi\left(\sigma_{p_y}\right) = a_3\phi_{p_y} + a_4\left[\frac{1}{\sqrt{2}}(\phi_3 - \phi_4)\right] \tag{17.8}$$

$$\psi(\sigma_{p_z}) = a_3\phi_{p_z} + a_4\left[\frac{1}{\sqrt{2}}(\phi_5 - \phi_6)\right] \tag{17.9}$$

Note that the three wave functions arising from the p_x, p_y, and p_z orbitals are identical except for the directional character as indicated by the subscripts. Therefore, they must represent a triply degenerate set (the t_{2g} set, which will be apparent later).

When we inspect the $d_{x^2-y^2}$ orbital, we see that the lobes have positive signs in the *x* direction and negative signs in the *y* direction as shown in Figure 17.12(f). The ligand group orbitals that match the symmetry of the $d_{x^2-y^2}$ orbital will be

$$\psi\left(\sigma_{d_{x^2-y^2}}\right) = a_5 \, \phi_{d_{x^2-y^2}} + a_6 \left[\frac{1}{2}(\phi_1 - \phi_2 + \phi_3 - \phi_4)\right] \qquad (17.10)$$

In looking at the d_{z^2} orbital, we note that it has positive lobes in the z direction with the ring in the xy plane having negative symmetry. Because the d_{z^2} orbital has positive lobes along the z-axis, the ligand orbitals in those directions will both have positive signs. With the "ring" being of negative symmetry, ligand orbitals in the xy plane will all be preceded by a negative sign. However, the orbital labeled as d_{z^2} is actually $d_{2z^2-x^2-y^2}$, which is needed in order the determine the coefficients of the orbitals in the wave function. The combination of ligand orbitals needed to match the symmetry of this orbital can be written as

$$\frac{1}{2\sqrt{3}}(2\phi_5 + 2\phi_6 - \phi_1 - \phi_2 - \phi_3 - \phi_4)$$

Therefore, the molecular orbital for this combination with the d_{z^2} orbital is written as

$$\psi(\sigma_{z^2}) = a_7 \, \phi_{z^2} + a_8 \left[\frac{1}{2\sqrt{3}}(2\phi_5 + 2\phi_6 - \phi_1 - \phi_2 - \phi_3 - \phi_4)\right] \qquad (17.11)$$

When constructing the molecular orbital energy level diagram, we make use of the fact that the energy of the atomic orbitals on the metal (assuming a first row metal) vary in the order $3d < 4s < 4p$. Moreover, the orbitals on ligands such as H_2O, NH_3, or F^- that hold the electrons being donated to the metal are usually lower in energy than any of those metal orbitals listed. For the ligands listed, the electrons reside in $2p$ orbitals or in hybrid orbitals made up of $2s$ and $2p$ orbitals (sp^3 in the case of H_2O and NH_3). As a result, when the energy level diagram is constructed, the orbitals on the metal side will be in the order given, but the ligand orbitals on the other side will lie lower than those of the metal. The result is a molecular orbital diagram like that shown in Figure 17.13.

Several things are apparent from Figure 17.13. First, the crystal field splitting, Δ_o, is represented as the difference in energy between a triply degenerate set of orbitals, t_{2g}, and a doubly degenerate set, e_g. However, according to the molecular orbital approach those orbitals are the three nonbonding $3d_{xy}$, $3d_{yz}$, and the $3d_{xz}$ and the antibonding e_g^* arising from the $3d_{x^2-y^2}$ and the $3d_{z^2}$ orbitals. Second, if six ligands donate six pairs of electrons to the metal ion, they will occupy the a_{1g}, e_g, and t_{1u} *molecular* orbitals. Because the bonding molecular orbitals lie closer in energy to the ligand orbitals, they are more like ligands orbitals than metal orbitals. The nonbonding t_{2g} and antibonding e_g^* have energies that are closer to those of the valence shell orbitals of the metal. From the molecular orbital diagram, we can see that for complexes containing metal ions that have 1, 2, or 3 electrons, the nonbonding orbitals are available (after the six bonding orbitals are filled by electrons donated from the ligands) so the configurations will be $(t_{2g})^1$, $(t_{2g})^2$, or $(t_{2g})^3$, respectively. With the configuration d^4 the electron configuration will be $(t_{2g})^4$ or $(t_{2g})^3(e_g^*)^1$ depending on the magnitude of Δ_o. This was also the case when bonding was described using the ligand field approach. It is also interesting to note that without placing electrons in any of the antibonding orbitals, the maximum number of electrons that can be accommodated is 18. It is thus apparent that the 18-electron rule corresponds to a closed shell arrangement of the bonding molecular orbitals.

As was shown earlier in this chapter, when four ligands are placed around a metal in a tetrahedral arrangement, the d orbitals are split into two sets, the t_2 and e subgroups (which are the d_{xy}, d_{yz}, and d_{xz} and the $3d_{x^2-y^2}$ and $3d_{z^2}$ orbitals, respectively). The interactions of the orbitals on the metal (which we

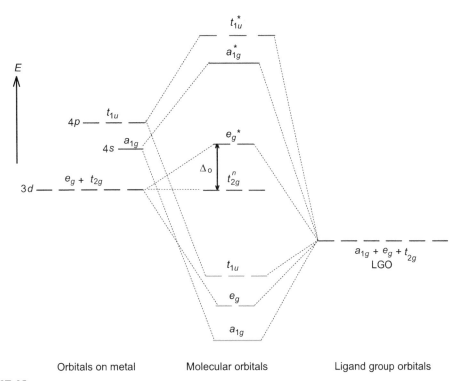

FIGURE 17.13
The molecular orbital energy level diagram for an octahedral complex.

will consider to be a first row metal) and ligands combine with different requirements on the symmetry of the ligand group orbitals. The four ligands can be considered on alternate corners of a cube containing the metal in the center. This structure is shown in Figure 17.14. Also shown in the figure are the orbitals on the four ligands that interact with the metal orbitals. The ligand orbitals are represented by positive and negative lobes as if they might be p orbitals, but they can be σ bonding orbitals of other types such as sp^3 orbitals on ammonia molecules. By visualizing in turn each of the d orbitals it can be seen how the combinations of metal orbitals and ligand group orbitals arise.

As was shown in Chapter 5 (see Table 5.2), the s and p orbitals in tetrahedral symmetry become a_1 and t_2, respectively. Because the $4s$ orbital on the metal is symmetric in all directions, it gives σ orbitals with the four ligand group orbitals (which give an a_1 orbital and a set of t_2 orbitals). However, the $3d_{x^2-y^2}$ and $3d_{z^2}$ orbitals, which have lobes lying along the axes, are pointing between the ligand orbitals. For example, in Figure 17.14, the positive orbital on ligand 1 points toward the metal and hence toward the d_{z^2} orbital. This is canceled by the negative lobe of ligand 3, which points toward the d_{z^2}. The result is that there is no *net* overlap between the ligands and the d_{z^2} orbital so it is nonbonding in character. By examining Figure 17.14 in terms of the interaction of the ligand metal orbitals we would find that the $3d_{x^2-y^2}$ orbital is also nonbonding. The $3d_{x^2-y^2}$ and $3d_{z^2}$ orbitals are identified as the e orbitals. Note that a tetrahedron does not have a center of symmetry so the "g" subscript is not used.

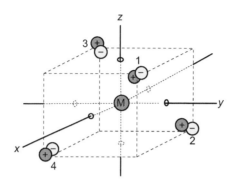

FIGURE 17.14
A tetrahedral complex with the lobes of the $3d_{x^2-y^2}$ and $3d_{z^2}$ orbitals directed between the ligands. These are nonbonding orbitals in a tetrahedral complex. The ligand orbitals have the positive lobes on ligands 1 and 2 pointed toward the metal, but ligands 3 and 4 have the negative lobes pointed toward the metal.

When we consider the $4p_x$ orbital, we see that the positive lobe projects in the x direction along the axis. Therefore, the appropriate combination of ligand orbitals to match the symmetry of that lobe requires a positive sign on the combination. The lobe that projects along the negative x-axis projects between two ligands so the combination of ligands on the two nearest corners of the cube (3 and 4) must be negative. We could progress in a similar way to find that the d_{xy}, d_{yz}, and d_{xz} orbitals, which were nonbonding in an octahedral complex are now σ bonding, but they also have the ability to combine with ligand orbitals to form π bonds. Although we will not write out the molecular wave functions as was done for the octahedral case, a qualitative molecular orbital diagram that results is shown in Figure 17.15.

From a consideration of the combination of ligand and metal orbitals, it should be apparent that the overlap is much more effective in an octahedral complex (where orbitals are directed at ligands) than in a tetrahedral complex (where orbitals are directed between ligands). The result is that the energy difference between the e and t_2^* orbitals in a tetrahedral complex is much smaller than that between the t_{2g} and e_g orbitals in an octahedral complex. As we saw when considering the two types of complexes by means of ligand field theory, Δ_t is only about half as large as Δ_o in most cases.

In addition to complexes having octahedral and tetrahedral geometry, there are numerous examples of square planar complexes. As we saw in Chapter 16, such cases are most prevalent when the metal is a d^8 ion and the ligands generate strong ligand fields. When considering orbital overlap between the metal and ligand orbitals, it is sometimes convenient to consider a square planar complex as an octahedral complex in which the ligands on the z-axis have been removed. That approach is useful when considering how the metal orbitals interact with the four ligands in a plane. Note that the four ligands are numbered as shown in Figure 17.11 except for positions 5 and 6 being omitted. In D_{4h} symmetry the d orbitals are split into the b_{1g} ($3d_{x^2-y^2}$), b_{2g} (d_{xy}), $a_{1g}(d_{z^2})$, and e_g (d_{xz} and d_{yz}) subsets. The p_x and p_y orbital points directly at ligands and in D_{4h} symmetry they constitute an e_u set. The p_z orbital has a_{2u} symmetry, and the four σ orbitals from the ligands give a_{1g}, e_u, and b_{1g} symmetry. Therefore, the b_{2g}, e_g, and a_{2u} behave as nonbonding orbitals.

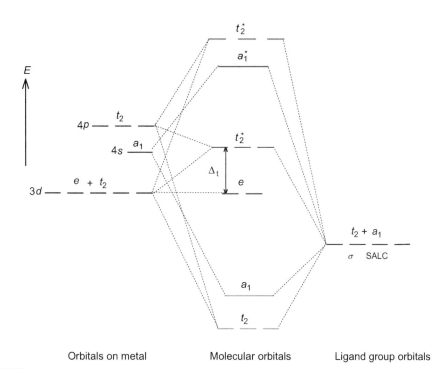

FIGURE 17.15
A qualitative molecular orbital diagram for a tetrahedral complex.

The pictorial approach that was taken with the octahedral complex can be used in this case to determine the symmetry character of the ligand group orbitals. We first consider the ligands interacting with the s orbital on the metal as shown in Figure 17.16(a). The combination of ligand group orbitals matching the symmetry of the s orbital is

$$(\phi_1 + \phi_2 + \phi_3 + \phi_4)$$

Although not shown, the p_x and p_y orbitals are directed toward the ligands in the same way as are the lobes of the $3d_{x^2-y^2}$ orbital. The p orbitals have positive and negative lobes along *each* axis rather than positive on the x-axis and negative on the y-axis as in the case of the $3d_{x^2-y^2}$ orbital.

When considering the interaction between the four ligands and the d_{z^2} metal orbital, it is seen that the positive lobes along the z-axis do not interact with the ligand σ orbitals, but the "ring" that has negative symmetry interacts with the four ligands. This is illustrated in Figure 17.16(b). Given the symmetry of the metal orbital, the appropriate combination of ligand orbitals can be written as

$$(-\phi_1 - \phi_2 - \phi_3 - \phi_4) = -(\phi_1 + \phi_2 + \phi_3 + \phi_4)$$

When we consider the four ligands lying on the x and y axes (as shown in Figure 17.16), it becomes apparent that the $3d_{x^2-y^2}$ orbital has positive lobes along the x-axis and negative lobes along the

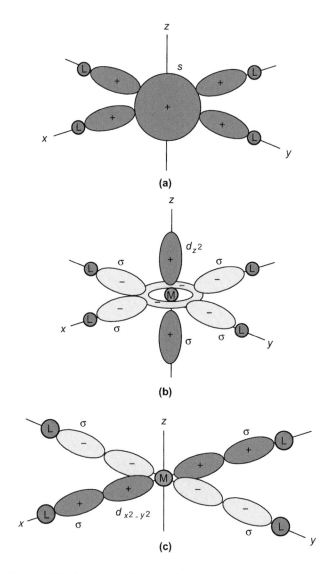

FIGURE 17.16

The combinations of metal and ligand orbitals in a square planar complex.

y-axis. As a result, the combination of ligand orbitals that matches the symmetry of the $3d_{x^2-y^2}$ orbital is

$$\left[\frac{1}{2}(\phi_1 - \phi_2 + \phi_3 - \phi_4)\right]$$

Although we will not write the complete wave functions as we did for the case of an octahedral complex, the molecular orbitals give rise to the energy level diagram shown in Figure 17.17. From the

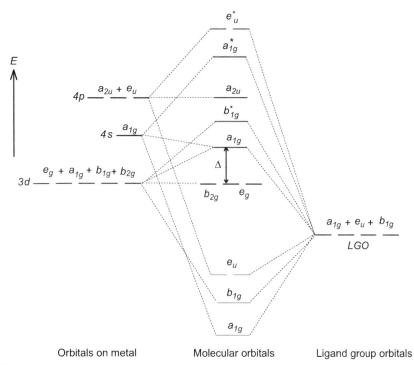

FIGURE 17.17

A molecular orbital energy level diagram for a square planar complex.

molecular orbital energy level diagram shown in Figure 17.17 it can be seen that the orbitals designated as e_g, a_{1g}, and b_{1g}^* correspond to the d_{xz}, d_{yz}, d_{z^2}, and d_{xy} in the crystal field diagram shown in Figure 17.6. In the crystal field model, Δ represented the difference in energy between the d_{xy} and the $d_{x^2-y^2}$ orbitals. In the molecular orbital model, Δ represents the difference in energy between the e_g and a_{1g} orbitals. The arrangement of molecular orbitals in the diagram also shows that d^8 ions would be likely candidates for forming square planar complexes because they would represent a closed shell arrangement with the a_{1g} orbital and all of the molecular orbitals of lower energy being filled. Although the same conclusion was reached in a different way earlier in this chapter, it is gratifying to see that even on an elementary level, the ligand field and molecular orbital approaches lead to similar conclusions. However, it should be mentioned that the molecular orbital methodology has evolved to a much higher level than the rudimentary overview that is presented here. High level calculations are now considered routine.

References for Further Study

Ballhausen, C.J., 1962. *Introduction to Ligand Field Theory*. McGraw-Hill, New York. One of the early classics on crystal and ligand field theory.

Drago, R.S., 1992. *Physical Methods for Chemists*. Saunders College Publishing, Philadelphia. This book presents high level discussion of many topics in coordination chemistry. Highly recommended.

Figgis, B.N., 1987. *Ligand field theory*. In: Wilkinson, G. (Ed.), *Comprehensive Coordination Chemistry*. Pergamon, Oxford. Higher level coverage of ligand field theory.

Figgis, B.N., Hitchman, M.A., 2000. *Ligand Field Theory and Its Applications*. John Wiley, New York. An advanced treatment of ligand field theory.

Jørgensen, C.K., 1971. *Modern Aspects of Ligand Field Theory*. North Holland, Amsterdam. An excellent, high level book.

Kettle, S.F.A., 1969. *Coordination Chemistry*, Appleton, Century, Crofts, New York. A good introductory book that presents crystal field theory in a clear manner.

Kettle, S.F.A., 1998. *Physical Inorganic Chemistry: A Coordination Approach*. Oxford University Press, New York. An excellent book on coordination chemistry that gives good coverage to many areas, including ligand field theory.

Porterfield, W.W., 1993. *Inorganic Chemistry: A Unified Approach*, 2nd ed. Academic Press, San Diego. Chapter 11 presents a good introduction to ligand field theory.

QUESTIONS AND PROBLEMS

1. Consider a linear complex with the two ligands lying on the z-axis. Sketch the splitting pattern for the d orbitals that would result.

2. For each of the following complexes, give the hybrid orbital type and the number of unpaired electrons.
   (a) $[Co(H_2O)_6]^{2+}$; (b) $[FeCl_6]^{3-}$; (c) $[PdCl_4]^{2-}$; (d) $[Cr(H_2O)_6]^{2+}$; (e) $[Mn(NO_2)_6]^{3-}$

3. If the complex $[Ti(H_2O)_4]^{3+}$ existed, what would be the approximate value for Dq?

4. For each of the following, sketch the arrangement of the d orbitals and put in electrons appropriately.
   (a) $[Cr(CN)_6]^{3-}$; (b) $[FeF_6]^{3-}$; (c) $[Co(NH_3)_6]^{3+}$; (d) $[Ni(NO_2)_4]^{2-}$; (e) $[Co(CN)_5]^{3-}$

5. For each of the complexes listed in question 4, write the formula for a specific ligand that would give a different result for the same metal ion.

6. For each of the following, sketch the arrangement of the d orbitals and put in electrons appropriately.
   (a) $[Mn(NO_2)_6]^{3-}$; (b) $[Fe(H_2O)_6]^{3+}$; (c) $[Co(en)_3]^{3+}$; (d) $[NiCl_4]^{2-}$; (e) $[CoF_3]^{3-}$

7. For each of the complexes listed in question 6, write the formula for a specific ligand that would give a different result for the same metal ion.

8. For each of the following, show the d orbital splitting pattern and the electron population in the orbitals. Give the hybrid bond type and predict the magnetic moments.
   (a) $[Mn(ox)_3]^{3-}$; (b) $[Ti(NH_3)_6]^{2+}$; (c) $[Co(CN)_4]^{2-}$; (d) $[Pd(H_2O)_4]^{2+}$; (e) $[NiF_4]^{2-}$

9. Which of the following complexes would undergo Jahn–Teller distortion?
   (a) $[FeCl_6]^{3-}$; (b) $[MnCl_6]^{3-}$; (c) $[CuCl_6]^{4-}$; (d) $[CrCl_6]^{3-}$; (e) $[VCl_6]^{4-}$

10. (a) Complexes of Fe^{3+} with several weak field ligands are only weakly colored and the color is not very different for any of the ligands. Explain in terms of crystal field theory why this is so.
    (b) Complexes of Fe^{3+} with strong field ligands are deeply colored and the color depends greatly on the nature of the ligand. Explain this observation in terms of crystal field theory.

11. Make a sketch of a cubic complex (start with a tetrahedral complex and add four ligands). Analyze the repulsion of each of the d orbitals and sketch the splitting pattern that would exist in a cubic field.

12. Would Jahn–Teller distortion be as significant for tetrahedral complexes as it is for octahedral complexes? For which of the electron configurations would Jahn–Teller distortion occur?

13. Determine the crystal field stabilization energy for $d^0 - d^{10}$ ions in tetrahedral complexes. Although there are no low spin tetrahedral complexes, assume that there are.

14. Explain why tetrahedral complexes of Co^{2+} are more stable than those of Ni^{2+}.

15. Although Ni^{2+} forms some tetrahedral complexes, Pd^{2+} and Pt^{2+} do not. Explain this difference.

16. The magnetic moment of $[Co(py)_2Cl_2]$ is 5.15 BM. Describe the structure of this compound.

CHAPTER 18

Interpretation of Spectra

In the previous chapter, the topic of spectral studies on coordination compounds was introduced only briefly in connection with ligand field theory and some of the attendant problems that are associated with interpreting the spectra were described. In this chapter, a more complete description will be presented of the process of interpreting spectra of complexes. It is from the analysis of spectra that we obtain information about energies of spectroscopic states in metal ions and the effects produced by different ligands on the d orbitals. However, it is first necessary to know what spectroscopic states are appropriate for various metal ions. The analysis then progresses to how the spectroscopic states for the metal ions are affected by the presence of the ligands and how ligand field parameters are determined from spectral data.

18.1 SPLITTING OF SPECTROSCOPIC STATES

As we have seen, an understanding of spin–orbit coupling is necessary to determine the spectroscopic states that exist for various electron configurations, d^n (see Section 2.6). Because they will be needed frequently in this chapter, the spectroscopic states that result from spin–orbit coupling in d^n ions that have degenerate d orbitals are summarized in Table 18.1.

The spectroscopic states shown in Table 18.1 are those that arise for the so-called *free* or *gaseous* ion. When a metal ion is surrounded by ligands in a coordination compound, those ligands generate an electrostatic field that removes the degeneracy of the d orbitals. The result is that e_g and t_{2g} subsets of orbitals are produced. Because the d orbitals are no longer degenerate, spin–orbit coupling is altered so that the states given in Table 18.1 no longer apply to a metal ion *in a complex*. However, just as the d orbitals are split in terms of their energies, the *spectroscopic states* are split in the ligand field. The spectroscopic states are split into components that have the same multiplicity as the free ion states from which they arise. A single electron in a d orbital gives rise to a ^2D term for the gaseous ion, but in an octahedral field, the electron will reside in a t_{2g} orbital, and the spectroscopic *state* for the t_{2g}^1 configuration is $^2T_{2g}$. If the electron were excited to an e_g orbital, the spectroscopic state would be ^2E_g. Thus, transitions between $^2T_{2g}$ and ^2E_g states would not be spin-forbidden because both states are doublets. Note that lower case letters are used to describe *orbitals*, whereas capital letters describe spectroscopic *states*.

A gaseous ion having a d^2 configuration gives rise to a ^3F ground state as a result of spin–orbit coupling. Although they will not be derived, in an octahedral ligand field, the t_{2g}^2 configuration gives three different spectroscopic states that are designated as $^3A_{2g}$, $^3T_{1g}$, and $^3T_{2g}$. These states are often referred to as *ligand field states*. The energies of the three states depend on the strength of the ligand field, **617**

Inorganic Chemistry. DOI: http://dx.doi.org/10.1016/B978-0-12-385110-9.00018-2

Table 18.1 Spectroscopic States for Gaseous Ions Having d^n Electron Configurations[a]

Ion	Spectroscopic States
d^1, d^9	^2D
d^2, d^8	$^3F, {}^3P, {}^1G, {}^1D, {}^1S$
d^3, d^7	$^4F, {}^4P, {}^2H, {}^2G, {}^2F, 2{}^2D, {}^2P$
d^4, d^6	$^5D, {}^3H, {}^3G, 2{}^3F, {}^3D, 2{}^3P, {}^1I, 2{}^1G, {}^1F, 2{}^1D, 2{}^1S$
d^5	$^6S, {}^4G, {}^4F, {}^4D, {}^4P, {}^2I, {}^2H, 2{}^2G, 2{}^2F, 3{}^2G, 3{}^2D, {}^2P, {}^2S$

[a] $2{}^3F$ means two distinct ^3F terms arise, and so on.

but the relationship is not a simple one. The larger the ligand field splitting, the greater the difference between the ligand field states of the metal ion. For the time being, we will assume that the function is linear, but it will be necessary to refine this view later. Table 18.2 shows a summary of the states that result from splitting the gaseous state terms of metal ions in an octahedral field produced by six ligands.

Figure 18.1 shows the approximate energies of the ligand field spectroscopic states as a function of the field strength for all d^n ions. In the drawings, the states are assumed to be linear functions of Δ_o, but this is not correct over a wide range of field strength. For the d^1 ion, the ^2D ground state is split into $^2T_{2g}$ and ^2E_g states in the ligand field. As the field strength, Δ_o, increases, a center of energy is maintained for the energies of the $^2T_{2g}$ and ^2E_g states in exactly the same way that a center of energy is maintained by the t_{2g} and e_g orbital subsets. Therefore, to give no *net* change in energy, the slope of the line for the ^2E_g state is $+(3/5)\Delta_o$, whereas that of the $^2T_{2g}$ state is $-(2/5)\Delta_o$.

Note that the ground state terms for d^n ions (except for d^5 and d^{10}) are all either D or F terms and that the state splitting occurs so that the center of energy is maintained. For the ligand field states that are produced by splitting the ^3F term (which results from a d^2 configuration), the center of energy is also preserved, even though there are three states in the ligand field. By looking at Table 18.2, it can be seen

Table 18.2 Splitting of Spectroscopic States in a Ligand Field[a]

Gaseous Ion Spectroscopic State	Components in an Octahedral Field	Total Degeneracy
S	A_{1g}	1
P	T_{1g}	3
D	$E_g + T_{2g}$	5
F	$A_{2g} + T_{1g} + T_{2g}$	7
G	$A_{1g} + E_g + T_{1g} + T_{2g}$	9
H	$E_g + 2 T_{1g} + T_{2g}$	11
I	$A_{1g} + A_{2g} + E_g + T_{1g} + 2 T_{2g}$	13

[a] Ligand field states have the same multiplicity as the spectroscopic state from which they arise.

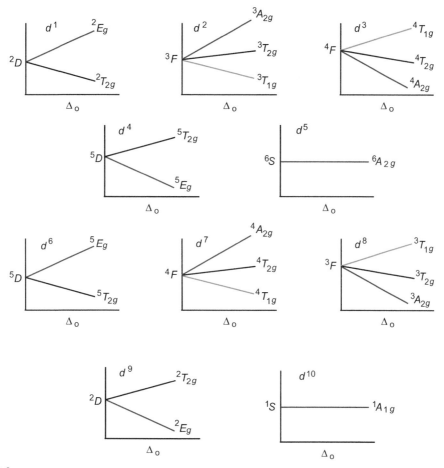

FIGURE 18.1
The splitting patterns for ground state D and F terms in an octahedral field.

that all the states that arise from splitting the D and F ground states for the gaseous ions have T, E, or A designation. When describing the splitting of the d orbitals in an octahedral field (see Chapter 17), the "t" orbitals were seen to be triply degenerate, whereas the "e" orbitals were doubly degenerate. We can consider the *spectroscopic states* in the ligand field to have the same degeneracies as the *orbitals*, which makes it possible to preserve the center of energy. For example, the lines representing the $^2T_{2g}$ and ^2E_g states have slopes of $-(2/5)\Delta_o$ and $+(3/5)\Delta_o$, respectively. Taking into account the degeneracies of the states, the energy of *sum* of the two subsets is

$$3[-(2/5)\Delta_o] + 2[+(3/5)\Delta_o] = 0$$

This result shows that even though the two states have energies that depend on the magnitude of the crystal field splitting, the *overall* energy change is 0. For purposes of determining the center of energy, an

"A" state can be considered as being singly degenerate. When the ^3F state corresponding to the d^2 configuration is split into $^3A_{2g}$, $^3T_{1g}$, and $^3T_{2g}$ components, the slopes of the lines must be such that the overall energy change is 0. Thus, the slopes of the lines and the multiplicities are related by

$$3[-(3/5)\Delta_o] + 3[+(1/5)\Delta_o] + 1[+(6/5)\Delta_o] = 0$$
$$(T) \qquad\qquad (T) \qquad\qquad (A)$$

which shows that the three ligand field states yield the energy of the ^3F gaseous ion state as the center of energy.

From the diagrams shown in Figure 18.1, it can be seen that the splitting pattern for a d^4 ion is like that for d^1 ion except for being inverted and the states having the appropriate multiplicities. Likewise, the splitting pattern for a d^3 ion is like that for d^2 except for being inverted and the multiplicity being different. The reason for this similarity is the "electron-hole" behavior that is seen when the spectroscopic states for configurations such as p^1 and p^5 are considered. Both give rise to a ^2P spectroscopic state, and only the J values are different. It should be apparent from the diagrams shown in Figure 18.1 that the ligand field splitting pattern is the same for d^1 and d^6, d^2 and d^7, and so on except for the multiplicity. In fact, it is easy to see that this will be true for any d^n and d^{5+n} configurations. It is also apparent that the singly degenerate "A" states that arise from the S states for d^5 and d^{10} configurations are not split in the ligand field. All the ligand field components and their energies in an octahedral field are summarized in Table 18.3.

When a transition metal ion is surrounded by ligands that generate a tetrahedral field, the splitting pattern of the d orbitals is inverted when compared with that in an octahedral field. As a result, the e orbitals lie lower than the t_2 set (note that there is no subscript "g" because a tetrahedron does not have a center of symmetry). A further consequence is that the energies of the ligand field spectroscopic states shown in Table 18.3 are also reversed in order compared with their order in octahedral fields. For example, in an octahedral field, the d^2 ion gives the states (in order of increasing energy) $^3T_{1g}$, $^3T_{2g}$, and

Table 18.3 Energies of Octahedral Crystal Field States in Terms of Δ_o

Ion	State	Octahedral Field States	Energies in Octahedral Field
d^1	^2D	$^2T_{2g} + {}^2E_g$	$-(2/5)\Delta_o, +(3/5)\Delta_o$
d^2	^3F	$^3T_{1g} + {}^3T_{2g} + {}^3A_{2g}$	$-(3/5)\Delta_o, +(1/5)\Delta_o, +(6/5)\Delta_o$
d^3	^4F	$^4A_{2g} + {}^4T_{2g} + {}^4T_{1g}$	$-(6/5)\Delta_o, -(1/5)\Delta_o, +(3/5)\Delta_o$
d^4	^5D	$^5E_g + {}^5T_{2g}$	$-(3/5)\Delta_o, +(2/5)\Delta_o$
d^5	^6S	$^6A_{1g}$	0
d^6	^5D	$^5T_{2g} + {}^5E_g$	$-(2/5)\Delta_o, +(3/5)\Delta_o$
d^7	^4F	$^4T_{1g} + {}^4T_{2g} + {}^2A_{2g}$	$-(3/5)\Delta_o, +(1/5)\Delta_o, +(6/5)\Delta_o$
d^8	^3F	$^3A_{2g} + {}^3T_{2g} + {}^3T_{1g}$	$-(6/5)\Delta_o, -(1/5)\Delta_o, +(3/5)\Delta_o$
d^9	^2D	$^2E_g + {}^2T_{2g}$	$-(3/5)\Delta_o, +(2/5)\Delta_o$
d^{10}	^1S	^1A_g	0

Crystal field state of lowest energy given first, and energies are listed in the same order.

$^3A_{2g}$ as shown in Table 18.3. In a tetrahedral field, the order of increasing energy for the states arising from a d^2 ion would be $^3A_2$, $^3T_2$, and $^3T_1$.

For a specific d^n electron configuration, there are usually several spectroscopic states that correspond to energies above the ground state term. However, they may not have the same *multiplicity* as the ground state. When the spectroscopic state for the free ion becomes split in an octahedral field, each ligand field component has the same multiplicity as the ground state (see Table 18.3). *Transitions between spectroscopic states having different multiplicities are spin-forbidden.* Because the T_{2g} and E_g spectroscopic states in a ligand field have the same multiplicity as the ground states from which they arise, it can be seen that the $T_{2g} \rightarrow E_g$ transition is the only spin-allowed transition for ions that have D ground states. In cases where the ground state is an F term, there is a P state of higher energy that has the same multiplicity. That state gives a T_{1g} state [designated as $T_{1g}(P)$] having the same multiplicity as the ground state T_{1g} term. Accordingly, spectroscopic transitions are possible from the ground state in the ligand field to the T state that arises from the P term. Therefore, for ions having T and A ground states, the spin-allowed transitions in an octahedral field are as follows.

Octahedral Field			
For *T* ground states		**For *A* ground states**	
ν_1	$T_{1g} \rightarrow T_{2g}$	ν_1	$A_{2g} \rightarrow T_{2g}$
ν_2	$T_{1g} \rightarrow A_{2g}$	ν_2	$A_{2g} \rightarrow T_{1g}$
ν_3	$T_{1g} \rightarrow T_{1g}(P)$	ν_2	$A_{2g} \rightarrow T_{1g}(P)$

The spin-allowed transitions for ions having T and A ground states in tetrahedral fields are as follows.

Tetrahedral Field			
For *T* ground states		**For *A* ground states**	
ν_1	$T_1 \rightarrow T_2$	ν_1	$A_2 \rightarrow T_2$
ν_2	$T_1 \rightarrow A_2$	ν_2	$A_2 \rightarrow T_1$
ν_3	$T_1 \rightarrow T_1(P)$	ν_2	$A_2 \rightarrow T_1(P)$

As shown above, for both octahedral and tetrahedral complexes, *three* absorption bands are expected. Many complexes do, in fact, have absorption spectra that show three bands. However, charge transfer (CT) absorption makes it impossible in some cases to see all three bands, so spectral analysis must often be based on only one or two observed bands.

18.2 ORGEL DIAGRAMS

The splitting patterns of the spectroscopic states that are shown in Table 18.3 can be reduced to graphical presentation in two diagrams. These diagrams were developed by L.E. Orgel, and they have since become known as *Orgel diagrams*. Figure 18.2 shows the Orgel diagram that is applicable for ions that give D ground states.

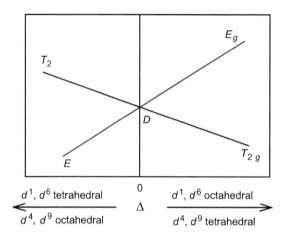

FIGURE 18.2

An Orgel diagram for metal ions having D spectroscopic ground states. The multiplicity of the D state is not specified because it is determined by the number of electrons in the d orbitals of the metal ion.

In an Orgel diagram, energy is represented as the vertical dimension, and the vertical line in the center of the diagram represents the gaseous ion where there is no ligand field ($\Delta = 0$). Note that the right-hand side of the diagram applies to d^1 and d^6 ions in *octahedral* fields or d^4 and d^9 ions in *tetrahedral* fields. This situation arises because the ligand field states are inverted for the two cases, and the electron-hole formalism also causes the orbitals to be inverted. As a result, the ligand field states are the same for a d^1 ion in an octahedral field as they are for a d^4 ion in a tetrahedral field.

The left-hand side of the Orgel diagram shown in Figure 18.2 applies to d^4 and d^9 ions in *octahedral* fields or to d^1 and d^6 ions in *tetrahedral* fields. Note that the "g" subscripts have been deleted on the left-hand side of the diagram. Although *both* sides of the diagram can apply to tetrahedral or octahedral complexes in some cases, it is customary to show the "g" on one side of the diagram but not on the other. This does *not* mean that one side of the diagram applies to tetrahedral complexes and the other to octahedral complexes. *Both* sides apply to *both* types of complexes. It should also be remembered that the magnitude of the splitting is quite different for the two types of complexes because Δ_t is approximately $(4/9)\Delta_o$.

The Orgel diagram that applies to ions that have F spectroscopic ground states is shown in Figure 18.3. This diagram also includes the P state, which has higher energy because in an octahedral ligand field that state becomes a T_{1g} state (T_1 in a tetrahedral field) that has the same multiplicity as the ground state. Therefore, spectral transitions are spin allowed between the ground state and the excited state arising from the P state.

For octahedral complexes containing d^2 and d^7 ions, the ground state is T_{1g}, which we will designate as $T_{1g}(F)$ to show that it arises from the F ground state of the free ion rather than the P state. Note that the lines representing the $T_{1g}(F)$ and $T_{1g}(P)$ in the Orgel diagram are curved with the curvature increasing as the magnitude of Δ increases. These states represent quantum mechanical states that have identical designations, and it is a characteristic of such states that they cannot represent the same energy. Therefore, these states interact strongly and repel each other. This phenomenon is often referred to as the *noncrossing rule*.

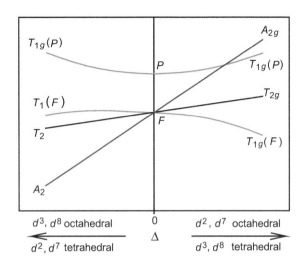

FIGURE 18.3

An Orgel diagram for metal ions having F spectroscopic ground states. The multiplicity of the F state is not specified. The P state having the same multiplicity as the ground state is also shown.

As has been stated, the multiplicity of the ligand field states is the same as the multiplicity of the ground state term from which they arise. Therefore, a d^2 ion gives a ^3F ground state that is split into the triplet $^3T_{1g}$, $^3T_{2g}$, and $^3A_{2g}$ states in an octahedral field. From the diagram, we can see that the possible spectral transitions are those from the $^3T_{1g}$ state to the three ligand field states of higher energy. On this basis, the Orgel diagram allows us to predict that three bands should be observed in the spectrum for a d^2 ion in an octahedral field. Frequently, the spectra of such complexes contain fewer than three bands, and there may be some ambiguity in assigning the bands to specific transitions. One reason why the assignment of some bands may be uncertain is that there may be charge transfer bands that occur in the same region of the spectrum (see Section 18.8). These bands are often intense, and they may mask the so-called $d-d$ transitions. Another factor that may complicate band assignment is that some of the bands are quite weak so they may be difficult to identify in the presence of other strong bands. The point is that for many complexes, matching the observed bands in the spectrum is not necessarily a simple straightforward matter. Moreover, the Orgel diagrams are only qualitative so we need a more quantitative method if we are to be able to perform calculations to obtain Δ and other ligand field parameters.

In the previous discussion, only the ligand field states that arise from the free ion ground state and excited states having the same multiplicity were considered. However, other spectroscopic terms exist for the free ion (see Table 18.1). For example, if a d^2 ion is considered, the ground state is ^3F, but the other states are ^3P, ^1G, ^1D, and ^1S. When spectral transitions are considered, only the ^3F and ^3P states concern us. We do need to understand that because of the noncrossing rule, some of the energies of the states will deviate from linear functions of Δ. A d^2 electron configuration corresponds to Ti^{2+}, V^{3+}, and Cr^{4+} ions, but even for the free ions, the spectroscopic states have different energies. For the free ions, the energies of the spectroscopic states are shown in Table 18.4.

The energies shown in Table 18.4 make it clear that a *complete* energy level diagram for a d^2 ion in an octahedral ligand field will be dependent on the specific metal ion being considered. Furthermore, the

Table 18.4 Energies of Spectroscopic Terms for Gaseous d^2 Ions

Term	Spectroscopic State Energies, cm^{-1}		
	Ti^{2+}	V^{3+}	Cr^{4+}
^3F	0	0	0
^1D	8,473	10,540	13,200
^3P	10,420	12,925	15,500
^1G	14,398	17,967	22,000

349.8 cm^{-1} = 1 kcal mol^{-1} = 4.184 kJ mol^{-1}

energies of states that have identical designations will obey the noncrossing rule so that they will vary in a nonlinear way with the field strength.

18.3 RACAH PARAMETERS AND QUANTITATIVE METHODS

From the discussion above, it is clear that we should expect a maximum of three spin-allowed transitions regardless of the d^n configuration of the metal ion. Because of spin–orbit coupling, the interelectronic repulsion is different for the various spectroscopic states in the ligand field. The ability of the electrons to be permuted among a set of degenerate orbitals and interelectron repulsion are important considerations (see Chapter 2). By use of quantum mechanical procedures, these energies can be expressed as integrals. One of the methods makes use of integrals that are known as the Racah parameters. There are three parameters, A, B, and C, but if only *differences* in energies are considered, the Parameter A is not needed. The Parameters B and C are related to the coulombic and exchange energies, respectively, that are a result of electron pairing. For an ion that has a d^2 configuration, it can be shown that the ^3F state has an energy that can be expressed as $(A - 8B)$, whereas that of the ^3P state is $(A + 7B)$. Accordingly, the difference in energy between the two states can be expressed in terms of B only because the Parameter A cancels. When states having a different multiplicity than the ground state are considered, the difference in energy is expressed in terms of both B and C.

Although the energies of the spectroscopic states of the first-row d^2 ions were shown in Table 18.4, compilations exist for all gaseous metal ions. The standard reference is a series of volumes published by the National Bureau of Standards (C.E. Moore, *Atomic Energy Levels*, National Bureau of Standards Circular 467, Vol. I, II, and III). Table 18.5 shows the energies for the spectroscopic states in d^n ions in terms of the Racah B and C parameters.

For the cases where the ground state of a free ion is an F term, the difference in energy between the ground state and the first excited state having the same multiplicity (a P term) is defined as $15B$, which is $(A + 7B) - (A - 8B)$. This is analogous to the case of splitting of d orbitals in a ligand field where the difference in energy is represented as 10Dq. The Parameter B is simply a unit of energy whose numerical value depends on the particular ion considered. For d^2 and d^8 ions,

$$15B = {}^3F \rightarrow {}^3P \tag{18.1}$$

Table 18.5 Energies of Spectroscopic States for Free Ions in Terms of Racah Parameters

d^2, d^8		d^3, d^7		d^5	
^1S	$22B + 7C$	^2H	$9B + 3C$	^4F	$22B + 7C$
^1G	$12B + 2C$	^2P	$9B + 3C$	^4D	$17B + 5C$
^3P	$15B$	^4P	$15B$	^4P	$7B + 7C$
^1D	$5B + 2C$	^2G	$4B + 3C$	^4G	$10B + 5C$
^3F	0	^4F	0	^6S	0

where $^3F \rightarrow {}^3P$ means the energy difference between the two states. For d^3 and d^7 ions,

$$15B = {}^4F \rightarrow {}^4P \tag{18.2}$$

It can be seen from Table 18.5 that all excited spectroscopic states having a multiplicity that is *different* from the ground state have energies that are expressed in terms of both B and C. As we have seen from the previous discussion, spin-allowed transitions occur only between states having the *same* multiplicity. Therefore, in the analysis of spectra of complexes, only B must be determined. It is found for some complexes that $C \approx 4B$, and this approximation is adequate for many uses.

The discussion so far has concerned the Racah parameters for *gaseous* ions. For example, the d^2 ion Ti^{2+} has a ^3P state that lies 10,420 cm^{-1} above the ^3F ground state so $15B = 10{,}420$ cm^{-1} and B is 695 cm^{-1} for this ion. Using this value for B results in an approximate value of 2780 cm^{-1} as the value of C for the gaseous ion. For many complexes of $+2$ ions of the first-row transition metals, B is in the range of 700–1000 cm^{-1} and C is approximately 2500–4000 cm^{-1}. For $+3$ ions of the first-row metals, B is normally in the range 850–1200 cm^{-1}. For the second- and third-row metal ions, the value of B is usually only about 600–800 cm^{-1} because interelectron repulsion is smaller for the larger ions.

When *complexes* of the metal ions are considered, the situation is considerably more complicated. The differences between energy states in the ligand field are related not only to the Racah parameters but also to the magnitude of Δ (or Dq). As a result, the energies for the three spectral bands must be expressed in terms of both the Dq and the Racah parameters. Because the observed spectral bands represent differences in energies between states having the same multiplicity, only the Racah B parameter is necessary. Even so, B is not a constant because it varies with the magnitude of the effect of the ligands on the d orbitals of the metal (the ligand field splitting). Analysis of the spectrum for a complex involves determining the value of Dq and B for *that complex*. Of course, $\Delta = 10$Dq, and we have used Δ to describe orbital splitting more frequently up to this point. When dealing with spectral analysis, the discussion can also be presented with regard to Dq.

For metal ions having d^2, d^3, d^7, and d^8 configurations, the ground state is an F state, but there is an excited P state that has the same multiplicity. For d^2 and d^7 ions in an *octahedral* field, the spectroscopic states are the same (except for the multiplicity) as they are for d^3 and d^8 ions in *tetrahedral* fields. Therefore, the expected spectral transitions will also be the same for the two types of complexes. The three spectral bands are assigned as follows [$T_{1g}(F)$ means the T_{1g} state arising from the F spectroscopic state]:

$$\nu_1 = T_{1g}(F) \rightarrow T_{2g}$$

$$\nu_2 = T_{1g}(F) \rightarrow A_{2g}$$

$$\nu_3 = T_{1g}(F) \rightarrow T_{1g}(P)$$

For octahedral complexes of d^2 and d^7 metal ions (or d^3 and d^8 ions in tetrahedral complexes), the energies corresponding to ν_1, ν_2, and ν_3 can be shown to be

$$E(\nu_1) = 5Dq - 7.5B + (225B^2 + 100Dq^2 + 180DqB)^{1/2} \tag{18.3}$$

$$E(\nu_2) = 15Dq - 7.5B + (225B^2 + 100Dq^2 + 180DqB)^{1/2} \tag{18.4}$$

$$E(\nu_3) = (225B^2 + 100Dq^2 + 180DqB)^{1/2} \tag{18.5}$$

The third band corresponds to the difference in energy between the $T_{1g}(F)$ and $T_{1g}(P)$ states that arise from splitting the F and P states. In the limit where $Dq = 0$ (as it is for a gaseous ion), the energy reduces to $(225B^2)^{1/2}$, which is $15B$. That is precisely the difference in energy between the F and P spectroscopic states.

The corresponding energies for the spectral transitions that occur for d^3 and d^8 ions in octahedral fields (or d^2 and d^7 ions in tetrahedral fields) are as follows.

$$E(\nu_1) = 10Dq \tag{18.6}$$

$$E(\nu_2) = 15Dq - 7.5B - 1/2(225B^2 + 100Dq^2 - 180DqB)^{1/2} \tag{18.7}$$

$$E(\nu_3) = 15Dq + 7.5B + 1/2(225B^2 + 100Dq^2 - 180DqB)^{1/2} \tag{18.8}$$

For these complexes, the first band corresponds directly to $10Dq$. If all three of the spectral bands are observed in the spectrum, it can be shown that

$$\nu_3 + \nu_2 - 3\nu_1 = 15B \tag{18.9}$$

The discussion above applies to complexes that are weak field cases. Spectral analysis for strong field cases is somewhat different and will not be discussed here. For complete analysis of the spectra of strong field complexes, see the book by A.B.P. Lever, *Inorganic Electronic Spectroscopy* listed in the references at the end of this chapter.

It is possible to solve Eqns (18.3–18.8) to obtain Dq and B for situations where two or more of the spectral bands can be identified. The resulting equations giving Dq and B as functions of frequencies of the spectral bands are shown in Table 18.6.

Although the equations shown in Table 18.6 can be used to evaluate Dq and B, the positions of at least two spectral bands must be known and correctly assigned. Historically, other approaches to evaluating ligand field parameters have been developed, and they are still useful. In the remainder of this chapter, some of the other approaches will be described. In many cases, unambiguous interpretations of spectra are difficult because of factors such as charge transfer bands, absorptions by ligands, and lack of ideal symmetry caused by Jahn–Teller distortion (see Chapter 17). The discussion that follows will amplify the introduction presented so far, but for complete details of spectral interpretation, consult the reference works by Ballhausen, Jørgensen, and Lever listed in the references.

Table 18.6 Equations for Calculating Dq and B from Spectra	
Spectral Bands Known	**Equations**
Ions having "T" ground states	
ν_1, ν_2, ν_3	$Dq = (\nu_2 - \nu_1)/10$ $B = (\nu_2 + \nu_3 - 3\nu_1)/15$
ν_1, ν_2	$Dq = (\nu_2 - \nu_1)/10$ $B = \nu_1(\nu_2 - 2\nu_1)/(12\nu_2 - 27\nu_1)$
ν_1, ν_3	$Dq = [(5\nu_3^2 - (\nu_3 - 2\nu_1)^2)^{1/2} - 2(\nu_3 - 2\nu_1)]/40$ $B = (\nu_3 - 2\nu_1 + 10Dq)/15$
ν_2, ν_3	$Dq = [(85\nu_3^2 - 4(\nu_3 - 2\nu_2)^2)^{1/2} - 9(\nu_3 - 2\nu_2)]/340$ $B = (\nu_3 - 2\nu_2 + 30Dq)/15$
Ions having "A" ground states	
ν_1, ν_2, ν_3	$Dq = \nu_1/10$ $B = (\nu_2 + \nu_3 - 3\nu_1)/15$
ν_1, ν_2	$Dq = \nu_1/10$ $B = (\nu_2 - 2\nu_1)(\nu_2 - \nu_1)/(15\nu_2 - 27\nu_1)$
ν_1, ν_3	$Dq = \nu_1/10$ $B = (\nu_3 - 2\nu_1)(\nu_3 - \nu_1)/(15\nu_3 - 27\nu_1)$
ν_2, ν_3	$Dq = [(9\nu_2 + \nu_3) - (85(\nu_2 - \nu_3)^2 - 4(\nu_2 + \nu_3)^2)^{1/2}]/340$ $B = (\nu_2 + \nu_3 - 30Dq)/15$

Adapted from Y. Dou, J. Chem. Educ. **1990**, 67, 134.

18.4 THE NEPHELAUXETIC EFFECT

When a metal ion is surrounded by ligands in a complex, the ligand orbitals being directed toward the metal ion produce changes in the total electron environment of the metal ion. One consequence is that the energy required to force pairing of electrons is altered. Although the energy necessary to force electron pairing in the gaseous ion can be obtained from tabulated energy values for the appropriate spectroscopic states, those values are not applicable to a metal ion that is contained in a complex. When ligands bind to a metal ion, the orbitals on the metal ion are "smeared out" over a larger region of space. The molecular orbital terminology for this situation is that the electrons become more delocalized in the complex than they are in the free ion. The expansion of the electron cloud is known as the *nephelauxetic effect*.

As a result of the nephelauxetic effect, the energy required to force pairing of electrons in the metal ion is somewhat smaller than it is for the free ion. When ligands such as CN^- are present, the nephelauxetic effect is quite large owing to the ability of the ligands to π bond to the metal as a result of back donation. As discussed in Chapter 17, antibonding orbitals on the CN^- have appropriate symmetry to form π bonds to nonbonding d orbitals on the metal ion. As a result, the Racah Parameters B and C for a metal ion are variables whose exact values depend on the nature of the ligands attached to the ion. The change in B from the free ion value is expressed as the *nephelauxetic ratio*, β, which is given by

$$\beta = \frac{B'}{B} \tag{18.10}$$

where B is the Racah parameter for the free metal ion and B' is the same parameter for the metal ion in the complex. C.K. Jørgensen devised a relationship to express the nephelauxetic effect as the product of a parameter characteristic of the ligand, h, and another characteristic of the metal ion, k. The mathematical relationship is

$$(1 - \beta) = hk \tag{18.11}$$

After substituting B'/B for β, the equation can be written as

$$B' = B - Bhk \tag{18.12}$$

From this equation, we see that the Racah parameter for the metal ion in a complex, B', is the value for the gaseous ion *reduced* by a correction (expressed as Bhk) for the cloud expanding effect. As should be expected, the effect is a function of the nature of the particular metal ion and ligand. As a result, analysis of the spectrum for a complex must be made to determine both Dq and B (which is actually B' for the complex). Table 18.7 shows values of the nephelauxetic parameters for ligands and metal ions.

The data shown in Table 18.7 indicate that the ability of a ligand to produce a nephelauxetic effect increases as the softness of the ligand increases. Softer ligands such as N_3^-, Br^-, CN^-, or I^- show a greater degree of covalency when bonded to metal ions, and they can more effectively delocalize electron density.

Table 18.7 Nephelauxetic Parameters for Metal Ions and Ligands*

Metal Ion	k Value	Ligand	h Value
Mn^{2+}	0.07	F^-	0.8
V^{2+}	0.1	H_2O	1.0
Ni^{2+}	0.12	$(CH_3)_2NCHO$	1.2
Mo^{3+}	0.15	NH_3	1.4
Cr^{3+}	0.20	en	1.5
Fe^{3+}	0.24	ox^{2-}	1.5
Rh^{3+}	0.28	Cl^-	2.0
Ir^{3+}	0.28	CN^-	2.1
Co^{3+}	0.33	Br^-	2.3
Mn^{4+}	0.5	N_3^-	2.4
Pt^{4+}	0.6	I^-	2.7
Pd^{4+}	0.7		
Ni^{4+}	0.8		

*From C.K. Jørgensen, Oxidation Numbers and Oxidation States, *Springer Verlag*, New York, 1969, p. 106.

It is also apparent that the nephelauxetic effect is greater for metal ions that are more highly charged. This is to be expected because these smaller harder metal ions will experience a greater reduction in inter-electronic repulsion by expanding the electron cloud than will a larger metal ion of lower charge. As the data show, the nephelauxetic effect correlates well with the hard–soft interaction principle.

18.5 TANABE–SUGANO DIAGRAMS

Qualitatively, the Orgel diagrams are energy level diagrams in which the vertical distances between the lines representing the energies of spectroscopic states are functions of the ligand field splitting. Although they are useful for summarizing the possible transitions in a metal complex, they are not capable of being interpreted quantitatively. The problems associated with B having values that depend on the ligand field splitting make it difficult to prepare a specific energy level diagram for a given metal ion. Y. Tanabe and S. Sugano circumvented part of this problem by preparing energy level diagrams that are based on spectral energies and Δ, but they prepared the graphs with E/B and Δ/B as the variables. When plotted in this way, the energy of the ground state becomes the horizontal axis and the energies of all other states are represented as curves above the ground state. Figure 18.4 shows a complete Tanabe–Sugano diagram for a d^2 ion. Even though the axes on Tanabe–Sugano diagrams have numerical scales (unlike an Orgel diagram), they are still not exactly quantitative. One reason is that B is

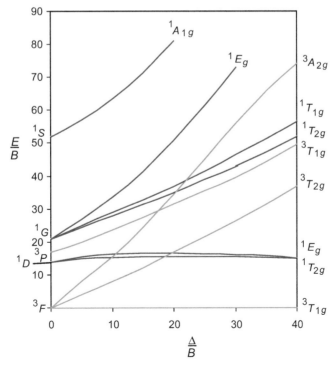

FIGURE 18.4
A complete Tanabe–Sugano diagram for a d^2 metal ion in an octahedral field. When $\Delta = 0$, the spectral terms (shown on the vertical axis) are those of the free gaseous ion. Spectroscopic transitions occur between states having the same multiplicity (shown in green).

not a constant. The nephelauxetic ratio, B'/B, varies from approximately 0.6 to 0.8 depending on the nature of the ligands. Some of the spectroscopic states have multiplicities that are different from that of the ground state as a result of electron pairing. The energies of those states are functions of both the Racah B and C parameters. However, the ratio C/B is not strictly a constant, and the calculations that are carried out to produce a Tanabe–Sugano diagram are based on a specific C/B value, usually in the range 4–4.5.

By looking at Tables 18.2 and 18.3, we see that for most metal ions, there will be a sizeable number of spectroscopic states in a ligand field. A *complete* Tanabe–Sugano diagram shows the energies of *all* these states as plots of E/B versus Δ/B. However, in the analysis of spectra of complexes, it is usually necessary to consider only states having the same multiplicity as the ground state. Therefore, only a small number of states must be considered when assigning spectral transitions. The Tanabe–Sugano diagrams shown in this chapter do not include some of the excited states but rather show only the states involved in d–d transitions. The result is a less cluttered diagram that still presents the details that are necessary for spectral analysis. The simplified Tanabe–Sugano diagrams for d^3–d^8 ions are shown in Figure 18.5. Note that the diagrams for d^4, d^5, d^6, and d^7 have a vertical line that typically occurs in the Δ/B range of 20–30. This is the result of electron spin pairing when the ligand field becomes sufficiently strong. Note that the ground state continues to be the horizontal axis, but the multiplicity is different and corresponds to that of the complex with fewer unpaired electrons.

The use of a Tanabe–Sugano diagram in spectral analysis is relatively simple. Figure 18.6 shows a simplified and enlarged Tanabe–Sugano diagram for a d^3 metal ion. Transitions between the $^4A_{2g}$ ground state and excited states having the same multiplicity can be represented as vertical lines between the horizontal axis and the lines representing the excited states. Suppose a complex CrX_6^{3-} has two absorption bands $\nu_1 = 11{,}000$ cm^{-1} and $\nu_3 = 26{,}500$ cm^{-1} and that Dq and B are to be determined from the diagram shown in Figure 18.6. For this complex, $\nu_3/\nu_1 = 2.4$, so the ratio of the lengths of the lines representing the third and first transitions must have this value. Because the lines representing the excited states *diverge* as Δ changes, there is only one point on the horizontal axis where this ratio of 2.4 is satisfied. That value of Δ/B is the one where the distances from the ground state (the horizontal axis) up to the first and third excited states have the ratio 2.4. The problem is to find the correct point on the Δ/B axis.

On Figure 18.6, a trial value of $\Delta/B = 30$ is shown that yields the vertical lines representing ν_3 and ν_1. The measured lengths of these lines give a ratio of 2.19, which is somewhat lower than the ratio of the experimental spectral energies. At a trial value of 10 for Δ/B (also shown in the figure), the lengths of the lines give a ratio of 3.11, which is higher than the ratio of the spectral energies. By carefully measuring the lengths of the lines corresponding to the transitions, it is found that the correct value of Δ/B is approximately 16. This value corresponds to values of E_1/B of approximately 14 and E_3/B of about 35 (by reading on the vertical axis). Because it is known that the energy of the ν_1 band is $E_1 = 11{,}000$ cm^{-1},

$$\frac{E_1}{B} = 14 = \frac{11{,}000 \text{ cm}^{-1}}{B} \qquad (18.13)$$

so B is approximately 780 cm^{-1}. We have already determined that Δ/B is approximately 16, so it is now possible to evaluate Δ, which we find to be approximately 12,000–13,000 cm^{-1}. If the calculations are repeated using the data for $E_3 = 26{,}500$ cm^{-1} and $\Delta/B = 16$, values of approximately 760 and 12,100 cm^{-1} are indicated for B and Δ, respectively. For Cr^{3+}, the free ion B value is 1030 cm^{-1} so the value indicated for B in the CrX_6^{3-} complex is, as expected, approximately 75% of the free ion value.

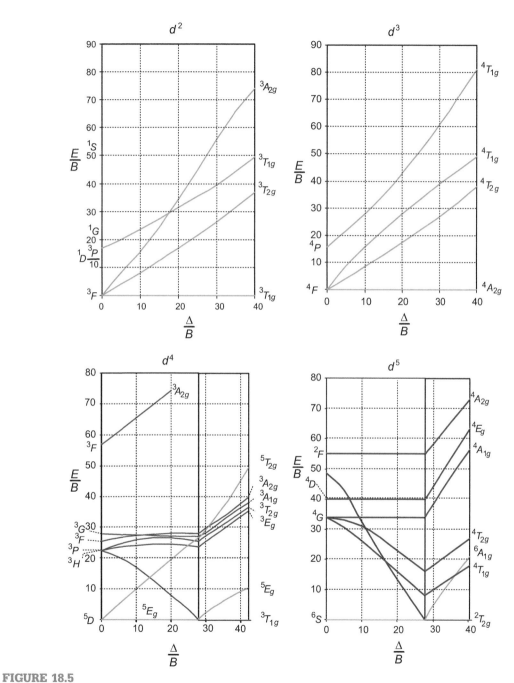

FIGURE 18.5

Simplified Tanabe–Sugano diagrams for d^n metal ions in octahedral fields. The drawings have been simplified by omitting several states that have multiplicities that do not permit spin-allowed transitions.

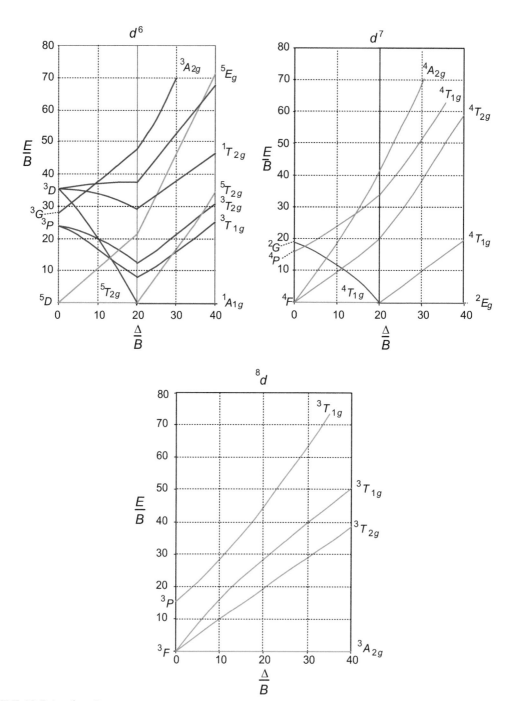

FIGURE 18.5 (continued).

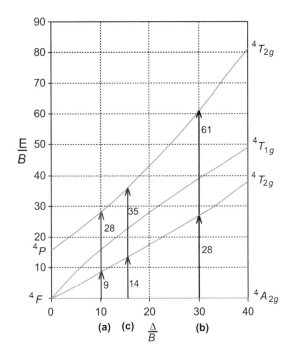

FIGURE 18.6
The simplified Tanabe–Sugano diagram for a d^3 ion in an octahedral field and its use for determining crystal field parameters (see text). At point (a) ($\Delta/B = 10$) on the figure, $\nu_3/\nu_1 = 28/9 = 3.11$. At point (b) ($\Delta/B = 30$) on the figure, $\nu_3/\nu_1 = 61/28 = 2.19$. At point (c) ($\Delta/B = 16$), $\nu_3/\nu_1 = 35/14 = 2.50$, which is approximately the correct value. Therefore, 16 is approximately the correct value for Δ/B.

In the foregoing illustration, we have assumed that ν_2 is either not observed or has an unknown energy. Having found that $\Delta/B = 16$, we can read from the Tanabe–Sugano diagram the energy of the $^4T_{1g}$ state corresponding to that value of Δ/B to find $E_2/B = 23$, which shows that the second spectral band should be at approximately 16,500 cm^{-1}.

Because a value of E/B can be read directly from the graph, B can be determined because the energies of the bands in the spectrum are already known. With B and Δ/B having been determined, Δ is readily obtained. Note, however, that this measurement is not particularly accurate, which means that Δ/B is not known very accurately. In fact, Tanabe–Sugano diagrams are not normally used to determine Δ and B quantitatively. They are useful as a basis for interpreting spectra, but the equations shown in Table 18.6 or other graphical or numerical procedures are generally used for determining Δ and B. One such graphical procedure will now be described.

18.6 THE LEVER METHOD

A method described by A.B.P. Lever (1968) provides an extremely simple and rapid means of evaluating Dq and B for metal complexes. In this method, the equations for $E(\nu_1)$, $E(\nu_2)$, and $E(\nu_3)$ are used with a wide range of Dq/B values to calculate the ratios ν_3/ν_1, ν_3/ν_2, and ν_2/ν_1. These values are shown as

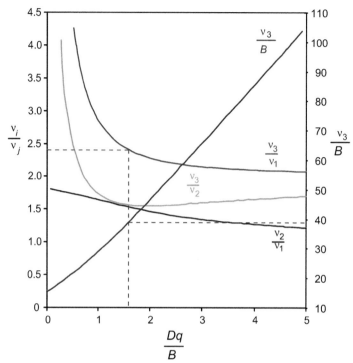

FIGURE 18.7
Diagram for using the Lever method to determine Dq and *B* for ions having *A* spectroscopic ground states. Drawn using data presented by Lever (1968).

extensive tables along with ν_3/B ratios in the original literature, but they can also be shown graphically as in Figure 18.7 (for ions having *A* ground states) and Figure 18.8 (for ions having *T* ground states).

When using the Lever method, experimental band maxima are used to calculate the ratios ν_3/ν_1, ν_3/ν_2, and ν_2/ν_1. The ratios are then located on the appropriate line on the graph, and by reading *downward* to the horizontal axis, the value of Dq/*B* is determined. Once the ν_i/ν_j ratio is located, the ν_3/B value can also be identified by reading to the right-hand axis, whether or not ν_3 is available from the spectrum. Therefore, because Dq/*B* and ν_3 are known, it is a simple matter to calculate Dq and *B*. When using the Lever method, it is best to use either the original tables of values or a large-scale graph with grid lines to be able to read values more accurately.

The procedure for using the Lever method will be illustrated by considering the CrX_6^{3-} complex described earlier in using the Tanabe–Sugano diagram. In this case, absorption bands were presumed to be seen at 11,000 and 26,500 cm^{-1} and that they represent ν_1 and ν_3, respectively. Therefore, $\nu_3/\nu_1 = 2.41$. Because Cr^{3+} is a d^3 ion, the ground state is $^4A_{2g}$ so Figure 18.7 is the appropriate one to use. Reading across to the vertical axis opposite the value $\nu_3/\nu_1 = 2.41$ gives a value of Dq/*B* equal to about 1.6. Similarly, the value ν_3/B is found to be about 38.5. Therefore,

$$Dq/B = 1.6 \tag{18.14}$$

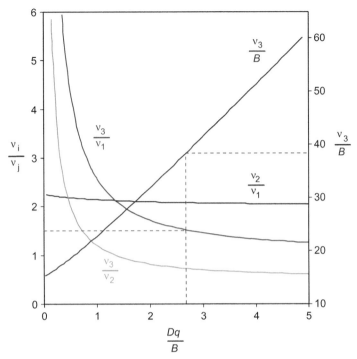

FIGURE 18.8

Diagram for using the Lever method to determine Dq and B for ions having T spectroscopic ground states. Drawn using data presented by Lever (1968).

$$\nu_3/B = 38.5 \qquad (18.15)$$

so that $B = Dq/1.6$ and $B = \nu_3/38.5$. From Eqns (18.14) and (18.15) we find that

$$Dq = \nu_3 \times 1.6/38.5 = 26,500 \text{ cm}^{-1} \times 1.6/38.5 = 1100 \text{ cm}^{-1}$$

which leads to a value of approximately 690 cm^{-1} for B. These values are in fair agreement with the values of 1200 and 760 cm^{-1} estimated for Dq and B, respectively, from the Tanabe–Sugano diagram. Because B for the gaseous Cr^{3+} ion is 1030 cm^{-1}, the nephelauxetic ratio for this complex is $B = 690$ cm^{-1}/1030 cm^{-1} = 0.67, which is a typical value for this parameter.

When using the Lever method, the question naturally arises regarding the situation in which the identities of the bands in the spectrum are unknown. Assuming that we do not know that the observed bands are actually ν_1 and ν_3, we calculate from the previous example that $\nu_i/\nu_j = 2.41$. It is readily apparent that this ratio could not be ν_2/ν_1 because the entire range of values presented for this ratio is approximately 1.2–2.4. If the 2.41 value actually represented ν_3/ν_2, a value of Dq/B of about 0.52 is indicated. From our previous discussions, it has been seen that for the first-row +3 metal ions, Δ_o is about 15,000–24,000 cm^{-1} (Dq approximately 1500–2400 cm^{-1}) depending on the ligands. It has also been shown that B values for the first-row metals are typically about 800–1000 cm^{-1}. Therefore, for this situation, a ratio of Dq/B of 0.52 is clearly out of the realm of possibility for an octahedral complex of this type. For that reason, a ratio of band energies of 2.41 is consistent only with the bands being assigned as v_3 and v_1. It can be shown that this is the case for other complexes as well. It is not actually necessary to know the assignments of the bands in most cases to use the Lever method to determine Dq and B. Assuming that the complexes have realistic values for Dq and B leads to the conclusion that only one set of assignments is possible. Also, only two band positions need to be known because the ratio of these gives both the Dq/B and v_3/B values from the figures.

Let us consider the case of VF_6^{3-}, which has bands at 14800 and 23,000 cm^{-1}. Because V^{3+} is a d^2 ion, the ground state term is $^3T_{1g}$, and in this case, $v_3/v_1 = 1.55$. From Figure 18.8, it can be seen (from the dotted lines) that a value of 1.55 for the ratio v_3/v_1 corresponds to Dq/$B = 2.6$ and $v_3/B = 38$. Therefore, using these values, it is found that

$$B = v_3/38 = Dq/2.6 \qquad (18.16)$$

so that Dq = 23,000 cm^{-1} × 2.6/38 = 1600 cm^{-1} and B is 600 cm^{-1}. These values for Dq and B are in agreement with those obtained for this complex using other procedures.

From the discussion earlier in this chapter, we know that a value of 16,000 cm^{-1} for Δ_o is typical of most complexes of a +3 first-row transition metal ion. For V^{3+}, the free ion B value is 860 cm^{-1} so the value 600 cm^{-1} found for V^{3+} in the complex indicates a value of 0.70 for the nephelauxetic ratio, β. All these values are typical of complexes of the first-row transition metal ions. Therefore, even though the identity of the bands may be uncertain, performing the analysis will lead to B and Dq values that will be reasonable only when the correct assignment of the bands has been made.

In the previous example, if an incorrect assumption is made that the bands represent v_2 and v_1, the ratio of energies having a value of 1.49 would indicate a value of Dq/B of only about 0.90. Using this value and the experimental spectral band energies leads to unrealistic values for Dq and B for a complex of a +3 first-row metal ion. Thus, even if the assignment of bands is uncertain, the calculated ligand field parameters will be consistent with the typical ranges of values for Dq and B for such complexes only for one assignment of bands. The Lever method is a simple, rapid, and useful method for analyzing spectra of transition metal complexes to extract values for ligand parameters. It is fully applicable to tetrahedral complexes as well by using the diagram that corresponds to the correct ground state term for the metal ion.

18.7 JØRGENSEN'S METHOD

An interesting approach to predicting the crystal field splitting for a given metal ion and ligand has been given by Christian Klixbüll Jørgensen. In Jørgensen's approach, the equation that has been developed to predict the ligand field splitting in an octahedral field, Δ_o, is

$$\Delta_o(cm^{-1}) = fg \qquad (18.17)$$

where f is a parameter characteristic of the ligand and g is a parameter characteristic of the metal ion. The values for parameters in this equation are based on the assignment of a value of $f = 1.00$ for water as

Table 18.8 Selected Values for the *f* and *g* Parameters for Use in Jørgensen's Equation

Metal Ion	g	Ligand	f
Mn^{2+}	8,000	Br^-	0.72
Ni^{2+}	8,700	SCN^-	0.73
Co^{2+}	9,000	Cl^-	0.78
V^{2+}	12,000	N_3^-	0.83
Fe^{3+}	14,000	F^-	0.9
Cr^{3+}	17,400	H_2O	1.00
Co^{3+}	18,200	NCS^-	1.02
Ru^{2+}	20,000	py	1.23
Rh^{3+}	27,000	NH_3	1.25
Ir^{3+}	32,000	en	1.28
Pt^{4+}	36,000	CN^-	1.7

From C. K. Jørgensen, Modern Aspects of Ligand Field Theory, *North Holland Publishing Co., Amsterdam,* 1971, pp. 347–8. York, 1969, p. 106.

a ligand, and values for other ligands were determined by fitting the spectral data to known crystal field splittings. Table 18.8 shows representative values for *f* and *g* parameters for several metal ions and ligands.

Equation (18.17) predicts values for Δ_o that are in reasonably good agreement with the values determined by more robust methods. In many instances, an approximate value for the ligand field splitting is all that is required, and this approach gives a useful approximation for Δ_o rapidly with a minimum of effort.

18.8 CHARGE TRANSFER ABSORPTION

Up to this point, the discussion of spectra has been related to the electronic transitions that occur between spectroscopic states for the metal ion in the ligand field. However, these are not the only transitions for which absorptions occur. Ligands are bonded to metal ions by donating electron pairs to orbitals that are essentially metal orbitals in terms of their character. Transition metals also have nonbonding e_g or t_{2g} orbitals (assuming an octahedral complex) arising from the *d* orbitals that may be partially filled, and the ligands may have empty nonbonding or antibonding orbitals that can accept electron density from the metal. For example, both CO and CN^- have empty π^* orbitals that can be involved in this type of interaction (see Chapter 16). Movement of electron density from metal orbitals to ligand orbitals and vice versa is known as charge transfer. The absorption bands that accompany such shifts in electron density are known as *charge transfer bands*.

Charge transfer (CT) bands are usually observed in the ultraviolet region of the spectrum, although in some cases, they appear in the visible region. Consequently, they frequently overlap or mask transitions

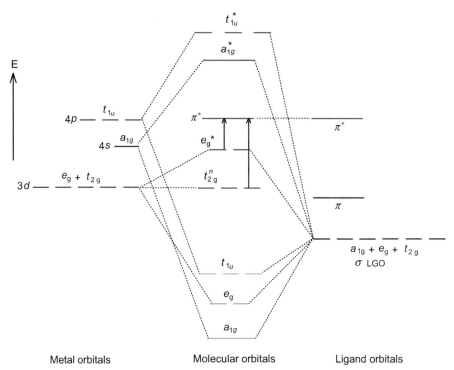

FIGURE 18.9

Interpretation of M → L charge transfer absorption in an octahedral complex using a modified molecular orbital diagram. The transitions are from e_g^* or t_{2g} orbitals on the metal to π^* orbitals on the ligands.

of the $d–d$ type. Charge transfer bands are of the spin-allowed type so they have high intensity. If the metal is in a low oxidation state and easily oxidized, the charge transfer is more likely to be of the metal to ligand type, indicated as M → L. A case of this type occurs in $Cr(CO)_6$ where it is easy to move electron density from the metal atom (which is in the zero oxidation state) especially because the CO ligands have donated six pairs of electrons to the Cr. The empty orbitals on the CO ligands are π^* orbitals. In this case, the electrons are in nonbonding t_{2g} orbitals on the metal so the transition is designated as $t_{2g} \to \pi^*$. In other cases, electrons in the e_g^* orbitals are excited to the empty π^* orbitals on the ligands. Figure 18.9 shows these cases on a modified molecular orbital diagram for an octahedral complex. Although CO is a ligand that has π^* acceptor orbitals, other ligands of this type include NO, CN^-, olefins, and pyridine.

The intense purple color of MnO_4^- is due to a charge transfer band that occurs at approximately 18,000 cm^{-1} and results from a transfer of charge from oxygen to the Mn^{7+}. In this case, the transfer is indicated as L → M, and it results from electron density being shifted from filled p orbitals on oxygen atoms to empty orbitals in the e set on Mn. As a general rule, the charge transfer will be M → L if the metal is easily oxidized, whereas the transfer will be L → M if the metal is easily reduced. Therefore, it is not surprising that Cr^0 in $Cr(CO)_6$ would have electron density shifted *from the metal to the ligands* and Mn^{7+} would have electron density shifted *from the ligands to the metal*. The ease with which electron density can be shifted from the ligands to the metal is related in a general way to how difficult it is to ionize or polarize the ligands.

We have not yet addressed the important topic of absorption by the ligands in complexes. For many types of complexes, this type of spectral study (usually infrared spectroscopy) yields useful information regarding the structure and details of the bonding in the complexes. This topic will be discussed later in connection with several types of complexes containing specific ligands (e.g. CO, CN^-, NO_2^-, and olefins).

18.9 SOLVATOCHROMISM

For reactions carried out in solutions, it has been known for many years that the nature of the solvent can have a great effect on the reaction rate. Although much more extensively studied in organic chemistry, solvent effects can also be pronounced in inorganic reactions. A few cases will be summarized to show the nature of the effects.

When a substance exhibits a different color when dissolved in different solvents, it is the result of *solvatochromism*. This term refers to the change in color that results from the solvent interacting with the solute. Such shifts in spectral bands have been used to correlate solvent effects on reaction rates (Reichardt, 2003; House, 2007). In general, complex dyes have been utilized to establish numerical scales that represent solvent effects, but simple molecules such as I_2 also exhibit solvatochromism (see Chapter 6). There is also a substantial body of literature dealing with solvatochromism of metal complexes. One such compound is $CoCl_2$ for which tetrahedral coordination (blue solution) results in organic solvents that do not coordinate strongly but octahedral coordination (characterized by a pink solution) occurs in water. Wagner-Czauderna, et al. (2004) have investigated the solvatochromism of $CoCl_2$ in mixed solvents that contain water and acetone, dimethylformamide (DMF), dimethylacetamide (DMA), or methylsulfoxide (DMSO). The equilibrium constants,

$$K = \frac{[\text{octahedral}]}{[\text{tetrahedral}]} \tag{18.18}$$

were obtained for several H_2O/organic solvent mixtures. For each of the organic solvents used (acetone, DMF, DMA, and DMSO), a linear relationship was obtained for the log K versus mole fraction of water (χ_w) in the mixed solvent. From analysis of the values for the equilibrium constants as functions of χ_w and temperature, it was observed that except for water/DMSO mixed solvent both ΔH and ΔS for equilibrium become more negative as χ_w increases. Thus, octahedral coordination is favored by the enthalpy change, but it is more unfavorable on the basis of entropy change.

Many other types of complexes also exhibit solvatochromism. Among them are complexes such as those shown in Figure 18.10 (Burgess, et al., 1998). The solvatochromism effect arises from the shift in a charge transfer band that is due to metal to ligand (M → L) electron migration. Moreover, there is also a rather small shift of the charge transfer band to higher wavenumbers as pressure increases in the range 1−1250 bar. A change in spectral properties resulting from a change in pressure is referred to as *piezochromism*.

The M → L charge transfer band for complexes having the general formula $[Fe(CN)_5L]^{3-}$ also exhibit solvatochromism. Accordingly, complexes of this type (where L = 3,5-dimethylpyridine or 3-cyanopyridine) have also been studied with regard to solvatochromatic behavior and the rates of substitution for L in water/organic mixed solvents (Alshehri, 1997). The organic solvents utilized were methanol, ethanol, 1,2-ethanediol, and 1,2-butanediol. In most cases, it was reported that there is a shift to lower

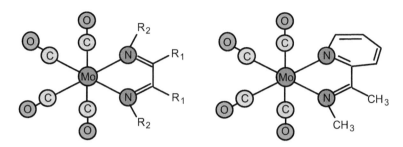

FIGURE 18.10
Two molybdenum complexes that exhibit solvatochromism.

wavenumber for the M → T charge transfer band as the mole percent of the organic constituent in the mixed solvent increases. Moreover, the rate constants for the reaction

$$[Fe(CN)_5L]^{3-} + CN^- \rightarrow [Fe(CN)_6]^{4-} + L \tag{18.19}$$

were observed to depend on the composition of the solvent. When the solvent was a mixture of water and 1,2-ethanediol, the rate constant was found to decrease for replacement of L (L = 4-cyanopyridine, 3-cyanopyridine, 4,4'-dipyridyl, 3,5-dimethylpyridine, or 4-phenylpyridine). On the other hand, for the case where L = 3,5-dimethylpyridine, the rate of substitution was generally found to increase as the mole fraction of alcohol (CH_3OH or C_2H_5OH) in the mixed solvent increases, but for certain leaving groups, the opposite trend was observed. For mixed solvents containing 1,2-ethanediol and 1,2-butanediol, there was a rather general relationship between the extent to which the charge transfer band was shifted as the percent of organic component in the solvent increased. There was also reported a decrease in the rate constant for substitution when L = 3-cyanopyridine or 3,5-dimethylpyridine.

The discussion to this point gives rise to more questions than answers. It is apparent that there is no clear understanding of the effects of the solvent on spectral properties and reaction rates. However, it is clear that the solvent is not simply an innocent bystander in coordination chemistry. A great deal of additional work is needed in this interesting and important area.

References for Further Study

Alshehri, S., 1997. *Transition Met. Chem.* 22, 553–556. An article dealing with solvatochromism.

Burgess, J., Maguire, S., McGranahan, A., Parsons, S.A., 1998. *Transition Met. Chem.* 23, 615–618. An article dealing with the solvatochromism in molybdenum carbonyl complexes.

Cotton, F.A., Wilkinson, G., Murillo, C.A., Bochmann, M., 1999. *Advanced Inorganic Chemistry*, 6th ed. John Wiley, New York. This reference text contains a large amount of information on the entire range of topics in coordination chemistry.

Douglas, B., McDaniel, D., Alexander, J., 2004. *Concepts and Models of Inorganic Chemistry*, 3rd ed. John Wiley, New York. A respected inorganic chemistry text.

Drago, R.S., 1992. *Physical Methods for Chemists*. Saunders College Publishing, Philadelphia. This book presents high level discussion of many topics in coordination chemistry.

Figgis, B.N., Hitchman, M.A., 2000. *Ligand Field Theory and Its Applications.* John Wiley, New York. An advanced treatment of ligand field theory and spectroscopy.

Harris, D.C., Bertolucci, M.D., 1989. *Symmetry and Spectroscopy.* Dover Publications, New York. A good text on bonding, symmetry, and spectroscopy.

House, J.E., 2007. *Principles of Chemical Kinetics,* 2nd ed. Elsevier, New York. Chapters 5 and 9 deal with solvent effects on reaction rates.

Kettle, S.F.A., 1969. *Coordination Chemistry.* Appleton, Century, Crofts, New York. A good introduction to interpreting spectra of coordination compounds is given.

Kettle, S.F.A., 1998. *Physical Inorganic Chemistry: a Coordination Approach.* Oxford University Press, New York. An excellent book on coordination chemistry that gives good coverage to many areas, including ligand field theory and spectroscopy.

Lever, A.B.P., 1968. *J. Chem. Educ.,* 45, 711. An article describing a novel method for determining Dq and B.

Lever, A.B.P., 1984. *Inorganic Electronic Spectroscopy,* 2nd ed. Elsevier, New York. A monograph that treats all aspects of absorption spectra of complexes at a high level. This is perhaps the most through treatment available in a single volume. Highly recommended.

Reichardt, C., 2003. *Solvents and Solvent Effects in Organic Chemistry,* 3rd ed. VCH, Weinheim. The standard reference on solvent effects. Highly recommended.

Solomon, E.I., Lever, A.B.P. (Eds.), 2006, *Inorganic Electronic Spectroscopy and Structure,* vols. I and II. Wiley, New York Perhaps the ultimate resource on spectroscopy of coordination compounds. Two volumes total 1424 pages on the subject.

Szafran, Z., Pike, R.M., Singh, M.M., 1991. *Microscale Inorganic Chemistry: a Comprehensive Laboratory Experience.* John Wiley, New York. Several sections of this book deal with various aspects of synthesis and study of coordination compounds. A practical, "hands on" approach.

Wagner-Czauderna, E., Boron-Cegielkowska, A., Orlowska, E., Kalinowski, M.K., 2004. *Transition Met Chem.* 29, 61–65. A report on solvatochromism in $CoCl_2$.

QUESTIONS AND PROBLEMS

1. For the following high-spin ions, describe the nature of the possible electronic transitions and give them in the order of increasing energy.

   (a) $[Ni(NH_3)_6]^{2+}$; (b) $[FeCl_4]^-$; (c) $[Cr(H_2O)_6]^{3+}$; (d) $[Ti(H_2O)_6]^{3+}$; (e) $[FeF_6]^{4-}$; (f) $[Co(H_2O)_2]^{2+}$

2. Use Jørgensen's method to determine Δ_o for the following complexes.

   (a) $[MnCl_6]^{4-}$; (b) $[Rh(py)_6]^{3+}$; (c) $[Fe(NCS)_6]^{3-}$; (d) $[Co(NH_3)_6]^{2+}$; (e) $[PtBr_6]^{2-}$

3. Spectral bands are observed at 17,700 and 32,400 cm^{-1} for $[Cr(NCS)_6]^{3-}$. Use the Lever method to determine Dq and B for this complex. Where would the third band be found? What transition does it correspond to?

4. The spectroscopic ground state for a certain first-row metal atom is $^6S_{5/2}$. What is the atom? What would be the spectroscopic ground state for the +2 and +3 ions of this metal?

5. Some values for the Racah B parameter are 918, 766, and 1064 cm^{-1}. Match these values to the ions Cr^{3+}, V^{2+}, and Mn^{4+}. Explain your assignments.

6. Spectral bands are observed at 8350 and 19,000 cm^{-1} for the complex $[Co(H_2O)_6]^{2+}$. Use the Lever method to determine Dq and B for this complex. Determine where the missing band should be observed. What transition does this correspond to?

7. A complex VL_6^{3+} has the two lowest energy transitions at 11,500 and 17,250 cm^{-1}.

   (a) What are the designations for these transitions?

   (b) Use the Lever method to determine Dq and B for this complex.

   (c) Where should the third spectral band be observed?

8. Use Jørgensen's method to determine Dq and B for $[Co(en)_3]^{3+}$. What would the approximate positions of the three spectral bands?

9. Peaks are observed at 264 and 378 nm for $[Cr(CN)_6]^{3-}$. What are Dq and B for this complex?

10. For the complex $[Cr(acac)_3]$ (where acac = acetylacetonate ion), v_1 is observed at 17,860 cm^{-1} and v_2 is at 23,800 cm^{-1}. Use these data to determine Dq and B for this complex.

11. For the complex $[Ni(en)_3]^{2+}$ (where en = ethylenediamine), v_1 is observed at 11,200 cm^{-1}, v_2 is at 18,450 cm^{-1}, and v_3 is at 29,000 cm^{-1}. Use these data to determine Dq and B for this complex.

12. For the complex $[Cr(NCS)_6]^{3-}$, the first two absorption bands are observed at 17,800 and 23,800 cm^{-1}. Where would the third band be expected to be seen?

13. For each of the following, tell whether you would or would not expect to see charge transfer absorption. For the cases where you would expect charge transfer absorption, explain the type of transitions expected.
(a) $[Cr(CO)_3(py)_3]$; (b) $[Co(NH_3)_6]^{3+}$; (c) $[Fe(H_2O)_6]^{2+}$; (d) $[Co(CO)_3NO]$; (e) CrO_4^{2-}.

CHAPTER 19

Composition and Stability of Complexes

Coordinate bonds between metals and ligands result in the formation of complexes under many different types of conditions. In some cases, complexes form in the gas phase, and the number of known solid complexes is enormous. However, it is in solutions that many of the effects of complex formation are so important. For example, in qualitative analysis, AgCl precipitates when a solution of HCl is added to one containing Ag^+. When aqueous ammonia is added, the precipitate dissolves as a result of the formation of a complex,

$$AgCl(s) + 2\,NH_3(aq) \;\rightarrow\; Ag(NH_3)_2{}^+(aq) + Cl^-(aq) \tag{19.1}$$

It is possible to determine the concentration of certain metal ions by performing a titration in which the complexation of the metal is the essential reaction. Typically, a chelating agent such as ethylenediaminetetraacetic acid (EDTA) is used because the complexes formed are so stable. The specific composition of complexes formed in solutions often depends on the concentrations of the reactants. As a part of the study of the chemistry of coordination compounds, some attention must be given to the systematic treatment of topics related to the composition and stability of complexes in solution. This chapter is devoted to these topics.

19.1 COMPOSITION OF COMPLEXES IN SOLUTION

Although it is well known that the most frequently encountered coordination numbers are 2, 4, and 6, we should expect that for some ligands and metal ions, different or unknown ratios of ligands to metal ion might occur under certain conditions. If several different complexes are formed, the problem of determining composition may be quite complicated. For simplicity, we shall assume that only one complex is formed between the metal and ligand or that the amounts of all other complexes formed can be neglected compared with the amount of the dominant complex.

The equation for the formation of a complex by Metal A and Ligand B is written as

$$n\,A + m\,B \;\rightleftarrows\; A_nB_m \tag{19.2}$$

for which the equilibrium constant is written as

$$K = \frac{[A_nB_m]}{[A]^n[B]^m} \tag{19.3}$$

643

Inorganic Chemistry. DOI: http://dx.doi.org/10.1016/B978-0-12-385110-9.00019-4

Solving Eqn (19.3) for the concentration of the complex leads to

$$[A_nB_m] = K[A]^n[B]^m \tag{19.4}$$

Taking the logarithm of both sides of this equation yields

$$\log[A_nB_m] = \log K + n\log[A] + m\log[B] \tag{19.5}$$

To obtain data for use with this equation, a series of solutions can be prepared in which the concentration of B is kept constant but the concentration of A is varied. For each of these solutions, the concentration of the complex $[A_nB_m]$ is measured as a function of [A]. For many complexes, the concentration of the complex can be measured by spectrometry because many complexes absorb at a wavelength that is different from that of the metal ion or ligand alone (see Chapter 18). When [B] is kept constant, Eqn (19.5) reduces to

$$\log[A_nB_m] = n\log[A] + C \tag{19.6}$$

where $C = \log K + m\log[B] =$ a constant. As is illustrated in Figure 19.1, a graph can then be made of $\log[A_nB_m]$ versus $\log[A]$ that yields a straight line of slope n.

This procedure can be repeated keeping the concentration of A fixed and varying the concentration of B. Under these conditions, we obtain the concentration of $[A_nB_m]$ as the concentration of B varies, and the graph of $\log[A_nB_m]$ versus $\log[B]$ yields a straight line of slope m. In this way, we can determine n and m, the numbers of metal ions and ligands, respectively, in the formula for the complex. Once the values of m and n are known, the value of K can be obtained because the intercepts of the two plots are $\log K + m\log[B]$ and $\log K + n\log[A]$, respectively.

Although this method is simple in principle, problems arise if more than one complex is formed. As the relative concentrations of A and B are varied, it is quite possible that such a situation could occur due to mass action effects. The plots might not be linear in this case, and it is possible that the complex(es) produced might absorb at a different wavelength. However, it should be pointed out that this method is not restricted to coordination compounds containing metal ions and ligands. It is applicable to cases where species of almost any type are forming aggregates as long as a property proportional to the concentration of the complex can be measured. Molecular association of the types discussed in Chapter 6 can often be studied in this way.

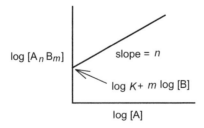

FIGURE 19.1
An illustration of the logarithmic method.

19.2 JOB'S METHOD OF CONTINUOUS VARIATIONS

Another method for determining the composition of a complex in solution is that known as *Job's method* or the *method of continuous variations*. Suppose that the formation of a complex between Metal A and m ligands B can be shown as

$$A + m\,B \;\rightleftarrows\; AB_m \tag{19.7}$$

for which the equilibrium constant can be written as

$$K = \frac{[AB_m]}{[A][B]^m} \tag{19.8}$$

A series of solutions can now be prepared such that the *total* concentration of metal and ligand is a constant, C,

$$[A] + [B] = C, \tag{19.9}$$

but we allow the concentration of each component to vary. For example, we might make solutions that have a total concentration of 1 M and that have compositions as follows: 0.1 M (A) and 0.9 M (B), 0.2 M (A) and 0.8 M (B), 0.3 M (A) and 0.7 M (B), etc. If as a result of complex formation the ionic strength of the solution changes, it is possible to keep the ionic strength almost constant by dissolving a salt such as $KClO_4$. Generally, ClO_4^- does not compete with other ligands, and the large concentration of ions from the salt means that small changes in the total concentration of ions due to forming complexes do not change the overall ionic strength significantly. Next, we select some physical property that is exhibited by the complex but which is not exhibited to a significant extent by the metal or unbound ligands. It is much simpler if the property is one that is linearly related to the concentration of the complex. For this reason and because many transition metal complexes are highly colored, absorbance of visible light by the solutions at a particular wavelength where the complex absorbs is frequently used as the property.

A graph (known as a Job's plot) is now made of the physical property (in effect, the concentration of the complex) as a function of the ratio $[B]/([B]+[A])$ (or sometimes $[B]/[A]$). A maximum or minimum (depending on the type of property measured) in the curve will be found when the ratio of the constituents in the solution is the same as it is in the formula for the complex. Figure 19.2 shows a graph constructed as described above.

It is easy to verify that the curve representing the measured property will exhibit a maximum or minimum at the same ratio of concentrations as the constituents are present in the complex. We can rearrange the expression for the equilibrium constant as

$$[AB_m] = K[B]^m[A] \tag{19.10}$$

but because $[A] + [B] = C$, we can write $[A] = C - [B]$, so substituting for $[A]$ we obtain

$$[AB_m] = K[B]^m\{C - [B]\} = KC[B]^m - K[B]^{m+1} \tag{19.11}$$

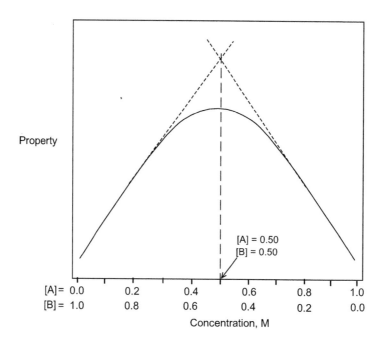

FIGURE 19.2
An illustration of the use of Job's method. In this case, the complex of A and B has a 1:1 composition.

Because we want to find out if the change in concentration of the complex with the change in concentration of B goes through a maximum or minimum, we need to take the derivative and set it equal to 0.

$$\frac{d[AB_m]}{d[B]} = 0 = mKC[B]^{m-1} - (m+1)K[B]^m \tag{19.12}$$

The sum of the two terms on the right-hand side of this equation is zero, so we can equate them. After dividing by $K[B]^{m-1}$, we obtain

$$(m+1)[B] = mC \tag{19.13}$$

By expanding the left hand side and letting $C = [A] + [B]$, we find that

$$m[B] + [B] = m\{[B] + [A]\} = m[B] + m[A] \tag{19.14}$$

from which we obtain

$$[B] = m[A] \tag{19.15}$$

or

$$m = \frac{[B]}{[A]} \tag{19.16}$$

This result signifies that the maximum in the amount of complex formed from a fixed total concentration of A plus B will occur when A and B are present in the solution in precisely the same ratio as they are in the complex.

This procedure has been used successfully to determine the composition of many complexes in solution. It is possible to extend this method to cases where more than one complex is formed but the application is quite difficult. Like the logarithmic method, Job's method can be applied to other cases of molecular interaction and is not limited to the formation of coordination compounds. Both methods are based on the assumption that one complex is dominant in the equilibrium mixture. Numerous other methods for determining the number of metal ions and ligands in complexes have been devised, but they are beyond the introduction to the topic presented here.

19.3 EQUILIBRIA INVOLVING COMPLEXES

In the preceding section, we have described two methods that are frequently used to determine the composition of complexes in solution. We will now turn our attention to a consideration of the simultaneous equilibria that are involved in complex formation. The widely used approach described here is known as Bjerrum's method, and it was described by Jannik Bjerrum many years ago.

If we consider a complex such as that represented by the formula AB_m, we can represent the formation of this complex as if it occurred in the m steps shown below.

$$\text{Step 1:} \quad A + B \rightleftarrows AB \tag{19.17}$$

$$\text{Step 2:} \quad AB + B \rightleftarrows AB_2 \tag{19.18}$$

$$\vdots \qquad \vdots$$

$$\text{Step } m: \quad AB_{m-1} + B \rightleftarrows AB_m . \tag{19.19}$$

For these reactions, we can write the equilibrium constants (known as stepwise formation or stability constants) as

$$K_1 = \frac{[AB]}{[A][B]} \tag{19.20}$$

$$K_2 = \frac{[AB_2]}{[AB][B]} \tag{19.21}$$

$$\vdots$$

$$K_m = \frac{[AB_m]}{[AB_{m-1}][B]} \tag{19.22}$$

In aqueous solutions, however, the complexes are usually present in compositions in which water molecules complete the coordination sphere of the metal up to a definite number, usually 2, 4, or 6. Thus, the complexes are formed as represented by the equations

$$A(H_2O)_m + B \rightleftarrows AB(H_2O)_{m-1} + H_2O \tag{19.23}$$

$$AB(H_2O)_{m-1} + B \rightleftarrows AB_2(H_2O)_{m-2} + H_2O \tag{19.24}$$

$$\vdots$$

$$AB_{m-1}(H_2O) + B \rightleftarrows AB_m + H_2O \tag{19.25}$$

We shall ignore for the moment the fact that the solvent plays a role and will represent the formation of the successive complexes as shown in Eqns (19.17–19.19). However, we should not lose sight of the fact that in aqueous solutions, the total coordination number of the metal is m, and if x sites are bonded to water molecules and y sites are where ligands are attached, then $x + y = m$. Because the constants K_1, K_2, ..., K_m represent the *formation* of complexes, they are called *formation constants*. The larger the value of a formation constant, the more stable the complex. Consequently, these constants are usually called *stability constants*.

It is important to note that the equilibrium constants for the formation of the complex AB_3, for example, can be represented as a product such that β_3 is given by

$$\beta_3 = K_1 K_2 K_3 = \frac{[AB]}{[A][B]} \times \frac{[AB_2]}{[AB][B]} \times \frac{[AB_3]}{[AB_2][B]} = \frac{[AB_3]}{[A][B]^3} \tag{19.26}$$

Thus, the *overall* stability constants for the various steps, β_i, can be expressed in terms of the product of several *stepwise* stability constants, K_i (e.g. $\beta_2 = K_1 K_2$; $\beta_4 = K_1 K_2 K_3 K_4$).

Frequently, it is of interest to consider the *dissociation* of complexes rather than their formation. For example,

$$AB_m \rightleftarrows AB_{m-1} + B \tag{19.27}$$

$$AB_{m-1} \rightleftarrows AB_{m-2} + B \tag{19.28}$$

$$\vdots$$

$$AB \rightleftarrows A + B \tag{19.29}$$

It should be apparent that the larger the equilibrium constants for these reactions, the more *unstable* the complexes. Consequently, the equilibrium constants for these reactions are called *dissociation* or *instability constants*. One must remember that the first step in the dissociation of AB_m corresponds to the reverse of the last step in the formation of the complex. Thus, if we represent the instability constants as k_i and a total of six steps are involved, the relationships between the k_i and K_i values are as follows.

$$K_1 = 1/k_6, \quad K_2 = 1/k_5, \dots, \quad K_6 = 1/k_1 \tag{19.30}$$

To determine the stability constants for a series of complexes in solution, we must determine the concentrations of several species. Moreover, we must then solve a rather complex set of equations to evaluate the stability constants. There are several experimental techniques that are frequently employed for determining the concentrations of the complexes. For example, spectrophotometry, polarography, solubility measurements, potentiometry, etc. may be used, but the choice of experimental method is based on the nature of the complexes being studied. Basically, however, we proceed as follows. A parameter is defined as the average number of bound ligands per metal ion, N, which is expressed as

$$N = \frac{(C_B - [B])}{C_A} \tag{19.31}$$

where C_B is the total concentration of ligand present in the solution, $[B]$ is the concentration of free or unbound ligand, and C_A is the total concentration of metal ion, A. The quantity $C_B - [B]$ represents the concentration of B that is bound in all the complexes present in the equilibrium mixture. Consequently, because one mole of $[AB_2]$ contains two moles of B and one mole of $[AB_3]$ contains three moles of B, etc, the concentration of bound B can be written as

$$C_B - [B] = [AB] + 2[AB_2] + 3[AB_3] + \ldots \tag{19.32}$$

The total concentration of metal will be the concentration of free metal plus all that bound in the various complexes.

$$C_A = [A] + [AB] + [AB_2] + [AB_3] + \ldots \tag{19.33}$$

Therefore, substituting the expression shown in Eqn (19.32) for $C_B - B$ and the expression above for C_A in Eqn (19.31) gives

$$N = \frac{[AB] + 2[AB_2] + 3[AB_3] + \cdots}{[A] + [AB] + [AB_2] + [AB_3] + \cdots} \tag{19.34}$$

The next step in the procedure for calculating the stability constants is to represent the concentrations of the several complexes in terms of the equilibrium constants for their formation. We do this by solving each equilibrium constant expression for the concentration of the complex. For example, the first step has an equilibrium constant

$$K_1 = \frac{[AB]}{[A][B]} \tag{19.35}$$

so solving for the concentration of the complex yields

$$[AB] = K_1[A][B] \tag{19.36}$$

For the complex AB_2,

$$K_2 = \frac{[AB_2]}{[AB][B]} \tag{19.37}$$

so that

$$[AB_2] = K_2[AB][B] = K_1 K_2[A][B]^2 \tag{19.38}$$

In the same way, it can be shown that

$$[AB_3] = K_1 K_2 K_3[A][B]^3 \tag{19.39}$$

Expressions for the concentrations of the other complexes can be found in a similar way. These expressions can now be substituted in Eqn (19.34) to obtain

$$N = \frac{K_1[A][B] + 2K_1 K_2[A][B]^2 + 3K_1 K_2 K_3[A][B]^3 + \cdots}{[A] + K_1[A][B] + K_1 K_2[A][B]^2 + K_1 K_2 K_3[A][B]^3 + \cdots} \tag{19.40}$$

A more useful equation is obtained by dividing each term in the numerator and denominator of this equation by [A],

$$N = \frac{K_1[\text{B}] + 2K_1K_2[\text{B}]^2 + 3K_1K_2K_3[\text{B}]^3 + \cdots}{1 + K_1[\text{B}] + K_1K_2[\text{B}]^2 + K_1K_2K_3[\text{B}]^3 + \cdots} \tag{19.41}$$

From this equation, it can be seen that the average number of ligands bound per metal ion depends only on the values of the stability constants for the complexes and the concentration of free or unbound ligand in the solution, [B]. To calculate m stability constants $(K_1, K_2, \ldots, K_m)$, we need m values of N each determined at a different but known value of [B]. What results is a set of m equations in m unknowns (the stability constants) that may be solved by the standard numerical, graphical, or computer techniques. Historically, this analysis of data presented a considerable challenge, especially in view of the limitations imposed by the use of experimental data. Solving such a series of equations when highly accurate data are available is readily handled by an advanced programmable calculator today, but it has not always been so.

Before a series of stability constants can be interpreted as indicating the correct order of the stabilities of the complexes, one further point must be considered. This factor arises from the fact that if one considers a complex such as $M(H_2O)_N$, there are N potential sites where the first ligand B may enter. Therefore, the probability that B will enter a site is proportional to N. That is, the first ligand to enter has N sites at which it may attach. The second ligand has only $N - 1$ sites available, and at a later point in the substitution process, the complex can be written as $MB_m(H_2O)_{N-m}$. The probability that another ligand will enter the coordination sphere of the metal is proportional to $N - m$ if the probability of a ligand entering is equal at all sites (i.e. there are no steric factors). The relationship between the probabilities for the first, second, third, ..., and last ligand to enter the coordination sphere can be expressed as the fractions

$$\frac{N}{1}, \frac{N-1}{2}, \cdots, \frac{N-m+1}{m}, \frac{N-m}{m+1}, \cdots, \frac{2}{N-1}, \frac{1}{N} \tag{19.42}$$

Keep in mind that in this case, the total number of sites is N. The relationship between the values of two successive K values must reflect the difference in probabilities of formation of the corresponding complexes. Therefore, we can write

$$\frac{K_m}{K_{m+1}} = \frac{(N-m+1)(m+1)}{m(N-m)} \tag{19.43}$$

The equilibrium constants for formation of the various complexes should be identical after the experimental values are corrected for the differences in probability for their formation. Also, the stability of the last complex as reflected by K_m should be independent of the fact that some of the intermediate complexes do not have equal probabilities for forming. If we represent the equilibrium constants that have been corrected for the differences in probability as K', we should find that

$$K_1K_2 \ldots K_m = K_1'K_2' \ldots K_m' \tag{19.44}$$

From this equation, we should find that the average equilibrium constant, K, may be represented as the mth root of either of the products of equilibrium constants.

$$K = (K_1 K_2 ... K_m)^{1/m} = (K_1' K_2' ... K_m')^{1/m} \tag{19.45}$$

Theoretically, however, the statistically corrected values, K', are all the same so that

$$(K_1 K_2 ... K_m)^{1/m} = K_1' = K_2' = ... = K_m' \tag{19.46}$$

We now have a sufficient number of conditions to determine the relationship between the measured equilibrium constants, K_1, K_2, ..., K_m, and the statistically corrected constants, K_1', K_2', ..., K_m'. The statistical corrections to be applied to the experimental stability constants are shown in Table 19.1 for the most common coordination numbers.

Many correlations have been made between the stability constants for a series of complexes and other properties. For example, basicity of the ligands, ionic radii, dipole moments, and other properties have been correlated with the stability constants of the complexes. However, before comparisons such as these are made, the stability constants should be corrected statistically to take into account the fact that successive complexes do not have the same probability of forming.

There is another problem that does not have such a simple solution. Consider the following reactions that represent the first and last step in the reaction of a metal ion, M^{3+}, in the form of an aqua complex, $[M(H_2O)_6]^{3+}$, with Cl^-.

$$[M(H_2O)_6]^{3+} + Cl^- \rightarrow [M(H_2O)_5 Cl]^{2+} + H_2O \tag{19.47}$$

$$[M(H_2O)Cl_5]^{2-} + Cl^- \rightarrow [MCl_6]^{3-} + H_2O \tag{19.48}$$

Note that in the first case, the Cl^- ion is approaching an aqua complex that carries a +3 overall charge. In the second case, the Cl^- is approaching an aquapentachloro complex that already carries a *negative* charge (-2), which is electrostatically unfavorable. Therefore, even after statistical correction of the stability constants is made, there is a great deal of difference in the likelihood that $[M(H_2O)_5Cl]^{2+}$ and $[MCl_6]^{3-}$ will form (the values of K_1 and K_6).

Table 19.1 Statistically Corrected Stability Constants

Coordination Number, N	Corrected Stability Constants, K'					
	K_1'	K_2'	K_3'	K_4'	K_5'	K_6'
2	$K_1/2$	$2K_2$				
4	$K_1/4$	$2K_2/3$	$3K_3/2$	$4K_4$		
6	$K_1/6$	$2K_2/5$	$3K_3/4$	$4K_4/3$	$5K_5/2$	$6K_6$

This is a difficult problem. First, the charged species are separated by water molecules. The force between charged particles that are separated by a medium having a dielectric constant ε is given by

$$F = \frac{q_1 q_2}{\varepsilon r^2} \tag{19.49}$$

where q_1 and q_2 are the charges on the particles, ε is the dielectric constant of the medium separating them, and r is the distance of separation. However, the metal ion is surrounded by water molecules that are coordinated to it. Those water molecules do not have the same dielectric constant as those of the bulk solvent because they are oriented in a particular way. Moreover, the high charge density of the metal ion induces additional charge separation in water molecules that are coordinated. The result is that we now have a very difficult task in trying to compensate for the difference in the tendencies of the complexes $[M(H_2O)_5Cl]^{2+}$ and $[MCl_6]^{3-}$ to form. In fact, although it is necessary to know that such differences exist, it is beyond the scope of this book to deal with the details of this problem.

19.4 DISTRIBUTION DIAGRAMS

If we know the stability constants for a series of complexes formed between a metal ion and several ligands, we have the information necessary to calculate the concentrations of all the species present at equilibrium. For example, we will suppose that a metal ion, M, forms several complexes with ligand X that are in equilibrium. In this example, we will assume that the metal ion has a coordination number of 4. The general procedure for dealing with complexes having other coordination numbers is the same as illustrated here. We can obtain the concentration of each complex following the procedure used earlier, which gives

$$[MX] = K_1[M][X] \tag{19.50}$$

$$[MX_2] = K_1 K_2 [M][X]^2 \tag{19.51}$$

$$[MX_3] = K_1 K_2 K_3 [M][X]^3 \tag{19.52}$$

$$[MX_4] = K_1 K_2 K_3 K_4 [M][X]^4 \tag{19.53}$$

The total concentration of metal ion, C_M, is the sum of the concentrations of the free (uncomplexed) metal added to the concentration of the metal ion that is bound in all the complexes present. This can be expressed as

$$C_M = [M] + [MX] + [MX_2] + [MX_3] + [MX_4] \tag{19.54}$$

When the substitutions are made for the concentrations of the complexes given in Eqns (19.50−19.53), we find that

$$C_M = [M] + K_1[M][X] + K_1 K_2 [M][X]^2 + K_1 K_2 K_3 [M][X]^3 + K_1 K_2 K_3 K_4 [M][X]^4 \tag{19.55}$$

The overall stability constants are given as the products of the stability constants for all steps up to the one being considered. These are represented as β_i, and they are related to the K values by the equations

$$\beta_1 = K_1 \tag{19.56}$$

$$\beta_2 = K_1 K_2 \tag{19.57}$$

$$\beta_3 = K_1 K_2 K_3 \tag{19.58}$$

$$\beta_4 = K_1 K_2 K_3 K_4 \tag{19.59}$$

Substituting for the products of the stability constants and factoring [M] from Eqn (19.55) gives

$$C_M = [M]\left(1 + \beta_1[X] + \beta_2[X]^2 + \beta_3[X]^3 + \beta_4[X]^4\right) \tag{19.60}$$

The fraction of the metal ion that is bound in each type of complex can now be calculated as well as the fraction of the metal that is free. The free metal has no ligands attached, so we will represent it as α_0, whereas the fraction bound to one ligand is α_1, etc. Therefore,

$$\alpha_0 = \frac{\text{Concentration of free metal}}{\text{Total concentration of metal}} = \frac{[M]}{C_M} \tag{19.61}$$

and by substitution, we find that

$$\alpha_0 = \frac{1}{1 + \beta_1[X] + \beta_2[X]^2 + \beta_3[X]^3 + \beta_4[X]^4} \tag{19.62}$$

Although this gives the fraction of the metal ion that is not contained in complexes, the fractions bound in complexes are given by

$$\alpha_1 = \frac{[MX]}{C_M} = \beta_1[X]\alpha_0 \tag{19.63}$$

$$\alpha_2 = \frac{[MX_2]}{C_M} = \beta_2[X]^2\alpha_0 \tag{19.64}$$

$$\alpha_3 = \frac{[MX_3]}{C_M} = \beta_3[X]^3\alpha_0 \tag{19.65}$$

$$\alpha_4 = \frac{[MX_4]}{C_M} = \beta_4[X]^4\alpha_0 \tag{19.66}$$

In these equations, [X] is the concentration of free ligand. With these five expressions that give the fraction of the free metal ion and the fraction in each complex, it is possible to vary [X] and calculate the fraction of the metal ion in each complex environment.

The graph that shows the fraction of each metal-containing species as a function of the concentration of free ligand is called a *distribution diagram*. As an example to illustrate the construction of a distribution diagram, the equilibrium that results when Cd^{2+} reacts with NH_3 will be considered. When water is omitted from the reactions, the steps can be shown as follows:

$$Cd^{2+} + NH_3 \rightleftharpoons Cd(NH_3)^{2+} \tag{19.67}$$

$$Cd(NH_3)^{2+} + NH_3 \rightleftharpoons Cd(NH_3)_2^{2+} \tag{19.68}$$

$$Cd(NH_3)_2^{2+} + NH_3 \rightleftharpoons Cd(NH_3)_3^{2+} \tag{19.69}$$

$$Cd(NH_3)_3^{2+} + NH_3 \rightleftharpoons Cd(NH_3)_4^{2+} \tag{19.70}$$

The equilibrium constants for the four steps in the formation of the cadmium ammine complexes are $K_1 = 447$, $K_2 = 126$, $K_3 = 27.5$, and $K_4 = 8.51$. Using these values, the overall stability constants are $\beta_1 = 447$, $\beta_2 = 5.63 \times 10^4$, $\beta_3 = 1.55 \times 10^6$, and $\beta_4 = 1.32 \times 10^7$. Equations (19.63–19.66) can be employed to calculate the concentrations of the fractions that contain the metal, α_i, by assuming a series of known values for the concentration of free NH_3. The range of ligand concentrations must be chosen with care based on the values of the stability constants. In this case, the range of NH_3 concentrations is from 10^{-5} M to 1 M and is expressed as pNH_3, which ranges from 5 to 0. These calculations give the values shown in Table 19.2. The distribution diagram for the complexes formed in the equilibria represented by the $Cd^{2+}–NH_3$ system is shown in Figure 19.3.

Table 19.2 Equilibrium Concentrations of Cd^{2+} Complexes in Solutions Containing NH_3

pNH$_3$	α_0	α_1	α_2	α_3	α_4
5.00	0.996	0.004	0.0	0.0	0.0
4.50	0.986	0.014	0.0	0.0	0.0
4.00	0.957	0.043	0.0	0.0	0.0
3.50	0.872	0.123	0.005	0.0	0.0
3.00	0.665	0.297	0.037	0.001	0.0
2.75	0.505	0.401	0.090	0.004	0.0
2.50	0.330	0.467	0.186	0.016	0.0
2.40	0.265	0.472	0.237	0.026	0.001
2.25	0.179	0.450	0.319	0.049	0.002
2.00	0.079	0.350	0.441	0.121	0.010
1.75	0.027	0.217	0.485	0.237	0.036
1.60	0.013	0.145	0.458	0.317	0.068
1.50	0.007	0.106	0.421	0.367	0.099
1.25	0.002	0.041	0.291	0.451	0.216
1.00	0.0	0.013	0.162	0.446	0.379
0.75	0.0	0.0	0.075	0.367	0.555
0.50	0.0	0.0	0.030	0.262	0.707
0.25	0.0	0.0	0.011	0.171	0.818
0.00	0.0	0.0	0.004	0.105	0.891

$pNH_3 = -log\ [NH_3]$.
If $pNH_3 = 2.00$, $[NH_3] = 10^{-2}$ M.
If $pNH_3 = 0.50$, $[NH_3] = 0.316$ M.

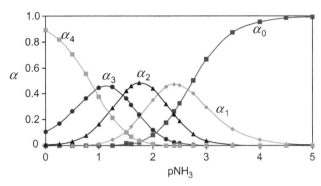

FIGURE 19.3
A distribution diagram for Cd^{2+} complexes with NH_3 constructed using data shown in Table 19.2.

The distribution diagram shows that at very low concentrations of NH_3, the Cd^{2+} is present almost entirely as the free ion. On the other hand, when the concentration of NH_3 is 1 M, the dominant species is $[Cd(NH_3)_4]^{2+}$. At intermediate concentrations of NH_3, several species are present, but the fractions must total 1, the entire fraction of Cd^{2+} because all the Cd^{2+} must be accounted for in the various complexes. By Le Chatelier's principle, increasing the concentration of NH_3 drives the system represented as shown in Eqns (19.67–19.70) to the right. The procedures described above can also be employed to construct a distribution diagram for the dissociation of a polyprotic acid such as H_3PO_4 by determining the concentrations of H_3PO_4, $[H_2PO_4^-]$, $[HPO_4^{2-}]$, and PO_4^{3-} as functions of $[H^+]$ or pH. It should be pointed out that determining the stability of complexes has been a significant area of research for many years, and an enormous body of literature on the subject exists. One of the classic research monographs in this area for many years has been the book by Rossotti and Rossotti (1961) that is cited in the references at the end of this chapter.

19.5 FACTORS AFFECTING THE STABILITY OF COMPLEXES

The formation of a coordinate bond is the result of the donation and acceptance of a pair of electrons. This in itself suggests that if a specific electron donor interacts with a series of metal ions (electron acceptors), there will be some variation in the stability of the coordinate bonds depending on the acidity of the metal ion. Conversely, if a specific metal ion is considered, there will be a difference in stability of the complexes formed with a series of electron pair donors (ligands). In fact, there are several factors that affect the stability of complexes formed between metal ions and ligands, and some of them will now be described.

Establishing correlations between the stability of complexes and factors related to the characteristics of the metal ion and ligands involved is not a new endeavor. One of the earliest correlations established showed that for many types of ligands, the stability of the complexes that they form with +2 ions of first-row transition metals varies in the order

$$Mn^{2+} < Fe^{2+} < Co^{2+} < Ni^{2+} < Cu^{2+} > Zn^{2+}$$

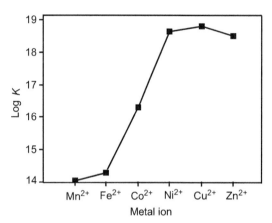

FIGURE 19.4

The effect of the nature of the metal ion on the stability of $EDTA^{4-}$ complexes.

Figure 19.4 shows a graph of the logarithm of K_1 for the formation of the complexes between these metal ions and $EDTA^{4-}$. The graph shows that the general order of stability of the complexes is in accord with the series of metal ions shown above. The order of stability for complexes containing the metal ions shown above is known as the *Irving–Williams series*. Complexes of these metals containing many other types of ligands show a similar trend in stability.

A measure of the Lewis acidity of a metal ion is determined by its affinity for a pair of electrons, and the greater this affinity, the more stable the complexes the metal ion will form. However, removing electrons from a metal to produce an ion is also related to the attraction the metal atom has for electrons. Therefore, it seems reasonable to seek a correlation between the stability constants for complexes of several metals with a given ligand and the total energy necessary for ionization to produce the metal ions. The first-row transition metal ions react in solution with ethylenediamine, en, to form stable complexes. We will consider only the first two steps in complex formation, which can be shown as follows:

$$M^{2+} + en \rightleftarrows M(en)^{2+} \qquad K_1 \qquad (19.71)$$

$$M(en)^{2+} + en \rightleftarrows M(en)_2^{2+} \qquad K_2 \qquad (19.72)$$

When a graph is made of the log K_1 versus the sum of the first and second ionization potentials for the metals, the result is shown in Figure 19.5. Clearly, the relationship is linear for the metal ions except Zn^{2+}, for which the complex is considerably less stable than that with Cu^{2+}. These results are in accord with the Irving–Williams series described earlier. The graph obtained when log K_2 is plotted against the total ionization potential has exactly the same characteristics as is shown in Figure 19.6.

Another type of correlation involving a metal ion with a series of ligands has been established for complexes between Ag^+ and numerous amines, for which the reactions can be shown as

$$Ag^+ + :B \rightleftarrows Ag^+:B \qquad K_1 \qquad (19.73)$$

$$Ag^+:B + :B \rightleftarrows B:Ag^+:B \qquad K_2 \qquad (19.74)$$

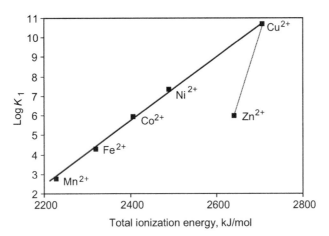

FIGURE 19.5
Variation in log K_1 for complexes of ethylenediamine with total ionization energy of the metal for +2 ions of first-row metals.

where B is an amine. The availability of the electron pair on a base such as an amine determines its basicity toward H^+,

$$H^+ + :B \rightleftarrows H^+:B \qquad K_B \qquad (19.75)$$

so there should be some correlation between K_B for the base and the stability constants for the silver complexes. When the basicities of a series of amines having similar structures are considered, there is a rather good correlation between K_B and K_1. This indicates that the stronger a base is toward H^+, the more stable the complex it will form with Ag^+. Primary and secondary amines generally follow the relationship well. However, if a different type of ligand is considered, such as pyridine, there is no correlation between its basicity toward H^+ and the stability of the complex it forms with Ag^+.

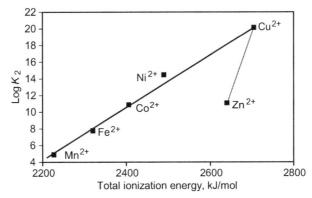

FIGURE 19.6
Variation in log K_2 for $M(en)_2^{2+}$ complexes with total ionization potential of the metal for +2 ions of first-row metals.

One reason for this is that although H^+ is a hard Lewis acid, Ag^+ is soft, so different characteristics of the base are being measured by the two types of equilibrium constants. In fact, the bonds between pyridine and Ag^+ have a substantial amount of covalency and π bond character that is impossible toward H^+. Although we have discussed correlations that involve equilibrium constants only for the formation of complexes by Ag^+, such correlations have also been made (or attempted) for complexes of other metals.

Another factor that affects trends in the stability constants of complexes formed by a series of metal ions is the crystal field stabilization energy. As was shown in Chapter 17, the aqua complexes for $+2$ ions of first-row transition metals reflect this effect by giving higher heats of hydration than would be expected on the basis of sizes and charges of the ions. Crystal field stabilization, as discussed in Section 17.4, would also lead to increased stability for complexes containing ligands other than water. It is a pervasive factor in the stability of many types of complexes. Because ligands that form π bonds through back donation are also generally strong field ligands, this becomes an additional consideration in regard to the overall stability of complexes.

In Chapter 9, the hard–soft acid–base (HSAB) principle was discussed, and numerous applications of the principle were presented. This principle is also of enormous importance in coordination chemistry. First-row transition metals in high oxidation states have the characteristics of hard Lewis acids (small size and high charge). Consequently, ions such as Cr^{3+}, Fe^{3+}, and Co^{3+} are hard Lewis acids that bond best to hard Lewis bases. When presented with the opportunity to bond to NH_3 or PR_3, these metal ions bond better to NH_3, which is the harder base. On the other hand, Cd^{2+} bonds better to PR_3 because of the more favorable soft acid–soft base interaction.

The thiocyanate ion provides an interesting test of these ideas. In the SCN^- ion, the sulfur atom is considered to be a soft electron donor, whereas the nitrogen atom is a much harder electron donor. Accordingly, Pt^{2+} bonds to SCN^- at the sulfur atom, whereas Cr^{3+} bonds to the nitrogen atom. Uncharged metal atoms are considered to be soft electron acceptors, and they form complexes with soft ligands such as CO, H^-, and PR_3. We will see many examples of such complexes in later chapters. On the other hand, we would not expect complexes between uncharged metal atoms and NH_3 to be stable.

Many observations concerning these trends had been made over the years, and in the 1950s, S. Ahrland, J. J. Chatt, and M. Davies presented a classification of metals based on their preferred interaction with donor atoms. Class A metals interact preferentially with ligands in which the donor atom is in the first row of the periodic table. For example, they prefer to bond to N rather than P donor atoms. Class B metals are those which interact better when the donor atom is in the second row of the periodic table. For example, a Class B metal would bond better to P than to N. The following table summarizes the behavior of metal atoms according to this classification.

	Strength as donor atom
Class A metals	$N >> P > As > Sb > Bi$
	$O >> S > Se > Te$
	$F > Cl > Br > I$
Class B metals	$N << P > As > Sb > Bi$
	$O << S \approx Se \approx Te$
	$F < Cl < Br < I$

Keep in mind that this classification, which so clearly now is a manifestation of the HSAB principle, predated that principle by approximately a decade. In retrospect, it seems almost obvious now.

One apparent contradiction to the HSAB principle involves the stability of the complexes $[Co(NH_3)_5NCS]^{2+}$ and $[Co(CN)_5SCN]^{3-}$. In the first of these, the thiocyanate ion is bonded to Co^{3+} through the nitrogen atom, as expected. However, in the second complex, SCN^- is bonded to Co^{3+} through the sulfur atom, and this arrangement is the stable one. The difference between these complexes lies in the effect that the other five ligands have on the metal ion. In $[Co(NH_3)_5NCS]^{2+}$, the Co^{3+} is a hard Lewis acid and the five NH_3 molecules are hard bases. Therefore, the hard character of the Co^{3+} ion as an electron pair acceptor is not altered. In $[Co(CN)_5SCN]^{3-}$, the five CN^- ligands are soft bases, so that the combination of a Co^{3+} ion with these soft bases alters the character of the Co^{3+} to the point where it behaves as a soft electron pair acceptor toward SCN^-. In other words, Co^{3+} *bonded to five soft ligands* is different from Co^{3+} *bonded to five hard ligands*. This effect is known as the *symbiotic effect*.

It is interesting to note that although NH_3 ($K_b = 1.8 \times 10^{-5}$) is a stronger base than pyridine ($K_b = 1.7 \times 10^{-9}$), there is often not as much difference in the stability constants of complexes containing these ligands as would be suggested on the basis of their interaction with H^+. This is especially true when the metal is from the second or third transition series. It appears that in the case of pyridine, which is a soft electron pair donor, there is some tendency for the ligand to form double bonds to the metal in which the electrons in the ring are arranged as in a quinone type of structure. This tendency is not possible for NH_3, which has no other pairs of electrons to use in multiple bonding or empty orbital of low energy in which to accept electrons.

Although the basicities of NH_3 and ethylenediamine, $H_2NCH_2CH_2NH_2$, are similar, en forms much more stable complexes. This means that the chelate formed in the reaction

$$M^{2+} + 2 \text{ en} \rightleftarrows M(\text{en})_2^{2+} \tag{19.76}$$

is more stable than the complex that is formed in the reaction

$$M^{2+} + 4 NH_3 \rightleftarrows M(NH_3)_4^{2+}. \tag{19.77}$$

Because the ethylenediamine forms chelate rings, the increased stability compared with NH_3 complexes is called the *chelate effect*. For both ligands, the atoms donating the electron pairs are nitrogen atoms. The difference in stability of the complexes is not related to the strength of the bonds between the metal ion and nitrogen atoms.

The equilibrium constant and free energy for a reaction are related by the equation

$$\Delta G = \Delta H - T\Delta S = -RT \ln K \tag{19.78}$$

From this equation, it is clear that a larger value for K results when ΔG is more negative, which could result if ΔH is more negative. As a result of the bonds between the metal ion and ligand being equal in number and about the same strength in both types of complexes, ΔH is approximately the same regardless of whether the complex contains NH_3 or en. The factor that is different is ΔS, the entropy change. When four NH_3 molecules enter the coordination sphere of the metal ion, four H_2O molecules leave the coordination sphere. As a result, there is an equal number of "free" molecules before and after the complex forms, and ΔS is approximately 0. On the other hand, when *two* en molecules enter the coordination sphere of the metal ion, *four* H_2O molecules are displaced. There is an increase in the number of free molecules, and the disorder of the system (entropy) increases. This *positive* value for ΔS (which occurs in the $-T\,\Delta S$ term) results in a more *negative* value for ΔG (larger value for K). Therefore, the chelate effect arises as a result of a positive ΔS when chelate rings form by displacing monodentate ligands. It is an *entropy effect*. As a result of the chelate effect, the equilibrium constant for the reaction

$$\left[Ni(NH_3)_6\right]^{2+} + 3\,en \; \rightleftarrows \; \left[Ni(en)_3\right]^{2+} + 6\,NH_3 \tag{19.79}$$

is approximately 10^9 even though ΔH is small.

Because of the chelate effect, ligands that can displace two or more water molecules from the coordination sphere of the metal generally form stable complexes. One ligand that forms very stable complexes is the anion ethylenediaminetetraacetate ($EDTA^{4-}$),

This ion can bond at six sites, so one $EDTA^{4-}$ ion replaces six water molecules when the reaction is carried out in aqueous solution. The result is the formation of complexes that have very large stability constants. This ligand is widely used in analytical chemistry in complexiometric titrations to determine concentrations of metal ions. Because it holds metal ions so securely, $EDTA^{4-}$ (in the form of Na_4EDTA, $Na_2CaEDTA$, or Ca_2EDTA) is added to salad dressings. Traces of metal ions catalyze oxidation reactions that lead to spoilage, but when $EDTA^{4-}$ is added, it binds to the metal ions so effectively that they cannot act as catalysts for the undesirable oxidation reactions. Many metal ions are effectively complexed (or sequestered) by $EDTA^{4-}$ or H_2EDTA^{2-} including main group ions such as Mg^{2+}, Ca^{2+}, and Ba^{2+}.

Not only is the presence of chelate rings a factor in determining the stability of complexes, but also ring size is important. Studies have shown that chelate rings having five or six members are generally more stable than those having other numbers of atoms. For example, when the series of ligands having the formula $H_2N(CH_2)_nNH_2$ (where $n = 2$, 3, or 4) forms complexes with the same metal ion, the most stable complex is with ethylenediamine ($n = 2$), which results in a five-membered chelate ring. If $n = 3$, which corresponds to 1,3-diaminopropane, the complexes, which have six-membered rings, are less stable than are those of en. The complexes with the ligand having $n = 4$ (1,4-diaminobutane) are even less stable. A similar situation exists for complexes of the anions of dicarboxylic acids, $^-OOC-(CH_2)_n-COO^-$ (where $n = 0$, 1,...).

Although ring size is important, so is ring structure. Acetylacetone (2,4-pentadione) undergoes the enolization reaction

$$
\text{CH}_3-\overset{\overset{\displaystyle :O:}{\|}}{\text{C}}-\text{CH}_2-\overset{\overset{\displaystyle :O:}{\|}}{\text{C}}-\text{CH}_3 \rightleftharpoons \text{CH}_3-\overset{\overset{\displaystyle :O:}{\|}}{\text{C}}-\text{CH}=\overset{\overset{\displaystyle \cdots\text{H}-\overset{..}{\text{O}}:}{|}}{\text{C}}-\text{CH}_3
\tag{19.80}
$$

A proton is easily removed to produce the acetylacetonate anion (abbreviated acac).

$$
\text{CH}_3-\overset{\overset{\displaystyle :O:}{\|}}{\text{C}}-\text{CH}=\overset{\overset{\displaystyle :\overset{..}{\overset{..}{O}}:^{-}}{|}}{\text{C}}-\text{CH}_3
$$

When this anion bonds to a metal ion through both oxygen atoms, a planar six-membered ring is formed. There are two double bonds as shown, and if there is some double bond character to one of the bonds to the metal ion, a conjugated system is formed, which can be represented as

$$\tag{19.81}$$

In these structures, only one of the chelate rings has been shown, but there would normally be two or three rings depending on whether the metal is +2 or +3 and has a coordination number of 4 or 6. The rings have considerable aromatic character, which leads to stability of the complexes. Moreover, such complexes are usually neutral in charge and have no other anion present. This also contributes to stability because there is no anion (usually a potential ligand, and in some cases, an oxidizing or reducing agent) to compete with the acac ligand. All these factors lead to complexes that are so stable that some of them can even be vaporized without extensive decomposition. Also, the chelate rings are stable enough to permit reactions to be carried out on them without disrupting the complex. One such reaction is bromination, which can be shown as

$$\tag{19.82}$$

Reactions such as this are electrophilic substitutions, and other electrophiles such as NO_2^-, CH_3CO^-, or CHO^- can be substituted on the rings without destroying the complex. Such chemical behavior illustrates the extreme stability of complexes of this type. It is apparent that the nature of the bonding in the chelate rings is as important as is their size. Other reactions of coordinated ligands will be shown in later chapters.

References for Further Study

Christensen, J.J., Izatt, R.M., 1970. *Handbook of Metal Ligand Heats*, Marcel Dekker, New York. A compendium of thermodynamic data for the formation of an enormous number of complexes.

Connors, K.A., 1987. *Binding Constants: The Measurement of Molecular Complex Stability*, Wiley, New York. An excellent discussion of the theory of molecular association as well as the experimental methods and data treatment. Deals with the association of many types of species in addition to metal complexes. Highly recommended.

Furia, T.E., 1972. *Sequestrants in foods in CRC Handbook of Food Additives*, 2nd ed., CRC Press, Boca Raton, FL. Chapter 6, Comprehensive tables of stability constants for many complexes of organic and biochemical ligands of metals.

Leggett, D.J. (Ed.), 1985. *Computational Methods for the Determination of Formation Constants*, Plenum Press, New York. A high level presentation of the theory of complex equilibria and computer programs for mathematical analysis.

Martell, A.E. (Ed.), 1971. *Coordination Chemistry*, vol. 1, Van Nostrand-Reinhold, New York. This is one of the American Chemical Society Monograph Series. Chapter 7 by Hindeman, J.C., and Sullivan, J.C. and Chapter 8 by Anderegg, G., deal with equilibria of complex formation.

Martell, A.E., Motekaitis, R.J., 1992. *Determination and Use of Stability Constants*, 2nd ed., John Wiley, New York. An excellent treatment of stability constants including methods of calculation and computer programs.

Pearson, R.G., 1968. *J. Chem. Educ.* 45, 581, 643. Two elementary articles on applications of the hard-soft interaction principle by its originator.

Rossotti, F.J.C., Rossotti, H., 1961. *Determination of Stability Constants*, McGraw-Hill, New York. Probably the most respected treatise on stability constants and experimental methods for their determination.

Rossotti, H., 1978. *The Study of Ionic Equilibria in Aqueous Solutions*, Longmans, New York. A wealth of information on the equilibria of complex formation.

www.adasoft.co.uk/scdbase/scdbase.htm An IUPAC data base of stability constants and distribution diagrams. This is just one of many web sites where tables of stability constants can be found.

QUESTIONS AND PROBLEMS

1. For each metal ion listed, predict which of the ligands would give the more stable complex and explain your choice.
   (a) Cr^{3+} with NH_3 or CO
   (b) Hg^{2+} with $(C_2H_5)_2S$ or $(C_2H_5)_2O$
   (c) Zn^{2+} with $H_2NCH_2CH_2NH_2$ or $H_2N(CH_2)_3NH_2$
   (d) Ni^0 with $(C_2H_5)_2O$ or PCl_3
   (e) Cd^{2+} with Br^- or F^-
2. Although acetylacetonate ion forms very stable complexes with many metal ions, acetate ion does not. Explain the difference in complexing behavior.
3. How would you expect $S_2O_3^{2-}$ to bond to Cr^{3+} assuming that all other ligands are H_2O? What would be different if the other ligands were CN^-? Explain your answer.
4. For the formation of ammonia complexes with Ag^+, $K_1 = 6.72 \times 10^3$ and $K_2 = 2.78 \times 10^3$. When the concentration of free NH_3 is 3.50×10^{-4} M, what is the average number of ligands bound per metal ion?

5. Data are given for the complexes formed between Ag^+ and NH_3 in Question 4. Using this information, determine the concentration of free NH_3 when the average number of ligands bound per metal ion is 1.65.

6. Consider the stability constants shown below for the ethylenediamine complexes of the metals listed.

Metal Ion	log K_1	log K_2	log K_3
Co^{2+}	5.89	4.83	3.10
Ni^{2+}	7.52	6.28	4.27
Cu^{2+}	10.55	9.05	−1.0

Provide an explanation the value for log K_3 for the complex of Cu^{2+}.

7. For the following pairs of ligands, predict which would form the more stable complex with the metal indicated.
   (a) Cr^{3+} with $C_6H_5-O^-$ or $o\text{-}O^--C_6H_4-CHO$
   (b) Fe^{2+} with $(C_2H_5)_2O$ or $(C_2H_5)_2S$
   (c) Zn^{2+} with $CH_3C(O)CH_2C(O)CH_3$ or $CH_3C(O)CH_2CH_2C(O)CH_3$
   (d) Cr^{3+} with NR_3 or PR_3 (R is an alkyl group)
   (e) Ni^0 with C_5H_5N or PR_3

8. Tell which of the following pairs of ligands forms more stable complexes with a first-row metal ion such as Cr^{3+}.
   (a) $H_2NCH_2CH_2NH_2$ or $NH_2CH_2CH_2CH_2NH_2$
   (b) $(CH_3)_3N$ or $H_2NCH_2CH_2NH_2$
   (c) $H_2PCH_2CH_2PH_2$ or $H_2NCH_2CH_2NH_2$
   (d) $CH_3C(O)CH_2C(O)CH_3$ or $H_2NCH_2CH_2NH_2$

9. Pyridine (C_5H_5N) forms complexes with Ag^+ that can be shown as

$$Ag^+ + py \rightleftarrows Ag(py)^+ \quad K_1 = 2.40 \times 10^2$$
$$Ag(py)^+ + py \rightleftarrows Ag(py)_2^+ \quad K_2 = 9.33 \times 10^1$$

When the concentration of free py is 2.50×10^{-4} M, what is the average number of ligands bound per metal ion?

10. For the formation of a complex according to the equation

$$x\,A + y\,B \rightleftarrows A_xB_y$$

the concentration of complex was determined at different concentrations of A and B as follows:

[A] = 0.200 M	[B] = 0.250 M	[A$_x$B$_y$] = 0.0467 M
[A] = 0.200 M	[B] = 0.400 M	[A$_x$B$_y$] = 0.191 M
[A] = 0.200 M	[B] = 0.750 M	[A$_x$B$_y$] = 1.26 M
[A] = 0.240 M	[B] = 0.220 M	[A$_x$B$_y$] = 0.0459 M
[A] = 0.420 M	[B] = 0.220 M	[A$_x$B$_y$] = 0.141 M
[A] = 0.700 M	[B] = 0.220 M	[A$_x$B$_y$] = 0.391 M

Using these data, determine the values of x and y for the complex.

11. When Cd^{2+} forms complexes with Cl^-, the stability constants are as follows:
$K_1 = 20.9$; $K_2 = 7.94$; $K_3 = 1.23$; $K_4 = 0.355$. Use the procedures described in this chapter to determine $[Cd^{2+}]$, $[CdCl^+]$, $[CdCl_2]$, $[CdCl_3^-]$, and $[CdCl_4^{2-}]$ for chloride ion concentrations in the range 10^{-4} M-10 M. Make a graph of the results in the form of a distribution diagram for the complexes

12. For the formation of two complexes AgL^+ and AgL_2^+, the average number of ligands bound per metal ion is 0.495 when the concentration of free ligand is 1.50×10^{-4} M and the number is 1.475 when the concentration of free ligand is 5.75×10^{-4} M. Determine the stability constants for the complexes.

CHAPTER 20

Synthesis and Reactions of Coordination Compounds

Preparation of coordination compounds and the transformation of one coordination compound into another form the basis for a vast amount of synthetic inorganic chemistry. In some cases, the reactions involve replacing one or more ligands with others in substitution reactions. In others, reactions may be carried out on ligands that are attached to metal ions without breaking the metal–ligand bond. These reactions, which often involve organic compounds, constitute the area of reactions of coordinated ligands. The fact that the organic molecule is bonded to a metal ion may make it easier to carry out certain reactions. Another type of reaction involving coordination compounds is that in which electrons are transferred between metal ions. This process is often influenced by the nature of the ligands surrounding the metal ions. Finally, it is important to have a general knowledge of how coordination compounds are prepared.

In this chapter, a survey of the enormously broad area of reactions of coordination compounds will be presented, and some of the basic mechanisms of the reactions will be presented. However, reactions of coordination compounds is such a very broad area that this chapter (as would be the case of any chapter) can present only the basic concepts and an elementary introduction to the field. More detailed coverage will be found in the references listed at the end of the chapter. The classic books in the field are Basolo and Pearson (1974) and Wilkins (1991), which present excellent and detailed reviews of the literature. We begin the chapter by illustrating some of the synthetic methods that have been useful for synthesizing coordination compounds.

20.1 SYNTHESIS OF COORDINATION COMPOUNDS

Coordination compounds have been produced by a variety of techniques for at least two centuries. Zeise's salt, $K[Pt(C_2H_4)Cl_3]$, dates from the early 1800s and Werner's classic syntheses of cobalt complexes were described over a century ago. Synthetic techniques used to prepare coordination compounds range from simply mixing the reactants to employing nonaqueous solvent chemistry. In this section, a brief overview of some types of general synthetic procedures will be presented. In Chapter 21, a survey of the organometallic chemistry of transition metals will be presented, and additional preparative methods for complexes of that type will be described there.

20.1.1 Reaction of a Metal Salt with a Ligand

One of the techniques for producing coordination compounds is to simply combine the reactants. Some reactions may be carried out in solution, but others may involve adding a **665**

Inorganic Chemistry. DOI: http://dx.doi.org/10.1016/B978-0-12-385110-9.00020-0

liquid or gaseous ligand directly to a metal compound. Some reactions of this type are the following:

$$NiCl_2 \cdot 6H_2O(s) + 3 \ en(l) \ \rightarrow \ \left[Ni(en)_3\right]Cl_2 \cdot 2H_2O(s) + 4 H_2O \tag{20.1}$$

$$Cr_2(SO_4)_3(s) + 6 \ en(l) \ \rightarrow \ \left[Cr(en)_3\right]_2(SO_4)_3(s) \tag{20.2}$$

The product obtained during the second reaction, $[Cr(en)_3]_2(SO_4)_3(s)$, is a solid mass. It has been found advantageous to dissolve ethylenediamine in an inert liquid that has a high boiling point such as toluene. Refluxing this solution while adding solid $Cr_2(SO_4)_3$ slowly gives a finely divided product that is easier to separate by filtration and is easy to purify. This technique can be applied to the preparation of numerous other types of complexes. In this case, changing the reaction medium gives a product that is more convenient to use in subsequent work.

An example of a synthesis utilizing a nonaqueous solvent is a common procedure that is used to prepare $[Cr(NH_3)_6]Cl_3$. The reaction is

$$CrCl_3 + 6 \ NH_3 \ \xrightarrow{\text{liq. NH}_3} \ \left[Cr(NH_3)_6\right]Cl_3 \tag{20.3}$$

It has been found that this reaction is catalyzed by sodium amide, $NaNH_2$. The function of the catalyst appears to involve the replacement of Cl^- by NH_2^-, which is the stronger nucleophile. Once the NH_2^- ion is attached to Cr^{3+}, it quickly removes a proton from a solvent molecule to be transformed into a coordinated NH_3 molecule. As will be shown later in this chapter, this type of behavior is also characteristic of coordinated OH^- in aqueous solutions.

Some alkenes will react with metal salts to give complexes that involve electron donation from the double bond. A classic case of this type is the formation of Zeise's salt. However, if the alkene has more than one potential donor site, bridged complexes may result. One such ligand is butadiene, which forms an interesting bridged structure with CuCl. The reaction is carried out at $-10\,°C$, with CuCl being added directly to liquid butadiene.

$$2 \ CuCl + C_4H_6 \ \rightarrow \ [ClCuC_4H_6CuCl] \tag{20.4}$$

Acetylacetone (2,4-pentadione) undergoes a tautomerization that can be shown as

$$\tag{20.5}$$

and the equilibrium is strongly solvent dependent. When a small amount of ammonia is present, the proton in the OH group is removed to form the acetylacetonate anion

which is an excellent chelating agent. The anion, abbreviated as acac, will react with many metal ions to form stable complexes. An example of this type of reaction is

$$CrCl_3 + 3 \ C_5H_8O_2 + 3 \ NH_3 \ \rightarrow \ Cr(C_5H_7O_2)_3 + 3 \ NH_4Cl \tag{20.6}$$

Complexes containing acac are especially stable because they exist as neutral complexes such as $M(acac)_3$ and $M(acac)_2$ when the metal ion is +3 and +2, respectively. Many other reactions have been carried out in which the ligand reacts with a metal compound. In many cases, it is preferable to start with an anhydrous metal compound, and dehydration can sometimes be accomplished by a reaction with $SOCl_2$.

20.1.2 Ligand Replacement Reactions

The replacement of one ligand by another is the most common type of reaction of coordination compounds, and the number of reactions of this type is enormous. Some are carried out in aqueous solutions, some in nonaqueous media, and others can be carried out in the gas phase. One reaction of this type is

$$Ni(CO)_4 + 4\,PCl_3 \ \rightarrow \ Ni(PCl_3)_4 + 4\,CO \tag{20.7}$$

When an octahedral complex such as $[Cr(CO)_6]$ reacts with pyridine, only three CO ligands are replaced. In the product $[Cr(CO)_3(py)_3]$, the CO and py ligands are *trans* to each other. Replacement reactions are important because frequently one type of complex is easily prepared and then it can be converted to another that cannot be obtained easily. Gaseous butadiene will react with an aqueous solution of $K_2[PtCl_4]$ to give a bridged complex in which Cl ligands are displaced.

$$2\,K_2[PtCl_4] + C_4H_6 \ \rightarrow \ K_2[Cl_3PtC_4H_6PtCl_3] + 2\,KCl \tag{20.8}$$

A large number of syntheses involve replacement reactions.

20.1.3 Reaction of Two Metal Compounds

Several synthetic processes involve the reaction of two metal salts. A well-known example of this type is the following:

$$2\,AgI + HgI_2 \ \rightarrow \ Ag_2[HgI_4] \tag{20.9}$$

This reaction has been induced in several ways, but one novel procedure involves the action of ultrasound on a suspension of the two solid reactants in dodecane. The ultrasonic vibrations create cavities in the liquid that implode driving particles of the solid together at high velocity. Under these conditions, the solids react in much the same way as if they were heated, but in cases in which an unstable product is formed, it is not thermally decomposed.

A variation of this type of reaction occurs when a metal complex already containing ligands reacts with a simple metal salt to give a redistribution of the ligands. For example,

$$2\,[Ni(en)_3]Cl_2 + NiCl_2 \ \rightarrow \ 3\,[Ni(en)_2]Cl_2 \tag{20.10}$$

20.1.4 Oxidation–Reduction Reactions

Many coordination compounds can be prepared when a compound of the metal is either reduced or oxidized in the presence of a ligand. Oxalic acid is a reducing agent, but it also serves as a source of the oxalate ion, which is a good chelating agent. An interesting reaction of this type involves the reduction of dichromate by oxalic acid as shown by the following equation:

$$K_2Cr_2O_7 + 7\,H_2C_2O_4 + 2\,K_2C_2O_4 \rightarrow 2\,K_3\left[Cr(C_2O_4)_3\right]\cdot 3H_2O + 6\,CO_2 + H_2O \qquad (20.11)$$

In other reactions, the metal may be oxidized as the complex is formed. Numerous complexes of Co(III) have been prepared by the oxidation of solutions containing Co(II). This technique is particularly useful in the case of cobalt because Co^{3+} is a strong oxidizing agent that reacts with water if it is not stabilized by complexation. In the following reactions, en is ethylenediamine, $H_2NCH_2CH_2NH_2$.

$$4\,CoCl_2 + 8\,en + 4\,en\cdot HCl + O_2 \rightarrow 4\left[Co(en)_3\right]Cl_3 + 2\,H_2O \qquad (20.12)$$

$$4\,CoCl_2 + 8\,en + 8\,HCl + O_2 \rightarrow 4\,trans\text{-}\left[Co(en)_2Cl_2\right]Cl\cdot HCl + 2\,H_2O \qquad (20.13)$$

Heating the product of the second reaction yields $trans$-$[Co(en)_2Cl_2]Cl$ by the loss of HCl. Dissolving $trans$-$[Co(en)_2Cl_2]Cl$ in water and evaporating the solution by heating result in the formation of cis-$[Co(en)_2Cl_2]Cl$ as the result of an isomerization reaction.

20.1.5 Partial Decompositions

Reactions in which volatile ligands such as H_2O and NH_3 are lost as a result of heating lead to the formation of other complexes. Generally, these reactions involve solids, and some such processes will be described later in this chapter. As a volatile ligand is driven off, another group can enter the coordination sphere of the metal. In other cases, there may be a change in the bonding mode of a ligand that is already present. For example, SO_4^{2-} may become bidentate to complete the coordination sphere of the metal. The synthesis of $[Co(NH_3)_5H_2O]Cl_3(s)$ is rather straightforward, and, after it is obtained, it is converted when heated to $[Co(NH_3)_5Cl]Cl_2(s)$ by the reaction

$$\left[Co(NH_3)_5H_2O\right]Cl_3(s) \rightarrow \left[Co(NH_3)_5Cl\right]Cl_2(s) + H_2O(g) \qquad (20.14)$$

Other compounds containing different anions are obtained by the following reaction:

$$\left[Co(NH_3)_5Cl\right]Cl_2 + 2\,NH_4X \rightarrow \left[Co(NH_3)_5Cl\right]X_2 + 2\,NH_4Cl \qquad (20.15)$$

This metathesis reaction can be carried out in aqueous solutions as a result of differences in solubility.

Two additional reactions representing partial decomposition are the following:

$$\left[Cr(en)_3\right]Cl_3(s) \rightarrow cis\text{-}\left[Cr(en)_2Cl_2\right]Cl(s) + en(g) \qquad (20.16)$$

$$\left[Cr(en)_3\right](SCN)_3(s) \rightarrow trans\text{-}\left[Cr(en)_2(NCS)_2\right]SCN(s) + en(g) \qquad (20.17)$$

These reactions have been known for almost 100 years, and they have been extensively studied. The reactions are catalyzed by the corresponding ammonium salt in each case, although other protonated amines function as catalysts. It appears that the function of the catalyst is to supply H^+, which helps to force an end of the ethylenediamine molecule away from the metal.

20.1.6 Precipitation Making use of the Hard–Soft Interaction Principle

As a consequence of the hard–soft interaction principle, ions of similar size and magnitude of charge interact best. That interaction includes the formation of precipitates. It is possible to make use of this principle when isolating complex ions that are relatively unstable. A well-known case of this type is the isolation of the $[Ni(CN)_5]^{3-}$ ion, which results when aqueous solutions containing Ni^{2+} also contain an excess of CN^-. Attempts to isolate $[Ni(CN)_5]^{3-}$ by adding K^+ were not successful because what was obtained was $K_2[Ni(CN)_4]$ and KCN. It was only when a large cation having a +3 charge was utilized that the pentacyanonickelate(II) ion was obtained in a solid product. The large +3 cation used was $[Cr(en)_3]^{3+}$. With that cation, the solid product was $[Cr(en)_3][Ni(CN)_5]$. The following equations describe the process:

$$Ni^{2+} + 4\,CN^- \rightarrow \left[Ni(CN)_4\right]^{2-} \tag{20.18}$$

$$[Ni(CN)_4]^{2-} + CN^- \longrightarrow [Ni(CN)_5]^{3-}$$
$$\overset{K^+}{\swarrow} \qquad \overset{[Cr(en)_3]^{3+}}{\searrow}$$
$$K_2[Ni(CN)_4] + KCN \qquad\qquad [Cr(en)_3]\,[Ni(CN)_5] \tag{20.19}$$

In many parts of this book, the utility of the hard–soft interaction principle has been described. In the situation described, the choice of an appropriate cation makes possible the isolation of a relatively unstable complex ion by providing a crystal environment that helps to stabilize the complex. The application of basic principles that relate to structure and stability can also be useful in synthetic chemistry.

20.1.7 Reactions of Metal Compounds with Amine Salts

Many years ago, L.F. Audrieth studied numerous reactions of amine hydrochloride salts. These compounds contain a cation that is a protonated amine that can function as a proton donor. Consequently, the molten salts are acidic, and they undergo many reactions in which they function as acids. This behavior is also characteristic of ammonium chloride as well as pyridine hydrochloride (or pyridinium chloride). Numerous metal oxides and carbonates react readily with the molten amine salts as illustrated by the following equation:

$$NiO(s) + 2\,NH_4Cl(l) \rightarrow NiCl_2(s) + H_2O(l) + 2\,NH_3(g) \tag{20.20}$$

Hydrothiocyanic acid (also known by the archaic name rhodanic acid), HSCN, is a strong acid, so it is easy to prepare stable amine hydrothiocyanate salts such as that of piperidine, $C_5H_{11}N$ (abbreviated as pip). In this reaction, the less volatile base, piperidine, replaces the volatile weak base, NH_3.

$$NH_4SCN(s) + pip(l) \rightarrow pipH^+SCN^-(s) + NH_3(g) \tag{20.21}$$

The salt pipHSCN, known as piperidinium thiocyanate or piperidine hydrothiocyanate, has a melting point of 95 °C. When metal compounds are added to this molten salt, thiocyanate complexes of the metals are produced. For example, the following reactions can be carried out at 100 °C in the presence of an excess of the amine hydrothiocyanate:

$$NiCl_2 \cdot 6H_2O(s) + 6 \text{ pipHSCN}(l) \rightarrow (\text{pipH})_4[Ni(SCN)_6](s) + 2 \text{ pipHCl}(l) + 6 H_2O(g) \quad (20.22)$$

$$CrCl_3 \cdot 6H_2O(s) + 6 \text{ pipHSCN}(l) \rightarrow (\text{pipH})_3[Cr(NCS)_6](s) + 3 \text{ pipHCl}(l) + 6 H_2O(g) \quad (20.23)$$

In some cases, the metal complex contains both piperidine and thiocyanate as ligands as illustrated by the following equation:

$$CdCO_3(s) + 2 \text{ pipHSCN}(l) \rightarrow [Cd(pip)_2(SCN)_2](l) + CO_2(g) + H_2O(g) \quad (20.24)$$

In addition to the reactions described above using molten pipHSCN, several reactions have been carried out at low temperature by sonicating mixtures of metal salts and pipHSCN. The use of ultrasound results in products of higher purity than when the molten salt is used. This is probably because some of the products are not very stable at the temperature of the molten salt (100 °C) and mixtures result under those conditions. In carrying out the reactions, the amine hydrothiocyanate and the metal compound were suspended in dodecane and pulsed ultrasound was applied. The following reactions are typical of preparations of this type:

$$MnCO_3(s) + 6 \text{ pipHSCN}(s) \rightarrow (\text{pipH})_4[Mn(NCS)_6](s) + 2 \text{ pip}(l) + CO_2(g) + H_2O(l) \quad (20.25)$$

$$MnCl_2 \cdot 4H_2O(s) + 4 \text{ pipHSCN}(s) \rightarrow (\text{pipH})_2[Mn(NCS)_4](s) + 2 \text{ pipHCl} + 4 H_2O(l) \quad (20.26)$$

$$Fe_2O_3(s) + 12 \text{ pipHSCN}(s) \rightarrow 2 (\text{pipH})_3[Fe(NCS)_6](s) + 6 \text{ pip}(l) + 3 H_2O(l) \quad (20.27)$$

$$NiCO_3(s) + 4 \text{ pipHSCN}(s) \rightarrow (\text{pipH})_2[Ni(NCS)_4](s) + 2 \text{ pip}(l) + CO_2(g) + H_2O(l) \quad (20.28)$$

In every case, the products were obtained in high purity in this convenient one-step synthesis. Although ultrasound has not been used extensively in inorganic chemistry, it is a well-known technique in organic synthesis.

In addition to heat, light, and ultrasound as sources of energy, microwaves have been employed in the synthesis of coordination compounds. Although this technique has been widely utilized in carrying out organic reactions, it can also be useful in inorganic chemistry. Suitable examples of this technique are the syntheses reported by Dopke and Oemke (2011) of platinum complexes containing the ligands $(C_6H_5)_2PCH_2CH_2P(C_6H_5)_2$, bis(diphenylphosphino)ethane (dppe) and triphenylphosphine. The complexes [Pt(dppe)Cl$_2$] and cis-[Pt(P(C$_6$H$_5$)$_3$)Cl$_2$] were prepared by subjecting solutions containing K$_2$[PtCl$_4$] and the ligands to microwaves. Compared to reactions induced by heating the solutions, reaction time was decreased and product purity was enhanced when microwaves were employed. Although microwave-assisted syntheses are not commonly used in inorganic chemistry, the technique undoubtedly deserves to be more widely used.

We have barely introduced the vast area of synthetic coordination chemistry. The methods described show that a wide variety of techniques have been employed, and they may be employed in combination in sequential steps to develop creative synthetic routes. The comprehensive compilations listed in the suggested readings should be consulted for further study. Several journals in inorganic chemistry are at least partially devoted to synthesis of materials, so the field is growing at a rapid rate.

20.2 SUBSTITUTION REACTIONS IN OCTAHEDRAL COMPLEXES

Coordination compounds undergo many reactions, but a large number of reactions can be classified into a small number of reaction types. When one ligand replaces another, the reaction is called a *substitution reaction*. For example, when ammonia is added to an aqueous solution containing Cu^{2+}, water molecules in the coordination sphere of the Cu^{2+} are replaced by molecules of NH_3. Ligands are held to metal ions because they are electron pair donors (Lewis bases). Lewis bases are nucleophiles (see Chapter 9), so the substitution of one nucleophile for another is a *nucleophilic substitution* reaction. Such a reaction can be illustrated as

$$L_nM-X + L' \rightarrow L_nM-L' + X \qquad (20.29)$$

where X is the leaving group and L' is the entering group. It is also possible to have one metal ion (Lewis acid) replace another in a different type of reaction. Lewis acids are electrophiles, so this type of reaction is an *electrophilic substitution*, which can be represented as

$$ML_n + M' \rightarrow M'L_n + M \qquad (20.30)$$

Because all the ligands are leaving one metal ion and attaching to another, this type of reaction is sometimes known as ligand scrambling.

Nucleophilic substitution reactions have rates that vary enormously. For example, the reaction

$$\left[Co(NH_3)_6\right]^{3+} + Cl^- \rightarrow \left[Co(NH_3)_5Cl\right]^{2+} + NH_3 \qquad (20.31)$$

is extremely slow, whereas the reaction

$$\left[Ni(CN)_4\right]^{2-} + {}^{14}CN^- \rightarrow \left[Ni(CN)_3({}^{14}CN)\right]^{2-} + CN^- \qquad (20.32)$$

is extremely fast. Complexes that undergo substitution very slowly are said to be *inert*; those that undergo rapid substitution are called *labile*. These qualitative terms have been used as descriptors for substitution reactions for many years. The fact that a complex undergoes substitution very slowly may indicate that there is no low-energy pathway for the reaction even though the product of the substitution reaction may be very stable.

Although the terms labile and inert have been in use for more than 50 years, they are only qualitative descriptions of substitution rates. A more appropriate way to describe the rates has been given by Gray and Langford (1968), which categorizes metal ions according to the rate of exchange of coordinated water with water in the bulk solvent. The four classes of metal ions are shown in Table 20.1.

Table 20.1 Classification of Metal Ions on the Basis of H_2O Exchange Rate

Class	k, sec^{-1}	Examples
I	~10^8	Li^+, K^+, Na^+, Ca^{2+}, Ba^{2+}, Cu^+, Hg^{2+}, Cd^{2+}
II	10^4–10^8	Mg^{2+}, Fe^{2+}, Mn^{2+}, Zn^{2+}
III	10^0–10^4	Be^{2+}, Al^{3+}, Fe^{3+}
IV	10^{-1}–10^{-9}	Co^{3+}, Cr^{3+}, Pt^{2+}, Pt^{4+}, Rh^{3+}, Ir^{3+}

In a general way, the ions in the first two classes would be considered labile, whereas those in the last two classes would be considered inert. Labile complexes are regarded as those in which the reaction is complete on a time scale that would be comparable to the time necessary to mix the solutions of the reacting species. Such reactions can be studied by flow techniques or by NMR line broadening. Inert complexes are those that can be followed by conventional kinetic techniques.

It is sometimes inferred that the ions in at least the first two categories attach solvent by electrostatic forces, whereas those in Class IV bond by predominantly covalent forces. This concept is not very realistic even though the rates of the exchange reactions may be compatible with that view. When substitution is viewed in terms of transition states, it can be seen that metal ions having empty d orbitals can readily form a hybrid orbital set that will accommodate an additional ligand. Therefore, a reaction can proceed by a pathway in which the transition state requires expansion of the coordination sphere, and such reactions should be rapid. In agreement with this view, complexes of Sc^{3+} (d^0), Ti^{3+} (d^1), and V^{4+} (d^2) are labile. Complexes containing d^3, d^4, d^5, and d^6 ions would require a ligand to leave before the entering ligand attaches (S_N1) or require the use of outer d orbitals to form a seventh bond. In either case, the reactions would likely be relatively slow compared to those of labile complexes. In agreement with this simple view, complexes of V^{2+}, Cr^{3+}, and Mn^{4+} (all of which are d^3 ions), low spin complexes of Co^{3+}, Fe^{2+}, Ru^{2+}, Rh^{3+}, Ir^{3+}, Pd^{4+}, and Pt^{4+} (all of which are d^6 ions), and low spin complexes of Mn^{3+}, Re^{3+}, and Ru^{4+} are all inert.

20.2.1 Mechanisms of Substitution Reactions

Mechanisms of substitution reactions can sometimes be described in terms of two limiting cases. In the first, the leaving group leaves the coordination sphere of the metal prior to the entering group attaching to the metal. Therefore, the transition state involves a complex that has a coordination number lower than that of the reactant. In such cases, the concentration of the entering group does not affect the rate of the reaction over a wide range of concentrations. The transition state involves only the complex, so the reaction is first order in complex, and the coordination number of the complex in the transition state is one less than it is in the reactant. This process can be illustrated as

$$ML_nX \underset{\text{Slow}}{\overset{\text{Slow}}{\rightleftharpoons}} [ML_n + X]^{\ddagger} \xrightarrow{\text{fast, +Y}} ML_nY + X \qquad (20.33)$$

In order to show the kinetic analysis of this process, we can separate the steps and assign rate constants as follows:

$$ML_nX \underset{k_{-1}}{\overset{k_1}{\rightleftharpoons}} ML_n + X \qquad (20.34)$$

$$ML_n + Y \xrightarrow{k_2} ML_nY \qquad (20.35)$$

The rate of formation of ML_nY can be expressed as

$$\text{Rate} = \frac{d[ML_nY]}{dt} = k_2[ML_n][Y] \qquad (20.36)$$

In the steady state approximation, the rate of formation of the transition state (sometimes called an intermediate if it has a sufficiently long lifetime) is the same as the rate at which it disappears. Therefore the concentration of the transition state as a function of time can be expressed as

$$\frac{d[ML_n]}{dt} = k_1[ML_nX] - k_{-1}[ML_n][X] - k_2[ML_n][Y] = 0 \tag{20.37}$$

From this equation we see that

$$k_1[ML_nX] = k_{-1}[ML_n][X] + k_2[ML_n][Y] \tag{20.38}$$

Solving for $[ML_n]$ gives

$$[ML_n] = \frac{k_1[ML_nX]}{k_{-1}[X] + k_2[Y]} \tag{20.39}$$

Substituting this expression for $[ML_n]$ in Eqn (20.36) gives

$$\text{Rate} = k_2[ML_n][Y] = \frac{k_1 k_2[ML_nX][Y]}{k_{-1}[X] + k_2[Y]} \tag{20.40}$$

In the dissociative mechanism, the concentration of the transition state ML_n is usually quite low and recombination with X to form the starting complex is insignificant compared to the reaction with Y. This means that $k_2[Y] \gg k_{-1}[X]$ and the first term in the denominator of the right hand side of Eqn (20.40) can be ignored to give

$$\text{Rate} \approx \frac{k_1 k_2[ML_nX][Y]}{k_2[Y]} \approx k_1[ML_nX] \tag{20.41}$$

Because it is the rate of *dissociation* of the M—X bond that determines the rate of substitution, the rate law involves only the concentration of the starting complex, ML_nX.

A reaction that has these characteristics is said to follow a *dissociative* or S_N1 pathway. This mechanism is distinguished by the bond to the leaving group being broken before the bond to the entering ligand starts to form. Some authors (such as Basolo and Pearson, 1974) define two types of S_N1 substitution. If the existence of the transition state having lower coordination number is verified, the mechanism is called $S_N1(\text{lim})$, the so-called limiting case. This could also be called the "strict" or "perfect" S_N1 case. The label S_N1 is used for a reaction in which the transition state of lower coordination number is not demonstrated, but all other factors agree with an S_N1 mechanism. The energy profile for the dissociative pathway is shown in Figure 20.1. It is interesting to note that for the reaction

$$[Co(NH_3)_5H_2O]^{3+} + X^- \rightarrow [Co(NH_3)_5X]^{2+} + H_2O \tag{20.42}$$

the rate constants have been determined at 25 °C for a wide range of ligands, X. For several ligands, the rate constants are as follows.

$X^- =$	Cl^-	Br^-	NO_3^-	NCS^-	NH_3
$10^6 \times k$, $M^{-1}\,sec^{-1}$	2.1	2.5	2.3	1.3	2

The essential feature of these data is that the rate is independent of the nature of the entering ligand. This behavior is characteristic of an S_N1 substitution mechanism.

If the entering group starts to bond to the metal before the leaving group exits the coordination sphere, the substitution reaction can be shown as

$$ML_nX + Y \underset{}{\overset{Slow}{\rightleftharpoons}} [Y\cdots ML_n\cdots X]^{\ddagger} \xrightarrow{fast,\ -X} ML_nY + X \qquad (20.43)$$

When written in terms of the elementary reaction steps, the process can be represented as

$$ML_nX + Y \underset{k_{-1}}{\overset{k_1}{\rightleftharpoons}} ML_nXY \qquad (20.44)$$

$$ML_nXY \xrightarrow{k_2} ML_nY + X \qquad (20.45)$$

The rate of formation of the product can be described in terms of the concentration of the transition state as follows:

$$Rate = \frac{d[ML_nY]}{dt} = k_2[ML_nXY] \qquad (20.46)$$

It is now necessary to express the concentration of the transition state as a function of time. The transition state is formed in the first reaction and consumed in the reverse of the first step and in the second step. Therefore, the steady state approximation yields

$$\frac{d[ML_nXY]}{dt} = k_1[ML_nX][Y] - k_{-1}[ML_nXY] - k_2[ML_nXY] = 0 \qquad (20.47)$$

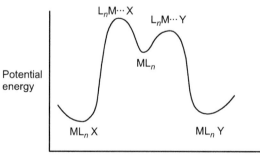

FIGURE 20.1
The energy profile for substitution of Y for X in a dissociative mechanism.

from which we obtain

$$k_1[\text{ML}_n\text{X}][\text{Y}] = k_{-1}[\text{ML}_n\text{XY}] + k_2[\text{ML}_n\text{XY}] \tag{20.48}$$

Solving this equation for the concentration of the intermediate gives

$$[\text{ML}_n\text{XY}] = \frac{k_1[\text{ML}_n\text{X}][\text{Y}]}{k_{-1} + k_2} \tag{20.49}$$

Substituting this result into Eqn (20.46) gives the rate law

$$\text{Rate} \approx \frac{k_1 k_2[\text{ML}_n\text{X}][\text{Y}]}{k_{-1} + k_2} \approx k_{\text{obs}}[\text{ML}_n\text{X}][\text{Y}] \tag{20.50}$$

Formation of the transition state in this case involves both [ML$_n$X] and Y, so the rate law contains the concentrations of both species.

In the process for which the rate law is a function of the concentration of *both* the complex and the entering ligand, the rate law describes an *associative* or S$_N$2 process. The associative pathway is characterized by the bond to the entering group being formed while the bond to the leaving group is still intact. In the transition state, the coordination number of the metal is larger than it is in either the reactant or product. In an S$_N$2 mechanism, the transition state is a complex in which the coordination number increases as the entering group attaches *before* the leaving group is completely detached. It is sometimes considered that these two processes can take place to a more or less equal extent (simultaneously in the transition state of the interchange mechanism as discussed later). In this case, the mechanism is described as S$_N$2, but this label is also used to describe a second type of associative process. If bond formation to Y is much more important than breaking the bond to X as the transition state forms, the mechanism is described as S$_N$2(lim) or as a "strictly" S$_N$2 process. In this case, the transition state is a complex in which *both* the entering and leaving ligands reside in the coordination sphere of the metal.

In aqueous solutions, a water molecule can bind to the activated complex to complete the coordination sphere. The complex formed, [ML$_n$H$_2$O], has some stability, so it represents a lower energy than the transition state, [ML$_n$]. Therefore, the complex [ML$_n$H$_2$O] is known as an *intermediate* because it is more stable than the transition state either before the H$_2$O enters or after it leaves. Figure 20.2 shows the energy profile for a substitution reaction that follows an associative pathway.

If the bond to the entering ligand starts to form as the bond to the leaving group is being broken, the pathway is referred to as an *interchange* mechanism, I. This is an intermediate process in which the bonds to both the entering and leaving groups exist simultaneously. The formation of the new bond may be more important than breaking the bond to the leaving group in some cases. Therefore, the interchange mechanism is a sort of one-step process in which the entering ligand is bonding to the metal as the leaving group moves out of the coordination sphere. If bond formation occurs essentially *before* the bond to the leaving group is ruptured, the interchange is known as *associative* interchange, I_a, and this is referred to as S$_N$2(lim) in older literature. However, if bond breaking to the leaving group is essentially complete *before* the bond forms to the entering group, the interchange is referred to as *dissociative* interchange, I_d, which is analogous to S$_N$2 in older terminology. Both interchange

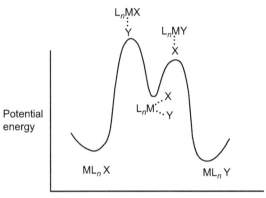

FIGURE 20.2

The energy profile for substitution of Y for X in an associative mechanism.

mechanisms can be regarded as S_N2 but with different degrees of overlap in the time scale in which bond making and bond breaking occur. The energy profile for this process is shown in Figure 20.3.

20.2.2 Some Factors Affecting Rates of Substitution

Substitution reactions occur as a metal−ligand bond is broken and a different ligand bonds to the metal. That such processes would be related to the properties of the metal and ligand should be expected. Therefore, when substitution occurs, there are some (largely predictable) effects as a result of different sizes and charges of the entering and leaving ligands and the size and charge of the metal ion. For example, if a substitution reaction proceeds by an S_N1 mechanism, increasing the sizes of the other ligands in the complex helps "pressure" the leaving group, making it more labile. On the other hand, in an S_N2 process, bulky ligands hinder the formation of the bond to the entering ligand, so the effect will be to decrease the rate.

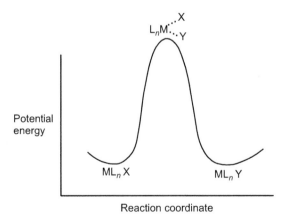

FIGURE 20.3

The interchange mechanism for a substitution reaction.

Increasing the charge on the metal ion makes it more difficult for the leaving group to depart, so the rate would be expected to decrease when the charge on the metal is higher. In contrast, a more highly charged metal ion attracts the entering group more strongly, so in $S_N2(lim)$ cases (where bond formation to the entering ligand is a dominant factor) the rate should increase. In an S_N2 mechanism, both bond breaking (which is hindered when the metal ion has a higher charge) and bond making are affected and in opposite ways, so no unambiguous conclusion can be given.

The size of the ligand that is being replaced influences the rate of substitution, and the effect can be rationalized in the following way. A larger leaving group makes it easier to depart, so the rate of an S_N1 reaction will increase as ligand size increases. A larger ligand generally hinders attachment of the entering group in an $S_N2(lim)$ process, so the rates are generally lower for larger leaving groups. In an S_N2 process, the effects of the size of the leaving group are opposing because both the entering and leaving groups are involved.

The size of the metal ion should also influence the rate of substitution reactions. The larger the metal ion, the less strongly the leaving group is attached, so an S_N1 process should have a higher rate for larger metal ions. In either an S_N2 or $S_N2(lim)$ process, there will be less steric crowding as the bond is formed to the entering group, so the rate should increase for larger metal ions.

Increasing the charge on the ligand that is leaving should hinder the formation of the transition state in an S_N1, S_N2, or $S_N2(lim)$ mechanism. On the other hand, increasing the charge (or polarity) of the entering ligand should have no effect in an S_N1 mechanism but would assist bond formation in either S_N2 or $S_N2(lim)$ mechanisms (which would increase the rate of substitution).

The effects predicted are qualitative at best. There are other factors that must be taken into account when predicting how various characteristics of the metal and ligand affect substitution reactions. For example, increasing the size of the metal ion is predicted to assist the formation of the transition state in S_N1, S_N2, or $S_N2(lim)$ cases. However, rates of reactions of Co^{3+}, Rh^{3+}, and Ir^{3+} complexes show that the second and third row metals react much more slowly than the first row metals. Realizing that formation of the transition state requires some sacrifice of ligand field stabilization energy, it is expected that forming the transition state of the first row (but smaller) ion would require less energy because for a given ligand, Dq is much smaller for the first row metal ions. Even this does not fully address all the problems because it is known that the ability of the ligands to form π bonds is also a factor. If, as is usually the case, the metal ion has nonbonding d orbitals, they may be involved in bonding to an entering ligand, which can facilitate bond formation during an S_N2 or $S_N2(lim)$ process. As a result of all these factors, the trends predicted above have some utility, but they should not be taken too literally. Neither do the effects of ligand field stabilization energy always agree with experience because the ligand field approach to bonding is from an inadequate description of the bonding in complexes.

Another troublesome factor is the influence of the solvent. In many cases, if two processes have comparable activation energies, it is possible for the solvent to affect the stability of the transition state to such an extent that the mechanism is controlled by these subtle differences in $\Delta H^{\ddagger}$ or $\Delta S^{\ddagger}$ brought about by the solvent. In general, solvents having high cohesion (large solubility parameter) favor the formation of transition states in which there is charge generation or separation. Solvents having low cohesion (small solubility parameter) generally favor formation of transition states in which charge is dispersed. Ion pairing (and dipole association) is more extensive in solvents that have lower cohesion, but solvents of high cohesion (and typically high polarity) generally solvate ions more strongly, which helps prevent ion pairing. The whole area of making predictions about rates of reaction of coordination compounds based on properties of ligands, metal ions, and solvents is a thorny problem in kinetics, with only some general trends being observable.

20.3 LIGAND FIELD EFFECTS

The ligand field stabilization energy associated with several geometrical arrangements of ligands was shown in Chapter 17. In the course of a substitution reaction in an octahedral complex, the transition state may involve either 5 bonds to the metal as in an S_N1 mechanism or 7 bonds in an S_N2 pathway. In either case, the ligand field stabilization energy is different from that of the starting complex. Therefore, part of the activation energy required can be attributed to the amount of ligand field energy lost in forming the transition state. The basis for calculating the ligand field energy in various possible transition states is provided by the orbital energies shown in Table 17.1.

Let us first consider the case of a substitution reaction in a complex of a d^6 ion such as Co^{3+} in a strong field. If the process takes place by an S_N1 process, the 5-bonded transition state may be presumed to have either a trigonal bipyramid or square-based pyramid structure. The orbital energies will be determined as follows:

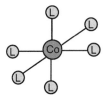 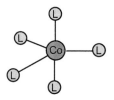 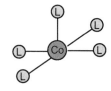

	Starting O_h Complex	Trigonal Bipyramid	Square-Based Pyramid
d_{z^2}	$6.00 \times 0 = 0$	$7.07 \times 0 = 0$	$0.86 \times 0 = 0$
$d_{x^2-y^2}$	$6.00 \times 0 = 0$	$-0.82 \times 1 = -0.82$	$9.14 \times 0 = 0$
d_{xy}	$-4.00 \times 2 = -8.00$	$-0.82 \times 1 = -0.82$	$-0.86 \times 2 = -1.72$
d_{yz}	$-4.00 \times 2 = -8.00$	$-2.72 \times 2 = -5.44$	$-4.57 \times 2 = -9.14$
d_{xz}	$-4.00 \times 2 = -8.00$	$-2.72 \times 2 = -5.44$	$-4.57 \times 2 = -9.14$
Total	$= -24.00$ Dq	$= -12.52$ Dq	$= -20.00$ Dq
Loss of CFSE		$= 11.48$ Dq	$= 4$ Dq

From this example, it can be seen that forming a square-based pyramid transition state results in a loss of only 4 Dq, whereas forming the trigonal bipyramid transition state involves the loss of 11.48 Dq. In cases in which Dq is rather large, this difference is considerable, which means that the square-based pyramid transition state is energetically more favorable than is a trigonal bipyramid. Therefore, the loss of one ligand and its replacement by another resulting in the product having the same configuration as the starting complex is consistent with the transition state being a square-based pyramid. Although the value of Dq may not be great enough to overcome other factors that *may* favor a trigonal bipyramid transition state for a first row d^6 ion, it is almost certainly large enough to do so for a second or third row d^6 ion. Thus, the loss of crystal field stabilization is so high that it forces

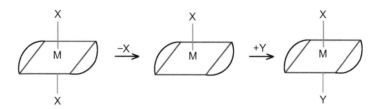

FIGURE 20.4
Substitution in which the transition state is a square-based pyramid.

complexes of these ions to undergo S_N1 substitution through a square-based pyramid transition state so the substitution reactions occur with retention of configuration. This is true for complexes of Pt^{4+}, Rh^{3+}, and Ir^{3+} as typified by the reactions

$$trans\text{-}[Pt(en)_2Cl_2]^{2+} + 2\ Br^- \rightarrow trans\text{-}[Pt(en)_2Br_2]^{2+} + 2\ Cl^- \tag{20.51}$$

$$cis\text{-}[Ru(en)_2Cl_2]^+ + 2\ I^- \rightarrow cis\text{-}[Ru(en)_2I_2]^+ + 2\ Cl^- \tag{20.52}$$

Thus, the fact there is no isomerization during substitution is consistent with the transition state being a square-based pyramid for second and third row metals. As shown in Figure 20.4, substitution would give a product having the same configuration as the starting complex.

Having rationalized that the transition state should be a square-based pyramid, it should be mentioned that there are numerous cases in which the transition state appears to be a trigonal bipyramid. We know that because the substitution occurs with a change in configuration. From the foregoing discussion, we would expect this to occur with first row transition metals because if 11.48 Dq must be sacrificed, this would be more likely if Dq is smaller (which is the case for first row metals). If a trigonal bipyramid transition state forms, there would be more than one product possible. This can be illustrated by the sequence shown in Figure 20.5. However, it should be kept in mind that many trigonal bipyramid structures are not rigid so some rearrangement may occur.

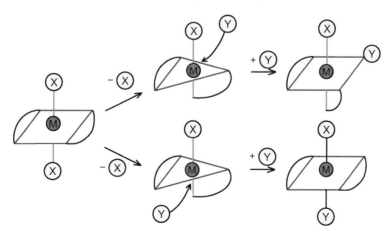

FIGURE 20.5
Attack on a trigonal bipyramid transition state during S_N1 substitution.

For first row metals, the number of Dq units lost is the same as for second and third row metals, but the magnitude of Dq is smaller. As a result, it is possible to get some rearrangement as substitution occurs. Therefore, the product can be a mixture of both *cis* and *trans* isomers.

$$\text{trans-}[ML_4AB] + Y \;\rightarrow\; \text{cis, trans-}[ML_4AY] + B \tag{20.53}$$

If this reaction were to take place by formation of a square-based pyramid transition state, the product would have a *trans* configuration. However, if the transition state is a trigonal bipyramid, the incoming ligand, Y, could enter either *cis* or *trans* to A.

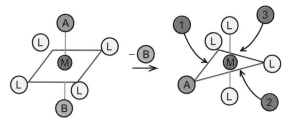

In this scheme, it is assumed that there is more room for the entering ligand to attach along an edge of the trigonal plane. From this scheme, it can be seen that attack by Y at positions 1 and 2 would lead to Y being *cis* to A, but attack at position 3 would leave Y *trans* to A. Thus, if A is in an equatorial position, there are two ways to get a *cis* product but only one way to get a *trans* product. As shown in the following scheme, the situation is different if A is in an axial position in the transition state.

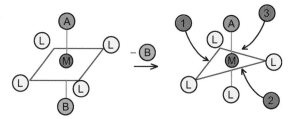

If we again assume that attachment of the entering ligand would occur along an edge of the trigonal plane, all three points of attack would complete the square planar arrangement, but all would place Y in positions that are *cis* to A. Therefore, we would expect that if A is in an axial position in the transition state, the product should all be the *cis* isomer. In the situation in which A occupies an equatorial position in the transition state, the product would be expected to be two-thirds *cis* and one-third *trans*. Next, we must examine the situation that would result if it were equally probable for A to be in an axial or an equatorial position. In that case, the transition state would be an equal mixture of the structures having A axial or equatorial. Once again we assume that attack is equally probable at all positions along the edges of the trigonal bipyramid. If this is true, we should expect the product to consist of five-sixths *cis* and one-sixth *trans* isomer. It must be emphasized that these are ideal cases and that in real complexes there are two complications. First, it is unlikely for it to be equally probable for A to be in axial and equatorial positions in the transition state. Second, because A and L are different ligands, it is unlikely that attack would be equally probable at any side of the trigonal plane. Steric differences would likely cause Y to have a different probability of entering at the three positions along the edges.

The ligand field stabilization energy is only one aspect of the formation of a transition state. Because the reactions are carried out in solutions, solvation of the transition state and the entering ligand may have enough effect to assist in the formation of a particular transition state. Also, the fact that some of the ligands (either those not being replaced or the ligands involved in the substitution) can form π bonds can affect the activation energy. As a result of these factors and the fact that the ligand field model is not really an adequate description of bonding, the discussion above is surely approximate. It is true, however, that the simple approach described above does agree with observations on substitution reactions.

20.4 ACID-CATALYZED REACTIONS OF COMPLEXES

As in many areas of chemistry, there are reactions of coordination compounds that are not what they appear. For example, the reaction

$$\textit{trans-}\left[Co(en)_2F_2\right]^+ + H_2O \ \rightarrow \ \left[Co(en)_2FH_2O\right]^{2+} + F^- \tag{20.54}$$

appears to be a substitution reaction in which H_2O replaces F^-. However, this reaction has a rate that is strongly pH dependent, suggesting that the reaction does not follow the usual pattern for complexes of Co^{3+}. Most substitution reactions of complexes of Co^{3+} are independent of the nature of the entering ligand, which indicates that they are formally S_N1 processes. The fact that H^+ is involved suggests that the reaction does not proceed by a dissociative pathway. Instead, the reaction is believed to proceed as follows:

$$\left[Co(en)_2F_2\right]^+ + H^+ \ \leftrightarrows \ \left[Co(en)_2F(HF)\right]^{2+} \tag{20.55}$$

In the first step, the reaction produces the conjugate acid of the complex by protonation of a coordinated F^- ion. That conjugate acid then undergoes substitution in a dissociative pathway.

$$\left[Co(en)_2F(HF)\right]^{2+} + H_2O \ \rightarrow \ \left[Co(en)_2FH_2O\right]^{2+} + HF \tag{20.56}$$

The concentration of the conjugate acid is the rate-determining factor because it is the dissociation of that species that leads to the formation of the product. Therefore, the rate law can be written as

$$\text{Rate} = k_1[\text{complex}] + k_2[\text{conjugate acid}] \tag{20.57}$$

From Eqn (20.55) it can be seen that the equilibrium constant is

$$K_{eq} = \frac{[\text{conjugate acid}]}{[H^+][\text{complex}]} \tag{20.58}$$

from which the concentration of the conjugate acid can be expressed as

$$[\text{conjugate acid}] = K_{eq}[H^+][\text{complex}] \tag{20.59}$$

Substituting this expression in the rate law shown in Eqn (20.57) gives

$$\text{Rate} = k_1[\text{complex}] + k_2 K_{eq}[\text{H}^+][\text{complex}] \qquad (20.60)$$

which shows the dependence on the concentration of acid.

Acid catalysis is observed most often in reactions of complexes containing ligands that are basic (so they can accept protons) or ligands that can form hydrogen bonds. Some complexes of this type include $[\text{Co}(\text{NH}_3)_5\text{CO}_3]^+$, $[\text{Fe}(\text{CN})_6]^{4-}$, and $[\text{Co}(\text{NH}_3)_5\text{ONO}]^{2+}$. Other complexes that are susceptible to acid-catalyzed reactions are those that contain coordinated *basic* ligands that can be dislodged by their bonding to H^+ instead of to the metal ion. Characteristic of this type of reaction is the process that takes place at 130 °C in the solid state.

$$[\text{Cr}(\text{en})_3](\text{SCN})_3(s) \xrightarrow{\text{NH}_4\text{SCN}(s)} \textit{trans-}[\text{Cr}(\text{en})_2(\text{NCS})_2]\text{SCN}(s) + \text{en}(g) \qquad (20.61)$$

Although this reaction will be discussed later, it is mentioned here because it is catalyzed by a solid acid such as NH_4SCN, which provides H^+ ions that bond to the pairs of electrons after breaking them loose from the metal. Over a rather wide range of catalyst concentrations, the rate is linearly dependent on the amount of solid acid. Once the NH_4^+ ion donates a proton, NH_3 is lost and the protonated ethylenediamine molecule is the acid that remains and continues to catalyze the reaction. Although base-catalyzed reactions of complexes may be better known, there are many acid-catalyzed reactions as well.

20.5 BASE-CATALYZED REACTIONS OF COMPLEXES

The reaction

$$[\text{Co}(\text{NH}_3)_5\text{Cl}]^{2+} + \text{OH}^- \rightarrow [\text{Co}(\text{NH}_3)_5\text{OH}]^{2+} + \text{Cl}^- \qquad (20.62)$$

appears to be a typical substitution reaction. However, unlike the majority of substitution reactions of Co^{3+} complexes, the reaction takes place rapidly and with a rate that is linearly dependent on OH^- concentration.

The mechanism that has been established for this reaction involves the reaction of OH^- with the complex to produce its conjugate base by removing a proton from a coordinated NH_3 molecule in a rapid step involving the equilibrium represented by the following equation:

$$[\text{Co}(\text{NH}_3)_5\text{Cl}]^{2+} + \text{OH}^- \underset{}{\overset{\text{fast}}{\rightleftharpoons}} \underset{\text{Conjugate base}}{[\text{Co}(\text{NH}_3)_4\text{NH}_2\text{Cl}]^+} + \text{H}_2\text{O} \qquad (20.63)$$

The next step is the loss of Cl^- in a dissociation step,

$$[\text{Co}(\text{NH}_3)_4\text{NH}_2\text{Cl}]^+ \overset{\text{slow}}{\rightleftharpoons} [\text{Co}(\text{NH}_3)_4\text{NH}_2]^{2+} + \text{Cl}^- \qquad (20.64)$$

followed by the rapid addition of water, which donates a proton to the coordinated NH_2^- and provides OH^- to complete the coordination sphere of the cobalt.

$$[\text{Co}(\text{NH}_3)_4\text{NH}_2\text{Cl}]^+ + \text{H}_2\text{O} \overset{\text{fast}}{\rightleftharpoons} [\text{Co}(\text{NH}_3)_5\text{OH}]^{2+} + \text{Cl}^- \qquad (20.65)$$

In this mechanism, the rate-determining step involves the dissociative reaction of the conjugate base. Because of this, the mechanism is known as the S_N1CB mechanism, in which the substitution is first order but with respect to the conjugate base. The overall rate is proportional to the concentration of the conjugate base so

$$\text{Rate} = k[\text{conjugate base}] \tag{20.66}$$

The equilibrium constant for the formation of the conjugate base can be written as

$$K_{eq} = \frac{[\text{conjugate base}]}{[\text{OH}^-][\text{complex}]} \tag{20.67}$$

from which the concentration of conjugate base is found to be

$$[\text{conjugate base}] = K_{eq}[\text{OH}^-][\text{complex}] \tag{20.68}$$

The rate can be expressed as

$$\text{Rate} = k\, K_{eq}[\text{OH}^-][\text{complex}] \tag{20.69}$$

This expression shows that the rate is proportional to $[\text{OH}^-]$. If the concentration of OH^- is held constant, the rate law can be written as

$$\text{Rate} = k'[\text{complex}] \tag{20.70}$$

where $k' = kK_{eq}[\text{OH}^-]$. This is the pseudo first-order rate law.

The S_N1CB mechanism has also been verified for the reaction

$$[\text{Pd}(\text{Et}_4\text{dien})\text{Cl}]^+ + \text{OH}^- \rightarrow [\text{Pd}(\text{Et}_4\text{dien})\text{OH}]^+ + \text{Cl}^- \tag{20.71}$$

where Et_4dien is $(\text{Et})_2\text{NCH}_2\text{CH}_2\text{NHCH}_2\text{CH}_2\text{N}(\text{Et})_2$, tetraethyldiethylenediamine, which contains three nitrogen atoms having unshared pairs of electrons and functions as a tridentate ligand. The first step in the reaction is the removal of a proton from a coordinated ligand in an equilibrium that is rapid.

$$[\text{Pd}(\text{Et}_4\text{dien})\text{Cl}]^+ + \text{OH}^- \underset{}{\overset{fast}{\rightleftharpoons}} [\text{Pd}(\text{Et}_4\text{dien-H})\text{Cl}] + \text{H}_2\text{O} \tag{20.72}$$

In this process, $\text{Et}_4\text{dien-H}$ represents a molecule of ligand that has had a proton removed. In this step, the reaction of the ligand can be shown as

$$(\text{Et}_2\text{NCH}_2\text{CH}_2)_2\text{NH} + \text{OH}^- \rightleftharpoons (\text{Et}_2\text{NCH}_2\text{CH}_2)_2\text{N}^- + \text{H}_2\text{O} \tag{20.73}$$

in which the only place where there is a hydrogen atom that can be removed by a base is on the middle nitrogen atom. Following the removal of H^+, the Cl^- is lost in a dissociative step.

$$[\text{Pd}(\text{Et}_4\text{dien-H})\text{Cl}] \underset{}{\overset{slow}{\rightleftharpoons}} [\text{PdEt}_4\text{dien-H}]^+ + \text{Cl}^- \tag{20.74}$$

Removal of the proton produces a negative charge in the position *trans* to where the Cl^- leaves, which enhances the process by a *trans* effect (see Section 20.9). After the dissociation of Cl^- from the transition state, the reaction with H_2O is rapid. A proton from H_2O replaces the one that was lost from the central nitrogen atom, and the remaining OH^- completes the coordination sphere of Pd^{2+}.

$$[Pd(Et_4dien-H)]^+ + H_2O \xrightarrow{\text{fast}} [Pd(Et_4dien)OH]^+ \tag{20.75}$$

The mechanism of the reaction is proved by studying the reaction when the ligand is one in which there is no hydrogen atom on the middle nitrogen atom. That ligand is $(Et_2NCH_2CH_2N)_2NC_2H_5$, in which the hydrogen atom has been replaced by an ethyl group. When the reaction of the complex $[Pd(Et_2dienC_2H_5)Cl]^+$ is studied, the rate of Cl^- replacement is independent of OH^- concentration. In this case, there is no possibility of forming the conjugate base.

20.6 THE COMPENSATION EFFECT

Although it is $\Delta G^{\ddagger}$ that determines the concentration of the transition state and hence the rate of a reaction, it is possible that reactions can proceed at considerably different rates, even though the $\Delta G^{\ddagger}$ values are essentially the same. This situation can arise because $\Delta G^{\ddagger}$ is made up of contributions from both $\Delta H^{\ddagger}$ and $\Delta S^{\ddagger}$ as illustrated by the equation

$$\Delta G^{\ddagger} = \Delta H^{\ddagger} - T\Delta S^{\ddagger} \tag{20.76}$$

If a series of reactions are carried out, it may happen that $\Delta H^{\ddagger}$ and $\Delta S^{\ddagger}$ for the two processes are different but that their values *compensate* so that $\Delta G^{\ddagger}$ is essentially constant. This situation can be explained by means of an example. Suppose a reaction is being carried out in a solvent and that solvation of the reactants and transition state occurs. For two different reactions, there will be two different transition states, TS_1 and TS_2, and we will suppose that transition states are charged or polar with the charges related by $TS_2 > TS_1$. If the solvent is polar, it will solvate TS_2 more strongly than it will solvate TS_1, which will be reflected in the heats of solvation of the transition states such that $\Delta H_2^{\ddagger}$ will be more negative than $\Delta H_1^{\ddagger}$. As a result of being attracted to the charged transition states, the solvent will become ordered or structured (lower or negative $\Delta S^{\ddagger}$) in the vicinity of both transition states. However, this will be more pronounced in the vicinity of TS_2 because it is more highly charged. Thus, $\Delta S_2^{\ddagger}$ will be more negative than $\Delta S_1^{\ddagger}$. Therefore, if $\Delta H_2^{\ddagger}$ is more negative than $\Delta H_1^{\ddagger}$ and $\Delta S_2^{\ddagger}$ is more negative than $\Delta S_1^{\ddagger}$, it is possible that $\Delta G^{\ddagger}$ may be approximately constant for the two cases. In other words, the effects of $\Delta H^{\ddagger}$ and $\Delta S^{\ddagger}$ offset each other because of the equation that relates the two quantities. This is known as the *compensation effect*. For a series of reactions (such as substitution by a series of ligands), it is possible that $\Delta G^{\ddagger}$ is approximately constant so that

$$\Delta H_1^{\ddagger} - T\Delta S_1^{\ddagger} = \Delta H_2^{\ddagger} - T\Delta S_2^{\ddagger} = C \tag{20.77}$$

For a series that includes several reactions,

$$\Delta H_i^{\ddagger} = T\Delta S_i^{\ddagger} + C \tag{20.78}$$

and a graph of the $\Delta H_i^{\ddagger}$ versus the $\Delta S_i^{\ddagger}$ values should be linear with a slope of T. This relationship is known as an *isokinetic relationship* and T is a temperature known as *isokinetic temperature*.

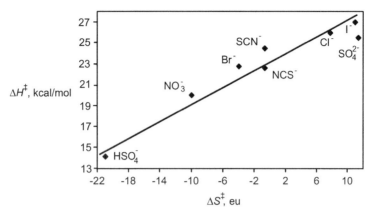

FIGURE 20.6
An isokinetic plot for the formation of $[Cr(H_2O)_5X]^{2+}$ by replacement of OH^-. *Constructed from the data given in D. Thusius,* Inorg. Chem., *1971, 10, 1106.*

The series of reactions

$$\left[Cr(H_2O)_5OH\right]^{2+} + X^- \rightarrow \left[Cr(H_2O)_5X\right]^{2+} + OH^- \tag{20.79}$$

where $X = Cl^-$, Br^-, I^-, SCN^-, etc. has been studied by Thusius (1971). The graph illustrating the isokinetic relationship is shown in Figure 20.6. The relationship between $\Delta H_i^{\ddagger}$ and $\Delta S_i^{\ddagger}$ shown in the graph is satisfactory considering that the nature of the ligands varies widely. Ligand substitution in this case follows a mechanism in which there is loss of OH^- followed by X^- entering the coordination sphere of the metal. The fact that the graph is linear is considered to be indicative of a common mechanism for all the substitution reactions.

For simplicity, we assumed that the transition states are charged. However, it is not necessary to do so because the only requirement is that the difference in entropy of forming the transition states be offset by the difference in enthalpy of activation. The transition states could have different polarities and the same result could be obtained. In fact, the transition states need not have high polarity. Forming a transition state in which there is a reduction in charge separation could result in more favorable solvation when the solvent is nonpolar. For there to be an isokinetic relationship for a series of reactions, it is required only that $\Delta H^{\ddagger}$ and $\Delta S^{\ddagger}$ be related in such a way that $\Delta G^{\ddagger}$ be approximately constant.

20.7 LINKAGE ISOMERIZATION

Linkage isomerization processes involve the change in bonding mode of one or more ligands that can bond to metal ions in more than one way. Some ligands that have this ability are CN^-, SCN^-, NO_2^-, SO_3^{2-}, and SO_4^{2-}. The first case of linkage isomerism involved pentaamminenitrocobalt(III) and pentaamminenitritocobalt(III) ions. In this case, the nitro form is the more stable of the two. The conversion of the nitrito linkage to the nitro bonding mode has been the subject of several studies, and a great deal is known about the process. In this case, the results of studying the reaction at high pressure have been especially important.

The activation energy for a reaction represents the energy barrier over which the reactants must pass in being transformed into products. Determining the rate constant (k) as a function of temperature allows a plot of ln k versus $1/T$ to be made, and the slope of the line $(-E/R)$ allows E to be determined. The formation of the transition state can be viewed as an equilibrium between the reactants and the transition state. The transition state is higher in energy than the reactants, so increasing the temperature increases the concentration of the transition state and thereby the rate of the reaction. Much less frequently studied than the temperature effect on the rate of a reaction is the effect of pressure. As a transition state forms from reactants, there is a volume change in most cases. By Le Chatelier's principle, increasing the pressure increases the concentration of the species occupying the smaller volume. If the transition state occupies a smaller volume than the reactants, increasing the pressure (at constant temperature) leads to an increase in the rate of the reaction. If increasing the pressure decreases the rate of the reaction, the transition state occupies a larger volume than the reactants. Generally, reactions that involve bond-breaking–bond-making steps have transition states that occupy a larger volume than the reactants. Intramolecular processes often pass through a transition state that has a smaller volume than the reactant.

The linkage isomerization

$$[Co(NH_3)_5ONO]Cl_2 \xrightarrow{\Delta,\ h\nu} [Co(NH_3)_5NO_2]Cl_2 \tag{20.80}$$

takes place both in solutions and the solid state when the starting isomer is heated or subjected to ultraviolet radiation. In regard to how the reaction takes place, it might be presumed that the metal–ligand bond is broken and the ligand then attaches in the more stable nitro form:

$$[Co(NH_3)_5ONO]^{2+} \leftrightarrows \left[Co(NH_3)_5^{3+} + ONO^-\right]^{\ddagger} \rightarrow [Co(NH_3)_5NO_2]^{2+} \tag{20.81}$$

However, if the reaction follows this pathway, the transition state should occupy a larger volume than the reactants, so an increase in pressure would decrease the rate of the reaction. Instead, when this reaction was studied at a series of high pressures, it was found that the rate increased with pressure.

It is possible to derive the following equation that relates the rate constant of a reaction to the applied pressure.

$$\ln k = \frac{\Delta V^{\ddagger}}{RT} P + \text{Constant} \tag{20.82}$$

By making a plot of ln k versus P, it is possible to determine the volume of activation, ΔV^*. For the linkage isomerization reaction shown in Eqn (20.81), the volume of activation is -6.7 ± 0.4 cm^3 mol^{-1}. Therefore, it can be concluded that the mechanism shown in Eqn (20.81) is not correct. In fact, the negative volume of activation indicates that in the transition state, the NO_2^- does not become detached from the metal ion. The mechanism has been shown to involve the movement of the nitrite ion to form a transition state in a process that can be described as shown in Figure 20.7.

The transition state (II) occupies a smaller volume than that of the starting complex, so an increase in pressure causes an increase in the rate of the reaction. In general, reactions that pass through transition

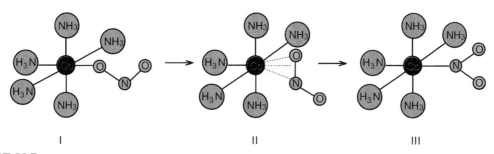

FIGURE 20.7
The mechanism of linkage isomerization of −ONO to −NO$_2$.

states that have smaller volumes than the reactants are enhanced by an increase in pressure. On the other hand, dissociation reactions that lead to two separate species in the transition state are retarded by increasing the pressure.

When the linkage isomerization is carried out in the solid state and the reaction quenched to very low temperature, it is observed that the infrared spectrum of the mixture contains bands that are not attributable to either Co−NO$_2$ or Co−ONO linkages. The bands are consistent with a transition state with the nitrite bonded as shown in the transition state above. In addition to the cobalt complex described above, the corresponding complexes of Rh^{3+} and Ir^{3+} also undergo linkage isomerization with the volumes of activation being -7.4 ± 0.4 cm^3 mol^{-1} and -5.9 ± 0.6 cm^3 mol^{-1}, respectively. This indicates that the linkage isomerization in these complexes also occurs without a bond-breaking step.

Additional experiments on the reaction of the cobalt complex have involved the reaction

$$\left[\text{Co(NH}_3)_5{}^{18}\text{OH}\right]^{2+} + \text{N}_2\text{O}_3 \rightarrow \left[\text{Co(NH}_3)_5{}^{18}\text{ONO}\right]^{2+} + \text{HNO}_2 \tag{20.83}$$

Because the ^{18}O is attached to the cobalt ion both before and after the reaction, it is reasonable to conclude that the Co−O bond is never broken. The reaction is believed to involve a transition state that can be shown

$$(\text{NH}_3)\text{Co}\!-\!{}^{18}\text{O}\!-\!\text{H}$$
$$\text{O}\!-\!\text{N}\!-\!\text{ONO}$$

which leads to the product being [Co(NH$_3$)$_5^{18}$ONO]$^{2+}$. The ^{18}O never leaves the coordination sphere of the cobalt. On the other hand, in the reaction shown in Eqn (20.83), the Co−^{18}ONO can be induced to undergo linkage isomerization in the same way that the nitrite that does not contain ^{18}O does.

20.8 SUBSTITUTION IN SQUARE PLANAR COMPLEXES

The majority of square planar complexes are those that contain d^8 metal ions, of which the most common examples are Ni^{2+}, Pd^{2+}, and Pt^{2+}, although some complexes containing Au^{3+} have also been studied. As a general trend, the rate of substitution in these complexes is Ni$^{2+} >$ Pd$^{2+} >$ Pt$^{2+} <$ Au^{3+}.

For the first three metal ions in the series, the rate is inversely related to the ligand field energy, as would be expected. The fact that Au^{3+} complexes undergo substitution much faster than those of Pt^{2+} comes from the fact that it has a higher charge, which gives rise to greater attraction for a potential ligand. Complexes of Pt^{2+} have been extensively studied, and much of the discussion will be concerned with those complexes.

Because most complexes of platinum are quite stable, substitution reactions are generally slow. For the reaction

$$[PtL_2XY] + A \rightarrow [PtL_2XA] + Y \tag{20.84}$$

the rate law has the form

$$Rate = k_1[\text{complex}] + k_2[\text{complex}][A] \tag{20.85}$$

Therefore, the observed first-order rate constant, k_{obs}, is given by

$$k_{obs} = k_1 + k_2[A] \tag{20.86}$$

Although the first term in Eqn (20.85) appears to be first order in complex, it usually represents a second-order process in which the solvent (which is usually a nucleophile) is involved. The relationships show that if a plot is made of k_{obs} versus [A], the result is a straight line having a slope of k_2 and an intercept of k_1. Therefore, the substitution process can be viewed as if it occurs by two pathways. This situation can be described as illustrated in Figure 20.8.

It is often found that the rate of substitution in square planar complexes varies greatly depending on the nature of the solvent. The term in the rate law that appears to be independent of the concentration of the entering ligand involves the solvent (which has essentially constant concentration). Therefore,

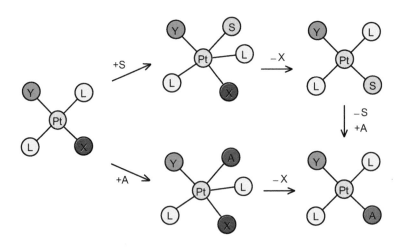

FIGURE 20.8
Substitution in a square planar complex in which the solvent participates in a second-order step.

the value of k_1 in the rate law depends on the solvent. Another way in which the solvent can affect the rate of substitution is related to how strongly the entering ligand is solvated. For example, if the entering ligand is solvated to different degrees in a series of solvents, the rate at which it enters the coordination sphere of the metal may show a decrease as the ability of the solvent to solvate the ligand increases. In order to be attached in the complex, the ligand must become partially "desolvated," and the more strongly the solvent is attached to the ligand, the more difficult that becomes.

There is a considerable difference in rate of substitution depending on the nature of the leaving group. For the reaction

$$[Pt(dien)X]^+ + py \rightarrow [Pt(dien)py]^{2+} + X^- \tag{20.87}$$

(where dien is diethylenetriamine, $H_2NCH_2CH_2NHCH_2CH_2NH_2$), the rate of the reaction was very fast when $X = NO_3^-$, and the value for k_{obs} was 1.7×10^{-8} sec^{-1} when $X = CN^-$. For a series of ligands, the rate of loss of X was found to vary in the following order.

$$NO_3^- > H_2O > Cl^- > Br^- > I^- > N_3^- > SCN^- > NO_2^- > CN^-$$

We have previously described the relationship known as the isokinetic plot in which $\Delta H^{\ddagger}$ is plotted against $\Delta S^{\ddagger}$ for the substitution reactions of a series of ligands. The reaction

$$[Pt(dien)X]^+ + L \rightarrow [Pt(dien)L]^+ + X \tag{20.88}$$

has been studied for a series of ligands, L. When the values for the enthalpy and entropy of activation are plotted, the result is Figure 20.9.

Although the fit of the data to the line is not perfect, it is adequate given the level of accuracy of the data. Because the point for H_2O falls far from the line through the other points, it is tempting to conclude that the mechanism is different in that case. The fact that the substitution reactions were

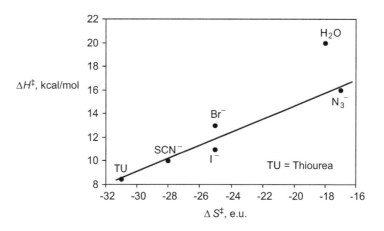

FIGURE 20.9
Isokinetic plot for the substitution of various ligands in [Pt(dien)Cl]Cl where dien is diethylenetriamine, $H_2NCH_2CH_2NHCH_2CH_2NH_2$. *Constructed from data given by Basolo and Pearson, 1974, p. 404.*

carried out in water makes this explanation even more plausible because the solvent is present in such an enormous excess. Therefore, in the rate law shown in Eqn (20.85), the large excess of water could be responsible for a substitution that proceeds with coordination of the solvent as the dominant feature. Having given some insight to the nature of substitution reactions in square planar complexes, we now turn to a dominant characteristic of these reactions.

20.9 THE TRANS EFFECT

One of the fascinating characteristics of substitution in square planar complexes is illustrated by the following equations:

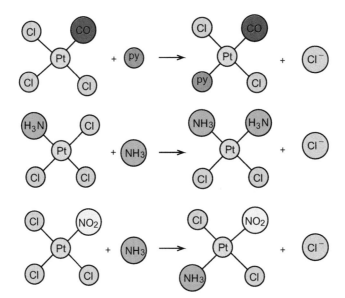

In the first of these reactions, all of the entering pyridine goes in the position *trans* to CO. If there were equal tendency for the three chloride ions to be replaced, the product should contain two-thirds *cis*- and one-third *trans*-[Pt(CO)(py)Cl$_2$]. The fact that the product consists only of the *trans* isomer indicates that CO is somehow directing substitution to go in the *trans* position to it. This influence is described as a *trans effect*. In the second reaction, none of the entering NH$_3$ goes in a position *trans* to the NH$_3$. Because all of the entering NH$_3$ goes opposite a coordinated Cl$^-$, we conclude that the Cl$^-$ must exert a stronger *trans* influence than does NH$_3$. In the last of these reactions, the fact that the product is *trans*-[Pt(NH$_3$)(NO$_2$)Cl$_2$] indicates that the NO$_2^-$ ligand is exerting an influence as to where the entering NH$_3$ goes. Otherwise, some of the Cl$^-$ being replaced would be coming from the positions *cis* to NO$_2^-$. Therefore, it can be concluded that NO$_2^-$ is exerting some sort of influence on the position *trans* to it in the complex. In some writing, the distinction is made between a *trans effect* and a *trans influence*. The former term is used to describe effects on rates of reactions (kinetics), whereas the latter is applied to static (thermodynamic) influences (such as bond lengths or stretching frequencies of metal–ligand bonds). The phenomena are interrelated to some extent, so no such distinction will be made here.

From the reactions shown, it becomes apparent that NH_3 must have less influence on the position *trans* to it than does Cl^-. Moreover, the first reaction shows that CO has a stronger influence than does Cl^-, and the last reaction shows that NO_2^- has a greater effect than Cl^-. By conducting reactions such as those shown, it is possible to determine the relative *trans* effect for a series of ligands. For several common ligands, the series can be shown as follows:

$$C_2H_4 \approx CO \approx N_3^- > (CH_3)_3P \approx H^- > NO_2^- > I^- > SCN^- > Br^- > Cl^- > py > NH_3 > OH^- > H_2O$$

In addition to the stereochemistry reaction products, there are numerous other manifestations of the *trans* influence of ligands in square planar complexes. One such factor is the lengths of bonds in the complexes. Consider the following structures:

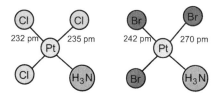

Here we can see that in $K[Pt(NH_3)Br_3]$ the length of the Pt–Br bond opposite the Pt–NH_3 bond is about 242 pm, whereas the Pt–Br bond opposite the other Br is 270 pm in length. A similar effect is seen in $K[Pt(NH_3)Cl_3]$, but the difference in bond lengths is smaller in this case because Cl^- exerts a weaker *trans* effect than Br^-. In the anion of Zeise's salt, $[Pt(C_2H_4)Cl_3]$,

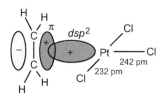

the Pt–Cl bond opposite C_2H_4 is 242 pm in length but that *trans* to Cl is 232 pm. These values show that the C_2H_4 molecule weakens and lengthens the Pt–Cl bond that is *trans* to it more than Cl^- does. This observation is in agreement with the order of the *trans* effect given above for numerous ligands.

Fifty years ago, Chatt et al. (1958) showed the effect that a ligand in the *trans* position has on the stretching frequency of a Pt–ligand bond. Although other complexes were also studied, the series of complexes where L was varied in *trans*-$[PtA_2LH]$ (where $A = PEt_3$) demonstrated the *trans* effect on the Pt–H stretching frequency in the infrared spectrum. The variation in Pt–H stretching frequency with the nature of L was found to vary as follows (band positions in cm^{-1}): CN^-, 2041; SCN^-, 2112; NO_2^-, 2150; I^-, 2156; Br^-, 2178; and Cl^-, 2183. These positions for the bands representing the stretching vibration of the Pt–H bond *trans* to the ligand being varied show that when $L = CN^-$, there is a substantial effect on the Pt–H bond. On the other hand, Cl^- has a much smaller effect when it is *trans* to the Pt–H bond. These observations are in agreement with the order of *trans* effect described earlier.

A series of reactions that can be represented as

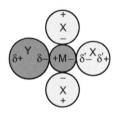

$$(20.89)$$

were studied, where L is the series of ligands Cl^-, Br^-, and NO_2^-. It was found that the activation energies were 79, 71, and 46 kJ mol^{-1}, and the rate constants were 6.3×10^{-3}, 18×10^{-3}, and 56×10^{-3} M^{-1} sec^{-1}, respectively, for these ligands. The effect of the ligand L on the position *trans* to it is clearly evident both in terms of rate of substitution and in the activation energy associated with the process.

One of the approaches to explaining the *trans* effect in a square planar complex is known as the polarization model. The underlying principle in this approach is that the metal ion and the ligands undergo mutual polarization that results in some small charge separation even in spherical ions and ligands. Suppose a complex has the composition MX_3Y and that Y is more polarizable than X. This situation is shown in Figure 20.10.

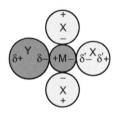

FIGURE 20.10
A square planar complex MX_3Y in which the ligands are polarizable.

Because Y is more polarizable than X, the charges separated in the ligands will be different. Based on the polarizabilities, $|\delta| > |\delta'|$. This causes the charge distribution in the metal to be such that the more positive side will be directed toward Y and the less positive side toward X. This has the effect of making X less tightly bound and easier to remove. Thus, the ease with which X is replaced should increase as the polarizability of Y increases. In general, this trend is observed with the *trans* effect of the halides varying in the order $I^- > Br^- > Cl^-$. Because the metal is also polarized somewhat, the observed *trans* effect for d^8 ions varies as $Pt^{2+} > Pd^{2+} > Ni^{2+}$.

Another explanation for the origin of the *trans* effect lies in the ability of the metal and ligand to form π bonds. The ability of a ligand to accept back donation is dependent on it having empty orbitals of suitable symmetry to match the d orbitals on the metal. Ligands such as CO, C_2H_4, and CN^- have such (antibonding) orbitals (see Chapter 16). Bonding to cyanide ion can be shown as

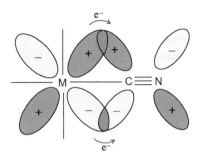

If back donation occurs to a ligand, the flow of electron density from the metal leaves less electron density to be donated in the opposite direction. It seems that this should have little effect on the donation of a pair of electrons on the ligand in the *trans* position to form a σ bond. Accordingly, the major factor appears to be the stabilization of a 5-bonded (trigonal bipyramid) transition state as a result of π bond formation. Ligands that readily form π bonds include some of those that generate the largest *trans* effect.

It is now apparent that ligands cause a *trans* effect by entirely different processes. For example, a large, soft ligand such as H⁻ generates a strong *trans* effect, but it cannot be as a result of π bonding. Basolo and Pearson (1974) have considered the *trans* effect to be made up of contributions of both a σ character and a π character. The result for several ligands can be shown as follows (S = strong; W = weak, M = moderate, etc.).

Ligand:	C_2H_4	CO	CN^-	PR_3	H^-	NO_2^-	I^-	NH_3	Br^-	Cl^-
σ Effect:	W	M	M	S	VS	W	M	W	M	M
π Effect:	VS	VS	S	M	VW	M	M	VW	W	VW

This series shows that some ligands give a *trans* effect that is made up of contributions from each type of interaction, but ligands that give large *trans* effects generally do so by only one dominant type of effect that is either σ or π in origin.

At this point, it is appropriate to mention some of the evidence that indicates a *trans* effect in octahedral complexes, but only a brief description will be given. In the reaction of $Mo(CO)_6$ with pyridine, only three CO ligands undergo replacement and the product has the structure

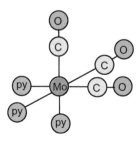

Clearly, this indicates that there is some sort of *trans* effect exerted by CO. Other observations of this type are also recorded.

Evidence of an octahedral *trans* effect also comes from the study of exchange of NH_3 in $[Co(NH_3)_5SO_3]^+$, which was followed using $^{15}NH_3$. It was found that only the NH_3 *trans* to SO_3^{2-} undergoes exchange to give *trans*-$[Co(NH_3)_4 (^{15}NH_3)SO_3]^+$. In a similar study that involved the exchange of NH_3 in *cis*-$[Co(NH_3)_4(SO_3)_2]^-$, it was found that the product had the structure

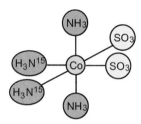

Kinetic analysis of the substitution reactions indicates that they follow a dissociative mechanism. It has also been shown that two water molecules in $[Cr(H_2O)_5I]^{2+}$ undergo exchange with labeled water. It is interesting that one exchange is rapid and occurs before I^- leaves. However, this is not true of the chloride compound. Therefore, it appears that the iodide ion labilizes the water *trans* to it, but the chloride does not.

In addition to the indications of an octahedral *trans* effect presented, there exists structural information in the form of bond lengths and spectral data similar to that described earlier for square planar complexes. Although the *trans* effect in octahedral complexes is not the dominant influence that is in square planar complexes, there is no doubt that there is such an effect.

20.10 ELECTRON TRANSFER REACTIONS

An aqueous solution containing complexes of two different metal ions may make it possible for a redox reaction to occur. In such cases, electrons are transferred from the metal ion being oxidized to the metal ion being reduced. For example,

$$Cr^{2+}(aq) + Fe^{3+}(aq) \rightarrow Cr^{3+}(aq) + Fe^{2+}(aq) \tag{20.90}$$

In this reaction, an electron is transferred from Cr^{2+} to Fe^{3+}, and such reactions are usually called electron transfer or electron exchange reactions. Electron transfer reactions may also occur in cases in which only one type of metal ion is involved. For example, the reaction

$$[^*Fe(CN)_6]^{4-} + [Fe(CN)_6]^{3-} \rightarrow [^*Fe(CN)_6]^{3-} + [Fe(CN)_6]^{4-} \tag{20.91}$$

represents an electron transfer from $[*Fe(CN)_6]^{4-}$ (where *Fe is a different isotope of iron) to Fe^{3+} in $[Fe(CN)_6]^{3-}$. This is an electron transfer in which the product differs from the reactants only in that a different isotope of Fe is contained in the +2 and +3 oxidation states.

Electron transfer between metal ions contained in complexes can occur in two different ways, depending on the nature of the metal complexes that are present. If the complexes are inert, electron transfer occurring faster than the substitution processes must occur without breaking the bond between the metal and ligand. Such electron transfers are said to take place by an *outer sphere* mechanism. Thus, each metal ion remains attached to its original ligands and the electron is transferred through the coordination spheres of the metal ions.

In the second case, the ligand replacement processes are more rapid than the electron transfer process. If this is the case (as it is with labile complexes), a ligand may leave the coordination sphere of one of the metal ions and may be replaced by forming a bridge utilizing a ligand already attached to a second metal ion. Electron transfer then occurs through a bridging ligand, and this is called an *inner sphere* mechanism.

For an outer sphere electron transfer, the coordination sphere of each complex ion remains intact. Thus, the transferred electron must pass through *both* coordination spheres. Reactions such as the following are of this type (where * represents a different isotope):

$$\left[^*Co(NH_3)_6\right]^{2+} + \left[Co(NH_3)_6\right]^{3+} \rightarrow \left[^*Co(NH_3)_6\right]^{3+} + \left[Co(NH_3)_6\right]^{2+} \tag{20.92}$$

$$\left[Cr(dipy)_3\right]^{2+} + \left[Co(NH_3)_6\right]^{3+} \rightarrow \left[Co(NH_3)_6\right]^{2+} + \left[Cr(dipy)_3\right]^{3+} \tag{20.93}$$

In the reaction shown in Eqn (20.93), the d^4 complex containing Cr^{2+} is inert owing to electron pairing giving the low spin state. There is an extreme variation in rates from very slow to very fast depending on the nature of the ligands present, and rate constants may vary from 10^{-6} to 10^8 $M^{-1} s^{-1}$.

The electron exchange between manganate (MnO_4^{2-}) and permanganate (MnO_4^{-}) takes place in basic solutions,

$$^*MnO_4^- + MnO_4^{2-} \rightarrow {}^*MnO_4^{2-} + MnO_4^- \tag{20.94}$$

and the reaction obeys the rate law

$$Rate = k\left[^*MnO_4^-\right]\left[MnO_4^{2-}\right] \tag{20.95}$$

When the solvent contains H_2O^{18}, no ^{18}O is incorporated into the MnO_4^- produced. Thus, the reaction is presumed not to proceed by forming oxygen bridges. However, the nature of the cations present greatly affects the rate. The rate of the reaction varies with the cation present in the order $Cs^+ > K^+ \approx Na^+ > Li^+$. This supports the belief that the transition state must involve a structure such as

$$^-O_4Mn\cdots M^+\cdots MnO_4^{2-}$$

Presumably, the function of M^+ is to "cushion" the repulsion of the two negative ions. The larger softer Cs^+ can do this more effectively than the smaller harder ions such as Li^+ or Na^+. Also, to form these bridged transition states, solvent molecules must be displaced from the solvation sphere of the cations. That process, because of their smaller sizes, would require more energy for the more strongly solvated Li^+ and Na^+. For the Cs^+ ion, which forms effective bridges, the rate of electron exchange has been found to be linearly related to Cs^+ concentration.

Similar results have been found for the electron exchange between $[Fe(CN)_6]^{3-}$ and $[Fe(CN)_6]^{4-}$. In that case, the acceleratory effects are found to vary with the nature of the cation in the order $Cs^+ > Rb^+ > K^+ \approx NH_4^+ > Na^+ > Li^+$ in accord with the size and solvation effects discussed above. For +2 ions, the order of effect on the rate is $Sr^{2+} > Ca^{2+} > Mg^{2+}$, in accord with the decrease in softness of these species. Exchange in these outer sphere cases is believed to involve the formation of bridged species containing cations that are probably less than fully solvated.

In aqueous solutions, Cr^{2+} is a strong reducing agent, and it reduces Co^{3+} to Co^{2+}. A number of electron transfer reactions involving complexes of these metals have been studied. High spin complexes of Cr^{2+} (d^4) are kinetically labile as are high spin complexes of Co^{2+} (d^7). However, complexes of Cr^{3+} (d^3) and low spin complexes of Co^{3+} (d^6) are kinetically inert. For the exchange reaction (O* represents ^{18}O)

$$\left[Co(NH_3)_5H_2O^*\right]^{3+} + \left[Cr(H_2O)_6\right]^{2+} \rightarrow \left[Co(NH_3)_5H_2O\right]^{2+} + \left[Cr(H_2O)_5H_2O^*\right]^{3+} \tag{20.96}$$

it was found that the rate law is

$$\text{Rate} = k\left[Co(NH_3)_5H_2O^{*3+}\right]\left[Cr(H_2O)_6^{2+}\right] \tag{20.97}$$

It was also found that the H_2O^* is quantitatively transferred to the coordination sphere of Cr^{3+}. Thus, the indication is that the electron is transferred from Cr^{2+} to Co^{3+}, but the H_2O^* is transferred from Co^{3+} to Cr^{2+} as reduction occurs. It appears that the electron transfer occurs through a bridged transition state that may have the structure

$$(NH_3)_5\ Co^{3+} \cdots \overset{\displaystyle H}{\underset{\displaystyle H}{\overset{\displaystyle |}{\underset{\displaystyle |}{O}}}} \cdots Cr\,(H_2O)\,_5^{2+}$$

The H_2O forming the bridge then ends up in the coordination sphere of the kinetically inert Cr^{3+} ion.

A large number of reactions similar to that above have been studied in detail. One such reaction is

$$\left[Co(NH_3)_5X\right]^{2+} + \left[Cr(H_2O)_6\right]^{2+} + 5\,H^+ + 5\,H_2O \rightarrow \left[Co(H_2O)_6\right]^{2+} + \left[Cr(H_2O)_5X\right]^{2+} + 5\,NH_4^+ \tag{20.98}$$

where X is an anion such as F^-, Cl^-, Br^-, I^-, SCN^-, or N_3^-. The Co^{2+} produced is written as $[Co(H_2O)_6]^{2+}$ because the high spin complexes of Co^{2+} (d^7) are labile and undergo rapid exchange with the solvent, which is present in great excess.

In these cases, it is found that X is transferred quantitatively from the Co^{3+} complex to the Cr^{2+} complex as electron transfer is achieved. Therefore, it is likely that electron transfer occurs through a bridging ligand that is simultaneously part of the coordination sphere of each metal ion and that the bridging group remains as part of the coordination sphere of the inert complex produced. The electron is thus "conducted" through that ligand. Rates of electron transfer are found to depend on the nature of X, and the rate varies in the order $I^- > Br^- > Cl^- > F^-$. However, for other reactions, an opposite trend is observed. There are undoubtedly several factors involved, which include F^- forming the strongest bridge but I^- being the best "conductor" for the electron being transferred because it is much easier to distort the electron cloud of I^- (it is much more polarizable and has a lower electron affinity). Therefore, in different reactions these effects may take on different weights leading to variations in the rates of electron transfer that do not follow a particular order with respect to the identity of the anion.

20.11 REACTIONS IN SOLID COORDINATION COMPOUNDS

To this point, several types of rate processes that occur in solutions have been described. However, the study of reactions of solid coordination compounds has yielded a large amount of information on behavior in these materials. Several types of reactions of solid complexes are known, but the discussion here will be limited to four common types of processes.

20.11.1 Anation

The most common reaction exhibited by coordination compounds is ligand substitution. Part of this chapter has been devoted to describing these reactions and the factors that affect their rates. In the solid state, the most common reaction of a coordination compound occurs when the compound is heated and a volatile ligand is driven off. When this occurs, another electron pair donor attaches at the vacant site. The donor may be an anion from outside the coordination sphere or it may be some other ligand that changes bonding mode. When the reaction involves an anion entering the coordination sphere of the metal, the reaction is known as *anation*. One type of anation reaction that has been extensively studied is illustrated by the equation

$$[Cr(NH_3)_5H_2O]Cl_3(s) \xrightarrow{\Delta} [Cr(NH_3)_5Cl]Cl_2(s) + H_2O(g) \tag{20.99}$$

in which the volatile ligand, H_2O, is lost and replaced by a Cl^- ion, which was originally an anion located elsewhere in the lattice (which is the origin of the term anation). Many reactions of this type have been studied for complexes containing several different metals.

In an interesting study on loss of water from pentaammineaquaruthenium(III) complexes, the kinetic analysis was performed by following the mass loss from the complexes as described in Chapter 8. The rate law used to model the process was

$$\log[M_0/(M_0 - M_t)] = kt \tag{20.100}$$

where M_0 is the initial mass of the sample and M_t is the mass of the sample at time t. It was found that the activation energies for the reaction

$$[Ru(NH_3)_5H_2O]X_3(s) \rightarrow [Ru(NH_3)_5X]X_2(s) + H_2O(g) \tag{20.101}$$

increase for several anions in the order $NO_3^- < Cl^- < Br^- < I^-$. As a result of the difference in rate when different anions are present, there is said to be an "anion effect" for this reaction. Table 20.2 shows some of the pertinent kinetic data for the reactions of the ruthenium complexes as well as those for the corresponding chromium and cobalt complexes.

The data shown in Table 20.2 were interpreted in terms of two possible mechanisms:
Mechanism I ("S_N1"):

$$[Ru(NH_3)_5H_2O]X_3(s) \xrightarrow{slow} [Ru(NH_3)_5]X_3(s) + H_2O(g) \tag{20.102}$$

$$[Ru(NH_3)_5]X_3(s) \rightarrow [Ru(NH_3)_5X]X_2(s) \tag{20.103}$$

Table 20.2 Kinetic Data for the Anation Reactions of Aqua Complexes[a]

Complex	E_a (kJ mol^{-1})	$\Delta S'$ (e.u.)	$10^4 k$ (sec^{-1}) (T, °C)
$[Cr(NH_3)_5H_2O]Cl_3$	110.5	−2.53	2.41 (65)
$[Cr(NH_3)_5H_2O]Br_3$	124.3	9.2	2.43 (76)
$[Cr(NH_3)_5H_2O]I_3$	136.8	15.4	1.61 (82)
$[Cr(NH_3)_5H_2O](NO_3)_3$	101.7	−2.49	1.38 (55)
$[Co(NH_3)_5H_2O]Cl_3$	79	−	4.27 (86)
$[Co(NH_3)_5H_2O]Br_3$	108	−	4.79 (85)
$[Co(NH_3)_5H_2O](NO_3)_3$	130	−	2.51 (85)
$[Ru(NH_3)_5H_2O]Cl_3$	95.0	−7.1	1.12 (43)
$[Ru(NH_3)_5H_2O]Br_3$	97.9	−5.2	0.77 (40)
$[Ru(NH_3)_5H_2O]I_3$	111.7	5.8	0.71 (40)
$[Ru(NH_3)_5H_2O](NO_3)_3$	80.8	−15.9	2.38 (41)

[a]Data from A. Ohyoshi, et al. (1975).

Mechanism II ("S$_N$2"):

$$[Ru(NH_3)_5H_2O]X_3(s) \xrightarrow{\text{slow}} [Ru(NH_3)_5H_2OX]X_2(s) \tag{20.104}$$

$$[Ru(NH_3)_5H_2OX]X_2(s) \rightarrow [Ru(NH_3)_5X]X_2(s) + H_2O(g) \tag{20.105}$$

In proposed Mechanism I, the loss of water from the complex is the rate-determining step, but removal of water from the coordination sphere of the metal ion should be independent of the nature of the anion that is not part of the coordination sphere of the metal ion. On the other hand, if Mechanism II is correct, the entry of X into the coordination sphere of the metal would be dependent on the nature of the anion because different anions would be expected to enter the coordination sphere at different rates. Because there is an observed anion effect, it was concluded that the anation reaction must be an S$_N$2 process. However, it is not clear how a process can be second order when both the complex cation *and* the anion are parts of the same formula. As discussed in Chapter 8, it is not always appropriate to try to model reactions in solids by the same kinetic schemes that apply to reactions in solutions.

For an ion to leave a lattice site in order to enter the coordination sphere of the metal would require the formation of a Schottky defect. The energy required to form this type of defect can be expressed by the equation

$$E_s = U \frac{1 - \left(1 - \dfrac{1}{\varepsilon}\right)}{A\left(1 - \dfrac{1}{n}\right)} \tag{20.106}$$

where U is the lattice energy, ε is the dielectric constant, A is the Madelung constant, and n is the exponent in the repulsion term in the Born–Landé equation. The energy predicted for creating the defect and the loss of crystal field stabilization energy in the complex would make the expected activation energy very high. Moreover, the lattice energy *decreases* as the size of the anion increases, but the activation energy is higher for larger anions.

A more realistic view of anation reactions is provided by considering the volatile ligand to be lost from the metal as the initial step. However, the ligand must go *somewhere*, and, in a crystal, the logical place is in *interstitial* positions. From there, the free volatile ligand must make its way out of the crystal. It is easy to show that for a solid lattice there is more free space as the difference between the sizes of the anions and cations increases. Therefore, given the large cation present, the activation energies would be expected to increase as the size of the anion increases, which is exactly the trend observed. The activation energy for loss of water from the nitrate complex does not follow the trend, but that is probably because nitrate is not a spherical ion. The planar NO_3^- ion should allow water to move through the crystal more easily than would spherical ions, which is consistent with the observed activation energies. In this scheme, the liberated volatile ligand creates a *defect* that *diffuses* through the crystal lattice. Numerous anation reactions in which a volatile ligand is lost have been found to give rates in accord with predictions based on the defect diffusion mechanism.

A unique type of anation occurs when the loss of water is followed by coordination of a ligand that is already part of the coordination sphere of another metal. A reaction of this type is

$$[Co(NH_3)_5H_2O][Co(CN)_6] \xrightarrow{140\ °C} [(NH_3)_5Co\text{-}NC\text{-}Co(CN)_5] + H_2O \qquad (20.107)$$

which leads to the formation of a cyanide bridge between two Co^{3+} ions. Because Co^{3+} with five NH_3 ligands is a hard Lewis acid and the same metal ion with cyanide ligands behaves as a soft Lewis acid, the Co−NC−Co linkage shown in the equation matches the hard−soft character of the ends of the cyanide ion. However, the reaction

$$[Co(NH_3)_5CN]^{2+} + [Co(CN)_5H_2O]^{2-} \rightarrow [(NH_3)_5Co\text{-}CN\text{-}Co(CN)_5] + H_2O \qquad (20.108)$$

leads to the formation of a CN bridge in which the softer end (C) of the ligand is bound to the harder of the Co^{3+} ions. Even though both cobalt ions have a +3 charge, the cobalt surrounded by CN^- ligands behaves as if it were soft due to the symbiotic effect. A reaction of this type has been investigated in which two different metals are involved. The reaction can be shown as

$$[Co(NH_3)_5H_2O][Fe(CN)_6](s) \rightarrow [(NH_3)_5Co\text{-}NC\text{-}Fe(CN)_5](s) + H_2O(g) \qquad (20.109)$$

Similar reactions have been carried out to produce complexes that contain thiocyanate bridges. The examples given are just a few among the large number of anation reactions.

20.11.2 Racemization

In Chapter 16, we described enantiomorphism in coordination compounds. An optically active compound of this type can sometimes be converted into the racemic mixture even in the solid state. In some cases the reaction is induced thermally, but racemization has also been brought about by high pressure. Although other types of complexes can exist as optical isomers, it has been the study of racemization in octahedral complexes that has been most extensive. Two types of processes have been identified: those in which ligands become detached from the metal and those in which there is no metal−ligand bond rupture (intramolecular).

If one end of a chelate ring on an octahedral complex is detached from the metal, the five-coordinate transition state can be considered as a fluxional molecule in which there is some interchange of positions. When the chelate ring reforms, it may be with a different orientation that could lead to racemization. If the chelate ring is not symmetrical (such as 1,2-diaminopropane rather than ethyl-enediamine), isomerization may also result. For reactions carried out in solvents that coordinate well, a solvent molecule may attach to the metal where one end of the chelating agent vacated. Reactions of this type are similar to those in which dissociation and substitution occur.

Numerous racemization reactions occur in which there is apparently no bond breaking as the transition state forms. Although there have been various schemes proposed to account for these observations, the two most likely and important models involve twist mechanisms. In a regular octahedral structure, C_3 axes pass through the structure and exit from the center of the triangular faces in the "upper" and "lower" halves of the structure. If one triangular face is rotated relative to the other around a C_3 axis, a trigonal prism results. However, for a trischelate complex there are two ways (actually there are four but most indications are that two are most important) in which the rings (that remain attached to the metal) can be oriented. In one way, the rings span the rectangular faces along the three longitudinal edges. This "trigonal twist" is known as the *Bailar twist* (Bailar, 1958). In a second type of trigonal twist, the transition state is a trigonal prism in which one of the chelating groups lies along the longitudinal edge of a rectangular face but the other two are attached along edges where the triangular faces join the rectangular faces. This mechanism is known as the *Ray and Dutt twist*. These mechanisms are illustrated in Figure 20.11.

If forming the transition state by a twist mechanism leads to some expansion of the lattice, it would be expected that a higher lattice energy would hinder racemization. In a very detailed study (Kutal and Bailar, 1972) of the racemization of $(+)$ - $[Co(en)_3]X_3 \cdot nH_2O$ (where $X = Cl^-$, Br^-, I^-, or SCN^-), it was found that the rate varied with the nature of the anion. The rate was found to decrease in the order $I^- > Br^- > SCN^- >> Cl^-$. It was also found that the degree of hydration of the complex affected the rate of racemization. Generally, the hydrated complexes reacted faster than anhydrous samples. It was also found that reducing the particle size increased the rate of racemization, but when the iodide compound was heated with water in a sealed tube, the racemization was *slower* than for the hydrated solid from which the water could escape. The fact that the hydrated samples racemized faster could indicate that an aquation–anation mechanism is involved, but the results obtained in the sealed tube experiments do not agree with that idea.

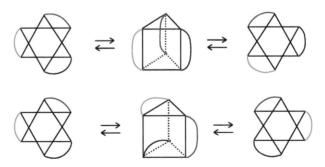

FIGURE 20.11
The Bailar (top) and Ray–Dutt (bottom) trigonal twist mechanisms. Note the difference in the orientation of the chelate rings in the trigonal prism transition states.

If the formation of the transition state constitutes a point defect that requires slight expansion of the lattice, the rate would be slowest for the chloride compound because the smaller ion would cause the lattice energy to be higher. The rate should be highest for the iodide compound because the large anion would lead to a lower lattice energy. The variation in the order of racemization with the nature of the anion agrees with this assessment. The fact that racemization is more rapid during the loss of water could be the result of slight lattice expansion as water diffuses through the crystal making it easier to form the transition state. Although the results will not be described here, the mechanism based on defects and diffusion is consistent with other racemization reactions (see O'Brien, 1983).

20.11.3 Geometrical Isomerization

Reactions in which isomerization of coordination compounds occur in solutions are common, and some reactions of this type in solid complexes have been studied. Generally, there is a change in color of the complex as the crystal field environment of the metal ion changes. Accordingly, some of the color changes that occur when complexes are heated *may* indicate isomerization, but very few geometrical isomerization reactions in solid complexes have been studied in detail. One such reaction is

$$\textit{trans-}\left[Co(NH_3)_4Cl_2\right]IO_3 \cdot 2H_2O(s) \xrightarrow{75-90°C} \textit{cis-}\left[Co(NH_3)_4Cl_2\right]IO_3(s) + 2 H_2O(g) \qquad (20.110)$$
$$\text{green} \qquad\qquad\qquad\qquad\qquad\qquad \text{violet}$$

Several interesting observations have been made on this reaction. The rate of isomerization was found to be the same as the rate of dehydration. All attempts to dehydrate the starting complex by conventional techniques were found to lead to isomerization. On the basis of this and other evidence, the mechanism proposed involves aquation in the complex followed by anation. In this process, water first displaces Cl^- in the coordination sphere and is then displaced by the Cl^-, possibly by an S_N1 mechanism. A trigonal bipyramid transition state could account for the Cl^- reentering the coordination sphere to give a *cis* product. The rate law for this reaction is of the form

$$-\ln (1 - \alpha) = kt + c \qquad (20.111)$$

(where c is a constant), and it is the first-order rate law described earlier (LeMay and Bailar, 1967).

Other compounds that undergo isomerization include *trans-*$[Co(pn)_2Br_2](H_5O_2)Br_2$, *trans-*$[Co(pn)_2Cl_2](H_5O_2)Br_2$, and *trans-*$[Co(pn)_2Br_2](H_5O_2)Cl_2$. All these contain the hydrated proton in the form of $H_5O_2^+$. When these compounds are heated, they lose water and hydrogen halide molecules as well as being converted to the *cis* form.

Isomerization involving a square planar complex is also known. Due to the *trans* effect, it easier to synthesize the *trans* isomer of many complexes than it is to prepare the *cis* complex. The following reactions lead to the formation of an unusual platinum complex:

$$[(C_2H_4)PtCl_3]^- + CO \rightarrow [PtCl_3CO]^- + C_2H_4 \qquad (20.112)$$

$$[PtCl_3CO]^- + RNH_2 \rightarrow \textit{trans-}[PtCl_2(CO)RNH_2] + Cl^- \qquad (20.113)$$

The *trans* compound melts at approximately 90 °C, and continued heating leads to isomerization to the *cis* structure. Geometrical isomerizations can also lead to a change in structure of the complex.

For example, a change from square planar to tetrahedral structure has been observed for the complex $[Ni(P(C_2H_5)(C_6H_5)_2)_2Br_2]$.

20.11.4 Linkage Isomerization

The oldest case of a linkage isomerization that has been studied in detail is

$$[Co(NH_3)_5ONO]Cl_2 \xrightarrow{\Delta} [Co(NH_3)_5NO_2]Cl_2 \qquad (20.114)$$
$$\underset{\text{red}}{} \qquad\qquad \underset{\text{yellow}}{}$$

which was investigated in the 1890s by Jørgensen and since that time by many other workers. This reaction takes place both in solutions and in the solid state, and it was discussed in some detail in Chapter 16 and earlier in this chapter.

Linkage isomerization reactions involving cyanide complexes have been known for many years. Generally, these reactions take place when a complex is prepared with a bonding mode that has the carbon end of CN^- bound to one metal ion and another metal ion is added that is forced to bond to the nitrogen end. If the metals have hard–soft characteristics that do not match the hard–soft character of CN^- (the carbon atom is the softer end), it is possible to induce a change in bonding mode, whereby the CN^- "flips" to give the appropriate match of hard–soft character. For example, when a solution containing $K_3[Cr(CN)_6]$ (the cyanide is C-bonded to Cr^{3+}) has added to it a solution containing Fe^{2+}, a solid product having the formula $KFe[Cr(CN)_6]$ is obtained. When this compound is heated at 100 °C for a short time, the color changes from brick red to green, but there is also another significant change. An absorption peak is observed at 2168 cm^{-1} in the infrared spectrum for the CN stretching vibration in $KFe[Cr(CN)_6]$. In fact, the CN stretching band is observed at approximately this position for almost all cyanide complexes in which the CN^- is bonded to a first row transition metal having a +3 charge. After heating the solid, the infrared band at 2168 cm^{-1} decreases in intensity, and a new band is seen at 2092 cm^{-1}. If the heating is continued, the band at 2168 cm^{-1} disappears, and only the band at 2092 cm^{-1} is observed. The position of this band is characteristic of CN stretching vibrations in which the CN is bonded to a metal ion with a +2 charge. What has happened is that the CN^- ions have changed bonding mode (referred to as a cyanide flip).

$$Cr^{3+} - CN - Fe^{2+} \rightarrow Cr^{3+} - NC - Fe^{2+} \qquad (20.115)$$

In this equation, only one of the CN^- ligands is shown. The product actually contains a three-dimensional network in which CN^- ions bridge between Cr^{3+} and Fe^{2+}. However, in this arrangement, the *harder* electron donor, nitrogen, is bonded to the *softer* metal ion, Fe^{2+}, and the *softer* electron donor, carbon, is bonded to the *harder* metal ion, Cr^{3+}. The fact that CN^- originally bonds to Cr^{3+} through the carbon end is not surprising because of the negative formal charge on that atom. However, when solid $KFe[Cr(CN)_6]$ is heated, the cyanide ions "flip" in order to give the arrangement that is in accord with the hard–soft interaction principle, which favors $Cr^{3+} - NC - Fe^{2+}$ linkages. Other reactions of this type have been observed in which a second row metal is involved. For example, when a solution containing Cd^{2+} is added to one containing $K_3[Cr(CN)_6]$, the solid product is $KCd[Cr(CN)_6]$. In that case, the Cd^{2+} ions are forced to bond to the nitrogen end of the cyanide ions because the carbon atoms are already bonded to Cr^{3+} and the solid contains $Cr^{3+} - CN - Cd^{2+}$ linkages. When heated, a cyanide flip occurs to give $Cr^{3+} - NC - Cd^{2+}$ linkages in accord with the hard–soft interaction principle. However, based on the intensities of the CN stretching bands, only two-thirds of the cyanide ions change bonding mode. Presumably, this is because Cd^{2+} has a coordination number of four, whereas that of Cr^{3+} is six, so two cyanides do not flip.

A somewhat unusual linkage isomerization occurs when $K_4[Ni(NO_2)_6] \cdot H_2O$ is heated. This complex is easily prepared by adding an excess of KNO_2 to a solution containing Ni^{2+}, and it has the NO_2^- ions coordinated to Ni^{2+} through the nitrogen atoms. Solid $K_4[Ni(NO_2)_6] \cdot H_2O$ exhibits two absorption bands in the infrared spectrum at 1325 cm^{-1} and 1347 cm^{-1} that are assigned to the N—O vibrations in $-NO_2$ linkages. When the compound is heated in the solid state to remove the water of hydration, new bands appear in the infrared spectrum at 1387 cm^{-1} and 1206 cm^{-1}. These absorptions are characteristic of M—ONO linkages. On the basis of the reduction in ligand field strength, which is determined spectroscopically, the product after dehydration is formulated as $K_4[Ni(NO_2)_4(ONO)_2]$. Therefore, the dehydration and linkage reaction can be written as

$$K_4[Ni(NO_2)_6] \cdot H_2O(s) \xrightarrow{\Delta} K_4[Ni(NO_2)_4(ONO)_2](s) + H_2O(g) \qquad (20.116)$$

The enthalpy change has been determined to be +14.6 kJ mol^{-1} for each Ni—NO_2 converted to Ni—ONO. This value is somewhat higher than the value of 8.62 kJ mol^{-1} required for conversion of $[Co(NH_3)_5NO_2]Cl_2$ to $[Co(NH_3)_5ONO]Cl_2$ in the solid state. Co^{3+} is a hard Lewis acid and attaching five NH_3 ligands leaves it as a hard acid. On the other hand, Ni^{2+} is a borderline acid, but the addition of six NO_2^- ligands, which are soft and capable of π bonding, indicates that the Ni^{2+} in the complex is a soft acid as a result of the symbiotic effect. Therefore, it would be expected that changing the bonding mode of NO_2^- to Ni^{2+} would require a greater amount of energy than it would when the soft NO_2^- ligand is bound to the hard Co^{3+}.

Adding a solution containing Ag^+ to one containing $[Co(CN)_6]^{3-}$ produces a solid in which there are $Ag^+-N\equiv C-Co^{3+}$ linkages, because the carbon end of the cyanide ions are attached to the Co^{3+} ion.

$$K_3[Co(CN)_6] + 3\,AgNO_3 \xrightarrow{H_2O} Ag_3[Co(CN)_6] \cdot 16H_2O + 3\,KNO_3 \qquad (20.117)$$

The product has a mismatch of hard—soft character because the Ag^+ is soft, but the nitrogen end of CN^- is the harder electron pair donor. For the $Ag^+-N\equiv C-Co^{3+}$ linkages the CN stretching band is observed at 2128 cm^{-1}. When $Ag_3[Co(CN)_6] \cdot 16H_2O$ is heated, dehydration occurs and a new band appears at 2185 cm^{-1} that is characteristic of $Ag^+-C\equiv N-Co^{3+}$ linkages. When $Ag_3[Co(CN)_6] \cdot 16H_2O$ is allowed to stand at room temperature for 24 h in ambient light, partial dehydration occurs, and the infrared spectrum shows that complete linkage isomerization takes place. A kinetic study of the isomerization showed that the isomerization follows a one-dimensional contraction, R1, rate law (see Chapter 8). The linkage isomerization reaction in $Ag_3[Co(CN)_6]$ can also be induced by ultrasound. The linkage isomerization reactions show the importance of the hard—soft interaction principle in predicting the stable bonding modes.

The chemistry of coordination compounds is a vast field that encompasses many kinds of work. So much of the field is concerned with solution chemistry that it is easy to forget that a great deal is known about solid state chemistry of coordination compounds. The brief survey presented shows that a great deal is known about some of the reactions, but there is much that needs to be done before many others will be understood.

References for Further Study

Atwood, J.E., 1997. *Inorganic and Organometallic Reaction Mechanisms*, 2nd ed. Wiley-VCH, New York. An excellent book on mechanistic inorganic chemistry.

Bailar, J.C., Jr., 1958. *J. Inorg. Nucl. Chem. 8*, 165. The paper describing trigonal twist mechanisms for racemization.

Basolo, F., Pearson, R.G., 1967. *Mechanisms of Inorganic Reactions*, 2nd ed. John Wiley, New York. A classic reference in reaction mechanisms in coordination chemistry.

Cosmano, R.J., House, J.E., 1975. *Thermochim. Acta, 13*, 127−131. Paper describing the thermally induced linkage isomerization in KCd[Fe(CN)$_6$].

Dopke, N.C., Oemke, H.E., 2011. *Inorg. Chim. Acta, 376*, 638−640. An article describing the use of microwaves in synthesis of platinum complexes.

Espenson, J.H., 1995. *Chemical Kinetics and Reaction Mechanisms*, 2nd ed. McGraw-Hill, New York. A book on chemical kinetics much of which is devoted to reactions of coordination compounds. Highly recommended.

Fogel, H.M., House, J.E., 1988. *J. Thermal Anal., 34*, 231−238. The paper describing the thermally induced changes in trans-dichlorotetramminecobalt(III) complexes.

Gray, H.B., Langford, C.H., 1968. *Chem. Eng. News*, April 1, p. 68. A excellent survey article, "Ligand Substitution Dynamics," that presents elementary concepts clearly.

House, J.E., 1980. *Thermochim. Acta, 38*, 59. A discussion of reactions in solids and the role of free space and diffusion.

House, J.E., 2007. *Principles of Chemical Kinetics*, 2nd ed. Academic Press/Elsevier, San Diego. A kinetics book that presents a discussion of several types of reactions in the solid state as well as solvent effects.

House, J.E., Bunting, R.K., 1975. *Thermochim. Acta, 11*, 357−360.

House, J.E., Kob, N.E., 1993. *Inorg. Chem., 32*, 1053. A report on the linkage isomerization in KCd[Fe(CN)$_6$] induced by ultrasound.

House, J.E., Kob, N.E., 1994. *Transition Metal Chemistry, 19*, 31. A report on isomerization in Ag$_3$[Co(CN)$_6$].

Kutal, C., Bailar, J.C., Jr., 1972. *J. Phys. Chem., 76*, 119. An outstanding paper on the racemization of a complex in the solid state.

LeMay, H.E., Bailar, Jr., J.C., 1967. *J. Am. Chem. Soc., 89*, 5577.

O'Brien, P., 1983. *Polyhedron, 2*, 223. An excellent review of racemization reactions of coordination compounds in the solid state.

Taube, H., 1970. *Electron Transfer Reactions of Complex Ions in Solution*, Academic Press, New York. One of the most significant works on the subject of electron transfer reactions.

Wilkins, R.G., 1991. *Kinetics and Mechanisms of Reactions of Transition Metal Complexes*, VCH Publishers, NY. Contains a wealth of information on reactions of coordination compounds.

QUESTIONS AND PROBLEMS

1. The reaction of two moles of P(C$_2$H$_5$)$_3$ with K$_2$[PtCl$_4$] produces a product having a different structure than does the reaction of two moles of N(C$_2$H$_5$)$_3$. Show the structure for the product in each case and explain the difference in the reactions.

2. The Pt−Cl stretching band is observed at 314 cm^{-1} in *cis*-[Pt((CH$_3$)$_3$As)$_2$Cl$_2$] but in the *trans* isomer it is seen at 375 cm^{-1}. Explain the difference in band position for the two isomers.

3. Explain why most substitution reactions involving tetrahedral complexes of transition metal ions take place rapidly.

4. Explain why substitution reactions of tetrahedral complexes of Be^{2+} are slow.

5. Predict the product for each of the following substitution reactions.
   (a) [PtCl$_3$NH$_3$]$^-$ + NH$_3$ →
   (b) [PtCl$_3$NO$_2$]$^{2-}$ + NH$_3$ →
   (c) [Pt(NH$_3$)$_3$Cl]$^+$ + CN$^-$ →

6. Suppose the series of complexes *trans*-[Pt(NH$_3$)$_2$LCl] is prepared where L = NH$_3$, Cl$^-$, NO$_2^-$, Br$^-$, or pyridine. If the position of the Pt–Cl stretching band is determined for each product, what will be the order of decreasing wave number? Explain your answer.

7. Consider the linkage isomerization Cr^{3+}–CN–Fe^{2+} → Cr^{3+}–NC–Fe^{2+} described in the text. Show by using ligand field theory (magnitudes of Dq) that this process should be energetically favorable.

8. The reaction *cis*-[Co(en)$_2$BrCl] with H$_2$O gives a product that is 100% *cis* isomer. Explain what this indicates about the mechanism.

9. Why is the rate of H$_2$O exchange in the aqua complex of Cr^{3+} about 100 times that for the aqua complex of Rh^{3+}?

10. The reaction of *trans*-[Ir Cl(CO)(P(p-C$_6$H$_4$Y)$_3$)$_2$] with H$_2$ at 30 °C depends on the nature of the group Y in the phenyl group. For several ligands, the $\Delta H^\ddagger$ and $\Delta S^\ddagger$ vary as follows:

Y =	OCH$_3$	CH$_3$	H	F	Cl
$\Delta H^\ddagger$, kcal mol^{-1}	6.0	4.3	10.8	11.6	9.8
$\Delta S^\ddagger$, e.u.	−39	−45	−23	−22	−28

Test these data to determine whether an isokinetic relationship exists and comment on the results.

11. It is found that the rate of substitution reaction between Mn(CO)$_5$Br and As(C$_6$H$_5$)$_3$ varies somewhat with the solvent. The rate constant at 40 °C when the solvent is cyclohexane is 7.44×10^{-8} sec^{-1} and when the solvent is nitrobenzene it is 1.08×10^{-8} sec^{-1}. In light of the principles described in Chapter 6, what does this observation indicate about the mechanism of the reaction? What would you expect a reasonable value for the rate constant to be if the solvent is chloroform? See Table 6.7.

12. The ligand diethylenetriamine, H$_2$NCH$_2$CH$_2$NHCH$_2$CH$_2$NH$_2$, forms stable complexes with Pt^{2+}. When the complexes [Pt(dien)X]$^+$ undergo reaction with pyridine, the rate constants for X = Cl$^-$, I$^-$, and NO$_2^-$ are 3.5×10^{-5}, 1.0×10^{-5}, and 5.0×10^{-8} sec^{-1}, respectively. What is the mechanism for the substitution reactions? Provide an explanation for the difference in rates of the reactions. What would be a reasonable estimate for the rate constant when X = Br$^-$? Explain your answer.

13. By referring to Table 17.1, calculate the ligand field activation energy for a square planar complex of Pd^{2+} undergoing substitution by an associative process that has a trigonal bipyramid transition state. If the complex in question has Dq = 1,200 cm^{-1}, what would be the activation energy in kJ/mol if only ligand field effects are considered?

14. When *trans*-[Rh(en)$_2$Cl$_2$]$^+$ undergoes substitution of the Cl$^-$ ions by two ligands Y, the rate constants when Y = I$^-$, Cl$^-$, or NO$_2^-$ are 5.2×10^{-5}, 4.0×10^{-5}, and 4.2×10^{-5}, respectively, with unspecified units. Discuss the mechanism of the substitution processes. What would be a reasonable value for k when Y = NH$_3$?

15. Compare the rates and products obtained for the following reactions.

    trans-[Pt(H$_2$O)$_2$ClI] + X →

    cis-[Pt(NH$_3$)$_2$ClBr] + X →

    cis-[Pt(NO$_2$)$_2$ClBr] + X →

    trans-[Pt(NH$_3$)(H$_2$O)Cl$_2$] + X →

16. Explain the difference with respect to the size of the neighboring groups on substitution in an octahedral complex by associative and dissociative mechanisms.

17. Although substitutions in many tetrahedral complexes are rapid, when the metal ion is Be^{2+} this is not the case. Explain why this is so.

18. How would you expect the rates of substitution reactions in complexes of Mg^{2+} and Al^{3+} to compare? Provide an explanation for your answer.

19. Suppose an octahedral complex of a d^3 ion undergoes substitution by a dissociative pathway. What are the two possible transition states? What is the difference in ligand field activation energy for the two possible pathways? Suppose the solvent interacts with the *less* energetically favorable transition state (on the basis of ligand field activation energy) with a solvation energy that is 25 kJ mol^{-1} *more* favorable than it does with the other transition state. If there are only ligand field and solvation energies involved, what value of Dq for the complex would make the two pathways have the same activation energy? Discuss the implications of this situation with regard to mechanisms of substitution reactions. How would the situation be different for complexes of third row metals compared to those of first row metals?

20. On the basis of ligand field effects, would it be easier to form a trigonal bipyramid transition state from a square planar or a tetrahedral starting complex?

21. Although the ligands contain nitrogen donor atoms in both cases, the rate of reaction of $[Ni(NH_3)_4]^{2+}$ is over 20 times as great as that of $[Ni(en)_2]^{2+}$. Explain this observation.

22. Why would the rate of exchange of chloride be different when the complex is $[Co(NH_3)_5Cl]^{2+}$ rather than $[Co(NH_2CH_3)_5Cl]^{2+}$? Which would undergo more rapid exchange of Cl^-?

23. How would the volume of activation and the entropy of activation be useful when deciding whether a substitution reaction follows a dissociative or interchange mechanism?

24. The aquation of $[Cr(H_2O)_5F]^{2+}$ is accelerated at lower pH, but the aquation of $[Cr(H_2O)_5NH_3]^{2+}$ is not. Explain this difference.

25. Why is the rate at which ligands leave the coordination sphere of low spin Fe^{2+} complexes different than from those of high spin Fe^{2+}?

26. The substitution reactions of $[ML_5A]^{3+}$, $[ML_5B]^{3+}$, and $[ML_5C]^{3+}$ gave $\Delta H^\ddagger$ and $\Delta S^\ddagger$ values as follows (kJ/mol and e.u.): 23.2 and -8; 30.3 and 9; 26.5 and -1. Is it likely that the reactions follow the same mechanism?

27. If *cis*-$[Co(NH_3)_4(H_2O)_2]_2(SO_4)_3 \cdot 3H_2O(s)$ is heated, what would be the most likely change to take place first? If heating is continued at higher temperature, what other changes could occur? Sketch structures of the product at each stage of decomposition.

28. Suppose the complexes having the formula *trans*-$[Pt(NH_3)_2LCl]$ (where $L = NH_3$, Cl^-, Br^-, NO_2^-, or py) are studied by infrared spectroscopy to determine the position of the Pt—Cl stretching bands. For the different ligands, L, what would be the order of increasing wave number for the bands?

29. When $KCd[Fe(CN)_6]$ is first prepared, there is an absorption band at 2155 cm^{-1}. After heating the solid at $100\,°C$ for a few minutes, a new band is seen at 2065 cm^{-1}. Explain what gives rise to the band in the starting material and what changes are indicated by the band at 2065 cm^{-1}.

30. Consider a complex $ML_6^{z\pm}$ in which M is a $+3$ metal ion. If substitution takes place in an S_N2 process, what would be the likely order of increasing rate for the following entering ligands: Cl^-, CN^-, Br^-, NH_3, and NCS^-? Explain your answer.

CHAPTER 21

Complexes Containing Metal–Carbon and Metal–Metal Bonds

In previous chapters, we have presented a great deal of information about structure and bonding in coordination compounds. This chapter will be devoted to describing some of the important chemistry in the broad areas of organometallic complexes and those in which there are metal–metal bonds. The body of literature on each of these topics is enormous, so the coverage here will include only basic concepts and a general survey.

It is important to realize that there is a great deal of overlap in the topics covered in this chapter. For example, the chemistry of metal carbonyls is intimately related to that of metal–alkene complexes because both types of ligands are soft bases and many complexes contain both carbonyl and olefin ligands. Also, both areas are closely associated with catalysis by complexes discussed in Chapter 22 because some of the best-known catalysts are metal carbonyls and they involve the reactions of alkenes. Therefore, the separation of topics applied is certainly not a clear one. Catalysis by metal complexes embodies much of the chemistry of both metal carbonyls and metal–alkene complexes.

21.1 BINARY METAL CARBONYLS

One of the very interesting series of coordination compounds consists of metal atoms bonded to carbon monoxide, the metal carbonyls. A remarkable characteristic of these compounds is that because the ligands are neutral molecules, the metals are present in the zero oxidation state. Although the discussion at first will be limited to the binary compounds containing only metal and CO, many mixed complexes are known that contain both CO and other ligands. Depending on the net charge on the other ligands, the metal may or may not be in the zero oxidation state. In any event, the metals in metal carbonyls occur in low oxidation states because of the favorable interaction of the soft ligands (Lewis bases) with the metals in low oxidation states (soft Lewis acids).

Generally, the metals that form stable carbonyl complexes are those in the first transition series from V to Ni, in the second row from Mo to Rh, and in the third row from W to Ir. There are several reasons for these being the metals most often found in carbonyl complexes. First, these metals have one or more d orbitals that are not completely filled so they can accept electron pairs from σ electron donors. Second, the d orbitals contain some electrons that can be involved in back donation to the π^* orbitals on the CO ligands. Third, the metals are generally in the zero oxidation state or at least a low oxidation state so they behave as soft Lewis acids making it favorable to bond to a soft Lewis base such as CO.

707

Inorganic Chemistry. DOI: http://dx.doi.org/10.1016/B978-0-12-385110-9.00021-2

The first metal carbonyl prepared was $Ni(CO)_4$, which was obtained by Mond in 1890. This extremely toxic compound was prepared by first reducing nickel oxide with hydrogen

$$NiO + H_2 \xrightarrow{\quad 400\,°C \quad} Ni + H_2O \tag{21.1}$$

then treating the Ni with CO.

$$Ni + 4\,CO \xrightarrow{\quad 100\,°C \quad} Ni(CO)_4 \tag{21.2}$$

Because $Ni(CO)_4$ is volatile (b. p. 43 °C) and cobalt will not react under these conditions, this process afforded a method for separating Ni from Co by the process now known as the *Mond process*. Although there are many complexes known that contain both carbonyl and other ligands (mixed carbonyl complexes), the number containing only a metal and carbonyl ligands is small. They are known as *binary metal carbonyls* and are listed in Table 21.1, and the structures of most of these compounds are shown in Figures 21.1–21.3.

The binary metal carbonyls are named by giving the name of the metal followed by the name "carbonyl" with the number of carbonyl groups indicated by the appropriate prefix. For example, $Ni(CO)_4$ is nickel tetracarbonyl, whereas $Cr(CO)_6$ is chromium hexacarbonyl. If more than one metal atom is present, the number is indicated by a prefix to indicate the number. Thus, $Co_2(CO)_8$ is dicobalt octacarbonyl, and $Fe_2(CO)_9$ has the name diiron nonacarbonyl.

The effective atomic number (EAN) rule (also known as the 18-electron rule) was described briefly in Chapter 16, but we will consider it again here because it is very useful when discussing carbonyl and olefin complexes. The composition of stable binary metal carbonyls is largely predictable by the EAN rule or the "18-electron rule," as it is also known. Stated in the simplest terms, the EAN rule predicts that a metal in the zero or other low oxidation state will gain electrons from a sufficient number of ligands so that the metal will achieve the electron configuration of the next noble gas. For the first row transition metals, this means the krypton configuration with a total of 36 electrons.

Table 21.1 Binary Metal Carbonyls

Mononuclear		Dinuclear		Polynuclear	
Compound	m. p., °C	Compound	m. p., °C	Compound	m. p., °C
$Ni(CO)_4$	−25	$Mn_2(CO)_{10}$	155	$Fe_3(CO)_{12}$	140 (d)
$Fe(CO)_5$	−20	$Fe_2(CO)_9$	100 (d)	$Ru_3(CO)_{12}$	−
$Ru(CO)_5$	−22	$Co_2(CO)_8$	51	$Os_3(CO)_{12}$	224
$Os(CO)_5$	−15	$Rh_2(CO)_8$	76	$Co_4(CO)_{12}$	60 (d)
$Cr(CO)_6$	subl.	$Tc_2(CO)_{10}$	160	$Rh_4(CO)_{12}$	150 (d)
$Mo(CO)_6$	subl.	$Re_2(CO)_{10}$	177	$Ir_4(CO)_{12}$	210 (d)
$V(CO)_6$	70 (d)	$Os_2(CO)_9$		$Rh_6(CO)_{16}$	200 (d)
$W(CO)_6$	subl.	$Ir_2(CO)_8$		$Ir_6(CO)_{16}$	−
				$Os_5(CO)_{16}$	
				$Os_6(CO)_{18}$	

(d) indicates decomposition.

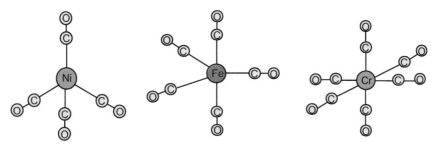

FIGURE 21.1
The structures of the mononuclear carbonyls of nickel, iron, and chromium.

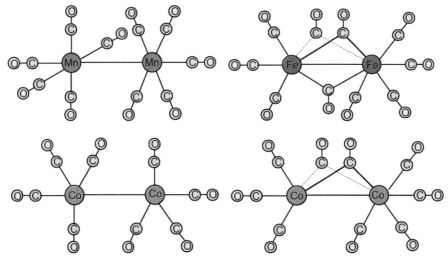

FIGURE 21.2
The structures of the binary metal carbonyls containing Mn, Fe, and Co.

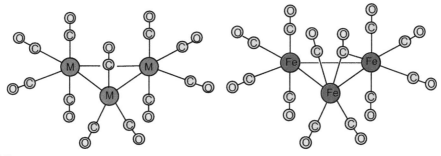

FIGURE 21.3
Structures of trinuclear metal carbonyls. In the structure on the left, M represents Ru or Os.

Actually, there is ample evidence to indicate that the rule is followed in many other cases as well. For example, Zn^{2+} has 28 electrons and receiving eight more electrons (two electrons each from four ligands) as in $[Zn(NH_3)_4]^{2+}$ results in 36 electrons being around the Zn^{2+}. Similarly, Co^{3+} has 24 electrons and six pairs of electrons from six ligands will raise the number to 36 (as in $[Co(CN)_6]^{3-}$). The vast majority of complexes containing Co^{3+} follow this trend. In the case of Cr^{3+}, the ion has 21 electrons and gaining six pairs results in there being only 33 electrons around the chromium ion. For complexes of metal ions, the EAN rule is followed much more frequently by those metals in the latter part of the transition series.

The example of Cr^{3+} mentioned above shows that the "rule" is by no means always true. However, for complexes of metals in the zero oxidation state containing soft ligands, such as CO, PR_3, alkenes, etc., there is a strong tendency for the stable complexes to be those containing the number of ligands predicted by the EAN rule. Because the Ni atom has 28 electrons, we should expect eight more electrons to be added from four ligands. Accordingly, the stable nickel carbonyl is $Ni(CO)_4$. For chromium(0), which contains 24 electrons, the stable carbonyl is $Cr(CO)_6$ as expected from the EAN rule. The six pairs of electrons from the six CO ligands bring the number of electrons around the chromium atom to 36.

The Mn atom has 25 electrons. Adding five carbonyl groups would raise the number to 35, leaving the atom one electron short of the krypton configuration. If the single unpaired electron on one manganese atom is then allowed to pair up with an unpaired electron on another atom to form a metal–metal bond, we have the formula $(CO)_5Mn–Mn(CO)_5$ or $[Mn(CO)_5]_2$, which is the formula for a manganese carbonyl that obeys the EAN rule.

Cobalt has 27 electrons and adding eight more electrons (one pair each from four CO ligands) brings the total to 35. Accordingly, forming a metal–metal bond using the remaining unpaired electrons would give $[Co(CO)_4]_2$ or $Co_2(CO)_8$ as the stable carbonyl compound. As we shall see, there are two different structures possible for $Co_2(CO)_8$, both agreeing with the EAN rule. Based on the EAN rule, it might be expected that $Co(CO)_4$ would not be stable, but the species $Co(CO)_4^-$ obtained by adding an electron to $Co(CO)_4$ would be stable. Actually, the $Co(CO)_4^-$ ion is well known and several derivatives containing it have been prepared.

The EAN rule also applies to carbonyls of metal clusters. For example, in $Fe_3(CO)_{12}$, there are eight electrons in the valence shells of the three Fe(0) atoms, which gives a total of 24. Then, the 12 CO ligands contribute an electron pair each, which gives another 24 electrons and brings the total to 48. This gives an average of 16 electrons per Fe atom, and the deficit must be made up by the metal–metal bonds. Because each atom contributes one electron to the bond, two bonds to other Fe atoms are required to give 18 electrons around each atom. As will be seen from the structure shown in Figure 21.3, the structure of $Fe_3(CO)_{12}$ has the three Fe atoms in a triangle with each bonded to two other Fe atoms. In a similar way, $Co_4(CO)_{12}$ has 12 CO ligands that contribute two each or a total of 24 electrons and four Co atoms that contribute 9 each or a total of 36 electrons. From both sources, the number of electrons available is 60, but for four metal atoms the number is 15 each. Therefore, to have 18 each would require each Co atom to form bonds to three others, and such a structure is shown in Figure 21.4, in which the four Co atoms are arranged in a tetrahedron.

Although the EAN rule is not always followed, it does offer a logical basis for predicting the composition of many complexes. The arguments presented here can, of course, be extended to second and third row transition elements. The EAN rule is also of use in predicting the formulas of stable complexes containing carbonyl groups and other ligands, as might result from ligand substitution reactions. This is particularly important when a ligand (such as an alkene containing several double bonds) can donate different numbers of electrons depending on the number needed by the metal to achieve the configuration of the next noble gas.

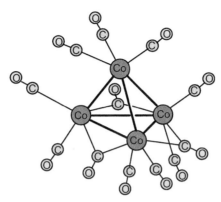

FIGURE 21.4
The structure of [Co$_4$(CO)$_{12}$], which is also the structure of the analogous rhodium and iridium compounds.

21.2 STRUCTURES OF METAL CARBONYLS

Mononuclear metal carbonyls contain only one metal atom, and they have comparatively simple structures. For example, nickel tetracarbonyl is tetrahedral. The pentacarbonyls of iron, ruthenium, and osmium are trigonal bipyramidal, whereas the hexacarbonyls of vanadium, chromium, molybdenum, and tungsten are octahedral. These structures are shown in Figure 21.1. The structures of the dinuclear metal carbonyls (containing two metal atoms) involve either metal–metal bonds or bridging CO groups, or both. For example, the structure of Fe$_2$(CO)$_9$, diiron nonacarbonyl, contains three CO ligands that form bridges between the iron atoms, and each iron atom also has three other CO groups attached only to that atom.

Carbonyl groups that are attached to two metal atoms simultaneously are called *bridging* carbonyls, whereas those attached to only one metal atom are referred to as *terminal* carbonyl groups. The structures of Mn$_2$(CO)$_{10}$, Tc$_2$(CO)$_{10}$, and Re$_2$(CO)$_{10}$ actually involve only a metal–metal bond, so the formulas are more correctly written as (CO)$_5$M–M(CO)$_5$. Two isomers are known for Co$_2$(CO)$_8$. One has a metal–metal bond between the cobalt atoms, and the other has two bridging CO ligands and a metal–metal bond. Figure 21.2 shows the structures of the dinuclear metal carbonyls. The structure for Mn$_2$(CO)$_{10}$ is sometimes drawn to appear as if four CO ligands form a square plane around the Mn atom. In fact, the four CO groups do not lie in the same plane as the Mn atom. Rather, they lie about 12 pm from the plane containing the Mn on the side opposite the metal–metal bond.

Structures of trinuclear, tetranuclear, and higher compounds are best thought of as metal clusters containing either metal–metal bonds or bridging carbonyl groups. In some cases, both types of bonds occur. The structures of Ru$_3$(CO)$_{12}$ and Os$_3$(CO)$_{12}$ are shown in Figure 21.3. The structure of Fe$_3$(CO)$_{12}$, also shown in Figure 21.3, contains a trigonal arrangement of Fe atoms, but it contains a bridging carbonyl group.

Carbonyls containing four metal atoms in the cluster are formed by Co, Rh, and Ir in which the metal atoms are arranged in a tetrahedron with 12 CO ligands, nine in terminal positions and three that are bridging. These structures are shown in Figure 21.4.

21.3 BONDING OF CARBON MONOXIDE TO METALS

Carbon monoxide has the valence bond structure shown as

$$|C\equiv O|$$

with a triple bond between C and O. The formal charge on the oxygen atom is +1, whereas that on the carbon atom is −1. Although the electronegativity of oxygen is considerably higher than that of carbon, these formal charges are consistent with the dipole moment being small (0.12 D), with the carbon being at the negative end of the dipole. The carbon end of the CO molecule is thus a softer electron donor, and it is the carbon atom that is bound to the metal. The molecular orbital energy diagram for CO is shown in Figure 3.9. The bond order (B.O.) in the molecule is given in terms of the number of electrons in bonding orbitals (N_b) and the number in antibonding orbitals (N_a) by

$$\text{B.O.} = \frac{N_b - N_a}{2} \tag{21.3}$$

For CO, this gives a bond order of $(8 - 2)/2 = 3$, a triple bond. For gaseous CO in which a triple bond exists, the C–O stretching band is observed at 2143 cm^{-1}. However, typically, in metal carbonyls the C–O stretching band is seen at 1850–2100 cm^{-1} for terminal CO groups. The shift of the CO stretching band upon coordination to metals reflects a slight reduction in the bond order, resulting from back donation of electron density from the metal to the CO. However, bridging carbonyl groups normally show an absorption band in the 1700 to 1850 range. The spectrum of $Fe_2(CO)_9$ shows prominent bands in these positions because both terminal and bridging CO ligands are present.

The molecular orbitals of CO are populated as shown in Figure 3.9, and the molecule is diamagnetic. However, the next higher unpopulated molecular orbitals are the π^* orbitals. By virtue of accepting several pairs of electrons from the ligands, the metal in a metal carbonyl acquires a negative formal charge. Thus, in $Ni(CO)_4$, the formal charge on the metal is −4. In order to remove part of this negative charge from the metal, electron density is donated back from the metal to the π^* orbitals on the ligands. The π^* orbitals on the CO have appropriate symmetry to effectively accept this electron density, resulting in a slight increase in the character of the metal–CO bond and a slight reduction from triple bond character in the CO ligand. The orbital interaction that permits back donation to occur can be shown as follows:

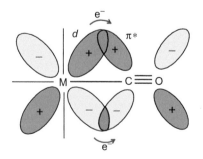

The valence bond approach to this multiple bonding can be shown in terms of the resonance structures.

$$M-C\equiv O \leftrightarrow M=C=O$$

In $Fe(CO)_5$, the formal charge on iron is -5, and in $Cr(CO)_6$, the formal charge on chromium is -6. We should expect that the back donation would be more extensive in either of these compounds than it is in the case of $Ni(CO)_4$. Because the greater back donation results in a greater reduction in C–O bond order, the infrared spectra of these compounds should show this effect. The positions of the CO stretching bands for these compounds are as follows:

$Ni(CO)_4$	2057 cm^{-1}
$Fe(CO)_5$	2034 cm^{-1}
$Cr(CO)_6$	1981 cm^{-1}

A similar reduction in C–O bond order occurs for an isoelectronic series of metal carbonyls as the oxidation states of the metals are reduced. Consider the positions of the C–O stretching bands in the following carbonyl species.

$Mn(CO)_6^+$	2090 cm^{-1}
$Cr(CO)_6$	1981 cm^{-1}
$V(CO)_6^-$	1859 cm^{-1}

In this series, the metal has a progressively greater negative charge that is partially relieved by back donation.

These data show that there is, in fact, more reduction in C–O bond order because the metal has a higher negative formal charge. There is a corresponding increase in metal–carbon stretching frequency, showing the increased tendency to form multiple metal–carbon bonds to relieve part of the negative charge.

Bonding of CO in a bridging position between two metal atoms is most appropriately described by considering the carbon to be double bonded to oxygen. If the carbon is considered to be hybridized sp^2, a σ bond between carbon and oxygen results from the overlap of a carbon sp^2 orbital with an oxygen atom p orbital. The π bond to oxygen results from the overlap of an unhybridized p orbital on the carbon atom with a p orbital on the oxygen atom. This leaves the remaining two sp^2 orbitals to bond simultaneously to the two metal atoms. As illustrated below, the C–O bond in a bridging CO is more like a double bond in a ketone than the triple bond in gaseous CO.

Stretching bands for bridging CO are found in the region characteristic of ketones, about 1700 to 1800 cm^{-1}. Accordingly, the infrared spectrum of Fe$_2$(CO)$_9$ shows absorption bands at 2000 cm^{-1} (terminal carbonyl stretching) and 1830 cm^{-1} (bridging carbonyl stretching). For most compounds containing carbonyl bridges, the C–O stretching band is seen around 1850 cm^{-1}. A great deal of what is known about the bonding in metal carbonyls has been determined by means of the various spectroscopic techniques, especially infrared spectroscopy. However, it is important to remember that the donation of a pair of electrons in forming the σ bond also has some effect on the carbonyl stretching vibration.

Although we have described bonding in terminal and bridging positions, CO is known to bond to metals in other ways. Some of the other types of linkages can be shown as follows:

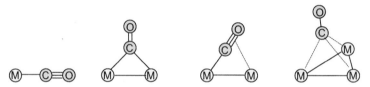

Further changes in C–O stretching vibrations occur when other ligands are present. For example, the CO stretching band in Cr(CO)$_6$ is found at 2000 cm^{-1}, whereas that in Cr(NH$_3$)$_3$(CO)$_3$ is found at approximately 1900 cm^{-1}. The structure of the latter compound has NH$_3$ *trans* to each CO ligand.

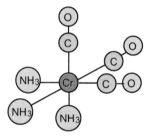

In this case, the product is the *fac* isomer in which all NH$_3$ ligands are *trans* to the CO molecules. Ammonia does not form π bonds to metals because it has no orbitals of suitable energy or symmetry to accept electron density. Thus, the back donation from Cr in Cr(NH$_3$)$_3$(CO)$_3$ goes to only three CO molecules, and the bond order is reduced even more than it is in Cr(CO)$_6$, in which back donation occurs equally to six CO molecules. There is, of course, an increase in Cr–C bond order and stretching frequency in Cr(NH$_3$)$_3$(CO)$_3$ compared to Cr(CO)$_6$. Based on the study of many mixed carbonyl complexes, it is possible to compare the ability of various ligands to accept back donation. When this is done, it is found that the ability to accept back donation decreases in the order

$$NO^+ > CO > PF_3 > AsCl_3 > PCl_3 > As(OR)_3 > P(OR)_3 > PR_3$$

The position of NO$^+$ should not be surprising because it has not only π^* orbitals of appropriate symmetry but also a positive charge.

The chemical behavior of metal carbonyls is influenced by the nature of other ligands present. A decrease in C–O CO BO MC BO results from an increase in M–C bond order. If other ligands are present that cannot accept electron density, more back donation to CO occurs so that the M−C bond will be stronger and substitution reactions leading to replacement of CO will be retarded. If other ligands are present that are good π acceptors, less back donation to the CO groups occurs. They will be labilized and substitution will be enhanced.

21.4 PREPARATION OF METAL CARBONYLS

As we have described previously, $Ni(CO)_4$ can be prepared directly by the reaction of nickel with carbon monoxide. However, most of the binary metal carbonyls listed in Table 21.1 cannot be obtained by this type of reaction. A number of preparative techniques have been used to prepare metal carbonyls, and a few general ones will be described here.

21.4.1 Reaction of a Metal with Carbon Monoxide
The reactions of carbon monoxide with Ni and Fe proceed rapidly at low temperature and pressure.

$$Ni + 4\,CO \rightarrow Ni(CO)_4 \tag{21.4}$$

$$Fe + 5\,CO \rightarrow Fe(CO)_5 \tag{21.5}$$

For most other metals, high temperatures and pressures are required to produce the metal carbonyl. By this direct reaction, $Co_2(CO)_8$, $Mo(CO)_6$, $Ru(CO)_5$, and $W(CO)_6$ have been prepared when suitable conditions are used.

21.4.2 Reductive Carbonylation
This type of reaction involves reducing a metal compound in the presence of CO. The reducing agents may include a variety of materials depending on the particular synthesis being carried out. For example, in the synthesis of $Co_2(CO)_8$, hydrogen is used as the reducing agent.

$$2\,CoCO_3 + 2\,H_2 + 8\,CO \rightarrow Co_2(CO)_8 + 2\,H_2O + 2\,CO_2 \tag{21.6}$$

Lithium aluminum hydride, $LiAlH_4$, has been used as the reducing agent in the preparation of $Cr(CO)_6$ from $CrCl_3$. Reduction by metals such as Na, Mg, or Al has also been used, as shown by the preparation of the $V(CO)_6^-$ ion:

$$VCl_3 + 4\,Na + 6\,CO \xrightarrow[\text{high pressure}]{\text{Diglyme, 100 °C}} [(diglyme)_2Na][V(CO)_6] + 3\,NaCl \tag{21.7}$$

From the reaction mixture, $V(CO)_6$ is obtained by hydrolyzing the product with H_3PO_4 and subliming $V(CO)_6$ at 45 to 50 °C. The reduction of CoI_2 can be used as a route to $Co_2(CO)_8$.

$$2\,CoI_2 + 8\,CO + 4\,Cu \rightarrow Co_2(CO)_8 + 4\,CuI \tag{21.8}$$

21.4.3 Displacement Reactions

Certain metal carbonyls have been prepared by the reaction of metal compounds directly with CO owing to the fact that CO is a reducing agent. For example,

$$2\ IrCl_3 + 11\ CO \rightarrow Ir_2(CO)_8 + 3\ COCl_2 \tag{21.9}$$

and

$$Re_2O_7 + 17\ CO \rightarrow Re_2(CO)_{10} + 7\ CO_2 \tag{21.10}$$

21.4.4 Photochemical Reactions

Photolysis of $Fe(CO)_5$ leads to partial removal of CO to produce diiron nonacarbonyl.

$$2\ Fe(CO)_5 \overset{h\nu}{\rightarrow} Fe_2(CO)_9 + CO \tag{21.11}$$

21.5 REACTIONS OF METAL CARBONYLS

Metal carbonyls undergo reactions with a great many compounds to produce mixed carbonyl complexes. A large number of these reactions involve the replacement of one or more carbonyl groups by a substitution reaction. Such reactions have also been studied kinetically in some cases.

21.5.1 Substitution Reactions

Many substitution reactions occur between metal carbonyls and other potential ligands. For example,

$$Cr(CO)_6 + 2\ py \rightarrow Cr(CO)_4(py)_2 + 2\ CO \tag{21.12}$$

$$Ni(CO)_4 + 4\ PF_3 \rightarrow Ni(PF_3)_4 + 4\ CO \tag{21.13}$$

$$Mo(CO)_6 + 3\ py \rightarrow Mo(CO)_3(py)_3 + 3\ CO \tag{21.14}$$

Substitution reactions of metal carbonyls frequently indicate differences in the bonding characteristics of ligands. In the case of $Mn(CO)_5Br$, radiochemical tracer studies have shown that only four CO groups undergo exchange with ^{14}CO.

$$Mn(CO)_5Br + 4\ ^{14}CO \rightarrow Mn(^{14}CO)_4(CO)Br + 4\ CO \tag{21.15}$$

The structure of $Mn(CO)_5Br$ can be shown as

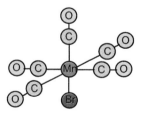

and the four CO molecules that undergo exchange reactions are those in the plane, which are all *trans* to each other. This indicates that the CO *trans* to Br is held more tightly because Br does not compete for π bonding electron density donated from Mn. In the case of the other four CO groups, competition between the groups, which are all good acceptors, causes the groups to be labilized. It was mentioned earlier that $Mn(CO)_3(py)_3$ results from the reaction of $Mn(CO)_6$ with py. The structure of $Mn(CO)_3(py)_3$ is

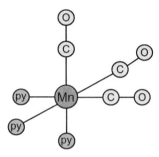

which has all three CO ligands *trans* to py. This structure maximizes the π donation to the three CO groups and reflects the differences in the ability of CO and py to accept electron density in π bonding with metal d orbitals. In cases in which good π acceptors are the entering ligands, all the CO groups may be replaced as shown in the reaction of $Ni(CO)_4$ with PF_3. These substitution reactions show that the π acceptor properties of CO influence the replacement reactions.

21.5.2 Reactions with Halogens

Reactions of metal carbonyls with halogens lead to the formation of carbonyl halide complexes by substitution reactions or breaking metal−metal bonds. The reaction

$$[Mn(CO)_5]_2 + Br_2 \rightarrow 2\,Mn(CO)_5Br \tag{21.16}$$

involves the rupture of the Mn−Mn bond, and one Br is added to each Mn. In the reaction

$$Fe(CO)_5 + I_2 \rightarrow Fe(CO)_4I_2 + CO \tag{21.17}$$

one CO from the iron is replaced by two iodine atoms so that the coordination number of the iron is increased to six. The formulas for these carbonyl halides obey the EAN rule.

The reaction of CO with some metal halides results in the direct formation of metal carbonyl halides as illustrated in the following examples:

$$PtCl_2 + 2\,CO \rightarrow Pt(CO)_2Cl_2 \tag{21.18}$$

$$2\,PdCl_2 + 2\,CO \rightarrow [Pd(CO)Cl_2]_2 \tag{21.19}$$

The structure of $[Pd(CO)Cl_2]_2$ is

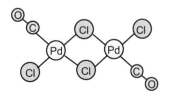

In this type of bridged structure, halogens serve as the bridging groups, and the bonding environment around each Pd atom is essentially square planar. Such compounds usually react by breaking the bridges followed by addition or substitution of ligands.

21.5.3 Reactions with NO

The nitric oxide molecule has one unpaired electron residing in an antibonding π^* molecular orbital. When that electron is removed, the bond order increases from 2.5 to 3, so in coordinating to metals, NO usually behaves as though it donates three electrons. The result is formally the same as if one electron were lost to the metal,

$$NO \rightarrow NO^+ + e^- \tag{21.20}$$

followed by coordination of NO^+, which is isoelectronic with CO and CN^-. Because NO^+ is the nitrosyl ion, the products containing nitric oxide and carbon monoxide are called *carbonyl nitrosyls*. The following reactions are typical of those producing this type of compound:

$$Co_2(CO)_8 + 2\,NO \rightarrow 2\,Co(CO)_3NO + 2\,CO \tag{21.21}$$

$$Fe_2(CO)_9 + 4\,NO \rightarrow 2\,Fe(CO)_2(NO)_2 + 5\,CO \tag{21.22}$$

$$[Mn(CO)_5]_2 + 2\,NO \rightarrow 2\,Mn(CO)_4NO + 2\,CO \tag{21.23}$$

It is interesting to note that the products of these reactions obey the EAN rule. Cobalt has 27 electrons and it acquires six from the three CO ligands and three from NO, which gives a total of 36. It is easy to see that $Fe(CO)_2(NO)_2$ and $Mn(CO)_4NO$ also obey the EAN rule. Because NO is considered as a donor of three electrons, two NO groups usually replace three CO ligands. This may not be readily apparent in some cases because metal–metal bonds are broken in addition to the substitution reactions.

21.5.4 Disproportionation

A number of metal carbonyls undergo disproportionation reactions in the presence of other coordinating ligands. For example, in the presence of amines, $Fe(CO)_5$ reacts as follows:

$$2\,Fe(CO)_5 + 6\,Amine \rightarrow [Fe(Amine)_6]^{2+}[Fe(CO)_4]^{2-} + 6\,CO \tag{21.24}$$

This reaction takes place because of the ease of formation of the carbonylate ions and the favorable coordination of the Fe^{2+} produced. The reaction of $Co_2(CO)_8$ with NH_3 is similar.

$$3Co_2(CO)_8 + 12\,NH_3 \rightarrow 2[Co(NH_3)_6][Co(CO)_4]_2 + 8\,CO \tag{21.25}$$

Formally, in each of these cases, the disproportionation produces a positive metal ion and a metal ion in a negative oxidation state. The carbonyl ligands will be bound to the softer metal species, the anion, and the nitrogen donor ligands (hard Lewis bases) will be bound to the harder metal species, the cation. These disproportionation reactions are quite useful in the preparation of a variety of carbonylate complexes. For example, the $[Ni_2(CO)_6]^{2-}$ ion can be prepared by the reaction

$$3\,Ni(CO)_4 + 3\,phen \rightarrow [Ni(phen)_3][Ni_2(CO)_6] + 6\,CO \tag{21.26}$$

The range of coordinating agents that will cause disproportionation is rather wide and includes compounds such as isocyanides, RNC:

$$Co_2(CO)_8 + 5\ RNC \rightarrow [Co(CNR)_5][Co(CO)_4] + 4\ CO \tag{21.27}$$

21.5.5 Carbonylate Anions

We have already seen that several carbonylate anions, such as $Co(CO)_4^-$, $Mn(CO)_5^-$, $V(CO)_6^-$, and $Fe(CO)_4^{2-}$, obey the EAN rule. One type of synthesis of these ions is that of allowing the metal carbonyl to react with a reagent that loses electrons readily, a strong reducing agent. Active metals are strong reducing agents, so the reactions of metal carbonyls with alkali metals should produce carbonylate ions. The reaction of $Co_2(CO)_8$ with Na carried out in liquid ammonia at $-75\ °C$ is one such reaction.

$$Co_2(CO)_8 + 2\ Na \rightarrow 2\ Na[Co(CO)_4] \tag{21.28}$$

Similarly,

$$Mn_2(CO)_{10} + 2\ Li \xrightarrow{THF} 2\ Li[Mn(CO)_5] \tag{21.29}$$

Although $Co(CO)_4$ and $Mn(CO)_5$ do not obey the EAN rule, the anions $Co(CO)_4^-$ and $Mn(CO)_5^-$ do.

A second type of reaction leading to the formation of carbonylate anions is the reaction of metal carbonyls with strong bases. For example,

$$Fe(CO)_5 + 3\ NaOH \rightarrow Na[HFe(CO)_4] + Na_2CO_3 + H_2O \tag{21.30}$$

$$Cr(CO)_6 + 3\ KOH \rightarrow K[HCr(CO)_5] + K_2CO_3 + H_2O \tag{21.31}$$

With $Fe_2(CO)_9$, the reaction is

$$Fe_2(CO)_9 + 4\ OH^- \rightarrow Fe_2(CO)_8^{2-} + CO_3^{2-} + 2\ H_2O \tag{21.32}$$

21.5.6 Carbonyl Hydrides

Some reactions of carbonyl hydrides will be illustrated in Chapter 22. Such species are involved in catalytic processes in which metal carbonyls function as hydrogenation catalysts. Generally, carbonyl hydrides are obtained by acidifying solutions containing the corresponding carbonylate anion or by the reactions of metal carbonyls with hydrogen. The following reactions illustrate these processes.

$$Co(CO)_4^- + H^+(aq) \rightarrow HCo(CO)_4 \tag{21.33}$$

$$[Mn(CO)_5]_2 + H_2 \rightarrow 2\ HMn(CO)_5 \tag{21.34}$$

The preparation of $Na[HFe(CO)_4]$ was shown in Eqn (22.30), and that compound can be acidified to give $H_2Fe(CO)_4$.

$$Na[HFe(CO)_4] + H^+(aq) \rightarrow H_2Fe(CO)_4 + Na^+(aq) \tag{21.35}$$

In general, the carbonyl hydrides are slightly acidic as is illustrated by the behavior of $H_2Fe(CO)_4$.

$$H_2Fe(CO)_4 + H_2O \rightarrow H_3O^+ + HFe(CO)_4^- \quad K_1 = 4 \times 10^{-5} \tag{21.36}$$

$$HFe(CO)_4^- + H_2O \rightarrow H_3O^+ + Fe(CO)_4^{2-} \quad K_2 = 4 \times 10^{-14} \tag{21.37}$$

As might be expected on the basis of the hard–soft interaction principle (see Chapter 9), large soft cations form insoluble compounds with these anions. Accordingly, the anions form precipitates upon reacting with Hg^{2+}, Pb^{2+}, and Ba^{2+}.

A few metal carbonyl hydrides can be prepared by direct reaction of the metal with CO and H_2, with the following being a typical reaction:

$$2\,Co + 8\,CO + H_2 \xrightarrow[\text{50 atm}]{150\,°C} 2\,HCo(CO)_4 \tag{21.38}$$

We now describe the structure and bonding in some of these compounds. As expected on the basis of HSAB, other ligands (and the metals) are soft. In general, two hydrogen atoms replace one CO ligand, or for metals that have an odd number of electrons, hydrogen contributes one bringing the number to 36 when the appropriate number of CO ligands is also present. As a result of manganese having 25 electrons, the addition of 10 electrons from five CO ligands and one from H brings the total to 36. Therefore, one stable carbonyl hydride of Mn is $HMn(CO)_5$, which has a structure in which the four carbonyl groups in the basal plane lie slightly below the manganese.

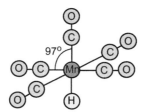

This compound behaves as a weak acid that dissolves in water to give $Mn(CO)_5^-$ by the reaction:

$$HMn(CO)_5 + H_2O \leftrightharpoons H_3O^+ + Mn(CO)_5^- \tag{21.39}$$

Other carbonyl hydrides include $H_2Fe(CO)_5$, $H_2Fe_3(CO)_{11}$, and $HCo(CO)_4$.

Some carbonyl hydrides, such as $[Cr_2(CO)_{10}H]^-$, have the hydrogen atom in bridging positions as shown in the following structure.

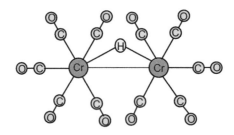

Two hydrogen bridges are present in some carbonyl hydride complexes. This is illustrated by the structure of the $[H_2W_2(CO)_8]^{2-}$ anion,

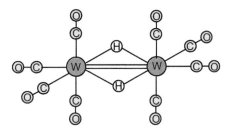

In this structure, the W—W distance is shorter than that of a typical single bond so it is shown as a double bond. As would be expected, this is also the structure of the $[H_2Re(CO)_8]$ complex, which is isoelectronic with $[H_2W_2(CO)_8]^{2-}$.

21.6 STRUCTURE AND BONDING IN METAL–ALKENE COMPLEXES

Alkene complexes of metals constitute one of the important classes of coordination compounds. In general, the metals are found in low oxidation states owing to the more favorable interaction with the soft π electron donors. The majority of metal–alkene complexes contain other ligands as well, although many complexes are known which contain only the metal and organic ligands. As we will see later, metal complexes containing olefins are important in explaining how some of the complexes behave as catalysts.

Probably the first metal–alkene complex was Zeise's salt, $K[Pt(C_2H_4)Cl_3]$ or the bridged compound $[PtCl_2(C_2H_4)]_2$. These compounds were first prepared by Zeise in about 1825. The palladium analogs of these compounds are also now known. A large number of metal–alkene complexes are known and some of the chemistry of these materials will be described here.

Because Zeise's salt has been known for so many years, it is not surprising that the structural features of the compound have been the subject of a great deal of study. Although this topic was introduced in Chapter 16, we will return to this important topic here because many of the ideas are useful when describing bonding in other organometallic compounds containing alkenes. In the anion of Zeise's salt, the ethylene molecule is perpendicular to the plane containing the Pt^{2+} and the three Cl^- ions as shown in Figure 21.5(a). As first proposed by Dewar in the early 1950s, a π orbital of the C_2H_4 molecule serves as the electron pair donor giving the usual σ bond to the metal as shown in

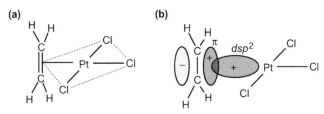

FIGURE 21.5
The bonding of C_2H_4 to Pt in Zeise's salt.

Figure 21.5(b). However, the π^* orbitals in C_2H_4 are empty and can accept electron density back donated from the metal. Relieving the negative formal charge on the metal is enhanced by this π^* orbital having suitable orientation and symmetry to interact effectively with the nonbonding d_{xz} orbital on the metal ion. Thus, there is a significant amount of multiple bond character to the metal–ligand bond. The π bond between the metal and ligand can be shown as the overlap of the π^* orbitals in C_2H_4 with a d orbital on the metal. The bonding in other olefin complexes is also described in this way.

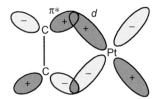

The typical C=C bond length is approximately 134 pm, and in the anion of Zeise's salt, the C–C bond length is 137.5 pm. Although there is *slight* lengthening of the bond in ethylene, it is not as much as would accompany the formation of two metal–ligand σ bonds. Therefore, when the bonding in this complex is considered in terms of resonance structures, it can be represented as

M ——|| I   M < II

Structure II contributes very little to the bonding. However, the C–C stretching band is shifted by about 100 cm^{-1} (to lower a wave number) as a result of back donation weakening the bond. Because alkenes are soft electron donors (and π acceptors), they exert a strong *trans* effect (see Chapter 20) that is manifested in the anion of Zeise's salt by causing the Pd–Cl bond *trans* to ethylene to be about 234 pm in length, whereas those in *cis* positions are 230 pm. Although there seems to be little contribution from a structure containing two σ bonds in the anion of Zeise's salt, that is not the case when the alkene is $(CN)_2C=C(CN)_2$, tetracyanoethylene and the other ligands are triphenylphosphine. In that case, the structure can be shown as

ϕ_3P Pt   152 pm
ϕ_3P 210 pm

In this case, the bond between the carbon atoms is about the same as it is for a C–C single bond. Moreover, unlike the anion of Zeise's salt, the carbon atoms are in the plane formed by platinum and the other ligands. Clearly, this represents a significant difference from the usual alkene complexes. In essence, a three-membered C–P–C ring is formed. It appears in this case that the ability of the triphenylphosphine ligands to form π bonds and the changes in the alkene π bond produced by four CN groups result in metal–carbon σ bonds being formed.

In $[PtCl_2(C_2H_4)]_2$, the olefin double bond also lies perpendicular to the plane of the other groups. In this bridged compound, two chloride ions function as the bridging groups and the ethylene molecules are *trans* to each other as shown in the following structure.

It is to be expected that two ligands that give very strong *trans* effect would be located opposite a ligand such as Cl$^-$, which has a weaker *trans* effect.

Some of the most interesting olefin complexes of metals are those containing CO and olefins as ligands. Moreover, metal carbonyls are frequently the starting complexes from which olefin complexes are obtained by substitution. The EAN rule enables predictions to be made regarding the total number of electrons used by polyene ligands in complexes. Essentially, each double bond that coordinates to the metal functions as an electron pair donor. A ligand that can exhibit several bonding sites is cyclohepta-1,3,5-triene (cht), which contains three double bonds that are potential electron pair donors.

In the complexes Ni(CO)$_3$(cht), Fe(CO)$_3$(cht), and Cr(CO)$_3$(cht) shown in Figure 21.6, the cht ligand bonds in different ways. In the first of these complexes, Ni has 28 electrons, so to satisfy the EAN rule, it

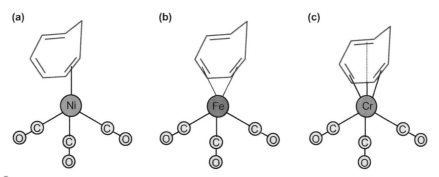

FIGURE 21.6
The ways in which cycloheptatriene functions as an electron donor.

needs to gain eight electrons. Because six electrons come from three CO molecules, only one double bond in cht will be bonded to Ni. In the Fe and Cr complexes, six electrons also come from the three CO molecules. However, according to the EAN rule, iron needs to gain a total of 10 electrons, so the cht will coordinate using two double bonds. In the case of the chromium complex, all three double bonds will be coordinated in order to satisfy the EAN rule (see Figure 21.6).

The "bonding capacity" of a ligand is called its *hapticity*. To distinguish between the three bonding modes of cht shown in Figure 21.6, the hapticity of the ligand is indicated by the term hapto, which is designated as h or η. When an organic group is bound to a metal by only one carbon atom by electrons in a σ bond, the bonding is referred to as monohapto, and it is designated as h^1 or η^1. In most cases, we will designate the hapticity of a ligand using η. When a π bond in ethylene functions as an electron pair donor, both carbon atoms are bonded to the metal and the bond is designated as h^2 or η^2. Figure 21.6(a) shows cht that is bonded to Ni by means of one double bond that connects two carbon atoms, so the bonding is η^2 in that case. In Figure 21.6(b), two double bonds spanning four carbon atoms are functioning as electron pair donors so the bonding is considered as η^4. Finally, in Figure 21.6(c) all three double bonds are electron pair donors, so the bonding of cht is η^6. Both the formula and name of the complex have the hapticity of the ligand indicated. For example, η^2-chtNi(CO)$_3$ is named as tricarbonyldihaptocycloheptatrienenickel(0). The formula η^4-chtFe(CO)$_3$ for the iron complex shown above is named as tricarbonyltetrahaptocycloheptatrieneiron(0).

Other organic ligands can be bound in more than one way also. For example, the allyl group can be bound by a σ bond between the metal and one carbon atom (η^1) or as a π donor encompassing all three carbon atoms (η^3).

$$CH_2 = CH - CH_2 - Mn(CO)_5$$

η^1-allyl

H$_2$C
/
HC —— Mn(CO)$_4$
\
H$_2$C

η^3-allyl

In Chapter 22, it will be shown that some change in bonding mode (probably η^1 to η^3 and the reverse) is believed to be one step in the isomerization of 1-butene catalyzed by metal complexes. In general, the change in bonding mode of the allyl group accompanies the loss of another ligand, frequently as a result of heating or irradiation. Interestingly, a complex having two allyl groups can exist as different isomers. An example of this is shown by the two forms of Ni(η^3-C$_3$H$_5$)$_2$.

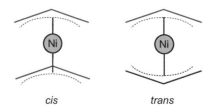

cis   trans

Butadiene can function as a donor of four electrons in which the bonding is delocalized. A complex of this type is $Fe(CO)_3(\eta^4\text{-}C_4H_8)$, which has the structure

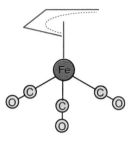

An interesting ligand is cyclooctatetraene (cot), C_8H_8, which is not aromatic because it does not follow the $4n + 2$ rule (see Chapter 5). This molecule has the structure

As a result of the iron atom having 26 electrons, gaining 10 from ligands would allow it to obey the 18-electron rule. Therefore, a stable complex between cot and iron is $Fe(cot)_2$ the structure of which can be shown as

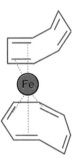

It should be noted that one cot ligand is η^4, whereas the other is η^6. So the formula is written as $Fe(\eta^6\text{-}cot)(\eta^4\text{-}cot)$. In addition to this complex that contains only one Fe atom, another example of the versatility of the cot ligand is in the complex $Fe_3(cot)_3$ in which the three Fe atoms form a triangle as in $Fe_3(CO)_{12}$. That complex has a structure in which each cot ligand is bonded in η^5 and η^3 fashion to each iron.

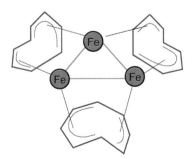

This compound, $Fe_3(\eta^5,\eta^3\text{-}C_8H_8)_3$, can be obtained from the $Fe(\eta^6\text{-}cot)(\eta^4\text{-}cot)$ shown above. Details of this structure and other complexes containing the cot ligand can be found in a recent article by Wang, Sun, Xie, King, and Schaefer (2011).

The bonding in organometallic compounds is of considerable interest and importance. If we wish to understand how the metal and ligand interact, it is appropriate to begin with a review of the bonding within the ligand and the types of orbitals that are available. In the case of the allyl group, the molecular orbitals were described in Chapter 5 by means of the Hückel molecular orbital method. The three molecular orbitals (bonding, nonbonding, and antibonding) are shown in Figure 21.7. It is with these orbitals that the metal orbitals must interact in forming a complex. With the z-axis taken as the vertical direction, the approach of the metal from that direction exposes the d_{z^2}, d_{yx}, and d_{xz} orbitals (those that project in the z direction) on the metal to the lobes of the ψ_1 orbital on the allyl group. The orbitals have same symmetry, so the bonding molecular orbital is made up of the lowest orbital on the allyl group and the d_{z^2} metal orbital. This interaction is shown in Figure 21.8(a).

There is a nodal plane in the orbital corresponding to ψ_2 at the middle carbon atom. The symmetry of the metal and ligand orbitals is appropriate for the interaction of the d_{yz} orbital on the metal to interact with ψ_2, and the interaction is shown in Figure 21.8(b). The combination of the d_{xz} orbital on the metal with the ψ_3 allyl orbital is shown in Figure 21.8(c). The bonding between other molecules with π systems and metals is similar.

Although there is a tendency for the olefin complexes to contain uncharged metals, a large number of complexes are known in which the metal ions are Pd^{2+}, Fe^{2+}, Cu^+, Ag^+, and Hg^{2+}. As we shall see, the formation of olefin complexes of these and other metals occurs as the metals catalyze certain reactions of the ligands.

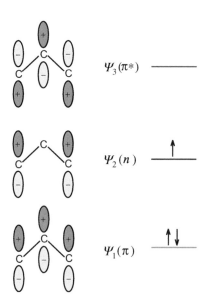

FIGURE 21.7
The molecular orbitals of the allyl species (see Section 5.6).

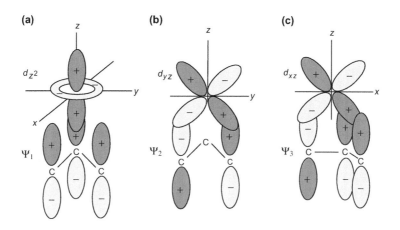

FIGURE 21.8
The interaction of metal d orbitals with the molecular orbitals on the allyl group. Note how the symmetry of the d orbital on the metal matches that of the allyl molecular orbital with which it interacts.

Since the early work dealing with Zeise's salt, many complexes have been prepared having the formula $[PtL(C_2H_4)X_2]$, where $L = $ quinoline, pyridine, ammonia, etc., and $X = Cl^-$, Br^-, I^-, NO_2^-, etc. Similar compounds have been prepared that contain olefins other than C_2H_4. Many of the complexes containing dienes, trienes, and tetraenes as ligands also contain carbonyl ligands. In fact, metal carbonyls are frequently starting complexes from which olefin complexes are obtained by substitution reactions.

21.7 PREPARATION OF METAL–ALKENE COMPLEXES

A number of synthetic methods are useful for preparing metal olefin complexes. A few of the more general ones will be described here, but the suggested readings given at the end of the chapter should be consulted for more details.

21.7.1 Reaction of an Alcohol with a Metal Halide

This method is one of those that can be used to prepare the ethylene complex known as Zeise's salt, $K[Pt(C_2H_4)Cl_3]$. Actually, the dimer $[Pt(C_2H_4)Cl_2]_2$ is obtained first, and the potassium salt is obtained by treating a concentrated solution of the dimer with KCl.

$$2\ PtCl_4 + 4\ C_2H_5OH \rightarrow 2\ CH_3CHO + 2\ H_2O + 4\ HCl + [Pt(C_2H_4)Cl_2]_2 \qquad (21.40)$$

$$[Pt(C_2H_4)Cl_2]_2 + 2\ KCl \rightarrow 2\ K[Pt(C_2H_4)Cl_3] \qquad (21.41)$$

Other metal complexes containing several types of unsaturated compounds as ligands can be prepared by similar reactions.

21.7.2 Reaction of a Metal Halide with an Alkene in a Nonaqueous Solvent

Reactions of this general type include the following examples:

$$2\,PtCl_2 + C_6H_5CH{=}CH_2 \xrightarrow[HC_2H_3O_2]{Glacial} [Pt(C_6H_5CH{=}CH_2)Cl_2]_2 \qquad (21.42)$$

$$2\,CuCl + CH_2{=}CH{-}CH{=}CH_2 \rightarrow [ClCuC_4H_6CuCl] \qquad (21.43)$$

21.7.3 Reaction of a Gaseous Alkene with a Solution of a Metal Halide

The classic synthesis of Zeise's salt, $K[Pt(C_2H_4)Cl_3]$ is an example of this type of reaction.

$$K_2PtCl_4 + C_2H_4 \xrightarrow[15\ days]{3\text{-}5\%\ HCl} K[Pt(C_2H_4)Cl_3] + KCl \qquad (21.44)$$

21.7.4 Alkene Substitution Reactions

The basis for this method lies in the fact that complexes of different olefins have different stabilities. For several common olefins, the order of stability of complexes analogous to Zeise's salt is

$$styrene > butadiene \approx ethylene > propene > butene$$

Thus, numerous replacement reactions are possible, of which the following is typical:

$$[Pt(C_2H_4)Cl_3]^- + C_6H_5CH{=}CH_2 \rightarrow [Pt(C_6H_5CH{=}CH_2)Cl_3]^- + C_2H_4 \qquad (21.45)$$

21.7.5 Reactions of a Metal Carbonyl with an Alkene

The reaction of $Mo(CO)_6$ with cyclooctatetraene is an example of this type of reaction.

$$Mo(CO)_6 + C_8H_8 \rightarrow Mo(C_8H_8)(CO)_4 + 2\,CO \qquad (21.46)$$

21.7.6 Reaction of a Metal Compound with a Grignard Reagent

An example of this type of reaction is that involving the preparation of $Ni(\eta^3\text{-}C_3H_5)_2$. This reaction can be shown as

$$2\,C_3H_5MgBr + NiBr_2 \rightarrow Ni(\eta^3\text{-}C_3H_5)_2 + 2\,MgBr_2 \qquad (21.47)$$

Similar reactions can be utilized to prepare allyl complexes of platinum and palladium. In this case, the product can exist in two isomers as described earlier. Analogous reactions can be used to prepare the tris-allyl complexes of several metals.

21.8 CHEMISTRY OF CYCLOPENTADIENYL AND RELATED COMPLEXES

In addition to complexes in which electron density is donated from localized σ bonds to metals, many complexes are known in which electrons in aromatic molecules are donated to metals. A brief survey of the chemistry of these interesting compounds will be presented here. However, the area is so vast that the interested reader will need to consult the references listed at the end of the chapter for further details.

Hydrogen atoms on the cyclopentadiene ring are very slightly acidic. Therefore, metallic sodium reacts with a solution of cyclopentadiene dissolved in tetrahydrofuran (THF) to liberate hydrogen, leaving sodium cyclopentadienide in solution.

$$2\,Na + 2\,C_5H_6 \rightarrow H_2 + 2\,Na^+C_5H_5^- \tag{21.48}$$

The reaction of this product with $FeCl_2$, which can be shown as

$$2\,NaC_5H_5 + FeCl_2 \rightarrow 2\,NaCl + Fe(C_5H_5)_2 \tag{21.49}$$

yields an orange solid (m. p., 173−174 °C) that is insoluble in water but soluble in organic solvents. The product, known as ferrocene, $Fe(C_5H_5)_2$, can also be produced in other ways. For example, when a solution of $FeCl_2 \cdot 4H_2O$ is added to a basic solution of cyclopentadiene, the following reaction occurs.

$$8\,KOH + 2\,C_5H_6 + FeCl_2 \cdot 4H_2O \rightarrow Fe(C_5H_5)_2 + 2\,KCl + 6\,KOH \cdot H_2O \tag{22.50}$$

As mentioned earlier, hydrogen atoms on the cyclopentadiene ring are very slightly acidic, and KOH facilitates the removal of H^+ leaving $C_5H_5^-$. Ferrocene was first produced in 1951, and it quickly became the focus of a great deal of organometallic chemistry.

Ferrocene is more correctly named as bis(η^5-cyclopentadienyl)iron(II), and the molecule has a structure that can be shown as follows:

This structure has been referred to as a "sandwich" structure. At low temperature in the solid state, the rings have the orientation shown in which they have a staggered configuration. However, the rotational barrier is very small (only about 4 kJ mol^{-1}), so there is free rotation at room temperature in the absence of crystal packing that hinders rotation. The fact that there is free rotation has been demonstrated by preparing the derivatives having one substituent on each ring and finding that only one product is obtained. If the molecule were "locked" in the staggered configuration so that free rotation did not occur, there would be three possible products having the structures shown in Figure 21.9.

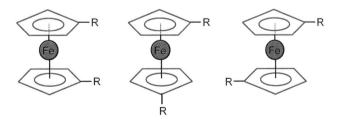

FIGURE 21.9
The products that should exist if ferrocene were prevented from having free rotation of the cyclopentadienyl ligands.

In addition to the compounds that contain two cyclopentadienyl rings, other compounds are also known in which only one ring is bonded to the metal and the other coordination sites are occupied by other groups. Such compounds are referred to as "half sandwich" structures and an example is the ion $[(\eta^5\text{-}C_5H_5)Fe(CO)_2(\text{olefin})]^+$, which has the structure

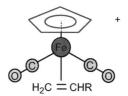

Brief mention was made in Chapter 7 regarding the application of lithium ferrite, $LiFeO_2$, in lithium batteries. The recent work by Khanderi and Schneider (2011) has involved the synthesis of lithium ferrite by decomposition of a half sandwich compound of iron that contains cyclopentadiene, cyclo-octadiene, dimethoxyethane, and lithium.

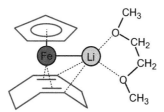

Thermal decomposition of this compound is the first step in the method to produce $LiFeO_2$. This type of study illustrates the important point that compounds synthesized in one area of chemistry often find applications in another. This will also be shown in Chapter 23 in which the use of organometallic compounds in medicine will be described.

The unusual ion $[Ni_2(C_5H_5)_3]^+$ illustrates another type of sandwich structure. This ion has the structure

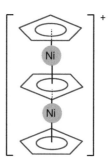

This structure is sometimes referred to as a "triple decker" sandwich.

In ferrocene and similar complexes, the $C_5H_5^-$ ion behaves as an aromatic electron donor. In order for an organic ring structure to achieve aromaticity, there must be 2, 6, or 10 electrons to form the π electron system ($4n + 2$ where $n = 0$, 1, 2, ... the number of atoms). Benzene is aromatic and forms complexes in which it behaves as if there were three double bonds that donate a total of six electrons. For cyclopentadiene, aromaticity is achieved when there is one additional electron as is the case of the anion $C_5H_5^-$. Formally, in ferrocene, the ligands are considered as negative ions and the iron as Fe^{2+}. The Fe^{2+} has 24 electrons and each ligand behaves as a donor of six electrons to Fe^{2+} giving a total of 36 electrons around Fe, in accordance with the 18-electron rule. If the iron was Fe^0 and each ligand was a donor of five electrons, the same total number of electrons around iron would result.

In ferrocene, the cyclopentadiene is bonded by the complete π system to the iron and, therefore, is η^5-C_5H_5. In other cases, $C_5H_5^-$ can bind to metals using a localized σ bond, so it is bonded as η^1. A compound of this type is $Hg(C_5H_5)_2$, which has the structure

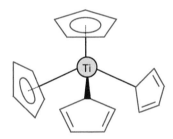

Another interesting compound that shows the different bonding ability of cyclopentadiene is $Ti(C_5H_5)_4$. In this compound, as in ferrocene, two of the cyclopentadienyl ions are coordinated as η^5-C_5H_5. The other two are bound through only one carbon atom in σ bonds to the metal (the η^1 bonding mode). This compound has the structure

Therefore, the formula for the compound is written as $(\eta^5$-$C_5H_5)_2(\eta^1$-$C_5H_5)_2Ti$, to show the difference in the bonding modes of the ligands.

Several complexes that are similar to ferrocene are known that contain other metals. They are generically referred to as metallocenes, having the general formula Mcp_2, and some of them are shown in Table 21.2.

Table 21.2 Some Common Metallocenes		
Compound	**m. p., °C**	**Color**
$Ticp_2$	—	Green
Vcp_2	167	Purple
$Crcp_2$	173	Scarlet
$Mncp_2$	172	Amber
$Fecp_2$	173	Orange
$Cocp_2$	173–174	Purple
$Nicp_2$	173–174	Green

A very large number of mixed complexes are known in which cyclopentadienyl and other ligands are present, with the EAN rule being obeyed in many cases. Cobaltocene does not obey the EAN rule, but it is easily oxidized to $(\eta^5\text{-}C_5H_5)_2Co^+$, which does obey the rule. Derivatives such as $(\eta^5\text{-}C_5H_5)Co(CO)_2$ and $(\eta^5\text{-}C_5H_5)Mn(CO)_3$ also follow the EAN rule. Because the cyclopentadienyl ion has a charge of -1, numerous complexes are known in which other negative ligands are also present. These include compounds such as $(\eta^5\text{-}C_5H_5)TiCl_2$, $(\eta^5\text{-}C_5H_5)_2CoCl$ (in which Co is considered as $+3$), and $(\eta^5\text{-}C_5H_5)VOCl_2$. A few compounds are also known in which the metal complex is an anion. For example,

$$Na^+C_5H_5^- + W(CO)_6 \rightarrow Na^+[(\eta^5\text{-}C_5H_5)W(CO)_5]^- + CO \tag{21.51}$$

Although there are several methods of preparing ferrocene and similar compounds, perhaps the most useful is that indicated in Eqn (22.49). In the second method, thallium cyclopentadienyl is prepared first by the reaction

$$C_5H_6 + TlOH \rightarrow TlC_5H_5 + H_2O \tag{21.52}$$

which shows that TlOH is a strong base. The reaction of TlC_5H_5 with $FeCl_2$ produces ferrocene.

$$FeCl_2 + 2\,TlC_5H_5 \rightarrow (\eta^5\text{-}C_5H_5)_2Fe + 2\,TlCl \tag{21.53}$$

A third method of preparing metallocenes involves the reaction of a metal halide with cyclopentadienylmagnesium bromide.

$$MX_2 + 2\,C_5H_5MgBr \rightarrow (\eta^5\text{-}C_5H_5)_2M + 2\,MgXBr \tag{21.54}$$

21.9 BONDING IN FERROCENE

A Hückel molecular orbital calculation for the cyclopentadiene system can be carried out as illustrated in Chapter 5. As shown in Figure 5.21, the Frost–Musulin diagram for cyclopentadiene indicates that the energies of the five molecular orbitals are $\alpha + 2\beta$, $\alpha + 0.618\beta$ (2), and $\alpha - 1.618\beta$ (2). Because the cyclopentadienyl anion has six electrons, only the three lowest energy levels are populated and they are the orbitals interacting with those on the iron. Figure 21.10 shows the orbitals of the cyclopentadienyl anion.

The interactions of the metal s and d_{z^2} orbitals with those of the cyclopentadienyl anion can be illustrated as follows. We begin by placing the metal orbital between the two ligands that have the lobes of their orbitals oriented to match the sign of the wave function of the metal orbital. Therefore, for the s and d_{z^2} orbitals of Fe^{2+}, the combinations with the ligand orbitals can be shown as (positive orbital lobes are shaded green and negative lobes are shaded yellow)

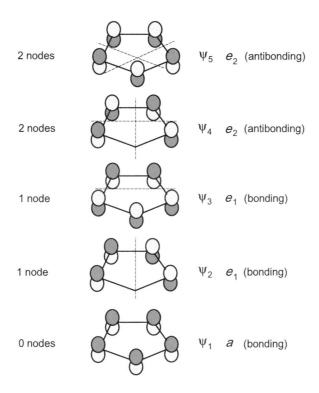

FIGURE 21.10
The five wave functions for the cyclopentadienyl anion. Green lobes indicate positive signs and yellow lobes indicate negative signs.

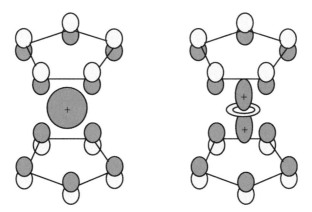

In these cases, the orbitals directed toward the ligands (along the z-axis) are positive, which requires the positive lobes of both ligand orbitals to be directed inward toward the metal. These combinations have a_{1g} symmetry and result in the combinations that are illustrated earlier. Because there are *five* molecular wave functions for *each* cyclopentadienyl ligand (those located above and below the metal ion), it is necessary to take into account the combinations of those wave functions with the metal orbitals. The procedure illustrated earlier can be followed with the remaining metal orbitals. They are only weakly bonded, which is not surprising because those orbitals have no directional component in the z direction where the ligands are located, and they require matching the symmetry character of ligand orbitals.

When the combinations of the two ψ_1 wave functions of the ligands and the p_z orbital of the metal are made, the signs of the lobes on the p_z orbitals require the negative lobes on the lower ligand wave function to be directed *upward* as shown in Figure 21.11. The combination of the d_{yz} orbital can be made using the ligand wave function ψ_2, which gives an e_{1g} molecular orbital. The combination of the d_{xz} and ψ_3 also gives an e_{1g} molecular orbital. However, although we will not show the diagrams, when the metal d_{xy} and $d_{x^2-y^2}$ orbitals are placed between the two ligands, it becomes evident that they must combine with the ligand orbitals having two nodes that are designated as e_{2g}. We now come to the point where the metal p_x and p_y orbitals must be combined with ligand orbitals. Because of the orientation of these orbitals, they form bonding molecular orbitals, but the bonding is not as effective (see Figure 21.11) as with the metal orbitals discussed so far. It is ψ_2 that interacts with the p_y orbital and ψ_3 that combines with the p_x orbital. Both these combinations give molecular orbitals having e_{1u} symmetry.

We have now determined the form of the molecular orbitals in ferrocene. A molecular orbital diagram for ferrocene is shown in Figure 21.12. For the ferrocene molecule, each ligand contributes six electrons and there are six from the Fe^{2+} ion, which gives a total of 18 electrons that must be placed in the molecular orbitals. The molecular orbital diagram for ferrocene shows that the 18 electrons populate the strongly bonding a_{1g} and a_{2u} states as well as the several e states that are less strongly bonding. Although two electrons populate the nonbonding a_{1g} state, no electrons are forced to populate the degenerate e_{1g}^* orbitals. That is not the case with cobaltocene, which has 19 electrons with one of them being forced to reside in the e_{1g}^* antibonding orbital. As a result, cobaltocene is much less stable and more easily oxidized than ferrocene, although $[Co(\eta^5\text{-}C_5H_5)_2]^+$ gives solids with some

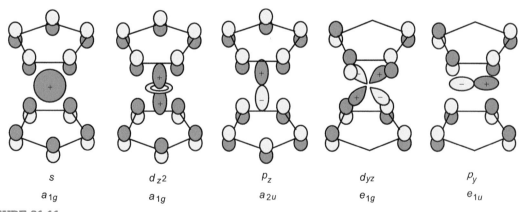

s	d_{z^2}	p_z	d_{yz}	p_y
a_{1g}	a_{1g}	a_{2u}	e_{1g}	e_{1u}

FIGURE 21.11
The overlap of s, p, and d orbitals on Fe with the molecular orbitals on the cyclopentadienyl ion. Positive regions of orbitals are shaded green and negative regions are shaded yellow. The orbitals on Fe should be considered as being located midway between the two cyclopentadienyl rings.

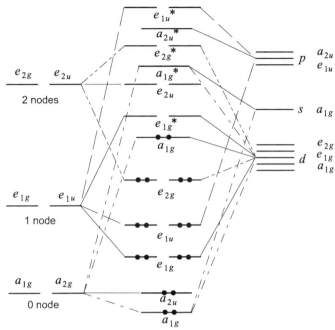

FIGURE 21.12
A qualitative molecular orbital diagram for ferrocene.

anions that are very stable. Nickelocene has a total of 20 electrons so two are forced to reside in the $e_{1g}{}^{*}$ orbitals, and compounds such as cobaltocene and nickelocene are easily oxidized. The chromium and vanadium compounds, $[Cr(\eta^5\text{-}C_5H_5)_2]$ and $[V(\eta^5\text{-}C_5H_5)_2]$, have 16 and 15 electrons, respectively, so they are highly reactive and readily bond to additional groups to conform to the 18-electron rule.

21.10 REACTIONS OF FERROCENE AND OTHER METALLOCENES

A great deal of chemistry has been carried out on complexes of the ferrocene type. Although most of the discussion here will deal with ferrocene itself, many reactions can be carried out with other metallocenes; however, the conditions might be different.

Ferrocene is a very stable compound that melts at 173 °C and can be sublimed without disruption of the metal complex. Many reactions exhibited by ferrocene are essentially those of an aromatic organic compound. For example, sulfonation of ferrocene can be achieved as follows:

$$\text{Fe} + H_2SO_4 \xrightarrow[\text{H}_2\text{SO}_4]{\text{glacial}} \text{Fe}\text{—SO}_3H + H_2O \qquad (21.55)$$

Acylation can be carried out by means of the Friedel–Crafts reaction,

$$(21.56)$$

Although the disubstituted product is shown in the equation above, some monosubstituted product is also obtained. In this process, $AlCl_3$ may react with CH_3COCl to generate the $CH_3C=O^+$ attacking species. It is also possible to introduce an acetyl group by the reaction of ferrocene with acetic anhydride, with phosphoric acid as the catalyst. In this reaction under less strenuous conditions, the dominant product is the monosubstituted derivative.

$$(21.57)$$

By a sequence of reactions, the two cyclopentadienyl rings can be connected. Using $C_2H_5OCOCH_2COCl$ as the starting acid chloride, it is possible to attach the acyl end to one of the C_5H_5 rings. Hydrolysis to remove the ethyl group and conversion of that end to an acid chloride followed by another Friedel–Crafts reaction and reduction of the product lead to

One of the most useful reactions of ferrocene with regard to synthesizing derivatives is *metalation*. In this reaction, the lithium derivative is first prepared by the reaction of ferrocene with butyllithium, a reaction showing the slight acidity of the hydrogen atoms on cyclopentadiene.

$$(21.58)$$

The derivative containing two lithium atoms can also be obtained by the subsequent reaction

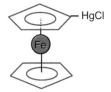

$$\text{(21.59)}$$

Better yields are obtained when polar solvents are utilized and an amine such as tetramethylethylene-diamine is present, which also facilitates the formation of the dilithium compound. The lithium derivatives undergo a large number of reactions that can be used to produce the enormous number of ferrocene derivatives. Rather than trying to show a great number of reactions, a few of the common reactants and the substituents that they introduce on the cyclopentadienyl rings are shown in Table 21.3.

Another derivative that is a useful intermediate for synthesizing other derivatives is the mercury compound that can be prepared from ferrocene by reaction of mercuric acetate in a solution that also contains chloride ion.

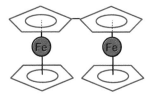

Borate esters react with the lithium compound to yield $(\eta^5\text{-cp})Fe(\eta^5\text{-cp-B(OR)}_2)$, which can be hydrolyzed to produce $(\eta^5\text{-cp})Fe(\eta^5\text{-cp-B(OH)}_2)$. This compound will undergo many reactions that lead to additional ferrocene derivatives. For example, the reaction with CH_3ONH_2 produces amino-ferrocene. The boric acid derivative will react with AgO in a coupling reaction that produces "diferro-cene" shown as

Table 21.3 Some Ferrocene Derivatives Obtained from the Lithium Derivatives

Reactant(s)	Substituent
N_2O_4	$-NO_2$
CO_2, H_2O	$-COOH$
$B(OR)_3$, H_2O	$-B(OH)_2$
$H_2NOCH_2C_6H_5$	$-NH_2$

This product can also be produced by the reaction of iodine with the mercury compound shown earlier, which yields the iodide. The iodide reacts with magnesium to produce a Grignard product, $(\eta^5$-cp) $Fe(\eta^5$-cp-MgI), which reacts in a coupling reaction to yield diferrocene. Although an enormous number of reactions can be carried out to produce derivatives of ferrocene, it is also possible by means of oxidizing agents to produce the +1 ferrocenyl carbocation by removing an electron.

Several ferrocene derivatives are produced from the dinuclear complex

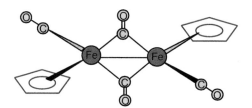

and heating this compound produces ferrocene. In this structure, the iron atoms and two bridging carbonyls lie in a plane and the cp rings are either *cis* or *trans* (as shown) with respect to the plane. Solutions of this compound appear to have a mixture of *cis* and *trans* forms as well as some of the compounds in which only a metal–metal bond holds the two halves together. When treated with an amalgam, reduction to $(C_5H_5)Fe(CO)_2^-$ occurs.

$$[(C_5H_5)Fe(CO)_2]_2 + 2\ Na \xrightarrow{\text{Na/Hg}} 2\ (C_5H_5)Fe(CO)_2^- + 2\ Na^+ \qquad (22.60)$$

The $(C_5H_5)Fe(CO)_2^-$ anion behaves as a nucleophile in many reactions, which leads to numerous novel products. For example, the reaction with CH_3I in THF can be shown as

$$(C_5H_5)Fe(CO)_2^- + CH_3I \rightarrow (C_5H_5)Fe(CO)_2CH_3 + I^- \qquad (22.61)$$

Acetyl chloride reacts with $(C_5H_5)Fe(CO)_2^-$ to produce an acyl derivative.

$$(C_5H_5)Fe(CO)_2^- + CH_3COCl \rightarrow (C_5H_5)Fe(CO)_2COCH_3 + Cl^- \qquad (22.62)$$

Numerous other derivatives are known that have the general formulas $(C_5H_5)Fe(CO)_2COR$ and $(C_5H_5)Fe(CO)_2X$ (where $X = Cl^-$, Br^-, I^-, SCN^-, OCN^-, CN^-, etc.).

A comprehensive chemistry of ferrocene and other metallocenes has been developed. The introduction presented here is meant to acquaint the reader with the general scope and nature of the field, and more details will be found in the references at the end of this chapter.

21.11 COMPLEXES OF BENZENE AND RELATED AROMATICS

Although we have considered compounds of $C_5H_5^-$ as though the ligand is a donor of six electrons, benzene is also a donor of six electrons. Therefore, two benzene molecules, each donating six electrons, would raise the total to 36 if the metal initially has 24. Such is the case if the metal is Cr^0. Thus, $Cr(\eta^6$-$C_6H_6)_2$ obeys the EAN rule, and its structure is

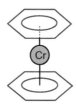

This compound has been prepared by several means including the following reactions:

$$3\ CrCl_3 + 2\ Al + AlCl_3 + 6\ C_6H_6 \rightarrow 3\ [Cr(\eta^6\text{-}C_6H_6)_2]AlCl_4 \tag{21.63}$$

$$2\ [Cr(C_6H_6)_2]AlCl_4 + S_2O_4^{2-} + 4\ OH^- \rightarrow 2\ [Cr(\eta^6\text{-}C_6H_6)_2] + 2\ H_2O + 2\ SO_3^{2-} + 2\ AlCl_4^- \tag{22.64}$$

Bis(benzene)chromium(0) is easily oxidized, and mixed complexes are obtained by means of substitution reactions. For example,

$$Cr(CO)_6 + C_6H_6 \rightarrow C_6H_6Cr(CO)_3 + 3\ CO \tag{21.65}$$

The structure of benzenetricarbonylchromium(0) can be shown as

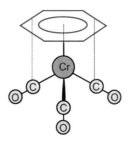

in which benzene is bonded in an η^6 manner. Note that in this case, the three carbonyl ligands are in staggered positions relative to the carbon atoms in the benzene ring (as indicated by the dotted vertical lines). Similar compounds have also been prepared containing Mo and W. Methyl-substituted benzenes such as mesitylene (1,3,5-trimethylbenzene), hexamethylbenzene, and other aromatic molecules have been used to prepare complexes with several metals in the zero oxidation state. For example, $Mo(CO)_6$ will react with $1,3,5\text{-}C_6H_3(CH_3)_3$, 1,3,5-trimethylbenzene, which replaces three carbonyl groups.

$$Mo(CO)_6 + 1,3,5\text{-}C_6H_3(CH_3)_3 \rightarrow 1,3,5\text{-}C_6H_3(CH_3)_3Mo(CO)_3 + 3\ CO \tag{21.66}$$

The product has a structure that can be shown as

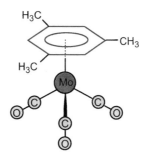

Although the chromium compound is the best known, other metals form similar complexes with benzene and its derivatives.

Hexamethylborazine, $B_3N_3(CH_3)_6$, is also a π electron donor. Because it donates six electrons, one molecule of $B_3N_3(CH_3)_6$ replaces three CO ligands. When the starting complex is $Cr(CO)_6$, the product, $(CO)_3Cr(\eta^6\text{-}B_3N_3(CH_3)_6)$, has the structure

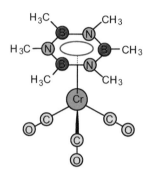

By using naphthalene as a ligand, it is possible to have both rings coordinated to metal atoms that have carbonyl groups as the other ligands. In fact, there is a novel complex that has each ring in naphthalene bonded to a $Cr(CO)_3$ group but on opposite sides of the plane of naphthalene. Thus, naphthalene is a bridging group between two $Cr(CO)_3$ fragments attached in *trans* positions.

Another aromatic molecule containing six π electrons is $C_7H_7^+$, the tropylium ion, derived from cycloheptatriene. This positive ion forms fewer complexes than benzene and these are less thoroughly studied. The molecule exhibiting an electron number of 10 and having an aromatic structure is the cyclooctatetraenyl ion, $C_8H_8^{2-}$. Some sandwich compounds containing this ligand are known as well as mixed carbonyl complexes. Although it will not be shown, the molecular orbitals on the $C_8H_8^{2-}$ anion match the symmetry of some of the f orbitals on heavy metals. Therefore, one of the known complexes that contain the $C_8H_8^{2-}$ ion is the uranium compound,

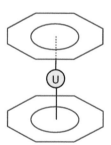

The $C_8H_8^{2-}$ ion behaves as a ligand having η^8. The uranium compound shown above can be prepared by the reaction

$$2\ C_8H_8^{2-} + UCl_4 \xrightarrow{\text{THF}} U(\eta^8\text{-}C_8H_8)_2 + 4\ Cl^- \qquad (21.67)$$

Being hybrid fields between organic and inorganic chemistry, the chemistry of metal–alkene complexes and organometallic compounds has developed at a rapid rate. There is no doubt that this type of chemistry will be the focus of a great deal of research for some time to come.

21.12 COMPOUNDS CONTAINING METAL—METAL BONDS

The development of the chemistry of compounds containing metal—metal bonds (often referred to as cluster compounds because they contain multiple metal atoms in close proximity) began about a century ago with the compound $Ta_6Cl_{14} \cdot 7H_2O$. Compounds having other "unusual" formulas were described, and, in the 1930s, the structure of $K_3W_2Cl_9$ was determined. The $[W_2Cl_9]^{3-}$ ion has two octahedra sharing a face, with the W—W distance being about 2.4 Å. That distance is shorter than that between atomic centers in metallic tungsten. As we shall see, one of the characteristics of many metal clusters is that there is multiple bonding between the metal atoms, which leads to short bonds. Although other clusters were reported in the interim, in 1963, the structure of $[Re_3Cl_{12}]^{3-}$ was elucidated by F. A. Cotton and coworkers (see Figure 21.13). This structure displays the common feature of clusters that contain three metal atoms, which is the trigonal planar arrangement for those atoms. The Re—Re bond length in $[Re_3Cl_{12}]^{3-}$ is 247 pm, which, although shorter than the single bond distance, is longer than the known triple bond length.

After a description of the structure of $[Re_3Cl_{12}]^{3-}$ was available, one of the most recognizable clusters was soon to be explained, that being $[Re_2Cl_8]^{2-}$ in which the Re-Re distance, is only 224 pm, indicating especially strong interaction between rhenium atoms. The structure of this interesting species can be shown as follows.

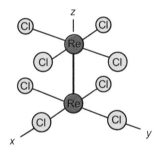

The interpretation of the bonding in this usual ion gives rise to some novel concepts. While describing the bonding in a species, it is important to keep in mind that only orbitals having the same symmetry can lead to positive overlap, and bond energies are dependent on the values of overlap integrals. Interpreting the bonding in the $[Re_2Cl_8]^{2-}$ (the structure of $[Tc_2Cl_8]^{2-}$ is similar) ion led to the concept of a *quadruple*

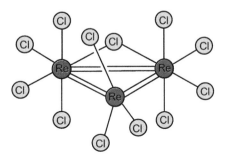

FIGURE 21.13
The structure of the $[Re_3Cl_{12}]^{3-}$ ion showing three double bonds.

bond, which can be visualized as occurring in the following way. The metal atoms lie along the z-axis, and the Cl ligands lie on the x- and y-axes on each end of the molecule. As shown in Figure 21.14, there is overlap of the d_{z^2} orbitals on the metal atoms to give a σ bond. On the basis of the overlap angle and geometry of the orbitals, the overlap is best when "end on," so the overlap of the d_{z^2} orbitals is the most effective. The d orbitals having spatial orientations that involve a component in the z direction, the d_{xz} and d_{yz}, overlap to give two π bonds. The interactions of the d_{xz} orbital are shown in Figure 21.14, and, although it is not shown, the interaction of the d_{yz} orbitals is similar to that of the d_{xz}. Overlap of these orbitals is somewhat less effective than that between the d_{z^2} orbitals.

Because the Cl atoms are located on the x- and y-axes, the lobes of the d_{xy} orbitals lie *between* the axes and are not used in bonding to the ligands. Therefore, the d_{xy} orbitals on the two Re atoms overlap slightly in *four* regions of space. Such a bond is (by extension of the σ and π notations for overlap in one and two regions, respectively) known as a δ bond. Although this overlap is less effective than either of the other types described above, it is responsible for the $[Re_2Cl_8]^{2-}$ ion having an eclipsed configuration. In fact, it can be shown that if one of the orbitals is rotated relative to the other, the value of the

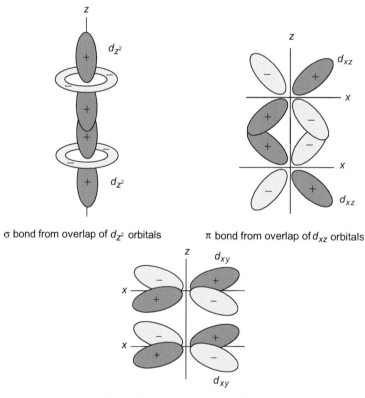

σ bond from overlap of d_{z^2} orbitals

π bond from overlap of d_{xz} orbitals

δ bond from overlap of d_{xy} orbitals

FIGURE 21.14

The formation of σ, π, and δ bonds by overlap of the d_{z^2}, and d_{xz}, and d_{xy} orbitals, respectively. The overlap of the d_{yz} orbitals is similar to that of the d_{xz} orbitals but is not shown.

overlap integral decreases and becomes equal to zero when the angle of rotation reaches 45°. This can be seen in Figure 21.15. Therefore, in the final analysis, the quadruple bond consists of one σ bond, two π bonds, and one δ bond. The configuration in which the Cl atoms are in eclipsed positions provides maximum overlap. From the rather intuitive analysis provided above, it is possible to arrive at a qualitative molecular orbital diagram like that shown in Figure 21.16 to describe the Re–Re bond.

On the basis of the molecular orbital diagram, the configuration of the metal–metal bond is $\sigma^2 \pi^4 \delta^2$, which shows a bond order of 4. The energy level scheme shown in Figure 21.16 is also useful for describing the bonding in other metal clusters. For example, Os^{3+} has a $5d^5$ configuration and it forms some compounds in which there is Os–Os bonding. In this case, there are 10 electrons to be accommodated and these give the configuration $\sigma^2 \pi^4 \delta^2 \delta^{*2}$ corresponding to a triple bond between the osmium ions. This is in agreement with the character of those bonds. In clusters of the type $Rh_2(RCOO)_4$ (in which the carboxyl groups form bridges similar to those shown below in the structure of $[Re(OOCCH_3)_4X_2]$), the d^7 configuration of Rh leads to 14 electrons being placed in the molecular orbitals. The resulting configuration, $\sigma^2 \pi^4 \delta^2 \delta^{*2} \pi^{*4}$ shows that the Rh–Rh bond is a *single* bond.

Following the interpretation of the structure of $[Re_2Cl_8]^{2-}$, further work led to "tying" the two "halves" of the structure together by introducing ligands that are simultaneously coordinated to both

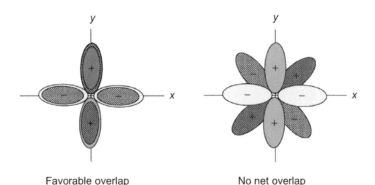

Favorable overlap   No net overlap

FIGURE 21.15

The overlap of d_{xy} orbitals in the $[Re_2Cl_8]^{2-}$ ion when the view is along the Re–Re bond. The lobes showing crosshatch are farther away from the viewer along the z-axis. On the left is the arrangement when the lobes having like signs are aligned, and, in this case, there is positive overlap of the d_{xy} orbitals. On the right is the arrangement when the ends have been rotated 45° relative to the other. In this configuration, there is a net overlap of zero for the d_{xy} orbitals.

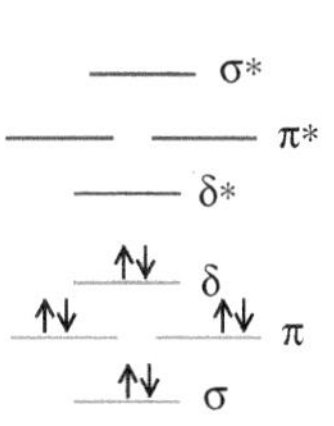

FIGURE 21.16

A simplified molecular orbital diagram for the $[Re_2Cl_8]^{2-}$ ion. Because Re^{3+} has the configuration $5d^4$, there are eight electrons all of which reside in bonding orbitals.

the Re atoms. One such structure, shown below, involves acetate ions (R represents CH_3 in the drawing) as the linking groups.

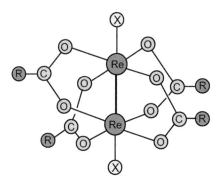

A compound of this type has been prepared in which X represents dimethylsulfoxide, $(CH_3)_2SO$, which is apparently bound to Re through the sulfur atom.

Although less numerous than complexes containing heavier transition metals, some cases are known in which there is a quadruple bond between the first row metals. One very interesting complex of this type is $[Cr_2(CO_3)_4(H_2O)_2]$, in which the carbonate ions form bridges between the two metal atoms to give a structure that can be shown as

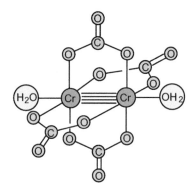

Numerous clusters are known in which the metal atoms are bound by single bonds. Earlier in this chapter, the discussion of metal carbonyls included $[Mn(CO)_5]_2$, which has this type of structure.

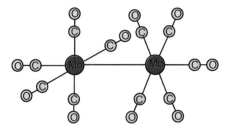

In general, clusters that contain only single bonds are larger units in which there are several metal atoms. The $[Mo_6Cl_8]^{4+}$ ion is a good example of this type of cluster.

Metal—metal bonds are responsible for the structures of complexes, such as $(OC)_5Mn-Mn(CO)_5$ and $Re_2Cl_8^{2-}$, that have no bridging groups between metal centers. These are classic complexes that represent the extremes in that the single $Mn-Mn$ bond present in $(OC)_5Mn-Mn(CO)_5$ is 292 pm in length and has a strength of less than 100 kJ/mol. In contrast, the quadruple bond in $Re_2Cl_8^{2-}$ is 224 pm in length and has a strength of several hundred kJ/mol. In numerous complexes of molybdenum, the $Mo-Mo$ bond has an average length of about 270 pm. However, in structures in which there is a triple bond, the bond length is only about 220 pm, and the quadruple bonds are only about 210 pm in length.

In this chapter, we have surveyed some of the most active and important areas of inorganic chemistry. The published literature in these fields is prodigious, so in keeping with a general textbook in inorganic chemistry that must introduce many fields, the coverage is far from complete. For further details on the material in this chapter, the interested reader should consult the references listed.

References for Further Study

Atwood, J.D., 1985. *Inorganic and Organometallic Reaction Mechanisms*. Brooks/Cole, Pacific Grove, CA. A good, readable book that contains a wealth of information.

Coates, G.E., 1960. *Organo-Metallic Compounds*. Wiley, New York. Chap. 6. This book is a classic in the field and Chapter 6 gives an introduction of over a hundred pages to this important field.

Cotton, F.A., Wilkinson, G., Murillo, C.A., Bochmann, M., 1999. *Advanced Inorganic Chemistry*, 6th ed. John Wiley, New York (Chapter 5). A 1300 page book that covers a great deal of organometallic chemistry of transition metals.

Crabtree, R.H., 1988. *The Organometallic Chemistry of the Transition Metals*. Wiley, New York. A standard book on organometallic chemistry.

Greenwood, N.N., Earnshaw, A., 1997. *Chemistry of the Elements*, 2nd ed. Butterworth-Heinemann, Oxford. This 1341 page book may well contain the most inorganic chemistry of any single volume. The wealth of information available makes it a good first reference.

Khanderi, J., Schneider, J.J., 2011. *Inorg. Chim. Acta* 370, 254—259. A report on a method to produce lithium ferrite.

Lukehart, C.M., 1985. *Fundamental Transition Metal Organometallic Chemistry*. Brooks/Cole, Pacific Grove, CA. This book is an outstanding text that is highly recommended.

Powell, P., 1988. *Principles of Organometallic Chemistry*. Chapman and Hall, London. A valuable resource in the field.

Purcell, K.F., Kotz, J.C., 1980. *An Introduction to Inorganic Chemistry*. Saunders College Pub., Philadelphia. This book provides an outstanding introduction to organometallic transition metal chemistry.

Wang, H., Sun, Z., Xie, Y., King, R.B., Schaefer III, H.F., 2011. *Inorg. Chem.* 50, 9256—9265.

QUESTIONS AND PROBLEMS

1. Which of the following should be most stable? Explain your answer.

   $Fe(CO)_2(NO)_3, Fe(CO)_6, Fe(CO)_3, Fe(CO)_2(NO)_2, Fe(NO)_5$

2. Draw the structure of the product in each of the processes indicated.
   (a) $Mo(CO)_6$ reacting with excess pyridine
   (b) The reaction of cobalt with CO at high temperature and pressure
   (c) $Mn_2(CO)_{10}$ reacting with NO
   (d) $Fe(CO)_5$ reacting with cycloheptatriene

3. Suppose a mixed metal carbonyl contains one Mn atom and one Co atom. How many CO molecules would be present in the stable compound? What would be its structure?

4. During a study of $Cr(NH_3)_3(CO)_3$, $Cr(CO)_6$, and $Ni(CO)_4$ by infrared spectroscopy, three spectra were obtained showing CO stretching bands at 1900, 2060, and 1980 cm^{-1}, but the spectra were not labeled. Match the spectra to the compounds and explain your answer.

5. Predict which of the following should be most stable and explain your answer.

$Fe(CO)_4NO, Co(CO)_3NO, Ni(CO)_3NO, Mn(CO)_6, Fe(CO)_3(NO)_2$

6. What is the structure (show all bonds clearly) of $Co(CO)_2(NO)(cot)$ where cot is cyclooctatriene?

7. How is benzene bound to Cr in the compound $Cr(CO)_3(C_6H_6)$? Explain your answer.

8. What is the structure (show all bonds clearly) of $Mn(CO)_2(NO)(cht)$?

9. For a complex $M(CO)_5L$, two bands are observed in the region 1900 to 2200 cm^{-1} Explain what this observation means. Suppose L can be NH_3 or PH_3. When L is changed from NH_3 to PH_3, one band is shifted in position. Will it be shifted to higher or lower wave numbers? Explain.

10. Describe the structure and bonding in the following where C_4H_6 is butadiene and C_6H_6 is benzene.
    (a) $Ni(C_4H_6)(CO)_2$
    (b) $Fe(C_4H_6)(CO)_4$
    (c) $Cr(C_6H_6)(CO)(C_4H_6)$
    (d) $Co(C_4H_6)(CO)_2(NO)$

11. Infrared spectra of $Ni(CO)_4$, $CO(g)$, $Fe(CO)_4^{2-}$, and $Co(CO)_4^-$ have bands at 1790, 1890, 2143, and 2060 cm^{-1}, respectively. Match the bands to the appropriate species and explain your line of reasoning.

12. Show structures (show all bonds clearly) for the following where C_8H_8 is cyclooctatetraene.
    (a) $Fe(CO)_3(C_8H_8)$
    (b) $Cr(CO)_3(C_8H_8)$
    (c) $Co(CO)(NO)(C_8H_8)$
    (d) $Fe(CO)_3(C_8H_8)Fe(CO)_3$
    (e) $Ni(CO)_2(C_8H_8)$
    (f) $Ni(NO)_2(C_8H_8)$
    (g) $Cr(CO)_4(C_8H_8)$
    (h) $Ni(CO)_3(C_8H_8)Cr(CO)_4$

13. Predict the products of the reactions indicated. More than one reaction may be possible.
    (a) $Fe_2(CO)_9 + NO \rightarrow$
    (b) $Mn_2(CO)_{10} + NO \rightarrow$
    (c) $V(CO)_6 + NO \rightarrow$
    (d) $Cr(CO)_6 + NO \rightarrow$

14. Starting with ferrocene, predict how you could prepare the following.
    (a) $(C_5H_5)Fe(C_5H_4CH_3)$;
    (b) $(C_5H_5)Fe(C_5H_4NO_2)$
    (c) $(C_5H_5)Fe(C_5H_4COOH)$;
    (d) $(C_5H_5)Fe(C_5H_4NH_2)$

15. When NO forms complexes with metals, it may form M–N–O bonds that are essentially linear in which the N–O stretching vibration gives rise to an absorption in the 1650–1900 cm^{-1} range. In other cases, the bonding is angular and the NO vibration is observed in the 1500–1700 cm^{-1} range. Explain this in terms of the difference in bonding.

16. Consider the positions of the CO stretching bands in the following complexes and provide an explanation.

$trans\text{-}[Ir(CO)Cl(P\phi_3)_2]$ 1967 cm^{-1} $trans\text{-}[Ir(CO)Cl(O_2)]$ 2015 cm^{-1}

CHAPTER 22

Coordination Compounds in Catalysis

In introductory chemistry courses, a catalyst is defined as a substance that alters the rate of a chemical reaction without being permanently altered itself. Reactions in which coordination compounds are involved in functioning as catalysts may begin and end with the same metal complex being present. During the course of the reaction, the catalyst may undergo numerous changes in coordination number, bonding mode, or geometry, and groups may enter or leave the coordination sphere of the metal. A complex that functions as a catalyst undergoes changes, and if the reaction were stopped at some instant in the process, the catalyst might be significantly different than at the initiation of the reaction. In fact, it might be recovered as a different complex. However, in a reaction scheme, the complex is regenerated at some point in its original form. The processes are actually series of reaction steps in which a metal complex facilitates the transformation of one substance into another. Because the catalyst is regenerated after a series of steps, the processes are often referred to as catalytic cycles.

Although a great deal is known about some of the processes, some catalytic processes or at least some of the steps in the process may not be completely understood. It is rather like a Born–Haber cycle that is used to represent the process by which a metal and a halogen are converted into a metal halide. The process is represented as a series of steps consisting of the sublimation of the metal, dissociation of the halogen, removal of the electron from the metal and placing it on the halogen, and combining the gaseous ions to form a crystal lattice. These steps lead from reactants to product, and we know the energies associated with them, but the reaction very likely does not literally follow these steps. Reaction schemes in which metal complexes function as catalysts are formulated in terms of known types of reactions, and in some cases the intermediates have been studied independent of the catalytic process. Also, the solvent may play a role in the structure and reactions of intermediates. In this chapter, we will describe some of the most important catalytic processes in which coordination chemistry plays such a vital role.

The other area of coordination chemistry that will be described in Chapter 23 is concerned with its role in some biochemical processes. There are a great number of such processes that depend on the presence of some metal ion for their effectiveness. The area of bioinorganic chemistry has grown enormously in recent years, and because a metal ion is involved, the topic is considered "fair game" for inorganic chemists as well as biochemists and organic chemists. However, because this area of study is extensively covered in courses in biochemistry, the survey presented in Chapter 23 will deal with only a few of the most common types of structures and reactions.

Even reactions that are energetically favorable may take place only slowly because there may be no low-energy pathway. The rate of a chemical reaction depends on the concentration of the transition state, which is determined by the energy required for it to form. In a general way, the function of a catalyst is to provide a way for the formation of a transition state that has a lower energy. Adsorption **747**

Inorganic Chemistry. DOI: http://dx.doi.org/10.1016/B978-0-12-385110-9.00022-4

of gaseous reactants on a metal surface is one way that alters the energy barrier to a reaction. For example, hydrogenation reactions can be carried out using a metal catalyst such as platinum, palladium, or nickel, and such reactions have been carried out industrially for many years. Even when a catalyst is used, such heterogeneous reactions require rather stringent conditions. In industrial processes that are carried out on a large scale, the energy cost may be significant, so the discovery of conditions that allow reactions to be carried out in a more cost-effective way is important. That is just the situation when reactions are carried out under homogeneous conditions in solutions in some cases at atmospheric pressure and room temperature.

The ability of complexes to catalyze several important types of reactions is of great importance, both economically and intellectually. For example, isomerization, hydrogenation, polymerization, and oxidation of olefins all can be carried out using coordination compounds as catalysts. Moreover, some of the reactions can be carried out at ambient temperature in aqueous solutions as opposed to more severe conditions when the reactions are carried out in the gas phase. In many cases, the transient complex species during a catalytic process cannot be isolated and studied separately from the system in which they participate. Because of this, some of the details of the processes may not be known with certainty.

Although not all facets of the reactions in which complexes function as catalysts are fully understood, some of the processes are formulated in terms of a sequence of steps that represent well-known reactions. The catalytic processes are represented as taking place through several steps that involve known reactions. The actual process may not be identical with the collection of proposed steps, but the steps represent chemistry that is well understood. It is interesting to note that developing kinetic models for reactions of substances that are adsorbed on the surface of a solid catalyst leads to rate laws that have exactly the same form as those that describe reactions of substrates bound to enzymes. In a very general way, some of the catalytic processes involving coordination compounds require the reactant(s) to be bound to the metal by coordinate bonds, so there is some similarity in kinetic behavior of all these processes. Before the catalytic processes are considered, we will describe some of the types of reactions that constitute the individual steps of the reaction sequences.

22.1 ELEMENTARY STEPS IN CATALYTIC PROCESSES

When a coordination compound functions as a catalyst, there are usually several steps in the process. The entire collection of steps constitutes the mechanism of the reaction. Before describing several of the important catalytic processes, we will describe the types of reactions that often constitute the elementary steps.

22.1.1 Ligand Substitution

The very essence of the catalytic schemes that involve coordination compounds is that reactants, other ligands, and solvent molecules must enter and leave the coordination sphere of the metal. In order for this to be rapidly accomplished, it is necessary that the complexes be of lower stability than one would associate with a complex such as $Cr(en)_3^{3+}$ or $Co(acac)_3$. It will be observed that in most cases, the transition metals are from the second or third transition series and that they generally start out in a low oxidation state. It will thus be concluded that the metal species are generally soft in character. In fact, d^8 ions of metals in low oxidation states are frequently encountered in these reactions. Complexes are often square planar species containing Pt(II), Pd(II), and Rh(I). One would not expect that a stable complex of a hard metal would be able to function in a way that would allow ligands to drift in and out of the coordination sphere as the process takes place. In many cases, hard metal ions such as Cr^{3+}, Al^{3+}, or Ti^{3+} have only one oxidation state (or at least a dominant one), and it is difficult for them to change oxidation states. Metals that function as catalysts in homogeneous processes are generally those in which it is easy to change their oxidation state. It is also a fact of life that some of these metals are extremely expensive. For example, as this is written the price of rhodium is over \$1000/oz! Approximately 60% of the rhodium produced comes from South Africa, but Russia is also a significant producer of rhodium. In some instances, the availability and price of precious metals are subject to political factors.

As a complex functions as a catalyst, it is often necessary for one ligand to enter the coordination sphere of the metal and another to leave (before or as the other ligand enters). These processes are substitution reactions, which were discussed in some detail in Chapter 20. As the catalytic processes are illustrated, it will be seen that some of the elementary steps are substitution reactions. A substitution reaction can be shown by the general equation

$$L_nM - X + Y \rightarrow L_nM - Y + X \tag{22.1}$$

in which L is a nonparticipating ligand and n is the number of those ligands.

In the various homogeneous catalytic schemes, the solvent may be coordinated to the metal or may simply be present as bulk solvent. When a ligand leaves the coordination sphere of a metal, it *may* be replaced by a molecule of solvent in a process that is either associative or dissociative. There is no general way to predict which type of mechanism is operative, so in some cases the substitution reactions will be described as they relate to specific processes. Because substitution reactions have been described in Chapter 20, several other types of reactions that constitute the steps in catalytic processes will be described in greater detail.

22.1.2 Oxidative Addition

Oxidative addition is a process by which an atom is simultaneously oxidized and the number of bonds to it is increased as groups are added. The term *oxad* is used to denote this type of reaction. This type of process is not limited to coordination compounds, and it is easy to find numerous examples in other areas of chemistry. The following examples and many others can be found in earlier chapters of this book.

$$PCl_3 + Cl_2 \rightarrow PCl_5 \tag{22.2}$$

$$2\,SOCl_2 + O_2 \rightarrow 2\,SO_2Cl_2 \tag{22.3}$$

$$ClF + F_2 \rightarrow ClF_3 \tag{22.4}$$

In each of these reactions, the oxidation state of the central atom is increased by 2 as the number of bonds is increased. In many cases, the number of bonds increases by two because of the oxidation states exhibited by the central atom. For example, in Eqn (22.2) phosphorus is changed from an oxidation state of $+3$ to $+5$ because the $+4$ oxidation state is not favorable for this atom. In Eqn (22.4), chlorine is oxidized from $+1$ to $+3$ because the $+2$ oxidation state is not normally accessible for that atom. Note that in each of these reactions, the number of bonds to the central atom is increased by two. In the reaction shown in Eqn (22.3), sulfur increases in oxidation from +4 to +6 and an additional bond (which has substantial double bond character) is formed.

An example of an oxad reaction of a coordination compound is

$$PtF_4 + F_2 \rightarrow PtF_6 \tag{22.5}$$

in which the platinum is oxidized from +4 to +6 with the formation of two additional bonds. One important complex that functions as a catalyst is *Wilkinson's catalyst*, $Rh(P\phi_3)_3Cl$ (where $\phi = $ phenyl, C_6H_5). A very significant property of this complex is that it undergoes an oxidative addition reaction with hydrogen.

$$Rh(P\phi_3)_3Cl + H_2 \rightarrow Rh(P\phi_3)_3ClH_2 \tag{22.6}$$

The product of the reaction has the structure

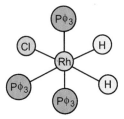

in which the rhodium is formally +3 with the hydrogen considered as H^-. In many oxad reactions, H_2 adds as two H^- ions in *cis* positions as if a hydrogen molecule approached the complex and added as atoms while the H–H bond simultaneously breaks. This is also true of gaseous HCl, which usually adds H and Cl in *cis* positions. On the other hand, in aqueous solutions where the HCl is largely dissociated, the addition is not restricted to *cis* positions, so the product contains H and Cl atoms in both *cis* and *trans* positions. The stereochemistry of the product depends on the solvent, and if the solvent has considerable polarity, the product consists of a mixture of *cis* and *trans* isomers. When the solvent is essentially nonpolar, the product contains hydrogen and halide in *cis* positions.

Another important complex is *trans*-[$Ir(P\phi_3)_2COCl$], which is known as *Vaska's compound*. This complex undergoes a large number of oxidative addition reactions because Ir^+ is easily oxidized to Ir^{3+} and that ion forms many stable octahedral complexes. When CH_3I is added, the octahedral product contains CH_3 and I in *trans* positions with the other ligands in a plane as shown below.

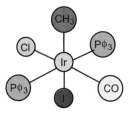

This oxad reaction is first order in both complex and CH_3I, and the entropy of activation is large and negative. It is believed that the reaction goes through an ionic transition state in which CH_3 adds to a pair of electrons on Ir followed by the attachment of I^- in the *trans* position.

The nature of the oxad reaction requires that the following conditions generally apply.

1. It must be possible to change the oxidation state of a metal. Normally, the metal changes oxidation state by 2 units, but two metal atoms can change by 1 unit each as shown in Eqns (22.10) and (22.11).
2. The metal must be able to increase its coordination number by 2 except in cases like that shown in Eqns (22.10) and (22.11).
3. Diatomic molecules such as H_2, Cl_2, or HCl add in *cis* positions when reacting as gases or as solutions in nonpolar solvents that do not assist in separating the molecules. In polar solvents, molecules that dissociate are not restricted to entering in *cis* positions.

One of the most interesting complexes that also figures prominently in catalytic processes is $Co(CN)_5^{3-}$. Because Co^{2+} is a d^7 ion, forcing the pairing of the electrons leaves one orbital occupied by only one electron and another empty. This makes possible the formation of dsp^3 hybrid orbitals, and the complex is trigonal bipyramidal as expected. It is the singly occupied orbital that gives the complex its unusual character, the ability to behave as a free radical. In fact, $Co(CN)_5^{3-}$ undergoes many reactions that are similar to those of other radicals. As expected, there is a coupling reaction, which can be shown as

$$2\ Co(CN)_5^{3-} \leftrightarrows Co_2(CN)_{10}^{6-} \tag{22.7}$$

Numerous other reactions of $Co(CN)_5^{3-}$ have been observed with a wide variety of substances. For example, the reaction with alkyl halides can be shown as follows:

$$RX + Co(CN)_5^{3-} \rightarrow Co(CN)_5X^{3-} + R\cdot \tag{22.8}$$

$$R\cdot + Co(CN)_5^{3-} \rightarrow RCo(CN)_5^{3-} \tag{22.9}$$

These reactions are typical oxad processes because the cobalt is formally converted from +2 to +3. Many other molecules such as H_2, H_2O_2, Br_2, O_2, C_2H_2, etc., are split in a homolytic manner during reactions with $Co(CN)_5^{3-}$. For example, the reaction with hydrogen is illustrated in the following equation:

$$H_2 + 2\ Co(CN)_5^{3-} \rightarrow 2\ HCo(CN)_5^{3-} \tag{22.10}$$

These reactions show the tendency of the metal to behave as a radical as a result of having an unpaired electron and a vacant position where a sixth ligand may be added.

The $[Co(CN)_5]^{3-}$ complex is an effective catalyst for some reactions, particularly the isomerization of alkenes. Newer and more efficient catalysts have been developed for some of the processes, but the catalytic behavior of the pentacyanocobalt(II) ion is also significant from a historical perspective. In reactions such as that shown in Eqn (22.10), *two* Co^{2+} ions increase *one* unit in oxidation state instead of the more common situation in which one metal ion increases by two units in oxidation state. The cobalt complex also reacts with CH_3I, Cl_2, and H_2O_2, which are indicated as X–Y in the equation

$$2\ [Co(CN)_5]^{3-} + X - Y \rightarrow [XCo(CN)_5]^{3-} + [YCo(CN)_5]^{3-} \tag{22.11}$$

Mechanistic studies have shown that the reaction represented in Eqn (22.10) is a one-step process that follows the rate law

$$\text{Rate} = k[H_2][Co(CN)_5^{3-}]^2 \tag{22.12}$$

whereas the reaction of $[Co(CN)_5]^{3-}$ with H_2O_2 follows a free radical mechanism. Other aspects of the chemistry of $[Co(CN)_5]^{3-}$ will be illustrated later in this chapter.

A different type of oxidative addition involves the transfer of a hydrogen atom from a coordinated group to the metal atom followed by (or simultaneous with) a change in bonding mode to the ligand from which the hydrogen atom came. This is illustrated in the following example, in which ϕ is a phenyl group, C_6H_5.

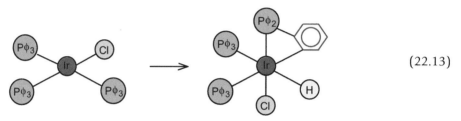

$$\tag{22.13}$$

A reaction of this type is sometimes referred to as *orthometallation* because it is from the *ortho* position that the hydrogen is transferred to the metal.

22.1.3 Mechanistic Considerations for Oxad Reactions

Much of Chapter 20 was devoted to the description of mechanisms of reactions of coordination compounds with emphasis on substitution reactions. Because there are numerous aspects of oxad reactions that are different from those involving substitution, we will address some of the mechanistic aspects of oxad reactions briefly in this section.

The mechanism of oxad reactions of Vaska's compound has been the subject of a considerable amount of study. It is known that the solvent plays a role in these reactions, but the nature of the solvent effect is not always clear. For example, when a series of *trans*-$[IrX(P\phi_3)_2(CO)]$ complexes (where $X = Cl$, Br, or I) undergo oxad reactions with H_2 in benzene, the rate decreases as X varies in the order $Cl^- > Br^- > I^-$. In contrast, when the oxad reaction is by CH_3I, the rate varies as $I^- > Br^- > Cl^-$. For both types of reactions, the entropy of activation is negative, which has been interpreted as indicating that association of the reactants occurs in the transition state. As described in Chapter 20, the compensation effect (as indicated by a linear isokinetic plot) is significant when considering a series of related reactions. By making use of the data presented by Atwood (1985), the isokinetic plot shown in Figure 22.1 was constructed.

It must be emphasized that a linear isokinetic plot does not *prove* that a common mechanism is operating, but it does give some credence to that conclusion. Neither does such a plot tell what the mechanism might be. Certainly the fit to the data is not perfect, but it is rather good in this case. One thing that is striking is the large negative values for the entropy of activation and the very low enthalpies of activation. The energy of the bond in the H_2 molecule is approximately 435 kJ/mol (104 kcal/mol), so there is a good indication that bond formation to the metal is at least as important as bond breaking

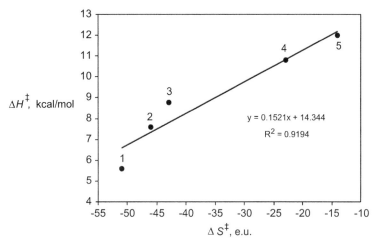

FIGURE 22.1

An isokinetic plot for oxidative addition of H_2 and CH_3I in $IrX(CO)(P\phi_3)_2$ in benzene. Points 1, 2, and 3 represent oxidative addition of CH_3I when X is Cl, Br, and I, respectively. Points 4 and 5 are for oxidative addition of H_2 when X is Cl and Br, respectively.

in the H_2 molecule. A plausible mechanism for oxad reactions of H_2 with metal complexes can be visualized as shown in Figure 22.2. This mechanism is usually referred to as a *concerted addition*.

The bonding of H_2 in metal complexes was described in Chapter 16. In connection with the oxad reaction in which the bonding is not static, it can be presumed that the σ orbital on the hydrogen molecule functions as an electron pair donor to an orbital on the metal atom. Simultaneously, the σ^* orbital on the H_2 molecule receives electron density from the populated d orbitals on the metal atom as a result of back donation. The result is that two M—H bonds form as the H—H bond is broken in a process that is accompanied by a very low activation energy.

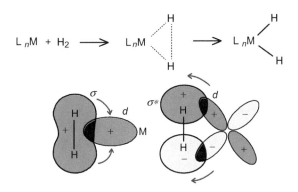

FIGURE 22.2

A possible mechanism for oxad of H_2 to a metal complex. The bonding of the hydrogen molecule before it dissociates is illustrated in color below the transition state. The red arrows indicate the direction of electron flow.

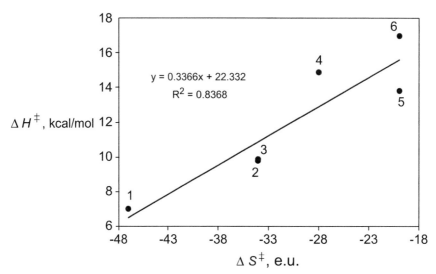

FIGURE 22.3

An isokinetic plot for oxidative addition of IrCl(CO)L$_2$ with CH$_3$I. The points 1 through 6 correspond to L = Pϕ_3, PEtϕ_2, PEt$_2\phi$, P(p-C$_6$H$_4$CH$_3$)$_3$, P(p-C$_6$H$_4$F)$_3$, or P(p-C$_6$H$_4$Cl)$_3$, respectively.

Kinetic data are available for the oxad reaction of CH$_3$I with a series of complexes that are derivatives of Vaska's compound, *trans*-IrCl(CO)L$_2$ where L = Pϕ_3, PEtϕ_2, PEt$_2\phi$, P(p-C$_6$H$_4$CH$_3$)$_3$, P(p-C$_6$H$_4$F)$_3$, or P(p-C$_6$H$_4$Cl)$_3$. When an isokinetic plot is made of the $\Delta H^{\ddagger}$ and $\Delta S^{\ddagger}$ values, the result obtained is as shown in Figure 22.3.

The points on the graph shown in Figure 22.3 show significant scatter, but the fit is sufficiently good to indicate that the reactions probably follow a common mechanism. In this case, the attacking species is CH$_3$I, but the ligands in the *trans* positions are varied. For the ligands listed in conjunction with Figure 22.3, the lowest rate constant (all values given are for reactions at 25 °C) was found when the group on P is p-C$_6$H$_4$Cl ($k = 3.7 \times 10^{-5}$ M^{-1} s^{-1}) and the highest occurred when p-C$_6$H$_4$CH$_3$ is present ($k = 3.3 \times 10^{-2}$ M^{-1} s^{-1}). Whether this is due to electronic or steric effects is not known, but it is interesting to note that the rate constant when the the group on P is p-C$_6$H$_4$F ($k = 1.5 \times 10^{-4}$ M^{-1} s^{-1}) was intermediate between those containing the chloride and methyl groups in the *para* position in the ligand. If the difference in rate constants is due to steric effects, it would be expected that the ligand containing Cl would have the smallest rate constant, and that is consistent with the observed k values. If the difference in rate constants is due to difference in electron density at the metal, the fact that CH$_3$ is electron releasing and fluorine is withdrawing would predict that greater electron density would be placed on the metal when the ligands contain CH$_3$ and lower when F is present. The observed rate constants show that trend, but the rate constant for the ligand containing Cl in the *para* position does not follow this trend because it is not as electron withdrawing as F.

In another study (see Atwood, 1985, p. 168), a series of oxad reactions was carried out at 30 °C with methyl iodide and *trans*-[IrCl(CO)(P(p-C$_6$H$_4$Y)$_3$)$_2$]. The identities of the groups in the *para* position of

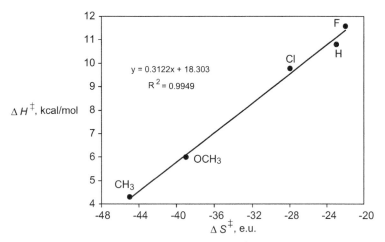

FIGURE 22.4

An isokinetic plot for the reaction of *trans*-[IrCl(CO)(P(*p*-C₆H₄Y)₃)₂]. The identity of Y is indicated for each point on the graph. *(Constructed from data given in Atwood (1985) p. 168).*

the phosphine ligands, Y, are shown in Figure 22.4, which shows graphically the values for the enthalpy and entropy of activation for reactions.

The isokinetic plot shown in Figure 22.4 illustrates a linear relationship that provides an excellent fit to the data. In this case, the effect produced by the ligand on the rate of oxidative addition is related to the electronic changes produced on the metal. The oxad rate is enhanced when Y is an electron-donating group that increases electron density at the metal. Such a relationship provides clear evidence that the mechanism is constant for the reactions with no significant differences due to steric effects. The kinetic data for this process are consistent with a nucleophilic attack by the ligand. The Hammett σ parameter is a measure of the electron withdrawing or releasing character of substituent groups. The kinetic data used to construct Figure 22.4 were originally correlated by means of the Hammett σ parameter, and a linear relationship was found between the logarithm of the rate constants and σ values for the substituents in *para* position. It was thus concluded that the effect produced by different substituents is related to their ability to change the electron density at the metal atom.

The oxidative addition of CH_3I to a series of rhodium(I) complexes containing substituted β-diketones and triphenylphosphite $[P(OC_6H_5)_3]$ ligands has been studied by Conradie, Erasums, and Conradie (2011). The results of this interesting study showed the effects that the nature of substituent groups located on the β-diketones had on the kinetics of the reaction. Also, the volume of activation was determined for the reactions. It was reported that the low values for $\Delta H^{\ddagger}$ and the negative values for $\Delta S^{\ddagger}$ and $\Delta V^{\ddagger}$ were consistent with a mechanism that involves association and simultaneous addition of CH_3 and I to the coordination sphere of the metal.

The rate of oxidative addition was found to be directly proportional to the concentration of CH_3I. However, it was also observed that the reactivity of the starting complex varied with the nature of the groups R_1 and R_2 on the β-diketones. This was attributed to the difference in electron density on the Rh atom caused by the effects of R_1 and R_2 in the ligand. The rates of the oxad reaction varied as follows where a β–diketone, R_1–COCHCO–R_2, is represented as R_1–β–R_2 and Fc is the ferrocenyl group:

$$CF_3-\beta-CF_3 \; < \; CF_3-\beta-C_6H_5 \; < \; CF_3-\beta-CH_3 \; < \; C_6H_5-\beta-C_6H_5 \; < \; CF_3-\beta-Fc$$
$$< \; CH_3-\beta-C_6H_5 \; < \; CH_3-\beta-CH_3 \; < \; CH_3-\beta-Fc \; < \; Fc-\beta-Fc$$

It was also reported that the addition of CH_3I to $[Rh(CF_3COCHCOFc)(P(OCH_3)_3)_2]$ takes place in two steps, with the first being the interaction of the metal with CH_3^+ followed by addition of I^-. Although negative values for $\Delta V^{\ddagger}$ were reported, they were smaller in magnitude than expected if a bond is being broken to give three species as a molecule of CH_3I is added to the metal. For the nine β-diketones, $\Delta H^{\ddagger}$ varied from 30 to 62 kJ mol^{-1} and $\Delta S^{\ddagger}$ varied from -51.6 to -145.4 J K^{-1} mol^{-1}, which was interpreted as being indicative of an associative mechanism. Thus, the mechanism was interpreted as being an intimate two-step S_N2 process. The authors report data from an earlier study by other workers and state that variations in the activation parameters did not have much effect on $\Delta G^{\ddagger}$, but the range stated is from 68 to 94 kJ mol^{-1}. A constant value for $\Delta G^{\ddagger}$ indicates that the compensation effect is applicable, which indicates a common mechanism for a series of reactions.

The data obtained Conradie, et al. in conjunction with additional data they cite provide the basis for a plot of $\Delta H^{\ddagger}$ versus $\Delta S^{\ddagger}$ to test the compensation effect as described in Section 20.6. When such a plot is made, the fit of the data is found to be good when the points corresponding to the ligands $CH_3-\beta-C_6H_5$ and $CF_3-\beta-CF_3$ are omitted. Therefore, a common mechanism of oxidative addition of CH_3I to the $[Rh(R_1COCHCOR_2)(P(OCH_3)_3)_2]$ complexes is indicated.

From the brief discussion of oxad reactions presented in this chapter, it is clear that a great deal is known about the mechanisms of such reactions, and it is also apparent that there is much left to learn.

22.1.4 Reductive Elimination

In the catalytic schemes to be described later, it will be seen that reactants become ligands and are frequently altered in some way (usually by adding or removing an atom or changing the position of a bond). After these changes, the altered molecule must be set free, and one important route of this type is known as reductive elimination. A reductive elimination reaction is the opposite of oxidative addition. The coordination number of the metal decreases (usually by two units) as the oxidation state decreases. The following reaction illustrates the reductive elimination of C_2H_6, which is one step in hydrogenation using Wilkinson's catalyst.

$$(22.14)$$

Note that in this case, the reductive elimination involves the transfer of a hydrogen atom from a *cis* position to the alkyl group. This is a typical condition for this type of reaction. A similar reaction takes place during the isomerization of alkenes as catalyzed by $RhCl(P\phi_3)_3$.

$$\text{(22.15)}$$

In catalytic processes, reductive elimination is essential in order to remove the product from the coordination sphere of the metal.

22.1.5 Insertion Reactions

In an insertion reaction, an entering group becomes bonded to a metal and a ligand while being positioned between them. Formally, such a reaction can be shown as

$$L_nM-X + Y \ \rightarrow \ L_nM-Y-X \tag{22.16}$$

in which the entering ligand, Y, becomes inserted between M and X and forms bonds to both. Molecules such as O_2, CS_2, CO, C_2H_4, SO_2, and $SnCl_2$ are among those that undergo insertion reactions.

When an insertion reaction occurs with the entering group containing more than one atom, there are two possible outcomes. For example, when a diatomic molecule enters the complex, it may do so in either 1,1-addition or 1,2-addition. These processes are illustrated in the following equations:

$$L_nM-X + AB \ \rightarrow \ L_nM-A-B-X \tag{22.17}$$

$$L_n M - X \ + \ AB \longrightarrow L_n M - \overset{\overset{\displaystyle B}{\displaystyle |}}{A}-X \tag{22.18}$$

In the first of these processes, both A and B (atoms 1 and 2 in the molecule) are bound to other groups, so this is known as 1,2-adddition. In the second process, only A (atom 1) is bound to the metal and the other ligand, so this is 1,1-addition.

The number of insertion reactions known to occur is very large. However, they can be considered to fall into a much smaller number of reaction *types* that arise as a result of the types of bonds between the metal and other ligand at the reactive site. The following equations illustrate these reaction types. One very simple insertion reaction is the formation of a Grignard reagent in which magnesium is inserted between R and X,

$$Mg + R-X \ \rightarrow \ RMgX \tag{22.19}$$

Reactions that are more typical of coordination compounds are the insertion reactions illustrated in the following equations:

$$M-H + H_2C{=}CH_2 \ \rightarrow \ M-CH_2-CH_2-H \tag{22.20}$$

$$HCo(CO)_4 + F_2C{=}CF_2 \ \rightarrow \ HCF_2CF_2Co(CO)_4 \tag{22.21}$$

These reactions are examples of 1,2-addition. In addition to insertion in the $M-H$ bond as shown above, this type of reaction also occurs with metal–halogen and metal–carbon bonds. The following equation shows an example of a reaction in which insertion occurs between a metal and a halogen:

$$(\eta_5\text{-}C_5H_5)Fe(CO)_2Cl + SnCl_2 \rightarrow (\eta_5\text{-}C_5H_5)Fe(CO)_2SnCl_3 \qquad (22.22)$$

This reaction involves insertion in which the Sn atom in $SnCl_2$ is bonded to both Fe and Cl as the result of 1,1-addition. One of the many insertion reactions given by CO is the following:

$$CO + RPtBr(PR_3)_2 \rightarrow RCOPtBr(PR_3)_2 \qquad (22.23)$$

In the product, the carbon atom of the CO ligand is bonded to both Pt and R, so this is an example of 1,1-addition.

Insertion reactions are not limited to those involving transition metals. Recent work by Steward, Dickie, Tang, and Kemp (2011) has shown that tin compounds such as $Sn(N(CH_3)_2)_2$ undergo insertion reactions with CO_2, CS_2, and SCO. The products may involve the formation of chelates in which only one tin atom is involved. When the molecule inserted is CS_2, the product has the structure

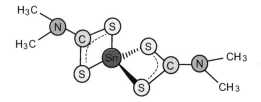

In other cases, more complex structures are obtained in which the S and O atoms in the inserted molecules form bridges between Sn atoms in adjacent units.

One of the best known and most extensively studied "insertion" reactions can be shown as

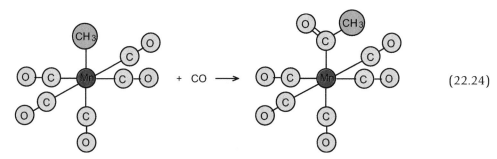

$$(22.24)$$

However, this reaction has been the subject of a great deal of study. Three of the transition states that could conceivably be formed during this reaction are illustrated in Figure 22.5. The transition states arise in the following ways:

1. The entering CO becomes inserted directly between Mn and CH_3.
2. Insertion of CO from an adjacent position occurs with replacement by incoming CO.
3. Migration of the CH_3 to an adjacent CO and attachment of CO at the site vacated by CH_3.

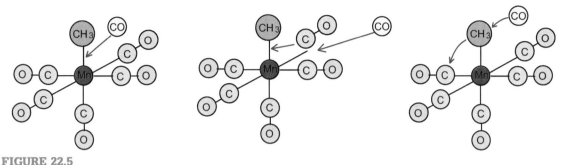

FIGURE 22.5
Possible transition states for the reaction shown in Eqn (22.24).

If all CO molecules have the same identity, it is impossible to tell one from another. What is needed is some way to distinguish the CO groups from each other, and isotopic labeling constitutes such a technique.

Many of the questions regarding the mechanism of the reaction shown in Eqn (22.24) have been answered by the use of ^{13}CO. The results have shown that when the entering carbon monoxide is ^{13}CO, there is no ^{13}CO in the acyl group. Instead, the labeled CO is in a position that is *cis* to the acyl group, so the reaction can be shown as

$$CH_3Mn(CO)_5 + {}^{13}CO \rightarrow cis\text{-}CH_3C(O)Mn(CO)_4({}^{13}CO) \qquad (22.25)$$

Therefore, it can be concluded that the reaction does not go through transition state I, which would result in the ^{13}CO being in the acyl group. However, this does not answer the question as to whether the CH_3 migrates to a CO in a *cis* position or whether the CO from a *cis* position migrates to the Mn−CH$_3$ bond and becomes inserted between Mn and CH_3. In accord with the principle of microscopic reversibility, the reaction for the loss of CO should proceed through the same transition state that was formed when the ligand was added. The structure of the molecule and the predicted products for decarbonylation are shown in Figure 22.6.

If the reaction forming the acyl group had taken place with migration of the methyl group, then decarbonylation should also occur by migration of the methyl group with the formation of the same transition state. Migration of the CH_3 group could move it to any of the four positions around the square planar part of the molecule. If the reaction proceeds as shown in structure I, the methyl group would displace ^{13}CO to form an Mn−CH$_3$ bond and there would be no ^{13}CO in the product. There would be a 25% probability for transfer of the methyl group to that position. Methyl group migration as indicated in structure II would lead to ^{13}CO remaining in the product, and there are two possibilities for the methyl group being in positions that are *cis* to the ^{13}CO and only one in which CH_3 would be *trans* to ^{13}CO. When all of the possible outcomes are considered, there should be twice as much product with ^{13}CO located *cis* to the CH_3 group as there is with it in the *trans* position. The product distribution from decarbonylation agrees with that assessment. Therefore, the "insertion" of CO in $CH_3Mn(CO)_5$ is actually the result of migration of the methyl group.

Studies have also been conducted to determine the effect of solvent on the rate of methyl migration. Removing the CH_3 group from a coordination site and placing it on a coordinated CO would result in

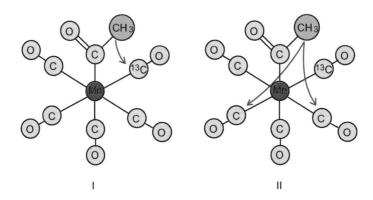

FIGURE 22.6
Possible paths of CH_3 transfer in decarbonylation. In structure II, only one of the two possible migrations of CH_3 to positions *cis* to the ^{13}CO is indicated.

a transition state having a smaller volume than the starting complex. Forming a transition state having a smaller volume than the reactants is favored by high pressure or solvents having high internal pressure. Accordingly, it would be expected that the rate would increase as the solubility parameter or cohesion energy increases for a series of solvents (see Chapter 6). Atwood (1985) shows relative rate data (for mesitylene, di-*n*-butylether, *n*-octylchloride, and DMF as solvents), which indicates that this is indeed the case. However, the data presented also show that the equilibrium constant varies with the solvent. This was interpreted as an indication that solvation of the alkyl and acyl complexes depends on the nature of the solvent.

A somewhat different approach can be taken by making use of the data presented by Basolo and Pearson (1967). The rate constants are given for formation of the intermediate, $CH_3COMn(CO)_4S$, in which it is assumed that a solvent molecule, S, occupies the position vacated by movement of the CH_3 group. A graph showing the $\ln k$ versus solubility parameter of the solvent is presented in Figure 22.7.

Although the agreement is far from perfect, there is a general increase in $\ln k$ as the solubility parameter of the solvent increases. Forming a transition of smaller volume is assisted when the solvent has high cohesion. In the early work, the trend was attributed to either a molecule of solvent occupying the vacant site or stabilization of a polar transition state by polar solvent molecules. However, the observed variation in rate with solvent is also consistent with the formation of a transition state having a smaller volume. This may be an explanation for the observation mentioned earlier that the equilibrium constant was also larger in solvents having high cohesion. It is interesting to note that for four of the six solvents represented by data points in Figure 22.7, the linear fit (the green line) is quite good. The fact that the points for tetrahydrofuran (THF) and DMF fall far from that relationship may indicate that they are involved in solvation of the transition state by forming a coordinate bond to the metal (see Chapter 6). Although the nature of the interactions is not known, it is likely that all six of the solvents do not play the same role in assisting the formation of the transition state. A THF complex of the type $CH_3Mn(CO)_4THF$ has been identified, so the conjecture that THF and DMF behave in a manner that is different from the other solvents is not without some justification. Additional work on the influence of the solvent could yield additional insight regarding the mechanisms of this and other reactions of coordinated ligands.

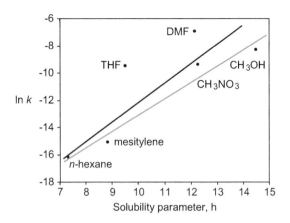

FIGURE 22.7
The variation in rate constant with solubility parameter for the reaction shown in Eqn (22.24). The green line shows the fit with the points for THF and DMF omitted. *(Constructed from data given in Basolo and Pearson 1967, p. 584).*

22.2 HOMOGENEOUS CATALYSIS

Few areas of chemistry have changed as rapidly as the field of homogeneous catalysis by coordination compounds. Not only is the chemistry fascinating and diverse, but also it is immensely practical. The ability to carry out hydrogenations, isomerizations, and other processes at or near ambient conditions is sufficient motivation to assure that knowledge will develop rapidly. That has certainly been the case. From a few rather sparse reactions has come a well-developed and rapidly growing field. Commensurate with that growth has been the expansion of the literature in the field. Numerous books have been written on the subject some of which are listed as references at the end of this chapter. It is impossible in a textbook to do more than describe a few of the most important types of processes and show the general nature of the field. The interested reader should consult additional sources to become familiar with the details of the catalytic processes. In the sections that follow, several types of reactions will be described, and the role of the complexes that function as catalysts will be shown. Keep in mind that each of the schemes is made up of a series of steps most of which are identical to some of the reactions described earlier.

22.2.1 Hydrogenation
In order for a complex to function as a hydrogenation catalyst, it is necessary for it to be able to bond to hydrogen and to the alkene. This requires the complex to be able to add hydrogen in an oxidative addition reaction, and that process is characteristic of a metal in a low oxidation state. After the hydrogen and alkene are part of the coordination sphere of the metal, a facile means must be available for them to react. The overall reaction can be written as

$$H_2 + \text{alkene} \;\rightarrow\; \text{alkane} \tag{22.26}$$

A complex that meets the requirements for a hydrogenation catalyst is $RhCl(P\phi_3)_3$ where ϕ is a phenyl group, C_6H_5. This complex is a coordinatively unsaturated 16-electron species that has become known

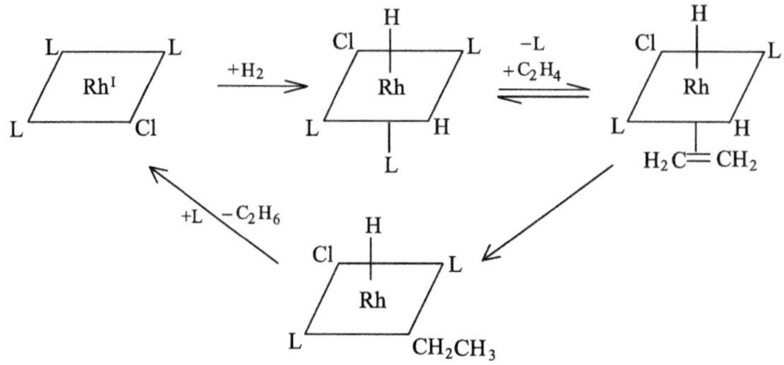

FIGURE 22.8
The hydrogenation of alkenes using Wilkinson's catalyst. L is $P(C_6H_5)_3$ or perhaps a solvent molecule.

as Wilkinson's catalyst because it was developed and utilized by the late Sir Geoffrey Wilkinson (who also received a Nobel Prize in 1973 for work on the structure of ferrocene). This versatile catalyst is effective in the hydrogenation of several types of compounds, which include alkenes and alkynes. Figure 22.8 shows the process that leads to hydrogenation of ethylene. Although the catalytic scheme is illustrated as if the hydrogenation of ethylene were occurring, more useful reactions involve hydrogenating longer chain alkenes. In fact, ethylene does not undergo hydrogenation very readily under these conditions, but replacing a hydrogen atom on the C_2H_4 group with an alkyl group, R, makes the scheme fit the general case.

The overall catalytic process can be considered as a series of steps. The first step involves the oxidative addition of hydrogen to give a complex having octahedral coordination. Addition of the alkene could occur by either an associative or dissociative pathway in which a molecule of $P\phi_3$ leaves. However, on the basis of NMR studies, it is believed that a small amount of this complex undergoes loss of one of the $P\phi_3$ ligands to give a five-bonded complex. It is also believed that it is possible for this intermediate to be interconverted between trigonal bipyramid and square base pyramid structures (as shown in Eqn (22.27)) because of its fluxional nature. It may be that the loss of the $P\phi_3$ ligand *trans* to H is facilitated as a result of a *trans* effect. It is to this intermediate that the alkene attaches. In order to form the reactive intermediate, a ligand must be lost and the large size of the $P\phi_3$ ligand seems to be one factor that enhances its dissociation.

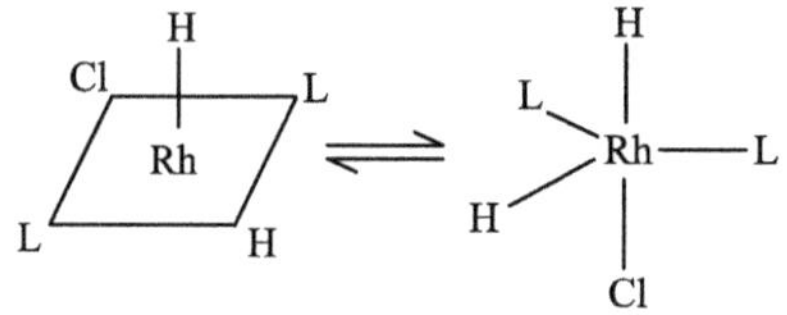

$$(22.27)$$

The reaction scheme shown in Figure 22.8 consists of a series of steps that represents the types of elementary reactions described earlier. The first step is the oxidative addition of hydrogen with the hydrogen atoms occupying *cis* positions. Subsequently, the entry of a molecule of triphenylphosphine into the coordination sphere occurs as a hydrogen transfer to the alkene occurs. This is followed by another hydrogen transfer that occurs in conjunction with reductive elimination of the alkane to regenerate the catalyst. Part of the interest and utility of this hydrogenation scheme is that it can be carried out at essentially ambient conditions. This represents a radical departure from the high temperature and pressure required when metal catalysts are used in hydrogenation. The mechanism shown in Figure 22.8 is only one of at least three that have been proposed, and it may deviate in some details from the actual mechanism (see structures I and II below). Keep in mind that positions indicated as being vacant may be occupied by a coordinated solvent molecule.

I   II

In the scheme shown in Figure 22.8, the rate-determining step is alkene insertion (viewed by some as hydrogen transfer). Because the hydride ion is a nucleophile, the reaction can be considered as a nucleophilic attack, which is influenced by the electron density in the alkene. A most interesting and important study dealing with this issue was reported by Nelson, Li, and Brammer (2005), in which the ionization potentials of a series of alkenes (a measure of electron density and availability) were correlated with the relative rate of reaction. The alkene that served as an index for comparing the rates was $CH_2=CHCH_2CH_2CH_2CH_3$, for which the k was set equal to 100. Rates for reactions of other alkenes (expressed by the k_{rel} values) were compared to that value, and they ranged from 1.4 for $(CH_3)_2C=C(CH_3)_2$ to 410 for $CH_2=CH-CH_2OH$. Plots of ionization potential versus log k_{rel} were generally linear and especially so for a series of terminal alkenes. The negative slope of the plots is consistent with nucleophilic attack of hydride on the alkene that causes the alkene to change from η^2 to η^1 bonding. Other good correlations were established for the relative rate of hydrogenation of alkenes and the energy of the lowest unoccupied molecular orbital. Although the correlations do not distinguish between mechanisms with regard to the possible transition states, they demonstrate that nucleophilic attack of hydride on the alkene is the rate-determining step in the process.

Some disagreement exists regarding the structure of the transition state (whether H or $P\phi_3$ is *trans* to the alkene) and whether solvent molecules occupy sites that are apparently vacant. In spite of some uncertainty regarding these details, the major issues regarding the catalyzed hydrogenation of alkenes using Wilkinson's catalyst are fairly well understood.

Following the successful use of Wilkinson's catalyst, other complexes having similar characteristics were studied. These include $RhH(P\phi_3)_3$, $RhHCl(P\phi_3)_3$, and $RhH(CO)(P\phi_3)_3$. The last two of these are 18-electron complexes, so it is necessary for a ligand to leave in order to form a bond to a molecule of the reacting alkene. Hydride complexes contain one hydrogen atom that can transfer to the alkene (or to which the H can transfer) to form an alkyl group. After the transfer, the reaction with gaseous hydrogen reforms the hydride complex, and a second hydrogen transfer during which reductive elimination of the alkane occurs completes the process. Some catalysts have been designed that contain a different ratio of $P\phi_3$ to rhodium, and others have been studied that contain different phosphine ligands having a different basicity than $P\phi_3$. These catalysts and those that contain more basic phosphines have not been found to be more effective than the intermediates shown in Eqn (22.27). The nature of the alkene is also a factor in the catalytic process.

Another important use for Wilkinson's catalyst is in the production of materials that are optically active (by what is known as enantioselective hydrogenation). When the phosphine ligand is a chiral molecule and the alkene is one that can complex to the metal to form a structure that has R or S chirality, the two possible complexes will represent two different energy states. One will be more reactive than the other, so hydrogenation will lead to a product that contains predominantly only one of the diastereomers.

Although there is a large amount of interesting chemistry associated with the $Co(CN)_5^{3-}$ ion, it is the ability to function as a catalyst in homogeneous systems that is the issue in this chapter. Because $Co(CN)_5^{3-}$ has the ability to split hydrogen (as shown in Eqn (22.10)), it is not surprising that it can function as a catalyst in hydrogenation reactions. The hydrogenation of 1,3-butadiene is illustrated in Figure 22.9.

As shown earlier, $Co(CN)_5^{3-}$ has the ability to split hydrogen molecules as a result of an oxidative addition reaction.

$$H_2 + 2\ Co(CN)_5^{3-}\ \rightarrow\ 2\ HCo(CN)_5^{3-} \tag{22.28}$$

Consequently, it is $HCo(CN)_5^{3-}$ that functions as a catalyst in hydrogenation processes. In the first step of the process shown in Figure 22.9, the alkene coordinates to $HCo(CN)_5^{3-}$ as one hydrogen atom is added to the molecule so that only one double bond remains. The monoene is bonded to the cobalt in η^1 fashion. In the second step, another $HCo(CN)_5^{3-}$ transfers hydrogen to the alkene, which undergoes reductive elimination and leaves having been converted to 1-butene.

One of the interesting aspects of this process is that the nature of the product depends on the concentration of CN^- in the solution. When the concentration of CN^- is high, the product is 1-butene, but at low concentrations of CN^-, the product is 2-butene. The explanation for this difference seems to be that when the cyanide ion concentration is low, the alkene changes bonding mode from η^1 to η^3 to complete the coordination sphere of the metal ion. The result is that the hydrogen atom from the second $HCo(CN)_5^{3-}$ is added to the terminal carbon atom in the η^3 arrangement and the product is 2-butene. In the presence of an excess of CN^-, the alkene maintains the η^1 bonding and the hydrogen

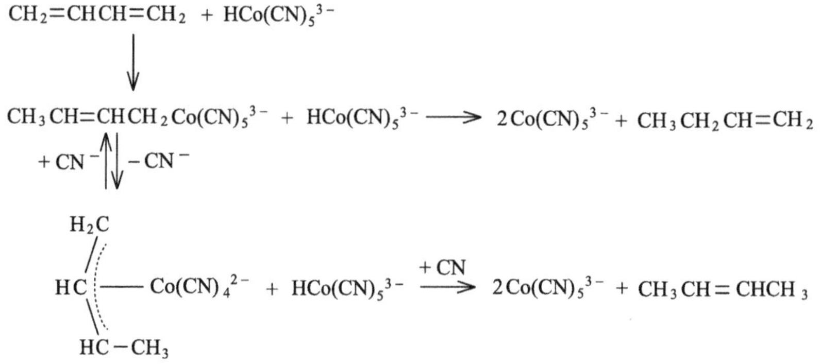

FIGURE 22.9
A plausible mechanism for hydrogenation of 1,3-butadiene catalyzed by $Co(CN)_5^{3-}$.

atom from the second $HCo(CN)_5^{3-}$ is added to carbon atom adjacent to the methyl group, which produces 1-butene.

22.2.2 Isomerization of Alkenes

The isomerization of alkenes is an industrially important process that can be used to prepare specific isomers used as monomers in polymerization. One step in the isomerization process involves a change in bonding mode of an alkene. For example, isomerization of 1-alkenes to produce 2-alkenes may take place as the alkene changes from η^2 to η^3 in the transition state. This mechanism can be shown as

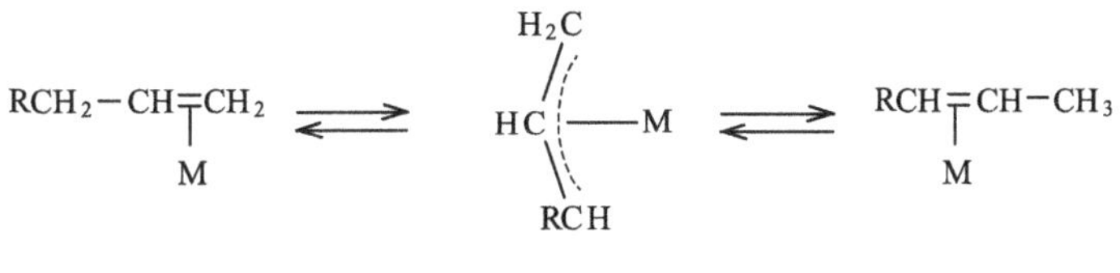

$$RCH_2-CH{=}CH_2 \; \rightleftarrows \; \underset{RCH}{\overset{H_2C}{HC{-}M}} \; \rightleftarrows \; RCH{=}CH-CH_3 \qquad (22.29)$$

One of the effective catalysts for isomerization of alkenes is a square planar complex of Rh^{1+}. That complex undergoes replacement of a ligand in the coordination sphere as shown in the reaction scheme presented in Figure 22.10. The isomerization process can be considered as taking place in a series of steps that involves a substitution reaction in which the alkene replaces a ligand L followed by the addition of H^+ and L to form a six-bonded complex. A hydrogen transfer changes the alkene from η^2 to η^1, which is reversed as the alkyl group is converted into a 2-ene as a ligand L enters the coordination sphere of the metal. Reductive elimination by loss of L and H^+ leads to a square planar complex in which the alkene is present as the 2-ene. Elimination of $RCH{=}CH_2CH_3$ as L enters the coordination sphere liberates the product and reforms the catalyst.

22.2.3 Polymerization of Alkenes (The Ziegler–Natta Process)

Ethylene and propylene are two monomers that are used in enormous quantities to produce poly-ethylene and polypropylene. These polymers are used for making a large variety of containers and other

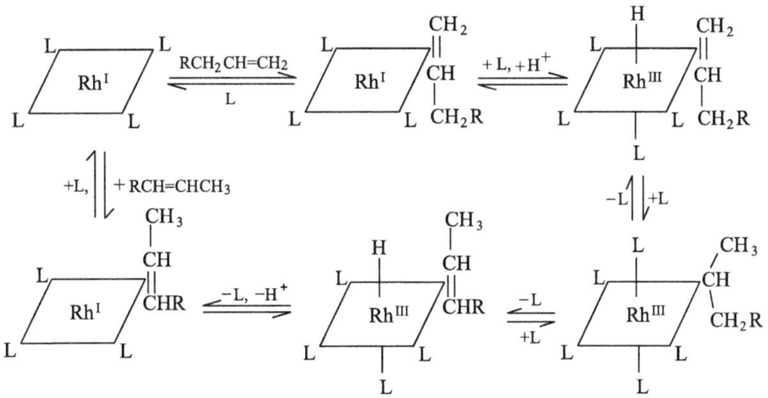

FIGURE 22.10
The reaction scheme for the isomerization of alkenes. L is a solvent molecule or Cl^-.

items. As a result, there has been a great deal of interest for many years in the reactions that yield these polymers.

The polymerization of alkenes is one of the important processes in which coordination chemistry is involved in what is known as the Ziegler–Natta process. Although polymerization of ethylene can be carried out at high temperature and pressure, a process by which this could be accomplished at room temperature and atmospheric pressure was discovered by Karl Ziegler in 1952. The process utilized $TiCl_4$ and $Al(C_2H_5)_3$ in a hydrocarbon solvent. It was later found by Giulio Natta that by using other catalyst combinations, polymers having stereoregularity could be produced. Ziegler and Natta shared the 1963 Nobel Prize for their enormously important work.

The polymerization reaction can be carried out using other combinations of a metal halide and a metal alkyl. Generally, a halide of titanium, vanadium, or chromium is used in combination with an alkyl of beryllium, aluminum, or lithium. Although the details of the process are not completely understood, the mechanism involves the replacement of a chloride ion by an ethyl group or an ethyl group attaching at a vacant site as the initial step. After a molecule of ethylene is coordinated to the titanium, it undergoes migration and insertion in the bond between Ti and C_2H_5. Another ethylene molecule becomes coordinated at the vacant site in the coordination sphere of the Ti and another insertion reaction occurs, which lengthens the chain and so on. The reaction scheme shown in Figure 22.11 illustrates this process. The crucial step in the process is the insertion reaction by which the hydrocarbon chain is lengthened.

Another reaction involving combining alkenes is the dimerization of ethene. That reaction is catalyzed by a rhodium complex, $RhClL_3$, as is illustrated in Figure 22.12. The first step in this process is an oxidative addition that is followed by a hydrogen transfer to a coordinated ethene molecule to convert it to an ethyl group. The third step involves the addition of a ligand as the C_2H_4 molecule undergoes an insertion reaction to give a coordinated butyl group.

22.2.4 Hydroformylation

Also referred to as the *oxo* process or *hydrocarbonylation*, *hydroformylation* is a route to producing an aldehyde from an alkene, hydrogen, and carbon monoxide. This process has been known for

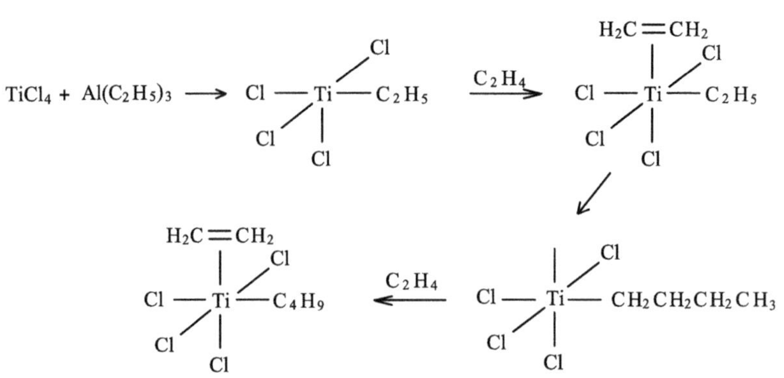

FIGURE 22.11
The catalytic scheme for polymerization by the Ziegler–Natta process.

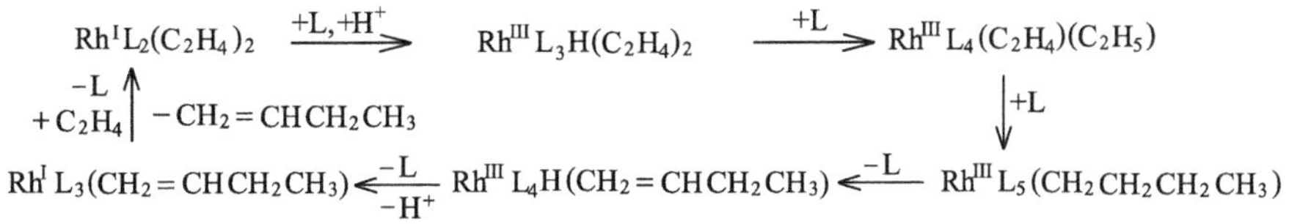

$$Rh^I L_2(C_2H_4)_2 \xrightarrow{+L, +H^+} Rh^{III} L_3H(C_2H_4)_2 \xrightarrow{+L} Rh^{III} L_4(C_2H_4)(C_2H_5)$$

$$\begin{array}{c} -L \uparrow \\ +C_2H_4 \end{array} \bigg| -CH_2=CHCH_2CH_3 \qquad\qquad\qquad \bigg| {+L} $$

$$Rh^I L_3(CH_2=CHCH_2CH_3) \xleftarrow[-H^+]{-L} Rh^{III} L_4H(CH_2=CHCH_2CH_3) \xleftarrow{-L} Rh^{III} L_5(CH_2CH_2CH_2CH_3)$$

FIGURE 22.12
Possible mechanism for the dimerization of ethene.

approximately 70 years, and it is still economically important because useful compounds are produced in enormous quantities by this means. The reaction is summarized by the following equation:

$$4\ RCH=CH_2 + 5\ H_2 + 4\ CO \rightarrow 2\ RCH_2CH_2CHO + 2\ RCH_2CH(O)CH_3 \qquad (22.30)$$

Figure 22.13 shows the scheme used to describe the hydroformylation process. The active catalyst is $HCo(CO)_3$, which is a 16-electron species that is coordinatively unsaturated. After that species is generated, the first step of the catalyzed process involves the addition of the alkene to the catalyst. In the next step, an insertion reaction occurs in which the alkene is inserted in the Co$-$H bond (nucleophilic attack by H^- on the alkene would accomplish the same result as described earlier), which changes the bonding from η^2 to η^1. This is followed by the migration of CO to the Co$-$alkyl bond, which generates the RC(O)$-$Co linkage. With the oxidative addition of H_2 and reductive elimination of the aldehyde, the process is complete. Although the cobalt carbonyl catalyst allows the reaction to be useful, it requires temperatures of approximately 150 °C and a pressure of 200 atmospheres. It is also generally the case that branched aldehydes are produced in greater quantity than straight chain products. This is a significant handicap because the linear aldehydes are converted to linear alcohols that are used in the manufacture of detergents. The oxo process is also important for the production of precursors for making polymers such as polyvinyl chloride.

In a more modern process, a rhodium complex having the structure

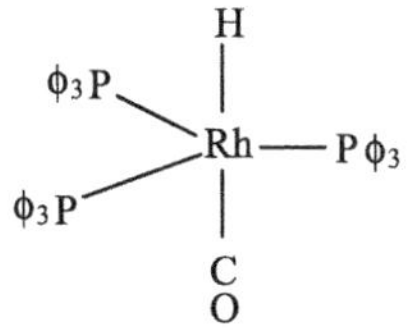

$$\begin{array}{c} H \\ \phi_3 P \diagdown \mid \\ {\diagup} Rh - P\phi_3 \\ \phi_3 P \diagup \mid \\ C \\ O \end{array}$$

is the starting complex for a catalytic cycle. The complete cycle is shown in Figure 22.14.

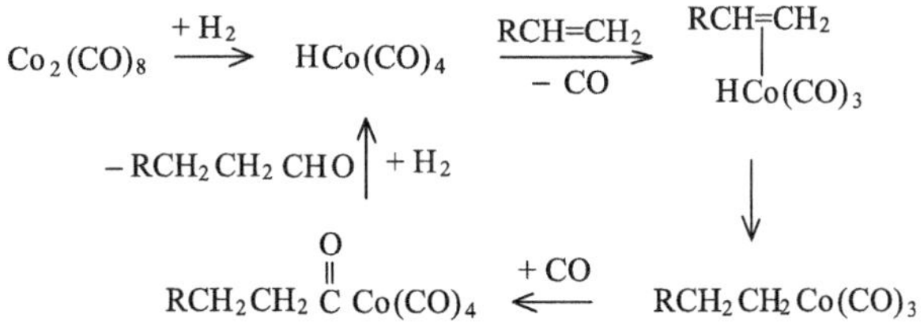

$$Co_2(CO)_8 \xrightarrow{+H_2} HCo(CO)_4 \xrightarrow[-CO]{RCH=CH_2} \begin{array}{c} RCH{=}CH_2 \\ \mid \\ HCo(CO)_3 \end{array}$$

$$-RCH_2CH_2CHO \uparrow\ +H_2 \qquad\qquad \downarrow$$

$$\underset{RCH_2CH_2 \overset{O}{\overset{\|}{C}} Co(CO)_4}{} \xleftarrow{+CO} RCH_2CH_2 Co(CO)_3$$

FIGURE 22.13
The reaction scheme for the hydroformylation process.

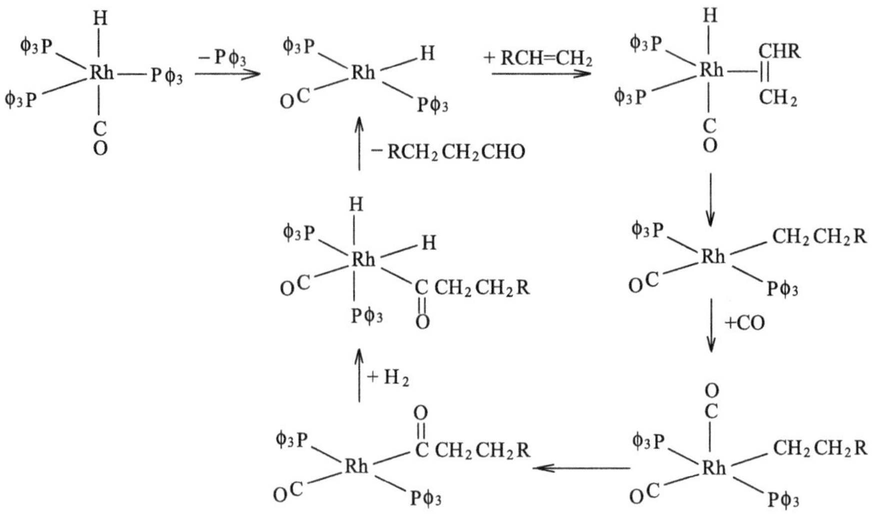

FIGURE 22.14
The hydroformylation process utilizing a rhodium catalyst.

The first step in hydroformylation with the rhodium catalyst involves the loss of a $P\phi_3$ ligand followed by coordination of the alkene. This is followed by a hydrogen transfer that is accompanied by a change in bonding mode of the alkene to the metal from η^2 to η^1. After adding CO to a vacant site to give a five-bonded complex, migration of CO and insertion in the metal–carbon bond leads to formation of $M–C(O)CH_2CH_2R$. An oxad reaction of H_2 followed by reductive elimination of RCH_2CH_2CHO completes the process. Hydroformylation is one of the most important processes in which a complex functions as a catalyst.

22.2.5 The Wacker Process
The Wacker process is one of the important industrial processes by which ethylene is converted to acetaldehyde. This process involves the addition of oxygen to an alkene, and the equation for the overall process can be shown as

$$C_2H_4 + H_2O + PdCl_2 \rightarrow CH_3CHO + Pd + 2\ HCl \tag{22.31}$$

It is believed that this reaction proceeds by the formation of a complex in which C_2H_4 is attached to the palladium.

$$PdCl_4^{2-} + C_2H_4 \rightarrow [PdCl_3C_2H_4]^- + Cl^- \tag{22.32}$$

In that complex, it may be that water reacts with the coordinated C_2H_4 to produce a σ-bonded CH_2CH_2OH group rather than an insertion reaction involving an OH group. The aldehyde is formed as H^+ is lost, and the palladium is produced as shown in Eqn (22.31). The palladium chloride catalyst can be recovered (the price of palladium is over \$500/oz as this is written) by the reaction with $CuCl_2$.

$$2\ CuCl_2 + Pd \rightarrow PdCl_2 + 2\ CuCl \tag{22.33}$$

In order to complete the cyclic process, cuprous chloride is oxidized:

$$4 \text{ CuCl} + 4 \text{ HCl} + O_2 \rightarrow 4 \text{ CuCl}_2 + 2 \text{ H}_2O \tag{22.34}$$

The Wacker process was developed in the late 1950s and is not widely used because other processes are more effective.

22.2.6 The Monsanto Process

Acetic acid is a major organic chemical that is manufactured in enormous quantities. It is used in numerous processes that include production of monomers as precursors for polymerization, as a solvent, and for many other industrial uses. One of the most important processes for the production of acetic acid is the Monsanto process. This method makes use of a rhodium catalyst that is obtained by a reaction involving RhI_3.

$$RhI_3 + 3 \text{ CO} + H_2O \rightarrow [RhI_2(CO)_2]^- + CO_2 + 2 \text{ H}^+ + I^- \tag{22.35}$$

The rate-determining step in the process is believed to be oxidative addition of CH_3I to the rhodium complex producing an octahedral rhodium +3 complex. The second step entails an insertion reaction in which CO is placed between the metal and the CH_3 group. This step could also involve migration of the CH_3 to a neighboring CO to give a $Rh-C(O)CH_3$ linkage or movement of the CO. After the addition of a molecule of CO, reductive elimination of CH_3COI generates the initial rhodium complex ion. The series of steps can be represented by the scheme shown in Figure 22.15. The acetyl iodide that is released in the last step of the process reacts with water to produce acetic acid:

$$CH_3COI + H_2O \rightarrow CH_3COOH + HI \tag{22.36}$$

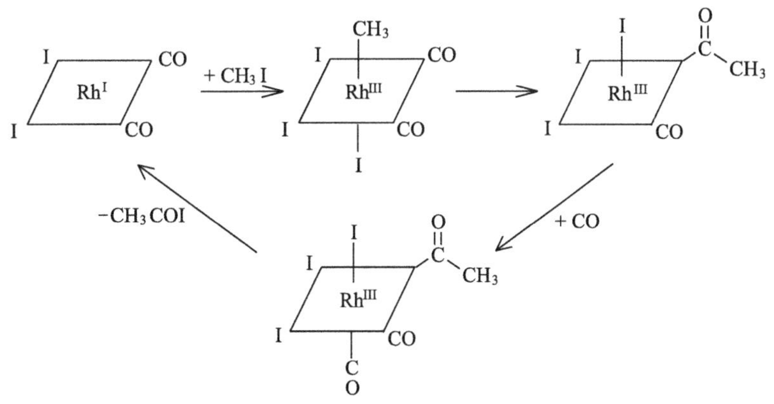

FIGURE 22.15
Reaction scheme showing the Monsanto process for making acetic acid.

In an alternate procedure, the acetyl iodide is converted to methyl acetate if methanol is the solvent for the process.

$$CH_3COI + CH_3OH \rightarrow CH_3COOCH_3 + HI \qquad (22.37)$$

Methyl iodide is regenerated by the reaction

$$CH_3OH + HI \rightarrow CH_3I + H_2O \qquad (22.38)$$

This reaction is efficient because of the soft—soft interaction of CH_3 and I and the hard—hard interaction that leads to the formation of water. As in the case of some other catalytic processes, the idealized scheme shown in Figure 22.15 does not necessarily represent the complete details of the process. Other products are known to include dimethyl ether and methyl acetate.

The applications of coordination compounds in catalysis that have been shown are by no means the only important cases. In fact, there are numerous reactions in which homogeneous catalysis forms the basis for a great deal of chemistry. From the examples shown, it should be apparent that this is a vast and rapidly developing field. It is also one that is important from an economic standpoint. Although the basic principles have been described in this chapter, the literature related to catalysis is extensive. For further details and more comprehensive reviews of the literature, consult the references listed.

References for Further Study

Atwood, J.D., 1985. *Inorganic and Organometallic Reaction Mechanisms*. Brooks/Cole, Belmont, CA. A good introductory survey of the field that contains many references. A second edition (1997) was published by Wiley-VCH, New York.

Basolo, F., Pearson, R. G. 1967. *Mechanisms of Inorganic Reactions*, 2nd ed., John Wiley, New York. A classic reference in reaction mechanisms in coordination chemistry.

Bhaduri, S., Mukesh, D., 2000. *Homogeneous Catalysis: Mechanisms and Industrial Applications*. Wiley-Interscience, New York.

Conradie, M.M., Erasmus, J.J.C., Conradie, J., 2011. *Polyhedron* 30, 2345—2353. An article dealing with oxidative addition of methyl iodide to rhodium complexes.

Cotton, F.A., Wilkinson, G., Gaus, P.L., 1999. *Advanced Inorganic Chemistry*, 6th ed. John Wiley, New York. One of the great books in inorganic chemistry. It also contains a great deal of material on organometallic chemistry and catalysis.

Gray, H.B., Stifel, E.I., Valentine, J.S., Bertini, I., 2006. *Biological Inorganic Chemistry: Structure and Reactivity*. University Science Books, Sausilito, CA.

Greenwood, N.N., Earnshaw, A., 1997. *Chemistry of the Elements*, 2nd ed. Butterworth-Heinemann, Oxford. A monumental reference work that contains a wealth of information about many types of coordination compounds.

Heaton, B., 2005. *Mechanisms in Homogeneous Catalysis: A Spectroscopic Approach*. Wiley-VCH, New York.

Jordan, R.B., 2007. *Reaction Mechanisms of Inorganic and Organometallic Systems*. Oxford University Press, New York.

Lukehart, C.M., 1985. *Fundamental Transition Metal Organometallic Chemistry*. Brooks/Cole, Belmont, CA. Contains good coverage on a broad range of topics.

Nelson, D.J., Li, R., Brammer, C., 2005. *J. Org. Chem.* 70, 761. Article dealing with the correlation of hydrogenation of alkenes with ionization potentials.

Stewart, C.A., Dickie, D.A., Tang, Y., Kemp, R.A., 2011. *Inorg. Chim. Acta* 376, 73—79. A report on insertion reactions in tin amides.

Szafran, Z., Pike, R.M., Singh, M.M., 1991. *Microscale Inorganic Chemistry: A Comprehensive Laboratory Experience*. John Wiley, New York. Chapters 8 and 9 provide procedures for syntheses and reactions of numerous transition metal complexes including Wilkinson's catalyst. This book also provides a useful discussion of many instrumental techniques.

Torrent, M., Solá, M., Frenking, G., 2000. *Chem. Rev.* 100, 439. A 54-page high level paper dealing theoretical aspects of processes in which metal complexes function as catalysts.

Van Leeuwen, P.W., 2005. *Homogeneous Catalysis: Understanding the Art.* Springer, New York.

QUESTIONS AND PROBLEMS

1. Explain how heating $Rh(CH_3)(P\phi_3)_3$ can produce CH_4.
2. Write equations for the following reactions.
   (a) The oxidative addition of CH_3I to $RhCO(P\phi_3)_3$
   (b) The reductive elimination of hydrogen from $HCo(CO)_4$
   (c) The oxidative addition of CH_3I to $Co(CN)_5^{3-}$
3. Write equations to show the hydrogenation of propene with the catalyst $H_3Co(P\phi_3)_3$.
4. Write equations for the following and draw structures for the products. In some cases, there may be more than one possible product.
   (a) Insertion of RNC in the $MnCH_3$ bond
   (b) Insertion of CS_2 in the M$-$C bond
   (c) Insertion of $F_2C=CF_2$ in the M$-$H bond
   (d) Insertion of RNC in the M$-$X bond
   (e) Insertion of SO_2 in a M$-$M bond
5. Explain why complexes containing Mn(I) are more likely to be good catalysts than those of Mn(III).
6. Consider the reaction (L = triphenylphosphine)

$$RuCl_2L_3 + H_2 \rightarrow HRuClL_3 + X$$

When triethylamine is present, this reaction is enhanced. Explain how and tell what the other product of the reaction, X, would be in that case.

7. Would the reaction

$$Fe(CO)_5 + I_2 \rightarrow cis\text{-}[Fe(CO)_4I_2] + CO$$

be oxidative addition, reductive elimination, or both? Explain.

8. In a study of the rate at which the ligand X is replaced by CO in the reaction

$$trans\text{-}[Cr(CO)_4XL] + CO \rightarrow [Cr(CO)_5L] + X$$

the rate was found to vary with the nature of L as follows:

L	$\Delta S^{\ddagger}$, e.u.	$\Delta H^{\ddagger}$, kcal/mol
$P(O\phi)_3$	0	32
CO	23	40
$P\phi_3$	12	36
$As\phi_3$	22	36

How would you interpret these results?

9. How would you explain to a person whose background in chemistry is limited the difference in behavior of $[Co(CN)_6]^{3-}$ and $[Co(CN)_5]^{3-}$ as potential catalysts?

10. Explain why the methyl migration reaction in $CH_3Mn(CO)_5$ has a rate when DMF is the solvent that is about 50 times that when mesitylene is the solvent. Give a reasonable estimate of the relative rate when the solvent is $CHCl_3$. Give a reasonable estimate of the relative rate when the solvent is n-C_6H_{14}. You may wish to consult Table 6.7.

11. For the reaction

$$L + CH_3Mn(cp)(CO)_3 \rightarrow CH_3C(O)Mn(cp)(CO)_2L$$

(where $L = PBu_3$, $P(OBu)_3$, $P\phi_3$, or $P(O\phi)_3$ in separate experiments), the rate is found to be almost independent of the nature of the entering group when the solvent is tetrahydrofuran (THF). Explain what this means in terms of the mechanism of substitution. When the solvent is toluene, the rates are much lower than when the solvent is THF, and there is a considerable difference in rate depending on the entering ligand (the rate when $L = PBu_3$ is about 16 times that when $L = P(O\phi)_3$). How would you interpret these observations?

12. When the reaction between $CH_3Mn(CO)_5$ and $AlCl_3$ takes place, the product has a structure in which there is a ring. Draw the structure of the product. When $AlCl_2C_2H_5$ is present, the rate of the reaction is less than 1% of what it is when $AlCl_3$ is present. Explain this observation.

13. In the reaction

$$L + cpRh(CO)_2 \rightarrow cpRh(CO)L + CO$$

the rate of substitution depends on nature of the entering ligand. For example, when L is PBu_3, $k = 4.3 \times 10^{-3}$ $M^{-1} s^{-1}$ and when L is $P(O\phi)_3$, $k = 7.5 \times 10^{-5}$ $M^{-1} s^{-1}$. Explain what this means in terms of the mechanism of substitution.

14. Explain why Ir is not as widely used as Rh in terms of the difference in coordination chemistry of the two metals.

15. For the oxidative addition of CH_3I to a series of rhodium(I) complexes containing β-diketones, Conradie, Erasmus, and Conradie, J. (2011) report data from their work and previous studies as follows. Test these data for the operation of a compensation effect. The β-diketones are designated as R_1–$COCHCO$–R_2, with R_1 and R_2 as indicated with Fc representing the ferrocenyl group.

R_1	R_2	$\Delta S^{\ddagger}$, J K^{-1} mol^{-1}	$\Delta H^{\ddagger}$, kJ mol^{-1}
CF$_3$	CF$_3$	−126	56
C$_6$H$_5$	CF$_3$	−123	51
CF$_3$	CH$_3$	−129	47
Fc	CF$_3$	−145.4	40
C$_6$H$_5$	CH$_3$	−127	30
CH$_3$	CH$_3$	−128	40
C$_6$H$_5$	C$_6$H$_5$	−115	49
Fc	CH$_3$	−51.6	62.3
Fc	Fc	−104	46.4

CHAPTER 23

Bioinorganic Chemistry

As analytical methods and techniques for determining molecular structure have become more sophisticated, so has the awareness that a small detail in a structure may have a major role in the function of that species. It has become increasingly apparent that such a small detail may be the presence of a metal ion. Thus, the relationship between inorganic chemistry and biochemistry has grown into the recognized discipline of bioinorganic chemistry with journals published specifically to report results of research in this hybrid area. Any separation between the fields of chemistry seems to have disappeared although numerous distinctions are sometimes still made. For example, adding an (inorganic) chlorine atom to a benzene molecule to produce chlorobenzene does not make the product part of "inorganic" chemistry. But if a zinc atom is contained in an enzyme having a molecular weight of 30,000, it suddenly becomes a "bioinorganic" structure because the presence of the zinc may cause extensive changes in the structure and function of the enzyme. To be sure, the molecule contains a metal, but the vast majority of the molecule is still "biochemical" in nature and most of its reactions are biochemical reactions. Nevertheless, it is appropriate to show at least some ways in which metal ions influence biochemical function even in an inorganic book.

From the outset, it is apparent that any attempt to cover a significant part of a discipline as vast as bioinorganic chemistry in a chapter of a general book on inorganic chemistry is doomed to failure. Entire volumes (even multiple volumes) have been written on the subject. Articles describing work in the field are published in numerous journals that deal with inorganic chemistry, biochemistry, and hybrid areas such as organometallics. Even organizing such a huge amount of information would be a daunting task. Although one may not be able to eat a whole pie, eating one piece can show the flavor and character of the pie. So it is when dealing with an enormous amount of material on a subject such as bioinorganic chemistry. In this chapter, the approach taken will be to concentrate on a few selected systems to show the flavor and character of the field but without any attempt to present a survey of the entire field. That task is addressed in some of the monographs listed in the suggested references at the end of the chapter.

23.1 WHAT METALS DO IN SOME LIVING SYSTEMS

When dealing with bioinorganic chemistry, it is ordinarily a metal that is the center of attention. In most instances, the metal is bound in some sort of complex in which the groups attached are large molecules that contain oxygen, nitrogen, or some other atom as the Lewis base. Consequently, this chapter will focus on the nature of the metal complexes rather than biochemical function.

773

Inorganic Chemistry. DOI: http://dx.doi.org/10.1016/B978-0-12-385110-9.00023-6

23.1.1 Role of Metals in Enzymes

When analyzing the role of metals (that term will also be used to denote metal ions), it is discovered that they perform functions that can be grouped in a few broad categories. One way in which metals function in biochemical structures is in enzyme activity. Enzymes are high molecular weight proteins (polypeptides) that facilitate reactions by providing a lower energy pathway for the reaction. It generally does that by binding to the reacting structure (known as the *substrate*) so that the enzyme–substrate complex is more reactive than the substrate alone. In fact, the substrate may not even undergo a reaction without the enzyme being present. The following categories summarize the ways in which enzymes may function as catalysts.

1. *Absolute specificity*, in which the enzyme catalyzes a single reaction.
2. *Group specificity*, in which the enzyme catalyzes reactions of only one type of functional group.
3. *Linkage specificity*, in which the enzyme alters the reactivity of a particular type of bond.
4. *Stereochemical specificity*, in which the enzyme catalyzes reactions of only one stereoisomer of a compound.

Enzymes may not function well or at all unless some other species known as a *cofactor* is present. An enzyme alone is referred to as the *apoenzyme*, and the combination of an enzyme and a cofactor is known as the *holoenzyme*. Among the species that function as cofactors are organic compounds that interact with the enzyme. If the organic moiety is strongly attached to the enzyme, it is called a *prosthetic group*, but if it is loosely bound to the enzyme, it is referred to as a *coenzyme*. For the purposes of this discussion, the most interesting cofactors are metal ions. Metal ions are required by approximately one-fourth of known enzymes for them to function. Depending on the type of enzyme, the appropriate metal ion cofactor may be Zn^{2+}, Mg^{2+}, Ca^{2+}, K^+, Fe^{2+}, or Cu^{2+}. A sizeable number of enzymes are sometimes called *metalloenzymes* because they have active sites that contain a metal. Vitamins are sources of metal ions for many enzymes.

The presence of some substances may hinder the action of an enzyme. Such substances are known as *inhibitors*, and in some cases, the inhibitor may be a metal. It is not necessary to describe all the ways in which an inhibitor may reduce the activity of an enzyme, but one way is by binding to the substrate. This is known as *competitive inhibition*, and it applies to cases in which the inhibitor competes with the substrate in binding to the enzyme. In *noncompetitive inhibition*, the inhibitor binds to the enzyme and alters its structure so it can no longer bind to the substrate. In some instances, certain metal ions function as inhibitors of enzyme action, which can be a cause of the toxicity described in Section 23.1.2.

One of the most important metals with regard to its role in enzyme chemistry is zinc. There are several significant enzymes that contain the metal, among which are carboxypeptidase A and B, alkaline phosphatase, alcohol dehydrogenase, aldolase, and carbonic anhydrase. Although most of these enzymes are involved in catalyzing biochemical reactions, carbonic anhydrase is involved in a process that is essentially of an inorganic nature. The apoenzyme (that does not contain the metal ion) does not function as a catalyst for the reaction. The conversion of carbon dioxide to bicarbonate can be shown as

$$CO_2 + H_2O \xrightarrow{\text{Carbonic anhydrase}} H^+ + HCO_3^- \tag{23.1}$$

This reaction is essential in maintaining a constant pH in blood by means of the bicarbonate buffer system. Carbonic anhydrase, which contains a single zinc atom in its structure, has a molecular weight of about 30,000. The enzyme contains a Zn^{2+} ion bound to histidine groups that have the structure

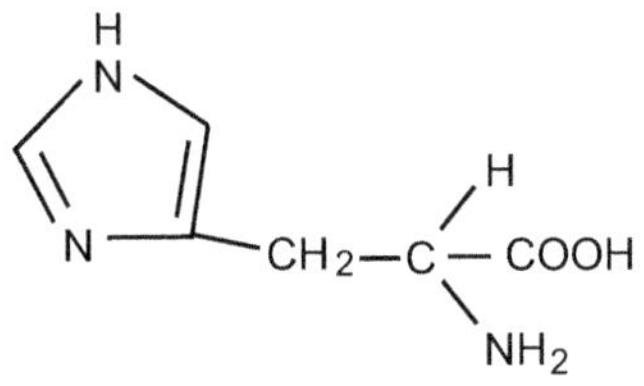

The exact role of the catalyst is not known, but it is believed to involve hydrolysis that can be represented by the equation

$$h_3ZnOH_2 + H_2O \rightarrow H_3O^+ + h_3ZnOH^- \tag{23.2}$$

where h represents a coordinated histidine group. With the product being more basic, its reaction with CO_2 (an acidic oxide) is facilitated:

$$h_3ZnOH^- + CO_2 \rightarrow HCO_3^- + h_3Zn \tag{23.3}$$

The structure representing the zinc complex in the enzyme is shown in Figure 23.1. The mechanism of the reaction involving the action of carbonic anhydrase is shown in Figure 23.2. The function of the zinc ion is to make the OH bond more polar, thereby enabling CO_2 to bind to the oxygen atom. Although

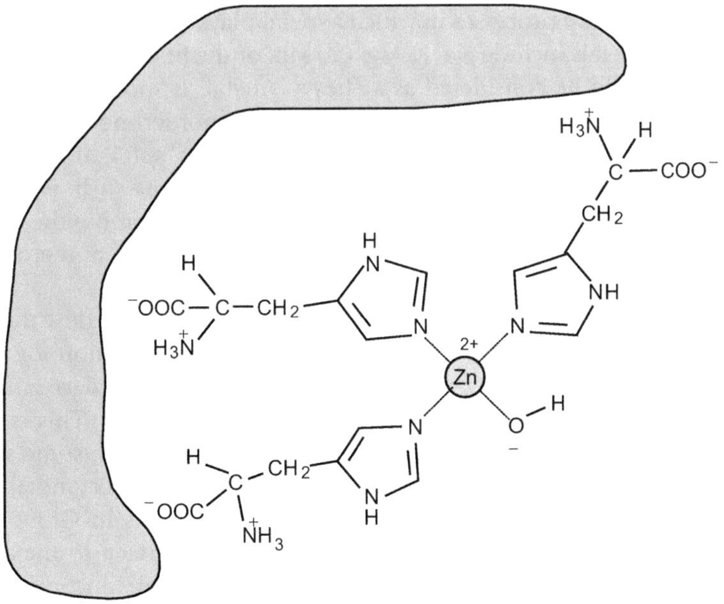

FIGURE 23.1
A representation of the structure of carbonic anhydrase. In this structure, zinc is surrounded tetrahedrally by three histidine molecules and one hydroxide ion or a water molecule.

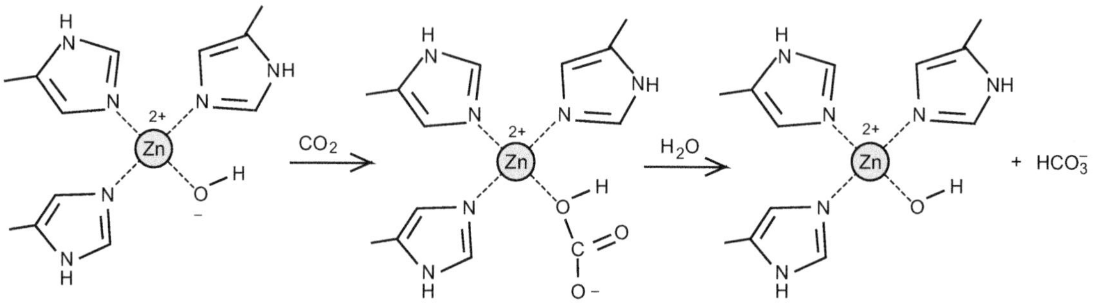

FIGURE 23.2

The mechanism by which carbonic anhydrase converts carbon dioxide into bicarbonate ions. Only the coordinated portions of the histidine residues are shown.

the mechanism is not completely understood, the rate of the reaction shown in Eqn (23.3) is increased by several orders of magnitude (up a factor of up to 10^9) by the enzyme. It is also been found that replacement of Zn^{2+} by Co^{2+} or Mn^{2+} results in a decrease in the effectiveness of the catalyst. Action of the enzyme is inhibited by species such as H_2S, CN^-, and others that bind strongly to the metal. Although the metal ion involved in the action of carbonic anhydrase is Zn^{2+}, other enzymes contain ions of Mg, Mn, K, Cu, Fe, Ni, and Mo.

23.1.2 Metals and Toxicity

The vast majority of biochemical processes in which a metal plays a role involve only a relatively small number of metals. Those metals include Na, K, Mg, Ca, Mo, or the first-row transition metals from V to Zn. Only molybdenum could be considered as a "heavy" metal. It should also be observed that the metal ions constitute those that can be considered as hard or borderline in hardness. It is a general property that ions of heavy metals having low charge (that is to say "soft") are toxic. These include Hg, Pb, Cd, Tl, and numerous others. Some heavy metals bind to groups such as the sulfhydryl (−SH) group in enzymes, thereby destroying the ability of the enzyme to promote the reaction in a normal way. In fact, a rather good relationship exists between the softness of numerous metals and their toxicity. However, beryllium is an extremely toxic metal that is not soft.

Metal toxicity arises from different biochemical functions that are affected by the presence of a toxic substance. In the case of metals, one way that toxicity arises is by substitution for another metal. One instance of this behavior relates to the toxicity of cadmium, which can replace zinc, a metal found in several enzyme systems. When this happens, the enzyme loses its activity. This is believed to be one factor related to the toxicity of beryllium, which can replace magnesium in some structures. There are numerous nonmetallic substances that are toxic with some of the most notable being cyanides, carbon monoxide, and hydrogen compounds of the heavier elements in Groups VA and VIA. The toxicity of these materials is related to their action as ligands that attach to metals in specific structures, which thereby prevents the metals from serving their normal functions. Perhaps, the best-known example of this type of behavior is carbon monoxide, which binds to iron in hemoglobin and myoglobin to prevent oxygen molecules from coordinating so that the oxygen-carrying ability of the iron is lost. In still other cases, toxic materials function by causing changes in the structure of some molecules to which the toxic substance is bound. Because there are several ways in which toxins function and many biochemical species with which they can interact, the list of toxic materials is extensive.

Of the metals that are listed above as the major participants in biochemical functions, some (such as Mg and Zn) rarely change oxidation states. Therefore, metals such as these are involved in processes in which there is no redox chemistry taking place. These metals function in some other way. On the other hand, metals such as Fe, Mn, Mo, and Cu can change oxidation states more easily so they are the metals that participate in redox reactions. For example, the role of iron in oxygen transport requires it to bond to oxygen and thereby, at least formally, to become oxidized in the process. There are other instances of this type of behavior. As mentioned earlier, the list of metals that are involved in the vast majority of biochemical processes is not a particularly long one.

23.1.3 Photosynthesis

The process by which plants convert water and carbon dioxide into carbohydrates and oxygen is the basis for the chemistry of life as we know it. This process produces the oxygen needed for respiration, and the carbohydrates produced are the foods for all forms of animal life. When glucose is the product, the overall reaction, which requires energy in the form of electromagnetic radiation, can be summarized by the equation

$$6\,CO_2 + 6\,H_2O \xrightarrow{h\nu} C_6H_{12}O_6 + 6\,O_2 \tag{23.4}$$

Because the reaction requires energy from light, it is known as *photosynthesis*. The equation looks simple, but photosynthesis is anything but that. The structures that are responsible for the absorption of light in order for its energy to be used are the chlorophylls, which contain porphyrin-type ligands. The porphyrin structure is derived from the basic unit known as *porphin*, which has the structure

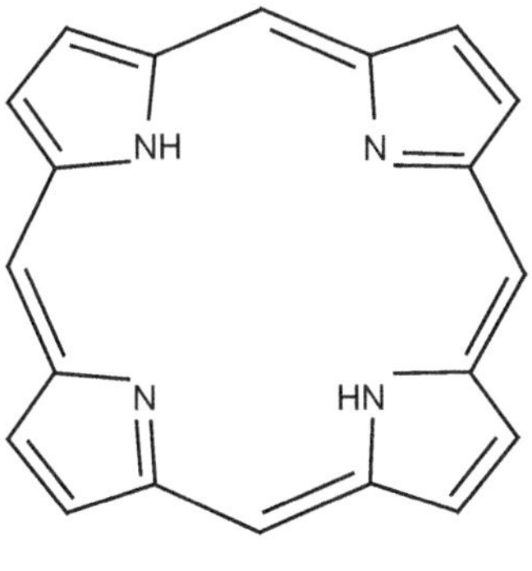

Porphin

This structure shows the four nitrogen atoms that provide the electron pairs for forming bonds to metals. The chlorophylls consist of several members of a series, three of which are chlorophyll *a*, *c*, and *d* that have structures derived from a molecule known as *chlorin*, which has the structure

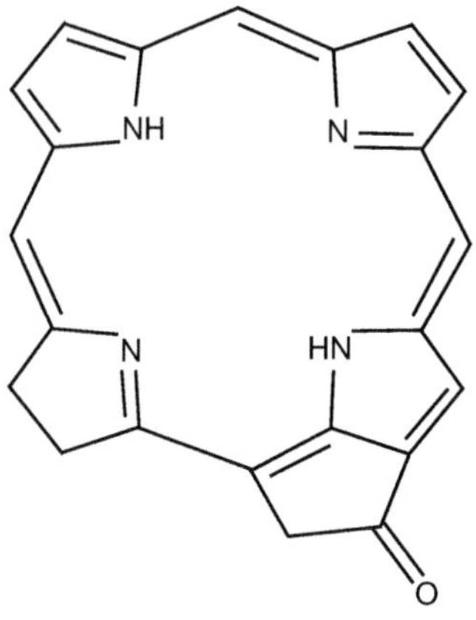

Chlorin

In chlorophyll *a*, which has the structure shown in Figure 23.3, the magnesium ion is not coplanar with the four nitrogen atoms to which it is bound but rather it is positioned approximately 30–50 pm above the plane of the nitrogen atoms, and it is usually bonded to at least one other ligand such as water.

Chlorophyll *a* units are arranged as if the molecules were layers that are held together by hydrogen bonding primarily by the water molecules between them.

The mechanism by which light is absorbed to provide the energy for photosynthesis is complex, but some steps are understood. Chlorophyll absorbs visible light most strongly in two regions that give rise to two absorption bands that are known as the Q band in the red region and the Soret band

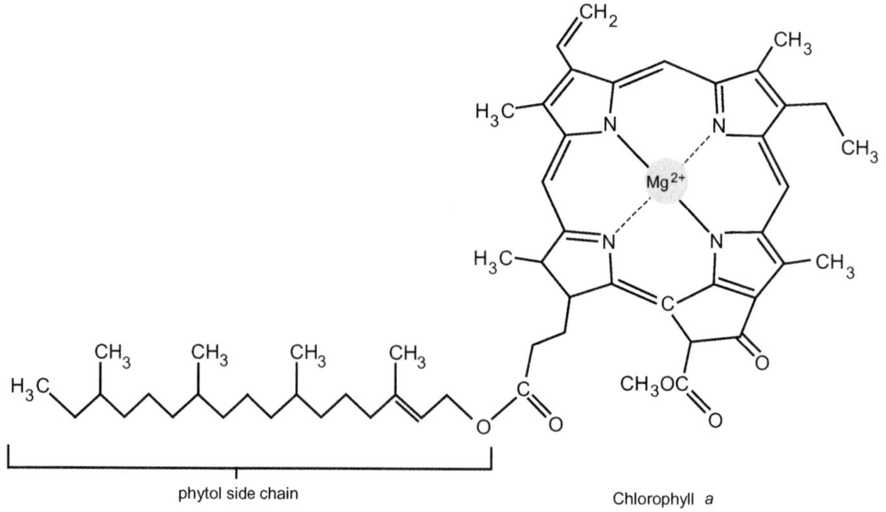

Chlorophyll *a*

FIGURE 23.3
The structure of chlorophyll *a*.

in the region near the blue to ultraviolet region. These bands are common absorptions for most structures that have porphyrin rings. These absorptions are associated with excitation of electrons in the porphyrin ring from π to π^* orbitals. Because the region near the middle of the visible spectrum (associated with a green color) represents a region of low absorption, the leaves appear green. Regions known as *organelles* are groups of molecules enclosed in a membrane that perform the function of absorbing the light. *Chloroplast* is the name applied to a region that is involved with photosynthesis. Two such centers are involved in photosynthesis, and they are referred to as photosystem I and photosystem II.

Chloroplasts are the organelles that contain the light-absorbing species that are contained in the thylakoid membrane within a leaf. There are two functional systems in the overall process. Photosystem I (which is sometimes referred to as P700) is the part of the process that represents the absorption of light having a wavelength of 600–700 nm by chlorophyll *a* and other pigments. Photosystem II (which is sometimes referred to as P680) involves the absorption of light having a wavelength of 680 nm by chlorophylls *a* and *b*. Plants have cells that function in both photosystems I and II, but only photosystem I is present in bacteria that support photosynthesis.

In photosystem I, absorption of a photon leads to an excited state that functions as a reducing agent. The electrons are passed from one species to another with several intermediate species that include ferredoxin (a protein containing iron and sulfur) before finally reducing CO_2. In photosystem II, electrons are transferred to a series of intermediates, of which a cytochrome *bf* complex is one entity. Ultimately, the transfer of electrons leads to the reaction

$$2\,H_2O \rightarrow O_2 + 4\,H^+ + 4\,e^- \tag{23.5}$$

The oxygen liberated during photosynthesis comes from water rather than CO_2.

23.1.4 Oxygen Transport

From the standpoint of the relationship of almost all animal life, the transport of oxygen by heme (also written as haem in some literature) is the basis for respiration. Heme is one of several proteins that contain iron. Others include materials such as myoglobin, ferritin, transferritin, cytochromes, and ferredoxins. In order to transport the oxygen required, the body of an average adult contains approximately 4 g of iron. In species such as mollusks, oxygen is transported by proteins that contain copper instead of iron. These are sometimes referred to as the copper blues. The structure of heme is shown in Figure 23.4.

In oxygen transport, the gas is absorbed in lung tissue and transported by heme to muscle tissue where it is transferred to myoglobin. Some oxygen is stored in myoglobin and released during exertion to oxidize glucose. After transferring an oxygen molecule to myoglobin, hemoglobin attaches a molecule of carbon dioxide and transports it to the lungs, from which it is exhaled. The oxygen transport occurs through the arterial network, and carbon dioxide is transported through the veins.

When an oxygen molecule is attached to the iron in the heme structure, the Fe^{2+} changes from high spin to low spin but the iron remains in the +2 oxidation state. In the heme structure, the Fe^{2+} ion resides above the plane of the four nitrogen atoms in the porphyrin structure and an imidazole group is attached to the iron on the side away from the plane of the four nitrogen atoms. This results in a structure that can be shown as

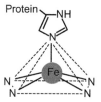

Heme

FIGURE 23.4
The structure of heme.

When an oxygen molecule is attached to the iron and it changes from high spin to low spin, the size of the Fe^{2+} changes. The ionic radius of Fe^{2+} in the high-spin state is approximately 78 pm, but in a low spin environment, it is about 61 pm. This reduction in size is sufficient to allow the Fe^{2+} ion to fit between the nitrogen atoms in the porphyrin ring and it resides in the same plane as the nitrogen atoms. The structure of heme with an oxygen molecule attached is represented as follows. This structure shows the angular orientation of the oxygen molecule and that the Fe^{2+} resides in the same plane as the four nitrogen atoms:

Hemoglobin is composed of four parts or subunits arranged in the form of a helix or spiral that is attached to a heme group. Each heme group contains Fe^{2+}, which is a potential site for attaching an oxygen molecule. The subunits are referred to as the α and β structures so that the four subunits consist of two α and two β structures. The α and β structures consist of different protein chains that are held together by bonds between $-NH_3^+$ in one chain and $-CO_2^-$ in another. The protein chains have a glutamic acid residue in normal hemoglobin, but there is a valine group in the hemoglobin if the individual is afflicted with sickle cell anemia. The subtle difference in structure causes the chains to be folded in a different manner that does not allow oxygen to bond as well, thereby affecting oxygen transport.

By the usual standards of coordinate bond strength, the binding of oxygen to iron in hemoglobin is quite weak. On the other hand, groups such as CN^-, CO, H_2S, and others bind strongly to Fe^{2+}, which makes it impossible for O_2 to attach. As a result, these substances are highly toxic, and they function by preventing the uptake of oxygen.

Myoglobin has a structure that is essentially one-fourth of that of hemoglobin. Therefore, it can bind to only one molecule of oxygen. If myoglobin containing no oxygen is oxidized, the iron is converted to Fe^{3+} and results in the structure known as *metmyoglobin*. This material is not an effective oxygen transport agent.

Oxygen is carried from the lungs to muscle tissue by the hemoglobin where it released to the myoglobin. In order for this to happen, it is necessary for the binding of oxygen by myoglobin to be stronger than that of hemoglobin. However, myoglobin has only one iron to bond to oxygen, whereas hemoglobin has four such sites. It is known that when one oxygen molecule bonds to hemoglobin, the other three sites attach oxygen more readily. This is known as the *cooperative effect*, and it is believed to arise as a result of changes in conformation of the chains. When the first oxygen molecule bonds to iron in hemoglobin and the iron moves into the plane of the four nitrogen atoms, the histidine ring is also drawn toward that plane. This results in subtle changes in conformation that make it easier for subsequent oxygen molecules to bind to the remaining iron atoms. Binding of oxygen in hemoglobin depends on the pH of the medium, and the ability to add oxygen decreases as the pH of the medium increases. This relationship is related to the fact that CO_2 and H^+ affect the cooperativity, and the concentrations of these species are pH dependent. As a result of being discovered by Christian Bohr, this behavior is referred to as the *Bohr effect*.

Hemoglobin and myoglobin differ in terms of their ability to add oxygen. Myoglobin has a greater attraction for oxygen, but hemoglobin has four times as many binding sites. Therefore, the curves showing the degree of oxygen saturation as a function of oxygen pressure are quite different. Figure 23.5 shows the variation of the fraction of available sites covered as a function of oxygen pressure.

If we represent the equilibrium for myoglobin reacting with oxygen as

$$Mb + O_2 \rightleftharpoons MbO_2 \tag{23.6}$$

the equilibrium constant can be written as

$$K = \frac{[MbO_2]}{[Mb] \times p} \tag{23.7}$$

In this expression, p represents the pressure of oxygen (which replaces concentration because oxygen is a gas). If we let f represent the fraction of the available sites that have oxygen attached, $(1-f)$ is the

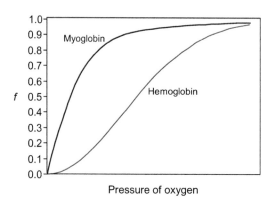

FIGURE 23.5
The qualitative relationship between the oxygen-binding curves for hemoglobin and myoglobin.

fraction of the sites that are still available where oxygen can attach. We can represent the rate of oxygen binding and the rate of oxygen detachment as being equal at equilibrium, which results in the following relationship:

$$k_a(1 - f)p = k_d f \tag{23.8}$$

In this equation, k_a is the rate constant for attaching oxygen and k_d is the rate constant for detaching oxygen. Solving for f gives

$$f = \frac{k_a p}{k_d + k_a p} \tag{23.9}$$

The equilibrium constant can be written in terms of the rate constants as

$$K = \frac{k_a}{k_d} \tag{23.10}$$

Solving for k_a, substituting the result in Eqn (23.9), and simplifying give

$$f = \frac{Kp}{1 + Kp} \tag{23.11}$$

It is interesting to note that this equation is of the same form as that which arises for adsorption of a reacting gas on a solid surface. In adsorption studies, the relationship is known as the Langmuir isotherm. The curve showing uptake of oxygen by hemoglobin fits a different relationship in which the pressure of oxygen is raised to a power that is equal to about 2.8. From Figure 23.5, it can be seen that at low pressure, the binding of oxygen by myoglobin is much greater than that of hemoglobin. At higher pressure of oxygen, the efficiency of oxygen binding is approximately the same.

23.1.5 Cobalamins and Vitamin B$_{12}$

It was recognized long ago that sheep were subject to a disease that was thought to be anemia. It was later found that the disease, known as pernicious anemia in humans, was the result of a deficiency in cobalt, not iron. Specifically, the deficiency is now recognized to be in vitamin B_{12}, which is a coenzyme that is required for making red blood cells. Cobalamins are derivatives of vitamin B_{12}. The vitamin B_{12} structure is similar to that of heme in that there are four nitrogen atoms bonded to the metal. However, the nitrogen atoms are contained in a structure known as the *corrin* ring, which can be shown as

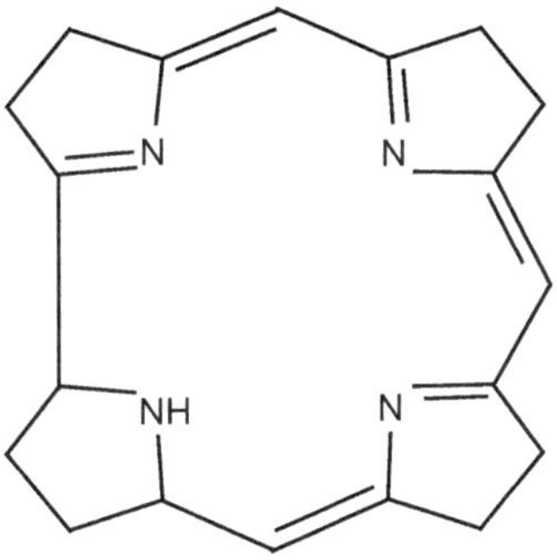

Corrin

In the corrin structure, the pyrrole rings are partially reduced and there is one hydrogen atom bonded to nitrogen. Also, there is one CH_2 group missing (on the left-hand side of the structure shown above). The simplified structure of vitamin B_{12} is shown in Figure 23.6. The cobalt bonded to the four nitrogen atoms is part of one five-membered chelate ring and one that contains six members in the equatorial positions.

The complete structure of vitamin B_{12} is shown in Figure 23.7. One of the most unusual features of the vitamin B_{12} structure is that cobalt is bonded directly to carbon. In heme, iron is bonded to five nitrogen atoms, but the sixth position is vacant until oxygen attaches. In vitamin B_{12}, the sixth position is that bonded to the group indicated by R in the structure shown. When contained in structures such as the cobalamins, cobalt can be reduced to +2 or even +1 as it is in vitamin B_{12}. There are several types of reactions for which vitamin B_{12} is a very effective catalyst. These include some that can be classified as exchange, oxidation–reduction, and methylation, and there are numerous derivatives known of

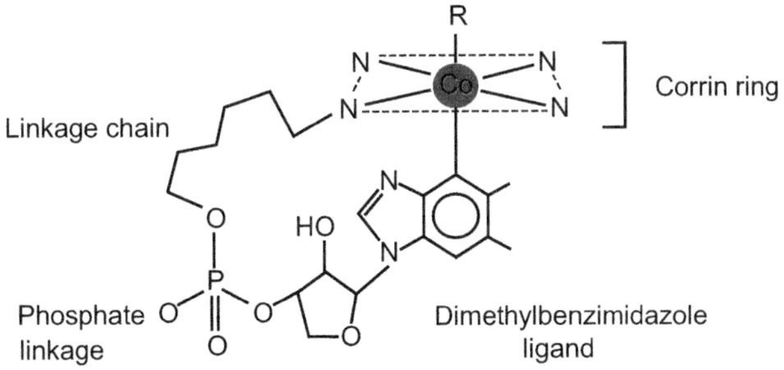

FIGURE 23.6
A simplified structure of vitamin B_{12}.

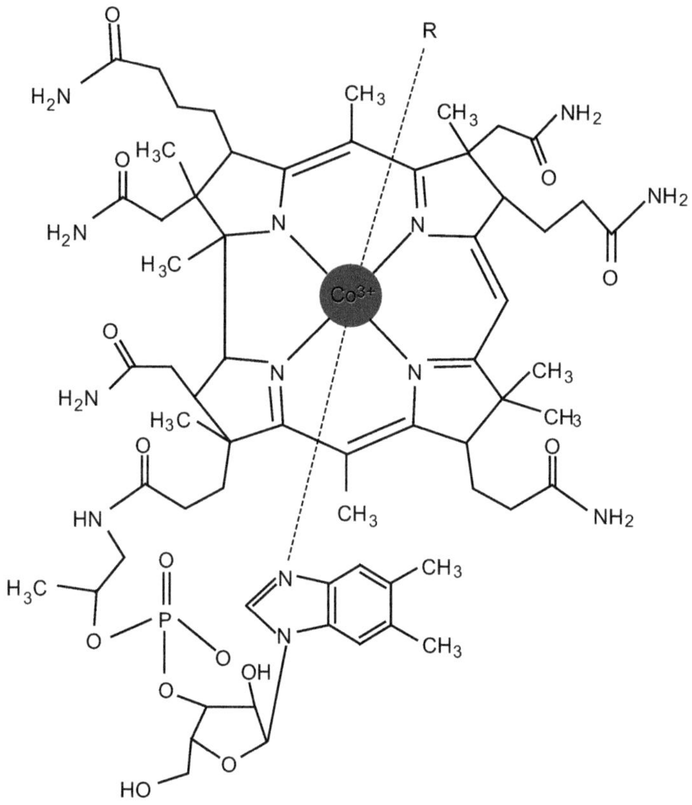

FIGURE 23.7
The structure of vitamin B_{12}.

vitamin B_{12}. One of the interesting reactions that is catalyzed by vitamin B_{12} involves moving a group from one atom to another as illustrated by the following:

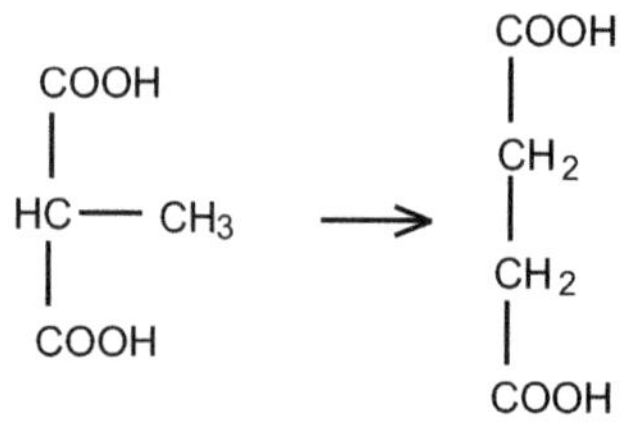

$$
\begin{array}{c}
\text{COOH} \\
|\\
\text{HC}\!-\!\text{CH}_3 \\
|\\
\text{COOH}
\end{array}
\quad\longrightarrow\quad
\begin{array}{c}
\text{COOH} \\
|\\
\text{CH}_2 \\
|\\
\text{CH}_2 \\
|\\
\text{COOH}
\end{array}
\tag{23.12}
$$

23.2 CYTOTOXICITY OF SOME METAL COMPOUNDS

In this section, some of the important metal complexes used in the treatment of cancer will be described. This is a field that is developing very rapidly, and the number of publications annually dealing with the topic is large. The interested reader should consult the current literature for recent developments. Especially useful will be the journals devoted specifically to bioinorganic chemistry. The references listed at the end of the chapter not only are useful in themselves but also give citations to earlier literature.

23.2.1 Platinum Complexes

The interface between inorganic chemistry and biochemistry has resulted in widespread study of the interaction of metal complexes with biological materials. As a result of developments in this area, there has been rapid growth in the use of metal complexes in the treatment of diseases and emergence of the field of metallotherapy or medicinal inorganic chemistry. It was mentioned in Chapter 16 that *cis*-diamminedichloroplatinum(II), *cis*-$[Pt(NH_3)_2Cl_2]$ (known as *cisplatin*), is effective in treating certain types of cancer. It provides perhaps the most dramatic development in the field of metallotherapy or metallodrugs. The relatively simple cisplatin complex has the following structure:

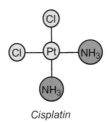

Cisplatin

Although the compound is effective in cancer treatment, such effectiveness is accompanied by considerable toxicity that can result in kidney damage and other side effects. The compound has a long history since it was first prepared in 1845 by Michel Peyrone, and its structure was explained in 1893 by Alfred Werner. Cisplatin has been used in the treatment of cancer since its approval in 1978 by the Food & Drug Administration. Following the development of cisplatin as a cancer treatment, over 3000 additional platinum complexes have been prepared and tested to determine their effectiveness (Alberto et al. 2009).

It is believed that cisplatin functions by bridging between adjacent base moieties in DNA, which the *trans* isomer cannot do. As a result of these effects, changes occur in the DNA structure so that its replication is prevented, which leads to apoptosis (changes in cell structure) and cell death. In this process, it is likely that cisplatin undergoes one or more aquation reactions in which the chloride ligands are replaced with the effective anticancer agent being *cis*-$[Pt(NH_3)_2(H_2O)(OH)]^+$. Cisplatin has been shown to be effective toward several types of cancer including those of the ovarian, bladder, and testicular types. However, side effects can lead to damage to kidneys. Another issue with the use of cisplatin is that cells (particularly in lung and ovarian cancer) can develop a resistance to the drug. Much is known about the specific interactions of cisplatin with cell components and how the resistance arises at the cellular level. An excellent and detailed review of this topic is available for the interested reader (Siddik, 2003).

One aspect that leads to difficulty in the use of cisplatin has to do with the rate of hydrolysis because it is the aqua complex that is the effective agent. Consequently, a great number of derivatives of cisplatin have been prepared that exhibit lower rates of hydrolysis. Three of those that are effective anticancer agents are *oxaliplatin*, *carboplatin*, and *nedaplatin*. These along with cisplatin are the four platinum compounds most often employed in cancer treatment, and they have the structures

Oxaliplatin   Carboplatin   Nedaplatin

Oxaliplatin is used in the treatment of colorectal cancers, and it has been shown to be effective on tumors that have become resistant to cisplatin. In addition to colorectal cancer, oxaliplatin is effective against ovarian, breast, pancreatic, and esophagogastric cancer. It appears that oxaliplatin forms linkages between purine base units in DNA.

Nedaplatin has been shown to be effective toward lung, gastric, testicular, and ovarian cancers, and it has less pronounced side effects than those produced by some other platinum complexes. A recent theoretical study of the mechanism of action by nedaplatin has been reported by Alberto et al. (2009). As in the case of cisplatin, it appears that the active agent in nedaplatin activity is cis-$[Pt(NH_3)_2(H_2O)(OH)]^+$, which forms bridges between adjacent base units in DNA. These bridges prevent replication and growth leading to apoptosis and cell death. The Lewis acid character of the platinum complex results in nedaplatin forming bridges between base sites in DNA leading to apoptosis and cell death. A significant advantage of nedaplatin over cisplatin is that the former exhibits less toxic side effects.

The mechanism by which cis-$[Pt(NH_3)_2(H_2O)(OH)]^+$ is produced has been addressed in the study by Alberto et al. Hydrolysis in both neutral and acidic solutions was addressed in that work. The formation of the aquahydroxo complex was shown to be more favorable energetically when the mechanism is S_N2, but two transition states were considered in which a fifth group, H_2O, was present. The first involves breaking the Pt—O bond in which the oxygen atom is bonded to the carbonyl group resulting in a transition state that has the structure

If the Pt—O bond to the oxygen atom attached to the CH_2 group is broken, the transition state has the structure

In either case, the second step in the aquation process leads to *cis*-$[Pt(NH_3)_2(H_2O)(OH)]^+$. The first step is rate-determining so the active aquation product is the aquahydroxo complex in either case. However, in acidic systems, the second step is not fast compared with the first so the partially aquated species may survive long enough for it to be involved in the formation of bridged structures with DNA.

Because cisplatin interacts with DNA by forming linkages between base units, it causes the DNA structure to become more rigid. The result is that the DNA becomes more compact, and there may be a relationship between the degree of compaction and the effectiveness of the bridging agent. Such interactions have been studied by X-ray and NMR techniques. These methods show distortion of the DNA in the regions where the linkages occur. It is believed that these structural changes result in interference with cell transcription and/or replication.

In addition to the platinum complexes already studied, recent work has shown that the two complexes shown in Figure 23.8 lead to DNA compaction by forming cross-linkages. However, the mechanism appears to be different from that exhibited by cisplatin (Yoshikawa et al. 2011). The compound shown in Figure 23.8(a) has a lower IC_{50} value (see below) than does cisplatin toward certain types of cancers. Rapid progress has been made in the development of platinum compounds for cancer treatment, but the quest for effective cancer drugs continues and metal complexes are a vital part of the effort.

23.2.2 Complexes of Other Metals

Although the first metal complex used in cancer treatment was cisplatin, an enormous number of studies have been conducted to assess the effectiveness of other metal complexes in cancer treatment. In this section, we will present a brief survey to show a few of the types compounds that have been employed.

Toxicity toward cancer cells or the inhibition of the action of an enzyme is expressed as a number known as the IC_{50} value. This value denotes the concentration of the inhibitor (usually expressed in micromoles per liter (μM)) that results in a decrease in the rate or activity to one-half of its value when no inhibitor is present. The lower the value, the more effective the inhibitor is at decreasing the function. Cisplatin has become the standard to which the effectiveness of other metal complexes against cancer is compared.

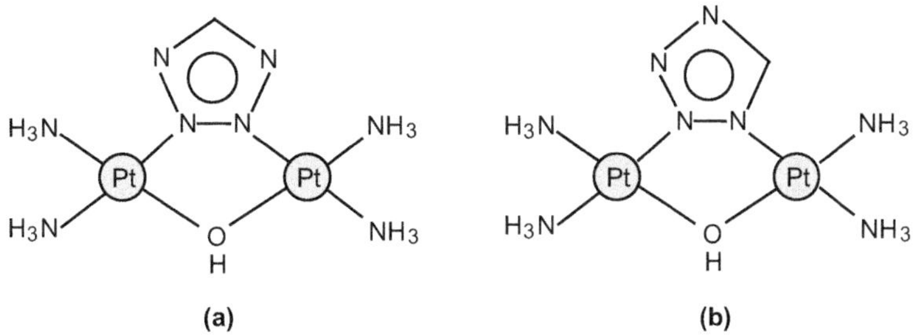

FIGURE 23.8

Two dinuclear platinum ions that have been found to have anticancer activity. Both complexes have an overall +2 charge, and they been used as perchlorate salts.

In recent years, it has been found that several titanium complexes also exhibit the property of cytotoxicity. Two such compounds are those shown in the following structures:

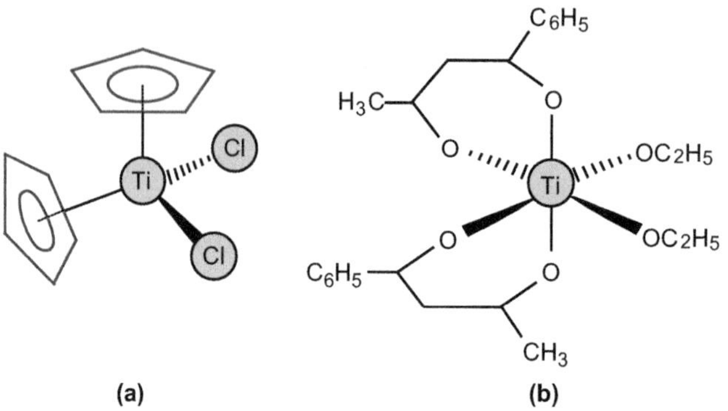

(a)   (b)

Compounds of this type are more effective than cisplatin at destroying cells in ovarian and colon cancers, but they have relatively short lifetimes in aqueous environments because they are not kinetically inert (see Chapter 20). Tzubery and Tshuva (2011) have extended such studies by preparing a series of titanium complexes containing the ligand

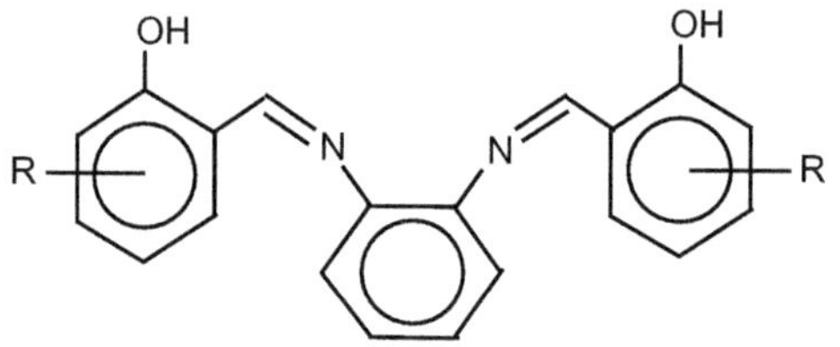

in which the R groups can represent 4-CH$_3$, 4-Cl, 2-Cl, or H. Thus, the structures of the titanium complexes can be shown as follows in which the positions of R on the ring may vary:

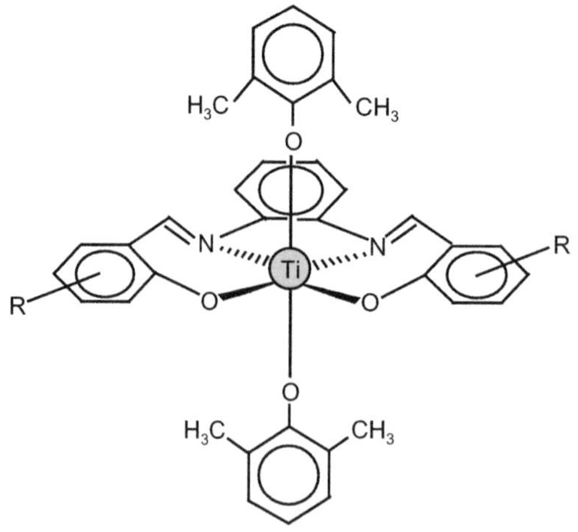

Such compounds are more resistant to destruction due to hydrolysis and most exhibit high antitumor activity. It was reported that when R = 4-CH$_3$, the activity was higher than when R = H and that when

R = 2-Cl, there was no antitumor activity. Titanium compounds are much less toxic than those that contain a heavy metal such as platinum. This is an enormously important and interesting area of chemistry that is developing rapidly. The article by Tzubery and Tshuva provides references to the literature that describe much of the potential medicinal chemistry of these titanium compounds.

In addition to the study of titanium complexes with cyclopentadiene as cytotoxic agents, several vanadium complexes (derivatives of vanadocene dichloride) have also been studied in this regard (Honzícek et al. 2011). Some of these complexes have the structures shown in Figure 23.9. Although $(C_5H_5)_2VCl_2$ hydrolyzes rapidly, the highly substituted compounds such as those containing $-CH_2C_6H_4OCH_3$ groups are much less soluble in water. This has resulted in speculation that the cytotoxicity of those compounds may arise from a different mode of action at the molecular level than is the case for $(C_5H_5)_2VCl_2$. In any event, it is extremely interesting and important that organometallic complexes containing first-row metals of relatively low toxicity are finding use in cancer research.

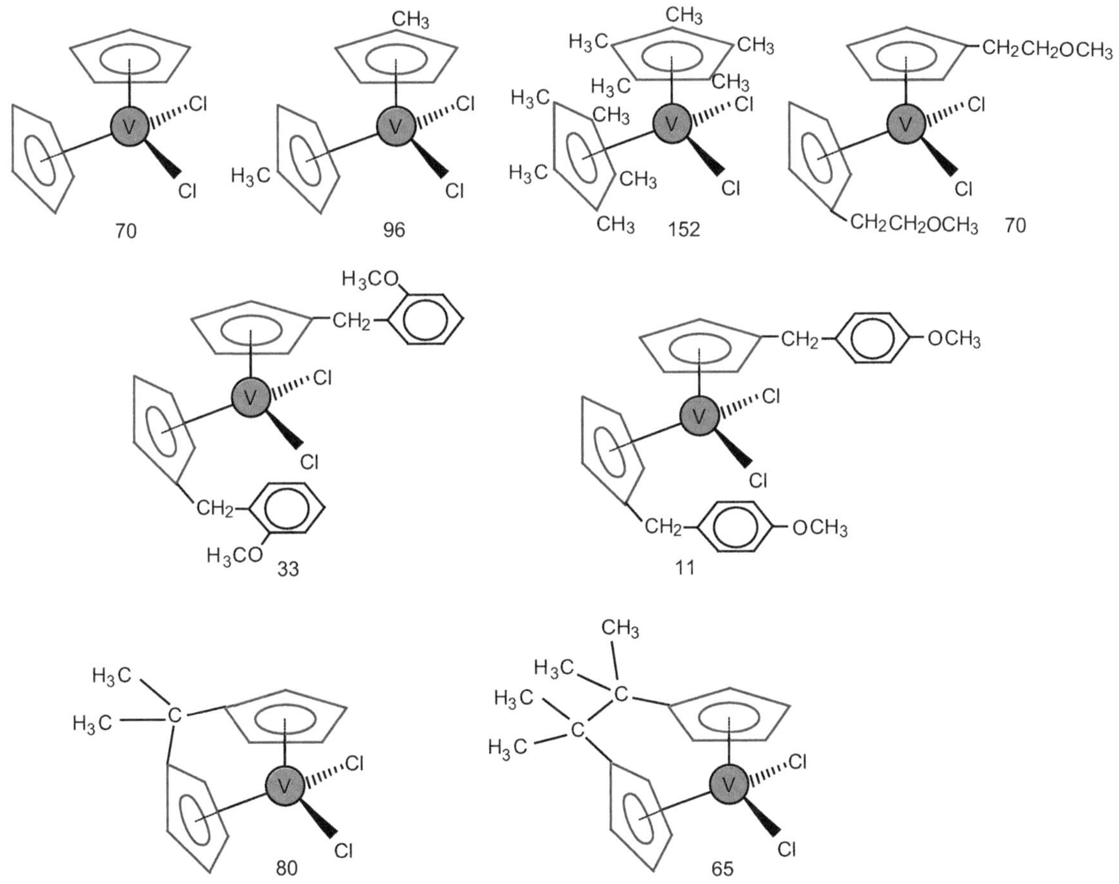

FIGURE 23.9
Some vanadium complexes that exhibit cytotoxic behavior. The numbers below each structure are the IC$_{50}$ values toward MOLT-4 human T-lymphocytic leukemia cells.

In addition to complexes containing Ti and V, several copper compounds have also been prepared that have anticancer activity. Some of them contain dimethyl derivatives of 4-nitropyridine-N-oxide, and they have the general formula $[CuL_2(NO_3)_2H_2O]$ except for the 3,5-dimethyl compound that has the formula $[CuL_2(NO_3)_2]$. The structures of these ligands that produce cytotoxic copper complexes are as follows:

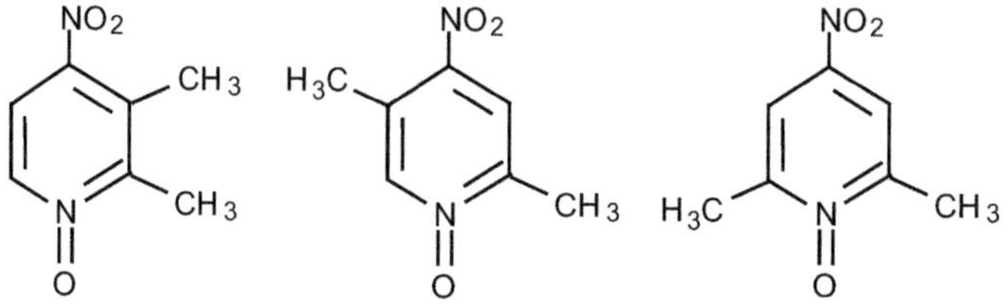

The five-bonded complexes containing the 2,3-dimethyl- and 2,5-dimethylpyridine-oxide have a structure that is essentially a square-based pyramid with the water molecule in the axial position. The two ligands L are in *trans* positions in the square base. Some of these complexes exhibit a rather high level of cytotoxic activity toward some types of colon and breast cancer and also toward mouse leukemia (Puszko et al., 2011).

Other copper complexes having cytotoxic properties that have been studied contain a Schiff base ligand, and they have structures shown below:

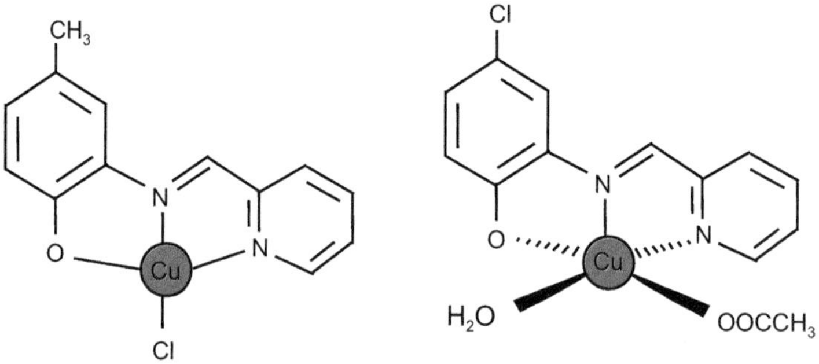

In addition to the complexes of the first-row metals described, complexes containing cobalt(II) have also been studied. Cytotoxic behavior is certainly not limited to the heavy transition metals even though some such compounds are effective.

Gold compounds have been presumed to have medicinal properties since ancient times. A very interesting account of the history of the use of gold compounds as pharmaceuticals has been given by Kean et al. (1997). One of the most studied and widely used complexes of gold is known as *auranofin*, and it has the structure

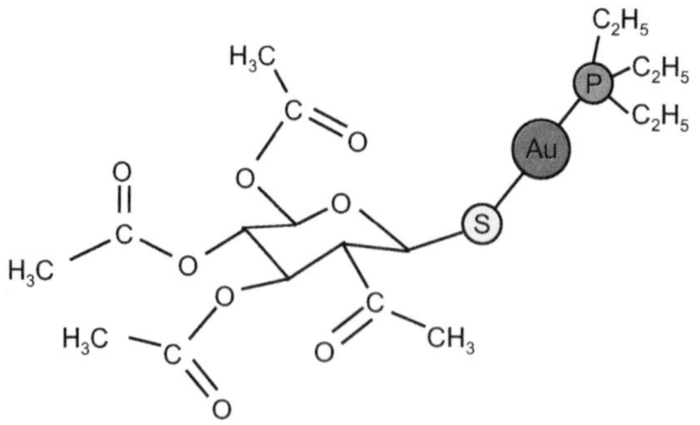

Auranofin is widely recognized as an effective drug for the treatment of rheumatoid arthritis, and it has been available for use in oral doses containing 3 mg of the compound for approximately 40 years. Not only is auranofin an effective agent for treatment of rheumatoid diseases but also it has been shown to have anticancer properties that are comparable to those of cisplatin. The action of auranofin is thought to involve ligand removal, which allows the gold ion to interact with sulfur linkages in enzymes. In terms of bond formation, this is a type of Lewis acid–base interaction that can be characterized as a soft acid bonding to a soft base. Another gold compound that releases gold *in vivo* is $Au(P(C_6H_5)_3)Cl$, and it has also been utilized as an anticancer agent. The details of the binding of auranofin to cysteine and selenocysteine have recently been described (Shoeib et al. 2010).

In addition to auranofin, numerous other gold compounds have been studied to determine their cytotoxic behavior. These include both gold(I) and gold(III) compounds. Among them are $[Au(en)_2]Cl_3$ and $[Au(1,10\text{-ophen})Cl_2]Cl$. Some of the gold compounds that have been studied (Gimeno et al., 2011) contain ferrocene groups, and an example of this type that exhibits anticancer activity has the structure

Platinum(II) is a d^8 ion as is gold(III). Accordingly, complexes containing gold(III) have been studied with regard to their effectiveness in treating cancers. Compounds containing gold(III) have exhibited anticancer activity as a result of their ability to inhibit mitochondrial processes. A wide variety of compounds have been studied, but those containing 1,10-phenanthroline and its derivatives have been of interest because studies have shown that the ligands have cytotoxicity themselves. One such complex that has been studied by Wein et al. (2011) contains 5,6-dimethylphenanthroline (5,6-dmp) as a chelating ligand. The compound is dichloro-5,6-dimethyl-1,10-phenanthrolinegold(III) tetrafluoroborate, and it has the structure

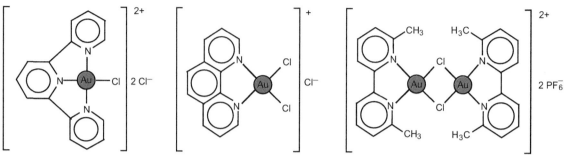

FIGURE 23.10
Compounds containing gold(III) that are effective against some types of cancer.

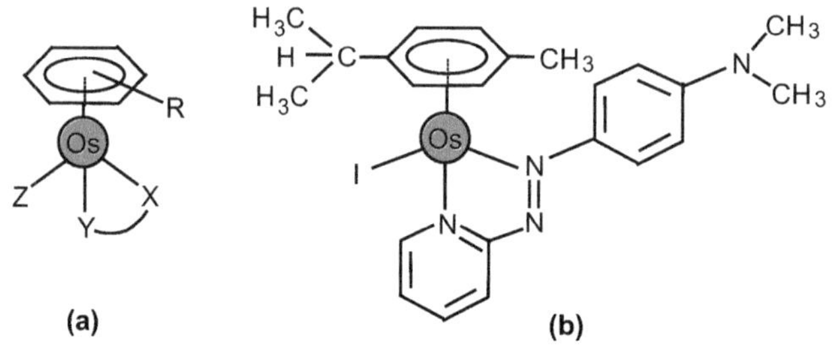

(a)   **(b)**

FIGURE 23.11
Osmium compounds utilized in anticancer treatment.

Toward certain types of cancers, the IC_{50} value for $[Au(5,6\text{-}dmp)Cl_2]BF_4$ is only about one-eighth of that of cisplatin. Cisplatin functions in cytotoxic activity by altering DNA structure, but gold(III) compounds are believed to be effective as a result of interaction with thio and imidazole groups in proteins. This causes anti-mitochondrial changes in an enzyme known as *thioredoxin reductase*. It appears that the mechanism of anticancer action by compounds containing platinum is quite different from those containing gold(III). Some of the recently studied gold(III) compounds that exhibit cyto-toxic activity have structures shown in Figure 23.10 (Casini et al., 2008; Vela et al., 2011).

The investigation of cyclopentadienyl and aryl compounds as effective anticancer agents has now spread to the use of some osmium compounds. Some of these have the "piano stool" configuration shown in Figure 23.11(a) with a specific example being shown in Figure 23.11(b) (Fu et al., 2010). This compound has shown activity toward several types of cancer cells including some types that are resistant to cisplatin.

23.3 ANTIMALARIAL METALLODRUGS

Although the discussion to this point has dealt primarily with antitumor action of complexes, this is by no means the only area of metallotherapy. Malaria is widespread throughout many parts of the world, and deaths from the disease are estimated to number from 700,000 to 1 million annually. It is caused

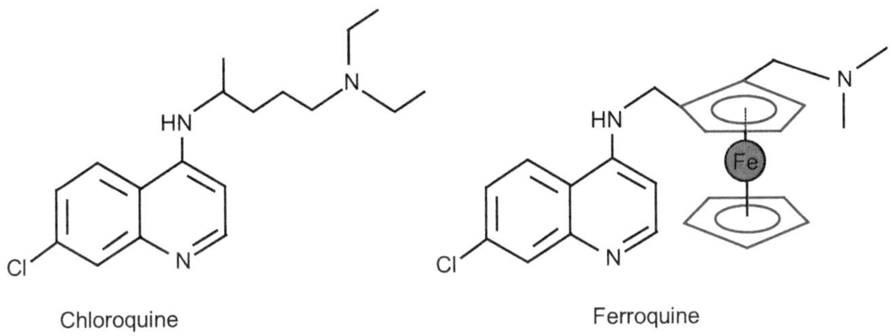

FIGURE 23.12
The structures of chloroquine and ferroquine.

by the parasite *Plasmodium falciparum* that is transmitted by the female *Anopheles* mosquito. One of the drugs that is employed in the treatment of malaria is known as *chloroquine*, and it has the structure shown in Figure 23.12.

A derivative of chloroquine that is being studied for its antimalarial activity is known as *ferroquine*, which is a ferrocene derivative of quinoline, that has the structure shown in Figure 23.12. Ferroquine has been found to be effective against malaria parasites that have developed resistance to chloroquine. Not only is ferroquine an effective drug for treating malaria but also are some "half sandwich" compounds that contain only one ring attached to the metal. One such compound contains the Cr(CO)$_3$ moiety, and it has the piano stool and half sandwich structure that can be shown as (Glans et al. 2011)

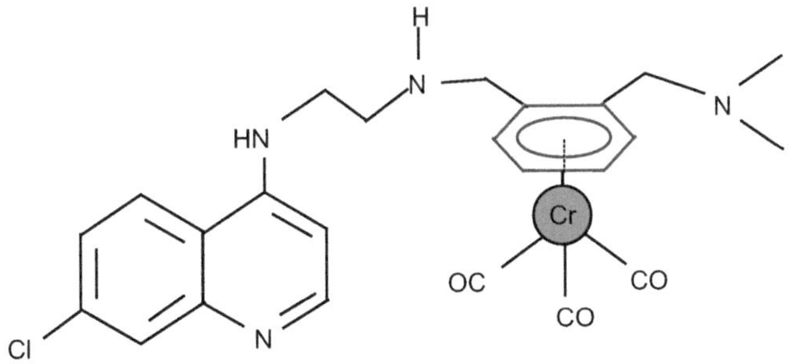

This compound is also effective against some strains of malarial parasites that have become resistant to chloroquine. It also provides an example of how a metal derivative of a therapeutic agent can enhance the effectiveness of known drugs.

As a result of there being journals specifically devoted to bioinorganic chemistry and numerous others devoting a large number of pages to publishing articles in the field, the literature is growing at a phenomenal rate. Moreover, the field is extremely broad, dealing with all types of living systems. A considerable number of both research monographs and textbooks have been published that deal with bioinorganic chemistry. Consequently, this chapter includes only a brief introduction to bioinorganic chemistry and shows the results of a few specific studies to give the flavor of the field. It should serve as an indication of the types of structures and reactions that constitute the field. Hopefully, the glimpses of bioinorganic chemistry given in this chapter will encourage the reader to pursue further study in this vast, interesting, and important field.

References for Further Study

Alberto, M.E., Lucas, M.F.A., Pauleka, M., Russo, N., 2009. *J. Phys. Chem.* B 113, 14473−14479. A study of nedaplatin effects on DNA.

Bailar Jr., J.C., 1971. *Am. Sci.* 59, 586. An older survey of bioinorganic chemistry by the late Professor Bailar, whose influence is still present in coordination chemistry.

Bonetti, A., Leone, R., Muggia, F., Howell, S.B. (Eds.), 2008. *Platinum and Other Heavy Metal Compounds in Cancer Chemotherapy.* Humana Press, New York. A book that contains chapters written by contributors who presented the material at a symposium.

Casini, A., Hartinger, C., Gabbiani, C., Mini, E., Dyson, P.J., Keppler, B.K., Messori, L., 2008. *J. Inorg. Biochem.* 102, 564−575. An article describing antitumor properties of gold(III) compounds.

Fu, Y., Habtemariam, A., Pizarro, A.M., van Rijt, S.H., Healey, D.J., Cooper, P.A., Shnyder, S.D., Clarkson, G.J., Sadler, P.J., 2010. *J. Med. Chem.* 53, 8192−8196.

Gasser, G., Ott, I., Metzler-Nolte, N., 2011. *J. Med. Chem.* 54, 3−25. A detailed review of the use of organometallic compounds in cancer treatment.

Gimeno, M.C., Goitia, H., Laguna, A., Luque, M.E., Villacampa, M.D., Sepúlveda, C., Meireles, M., 2011. *J. Inorg. Biochem.* 105, 1373−1382. Results of a study on antitumor properties of gold complexes.

Glans, L., Taylor, D., de Kock, C., Smith, P.J., Haukka, M., Moss, J.R., Nordlander, E., 2011. *J. Inorg. Biochem.* 105, 985−990. A description of antimalarial activity of chromium complexes.

Gray, H.B., Stifel, E.I., Valentine, J.S., Bertini, I. (Eds.), 2006. *Biological Inorganic Chemistry: Structure and Reactivity.* University Science Books, Sausilito, CA.

Honzícek, J., Klepalova, I., Vinklarek, J., Padelkova, Z., Cisarova, I., Siman, P., Rezacova, M., 2011. *Inorg. Chim. Acta* 373, 1−7. A report on the use of vanadium complexes in cancer research.

Kean, W.F., Hart, L., Buchanan, W.W., 1997. *Br. J. Rheumatol.* 36, 560−572. A review of the chemistry, uses, and effects of auranofin.

Kraatz, H.-R., Metzler-Nolte, N. (Eds.), 2006. *Concepts and Models in Bioinorganic Chemistry.* Wiley-VCH, New York.

Lippard, S.J., 1994. *Principles of Bioinorganic Chemistry.* University Science Books, Sausilito, CA.

Puszko, A., Brzuszkiewicz, A., Jezierska, J., Adach, A., Wietrzyk, J., Filip, B., Pelczynska, M., Cieslak-Golonka, M., 2011. *J. Inorg. Biochem.* 105, 1109−1114.

Sekhon, B.S., 2011. *J. Pharm. Educ. Res.* 2, 1−20. A survey of the uses of inorganic compounds in medicine.

Siddik, Z.H., 2003. *Oncogene* 22, 7265−7279. A review dealing with the mechanism of action of cisplatin and the development of resistance to the drug.

Shoeib, T., Atkinson, D.W., Sharp, B.L., 2010. *Inorg. Chim. Acta* 363, 184−192. A paper dealing with the structure of auranofin and its binding *in vivo*.

Tzubery, A., Tshuva, E.Y., 2011. *Inorg. Chem.* 50, 7946. An article that deals with the anticancer properties of titanium complexes. Extensive references to earlier work are presented.

Vela, L., Contel, M., Palomera, L., Azeceta, G., Marzo, I., 2011. *J. Inorg. Biochem.* 105, 1306−1313. The anticancer properties of gold(III) compounds are described in this article.

Wein, A.N., Stockhausen, A.T., Hardcastle, K.I., Saadein, M.R., Peng, S., Wang, D., Shin, D.M., Chem, Z., Eichler, J.F., 2011. *J. Inorg. Biochem.* 105, 663−668. Describes research on effects of gold(III) complexes on certain types of cancer.

Yoshikawa, Y., Komeda, S., Uemura, M., Kanbe, T., Chikuma, M., Yoshikawa, K., Imanaka, T., 2011. *Inorg. Chem.* 50, 11729−11735. A forthcoming article describing compaction of DNA and anticancer activity of dinuclear platinum complexes

QUESTIONS AND PROBLEMS

1. It has been suggested that when oxygen binds to iron in hemoglobin, an electron is transferred from iron to the O_2 molecule. Explain what this would mean in terms of oxidation states and charges on the species. What types of evidence would you need to be able to determine if this is what happens?

2. One of the absorption bands in chlorophyll *a* appears at about 700 nm. What energy does this correspond to in kJ mol^{-1}? Would this energy be sufficient to break an OH bond? See Table 4.1.

3. When an individual is being treated for exposure to carbon monoxide, oxygen is administered. Explain the basis for this procedure.

4. If accidental ingestion of a toxic heavy metal has occurred, one emergency procedure calls for administering egg white, but a more modern treatment involves administering ethylenediaminetetraacetate. Explain why this treatment is recommended.

5. Explain why complexes containing Ti—Cl bonds have short lifetimes *in vivo*.

6. If you were designing a synthetic oxygen carrier, what characteristics would be required of the metal ion? Suggest some alternative choices to Fe^{2+} and describe the characteristics of coordination compounds of the metal(s) you choose.

7. Describe why nitrites (oxidizing agents) diminish the oxygen-carrying capability of the blood.

8. Given that the effective agent resulting from cisplatin is a hydrolysis product, why would the phosphine analog not be as useful?

9. In addition to carbonic anhydrase, Zn^{2+} is present and active in several other enzymatic processes. Explain why some other metals might be less effective than Zn^{2+}.

Ionization Energies

Element	1st Ionization Potential (kJ mol^{-1})	2nd Ionization Potential (kJ mol^{-1})	3rd Ionization Potential (kJ mol^{-1})
Hydrogen	1,312.0	—	—
Helium	2,372.3	5,250.4	—
Lithium	513.3	7,298.0	11,814.8
Beryllium	899.4	1,757.1	14,848
Boron	800.6	2,427	3,660
Carbon	1,086.2	2,352	4,620
Nitrogen	1,402.3	2,856.1	4,578.0
Oxygen	1,313.9	3,388.2	5,300.3
Fluorine	1,681	3,374	6,050
Neon	2,080.6	3,952.2	6,122
Sodium	495.8	4,562.4	6,912
Magnesium	737.7	1,450.7	7,732.6
Aluminum	577.4	1,816.6	2,744.6
Silicon	786.5	1,577.1	3,231.4
Phosphorus	1,011.7	1,903.2	2,912
Sulfur	999.6	2,251	3,361
Chlorine	1,251.1	2,297	3,826
Argon	1,520.4	2,665.2	3,928
Potassium	418.8	3,051.4	4,411
Calcium	589.7	1,145	4,910
Scandium	631	1,235	2,389
Titanium	658	1,310	2,652
Vanadium	650	1,414	2,828
Chromium	652.7	1,592	2,987
Manganese	717.4	1,509.0	3,248.4
Iron	759.3	1,561	2,957
Cobalt	760.0	1,646	3,232
Nickel	736.7	1,753.0	3,393
Copper	745.4	1,958	3,554
Zinc	906.4	1,733.3	3,832.6

(Continued)

(Continued)

Element	1st Ionization Potential (kJ mol^{-1})	2nd Ionization Potential (kJ mol^{-1})	3rd Ionization Potential (kJ mol^{-1})
Gallium	578.8	1,979	2,963
Germanium	762.1	1,537	3,302
Arsenic	947.0	1,798	2,735
Selenium	940.9	2,044	2,974
Bromine	1,139.9	2,104	3,500
Krypton	1,350.7	2,350	3,565
Rubidium	403.0	2,632	3,900
Strontium	549.5	1,064.2	4,210
Yttrium	616	1,181	1,980
Zirconium	660	1,267	2,218
Niobium	664	1,382	2,416
Molybdenum	685.0	1,558	2,621
Technetium	702	1,472	2,850
Ruthenium	711	1,617	2,747
Rhodium	720	1,744	2,997
Palladium	805	1,875	3,177
Silver	731.0	2,073	3,361
Cadmium	867.6	1,631	3,616
Indium	558.3	1,820.6	2,704
Tin	708.6	1,411.8	2,943.0
Antimony	833.7	1,794	2,443
Tellurium	869.2	1,795	2,698
Iodine	1,008.4	1,845.9	3,200
Xenon	1,170.4	2,046	3,097
Cesium	375.7	2,420	—
Barium	502.8	965.1	—
Lanthanum	538.1	1,067	—
Cerium	527.4	1,047	—
Praseodymium	523.1	1,018	—
Neodymium	529.6	1,035	—
Promethium	535.9	1,052	—
Samarium	543.3	1,068	—
Europium	546.7	1,085	—
Gadolinium	592.5	1,167	—
Terbium	564.6	1,112	—
Dysprosium	571.9	1,126	—
Holmium	580.7	1,139	—
Erbium	588.7	1,151	—

(Continued)

Element	1st Ionization Potential (kJ mol^{-1})	2nd Ionization Potential (kJ mol^{-1})	3rd Ionization Potential (kJ mol^{-1})
Thulium	596.7	1,163	—
Ytterbium	603.4	1,176	—
Lutetium	523.5	1,340	—
Hafnium	642	1,440	—
Tantalum	761	(1,500)	—
Tungsten	770	(1,700)	—
Rhenium	760	1,260	—
Osmium	840	(1,600)	—
Iridium	880	(1,680)	—
Platinum	870	1,791	—
Gold	890.1	1,980	—
Mercury	1,007.0	1,809.7	—
Thallium	589.3	1,971.0	—
Lead	715.5	1,450.4	—
Bismuth	703.2	1,610	—
Polonium	812	(1,800)	—
Astatine	930	1,600	—
Radon	1,037	—	—
Francium	400	(2,100)	—
Radium	509.3	979.0	—
Actinium	499	1,170	—
Thorium	587	1,110	—
Protactinium	568	—	—
Uranium	584	1,420	—
Neptunium	597	—	—
Plutonium	585	—	—
Americium	578.2	—	—
Curium	581	—	—
Berkelium	601	—	—
Californium	608	—	—
Einsteinium	619	—	—
Fermium	627	—	—
Mendelevium	635	—	—
Nobelium	642	—	—

Numbers in parentheses are approximate values.

Character Tables for Selected Point Groups

C_2	E	C_2		
A	1	1	z, R_z	x^2, y^2, z^2
B	1	−1	x, y, R_x, R_y	yz, xz

C_s	E	σ_h		
A'	1	1	x, y, R_z	x^2, y^2, z^2, xy
A"	1	−1	z, R_x, R_y	yz, xz

C_i	E	i		
A_g	1	1	R_x, R_y, R_z	x^2, y^2, z^2, xy, xz, yz
A_u	1	−1	x, y, z	

C_{2v}	E	C_2	$\sigma_v(xz)$	$\sigma_v(yz)$		
A_1	1	1	1	1	z	x^2, y^2, z^2
A_2	1	1	−1	−1	R_z	xy
B_1	1	−1	1	−1	x, R_y	xz
B_2	1	−1	−1	1	y, R_x	yz

C_{3v}	E	$2C_3$	$3\sigma_v$		
A_1	1	1	1	z	$x^2 + y^2$, z^2
A_2	1	1	−1	R_z	
E	2	−1	0	(x, y), (R_x, R_y)	$(x^2 - y^2, xy)$, (xz, yz)

C_{4v}	E	$2C_4$	C_2	$2\sigma_v$	$2\sigma_d$		
A_1	1	1	1	1	1	z	$x^2 + y^2, z^2$
A_2	1	1	1	-1	-1	R_z	
B_1	1	-1	1	1	-1		$x^2 - y^2$
B_2	1	-1	1	-1	1		xy
E	2	0	-2	0	0	$(x, y), (R_x, R_y)$	(xz, yz)

C_{2h}	E	C_2	i	σ_h		
A_g	1	1	1	1	R_z	x^2, y^2, z^2, xy
A_u	1	1	-1	-1	z	
B_g	1	-1	1	-1	R_x, R_y	xz, yz
B_u	1	-1	-1	1	x, y	

$C_{\infty v}$	E	$2C_\infty$	$\infty \sigma_v$		
$A_1(\Sigma^+)$	1	1	1	z	$(x^2 + y^2, z^2)$
$A_2 (\Sigma^-)$	1	1	-1	R_z	
$E_1 (\Pi)$	2	$2\cos\phi$	0	$(R_x, R_x), (x, y)$	(xz, yz)
$E_2 (\Delta)$	2	$2\cos 2\phi$	0		$(x^2 - y^2, xy)$
$E_3 (\phi)$	2	$2\cos 3\phi$	0		

D_2	E	$C_2(z)$	$C_2(y)$	$C_2(x)$		
A	1	1	1	1		x^2, y^2, z^2
B_1	1	1	-1	-1	z, R_z	xy
B_2	1	-1	1	-1	y, R_y	xz
B_3	1	-1	-1	1	x, R_x	yz

D_3	E	$2C_3$	$3C_2$		
A_1	1	1	1		$x^2 + y^2, z^2$
A_2	1	1	-1	z, R_z	
E	2	-1	0	$(x, y), (R_x, R_y)$	(xz, yz), $(x^2 - y^2, xy)$

D_{2h}	E	$C_2(z)$	$C_2(y)$	$C_2(x)$	i	$\sigma(xy)$	$\sigma(xz)$	$\sigma(yz)$		
A_g	1	1	1	1	1	1	1	1		x^2, y^2, z^2
B_{1g}	1	1	−1	−1	1	1	−1	−1	R_z	xy
B_{2g}	1	−1	1	−1	1	−1	1	−1	R_y	xz
B_{3g}	1	−1	−1	1	1	−1	−1	1	R_x	yz
A_u	1	1	1	1	−1	−1	−1	−1		
B_{1u}	1	1	−1	−1	−1	−1	1	1	z	
B_{2u}	1	−1	1	−1	−1	1	−1	1	y	
B_{3u}	1	−1	−1	1	−1	1	1	−1	x	

D_{3h}	E	$2C_3$	$3C_2$	σ_h	$2S_3$	$3\sigma_v$		
A_1'	1	1	1	1	1	1		$x^2 + y^2, z^2$
A_2'	1	1	−1	1	1	−1	R_z	
E'	2	−1	0	2	−1	0	(x, y)	$(x^2 - y^2, xy)$
A_1''	1	1	1	−1	−1	−1		
A_2''	1	1	−1	−1	−1	1	z	
E''	2	−1	0	−2	1	0	(R_x, R_y)	(xz, yz)

D_{4h}	E	$2C_4$	C_2	$2C_2'$	$2C_2''$	i	$2S_4$	σ_h	$2\sigma_v$	$2\sigma_d$		
A_{1g}	1	1	1	1	1	1	1	1	1	1		$x^2 + y^2, z^2$
A_{2g}	1	1	1	−1	−1	1	1	1	−1	−1	R_z	
B_{1g}	1	−1	1	1	−1	1	−1	1	1	−1		$x^2 - y^2$
B_{2g}	1	−1	1	−1	1	1	−1	1	−1	1		xy
E_g	2	0	−2	0	0	2	0	−2	0	0	(R_x, R_y)	(xz, yz)
A_{1u}	1	1	1	1	1	−1	−1	−1	−1	−1		
A_{2u}	1	1	1	−1	−1	−1	−1	1	1	1	z	
B_{1u}	1	−1	1	1	−1	−1	1	−1	−1	1		
B_{2u}	1	−1	1	−1	1	−1	1	−1	1	−1		
E_u	2	0	−2	0	0	−2	0	2	0	0	(x, y)	

D_{5h}	E	$2C_5$	$2C_5^2$	$5C_2$	σ_h	$2S^5$	$2S_5^3$	$5\sigma_v$		
A_1'	1	1	1	1	1	1	1	1		$x^2 + y^2, z^2$
A_2'	1	1	1	-1	1	1	1	-1	R_z	
E_1'	2	$2\cos 72°$	$2\cos 144°$	0	2	$2\cos 72°$	$2\cos 144°$	0	(x,y)	
E_2'	2	$2\cos 144°$	$2\cos 72°$	0	2	$2\cos 144°$	$2\cos 72°$	0		$(x^2 - y^2, xy)$
A_1''	1	1	1	1	-1	-1	-1	-1		
A_2''	1	1	1	-1	-1	-1	-1	1	z	
E_1''	2	$2\cos 72°$	$2\cos 144°$	0	-2	$-2\cos 72°$	$-2\cos 144°$	0	(R_x, R_y)	(xz, yz)
E_2''	2	$2\cos 144°$	$2\cos 72°$	0	-2	$-2\cos 144°$	$-2\cos 72°$	0		

D_{2d}	E	$2S_4$	C_2	$2C_2'$	$2\sigma_d$		
A_1	1	1	1	1	1		$x^2 + y^2, z^2$
A_2	1	1	1	-1	-1	R_z	
B_1	1	-1	1	1	-1		$x^2 - y^2$
B_2	1	-1	1	-1	1	z	xy
E	2	0	-2	0	0	$(x, y), (R_x, R_y)$	(xz, yz)

D_{3d}	E	$2C_3$	$3C_2$	i	$2S_6$	$3\sigma_d$		
A_{1g}	1	1	1	1	1	1		$x^2 + y^2, z^2$
A_{2g}	1	1	-1	1	1	-1	R_z	
E_g	2	-1	0	2	-1	0	(R_x, R_y)	$(x^2 - y^2, xy), (xz, yz)$
A_{1u}	1	1	1	-1	-1	-1		
A_{2u}	1	1	-1	-1	-1	1	z	
E_u	2	-1	0	-2	1	0	(x, y)	

S_4	E	S_4	C_2	S_4^3		
A	1	1	1	1	R_z	$x^2 + y^2, z^2$
B	1	-1	1	-1	z	$x^2 - y^2, xy$
E	1	$\pm i$	-1	$\pm i$	$(x, y), (R_x, R_y)$	(xz, yz)

T_d	E	$8C_3$	$3C_2$	$6S_4$	$6\sigma_d$		
A_1	1	1	1	1	1		$x^2 + y^2 + z^2$
A_2	1	1	1	−1	−1		
E	2	−1	2	0	0		$(2z^2 − x^2 − y^2, x^2 − y^2)$
T_1	3	0	−1	1	−1	(R_x, R_y, R_z)	
T_2	3	0	−1	−1	1	(x, y, z)	(xz, yz, xy)

O_h	E	$8C_3$	$6C_2$	$6C_4$	$3C_2$	i	$6S_4$	$8S_6$	$3\sigma_h$	$6\sigma_d$		
A_{1g}	1	1	1	1	1	1	1	1	1	1		$x^2 + y^2 + z^2$
A_{2g}	1	1	−1	−1	1	1	−1	1	1	−1		
E_g	2	−1	0	0	2	2	0	−1	2	0		$(2z^2 − x^2 − y^2, x^2 − y^2)$
T_{1g}	3	0	−1	1	−1	3	1	0	−1	−1	(R_x, R_y, R_z)	
T_{2g}	3	0	1	−1	−1	3	−1	0	−1	1		(xz, yz, xy)
A_{1u}	1	1	1	1	1	−1	−1	−1	−1	−1		
A_{2u}	1	1	−1	−1	1	−1	1	−1	−1	1		
E_u	2	−1	0	0	2	−2	0	1	−2	0		
T_{1u}	3	0	−1	1	−1	−3	−1	0	1	1	(x, y, z)	
T_{2u}	3	0	1	−1	−1	−3	1	0	1	−1		

I_h	E	$12C_5$	$12C_5^2$	$20C_3$	$15C_2$	i	$12S_{10}$	$12S_{10}^3$	$20S_6$	15σ		
A_g	1	1	1	1	1	1	1	1	1	1		$x^2 + y^2 + z^2$
T_{1g}	3	x	y	0	−1	3	y	x	0	−1	(R_x, R_y, R_z)	
T_{2g}	3	y	x	0	−1	3	x	y	0	−1		
G_g	4	−1	−1	1	0	4	−1	−1	1	0		
H_g	5	0	0	−1	1	5	0	0	−1	1		$(2x^2 − x^2 − y^2, x^2 − y^2, xy, xy, yz)$
A_u	1	1	1	1	1	−1	−1	−1	−1	−1		
T_{1u}	3	x	y	0	−1	−3	−y	−x	0	1	(x, y, z)	
T_{2u}	3	y	x	0	−1	−3	−x	−y	0	1		
G_u	4	−1	−1	1	0	−4	1	1	−1	0		
H_u	5	0	0	−1	1	−5	0	0	1	−1		

where $x = \frac{1}{2}(1+\sqrt{5})$ and $y = \frac{1}{2}(1-\sqrt{5})$.

Index

Note: Page numbers with "f" denote figures; "t" denote tables.